C0-AUP-170

Teacher Education Programs and Online Learning Tools:

Innovations in Teacher Preparation

Richard Hartshorne
University of Central Florida, USA

Tina L. Heafner
University of North Carolina at Charlotte, USA

Teresa M. Petty
University of North Carolina at Charlotte, USA

Managing Director: Lindsay Johnston
Senior Editorial Director: Heather A. Probst
Book Production Manager: Sean Woznicki
Development Manager: Joel Gamon
Development Editor: Hannah Abelbeck
Assistant Acquisitions Editor: Kayla Wolfe
Typesetter: Alyson Zerbe
Cover Design: Nick Newcomer

Published in the United States of America by
Information Science Reference (an imprint of IGI Global)
701 E. Chocolate Avenue
Hershey PA 17033
Tel: 717-533-8845
Fax: 717-533-8661
E-mail: cust@igi-global.com
Web site: http://www.igi-global.com

Library of Congress Cataloging-in-Publication Data

Teacher education programs and online learning tools: innovations in teacher preparation / Richard Hartshorne, Tina Heafner and Teresa Petty, editors.
p. cm.
Includes bibliographical references and index.
Summary: "This book presents information about current online practices and research in teacher education programs, and explores the opportunities, methods, and issues surrounding technologically innovative opportunities in teacher preparation"--Provided by publisher.
ISBN 978-1-4666-1906-7 (hardcover) -- ISBN 978-1-4666-1907-4 (ebook) -- ISBN 978-1-4666-1908-1 (print & perpetual access) 1. Teachers--Training of--United States. 2. Teachers--Training of--Technological innovations--United States. 3. Teachers--Training of--Computer-assisted instruction. 4. Teachers--Training of--Research--United States. I. Hartshorne, Richard, 1971- II. Heafner, Tina. III. Petty, Teresa M.
LB1715.T4236 2013
370.71'1--dc23

2012010238

British Cataloguing in Publication Data
A Cataloguing in Publication record for this book is available from the British Library.

Editorial Advisory Board

List of Reviewers

Table of Contents

Detailed Table of Contents

Section 1
Issues and Trends in Online Education

Over the last two decades, online learning has emerged as a commonly used instructional platform in higher education. While online courses have become commonplace, this transitional time has been accompanied by growing pains. Adjusting to new ways of facilitating learning experiences required thoughtful reconsideration of epistemological beliefs and existing practices. The three chapters in this first section provide personal, programmatic, and unit views of how these changes were experienced and the challenges each faced through the process of migrating to online learning. As the trend for growth in online learning continues at an even faster pace, the insights shared from these authors are important lessons to consider as online learning becomes even more pervasive in teacher education programs. These chapters also provide a context for exploring the innovations of praxis that technology engenders.

This chapter details the experiences of a university professor whose perspectives shifted from one of initial dissent to eventual advocacy for online learning as a delivery mode for her reading/literacy courses. Spanning eight years, her distance education teaching practices were shaped by her personal ventures as an online student, the outcomes gained by enhancing the social presence of her online courses, collaboration with colleagues, and systematic examination of her online teaching practice relative to its rigor, quality, and effectiveness within a teacher preparation program. Insights gained while teaching online conclude with recommendations for faculty members, institutions, systems, and organizations with vested interest in the future of teacher education.

At present, there are very few examples of the preparation of teachers for the online environment in teacher education. Even more unfortunate is that less than 40% of all online teachers in the United States reported receiving any professional development before they began teaching online. While some virtual schools provide some training to their own teachers, in most instances, no such training is provided to the school-based personnel. This is unfortunate, as K-12 student success in online learning environments require support from both the online teacher and the local school-based teacher. Clearly, there is a need for teacher education programs to equip all teachers with initial training in how to design, deliver, and – in particular – support K-12 online learning. This chapter begins with an examination of the act of teaching online and how that differs from teaching in a face-to-face environment. Next, the chapter describes existing teacher education initiatives targeted to pre-service teachers (i.e., undergraduate students), and then in-service teachers (i.e., graduate students). This is followed by an evaluation of current state-based initiatives to formalize online teaching as an endorsement area. Finally, a summary of the unique aspects of teaching online and how some initiatives have attempted to address these unique skills, before outlining a course of action that all teacher education programs should consider adopting.

Online teacher programs are diverse in their models, expressing a variety of learning objectives, pedagogies, technological platforms, and evaluation methods. Promoting and assessing collaborative learning of an online teacher programs is one of the major challenges, in part because collaboration includes complex cognitive and social-emotional dimensions. This chapter focuses on teachers' academic programs in online learning environment and examines the conditions for effective teaching and assessing collaborative problem solving. The chapter provides readers a look at the rationale, implementation, and assessment of collaborative learning in online teacher program and presents the conditions for effective design of collaborative learning for pre- and in-service teachers. Examples from two empirical studies are provided on how the collaborative learning environment leverages teachers' constructivist teaching, ongoing feedback, and evaluation to prepare teachers for instruction in technology-rich environments.

This chapter reports a study which examined experiences of nine beginning teachers who completed their initial teacher education in the online mode. The study investigated reported perceptions during their first six months teaching. Participants found the content of the online program comprehensive, prepared them well to begin teaching, and provided an opportunity for Māori (New Zealand's indigenous people) to become high school teachers. Main advantages of studying online were: flexibility; saving time and money; developing skills and personal attributes such as independence; and, for some, the only way to become teachers. The major disadvantage was the difficulty of studying alone despite an interactive delivery platform. Also, participants were concerned learning online did not allow modeling of teaching

Teacher education courses offered online are becoming increasingly common. Unfortunately, few instructors of online teacher education courses have specific preparation for teaching adult learners or in teaching online courses, resulting in faltering attempts to transfer traditional methodology such as lectures to online platforms. Given this new reality, it is up to teacher educators to continue to refine the methods of online instruction. This methodology must not be simply transferring face-to-face methods of teaching into online teaching methods. Instead, teacher educators must become learners again, considering what makes online learners unique and what methods are most effective when utilized online. This chapter considers the background of distance education and examines relevant literature on adult learners. Links between differentiated learning, distance education, and adult learner research present a way of highlighting the possibilities and continuing the conversation about what makes online learning effective for teacher education students and for the K-12 students they will teach. Differentiated instruction is proffered as a means of meeting the needs of adult learners in online teacher education courses. Specific examples of differentiating content, process, and product are suggested.

Teacher candidates in online courses engage in authentic learning to foster 21st-Century practices similar to those of their K–12 students, namely information and technology literacy and media production. This chapter describes instructional practices used in six online literacy courses for pre-service and in-service teacher candidates. The instructor assumed multiple roles during online instruction, including pedagogue, technologist, and evaluator. Although the course designs were highly structured, the instructor incorporated multiple resources to support diverse learners, to foster independent learning, to promote critical thinking and reflection on how instructional strategies can be used in K–12 classrooms, and to facilitate small group collaboration through authentic problem-solving tasks. Online courses for teacher education programs can serve as a vehicle for supporting candidates' information and technology skills. Online instructors can assume the primary role of pedagogue to help candidates connect their content area with best practices in literacy and technology.

While technology has always played a role in teaching and learning, with the advent of Information Communication Technologies (ICTs), schools have struggled to keep pace with Web 2.0 tools available for teaching and learning. Multiliteracies, a term coined by scholars who published under the name The New London Group in 1996, have helped provide a theoretical foundation for applying new texts and tools to teaching and learning; however, much of the scholarship around Multiliteracies remains in the academic and theoretical domain. This chapter addresses a suggested pedagogic framework, or metastructure, for applying Multiliteracies to teacher education and by extension to P-12 classrooms. Ways in which Web 2.0 tools have been used in undergraduate and graduate education courses are documented and discussed.

Section 5
Comparing Delivery Options in Teacher Education

Online course delivery continues to evolve as it increases in popularity. A variety of delivery options exist and are used in teacher education courses. Instructors utilize both synchronous and asynchronous methods as well as hybrid and fully web-based approaches. In this section, one author discusses the level of critical thinking of pre-service teachers participating in an urban education course through online discussions utilizing Bloom's Revised Taxonomy. Another author examines student preferences in regards to a fully web-based course and a blended model of instruction. In the final chapter of this section, the author investigates an innovative approach to early field-based experiences. These chapters examine various course delivery options and explore their utility in online teacher education courses.

In this chapter, a teacher educator examines the level of critical thinking of her pre-service teachers participating in an urban education course through online discussions. The objective was to see if online discussions, which were the heart of the learning process, could be an effective strategy to promote critical thinking skills. Using the revised version of Bloom's Taxonomy as a guide, participants' posts and responses were assessed to determine the quality of thinking that occurred in the online discussion forum. Results show that utilizing online discussion forums can be an effective pedagogy for classes where complex, often controversial, issues such as social justice, equity, and white privilege are discussed.

This chapter summarizes the results of a quasi-experiment conducted to determine the relative effectiveness of preparing pre-teacher education university students using a fully web-based course conducted asynchronously versus a blended model of instruction using the same LMS for forty percent of instructional time. The project evaluated two large sections of SPED 2100 "Introduction to Students with Special Needs." Data was collected to evaluate the extent to which pre-teacher education students developed understanding of critical information related to human development factors, psychological, sociological, and policy foundations of teaching students with special needs. Further, data collection examined student preferences in learning and the extent to which students developed comparable perception of preparedness for the future teaching roles. Results indicated no significant differences regarding content knowledge, but varying perspectives on the potential for success in fully web-based courses dependent largely on learner profile and the point of development in university coursework.

The purpose of this study was to explore pre-service teachers' behaviors in and perceptions of traditional field-based and virtual models of early field experiences. Specifically, this study examined some of the strengths and limitations associated with each model. Fifty undergraduate students participated in either a traditional field-based or a virtual field experience and completed an online questionnaire that examines various behaviors and student perspectives related to each model of early field experiences. The virtual field experiences include activities in the Inquiry Learning Forum (ILF), a web-based environment where students can observe and discuss diverse pedagogical practices and conceptual issues captured in a collection of video-based classrooms. The results of this study suggest that a virtual field experience, which utilizes video-based cases, may promote reflective practices, which could be especially valuable to students early in their teacher education program. In addition, this study suggests that the strengths and limitations of each format need to be considered in relation to the goals and objectives of the early field experience, and discusses the possibility of a hybrid model of field experiences.

Section 6
Innovative Online Teaching and Learning Practices in Teacher Education

The five chapters presented in this section present research-based ideas of new ways of integrating technology into teacher education programs. The focus of the chapters include: presenting a framework for integrating online learning opportunities into an English teacher education program; discussing results and implications of a longitudinal study of a mathematics methods course for teacher candidates taught in both hybrid and 100% asynchronous online formats; introducing the theoretical foundation, implementation, implications, and future applications of a tool to provide an online scaffold for teachers to inquire about their technology integration practices, a mechanism to synthesize action research information from varied sources, and a method to gather evidence of student learning within technology integration inquiries; and illustrating methods in which faculty in a fully online teacher leadership Master's program utilized a design-based approach to systematically modify program courses to increase student learning outcomes.

In this chapter, a framework and methods employed to integrate online learning opportunities into an English teacher education program at a large, public university in the southeastern United States, is discussed. The authors focus on their efforts to extend pre-service secondary English language arts teachers' understandings of what constitutes literacy and what counts as text in the secondary English language arts classroom in a blended technology- and media literacy-focused methods course, a required component of a three-semester English education Master's degree program. Specifically, the authors document the ways they nudge pre-service teachers to consider the kinds of literacy events they might design and the types of literacy practices they might promote to support literacy learning with interactive online technologies and popular media in English language arts classrooms.

Foreword

As an Associate Editor-in-Chief of the *Journal of Technology and Teacher Education (JTATE)*, I have the pleasure of reviewing some of the top published articles that address the role of technology in preparing teachers. The foundation of research for the field has grown immensely. There are authored works that cover almost every aspect of technology, ranging from virtual reality and mobile computing to games and simulations for preparing teachers.

There is one area of research where we are still lacking—online education. It is naïve to suggest that such articles do not exist. I have seen such articles in JTATE, and many of the authors of the chapters in this book have themselves published research, theory, and practice articles about the use of Web 2.0 and 3.0 tools to improve teaching and teacher education. However, online learning as a field is relatively new. Although the construction of online programs at all educational levels began in the 90s, it never really gained traction until the mid 2000's. This has meant a relative dearth of research articles specifically on practices within said programs. Such a lack of foundational knowledge has correlated to a lack of teacher preparation in this important area.

Perhaps the best evidence of this argument is to examine such existing practices. Many colleges of teacher education still only require one or two courses in technology—with many of these courses having topics like desktop publishing and spreadsheets for in classroom use. One could also explore state and national standards for online education. Most states do not have online education as a part of their certification process. And national agencies are struggling to create standards for teaching and learning online.

The most troublesome findings come from Rice and Dawley, colleagues at Boise State. They found that most teachers who were teaching online received little to no instruction prior to their online experience. Some who did receive training were given access long after the course had started and through a face-to-face medium. Finally, many teachers who were teaching online had never taken an online course themselves.

This lack of participation does not, however, correlate with what is happening in the field. An estimated two million P-12 students took online classes last year in the United States. All fifty states have some sort of legislation regarding online learning. And, some sources suggest that in ten years, a majority of classes will be online.

In sum, you have an exponential growth of online practice mixed with a relatively poor response by many who are preparing teachers. This is one of two reasons this book is so important. A collection of research related to online education and teacher preparation will help form the foundation by which colleges and other preparatory in-service or pre-service programs can rethink their efforts and meet real world demands.

A second reason relates to policy efforts. Legislative conversations about online learning and teacher professional development are happening globally. For instance, in the state of Ohio, legislators spent the last year examining best practices related to online education. One topic of discussion was the research showing teachers matter when it comes to online success. If teachers matter, what policies should be put in place to reform how they are prepared? Legislators should be able to point to a foundation of best practice research that guides their thinking. This book will help support that research effort.

The term online education mixed with teacher preparation means at least four things.

1. Preparing in-service teachers through online professional development
2. Preparing pre-service teachers to teach online
3. Using online tools to prepare both in-service and pre-service teachers
4. Using online delivery to prepare both in-service and pre-service teachers to teach online

The importance of this book is that it touches on all four areas. It does so directly at times, and at other times indirectly while discussing the latest technology or a specific educational approach. The book is divided into seven sections, ranging from collaboration to delivery options. However, it is important to note that there are two other critical themes that run throughout this book.

The first theme is **innovation**. Some of the tools discussed and presented are on the cutting edge of educational technology research. More importantly, the editors have done an important task in getting authors to think about the present and the future. What is technology and teacher preparation going to look like in five years? Ten years? The second theme is **pedagogical content knowledge**. If knowing math is different than knowing how to teach, and both are different than knowing how to teach math, then surely teaching math online or with online tools is also different. This book recognizes that in covering various content areas.

This book is an important step in preparing colleges of education and teacher educators to think critically about the role of online tools and online education in preparing teachers. It is my hope that such a book leads to changes where teachers are prepared differently. I look forward to a present and future where pre-service teachers have opportunities for blended or online internships, and where pre-service teachers get just-in-time collaborative help from mentors far away. The research presented in this book suggests that this not only can happen and does exist, but that it can happen ubiquitously.

Richard E. Ferdig
Kent State University, USA
May 8, 2012

Richard E. Ferdig *is the RCET Research Professor and Professor of Instructional Technology at Kent State University. He works within the Research Center for Educational Technology and also the School of Lifespan Development & Educational Sciences. He earned his PhD in educational psychology from Michigan State University. At Kent State University, his research, teaching, and service focus on combining cutting-edge technologies with current pedagogic theory to create innovative learning environments. His research interests include online education, gaming, and what he labels a deeper psychology of technology. In addition to publishing and presenting nationally and internationally, Ferdig has also been funded to study the impact of emerging technologies.*

Preface

CONTEXT AND PURPOSE

The infiltration of technology in everyday life and its spillover to work have created new communication avenues as well as innovative opportunities to learn. As a result, technology has begun to redefine the nature of teaching and the scope of learning. The manner in which institutions of higher education serve their student populations has evolved with the advent of online learning, Web 2.0 tools, and emergent technologies. Likewise, PK-12 schools have continued to expand technology resources and availability, which has rapidly changed the workplace environment for teachers. How have teacher education programs adapted to meet the changing expectations for technology-mediated teaching and learning? The manner in which teacher education programs have navigated these new spaces for learning serves as an important dialogue for exploring new possibilities for teaching and learning in the 21st Century. Thus, the focus of this book is to create a forum for examining the new teacher preparation classroom.

Over the past decade, the World Wide Web has become a critical tool for preservice teachers to utilize in their teaching and learning experiences. National efforts have been made to encourage technology integration in teacher education with expectations for frequent and successful applications with K-12 learners. One consequence has been a growing trend in higher education to provide students with more online educational opportunities. While online learning has become pervasive in many fields in higher education, one area in which it has been somewhat slow to catch on is teacher education, resulting in fewer opportunities for technology-mediated learning experiences in K-12 classrooms. However, for a variety of reasons (e.g., technological advances, budgetary concerns, technological expectations of students), teacher education programs are increasingly implementing online components in their programs. While this trend is expanding, little research has empirically explored the effectiveness of online education in teacher education programs. In response, this book brings together a collection of research examining various forms of and experiences with online learning in teacher preparation. Furthermore, researchers explore the utility and versatility of web-based tools for training teachers in the use of technology mediated learning tools with K-12 students.

It is important to understand the theoretical, pedagogical, technological, financial, and logistical issues, as well as management approaches, instructional delivery options, and policy considerations needed to create quality online teacher education programs. One purpose of this book is to present information about the current practices and research in online teacher education programs, while also presenting opportunities, methods, and issues involved with implementing online learning opportunities. This book presents a discussion of issues that have arisen when traditionally face-to-face teacher education programs have modified their practices to include online components. Another objective for the book is to discuss issues related to evidence of student learning and assessment in the context of various online aspects of teacher education programs. Furthermore, the book is a resource for disseminating information about current research related to online practices in teacher education programs and effective technology tools that support learning outcomes. A final objective of this book is to present empirical evidence of preservice teacher learning and assessment in the context of various online aspects of teacher education programs.

IMPETUS

The editors of this book embarked several years ago on an effort to create meaningful online solutions for perceived barriers to online teacher education. When online courses were first introduced at the institution in which they were all employed, it was narrowly understood as a replicate tool for mimicking current coursework. Strict steps were taken to ensure the similarities between course delivery methods resulting in asynchronous only learning platforms for online courses. The epistemological stumbling block was how to provide observations of teaching, typically face to face in nature, in an online asynchronous format. Given this limitation, the prospect of developing a 100% online program seemed insurmountable. Concurrently, institutional decisions were made to restrict online coursework in teacher licensure to only lateral entry teachers. Internal contradictions of using online learning to expand programs surfaced as enrollment in the online program depended on a student's proximity to the university or employment status.

As state-wide and even international interest grew in a 100% online teacher preparation program, questions emerged as to whether enrollment restrictions should be maintained. To exacerbate the problem, state demands for increasing the number of highly qualified teachers to fill classroom shortages became a mandate. Thus, this book's editors sought to create a technology-mediated solution that might allow them to improve the quality of online learning and address aforementioned issues. They used as their guiding principle the view that technology could create new spaces and places for learning that did not mirror current processes. They initiated efforts to expand delivery modes to include both synchronous and asynchronous tools in addition to emergent, web-based technologies. The outcome of the editors' research and development has been *ROGI: The Remote Observation of Graduate Interns* and *WiTL: Windows into Teaching and Learning*. Through the work and study of many researchers whose ideas served to inform our efforts, they came to the understanding that others could benefit from a collection of research that would showcase the possibilities of technology in teacher education as well as address how challenges can be overcome.

As online learning becomes ubiquitous in higher education and in PK-12 schools, teacher preparation programs can be at the forefront of innovation rather than the traditional role of lagging behind with technology integration. Teaching and learning are at the core of education; therefore, leaders in pedagogy can pave new pathways for how technology can re-envision work in academic, schools, and in expanding, globally dynamic communities. The contributions of this book are timely ideas for guiding the development of these new learning spaces and in defining the role online learning can and should play in teacher education in the 21st Century.

HOW TO USE THIS BOOK

Each section contains several chapters supporting the section theme. Beginning with issues and trends in online teacher education and concluding with an exploration of technology tools for moving forward in the evolution of online teacher education, this book presents research and theoretical guidelines regarding online teaching and learning.

The Roadmap

The first section examines **Issues and Trends in Online Education.** These chapters provide a context for exploring the innovations of praxis that technology engenders and the growing pains that accompany these changes. As the trend for growth in online learning continues at an increasingly fast pace than in the last few decades, the insights shared from these authors are important lessons to consider as online learning becomes even more pervasive in teacher education programs. Terry Atkinson presents, in the opening chapter, her experiences as a university instructor who navigated the shift from traditional teaching to online instruction. *The Journey from Dissenter to Advocate: Insights Gained While Teaching Online* is an eloquent presentation of the evolution of pedagogical beliefs and practice for meaningful online instruction. Next, Teresa Petty, Tina Heafner, and Richard Hartshorne discuss the evolution of a remote online observation process and relevant research findings that have shaped the project over the course of four years. In *Remote Observation of Graduate Interns: A Look at the Process Four Years Later*, the authors describe how this process remains an active synchronous observation method for preservice teachers, how it has progressed as an evolving technological tool, and the resulting shifts in college policies to adopt its utility. This section concludes with a chapter titled, *Perceptions of Preparation of Online Alternative Licensure Teacher Candidates,* which examines a comparison of online and face to face teacher preparation programs. The authors of this chapter, Joyce Brigman and Teresa Petty, investigate how to most effectively produce teachers possessing a high sense of preparedness for the classroom, regardless of the licensure pathway and modes of delivery, including both face-to-face and online learning.

Section two, **Online Teaching and Learning Initiatives in Online Teacher Education**, brings together innovative ideas for addressing current challenges in teacher education. In the first chapter, *Virtually Unprepared: Examining the Preparation of K-12 Online Teachers*, Michael Barbour, Jason Siko, Elizabeth Gross, and Kecia Waddell begin by providing an examination of differences between teaching online and teaching in face-to-face environments, as well as current teacher education initiatives targeted at both preservice and inservice teachers, followed by an evaluation of current initiatives to formalize online teaching as an endorsement area. The chapter concludes by outlining, based on results of previous initiatives, a course of action that all teacher education programs should consider adopting. In the second chapter in this section, Yigal Rosen and Rikki Rimor provide readers with a look at the rationale, implementation, and assessment of collaborative learning in online teacher education programs and present the conditions for effective design of collaborative learning for pre- and in-service teachers. In *Teaching and Assessing Problem Solving in Online Collaborative Environment*, these authors describe examples of how the collaborative learning environment helps prepare teachers for instruction in technology-rich environments. In the third chapter, *Online Teacher Education: A Case Study from New Zealand*, Ann Yates reports research on a program in New Zealand which examined experiences of nine beginning teachers who completed their initial teacher education in the online mode, and also reported participant perceptions during their first sixth months of teaching. A major purpose of the program was to provide an opportunity for Māori (New Zealand's indigenous people) to become high school teachers. Furthermore, Yates addresses major advantages and disadvantages found, which aligned with much of the existing research related to online teaching and learning. The last chapter in this section, *Training Teachers for Virtual Classrooms: A Description of an Experimental Course in Online Pedagogy*, provides a starting point for those wishing to create their own courses in online pedagogy. Wayne Journell, Melissa Beeson, Jerad Crave, Miguel Gomez, Jayme Linton, and Mary Taylor describe an experimental course designed to train teachers for virtual instruction. Topics in the course included the history of online education, online learning theories, the creation of online communities, online assessments, and ways to differentiate online courses for learners with special needs. Students were also afforded opportunities to put theoretical knowledge into practice by experiencing various forms of synchronous and asynchronous communication and designing their own online course.

The third section, **Supporting Online Learning in Teacher Education through Collaboration**, examines the role and importance of collaboration in online learning. In *Promoting Collaborative Learning in Online Teacher Education*, Vassiliki Zygouris-Coe discusses the importance of collaborative learning in online teacher education. The author provides details on course design issues, instructional practices, benefits and challenges associated with collaborative learning in an online teacher education course. Zygouris-Coe also discusses implications for further development and evaluation of collaborative learning in teacher preparation programs. Next, David Dunaway discusses the various ways that Web 2.0 tools can be used to create virtual cooperative learning experiences for aspiring teachers. In *Creating Virtual Collaborative Learning Experiences for Aspiring Teachers*, Dunaway discusses the utilization of these tools referencing his own experiences with the implementation of these tools in an online Masters of Administration course. He explores ways in which colleges of education can support reflection and collaboration while diminishing feelings of isolation of classroom teachers. This section concludes with a chapter titled, *Cyber-Place Learning in an Online Teacher Preparation Program: Engaging Learning Opportunities through Collaborations and Facilitation of Learning*. Victoria Cardullo explores literature supporting collaborative learning in online teacher preparation programs. An investigative look at the implications of learner-centered and place-based approaches to teaching is taken. In this chapter, it is the author's intent to offer guidelines for transference of classroom best practices to a cyber-place learning environment that will align with teacher preparation programs.

The fourth section of the book, **Literacy Education in Online Learning**, provides a specialized view of how literacy learning is formalized in an online environment. This section begins with an examination of unique attributes of non-traditional learners. Dixie Massey describes in her chapter, *Differentiating Instruction for Adult Learners in an Online Environment*, the limited preparation of online instructors and the issues that arise when they encounter adult learners. She presents first the limitations of existing online instructional practices for meeting the needs of these diverse distance education learners and concludes with recommendations that differentiated instruction is a viable response for making content, processes, and learning accessible to adult learners. The second chapter in this section, *Teacher Education in Online Contexts: Course Design and Learning Experiences to Facilitate Literacy Instruction for Teacher Candidates*, describes instructional practices used in six online literacy courses for preservice and inservice teacher candidates. Salika Lawrence describes instructor roles as including pedagogue, technologist, and evaluator, all in an online literacy program. Although the course designs were highly structured, the instructor incorporated multiple resources to support diverse learners, to foster independent learning, to promote critical thinking and reflection on how instructional strategies can be used in

K–12 classrooms, and to facilitate small group collaboration through authentic problem-solving tasks. This section concludes with a chapter by Bruce Taylor and Lindsay Yearta. In *Putting Multiliteracies into Practice in Teacher Education: Tools for Teaching and Learning in a Flat World*, the authors present a pedagogic framework or metastructure for applying Multiliteracies to teacher education and by extension to P-12 classrooms. Multiliteracies, a term coined by scholars who published under the name The New London Group in 1996, provides a theoretical foundation for applying new texts and tools to teaching and learning; however, much of the scholarship around Multiliteracies remains in the academic and theoretical domain. Taylor and Yearta recommend practical and integrative online literacy course tasks that promote authentic applications of technology-mediated multiliteracies.

Comparing Delivery Options in Teacher Education, the fifth section of the book, examines the various models of instruction that online learning platforms offer. In *(Re)Assessing Student Thinking in Online Threaded Discussions*, Felicia Saffold utilizes Bloom's Revised Taxonomy to determine the quality of thinking that occurred in the online discussion forum of an urban education course offered to preservice teachers. The author discusses findings that indicate the utilization of online discussion forums can be an effective pedagogy for classes on complex, often controversial issues such as social justice, equity, and white privilege. In the second chapter, Chris O'Brien, Shaqwana Freeman, John Beattie, LuAnn Jordan, and Richard Hartshorne investigate web-based instructional methods versus blended instructional methods for delivery in an introductory course in Special Education. In this chapter, *Investigation of Blended vs. Fully Web-Based Instruction for Pre-Teacher Candidates in a Large Section Special Education Survey Course*, the authors present data that were collected to evaluate the extent to which pre-teacher education students developed understanding of critical information related to human development factors, psychological, sociological, and policy foundations of teaching students with special needs. Student preferences in learning and the extent to which students developed comparable perceptions of preparedness for the future teaching roles is also discussed. The last chapter in this section is *Examining Student Behaviors in and Perceptions of Traditional Field-Based and Virtual Models of Early Field Experiences.* In this chapter, Hyo-Jeong So and Emily Hixon examine student behaviors in and perceptions of traditional field-based and virtual models of early field experiences in teacher preparation programs. The authors discuss various behaviors and student perspectives related to each model of early field experiences, suggesting that a virtual field experience that utilizes video-based cases may promote reflective practices, which could be especially beneficial to preservice teachers.

The sixth section of the book, **Innovative Online Teaching and Learning Practices in Teacher Education**, presents research-based ideas of new ways of integrating technology. Luke Rodesiler and Barbara Pace co-author the first chapter, titled, *Preparing Pre-Service English Language Arts Teachers to Support Literacy Learning with Interactive Online Technologies*. The authors of this chapter discuss a framework for integrating online learning opportunities into an English teacher education program. Efforts focused on extending pre-service secondary English language arts teachers' understandings of what constitutes literacy and what counts as text in the secondary English language arts classroom in a blended technology- and media literacy-focused methods course. Specifically, methods of encouraging pre-service teachers to consider the kinds of literacy events they might design and literacy practices they might promote to support literacy learning with interactive online technologies and popular media were addressed. Next, Drew Polly presents results and implications of a longitudinal study of a mathematics methods course for teacher candidates taught in both hybrid and 100% asynchronous online formats. In *Designing and Teaching an Online Elementary Mathematics Methods Course: Promises, Barriers, and Implications*, he expounds upon findings that participants valued the amount of support provided by the instructor and communication with classmates, had mixed comments about having to take ownership of their learning, and disliked the amount of work in the course. Lastly, Polly shares participant work sample experiences during their clinical projects as examples of online learning outcomes. The third chapter in this section, *Taking Action Research in Teacher Education Online: Exploring the Possibilities*, explores online tools that can facilitate distance action research coaching. In an effort to provide teachers with a vehicle to engage in this professional development, as well as raise teacher voices in educational reform, Nancy Dana, Desi Krell, and Rachel Wolkenhauer discuss action research in teacher education. As the systematic, intentional study by teachers of their own praxis, these authors describe how critical and powerful research-based professional development can be for practicing teachers. The authors extend the coaching of action research to online environments and examine the implications. Furthermore, the authors provide possibilities for negotiating challenges of time and space while enhancing both the quantity and quality of the teacher educator's action research coaching opportunities. The fourth chapter in the section, authored by Kara Dawson, Cathy Cavanaugh, and Albert Ritzhaupt, provides an introduction to ARTI, (Action Research for Technology Integration). ARTI is an online tool designed to support the merger of action research and technology integration. In *ARTI: An Online Tool to Support Teacher Action Research for Technology Integration*, the authors address the theoretical foundation of ARTI, the conceptual design of ARTI as a tool to provide: (1) an online scaffold for teachers to inquire about how their technology integration practices, (2) a mechanism to synthesize action research information from multiple teachers, and (3) a mechanism to capture evidence of student learning within technology integration inquiries. Lastly, examples of the implementation and implications of ARTI, along with future possibilities, are discussed. The last chapter in this section, *Improving Student Learning in a Fully Online Teacher Leadership Program: A Design-Based Approach*, describes the methods in which faculty in a fully online Master's program in teacher leadership used a design-based approach, grounded in theory and informed by data, to iteratively improve core courses and student learning from them. Specifically, the authors, Scott Day, Leonard, Karen Swan, Daniel Matthews, and Emily Boles, discuss revisions of courses to meet Quality Matters (QM) standards for online course design, as well as incremental and ongoing revisions related to course implementation and based on student responses to the Community of Inquiry (CoI) survey.

The final and seventh section in the book, **Moving Forward**, describes innovative practices that have the potential to chart new pathways for teacher education programs and in schools through online and Internet tools. The first chapter, *Puttering, Tinkering, Building, and Making: A Constructionist Approach to Online Instructional Simulation Games*, addresses instructional simulation games as models of the real world that allow students to interact with events and objects that are normally inaccessible within a classroom setting. Joe Feinberg, Audrey Schewe, Christopher Moore, and Kevin Wood examine the critical role of the teacher in determining the success or failure of modeling in a classroom environment, as well as the importance of teacher educators to effectively model how to implement instructional simulation games in the classroom. The authors conclude with a discussion of the role of constructionism and the benefits of incorporating modeling in teacher education. The second chapter, by David Gibson and titled, *Teacher Education with simSchool*, introduces an innovative online learning platform for the preparation of teachers. Through simulations, which address some of the systemic challenges of teacher education in the U.S., Gibson contrasts traditional course-based online learning experiences with a simulation approach to four areas of teacher preparation: conceptions of teaching and learning, the organization of knowledge, assessment practices and results, and the engagement of communities of practice. He also outlines a rationale for the new approach based on self-direction and personal validation in a complex but repeatable practice environment, supported by emergent interdisciplinary knowledge concerning the unique affordances of digital media assessment and social media. The final chapter in this section and in the book describes one of the projects mentioned earlier in the preface. Tina Heafner and Michelle Plaisance present Windows into Teaching and Learning (WiTL), a project that captures the nuisance of online learning as a method for transforming school-based clinical experiences in teacher preparation programs. *Windows into Teaching and Learning: Uncovering the Potential for Meaningful Remote Field Experiences in Distance Teacher Education* outlines the theoretical context in which the project was conceptualized and developed. The authors overview the challenges facing colleges of education in providing meaningful and relevant clinical experiences to pre-service teachers enrolled in online distance education courses as the impetus of WiTL. They share the potential impact of WiTL as a practical tool for facilitating purposefully integrative clinically-based online coursework.

CONCLUSION

In the spirit of the words of contributing authors Taylor and Yearta, "conclusions" seem more like "intermission" rather than an ending to ideas shared this book. This collection of chapters captures past, current, and evolving applications of online learning tools and practices in teacher education. The contributing authors present research to guide their understanding of the effectiveness and utility of course designs, program organization and the overall effectiveness of online instruction. They question the viability and utility of online and technology-based tools as well as the methods employed for preparing teachers. While there are mixed results in their studies and applications, they overwhelming conclude that online learning is here to stay. The issue that they consistently grapple with is how can teacher educators embrace the possibilities of technology to enhance the quality and effectiveness of preparing teachers for the rapidly changing world preservice teachers will encounter in 21st Century schools. The insights they share can positively impact future directions for online learning and serve to pave new pathways for learning in higher education. Therefore, the intermission begins with the final chapter in this book. As readers look forward and learn from these innovations, they can begin to shape the new teacher preparation classroom for the challenges that have yet to be discovered in a ubiquitous technological society. With the advent of new technologies, teacher preparation programs can be at the forefront of innovation and the research surrounding their work can create new, effective, and engaging learning spaces as well as define the role online learning will continue to play in 21st Century teacher education.

Tina L. Heafner
University of North Carolina at Charlotte, USA

Richard Hartshorne
University of Central Florida, USA

Teresa M. Petty
University of North Carolina at Charlotte, USA

Section 1
Issues and Trends in Online Education

Chapter 1
The Journey from Dissenter to Advocate:
Insights Gained while Teaching Online

Terry S. Atkinson
East Carolina University, USA

ABSTRACT

This chapter details the experiences of a university professor whose perspectives shifted from one of initial dissent to eventual advocacy for online learning as a delivery mode for her reading/literacy courses. Spanning eight years, her distance education teaching practices were shaped by her personal ventures as an online student, the outcomes gained by enhancing the social presence of her online courses, collaboration with colleagues, and systematic examination of her online teaching practice relative to its rigor, quality, and effectiveness within a teacher preparation program. Insights gained while teaching online conclude with recommendations for faculty members, institutions, systems, and organizations with vested interest in the future of teacher education.

INTRODUCTION

While teaching teacher education courses during the past 12 years at two different universities, the impact of technology on my own teaching and learning has gained a far greater reach than I ever imagined. As a fixed-term instructor at my alma mater, I taught a full load of face-to-face courses for two years after earning my doctorate at a teacher preparation institution where online instruction was non-existent and personal interaction with students was greatly valued. Spending time as both a graduate student and an instructor/supervisor of preservice teachers in this environment left an

DOI: 10.4018/978-1-4666-1906-7.ch001

imprint on me that has not faded. It was for this reason, which I never considered, nor imagined, teaching students via any mode other than face-to-face as I accepted a tenure-track teaching position at an institution almost twice the size of my alma mater. While I was initially impressed with the access to and support for technology in my new setting, I was unaware as I penned my name on my contract of its long-standing reputation for distance education delivery within rural areas of the state and with the military across the globe. My experience with technology integration and distance education would encounter a shift into fast forward and lead me through a transformation from dissenter to advocate of teaching and learning online.

In the following chapter, the details of this personal journey will unfold in terms of examining the literature related to the evolution of online instruction in higher education settings. This literature frames a plea to develop more numerous and effective online course offerings issued in 2005 by our incoming university system President, Erskine Bowles (Durham, 2005). He discussed the inevitable growth of online learning in higher education and challenged our faculty members to realize that if we did not offer such courses, for-profit institutions would. He punctuated his message by saying that surely *we* could do this better than *they*. His call, as well as my own experiences as a student, as an instructor, and as a volunteer in surrounding public schools, impacted me deeply and fueled my thirst for challenging myself as a technology learner, an academic, and an online instructor. Last, I offer some reflections and recommendations based on what I have learned during this journey and suggest future possibilities for how online teaching and learning can become more rigorous and rewarding for teacher educators and their students.

ONLINE LEARNING: FACTS, FIGURES, CHALLENGES

The literature related to online distance education has a somewhat brief history since 1981 when online courses were first offered to adult education students through the Electronic Information Exchange System (EIES) (Harasim, 2006). Developed and managed with funding from the National Science Foundation at the New Jersey Institute of Technology, EIES was originally created for use in scientific research communities. Specialized communication systems that evolved within EIES later led to the initial development of courses delivered through university-based computer networks to undergraduates at the New Jersey Institute of Technology. By 1985, the first graduate courses offered in a similar manner premiered at the University of Toronto and New York's New School of Social Research which led to degree programs at these same institutions one year later and at the University of Phoenix in 1989 (Hiltz, Turoff, & Harasim, 2007).

The launch of the Internet in 1989 paired with the advent of the World Wide Web in 1992 led to increased opportunities and global access for learners to further develop their understandings online. Few settings have been impacted more dramatically by this increase than American higher education (Moller, Foshay, & Huett, 2008). Enrollment in online classes at the college and university level has skyrocketed since the Internet's debut and has most recently been further boosted by competition among for-profit, private, and public institutions. Evidence from the Sloan-C 2010 report, *Class Differences: Online Education in the United States* (Allen & Seaman, 2010), substantiates this significant and ongoing growth in online instruction since its early beginnings, designated by courses where at least 80 percent of content is delivered online.

- During the fall 2009 term, almost 5.6 million students took at least one online course; an increase of more than one million students than reported the previous year.
- While the overall higher education student population reported less than 2 percent growth in one year, these increases reflect a 21 percent growth rate for online enrollments.
- Based on 2009 data, approximately 30 percent of all U.S. higher education students took at least one online course (p. 2).

Increased offerings of online courses within higher education have raised concerns within the academic community. Many university faculty members devalue online learning and are reluctant to adopt such course experiences (Allen & Seaman, 2007). While reasons for this resistance vary, concern about rigor and quality in online education has led to major programs such as Quality Matters, which certifies K-12 and higher education online course quality at more than 400 colleges and universities across the United States. The Quality Matters (QM) non-profit initiative is an outcome of MarylandOnline's efforts to promote and improve the quality of online education using a faculty-centered, peer review process (MarylandOnline, 2010). Despite the fact that studies including a U. S. Department of Education (2009) meta-analysis and review of empirical online learning research have identified multiple favorable outcomes of online learning, some researchers note concern about specific shortcomings of online instructional delivery. Particular to the field of teacher education, Duffy, Webb, and Davis (2009) argue that literacy education is at a crossroads, impacted by factors that can marginalize teacher education's ability to develop professional teachers. They define such teachers as those who should be informed pedagogical decision-makers who can "diagnose student strengths and needs," "respond appropriately to student misunderstandings during lessons," "design problem-solving tasks in literacy," "decide how to re-teach when things go wrong," "adapt instruction for students whose background and language may be quite different from the teachers'," and "develop excited, engaged and inspired readers" (p. 190). Duffy and colleagues then point to specific factors that impede teachers from developing these professional abilities, one of which is online distance learning in reading or writing methods courses. They further elaborate that while online learning might be appropriate for learning declarative or procedural knowledge, virtual course delivery cannot adequately focus on conditional knowledge allowing students to think their way through the whys, whens, and under what conditions typical of challenging teaching situations. In sum, Duffy, Webb, and Davis challenge the notion that online distance education courses provide "the intellectual context necessary to foster an informed, reasoned, and passionate corps of teachers" (p. 192).

SHIFTING PERCEPTIONS OF ONLINE TEACHING ACROSS TIME

Continuing demand for online course delivery is a reality in today's higher education settings, including the large public university in which I currently teach. A dabbler in technology by nature, I had integrated multiple technology tools in my past teacher preparation courses. But, as a nascent academic, the idea that online delivery of my clinical reading diagnosis graduate courses could equal the rigor of learning and application attained in face-to-face delivery was beyond my comprehension. Such change would be challenging and came to involve more than four years of reflecting, soul-searching, and personal learning. A variety of learning, teaching, and service experiences served as important initial catalysts, leading to resulting actions and outcomes that have shaped me into the advocate for online teaching and learning that I am today.

Initial Catalysts

While the numbers of university students learning online increase yearly as documented by annual Sloan-C reports (Allen & Seaman, 2010), it was not this obvious growing presence of online education that altered my perceptions sufficiently to redesign my graduate level courses to be taught online. Actual experiences with my own students and in surrounding K-12 schools served as catalysts for this transformation. While earlier teaching at my alma mater in a large, cosmopolitan setting had involved work in high-need public schools, I was totally unprepared for the differences that I would encounter after accepting a tenure-track position ten years ago in a rural area of the same state.

My initial semester's teaching assignment included a face-to-face graduate level clinical reading diagnosis course that met one night a week for a three-hour block. What I was surprised to find was that it was to be delivered to a cohort of students at a school site more than two hours away. In my new role and across unknown territory, I typically drove my Mondays away from noon until nine to meet 20 voracious teachers who soaked up my every word. Leaving them was often a challenge as they stayed afterward, seeking help and guidance. While I began the semester with a plan to "take on the world," I ended it with the stark reality that driving hours away to deliver reading instruction to rural teachers was dreadfully inadequate in terms of the large scope of need in the counties surrounding our university. As I began my next semester, I reframed my efforts to polish and perfect this course by focusing on a more attainable goal, situating my graduate course within in local school setting.

With more than 15 years of experience as a classroom teacher, I was committed to maintaining a presence in K-12 schools while teaching at the university level. During the following two years, I scheduled my graduate level clinical diagnosis reading courses in local elementary schools where struggling students were recruited to receive free after school reading support. The challenges were many as my graduate students sought to establish relationships with students who were often unmotivated, repeatedly absent, and had little investment in an after-hours academic experience. These face-to-face courses were taught in two different schools, based on the teaching settings of my graduate students, most of whom were practicing teachers. I found that those who gained the most from the courses and their tutoring sessions supported students who were members of their own classrooms. While coming to this conclusion, I was also investing a significant amount of time in three surrounding high-poverty counties that had contacted our university seeking literacy support for their teachers. I observed and taught in some classrooms while I met with teachers, students, and administrators in an effort to better understand each school's culture. Outdated reading and literacy instruction that did little to meet the needs of student populations was common. In each of these settings, hard-working teachers with few resources shared their dreams of earning graduate degrees in Reading Education. Their hopes were dashed by the distance between our campuses and their responsibilities to jobs, families, and communities that made traveling for evening classes hugely challenging and often impossible. I vividly remember returning from all three of these counties with the thought that our university must certainly take graduate level reading instruction to these teachers, not through random professors driving miles to meet students in remote locations, but through other innovative delivery means. Teaching my course online could eliminate this problematic distance and ultimately allow many more teachers to learn about and apply the clinical diagnosis of reading difficulties with students in their own classrooms. How could I not attempt to deliver an online version of my face-to-face course, despite the fact that I had no idea about how to begin?

Resulting Actions

As any teacher might take on such a task, my quest to become an online instructor began with looking to others and learning all that I could. I reviewed online courses offered by other instructors in my department whose work I admired and spent the summer prior to my fourth year of university teaching engaged in course management system tutorials. As my first attempt at online teaching took flight, I quickly determined that the course content must involve far more than simply attempting to transcribe what might be said in a face-to face course into well-written narratives that could be accessed online. After describing the actions that have impacted my instructional practice such that I now promote teaching students from a distance, resulting outcomes, recommendations, and future research directions will be summarized in following sections.

Becoming an Online Student

Lack of satisfaction with my initial virtual course delivery led to one of the best moves I ever made in my journey to learn more about teaching online; I located an online teaching expert, who at that time served as an assistant professor in my university's Educational Technology program, and made an appeal for help. Dr. Susan Colaric, who now serves as Director of Instructional Technology at Florida's Saint Leo University, willingly took me under her wing and in addition to meeting with me individually, walked me through her online course titled, Instructional Strategies for Distance Education. Until this time, I never imagined that an online course even existed that could guide student learners through activities and experiences that were exemplary in online courses. Dr. Colaric offered me access to her online course and I studied it carefully, sending back questions via email and wanting to know more. Soon afterward, she contacted me to ask if I would like to enroll as a student in this same online course during the coming semester, which she planned to offer exclusively to interested faculty members. She sought to allow faculty to experience learning at a distance from the perspective of an online student, to read recommended articles and text excerpts, and to examine recommended online teaching practices within discussion forum conversations. I could not wait to get started.

As the semester progressed, I enriched my own knowledge, refined my current online course, and made plans for more extensive changes in my upcoming fall courses. In addition to learning about what online learning entailed according to experts in the field (Palloff & Pratt, 2001) and as defined by the Southern Association of Colleges and Schools (2000), course module content focused on instructional design, web searching and evaluation, online discussions, technology tools for teaching and learning, and online collaboration. As we learned, we applied, often through new eyes as we discussed possibilities and worked collaboratively with course colleagues who served as faculty members within many disciplines across our campus. This iterative professional work and dialogue was of immeasurable benefit and helped me begin to realize the power of online learning tools among faculty members on the same campus who otherwise might not have connected and interacted.

This initial online learning experience led me to enroll in additional instructional technology courses several years later and eventually add-on instructional technology to my current teaching license. The more I learned about instructional design and technology integration from excellent online instructors and related reading (Reiser & Dempsey, 2007), the more opportunities I had to experience what worked for me as a learner, gain perspectives from a learning community comprised of dozens of course colleagues in other parts of the state and country, and apply what I learned within the online courses that I was teaching.

Focusing on Social Presence

Remaining true to my belief that personal interaction with my students was paramount in terms of my teaching effectiveness, I searched for online teaching experts who shared this concern. Making online teacher-student and student-student interactions more like those occurring in a face-to-face setting led me to the work of Steven R. Aragon, a human resource professor at the University of Illinois at Urbana-Champaign. Aragon argues that establishing social presence is a key factor in building a sense of community among learners at a distance (2003). With roots in the communications literature, social presence is one component of the Community of Inquiry model (Garrison, Anderson, & Archer, 2000) involving the communication factors that make interpersonal relationships salient, intimate, and immediate (Gunawardena & Zittle, 1997; Short, Williams, & Christie, 1976). Aragon's specific social presence practices reinforced many of my beliefs about the need to know my students well and work with and among them. His work was influential in making a shift to using Moodle as a course management system. Developed for use in K-12 school settings, Moodle's grouping and communication features foster social presence by displaying students' pictures with all teacher-student and student-student communication and through optional automatic email notification about news items, grade postings, and group discussion board postings. As Aragon suggests, while not difficult to implement, establishing and maintaining social presence in an online course requires deliberate, conscious, and ongoing effort. Revisiting Aragon's work in an ongoing way continues to have a profound impact on my day-to-day online teaching. Due to their relevance for other teacher educators, Aragon's recommendations for bolstering social presence are summarized in Table 1.

Collaborating with Colleagues

As I focused on integrating all that I was learning about how to teach and learn more effectively at a distance, my confidence as an online instructor continued to grow in tandem with my competence. Indeed, taking what I learned as an online student with faculty members across our campus and applying what I gained from examining Aragon's work led not only to greater satisfaction as an instructor, but also to positive feedback from students. Taking the risk to collaborate with fellow faculty members while teaching online seemed a natural next step. Thus, one colleague and I agreed to spend a summer session mentoring, supporting, and challenging one another while teaching different sections of the same content area reading course online. This particular course offered specific teaching challenges as students from various content area disciplines (dance, theater, math, science, physical education) enrolled as a requirement for gaining teaching licensure. Past difficulty had been noted with establishing common ground among students with such a wide array of backgrounds. My colleague and I sought to overcome this problem by addressing course objectives around redesigned collaborative group work related to the reading of young adult novels. Students chose from multiple novels selected for their high potential for content area integration. They engaged in virtual conversations about their novels as they learned about content area reading practices and planned related student learning experiences. The additional communal bond that all group members shared was further strengthened by another course element. Our state's Department of Public Instruction had charted graduate faculty with the task of "revisioning" our graduate courses to align with the 21st Century Skills P-21 Framework (2004). As we continued to redesign and improve this online course during two following years, my colleague, our students, and I shared the experience of learning and sharing

Table 1. Factors contributing to social presence in online courses

Course Design
Create and send initial student welcome messages
Include student profiles meant to share personal and/or professional information
Integrate teacher-student and student-student audio messages
Limit enrollment to fewer than 30 course participants
Include opportunities for students to learn collaboratively
Instructor Participation
Contribute to online discussions
Answer questions sent via email or voicemail promptly
Provide substantive and frequent student feedback, particularly personalized comments sent to individual students
Initiate and participate in conversations with students about personal or professional topics of interest
Share appropriate personal experiences or stories with students
Include tasteful humor whenever possible
Embed emoticons as non-verbal cues
Address students by their preferred names
Provide opportunities for students to interact with their instructor
Student Participation (facilitated by clear expectations, instructor modeling, and accountability, as appropriate)
Contribute to online discussions
Respond promptly to email questions from colleagues and instructor
Initiate and participate in conversations with fellow students and/or course instructor about personal or professional topics of interest
Share appropriate personal experiences or stories with student colleagues and/or course instructor
Include tasteful humor whenever possible
Embed emoticons as non-verbal cues
Address students and course instructor by appropriate titles

From Aragon, S.R. (2003). Creating social presence in online environments. *New Directions for Adult and Continuing Education, 100, 57-68.*

through new technology tools. In addition to offering e-reader selections as alternatives to hard copy novels and communicating learning insights through Glogs (http://edu.glogster.com/) and VoiceThread (http://voicethread.com/products/k12/educator/), students built group WIKIs to create novel-related vocabulary and questioning activities, as well as multimodal companion text suggestions.

Throughout this online teaching collaboration, my colleague and I learned with and among our students, embracing the 21st century "New Literacies" mindset that what it means to be literate constantly changes as related technology transforms and emerges (Lankshear & Knobel, 2003; Leu, 2000). Enrolled as co-teachers in each other's online courses, we shared students' comments, learning outcomes, and addressed concerns collaboratively. Learning about and integrating technology tools with which we had little prior expertise or experience, strengthened our bond as colleagues and with all course participants, creating a learning community that promoted risk-taking and often encouraged students to share alternative technology tools or approaches with which they were familiar. This sort of "give and take" among all course participants encouraged my colleague and me to extend our understandings, both as technology learners and online facilitators, in ways that we had not previously experienced. In our opinion, dovetailing extensive technology integration as we did within this online course allowed us to create an optimal shared learning environment, reframing teachers' traditional consideration of classroom as "MySpace," to that of "OurSpace" (McClay & Mackey, 2009).

This online teaching collaboration was soon extended to multiple reading/literacy colleagues in our department who taught the same content area reading course. Four additional colleagues eventually joined in virtual sharing while teaching the same online course and this fact was made completely transparent to our students. We all agreed that benefits of working closely together while teaching the same online course have multiple positive outcomes in terms of collective problem-solving and sharing of student learning. Additionally, these benefits are of particular relevance in the field of teacher education as we sought to be an example for our students. Teaching has a long-standing reputation of isolationism, often cited as a contributing factor in teacher frustration, dissatisfaction, or in decisions to leave the profession (Carroll & Fulton, 2004). These feelings of solitude can be exacerbated by teaching online. Our efforts to support and coach one another connected us with members of a professional mentoring community who met both face-to-face and online (and has since led to a collaborative teaching model used for all of our online reading courses). As we collectively strengthened our own pedagogy, we served as models for our students, all of whom were veteran or preservice teachers. This notion of modeling by example was intentional, bolstered by studies documenting that peer coaching, mentoring, and professional learning communities combat or overcome teacher isolation and ultimately lead to better teaching (Darling-Hammond, 2003; Heider, 2005; NCTAF, 2002).

Studying Online Teaching Practice

The virtual mentorship model described in the previous section flourished as more and more reading faculty members began teaching online. One faculty member served as mentor for all online reading courses taught during a particular semester, involving tenure-track, fixed-term lecturer, and adjunct faculty members, some of whom taught in local public schools as literacy facilitators or at universities across the country. With multiple cohorts of students enrolling in our reading graduate program, I found my role as mentor for our reading diagnosis course grow from working with one colleague, to working with a second, a former doctoral colleague from my alma mater, who now teaches at a university in Washington state. The three of us knew one another through collaborations at a yearly literacy research conference and it was a face-to-face meeting at one of these events that led to a multi-year study examining the efficacy of our online pedagogy.

My colleagues and I had followed the work of Gerald Duffy closely during past years as he served as an Endowed Scholar at my alma mater. While we admired and valued his career-long contributions to the field of teacher effectiveness and its implications for reading/literacy (Duffy, 1993; Duffy, 1994; Duffy, 1998; Duffy, 2002) we were struck by his vocal disdain for online courses and had lengthy discussions with him about the subject at these yearly research meetings. Thus, when he published a book chapter in *Changing Literacies for Changing Times* (Duffy, Webb, & Davis, 2009), citing online instruction as a major concern in marginalizing teacher education's future potential to develop professional teachers, we planned a study to challenge this notion. As mentioned earlier in this chapter, Duffy, Webb, and Davis argue that while online instruction might be an appropriate format for delivering declarative or procedural knowledge, it lacks the ability to build the kinds of conditional knowledge needed by professional teachers. The presence of conditional knowledge is evidenced by what Duffy and colleagues define as "thoughtful adaptation," a teacher's ability to adjust the complexities of literacy instruction both in planning and "on the fly" in the classroom. Moreover, thoughtfully adaptive teachers make iterative decisions based on the needs of their students and the hows, whys, and whens of using particular instructional methods (Duffy, Miller, Kear, Parsons, Davis, & Williams, 2008).

Duffy's questions and concerns led to lengthy discussions about our clinical diagnosis course and online instructional practices. We designed a related study to specifically examine instructional tasks, learning outcomes, and follow-up interview responses of our 60 students for evidence of conditional knowledge and thoughtful adaptations. At the end of the following semester, study findings documented that students did learn to be adaptive in this online clinical diagnosis course and that the factors contributing most significantly to this outcome included the inherent qualities of the learning tasks involved, as well as the ongoing and iterative feedback that was provided by the instructors. Submission of assessment data and a proposed plan for initial literacy support of a small group of classroom students was discussed with each course participant in order to launch instruction. As subsequent lesson reflections were submitted, instructor-course participant conversations continued about plans and continued literacy instruction that were adjusted based on student success and understandings. Thus, the nature of the instruction, rather than the mode of delivery, determined what course participants learned, understood, and put into practice with the students they assessed and supported during the semester (Parsons, Massey, Vaughn, Scales, Faircloth, Howerton, Atkinson, & Griffith, 2011). While the particulars of data collection and analysis were of great value in documenting these findings, the overarching notion of aligning instruction with a focus on conditional knowledge had a profound impact on my thinking about online teaching and learning, particularly as a teacher educator. Indeed, my dilemmas about how to best frame online learning experiences for veteran or preservice teachers continue to be resolved through this lens. Students must be using what they learn to think their way through the whys, whens, and under what conditions typical of challenging teaching situations. If not challenged to extend their understandings to this degree, their potential to develop fully as teaching professionals is being shortchanged. Based on my experience, consideration of this maxim holds promise for all teacher educators, regardless of the delivery format they employ.

CONTINUING THE JOURNEY: WHAT POSSIBILITIES LIE AHEAD?

As I begin my eighth year of teaching online, I reflect on navigating the challenges inherent in teaching at a distance realizing that they could have led to a far different outcome. Facing the ubiquity of online teaching both nationally (Allen and Seaman, 2010) and at my own institution, I was indeed fortunate to teach on a campus that offered instructional technology courses where I not only gained understandings about best distance education teaching practice and discussed them with faculty colleagues, but also experienced learning from the perspective of an online student. Moreover, teaching with and among colleagues who were and continue to be interested in improving their teaching practice led to the highly beneficial online mentorship collaboration and clinical diagnosis course study described in previous sections. Gaining insights and improving practice within rich professional learning communities had a deep impact on my evolution as an online instructor and bears recommendation for either nascent instructors or those involved in current teaching who wish to examine and improve their effectiveness (Darling-Hammond, 2003; Heider, 2005; NCTAF, 2002). However, this retrospect describes my voluntary and self-initiated participation in these opportunities. As with the discovery and implementation of Steven R. Aragon's work related to social presence (2003), I happened upon his writing in an effort to foster more authentic teacher-student and student-student interactions in my online courses. Learning about appropriate and productive online teaching practices should not occur through happenstance. This is especially true in the field of teacher education where the development of conditional knowledge should frame most, if not all, declarative and procedural learning experiences (Duffy, Webb, & Davis, 2009).

So how might teacher education institutions make a more deliberate effort to support online instructors and quell the concerns of those who fear that online teaching involves little more than dispensing information? In the midst of strict budgetary times, it is most feasible to begin with what might happen within these institutions themselves or among pairs or small groups of institutions. If guidelines for online instruction or support workshops do not exist, groups of interested and hopefully experienced online faculty might form a learning community and seek to develop such tools. In addition to studying their own experiences in teaching from a distance, relevant readings selected from this volume or from other sources pertinent to teacher education might be examined and discussed in order to identify common findings, conclusions, and recommendations. Guideline drafts would benefit from comparison with rubrics designed to assess online course quality in higher education and K-12 education through initiatives such as Quality Matters (MarylandOnline, 2010) and the International Association for K-12 Online Learning (North American Council for Online Learning, 2011). Additionally, connections should be made in such online study ventures with instructional or educational technology faculty within or across institutions, who can share their expertise and relate it to multiple content areas within teacher education programs.

Beyond the efforts made by individual teacher education institutions to bolster the quality and rigor of distance education courses, larger entities should focus their energies on creating system-wide or organizational resources offering guidance to achieve this goal. Within the state of North Carolina, the 15 public teacher education institutions within the University of North Carolina General Administration (UNCGA) system are uniquely poised to benefit from online instructional support through The University of North Carolina-ONLINE portal (University of North Carolina System, 2007). This portal, which "offers comprehensive descriptions of and contact, application, admission, and tuition and fee information for more than 240 online programs in 22 fields of study offered by the 16 constituent universities of one of the world's most prestigious university systems" (University of North Carolina System, 2007, p. 1) to current and prospective University of North Carolina students, could potentially connect online instructors of teacher education courses with quality improvement initiatives or resources. On an even larger scale, launching discussions or ventures related to improving online teacher education quality would be noble ventures for organizations such as the American Association of Colleges for Teacher Education (AACTE) who have the means and infrastructure to impact multiple institutions across the country.

CONCLUDING THOUGHTS: ADVOCACY WITH A CAVEAT

Charting a successful journey across the unknown terrain of online instruction during the past eight years has involved significant challenges and a time commitment that I could have never anticipated. But as higher education continues to be transformed by the impact and explosive growth of distance education (Allen & Seaman, 2010; Moller, Foshay, & Huett, 2008), so are teacher education institutions. The need to reach greater numbers of students who are either non-traditional or who cannot travel to a university campus is essential within the field of teacher preparation. This is especially true in the rural locales surrounding my university where attracting quality teachers is an ongoing challenge. Pleas from teachers during my first semesters of work in surrounding rural schools served as a catalyst, convincing me that the need for online course offerings outweighed my ability to know how to deliver such instruction effectively. Thus, I initiated subsequent actions that enriched my understandings about planning and facilitating meaningful online learning experiences for and with my students.

Thus, as I conclude the story of my journey from dissenter to advocate, I must qualify my advocacy of distance learning with multiple caveats that may serve as recommendations for interested teacher educators. My promotion of online teacher education instruction rests on the inclusion of particular learning conditions. Learning outcomes must clearly connect with conditional knowledge and evidence of social presence should support and encourage productive communication among all stakeholders. Further, course facilitators must understand and empathize with the virtual challenges of their students. This perspective can be fostered by making an ongoing and conscious effort to assume a "New Literacies" mindset (Lankshear & Knobel, 2003; Leu, 2000) and embracing the ongoing changes inherent in a 21st century environment. Learning with, among, and from one's students must become the norm of teacher educators who understand that the nature of one's instruction determines the quality of learning outcomes, within online teaching environments and beyond (Parsons, Massey, Vaughn, Scales, Faircloth, Howerton, Atkinson, & Griffith, 2011).

A final caveat must be offered if quality distance education is expected to buoy teacher education institutions in a political and media landscape fraught with scrutiny, criticism, and competition from for-profit entities (Allington, 2002; Darling-Hammond & Bransford, 2005). Rather than relegating the outcomes of online instruction to faculty members who depend on their self-ascribed efforts to plan and implement effective courses, teacher education institutions must build in options for identifying, supporting, and perfecting exemplary distance education. In doing so, specifics must qualify what makes online learning experiences relevant and effective for veteran and preservice teachers and different from online learning for the populace in general. Such a focus offers a call for emerging and future research. Coordinating efforts among engaged faculty members, post-secondary institutions, university systems, and professional organizations offers the potential to put teacher educators in the position of determining how and if online teaching and learning practices can move us forward as the 21st century unfolds.

REFERENCES

Allen, E., & Seaman, J. (2011). *Class differences: Online education in the United States, 2010*. Newburyport, MA: Sloan Consortium (Sloan-C).

Allington, R. (2002). *Big brother and the national reading curriculum: How ideology trumped evidence*. Portsmouth, NH: Heinemann.

Aragon, S. R. (2003). Creating social presence in online environments. *New Directions for Adult and Continuing Education*, *100*, 57–68. doi:10.1002/ace.119

Carroll, T., & Fulton, K. (2004). The true cost of teacher turnover. *Threshold,* Spring, 16-17. Retrieved August 14, 2011, from http://www.nctaf.org/resources/news/nctaf_in_the_news/

Darling-Hammond, L. (2003). Keeping good teachers. *Educational Leadership*, *60*(8), 6–13.

Darling-Hammond, L., & Bransford, J. (2005). *Preparing teachers for a changing world: What teachers should learn and be able to do*. San Francisco, CA: Jossey-Bass.

Duffy, G. (1993). Teachers' progress toward becoming expert strategy teachers. *The Elementary School Journal*, *94*(2), 109–120. doi:10.1086/461754

Duffy, G. (1994). How teachers think of themselves: A key to mindfulness. In Mangieri, J., & Collins-Block, C. (Eds.), *Creating powerful thinking in teachers and students: Diverse perspectives* (pp. 3–26). Fort Worth, TX: Holt, Rinehart & Winston.

Duffy, G. (1998). Teaching and the balancing of round stones. *Phi Delta Kappan*, *79*, 777–780.

Duffy, G. (2002). Visioning and the development of outstanding teachers. *Reading Research and Instruction*, *41*, 331–344. doi:10.1080/19388070209558375

Duffy, G. G., Miller, S. D., Kear, K. A., Parsons, S. A., Davis, S. G., & Williams, J. B. (2008). Teachers' instructional adaptations during literacy instruction. In Y. Kim, V. J. Risko, D.L. Compton, D. K. Dickinson, M. K., Hundley, R. T. Jimenez, K. M. Leander, & D. W. Rowe (Eds.), *57th yearbook of the National Reading Conference* (pp. 160-171). Oak Creek, WI: National Reading Conference.

Duffy, G. G., Webb, S., & Davis, S. G. (2009). Literacy education at a crossroad: A strategy for countering the trend to marginalize quality teacher education. In J. V. Hoffman & Y. Goodman (Eds.), *Changing literacies for changing times* (pp. 189-197). New York, NY: Routledge, Taylor and Francis.

Durham, J. (2005). Bowles plans to help ECU meet goals. *Pieces of Eight*. Retrieved August 17, 2011, from http://www.ecu.edu/cs-admin/news/poe/0015/bowles.cfm

Garrison, D. R., Anderson, T., & Archer, W. (2000). Critical inquiry in a text-based environment: Computer conferencing in higher education. *The Internet and Higher Education*, *2*(2-3), 87–105. doi:10.1016/S1096-7516(00)00016-6

Gunawardena, C. N., & Zittle, F. J. (1997). Social presence as a predictor of satisfaction within a computer-mediated conferencing environment. *American Journal of Distance Education*, *11*(3), 8–26. doi:10.1080/08923649709526970

Harasim, L. (2006). A history of e-learning: Shift happened. In Weiss, J., Nolan, J., & Trifonas, P. (Eds.), *International handbook of virtual learning environments* (pp. 25–60). Dordrecht, The Netherlands: AA Dordrecht. doi:10.1007/978-1-4020-3803-7_2

Heider, K. L. (2005). Teacher isolation: How mentoring programs can help. *Current Issues in Education*, *8*(14). Retrieved August 18, 2011, from http://cie.ed.asu.edu/volume8/number14/

Hiltz, S. R., Turoff, M., & Harasim, L. (2007). Development and philosophy of the field of asynchronous learning networks. In Andrews, R., & Haythornthwaite, C. (Eds.), *The sage handbook of e-learning research* (pp. 55–72). London, UK: Sage, Ltd.

Lankshear, C., & Knobel, M. (2003). *New literacies: Changing knowledge and classroom learning*. Buckingham, UK: Open University Press.

Leu, D. J. Jr. (2000). Literacy and technology: Deictic consequences for literacy education in an information age. In Kamil, M. L., Mosenthal, P. B., Pearson, P. D., & Barr, R. (Eds.), *Handbook of reading research* (*Vol. 3*, pp. 743–770). Mahwah, NJ: Erlbaum.

MarylandOnline. (2010). *Quality matters (QM)*. Retrieved August 16, 2011, from http://www.qmprogram.org

McClay, J. M., & Mackey, M. (2009). Distributed assessment in OurSpace: This is not a rubric. In Burke, A., & Hammett, R. F. (Eds.), *Assessing new literacies: Perspectives from the classroom*. New York, NY: Peter Lang.

Moller, L., Foshay, W., & Huett, J. (2008). The evolution of distance education: Implications for instructional design on the potential of the Web. *TechTrends*, *52*(4), 66–70. doi:10.1007/s11528-008-0179-0

National Commission on Teaching and America's Future. (2002). *Unraveling the "teacher shortage" problem: Teacher retention is the key*. Washington, DC: National Commission on Teaching and America's Future.

North American Council for Online Learning. (2011). *iNACOL national standards of quality for online courses*. Retrieved August 8, 2011, from http://www.inacol.org/research/nationalstandards/index.php

Palloff, R. M., & Pratt, K. (2001). *Lessons from the cyberspace classroom: The realities of online teaching*. San Francisco, CA: Jossey-Bass.

Parsons, S. A., Massey, D., Vaughn, M., Scales, R. Q., Faircloth, B. S., & Howerton, S. (2011). Developing teachers' reflective thinking and adaptability in graduate courses. *Journal of School Connections*, *3*(1), 91–111.

Partnership for 21st Century Skills. (2004). *Framework for 21st century learning*. Retrieved August 2, 2011, from http://www.p21.org/index.php?option=com_content&task=view&id=254&Itemid=120

Reiser, R. A., & Dempsey, J. V. (2007). *Trends and issues in instructional design* (2nd ed.). Upper Saddle River, NJ: Pearson Education, Inc.

Short, J. E., Williams, E., & Christie, B. (1976). *The social psychology of telecommunications*. New York, NY: Wiley.

Southern Association of Colleges and Schools. (2000). *Best practices for electronically offered degree and certificate programs*. Retrieved March 15, 2005 from http://www.sacscoc.org/pdf/commadap.pdf

U. S. Department of Education. (2009). *Evaluation of evidence-based practices in online learning: A meta-analysis and review of online learning studies*. Retrieved August 10, 2011, from http://www.ed.gov/about/offices/list/opepd/ppss/reports.html

University of North Carolina System. (2007). *The University of North Carolina online*. Retrieved August 15, 2011, from http://online.northcarolina.edu/

ADDITIONAL READING

Allen, E., & Seaman, J. (2005). *Online nation: Five years of growth in online learning*. Needham, MA: Sloan Consortium (Sloan-C).

Allen, E., & Seaman, J. (2008). *Staying the course: Online education in the United States, 2008*. Needham, MA: Sloan Consortium (Sloan-C).

Allen, E., & Seaman, J. (2010). *Learning on demand: Online education in the United States, 2009*. Needham, MA: Sloan Consortium (Sloan-C).

Bender, T. (2003). *Discussion-based online teaching to enhance student learning*. Sterling, VA: Stylus Publishing.

Conrad, R. M., & Donaldson, J. A. (2004). *Engaging the online learner: Activities and resources for creative instruction*. San Francisco, CA: Jossey-Bass.

Fink, L. D. (2003). *Creating significant learning experiences: An integrated approach to designing college courses*. San Francisco, CA: Jossey-Bass.

Garrison, D. R., & Anderson, T. (2003). *E-learning in the 21st century: A framework for research and practice*. London, UK: RoutledgeFalmer. doi:10.4324/9780203166093

Hanna, D. E., Glowacki-Dudka, M., & Conceicai-Runlee, S. (2000). *147 practical tips for teaching online groups: Essentials of web-based education*. Madison, WI: Atwood Publishing.

National Center for Education Statistics. (2008). *Distance education at degree-granting postsecondary institutions*. Retrieved August 9, 2011, from http://www.inacol.org/research/research.php

North American Council for Online Learning. (2011). *iNACOL national standards for quality online programs*. Retrieved August 8, 2011, from http://www.inacol.org/research/nationalstandards/index.php

North American Council for Online Learning. (2011). *iNACOL national standards for quality online teaching*. Retrieved August 8, 2011, from http://www.inacol.org/research/nationalstandards/index.php

North American Council for Online Learning. (2011). *Promising practices in online learning*. Retrieved August 8, 2011, from http://www.inacol.org/research/promisingpractices/index.php

Palloff, R. M. (2003). *The virtual student: A profile and guide to working with online learners*. San Francisco, CA: Jossey-Bass.

Palloff, R. M., & Pratt, K. (2007). *Building online learning communities: Effective strategies from the virtual classroom* (2nd ed.). San Francisco, CA: Jossey-Bass.

Renes, S. L., & Strange, A. T. (2011). Using technology to enhance higher education. *Innovative Higher Education*, *36*, 203–213. doi:10.1007/s10755-010-9167-3

Rovai, A. (2002). Building sense of community at a distance. *International Review of Research in Open and Distance Learning*, *3*(1), 1–16.

Salmon, G. (2003). *E-tivities: The key to active online learning*. London, UK: Kogan Page Limited.

Southern Regional Education Board. (2009). *Guidelines for professional development of online teachers*. Retrieved August 8, 2011, from http://www.inacol.org/research/reports.php

Wenger, E. (1998). *Communities of practice: Learning, meaning, and identity*. Cambridge, UK: Cambridge University Press.

Wenger, E., McDermott, R., & Snyder, W. M. (2002). *Cultivating communities of practice: A guide to managing knowledge*. Boston, MA: Harvard University Press.

Young, S., & Bruce, M. A. (2011). Classroom community and student engagement in online courses. *Journal of Online Teaching and Learning*, *7*(2), 219–230.

KEY TERMS AND DEFINITIONS

Conditional Knowledge: Grounded in the notion of metacognition, conditional knowledge moves beyond facts and skills to consider the whens, whys, and under what conditions evident in authentic application.

Declarative Knowledge: Factual knowledge or information.

Face-To-Face Instruction: Learning experiences that take place in a physical location during regular meetings between students and instructor.

Learning Community: A collective group of individuals sharing the common goal of enhancing their understandings about a topic or idea of mutual interest.

Online Instruction: Learning experiences characterized by the fact that at least 80% of instructional delivery takes place online, typically though a course management system (for the purposes of this chapter, the terms *online instruction, online learning, distance instruction, distance learning*, and *learning at a distance* are used synonymously).

Procedural Knowledge: Knowledge or information related to how to perform or operate.

Social Presence: As defined in the communication literature by Gunawardena & Zittle (1997) and Short, Williams, & Christie (1976) social presence involves the communication factors that make interpersonal relationships salient, intimate, and immediate.

Teacher Education Institutions: Postsecondary colleges and universities whose enterprise involves the preparation, degree-granting, and licensure of professional teachers, typically for K-12 classrooms.

Thoughtful Adaptations: A teacher's ability to adjust the complexities of literacy instruction in response to student understanding and outcomes, both in planning and "on the fly" in the classroom.

Chapter 2
Remote Observation of Graduate Interns
A Look at the Process Four Years Later

Teresa M. Petty
University of North Carolina at Charlotte, USA

Tina L. Heafner
University of North Carolina at Charlotte, USA

Richard Hartshorne
University of Central Florida, USA

ABSTRACT

The Remote Observation of Graduate Interns (ROGI) is a method crafted by researchers at the University of North Carolina at Charlotte that allows graduate interns completing their student teaching experience to be observed remotely. Initially developed as a teacher shortage solution, ROGI remains an active method of observing interns geographically removed from the university through a virtual, synchronous format. Since its inception, ROGI has progressed as a technological tool, and college policies have evolved to adopt its utility. Authors describe the components of ROGI, its implementation, and ways in which the process has changed over the first four years of use. They present research to articulate how technology-mediated processes introduced new ways of thinking about traditional approaches to teacher education and new challenges that accompanied this innovation. Authors conclude with recommendations for future research and how other researchers might embrace the potential of emerging technologies in preparing teacher educators.

DOI: 10.4018/978-1-4666-1906-7.ch002

INTRODUCTION

In 2006, the state of North Carolina, like many other states across the U.S., was faced with a critical teacher shortage. This shortage was especially prevalent in high-need content areas across the state, such as mathematics, science, foreign language, middle grades, and special education. Certain geographic regions in the state also faced significant teacher shortages across the board, not just in the content areas that were considered high-need in other areas of the state. In response to this growing need to recruit more high quality teachers, the University of North Carolina at Charlotte's Department of Middle, Secondary, and K-12 Education, decided to open its teacher licensure program to anyone in the state pursuing a teaching license. While online courses had previously been utilized in teacher licensure programs prior to the decision to expand licensure opportunities, the program was limited to lateral entry teachers only. Lateral entry teachers are practicing teachers who have already achieved a Bachelor's degree and are concurrently teaching and taking courses to obtain a teaching license.

After careful consideration, the decision was made to offer a 100% online Graduate Certificate (licensure) program in middle grades and secondary education, addressing the teacher shortage in certain content areas while also providing outreach to areas of the state where teacher shortage was a growing concern. However, one logical barrier remained: how to facilitate a Graduate Student Teaching Internship online. The response was the creation of ROGI (the Remote Observation of Graduate Interns), a technology-mediated distance education observation process. In its inception, the project designers explored asynchronous evaluation tools (i.e. Adobe Premier) but found limitations in this delayed format. Thus, synchronous technological applications were sought to provide a replicable process that mirrored the observation experiences of being in the classroom. This chapter describes the evolutionary process of ROGI.

BACKGROUND

Initially, online offerings in the Graduate Certificate program were limited to lateral entry teachers. The rationale for offering online coursework to lateral entry teachers only was the supervision component required during the Graduate Student Teaching Internship. Supervision of lateral entry teachers took place formally in the schools through observations by administration and mentor teachers. For those interns who were not lateral entry teachers, supervision from the university was required. The intern was assigned a university supervisor who is responsible for observing the intern three times throughout the semester. It would, however, be problematic to supervise these individuals in their Graduate Internship. In putting together a 100% online program there was one missing component: how to offer a graduate internship in an online venue.

The faculty worked diligently to develop quality online courses that would be piloted and then implemented fully in fall 2007. The transition to a 100% online program, including the Graduate Student Teaching Internship, at first seemed like a doable task; however, design, development, and implementation of an online internship experience presented itself as a significant challenge. Thus, the dialogue shifted from whether or not to expand distance education to how to offer the Graduate Student Teaching Internship in an online environment. Much discussion surrounded a number of concerns: Could technology facilitate a teaching observation? If so, how could this be done most effectively? What tools would be necessary to facilitate the process? Will these technological applications be easy for university supervisors and graduate interns to use?

Literature Review

Impending changes within teacher education, whether motivated by national or state educational mandates, teacher shortages, shifting teacher education candidate needs, growth of second career professionals seeking employment, or tighter operating budgets, have encouraged teacher education programs to seek alternative methods for teacher preparation. Subsequent innovative changes to program platforms coincided with emerging technologies resulting in the introduction of new learning pathways. Although many education programs have investigated or implemented online coursework and teacher preparation experiences (Sharpe et al., 2003; Kent, 2007; Good et al., 2005), challenges of conducting teaching observations in a virtual setting posed limitations to the scope of online licensure opportunities. As new technology tools became available and users adapted to embrace their potential, challenges were no longer insurmountable. Research began to show evidence of this shifting paradigm and served to inform the evolution of ROGI.

Communication at a Distance

While asynchronous observation tools were part of the initial phase of ROGI, they were supplanted by synchronous observation methods that mimicked face to face observation processes. As the project progressed the use of asynchronous tools became useful in supporting reflection-based pedagogical growth and thus, ROGI became a composite of synchronous and asynchronous learning. The technology tools enabled participants to see at a distance and to bridge geographic barriers.

Asynchronous Communications

In the last decade of the 20th Century, distance education established a presence in higher education course offerings. Early research noted the prevalence of asynchronous learning and instructor preference for this format (Soo, & Bonk, 1998). In a project focused on defining the types of communication essential to online learning, Soo and Bonk (1998) surveyed eight veteran distance educators and concluded that their preference for asynchronous settings stemmed, in part, from the ability of asynchronous platforms to support the most important type of communication for effective online learning: learner-to-learner interaction. This type of interaction was consistently ranked highest in terms of importance, closely followed by other asynchronous settings, such as learner-teacher and learner-material. In an ongoing, online survey being conducted by Branon and Essex (2001), asynchronous instruction was found to be useful for the flexibility it affords, the capability of archiving discussions for future use, and the perception that it provides a means of eliciting more in-depth, reflective responses from participants.

Synchronous Communications

Mason (2000) explored the use of computer-mediated communication with a preservice teacher cohort during student teaching and found several benefits. Peer collaboration was enhanced, video conferencing allowed for more immediate feedback while the preservice teachers were able to engage in professional dialogue. Benbunan-Fich and Hilz (2003) determined that online courses could be improved when professors structured them to support the growth of a learning community. They noted that by using collaborative learning methods and being available to interact with students through synchronous and asynchronous tools, students became more engaged and experienced increased satisfaction with online learning. Jackobsson (2006) found that students were able to effectively utilize online collaborative environments to achieve beneficial critical reflection through the use of computer-mediated communication.

Seeing at a Distance

Recognizing the potential of digital tools, the Northwest Educational Technology Consortium released a report in 2005 defining technological best practices as the use of video conferencing and suggested this tool presented numerous possibilities for teacher education programs. Furthering this potential, Johnson and colleagues (2006) used video conferencing effectively in a cybermentoring project at Washington State University, with high degrees of satisfaction and learning effectiveness observed among participants. Work by O'Connor and colleagues (2006/2007) moved beyond observation of mentors and used video conferencing to observe and interact with student teachers. In this work it was found that the use of video conferencing was considered very effective in terms of both satisfaction with interactions and perceived learning. Another approach reported by Kent (2007) used interactive videoconferencing in lieu of field experience as placements of students in classrooms. Interactive video conferencing provided a common experience for all graduate interns and ensured that all students were exposed to teaching practices that were being taught in their university setting.

Bridging Geographic Distance

Technology tools connect remote geographic regions with higher education learning opportunities. Walizer, Jacobs, and Danner-Kuhn (2007) concluded that the challenges of candidate observations in remote field experience locations could be overcome by using web cameras and videoconferencing software. Gillis (2008) investigated student views of videoconferencing in teacher education, and found that technology opened opportunities to learn for students in rural areas. Another project, PT3 (Preparing Tomorrow's Teachers to use Technology), added yet another dimension to distance education. In addition to observing the classroom and the actions of the students and teacher, preservice teachers also interacted with the children and teacher, and prepared and presented a variety of enrichment activities (Lehman & Richardson, 2007). Through the use of video conferencing, Lehman and Richardson (2007) connected remote classrooms, students, and teachers to preservice teacher education preparation.

ROGI

ROGI (the Remote Observation of Graduate Interns) has been used for over four years now at the University of North Carolina at Charlotte. It first came as a solution to teacher shortages, allowing the recruitment of potential teachers from across the state. Its continued use has provided outreach to counties that still face a shortage in teaching; particularly in high need content areas such as mathematics, science, and special education. It has facilitated program growth from twelve original counties (see Figure 1) to thirty-four counties (see Figure 2).

What is ROGI?

ROGI is a technology-mediated performance-based assessment and reflection of teaching (Petty & Heafner, 2009). ROGI utilizes Wimba, a state of the art multimedia conferencing platform, to facilitate the Graduate Student Teaching Internship and to communicate with the Graduate Interns remotely. It allows for a 100% online internship semester.

Prior to the beginning of the semester of the Graduate Student Teaching Internship, the interns are placed in a classroom with a practicing teacher. If the intern is a lateral entry teacher, he/she is allowed to continue teaching in his/her own classroom. Graduate Interns must participate in monthly seminar sessions with the university supervisors and their classmates. Also, the university supervisor observes them three times during the semester, each time in their teaching environment. Following these observations, the interns participate in post-conference debriefing sessions with their university supervisor. They also have various assignments, such as lesson plans and weekly reflection statements, which they must

Figure 1. North Carolina counties served prior to the implementation of ROGI.

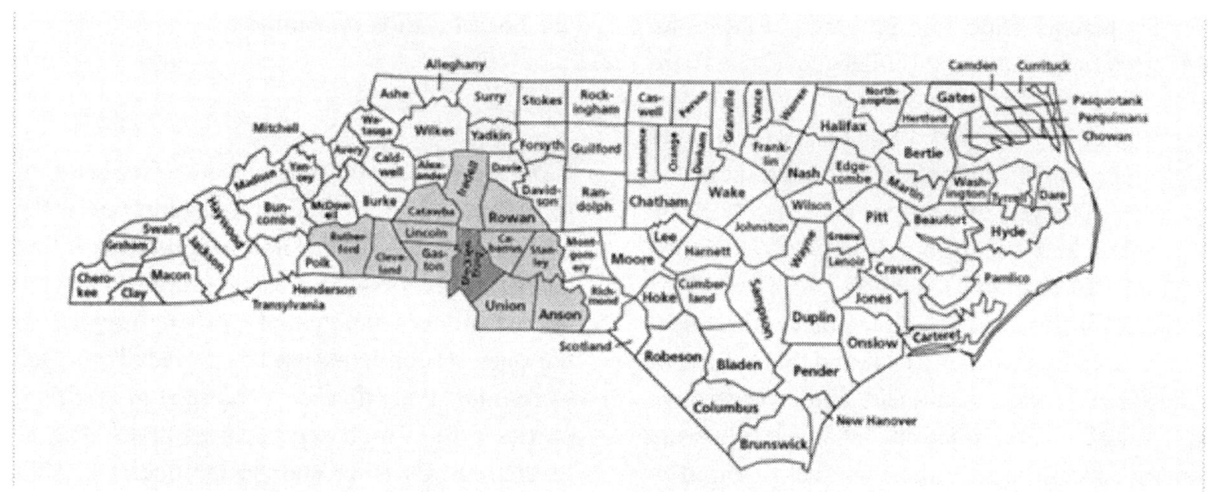

complete and submit to their university supervisor during the semester. Additional communications between supervisor and intern is further facilitated through email, texting, and Moodle, as needed.

Seminar Sessions

In a traditional face-to-face Graduate Student Teaching Internship, graduate interns meet monthly on campus to discuss various topics and issues related to teaching and learning. These monthly meetings are not feasible for graduate interns geographically removed from the university. ROGI is the solution to facilitate these seminars, allowing university supervisors and graduate interns the opportunity to hold virtual synchronous seminar sessions. These sessions occur synchronously using Wimba, the university-supported web conferencing application. In addition to the communicative aspect of the seminar sessions, requisite curriculum resources are collectively shared in Moodle, the university-supported learning management sys-

Figure 2. North Carolina Counties served since the implementation of ROGI in Fall 2008 (through Spring 2012)

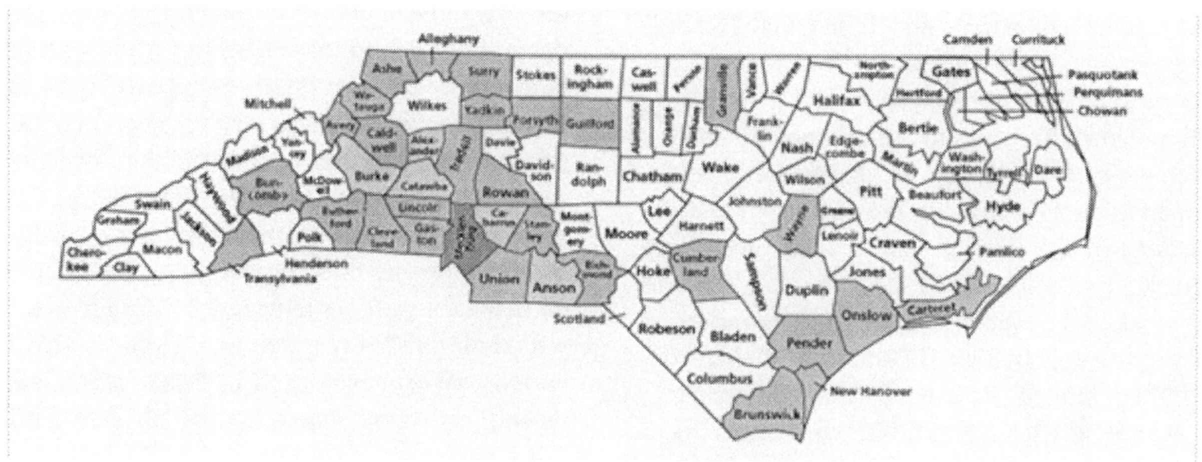

tem. In order to participate in the ROGI process and associated activities, graduate interns need a computer with Internet access, a headset with a microphone, and a webcam, each of which are bundled as a required student purchase, much like a textbook. The university supervisor leads the seminar session providing electronic materials for the interns relevant to the seminar topic. He/she is able to post the materials, including multimedia presentations, documents, and websites, in Wimba for the interns to view during the session. Interns participate as they would in a face-to-face environment; however, the utilization of Wimba allows participation across the state. Additional supporting documents, or materials that cannot be exchanged via Wimba, are posted in Moodle for asynchronous use.

Synchronous communication between the university supervisor and graduate interns or between interns can be text-based, video-based, and/or audible. Interns can individualize communications with each other and/or their university supervisor by using the text chat feature. This enables interns to ask questions and/or make comments throughout the seminar session. The university supervisor can respond to these comments via text or using their microphones. Interns may also communicate via the microphone. This works much like a face-to-face session when a supervisor calls on a student to speak. In Wimba, as well as many other web conferencing tools, there is a *hand raise* feature, which allows the intern to let the supervisor know that he/she has a question. The supervisor may then call on the intern, allowing him/her to speak. Finally, a webcam allows all participants to *see* each other. This helps to create a sense of community and allows the participants to associate a face with a voice. The inclusion of video in these sessions helps to diminish the feelings of isolation often felt by learners in online courses. Interns have commented that they enjoy this feature of the technology.

Graduate interns who have participated in these virtual seminar sessions convey an ease of use of the technology (Heafner, Petty, & Hartshorne, 2012). They enjoy the communicative aspects that the technology offers (Hartshorne, Heafner & Petty, 2011). They also report the added convenience of eliminating the drive to campus, especially for those located remotely from the university (Heafner & Petty, 2010). Furthermore, seminars create a professional network that enables interns to examine diverse school contexts in a socially dynamic way.

The seminar experience has evolved significantly over time. In the initial semesters of ROGI, a different conferencing tool was used, and interns and university supervisors did not use webcams. Additionally, due to the newness and lack of familiarity with all features of the web conferencing tool, university supervisors did not fully utilize all of the features; particularly, the sharing features, resulting in fewer electronic resources disseminated through the use of the conferencing tool. As supervisors have gotten more comfortable with the technology and ROGI has become a more common practice, the experience for both interns and supervisors has become more comprehensive, social, and rewarding.

Remote Teaching Observation

The Graduate Internship requires three teaching observations of the graduate intern by the university supervisor. In the typical face-to-face Graduate Internship, the university supervisor travels to the school to observe the intern; however, this is not realistic for interns located remotely from the university. Thus, remote observations are facilitated using a variety of technological applications and processes. Wimba, the videoconferencing tool utilized for the seminar experience, is also as a primary tool for facilitating the remote observation of graduate interns. However, in addition to software applications, hardware is also needed to make this experience work. To streamline the process, graduate interns are responsible for purchasing the required webcam and a wireless headset, both of which are necessary for the various activities in the ROGI process: the seminar sessions, teaching observations, and post-observation debriefings.

Prior to the initial intern observation, the graduate intern makes sure that the webcam and wireless headset are connected to the classroom computer/laptop that has Internet access. The graduate intern is responsible for making sure that the webcam is working properly and is positioned in a good location in the classroom so that the university supervisor can see the majority of the students during the teaching observation. The intern is also responsible for making sure the audio is working and that it works throughout the classroom in such a manner that, when he/she circulates, the audio quality is maintained. Interns who experience technical difficulty may contact the technical support resources at the university. Thus, it is important for the interns to plan ahead to ensure that everything is working properly on the day of the teaching observation.

Prior to the beginning of a remote observation, the university supervisor accesses Wimba from his/her office or some other remote location. He/she ensures that all of the technological tools, both hardware and software are working properly from his/her computer. For the teaching observation, other than a computer with Internet access, the only piece of hardware that is needed is a headset, so that the university supervisor can hear what is being said in the classroom. Because the teaching observation only requires one-way video, with the video camera being located in the classroom of the graduate intern, it is not necessary that the supervisor have a webcam accessible during the observations.

One feature of Wimba that is occasionally used during the teaching observation process is text chat. The university supervisor can communicate with the cooperating teacher during the observation if needed. For example, the university supervisor can ask the cooperating teacher to move the camera as needed to provide an optimal view of various students, students groups, or the intern. This allows the university supervisor the opportunity to see the entire classroom at various times. If the university supervisor misses something due to camera placement, the cooperating teacher can inform the supervisor of the situation. While the cooperating teacher is typically not present during the post-observation debriefing, this can be amended for a struggling intern, or if the cooperating teacher feels that their presence would be beneficial in the debriefing, perhaps to highlight concerns of the observation.

Graduate interns who have participated in the remote observation process report positive experiences (Hartshorne, Heafner & Petty, 2011). The idea of having the university supervisor observe remotely so that they do not have someone physically in the classroom is viewed positively by most graduate interns, as many report a greater feeling of ease knowing that they are not ‘performing’ but rather teaching (Heafner, Petty, & Hartshorne, 2012). Many interns have stated that they often forget the camera is even in the classroom (Heafner, Petty, & Hartshorne, 2011).

When ROGI was initially piloted, as with the seminar sessions, Centra rather than Wimba was the university-supported web conferencing tool, and was used to facilitate teaching observations. At this time, more technological equipment was also utilized during the observation process. As part of the ROGI process, this equipment was sent to the school prior to the observation semester. The equipment was usually driven to the school for the first observation, set up, and tested to make sure everything was working properly. If someone at the school had volunteered to complete the initial set up, the equipment was sometimes mailed to the participating school. The equipment included: net sharing camera, laptop computer, tripod, and wireless microphone, resulting in a significantly more costly process. Some schools preferred to use the classroom computer, which eliminated the need for the laptop computer, thus reducing the costs.

Figure 3. Screenshot of the remote observation process including text chat, video, and candidate lesson plan

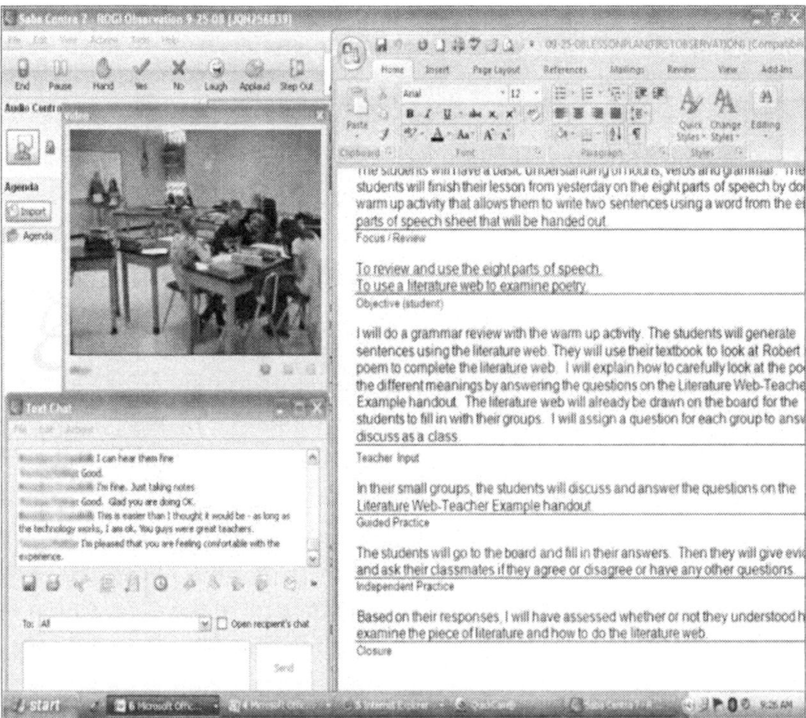

The net streaming camera was connected to the laptop computer and set on a tripod throughout the observation. The teacher candidate wore a wireless microphone to provide clear audio. A camera operator (cooperating teacher) was also used so that the camera could be spanned across the classroom. This provided the opportunity for all students to be seen by the university supervisor. Both the graduate intern and the university supervisor logged onto Centra (see Figure 3).

Post Conference Debriefing

Perhaps the most important component of the entire observation process is the post conference debriefing (see Figure 4). Following the teaching observation, the university supervisor and graduate intern meet to discuss the observation. During this debriefing, the university supervisor provides feedback to the intern regarding the teaching observation and the intern's progress. In the past, this post conference debriefing would typically take place in a face-to-face environment. ROGI, however, has afforded the interns and supervisors the opportunity to conduct this debriefing in a virtual, synchronous setting mimicking that of a face-to-face setting. For the post conference debriefing, the equipment used for the teaching observation is also used. The university supervisor adds a webcam to his/her computer so that there is two-way video, creating a stronger sense of personalization between the intern and supervisor. Wimba is used to facilitate the post conference debriefing, as an extension of the observation process, as both the intern and the supervisor remain logged onto Wimba after the teaching observation.

During the post conference debriefing, the university supervisor can post the teaching observation instruments in Wimba, allowing the intern to follow along as the supervisor talks through the various criteria used to rate the intern's teaching. The intern is able to ask questions, express any concerns, and gain clarity on what he/she needs to do to improve the teaching experience before the next observation. The supervisor and intern then develop an action plan for the candidate to follow before the intern is observed again.

Graduate interns and university supervisors have reported success with this method of post conference debriefing. The equipment is accessible immediately following the observation so that the intern can receive immediate feedback. The interns also describe the experience as a very helpful one in recognizing limitations of their teaching and learning of ways to correct deficiencies and improve their teaching. They describe the use of Wimba as effortless.

Initially, Centra was used to facilitate this post conference debriefing (see Figure 4). Since the intern wore a wireless microphone, he/she had to switch to a headset so he/she could both, hear and speak with the university supervisor. A headset also allowed the comments of the supervisor to be heard by only the intern. Since the webcam was not used early on, the tripod that situated the netstreaming camera was moved so that the intern was visible to the supervisor. When the conferencing could not take place in the same room as the teaching observation, all of the equipment had to be moved, which was a cumbersome process. With minimal equipment used in the updated version of ROGI, this problem has been minimized.

Electronic Teaching Portfolio

Another requirement of the Graduate Internship is the Teaching Portfolio. Interns in the online program create electronic teaching portfolios and websites that house these portfolios. The portfolio includes a variety of materials specific to the observations, including: lesson plans—which must be posted to the portfolio prior to an observation for university supervisor review and feedback, and any materials that might be used during a remote observation—so that the supervisor has access to the materials prior to the observation. Other materials that are also requirements of the portfolio include: daily reflections via an online journal, handouts from seminar, observation feedback, observation notes of other teachers observed by the candidate, video of a teaching session, lesson plan critiques, and a required assessment project.

Interns have expressed enjoyment in learning to create a website. Additionally, they have found that storing everything in one location and saving time and money on making hardcopies to be an added benefit of the portfolio development process. They also like not having to carry around a rather large notebook, which is typical of a traditional internship experience. The electronic teaching portfolio component of the observation experience has not changed much since the inception of ROGI. While various web-editing tools have been used and still vary among university supervisors, the electronic teaching portfolio consists of the same required elements as it did four years ago.

Electronic Observation Instruments

Several observation instruments are required during the Graduate Internship. First, the university supervisor must complete an Observation Feedback Form for each observation. This includes identified intern strengths, suggestions for improvement and other comments or questions. The supervisor must also complete a Student Teacher Assessment Rubric (STAR) for each of the three observations, as well as the summative rating at the conclusion of the Graduate Student Teaching Internship semester. Both the STAR and the summative rating of the Graduate Student Teaching Internship semester have several criteria based on current teaching standards. University supervisors can access these observation rubrics in TaskStream, the College of Education's student data management system. University supervisors

Figure 4. Screenshot of the post conferencing process including text chat, video, and observation instrument

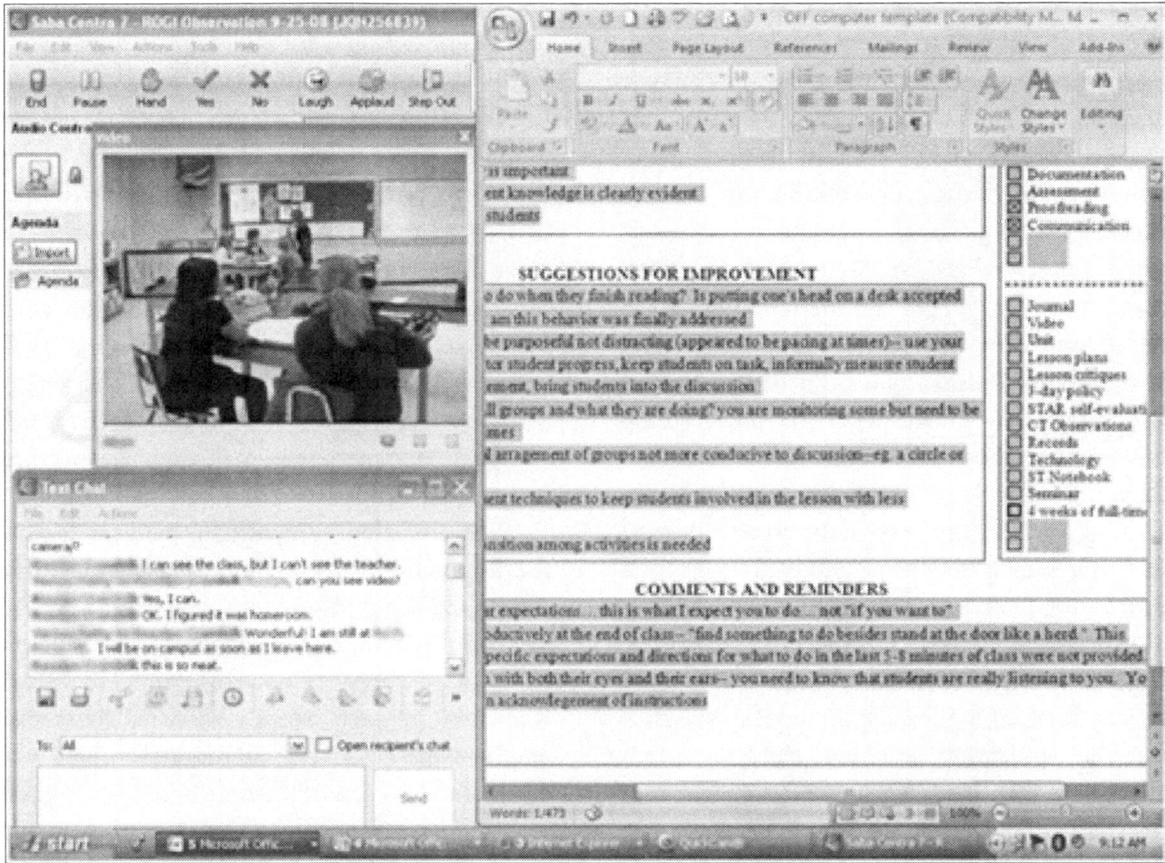

complete assessments of all of their graduate interns, and the graduate interns then receive their ratings electronically through TaskStream. This eliminates the constant exchange of papers and is also a cost saving measure, as it saves money on printing numerous observation documents.

Observation instruments are also completed by interns' administrators and/or cooperating teachers. These documents include the Observation Feedback Form, Capacity of Teaching Form, and Student Teacher Assessment Rubric. Additionally, a survey regarding intern dispositions is also completed by the administrators and cooperating teachers. As with other observation instruments and documentation, all of these forms are housed in TaskStream. Administrators and cooperating teachers are provided access to TaskStream, and can complete all of the observation forms electronically.

When ROGI first began, most of these forms were not completed electronically through a centralized system. Many forms that needed administrator and cooperating teacher completion were mailed to schools, completed, and returned. Whenever possible, forms were distributed via fax or e-mail. For example, university supervisors could complete observation forms and then email them to students. The new process of having all ROGI-related documentation stored in a central and easily accessible location has made the completion and distribution of observation instruments much less cumbersome, helping to streamline the overall ROGI process.

WHAT DO DATA SUGGEST ABOUT ROGI AFTER FOUR YEARS?

Since the pilot semester, authors have been collecting both quantitative and qualitative data to examine both the effectiveness of the project and perceptions of participants within the project. Interviews with graduate interns and university supervisors have been conducted each semester. Observation forms and summative internship assessments have also been complied as data sources. What follows is a brief description of what inferences learned from the project data analyses. For those seeking a more extensive evaluation of project outcomes, references are provided for authors' research. Seven themes data-based themes that summarize this collective research are described. The purpose in describing a compilation of findings is to present an overview of how technology-mediated processes introduce new ways of thinking about traditional approaches to teacher education and new challenges that accompany this innovation. Authors conclude with recommendations for future research and how other researchers might embrace and utilize the potential of new technologies in preparing teacher educators.

Technology, Specifically Online Observations, Can be Used to Measure Teaching Effectiveness

What is the purpose of the observation process? One university supervisor summarized her idea of what the process should be. She stated:

"It is the ability to be able to document what graduate interns do, and the impact that it has on their kids. So, I need to have both process and outcome data every time I do an observation. Over a series of observations, what I want to see is growth—and to the point where, by the final observation we've got, consistently, high-level performance. Um, but I think the feedback is probably the most important thing that I do in an observation and I use data and I record data, and I look at even what the children have done."

Regardless of the method used to observe graduate interns, either traditional face-to-face or remote, university supervisors have to be able to measure teaching effectiveness. To do this, the same observation instruments are utilized regardless of the type of teaching observation being conducted. When ROGI was first implemented, the researchers collected data from these observation assessment rubrics and analyzed them to insure a comparable experience for those observed face-to-face and remotely.

First, statistical comparisons were made using the observation assessment rubric to evaluate differences among observers (cooperating teacher or school administrator and university supervisors) and differences between graduate interns. Repeated measures were performed by teacher and between observers. Summative scores were calculated for each of the ten variables identified within the scoring rubric. Variables aligned with the ten Interstate New Teacher Assessment and Support Consortium (INTASC) standards for initial licensure (see http://www.ccsso.org/Projects/interstate_new_teacher_assessment_and_support_consortium/). Intra-class comparisons were made to determine if differences in observation scores were attributed to variance among observers or differences between graduate interns.

The remote observation process was useful in promoting increases in teaching knowledge, skills, and dispositions, as each graduate intern concluded the graduate internship with a rating between *Proficient* (3.00) and *Distinguished* (4.00) as defined in the observation assessment rubric. Learning outcomes aligned with expectations of growth over time during the internship semester. This is indicated in Figure 5.

As graduate interns honed their knowledge and skills, they improved in their teaching. Moreover, while variations in university supervisor

Figure 5. Average summative evaluation of teaching by observer over time

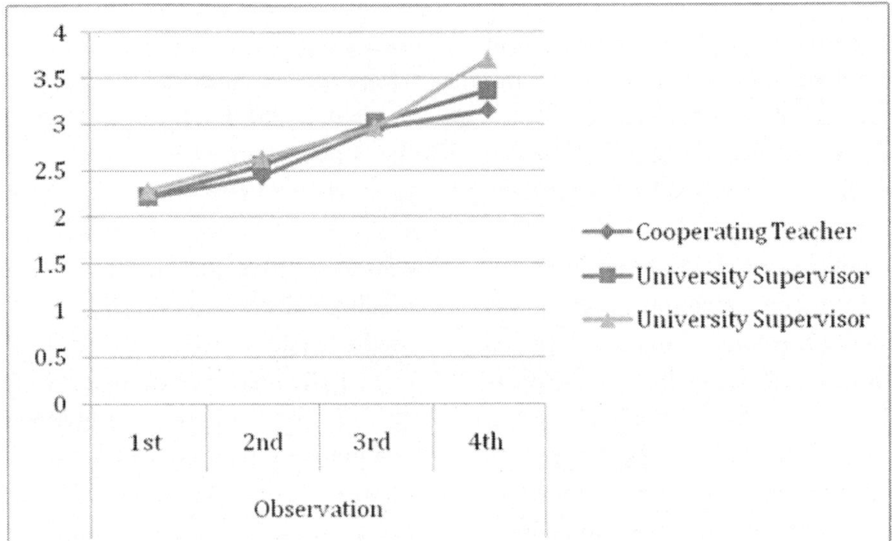

ratings existed, the observation assessment rubric ratings for the remote observation processes were consistent with those of the face-to-face observations. There were no significant differences in the overall teaching evaluations as a result of the remote observation methods used. Teaching evaluations were consistent across observation process.

For the graduate intern and the university supervisor, the observation experience, either remote or face-to-face, served the purpose of providing various perspectives on teaching and supporting reflective practice. Interns noted the role of the observer as "to see and to give you [the intern] some input on knowledge that they [the university supervisor] may have, ideas that may help you with things that went wrong with that class." University supervisors highlighted the role of observations as allowing interns "to take time to reflect on their teaching, classroom management, student behaviors, and overall impact on student learning outcomes." Both acknowledged the value of post-conferencing in improving practice and pedagogy, such as "I tried to use the suggestions and be better or improve on them the next time they came in or the next time I was observed" and "I would always use their feedback to try and come up with better lessons and improve by using their suggestions." While there are no differences in methods of observation in instructors' ability to measure teaching effectiveness or supporting interns in their professional growth, there in an added value of technology that extends evaluation and learning beyond the confines of traditional face to face processes.

Technology Tools Open New Spaces for Work and Collaboration

The technology afforded by ROGI offered new dimensions for both work and collaboration. University supervisors were no longer confined to long days spent driving to and from schools, waiting in school offices to visit graduate interns, and perhaps remaining at the school long after a teaching observation to conduct the post conference debriefing with the graduate intern. ROGI allowed university supervisors the flexibility to work from campus offices or even their home offices, cutting down on time spent traveling which allowed the supervisors to be more productive during days of teaching observations. One university supervisor highlighted the convenience that ROGI offered as shown in this comment, "Yes, I like the process because I could do it in my sweat pants, sweat shirts and baseball cap." In general, convenience enabled observers to create a "new office" and increase workplace satisfaction due to greater levels of productivity achieved with the use of technology.

With travel eliminated, ROGI offered the option of increasing the number of teaching observations if needed. Typically, a graduate intern is observed three times during the internship; however, there is occasionally the need to observe the intern more. This occurs if the intern is struggling or perhaps when the intern may want additional feedback. The face to face observation inhibits this option because travel funds are limited. One university supervisor confirmed this, stating "one thing we definitely found helpful last semester was being able to opt for more observations if needed which was a positive outcome of ROGI."

Not only did added observations benefit struggling interns, ROGI also offered the option of having multiple observers at one time, which provided the interns with additional feedback from a different perspective. Another comment of a university supervisor supported this notion. "It was also helpful that two supervisors could watch at the same time....Typically with face-to-face you would not have two people go out to the school because it is too tough on the intern and interferes with student behavior." Restrictive travel budgets make multiple observers a rare and discouraged practice.

Technology is Convenient and Flexible

Consistent with research in online teaching and learning, a major theme that emerged from interviews with graduate interns and university supervisors was the convenience and flexibility afford by the technological applications utilized. From the graduate interns' perspective, the benefits of the digital format and flexible access to feedback and processes often made the drawbacks typically associated with online learning obsolete. For example, one intern stated, "I liked having feedback electronically. It allowed me the convenience of looking at it and responding to it when I have time. I can go back and read it again." The additional benefits of flexible access to feedback, as well as the ability to revisit feedback, were seen as very beneficial by many graduate interns. Also, contrary to what one might expect, the increased number of interactions and correspondence between university supervisors and graduate interns afforded by the technology promoted the development of a personal rapport. For example, one intern commented,

"I thoroughly enjoyed uh reading the feedback that I received online, and uh, I actually used the feedback to fix the shortcomings that were found in my lesson," and "I really appreciated the comments on the lessons, I have read those. Like I said, I would always use those to try and come up with better lessons and improve by using their suggestions. They have been honest, which I appreciate."

In addition to access to feedback and increased communications and interactions between graduate interns and university supervisors, the convenience and flexibility afforded by the technology allowed for special innovative ways of addressing special circumstances that would be prohibitive in the traditional face-to-face observation process. These include the capacity to conduct additional observations, convenience of observing from the office or home, and the ability to overcome geographic and time limitations to complete multiple observations at remotely located schools during a single day. For example, one university supervisor commented, "it's very convenient. I mean, you know, we can observe people at another, umm, school that's, you know, two hours away." Another university supervisor getting ready to participate in the ROGI process commented,

"...and if I have a lot of flexibility like I think there is with that new remote observation program...I see teachers who would normally get more observation – who need that additional support – they are not going to get it 'cause someone is not point to want to walk out. They're not going to want to drive 45 minutes to get there. And I mean, even with just the conversation I have with other university supervisors, I'm sure that happens. The student needs additional support, but it's just too time-consuming to drive all the way out there."

Thus, the benefits of the convenience and flexibility that the technology associated with ROGI provided emerged as a theme through interviews with graduate interns as well as veteran and new university supervisors participating in ROGI. As the technological tools improve and the process becomes more streamlined, it is anticipated the new and unforeseen conveniences will emerge.

Technology Offers a Cost-Effective and Time-Saving Option to Constricting Budgets

Another benefit that emerged from ROGI was the cost-benefit and time savings. The figures for one semester of the project provided an example of travel expenditures incurred by the institution. In examining the travel time and costs for the pilot semester (Spring 2008), significant time and cost savings emerge. To conduct a single observation, Table 1 illustrates a savings of $414.85 in costs and 12 hours 46 minutes of time. If this is compounded for the required three observations, the cost savings increase to $1244.55 and the time savings increase to 38 hours and 18 minutes.

However, the time and cost savings mentioned previously are not indicative of the actual observation process for a normal semester. For example, typically a university supervisor is assigned to supervise six students, and schedules three or four observations for each. Thus, costs associated with a typical semester for a university supervisor, increased savings in time and overall expenditures. Table 2 illustrates data from schools that participate in a typical remote observation semester. Looking at these values more closely, $250.59 is saved in travel costs associated with conducting a single set of observations for each student. Additionally, time savings for a single observation of each student amounts to 9 hours and 10 minutes. Cost savings to conduct the required three to four observations ranges from $751.77 to $1002.36. Additionally, time savings range between 27 hours, 30 minutes and 36 hours, 40 minutes for travel time, which would be realized for each university supervisor.

Time savings also emerged as a major benefit identified through interviews with university supervisors, who consistently cited additional time benefits associated with ROGI, beyond travel time. The elimination of wait-time at schools before and between classes, time scheduling challenges to accommodate multiple observations within the confines of the school day, and reduced stress caused by the reduced travel time, were three other major time benefits themes that emerged. One university supervisor noted, "the amount of time that I wasted in doing the face-to-face, whether it was real or perceived was immense; the driving time and going and sitting in the (school) office to sitting in the classroom and waiting [...] It seemed like everything was just about twice as long as or more than it needed to be. I thought that from a time saving perspective, the remote observation was extremely beneficial." Another university supervisor added,

"Well, when we teach...and we have to run to the school; come to campus at 11 o'clock – which is a horrible time to find a parking spot – it's stressful. I think you can measure feedback forms and all

Table 1. Round trip cost for one observation at each school in one semester

School	Mileage	Cost*	Travel Time
School 1	321.48	$188.06	5 hours 18 minutes
School 2	81.74	$47.81	1 hour 48 minutes
School 3	57.84	$33.83	1 hour 10 minutes
School 4	97.04	$56.76	1 hour 48 minutes
School 5	151.10	$88.39	2 hours 42 minutes
Total	**709.20**	**$414.85**	**38 hours 18 minutes**

*Mileage reimbursement rate is $0.58

that, but I don't think – the travel budget, and all that – I don't think it ever takes into account the stress that it puts on a faculty member to run here and run there and not feel productive because I never finish anything (laugh), you know. And I'm running to a class, and I can't meet with students because I've got to run back out to school, so it's...it's something that you can't measure – you know, that productivity of a faculty member who wants to do supervision, but can't take the time to do it. And I value it for my teaching because when I go out and I observe student teachers, it enriches and I bring back to my classroom – when I teacher at the University, and I can tell stories and give examples, and tell them what it looks like out there. But, I...I can't...you know, I really have trouble doing that, you know, taking on this many – which are just the sheer numbers of our program with budget cuts, the real advantage would be to engage more faculty members and take less time to do it, that it really enhances your teaching."

Finally, another university supervisor commented,

"...it's less schedules to try to match to make work. So that, I think, would be easier to try to... when I think of – I've got four different schools, and can I get to this one, and this one, and travel time, and so on – if I was in one location, you could certainly knock out quite a few – faster and quicker because of that last travel time. That would be good."

In these current economic times, many teacher education programs are being asked to do more with reduced resources, while not sacrificing quality. ROGI affords a series of time and cost saving benefits to assist in addressing this charge.

Table 2. Round trip cost for one observation at each school (typical semester)

School	Mileage	Cost*	Travel Time
School A	38.54	$22.55	54 minutes
School B	77.52	$45.35	2 hours
School C	74.70	$43.70	1 hour 22 minutes
School D	79.24	$46.35	1 hour 34 minutes
School E	58.38	$34.15	1 hour 20 minutes
School F	99.98	$58.49	2 hours
Total	**428.36**	**$250.59**	**9 hours 10 minutes**

*Reimbursement per mile ($.585)

The benefits previously cited, coupled with the significant cost savings of the full implementation of the ROGI process holds significant promise, and can serve as a mechanism for augmenting traditional face-to-face observations (and programs) and addressing future teacher preparation needs through alternative methods of licensure and the recruitment of more geographically dispersed individuals.

Technology Cannot Fully Replicate Face-to-Face Experiences

ROGI is a comparable, not equivalent alternative to face-to-face observations. When technology is used to facilitate online observations, there will always be a physical distance between the observer and the intern being observed. From the vantage point of the observer, distance matters. A theme that was present in interviews with university supervisors was a resounding concern that something was lost by not being in the classroom. This distance was translated as impersonalizing the observation process. An experienced university supervisor commented, "I liked it [ROGI] okay because I could still give them some of the same information I was sharing with the face-to-face group, but it still seemed a bit impersonal." Another university observer shared similar views and in critiquing ROGI stated, "The interpersonal relationship is the only thing that is missing. And I just don't want the student that is doing the online to think, well if she had been here or something to that effect." These later sentiments reflected overall university supervisors' notions that online observations were different from face to face.

For observers the visual and sensory experiences of being in the classroom could not full be replicated remotely. A university supervisor described in his interview the importance of co-presence. He comment,

"There's a lot that you can't see, I think, from... that...would be challenging to see on camera. Um, I think you get a very unique perspective sitting at the back of the room; being able to look around and seeing what's going on. Sometimes a perspective that the teacher doesn't get to see. So I think that's important. Um, you know, um, sometimes, just that personal – just your presence being there."

University supervisors consistently articulated this co-presence difference between face-to-face and remote observations. The views expressed in this quote capture observers' sentiments:

"I prefer the face-to-face because you're more personal with them. You can go on a tangent with different topics. Sometimes when they are online as if you are going through what you need to go through and that's it. Whereas face-to-face, we could piggyback on things they would discuss; oh yeah I felt that way too. Not feel threaten to do that, and you would think that it would be the other way around that they wouldn't feel threatened because no one sees them online. I just kind of like that face-to-face interaction."

Another university observer commented that, "I don't think it [ROGI] should replace it [face to face], though. Just because I think there's... when you've got that time together, there's a lot more that comes out, and... and maybe you might feel a little -- not necessarily rushed, but maybe a little more scheduled with just the, um, with the cam... the WebCam instead of the face-to-face." He emphasized that "you can actually sit down and see facial expressions. Things come to you easier when you are face-to-face. Questions come...." University supervisors held to strong standing beliefs that relationships were best nurtured face to face.

Distance Can be Good

Distance was not always viewed negatively. Physical distance in the remote debriefing session allowed a less personal conversation and a more professional dialogue focused on what was observed and experienced in the lesson that was observed. The debriefing session was frequently described as: "less informal than face-to-face." Observer comments also noted a difference in expectations accompanied remote debriefing sessions. These views of "less informal" were positively described in the following quotes: "… when they [post-conferences] are online, you go through what you need to go through and that's it…" and "Sometimes when you have something for a student to work on, it is easier to tell them without the personal face-to-face. It is a little bit harder to tell someone face to face they are doing a bad job…"

Despite limitations, remote observations occurring "at the same time..." were described as having positive outcomes, "because then you're not always getting the cooperating teachers feedback about that particular lesson" when you are in the classroom. With ROGI, "you're getting, right then and there, something" in the form of text dialogue with the cooperating teacher who serves simultaneously as the camera operator. "So," in the words of this observer, "I think that's great." Interns too shared dualities of views of remote observations. As one intern expressed a preference for face-to-face because it was "more personable" and "right there on the spot." He also noted that "the same could be accomplished remotely."

ROGI also provided a sense of comfort to the graduate interns. After they had the camera in the classroom for a few days, both the interns and their students adjusted to it being there and their behavior was synonymous of a typical teaching day. One intern made the following comment regarding ROGI:

"I guess it (ROGI) made me feel more comfortable that someone wasn't sitting there, you know, looking over my shoulder per se. Like I said earlier, I liked the fact that someone would be there if I absolutely needed them, but remote observation made it feel like nobody was looking over your shoulder. The camera was there, and you knew someone was watching but it gave you a sense of disconnection, like you are doing the whole thing on your own, someone is not really there supervising you."

ROGI truly offered a different spin on the traditional observations for both the graduate interns and university supervisors. Interns were provided a comfortable environment for the sometimes stressful teaching observations while university supervisors redefined their work spaces from the K-12 classrooms to their own offices. As commented by one university supervisor, ROGI "was like being in two places at one time."

Technology, Despite Limitations, is Preferred

In describing face-to-face observations during interviews, observers noted a preference for the face-to-face observation over remote observations because, "A level of personalization exists in a face-to-face environment that does not remotely just because you are in the classroom." However, when comparing all the benefits and limitations of ROGI, overwhelmingly, observers expressed a preference for the flexibility, convenience, and time savings afforded by ROGI. In sum, these observers' reflections captured the challenge yet possibilities of ROGI.

"If I had to choose between doing an observation face-to-face or remotely, I would chose face-to-face because of personalization and rapport building that are afforded in that environment. However, considering all that I am expected to do in this job, I would select remote observations because of the convenience and time savings that are gained through the use of technology. I guess that makes the two processes equal."

"But there is still one significant difference that leads me to conclude that I prefer remote observations. This is the added value that technology offers. I can do things with technology that I never considered when I conducted observations onsite. I can hear teacher-student exchanges with more clarity. I can disappear from the classroom and see realistic teaching and student behavior. I can observe and text with the cooperating teacher and other colleagues simultaneously. In this sense, technology has helped me see new possibilities for the role of technology in teacher preparation."

University supervisors noted these as well as the following differences as justifications for their clear preference overall for remote observations: "from a work perspective and work load management, I would choose remote observations over face-to-face, because I think it is more feasible for balancing all of the other expectations of a tenure track faculty position" and "for me, benefits of the remote observation outweigh the benefits of face-to-face."

CONCLUSION

In the case of ROGI, remote observations despite limitations were not eliminated at this institution. The process continued to adapt and grow, but why? From its inception, ROGI has been an evolving process. In early implementation phases, project directors examined asynchronous tools for observations and then migrated to synchronous technologies when expectations of the observation experience were not met asynchronously. While initially technologies that enabled synchronous observations were selected, later archiving synchronously recorded teaching and post-conference debriefings were found to add the opportunity for deeper self-reflection in an asynchronous format. Allowing interns to review the lesson and teaching critique at a later time enabled project directors and participants to redefine the role of feedback and reflection in teacher preparation.

Over time, the synchronous observation process also changed. Instructors have sought additional technologies to build collaborative learning partnerships with remote interns that bridge physical distances. Experimentations with reflective blogs, interactive texting, and even keeping up with interns' Facebook posts have been integrated into the ROGI experience. New technologies such as: HD cameras, less expensive wireless microphones, and the use of a second camera have been examined in an attempt to improve the quality of what is seen and what is heard during the observation as well as how feedback is provided to interns. The philosophy of what can technology do to help university instructors and supervisors perform their work differently with meaningful outcomes has led project directors to explore the use of synchronous remote observations in early clinical experiences (see the last chapter in this book on WiTL).

ROGI initially served as an innovative method of addressing the teacher shortage issue in North Carolina. Through its use, the Graduate Certificate Program was made available online to anyone in the state wishing to pursue a teaching license. This outreach allowed many students the opportunity to pursue a teaching career that may otherwise have been impossible. ROGI currently serves the university in continuing to recruit potential and diverse teachers from areas of the state that still face teacher shortages and also in specific content areas that encounter shortages. ROGI also allows university supervisors to observe graduate interns from across the state rather than the thirteen county radius that was previously served in the traditional face-to-face observation experiences. Due to the implementation of ROGI, the number of counties across the state represented in the online program has almost tripled. As the use of ROGI continues in the Graduate Certificate Program, its effectiveness and efficiency will continue to be evaluated. With the ongoing changes and improvements in technology, ROGI is likely to look very different in the next four years.

REFERENCES

Allen, E., & Seaman, J. (2007). *Online nation five years of growth in online learning*. Needham, MA: The Sloan Consortium.

Arbaugh, J. B., & Benbunan-Fich, R. (2005). Contextual factors that influence ALNs. In Hiltz, S. R., & Goldman, R. (Eds.), *Learning together online: Research on asynchronous learning networks* (pp. 123–144). Mahwah, NJ: Lawrence Erlbaum Associates.

Barbour, M. (2010). Researching K-12 online learning. *Distance Learning*, *7*(2), 6–12.

Benbunan-Fich, R., & Hilz, S. R. (2003). Mediators of the effectiveness of online courses. *IEEE Transactions on Professional Communication*, *46*(4), 298–312. doi:10.1109/TPC.2003.819639

Branon, R. F., & Essex, C. (2001). Synchronous and asynchronous communication tools in distance education. *TechTrends*, *45*(1), 3642. doi:10.1007/BF02763377

Desai, M., Hart, J., & Richards, T. (2009). E-learning: Paradigm shift in education. *Education*, *129*(2), 327–334.

Gillis, D. (2008). Student perspectives on video conferencing in teacher education at a distance. *Distance Education*, *29*(1), 107–118. doi:10.1080/01587910802004878

Good, A. J., O'Connor, K. A., Greene, H. C., & Luce, E. F. (2005). Collaborating across the miles: Telecollaboration in a social studies methods course. *Contemporary Issues in Technology & Teacher Education*, *5*(3/4), 300–317.

Hampel, R., & Baber, E. (2003). Using internet-based audio-graphic and video conferencing for language teaching and learning. In Felix, U. (Ed.), *Learning language online towards best practice* (pp. 171–191). The Netherlands: Dripps, Meppel.

Hartshorne, R., Heafner, T., & Petty, T. (2011). Examining the effectiveness of the remote observation of graduate interns. *Journal of Technology and Teacher Education*, *19*(4), 395–422.

Hatch, T., & Grossman, P. (2009). Learning to look beyond the boundaries of representation. *Journal of Teacher Education*, *60*(1), 70–85. doi:10.1177/0022487108328533

Heafner, T. L., & Petty, T. (2010). Observing graduate interns remotely. *Kappa Delta Pi Record*, *47*(1), 39–43.

Heafner, T. L., Petty, T. M., & Hartshorne, R. (2011). Evaluating modes of teacher preparation: A comparison of face-to-face and remote observations of graduate interns. *Journal of Digital Learning in Teacher Education*, *27*(4), 154–164.

Heafner, T. L., Petty, T. M., & Hartshorne, R. (2012). Moving beyond four walls: Qualitative evaluation of ROGI (Remote Observation of Graduate Interns) for the expanding online teacher preparation classroom. In Alias, N. A., & Hashim, S. (Eds.), *Instructional technology research, design and development: Lessons from the field* (pp. 370–400). Hershey, PA: IGI Global Publishing.

Humphries, S. (2010). Five challenges for new online teachers. *Journal of Technology Integration*, *2*(1), 15–24.

Jakobsson, A. (2006). Students' self-confidence and learning through dialogues in a net-based environment. *Journal of Technology and Teacher Education*, *14*(2), 387–405.

Johnson, C. (2005). Lessons learned from teaching web-based courses: The 7-year itch. *Nursing Forum*, *40*(1), 11–17. doi:10.1111/j.1744-6198.2005.00002.x

Johnson, T., Maring, G., Doty, J., & Fickle, M. (2006). Cybormentoring: Evolving high-end video conferencing practices to support preservice teacher training. *Journal of Interactive Online Learning*, *5*(1), 59–74.

Kent, A. (2007). Powerful preparation of preservice teachers using interactive video conferencing. *Journal of Literacy and Technology*, *8*(2), 41–58.

Lehman, J. D., & Richardson, J. (2007, April). *Linking teacher preparation program with k-12 schools via video conferencing: Benefits and limitations*. Paper presented at the Annual Meeting of the American Educational Research Association, Chicago, IL. Retrieved from http://p3t3.education.purdue.edu/AERA2007_Videoconf_Paper.pdf.

Mason, C. (2000). Online teacher education: An analysis of student teachers' use of computer-mediated communication. *The International Journal of Social Education*, *15*(1), 19–38.

Moller, L., Foshay, W., & Huett, J. (2008). The evolution of distance education: Implications for instructional design on the potential of the Web. *TechTrends*, *52*(4), 66–70. doi:10.1007/s11528-008-0179-0

Northwest Educational Technology Consortium. (2005). *Videoconferencing to enhance instruction*. Retrieved from http://www.netc.org/digitalbridges/uses/useei.php

O'Connor, K. A., Good, A. J., & Greene, H. C. (2006). Lead by example: The impact of teleobservation on social studies methods courses. *Social Studies Research and Practice*, *1*(2), 165–178.

Olson, G. M., & Olson, J. S. (2000). Distance matters. *Human-Computer Interaction*, *15*(2), 139–178. doi:10.1207/S15327051HCI1523_4

Olson, G. M., Zimmerman, A., & Bos, N. (Eds.). (2008). *Scientific research on the Internet*. Cambridge, MA: MIT Press.

Petty, T., & Heafner, T. (2009). What is ROGI? *Journal of Technology Integration in the Classroom*, *1*(1), 21–27.

Sharpe, L., Hu, C., Crawford, L., Saravanan, G., Khine, M. S., Moo, S. N., & Wong, A. (2003). Enhancing multipoint desktop video conferencing (MDVC) with lesson video clips: Recent developments in pre-service teaching practice in Singapore. *Teaching and Teacher Education*, *19*(3), 529–541. doi:10.1016/S0742-051X(03)00050-7

Soo, K. S., & Bonk, C. J. (1998, June). *Interaction: What does it mean in online distance education?* Paper presented at The World Conference on Educational Multimedia and Hypermedia & World Conference on Educational Telecommunications, Freiburg, Germany.

Stodel, E. J., Thompson, T. L., & MacDonald, C. J. (2006). Learners' perspectives on what is missing from online learning: Interpretations through the community of inquiry framework. *International Review of Research in Open and Distance Learning*, *7*(3), 1–24.

Venn, M. L., Moore, R. L., & Gunter, P. L. (2001). Using audio/video conferencing to observe field-based practices of rural teachers. *Rural Educator*, *22*(2), 24–27.

Walizer, B. R., Jacobs, S. L., & Danner-Kuhn, C. L. (2007). The effectiveness of face-to-face vs. web camera candidate observation evaluations. *Academic Leadership*, *5*(3), 1–9.

Young, A., & Lewis, C. W. (2008). Teacher education programmes delivered at a distance: An examination of distance student perceptions. *Teaching and Teacher Education*, *24*(3), 601–609. doi:10.1016/j.tate.2007.03.003

Young, S. (2006). Student views of effective online teaching in higher education. *American Journal of Distance Education, 20*(2), 65–77. doi:10.1207/s15389286ajde2002_2

ADDITIONAL READING

Barnett, M., Harwood, W., Keating, T., & Saam, J. (2002). Using emerging technologies to help bridge the gap between university theory and classroom practice: Challenges and successes. *School Science and Mathematics, 102*(6), 299–313. doi:10.1111/j.1949-8594.2002.tb17887.x

Beck, R. J., King, A., & Marshall, S. K. (2002). Effects of video case construction on preservice teachers' observations of teaching. *Journal of Experimental Education, 70*(4), 345–361. doi:10.1080/00220970209599512

Benbunan-Fich, R., Hiltz, R., & Harasim, L. (2005). The online interaction learning model: An integrated theoretical framework for learning networks. In Hiltz, S. R., & Goldman, R. (Eds.), *Learning together online: Research on asynchronous learning networks* (pp. 19–37). Mahwah, NJ: Lawrence Erlbaum.

Brush, T., Igoe, A., Brinkerhoff, J., Glazewski, K., Ku, H., & Smith, T. C. (2001). Lessons from the field: Integrating technology into preservice teacher education. *Journal of Computing in Teacher Education, 17*(4), 16–20.

Knipe, D., & Lee, M. (2002). The quality of teaching and learning via videoconferencing. *British Journal of Educational Technology, 33*(3), 301–311. doi:10.1111/1467-8535.00265

Martin, M. (2005). Seeing is believing: The role of videoconferencing in distance learning. *British Journal of Educational Technology, 36*(3), 397–405. doi:10.1111/j.1467-8535.2005.00471.x

Phillion, J., Johnson, T., & Lehman, J. D. (2003). Using distance education technologies to enhance teacher education through linkages with K-12 schools. *Journal of Computing in Teacher Education, 20*(2), 63–70.

Pickering, L., & Walsh, E. (2011). Using videoconferencing technology to enhance classroom observation methodology for the instruction of preservice early childhood professionals. *Journal of Digital Learning in Teacher Education, 27*(3), 99–108.

KEY TERMS AND DEFINITIONS

Graduate Certificate Program: Program of study including professional education coursework that leads to the initial teacher licensure.

Graduate Intern: Someone participating in the Graduate Student Teaching Internship who is seeking teacher licensure.

Graduate Student Teaching Internship: The semester long student teaching experience for graduate interns.

Lateral Entry Teacher: Practicing teachers who have already achieved a Bachelor's degree and are concurrently teaching and taking courses to obtain a teaching license.

Post Conference Debriefing: Meeting between graduate interns and university supervisors that occurs after the teaching observation, and allows for reflection of the teaching experience by both the university supervisor and graduate intern. It also allows for collaboration to develop plans for action to improve future teaching.

ROGI: The Remote Observation of Graduate Interns is a technology mediated performance-based assessment and reflection of teaching that utilizes Wimba, a state of the art multimedia conferencing platform, to facilitate the Graduate Student Teaching Internship and to communicate with the Graduate Interns remotely.

Synchronous Learning: Learning environments occurring in real-time, typically with all stakeholders in geographically dispersed locations.

Chapter 3
Perceptions of Preparation of Online Alternative Licensure Teacher Candidates

M. Joyce Brigman
University of North Carolina at Charlotte, USA

Teresa M. Petty
University of North Carolina at Charlotte, USA

ABSTRACT

This chapter seeks to investigate the perceived sense of preparation for the classroom that leads to teacher effectiveness. The focus of this chapter is an exploration of the increasing role of alternative licensure and distance education in the preparation of teachers and results of a recent study concerning perceptions of a sense of preparedness espoused by alternative licensure teacher candidates after their online program completion.

INTRODUCTION

As observed by Labaree, "teaching is an enormously difficult job that looks easy" (2004, p. 39). Perhaps the same could be applied to the preparation of teachers as well. Colleges of education are called upon to provide preparation programs that include meaningful experiences conducive to teacher candidates' understanding of their future schools and students (Singer, Catapano, & Huisman, 2010). This call also applies to the current proliferation of more non-traditional avenues such as alternative licensure routes and distance learning. A crucial part of increasing teacher effectiveness is in the area of teacher preparation with the end result being, as United States Education Secretary Duncan has proposed, "to ensure that students exiting one level are prepared for success, without remediation, in the next" (US Department of Education, 2009, p. 207). This challenge

DOI: 10.4018/978-1-4666-1906-7.ch003

entails exploring the means of providing teachers with an efficacious sense of preparedness in their overall teaching abilities including those teachers prepared through alternative licensure avenues as well as through distance learning program delivery.

The importance of an effective teacher is hardly a novelty. Teachers, and by inference schools themselves, have tremendous impact on student achievement (Rivkin, Hanushek, & Kain, 2005). Teachers possessing proper skills and demonstration of care for their students remain our best means of improvement (DiGiulio, 2004). What teachers actually do in the classroom makes a great difference (Wenglinsky, 2000). Daily they face a myriad of decisions that impact their students' futures and require a wide range of knowledge (Darling-Hammond and Bransford, 2005). Teachers typically share a large portion of students' days for the majority of the year. What occurs in a shared year can impact students' educational foundation for ensuing instruction and solid grounding in basic educational knowledge. In educating students in our nation's classrooms to be capable in an increasingly competitive world, teachers must be able to deliver curriculum that is accurate, meaningful, and appropriate. Actions of such teachers encompass both what is taught and how it is taught (Fenstermacher & Richardson, 2005).

This chapter seeks to investigate the sense of preparation for the classroom that teachers require to become that effective entity. The focus of this chapter is an exploration of the increasing role of alternative licensure and distance education in the preparation of teachers and results of a recent study concerning perceptions of a sense of preparedness espoused by alternative licensure teacher candidates after their online program completion. Four themes for this chapter include:

- Growth and current role of alternative licensure
- Growth and current role of distance learning in schools of education
- A sense of preparedness and INTASC Standards
- Online alternative licensure teacher candidates' perceptions of preparedness

BACKGROUND

Growth and Current Role of Alternative Licensure

A few decades ago, the sole route to acquiring teaching licensure for public schools lay largely through a traditional on-campus undergraduate four-year program. Now the route to teaching is definitely multi-faceted. In fact, 35 percent of teachers responding to the most recent *Survey of the American Teacher* had experienced a previous career before entering the classroom (Met Life, 2010). During the 1980's, alternative programs serving as pathways to teacher licensure were created to a certain extent in answer to shortages in the teaching workforce (Richardson & Roosevelt, 2004). In 1983, fewer than nine states allowed alternative teacher education programs at American colleges or universities.

By 1993, however, there existed post-baccalaureate alternative teacher preparation programs in 40 states. With 80% of states offering alternative programs, it was believed that teacher shortages could be addressed (Wilson, Floden, & Ferrini-Mundy, 2002). To illustrate the growing proportions of alternatively licensed teachers among the work force by the year 2000, 45 of the 50 states possessed alternative licensure enactments (Hess, Rotherham, & Walsh, 2004). By the spring of 2010, there existed 125 alternative pathways to teaching preparation across all states preparing 62,000 new teachers annually with approximately 600 program sources (AACTE, 2010). Of the total number of AACTE reporting institutions, 540 offered a master's program or post baccalaureate program, respectively in the form of a Master of Arts degree, or alternative licensure without a final degree, for initial licensure (Ludvig, Kirshstein, Sidana, Ardila-Rey, & Bae, 2010).

Obviously, teachers who achieve their initial licensure through routes such as these comprise a discernable segment of the teaching force. Almost 60,000 teachers across the country received licensure through alternative programs in the 2008-09 academic year (National Center for Alternative Licensure, 2010). At the same time, according to the *Survey of the American Teacher: Collaborating for Student Success*, many who have moved to teaching from other careers are reportedly more likely to work in schools with students in low income families and high numbers of minority students (2010). These alternative pathways to licensure have attracted diverse potential educators in terms of age, ethnicity, and background (Hart, 2010; Wilson, et al., 2002).

Demographically, the National Center for Education Information reports higher numbers of males, minorities, and older candidates when compared to those who secure licensure by more traditional pathways to teaching. The center reports 37% male and 63% female participants in alternative programs with 68% White, 12% Black, and 19% other minorities among the nation's alternative certification routes (Feistritzer, 2005). Typically, these teacher candidates have the desire to be effective in the classroom and seek the benefits of quality programs that will enable them to reach that goal (Hart, 2008). At the same time, these multiple pathways must remain of such quality so as not to weaken the universal integrity of teacher preparation (Berry, Daughtrey, & Wieder, 2010).

The number and varying program requisites for alternative licensure programs in the United States are numerous (Boyd, et al., 2004; Richardson & Roosevelt, 2004). Because of the variations in these programs, it is difficult to make generalized statements that can be applied across the board to their effectiveness (Goe & Stickler, 2008; Buck & O'Brien, 2005). Existing alternative licensure pathways vary widely with some programs operating similarly to traditional preparation programs and others with far less rigorous requirements (Boyd, et al., 2004). Hence, the fundamental question is "not how new teachers are prepared but how well they are prepared and supported, whatever preparation pathway they may choose" (National Commission on Teaching and America's Future, 2003, p. 19). Recommendations for high-quality alternative programs leading to licensure include high program entrance requirements, mentoring and supervision of candidates, pedagogical instruction, classroom management, content curriculum, working with diverse learners, ongoing assessment, lesson design practice and high program exit standards (Wilson, Floden, & Ferrini-Mundy, 2002). These same characteristics could be applied to alternative licensure programs experienced through distance learning avenues of delivery as well.

Growth and Current Role of Distance Learning in Schools of Education

At the same time that the number of alternative licensure programs increased nationally, there has also been a transformation of teacher preparation in the United States in terms of delivery. No longer confined to a physical structure, the classroom dedicated to the preparation of future and current teachers has now become less physical and more digital in nature. The concept that a community of learning is solely applicable to a location of a four-walled structure is misleading in the context of the 21st century. Now a classroom of learners and what that means has evolved to include not only the traditional classroom but the less traditional classroom of cyberspace. The availability and possibilities of online learning has emerged as reaching far beyond the former days of correspondence courses (Maeroff, 2003).

By the 1990's, the everyday use of computers and the Internet was common enough to integrate into university students' studies (Mehrotra, Hollister, & McGahey, 2001). A generation later has seen instructional technology metamorphosing from the world of overhead projectors and slide carousels to interactive communication and asynchronous modalities (Johnson, 2003). With Web 2.0 technologies of far greater collaborative context, the world of distance learning has surpassed the initial generation of technological capabilities and has now been eagerly embraced by post-secondary educators. Advances in technology and communication techniques have made available the tools necessary in integrating "convenience and flexibility of time-and-place independent learning opportunities" (Eastman & Swift, 2001, p. 33). Furthermore, there is a simultaneous interest on the part of these institutions to develop less costly alternatives to face-to-face coursework (Hollister, McGahey, & Mehrotra, 2001).

As the general development and proliferation of distance learning coursework has seen phenomenal and widespread growth in recent decades in almost any area of study, it is possible to take courses and even entire programs of study online from respected institutions of learning. Based on data gathered for the Sloan Consortium's eighth annual report on higher education's implementation of online learning, degrees and certificate programs obtainable through distance learning have increased almost exponentially. As of the fall of 2009, 31% of the country's higher institutions reported having specific education programs which were fully implemented online in the United States. Of those which did so, public institutions were overwhelmingly in the forefront of planning and implementing these distance education programs and enrolling greater numbers of students (Allen & Seaman, 2010).

Growing segments of post-secondary students achieve degrees and certificates through distance education modes of program delivery. Based on the most recent information from the National Center for Educational Statistics, approximately two thirds of the nation's 4,160 two- and four-year institutions of higher learning included distance education coursework (Parsad, Lewis, & Tice, 2008). Of those reporting, 600 four year public institutions (88%) offered distance education coursework for college credit and 1,000 (97%) of public two year institutions did so. In the eighth annual report on the state of distance learning in the United States, more than 5.6 million students were enrolled in at least one distance education course indicating an increase of 21% from the previous reporting year and revealing at least 30% of post secondary students enrolled in at least one online course (Allen & Seaman, 2010).

Distance education opportunities are responsive to an upsurge of interest on the part of both learners and providers of that learning. Societal changes have helped to drive the transformation of distance education. Distance education is frequently upheld as a means of broadening access for a diversity of students (Eastman & Swift, 2001). More nontraditional students such as those who are older than the traditional college years, those who seek part-time educational opportunities, or those who are seeking career changes in an economic downturn gravitate towards courses and programs that are not bound by specific times and locations. As greater requirements are increasingly important for advanced career and lifelong opportunities, the needs for supple scheduling, access, and flexibility are essential. To serve this population, institutions of higher education have had to revise their traditional modes of service to meet these students' needs (Merton, 2002).

A definition of distance education can be construed as instruction taking place where teacher and student are physically distanced from one another (Hollister et al., 2001). In considering the provision of such instruction via distance education, certain challenges are faced by higher education in delivering distance education programs. The primary challenge is how to structure programs delivered via distance learning while simultaneously preserving the integrity and values of the content of the delivered program (Maeroff, 2003). In other words, the vehicle of the program's delivery should in no way be influential in diluting or revising the program itself. The program is the program. The system of delivery whether on-campus or online is simply that --- the system of delivery. The quality of the program is derived from its actual content, not from the technology used in its delivery (Eastman & Swift, 2001).

In face-to-face courses taking place on campus, the instructor manages the delivery of the content and makes pedagogical decisions regarding how this is done. Similarly, in distance learning, the instructor typically manages delivery of the course through modules or units which sub-divide the course's content in increments of time allowed for each course topic. Within each module/unit is usually an introduction to the topic, readings, some form of discussion, associated assignments for students, and is typically asynchronous in nature. According to Paquette (2004), the on-line training model, or distance learning model, possesses characteristics which emancipates students from place and time restrictions and provides an entrée to local as well as global knowledge, learner collaboration, and ongoing assistance. Some believe that this delivery model "maximizes learning efficiency" on the part of the learner (Paquette, 2004, p. 14).

Asynchronous online learning occurs when students access the course at a time convenient to them largely through discussion boards, email, or other communication systems whereas synchronous delivery takes place in real time where students virtually gather at one time often through video-conferencing or chat rooms (Hrastinski, 2008). In asynchronous coursework, remote students pace themselves as they are not necessarily working together at one time. They may collaborate and intersect with the provided material within the parameters set by the instructor. The instructor functions as advisor and course expert and maintains communication with students through email, discussion boards, and other exchanges (Paquette, 2004). More common is the asynchronous delivery where there is no requirement for a real-time class meeting.

With each passing year, distance learning grows in scope and delivery and is no longer considered an outgrowth program but a viable and valued part of institutions of higher learning. In fact, the bulk of recent growth in online enrollment appears to originate from expansions to institutions' existing programs (Allen & Seaman, 2010). Furthermore, it is possible to experience a variety of depth and breadth regarding distance learning experiences. In today's educational world, students may more readily select program offerings which more precisely fit their particular circumstances and idiosyncratic needs based on factors such as current employment hours, career goal aspirations and/or familial responsibilities. The flexibility afforded by distance learning offers students the means by which their individual educational needs might be met. In considering the quality of selecting one avenue of learning over the other, Daves and Roberts (2010) discovered in their study comparing education students' perceptions involving online and face-to-face programs, that students in both program settings had similar experiences and overall satisfaction in terms of social connectedness and learning.

Within this context of distance education, the field of educating the teaching profession has been impacted as well. In order to extend their teaching licensure programs, institutions have sought to integrate these programs into their catalog of program offerings (Ernst, 2008; Smith, Smith, & Boone, 2000). Among AACTE institutions, more distance learning courses were offered at colleges and universities with greater numbers of graduate students and larger student enrollment in schools of education (Ludvig, Kirshstein, Sidana, Ardila-Rey, & Bae, 2010). With increasing globalization as well as enhanced communication capabilities, institutional practices and programs have swiftly evolved to offer distance education course options as part of their offerings (Burbules & Callister, 2000). As distance education expanded so did competition among institutions of higher learning to provide both the more classroom-centered web-enhanced courses and the more totally web-based courses (Mehrotra, Hollister, & McGahey, 2001).

Furthermore, general methods of instruction can be transplanted effectively into distance learning settings including those of teacher preparation without changing actual program content (Olson & Werhan, 2005). Currently, colleges of education responsible for preparing the nation's teachers are confronted by two challenges. First, they are mandated to produce effective educators capable of teaching students of great diversity (Emihovich, 2008). Second, they are urged to provide increased numbers of online courses as well as blended courses in answer to demand (Schrum, Burbank, & Capps, 2007).

This movement certainly encompasses programs responsible for the preparation of future teachers. In fact, the viability of using distance education as a means of alternative teaching licensure certainly allows for those students who might not otherwise have access to such learning opportunities (Olson & Werhan, 2005).

A Sense of Preparedness and INTASC Standards

Teacher effectiveness is viewed even more closely in an age where our schools are progressively more scrutinized and held accountable for students' academic progress (Goe, 2007). Provision of these effective teachers is an essential challenge of current school reform (Hart, 2010). Upon entering the classroom, novice teachers face growing student populations who live in poverty, possess learning challenges, speak a language other than English, or are minorities (Darling-Hammond, 2006). The teacher has been found to make a greater difference in students' academic progress even when compared to other important classroom factors such as class size, student achievement level, or the demographic make-up of students in the classroom (Wright, Horn, & Sanders, 1997). Therefore, how a teacher has been prepared prior to entering such an environment becomes pivotal in the success of the novice teacher and the students that teacher encounters.

In establishing identifying criteria for effective novice teachers, among the most thorough standards were those introduced in 1992 by the Interstate New Teacher Assessment and Support Consortium (INTASC). This collection of standards reflected universal and identifiable expectations for teachers entering the classroom (Alban, Proffitt, & SySantos, 1998; Capraro, Capraro, & Helfeldt, 2010). Subsequently, many states adopted these standards as initial licensure criteria and they were subsequently integrated into many university teacher preparation programs as a means of training and producing effective teachers.

The standards established an identifiable collection of exact principles with which the novice teacher upon entering the classroom actually used to interact with students who would benefit from teacher possession of knowledge, skills, and dispositions or, reversely, does not benefit from them. These knowledge, skills, and dispositions of the INTASC standards could also provide a means of measuring teacher preparation program outcomes (Lang & Wilkinson, 2008).

A perceived sense of preparedness for the classroom and its accompanying challenges aid in promoting self-efficacious teaching behaviors. Teachers who recognized their own preparedness in their teaching programs were more likely to be self-efficacious (Darling-Hammond, 2003; Gallo & Little, 2003; Hoy & Woolfolk, 1990). Thus, in considering teachers' sense of preparedness, it is necessary to contemplate self-efficacious behavior and its general impact. Social cognitive theory focuses on the concept of agency, or how individuals are cognizant of their own capabilities in influencing their lives (Goddard, 2003). Indeed, it is this agency that is the driving force in individuals' pursuit of demanding and personally satisfying goals. Teacher efficacy could be defined as the self-evaluation of one's own abilities to causally impact student learning (Tschannen-Moran & Hoy, 2001). Therefore, it is necessary to develop a self-efficacy where teachers feel that they have the power to positively affect their students' learning outcomes (Ashton, 1984). Teacher efficacy is the springboard for future classroom success where teacher actions are built on the internal belief, or sense, that they are actually prepared to be successful with their students.

Possessing a sense of preparedness, or assurance in one's pre-service training and subsequent abilities necessary for the classroom, is the first step in becoming an effective teacher. Teachers professing that they had received sufficient preparation for the classroom were more likely to be self-efficacious in their classrooms while those professing a lower sense of preparation for the classroom were more likely to feel less prepared to successfully teach (Gallo & Little, 2003). Thus, the need to establish a high sense of preparedness is elemental to subsequent teacher development of self-efficacious classroom behaviors. Therefore, ascertaining teacher candidate perceptions of preparedness for the classroom becomes vital to ascertaining pre-service program effectiveness.

ONLINE ALTERNATIVE LICENSURE TEACHER CANDIDATES' PERCEPTIONS OF PREPAREDNESS

The Study's Purpose and Participants

The purpose of this study was to investigate teacher candidates' perceptions of preparedness after completion of an online alternative licensure program at a large urban university in the southeast United States. As part of this study, the following research question was posed: *Is there a significant difference in the reported sense of preparedness between completers of a graduate certificate enrolled in a distance education teacher preparation program and those in an on campus teacher preparation program?*

All participants of this study were graduate students who possessed undergraduate degrees and had been enrolled in the college's alternative licensure program known as the Graduate Certificate in Teaching Program. This program, although not degreed, leads to a recommendation for licensure and centers on knowledge, skills, and dispositions deemed crucial for a teacher commencing a career in the classroom. By the end of their program, participants would have completed 18 hours of graduate teaching courses, any content-deficiency coursework, extensive clinical experiences in schools, a student teaching internship, a portfolio, and PRAXIS II requirements. Program completers also had the option to apply and complete 21 additional hours to receive a Masters of Arts in Teaching degree from the university. A sample of convenience was used for this particular study. The determination of this particular pool of participants was due to their status as alternative licensure students who, upon program completion, would most likely possess the program knowledge and experiences necessary to provide an accurate picture of the current perception of candidate preparedness for the classroom.

Measurement

Upon completion of their alternative certificate program, participants submitted the College of Education's Initial Licensure Completers Exit Survey (hereafter referred to as the exit survey), which was the primary instrument used in the study. The exit survey was a created component of the College of Education's Comprehensive Candidate Assessment System (CCAS). In recognizing the importance of high teacher quality, the college adhered to the concept that "work of a College of Education must be grounded in evidence-based practice and professional wisdom and the work must be informed by ongoing assessments that lead to continuous improvement" (Anderson et al., 2004, p. 8). In emphasizing these attributes of effective professional educators, the college's assessment system also reflected the tenets of appropriate accrediting bodies and standards required by the state's Department of Public Instruction. This included strong consideration of INTASC standards in the development of the conceptual framework as well as the college's assessment system, including the exit survey (Anderson et al., 2004).

All initial licensure programs possessed common assessments and standards during the time frame of data collection for this study (e.g. fall 2008-fall 2010 semesters). This indicated INTASC standards being used commonly across courses and programs (College of Education Data Annual Report, 2009). The exit survey was used by the College of Education to measure teacher candidate perceptions of their preparation experiences upon program completion. Prior to students' final conferences with their supervisors, teacher candidates were asked to digitally complete the exit survey. Items included in the survey instrument were comparatively short and understandable and had been developed and field tested in the fall of 2003. Full implementation of this instrument began in the spring of 2004. Survey research is often used to obtain better understanding of current situations (Leedy & Ormomd, 2001).Therefore, utilizing survey research methodology for this study offered a viable vehicle for gaining insight into teacher candidates' perceptions of their preparedness. With survey research, it is possible to clearly describe a situation or scrutinize relationships among variables, as well as access information regarding participant opinions or feelings regarding identified topics (Muijs, 2004).

The survey possessed two sections. Section 1 measured teacher candidates' beliefs regarding relevancy of their preparation. Preparation areas included general education, major background deficiency courses, academic concentration courses, professional education courses, clinical experiences, and student teaching. Section 2 of the exit survey was based on the INTASC standards which reflected the four essential sections of core principles, knowledge, dispositions, and performance (INTASC Consortium, 1992). A 3-point Likert scale was used with the following ratings:

- 3 = I have developed exceptionally effective knowledge, skills, and dispositions in relationship to this standard.
- 2 = I have developed adequate effective knowledge, skills, and dispositions in relationship to this standard.
- 1 = I do not yet have the knowledge, skills, and dispositions in relationship to this standard (Initial Licensure Completers Exit Survey, n.d.).

Procedure

To measure program completers' sense of preparedness, the eleven INTASC-based survey items were used to develop a construct of preparedness resulting in a scale of 3-33. Scales with multiple items have been found to be viable in quantifying a construct that is not directly measureable (Gliem and Gliem, 2008). Likert scale items based on INTASC standards, which were combined to create a composite score for the construct of preparedness are displayed in Table 1.

As INTASC standards and their principles have been widely accepted by leaders in education, the instrument's 11 INTASC-based survey items could be considered valid measurements in the expected preparation of beginning teachers (Lang & Wilkinson, 2008). A predictive validity existed as INTASC-based questions would be generally expected to predict preparedness regarding the classroom. To measure participants' sense of preparedness, a scale factor, termed "preparedness," was created from the INTASC-based questions comprising the major portion of the exit survey instrument. Face validity was evident where, as described by Sirkin (2005), expert individuals have found such a measurement used

Table 1. Survey items used to create a construct of preparedness (INTASC Consortium, 1992, p. 14-33)

INTASC- Based Standard	Description
Content Pedagogy	Understands the central concepts, tools of inquiry, and structures of the discipline(s) she/he teaches and can create learning experiences that make these aspects of subject matter meaningful to students.
Student Development	Understands how children learn and develop and creates learning opportunities to support their intellectual, social, and personal development.
Diverse Learners	Understands how students differ in their approaches to instructional opportunities that are adapted to diverse learners.
Multiple Instructional Strategies	Understands and uses a variety of instructional strategies to encourage student development of critical thinking, problem solving, and performance skills
Motivation and Management	Understands individual and group motivation and behavior to create a learning environment that encourages social interaction, active engagement in learning, and self motivation.
Communications and Technology	Uses effective verbal, nonverbal, and media communication techniques to foster active inquiry, collaboration, and supportive interaction in the classroom.
Planning	Plans based upon knowledge of subject matter, students, the community, and curriculum goals.
Assessment	Understands and uses formal and informal assessment strategies to evaluate and ensure the continuous intellectual, social, and physical development of the learner.
Professional Development	As a reflective practitioner, continually evaluates the effects of choices and actions on students, parents, and other professionals in the learning community and actively seeks out opportunities to grow professionally.
School and Community Involvement	Fosters relationships with school colleagues, parents, and agencies in the larger community to support students' learning and well being.
Work with Families	Understands differences in families and creates opportunities to involve families in supporting student learning.

as appropriately addressing the concept being measured. To aid in drawing accurate conclusions regarding the data and measuring internal survey consistency, Cronbach's alpha was performed which indicated strong internal consistency and reliability of the factor. To determine item dimensionality, a factor analysis was also conducted.

It has been noted that the principles of INTASC "provide an appropriate standards-based construct that allows college personnel to do what they need to do -- make decisions that are less likely to be successfully challenged while, at the same time, providing data that can be aggregated to improve the outcomes of their programs. Valid, reliable, and fair measurement can lead to information about what students are learning and what they are not learning. It can also serve as a predictor of future behavior and a tool for both individual and program improvement" (Lang & Wilkinson, 2008, p.4).

Discussion of Results

Study results suggested that participants' sense of preparedness as measured by the INTASC standards-based construct was not affected by the mode of delivery –online or face-to face on the university's campus. Both groups of program completers rated their preparation similarly highly. One-way ANOVA was used to gain information to ascertain any difference between modes of delivery. Table 2 below displays the means and standard deviations comparing online and on-campus completers' sense of preparedness.

No significance between groups in their perception of preparedness was evident, $F(2, 440) = .145, p>.05$. No significant differences of the two modes of delivery were evident (p=.865). However, the extremely low numbers (N=47) of online completer respondents was a limiting factor that should definitely be considered. A further limita-

Table 2. Online and on-campus program completers

	Number	Mean	Standard Deviation
Online Program Completers	*N*=47	29.0000	3.45153
On-Campus Program Completers	*N*=387	29.1886	3.59160

tion involved whether participants were actually already working as a teacher while pursuing licensure or possessed previous work experience akin to teaching.

As both online and on-campus groups followed the same program requirements and shared the same course content, the comparable results of online and on-campus program completers were perhaps not surprising. Furthermore, the mode of delivery (e.g. online, on-campus) did not indicate a significant difference between middle and secondary program completers' sense of preparedness. It has been the case in other studies that students in both online and face-to-face modes of program delivery have reported similar experiences and satisfaction with their learning (Daves & Roberts, 2010; Steinweg, Davis & Thomson, 2005).

Distance learning is both a current and future vehicle of alternative teacher education (Olsen & Werhan, 2005). Additionally, it has been observed that both teacher candidates' state licensure and self-assessments were comparable to those who had completed their program in face-to-face coursework (Harrell & Harris, 2006). This may well be attributed to the common implementation of program coursework. As noted by Maeroff (2003), the integrity and values of a program must be preserved in distance education as it is in on-campus programs.

In considering the eleven original INTASC-based questions used in the exit survey, Table 3 below describes the online program completers' Likert responses based on the instrument's three point scale.

As can be seen, the only exit survey questions where the median deviated from a 3.0 result involved "Work with Families" (median of 2.0) and, to a lesser extent, "Student Development" (2.85). As the distribution of scores was highly skewed due to the majority of participants self-rating themselves at the high end of the Likert range, these median scores reveal distinctive differ-

Table 3. Description of online middle and secondary program completers

INTASC Standard-Based Question	Mean	Median	Standard Deviation
Content Pedagogy	2.85	3.0	.360
Student Development	2.66	2.85	.479
Diverse Learners	2.62	3.0	.491
Multiple Instructional Strategies	2.62	3.0	.534
Motivation and Management	2.62	3.0	.534
Communication and Technology	2.70	3.0	.507
Planning	2.62	3.0	.534
Assessment	2.57	3.0	.580
Professional Development	2.79	3.0	.463
School and Community Involvement	2.55	3.0	.583
Work with Families	2.40	2.0	.648

ences between these two areas as compared with other areas of preparedness within the construct. Most particularly, in an exit survey where respondents were highly likely to rate themselves as "exceptionally effective", a self-rating of "adequate" in preparedness for working with families is notable. This could well be an area of improvement for the preparation program itself. The actual percentages of responses of online program completers are reported below in Table 4.

Again, exit survey responders rated themselves highest in areas of "Content Pedagogy" (85.1%) and "Professional Development" (80.9%). "Work with Families" was rated lowest of the items with 8.5% actually professing a lack of knowledge, skills, and dispositions of this standard. Results thus indicate the value of looking more closely at that particular standard in teacher candidate preparation.

Effective teachers typically possess a degree of efficacy in their profession that is largely inspired by a sense of preparedness for the classroom. For this study, that sense of preparedness was reflected in respondents' self-reported perceptions regarding areas of content pedagogy, diverse learners, multiple instructional strategies, motivation and management, communications and technology, planning, assessment, professional development, school and community involvement, and work with families. Thus, their sense of preparedness has as a foundation of the principles of the INTASC standards which have been commonly accepted for novice teachers. The principles of the INTASC standards have constituted universal understandings for novice teachers since the early 1990's. As of April of 2011, INTASC standards have been revised in order to continue guidance of beginning teachers. They certainly re-validate and continue the function of the original INTASC standards and expressed what teachers should be able to do to foster the 21st century goal of producing career-and-college ready students (Council of Chief State School Officers, 2011).

SOLUTIONS AND RECOMMENDATIONS

Distance education is here to stay. In considering the use of distance education in teacher preparation, the implication is no longer viability. That has certainly been proven. Online learning is a reasonable alternative to more traditional face-to-face classroom settings especially as the majority of university leadership report that learning outcomes for their online students are similar to or better than those for on-campus students (Allen & Seaman, 2010). Growing online alternative licensure programs will continue to proliferate as demand allows (Olson & Werhan, 2005).

The major implications lie in how to most effectively produce these teachers possessing a high sense of preparedness for the classroom. Regardless of the licensure pathway selected, all teachers should experience high program standards from those who are responsible for their preparation (National Commission on Teaching and America's Future, 2003). Teacher candidates' sense of preparedness is nurtured largely by a program of preparation that, in turn, has the responsibility of producing effective teachers through various pathways and modes of delivery including both face-to-face and online learning. That sense of preparedness fosters the self-efficacy required to prepare students successfully and the strength of those seeds of teacher efficacy can have profound effects over the span of a career (Tschannen-Moran & Hoy, 2001). Most teachers who have entered the teaching profession rated their preparation program highly (89%) but also often rated those programs much lower (28%) in facing classroom challenges after they entered the classroom (Hart, 2010). Thus, further long-term scrutiny of teachers' sense of preparation is indicated.

Teacher Educators/Researchers should view alternative licensure program completers' high sense of preparedness in two ways. First, this is a validation of programming offered to

Table 4. Percentages of responses of online middle and secondary program completers

INTASC Standard-Based Question	Online Middle and Secondary Program Completers	Response Percentages
Content Pedagogy	1 = I do not yet have the knowledge, skills, and dispositions in relationship to this standard	0
	2 = I have developed adequate effective knowledge, skills, and dispositions in relationship to this standard.	14.9
	3 = I have developed exceptionally effective knowledge, skills, and dispositions in relationship to this standard.	85.1
Student Development	1 = I do not yet have the knowledge, skills, and dispositions in relationship to this standard	0
	2 = I have developed adequate effective knowledge, skills, and dispositions in relationship to this standard.	34.0
	3 = I have developed exceptionally effective knowledge, skills, and dispositions in relationship to this standard.	66.0
Diverse Learners	1 = I do not yet have the knowledge, skills, and dispositions in relationship to this standard	0
	2 = I have developed adequate effective knowledge, skills, and dispositions in relationship to this standard.	38.3
	3 = I have developed exceptionally effective knowledge, skills, and dispositions in relationship to this standard.	61.7
Multiple Instructional Strategies	1 = I do not yet have the knowledge, skills, and dispositions in relationship to this standard	2.1
	2 = I have developed adequate effective knowledge, skills, and dispositions in relationship to this standard.	34.0
	3 = I have developed exceptionally effective knowledge, skills, and dispositions in relationship to this standard.	63.8
Motivation and Management	1 = I do not yet have the knowledge, skills, and dispositions in relationship to this standard	2.1
	2 = I have developed adequate effective knowledge, skills, and dispositions in relationship to this standard.	34.0
	3 = I have developed exceptionally effective knowledge, skills, and dispositions in relationship to this standard.	63.8
Communication and Technology	1 = I do not yet have the knowledge, skills, and dispositions in relationship to this standard	2.1
	2 = I have developed adequate effective knowledge, skills, and dispositions in relationship to this standard.	25.5
	3 = I have developed exceptionally effective knowledge, skills, and dispositions in relationship to this standard.	72.3
Planning	1 = I do not yet have the knowledge, skills, and dispositions in relationship to this standard	2.1
	2 = I have developed adequate effective knowledge, skills, and dispositions in relationship to this standard.	34.0
	3 = I have developed exceptionally effective knowledge, skills, and dispositions in relationship to this standard.	63.8

continued on following page

Table 4. Continued

INTASC Standard-Based Question	Online Middle and Secondary Program Completers	Response Percentages
Assessment	1 = I do not yet have the knowledge, skills, and dispositions in relationship to this standard	4.3
	2 = I have developed adequate effective knowledge, skills, and dispositions in relationship to this standard.	34.0
	3 = I have developed exceptionally effective knowledge, skills, and dispositions in relationship to this standard.	61.7
Professional Development	1 = I do not yet have the knowledge, skills, and dispositions in relationship to this standard	2.1
	2 = I have developed adequate effective knowledge, skills, and dispositions in relationship to this standard.	17.0
	3 = I have developed exceptionally effective knowledge, skills, and dispositions in relationship to this standard.	80.9
School and Community Involvement	1 = I do not yet have the knowledge, skills, and dispositions in relationship to this standard	4.3
	2 = I have developed adequate effective knowledge, skills, and dispositions in relationship to this standard.	36.2
	3 = I have developed exceptionally effective knowledge, skills, and dispositions in relationship to this standard.	59.6
Work with Families	1= I do not yet have the knowledge, skills, and dispositions in relationship to this standard	8.5
	2 = I have developed adequate effective knowledge, skills, and dispositions in relationship to this standard.	42.6
	3 = I have developed exceptionally effective knowledge, skills, and dispositions in relationship to this standard.	48.9

their students enrolled in an alternative licensure program. That these students feel that they are prepared for the P-12 classroom is testament to a valuable learning experience. Second, further assessments of programs using student feedback would be most helpful in scrutinizing current programs and making possible future revisions. As programs across the nation vary in their graduates' perceptions, it would be beneficial to self-evaluate those programs in order to further develop their effectiveness (Darling-Hammond, Chung, & Frelow, 2002). In ascertaining the reality of respondents' reported perceptions, longitudinal studies which involve further inquiry into perceptions of preparation over time would be helpful. Additionally, integrating perspectives of students in distance education programs could increase providers' understanding and assessment of online programs (Zhu, 2006). The following recommendations for practice by teacher educators are proposed:

- Implement content commonality of programs across modes of delivery (e.g. distance education, on-campus).
- Implement ongoing scrutiny to ensure that modes of delivery possess common content with on-campus sections of the same coursework (including adjunct faculty).
- Consider modes of delivery to be equally valuable due to their common program content with on-campus sections of the same coursework.

P-12 Schools, Students and their Parents are the ultimate beneficiaries of programs whose teacher candidates have built a sense of preparedness for the classroom. Teacher preparation and experience play significant roles in student success (Aaronson, Barrow, & Sanders, 2007). Therefore, career and college readiness is certainly impacted by the caliber of effective teachers. College enrollment directly after high school graduation was 70% in 2009 with lower percentages reported for students of low income (55%) or Hispanic (62%) or Black families (63%) (Aud, Hussar, Kena, Bianco, Frohlich, & Kemp, 2011). A cumulative impact of effective teachers over time can most likely result in better prepared students. Therefore, it would be helpful to consider the long term effects of teacher preparation programs on student outcomes.

FUTURE RESEARCH DIRECTIONS

There are many possibilities for future research triggered by this study. As teaching, the role of teachers and education in the United States are currently in a state of flux, further research in teacher preparation, program assessments, and related research is paramount. The following summarizes a few recommendations for future research which could have impact on the field of education:

1. This study triggers curiosity regarding further examination of teacher candidates' perceptions from program completers' exit surveys and other measurement tools. Possibly, greater examination and cross-comparisons of other factors such as the quality of field experiences, content-related coursework, education-related coursework, or the value of advising could also be targeted. Program evaluations of those might prove informative as well.
2. Additional research on alternative licensure programs and their impact on student achievement is needed. These programs are not likely to fade. Therefore, scrutiny of their effectiveness as they continue is crucial. Longitudinal studies would deliver greater information on long-term effects of preparation and program completers' perceptions. In other words, does the highly reported sense of preparedness continue long past the point of program completion?
3. Research seeking information on perspectives from other sources such as faculty, school personnel, or P-12 students would aid in creating a fuller multi-dimensional picture of a sense of preparedness for the classroom. As outcome based data becomes more prevalent in twenty first century American education, it would be proactive to seek possible connections between teacher candidates' and teachers' sense of preparedness and their students' academic outcomes.
4. As greater numbers of online program students complete their program, it would be of definite value to continue to ascertain their perceptions of preparedness. Using other means of comparing their perceptions with on-campus program completers would have value as well. It would also be of benefit to discover further data on distance education program completers' subsequent pursuit of Master of Arts in Teaching.
5. As this study solely focused on middle and secondary teacher candidates, pursuing research on other program areas such as elementary, TESL, foreign languages, or special education would be informative. Pursuing similar data on programs on a state, regional or national level would produce a bigger picture of teacher preparation and preparedness for the classroom.
6. A longitudinal study of program completers' perspectives over the coming decade would encompass a larger sampling with more

definitive results. This would also aid in ascertaining teachers' sense of preparation as their classroom experiences increase.

7. Using program completer's sense of preparedness with student outcomes would garner valuable information on the connection between teacher preparation and their classroom performance.

Study Limitations

The small sample used in this study could hamper scrutiny of relationships among variables. Future studies with larger samples could further examine program completers' demographic variables' impact on perceptions of preparedness.

Results of the study pertain to a mid-sized urban university in the southeastern United States. Therefore, findings reflect the singular context and participants of this site and may not be akin to findings at other alternative licensure teacher preparation sites.

Participants' self-reporting of their perceptions formed the basis for the measurement instrument. Although the instrument itself is statistically valid, the personal perceptions of respondents may or may not reflect the same findings from other sources of perception such as faculty or school personnel. It may also be that respondents' beliefs and perceptions regarding their preparation program combine with personal beginning attitudes towards teaching coursework in general.

Participants' past or current experiences may well have had impact on their sense of preparedness. It was unknown based on the data collected for this study which alternative licensure participants were actually practicing in the classroom prior to a specific program's completion and which participants were not. This would include past teaching experiences, length of experiences, or even other experiences that would technically be deemed non-teaching and yet involved skills that could be applicable to the classroom.

CONCLUSION

A sense of preparedness can contribute to the development of self-efficacy. Collectively, whole school staffs of efficacious teachers may then have significant impact on students' learning (Goddard, Hoy, & Hoy, 2000; Tschannen-Moran & Barr, 2004). Schools with a core of such active teachers can accomplish much towards affecting student outcomes. They can be the catalysts for their students' attainment of the highest possible levels of academic achievement (Wenglinsky, 2002). In improving the future of effectiveness in the nation's teaching force, cumulative effects over time from a continual sequence of effective teachers will result in successful students who are indeed career and college ready.

In considering teacher educations programs, there will continue to be a range of avenues to teaching licensure and modes of delivery such as distance education. However, even though research has not shown evidence of dominance by any particular type of teacher preparation program over another, research does suggest that elements akin to program consistency and clarity of vision regarding learning and teaching are related to the quality of teachers and subsequent student achievement (AERA, 2005).

The future may well include additional totally web-based delivery modes, as well as blended configurations of face-to-face and online coursework. Discussion will continue to abound regarding the use of the Internet for teacher preparation (Zirkle, 2005). However, what is certain is that distance education and teacher preparation, including that of alternative licensure preparation, will continue to develop, evolve, and certainly remain a part of the fabric of schools of education. As envisioned by the American Association of the Colleges of Teachers of Education, "we have the opportunity to establish a vibrant vision for educator preparation, one that leverages the best of what has worked in the past, combined with what educators need now and in the future, in order to prepare all students for the future they deserve" (2010, p. 11).

REFERENCES

Aaronson, D., Barrow, L., & Sander, W. (2007). *Teachers and student achievement in the Chicago Public High Schools*. Retrieved from http://www.csa.com

Alban, T., Proffitt, T. D., & SySantos, C. (1998). *Defining performance based assessment within a community of learners: The challenge & the promise*. Retrieved from http://www.csa.com

Allen, E. I., & Seaman, J. (2010). *Class differences: Online education in the United States*, 2010. The Sloan Consortium. Babson Survey Research Group, USA. Retrieved from http://sloanconsortium.org/publications/survey/pdf/class_differences.pdf

American Association of Colleges of Teacher Education and the Partnership for 21st Century. (2010). *Evidence of teacher effectiveness by pathway to entry into teaching*. Retrieved from http://aacte.org/pdf/Publications/Reports%20_Studies/Evidence%20of%20Teacher%20Effectiveness%20by%20Pathway.pdf

American Association of Colleges of Teacher Education and the Partnership for 21st Century Skills. (2010). *Educator preparation: A vision for the 21st century*. Retrieved from http://aacte.org/email_blast/president_e-letter/files/02-16-2010/Educator%20Preparation%20and%2021st%20Century%20Skills%20DRAFT%2002151O.pdf

American Education Research Association. (2005). *The impact of teacher education: What do we know?* Retrieved from http://www.aera.net/uploadedFiles/News_Media/News_Releases/2005/STE-WhatWeKnow1.pdf

Anderson, K., Allen, L., Brooks, K., DiBiase, W., Finke, J., & Gallagher, S. Calhoun, M. (2004). *Rising to the challenge: Preparing excellent professionals: The conceptual framework for professional education programs at UNC Charlotte* (2nd ed.). Retrieved from http://education.uncc.edu/coe/Conceptual_Framework/conceptual%20framework.pdf

Ashton, P. (1984). Teacher efficacy: A motivational paradigm for effective teacher education. *Journal of Teacher Education, 35*(5), 28–32. doi:10.1177/002248718403500507

Aud, S., Hussar, W., Kena, G., Bianco, K., Frohlich, L., Kemp, J., & National Center for Education Statistics. (2011). *The condition of education 2011*. NCES 2011-033, National Center for Education Statistics.

Berry, B., Daughtrey, A., & Wieder, A. (2010). *Preparing to lead an effective classroom: The role of teacher training and professional development programs*. Center for Teaching Quality. Retrieved from http://www.eric.ed.gov/PDFS/ED509718.pdf

Burbules, N. C., & Callister, T. A. Jr. (2000). Universities in transition: The promise and the challenge of new technologies. *Teachers College Record, 102*(2), 271–293. doi:10.1111/0161-4681.00056

Capraro, M., Cpraro, R., & Helfeldt, J. (2010). Do differing types of field experiences make a difference in teacher candidates' perceived level of competence? *Teacher Education Quarterly, 37*(1), 131–154.

College of Education. (2009). *Annual data report: Introduction*. Retrieved from http://education.uncc.edu/assessment/secure/2009_Data_Report.htm

Council of Chief State School Officers. (2011, April). *Interstate Teacher Assessment and Support Consortium (INTASC) model core teaching standards: A resource for state dialogue*. Washington, DC: Author. Retrieved from http://www.ccsso.org/Documents/2011/_Model_Core_Teaching_Standards_2011.pdf

Darling-Hammond, L. (2000). How teacher education matters. *Journal of Teacher Education*, *51*(3), 166–173. doi:10.1177/0022487100051003002

Darling-Hammond, L. (2003). Keeping good teachers: Why it matters, what leaders can do. *Educational Leadership*, *60*(8), 6–13.

Darling-Hammond, L. (2006). *Powerful teacher education: Lessons from exemplary programs*. San Francisco, CA: John Wiley & Sons.

Darling-Hammond, L., & Bransford, J. (2005). *Preparing teachers for a changing world: What teachers should learn and be able to do*. San Francisco, CA: Jossey-Bass.

Darling-Hammond, L., Chung, R., & Frelow, F. (2002). Variation in teacher preparation: How well do different pathways prepare teachers to teach. *Journal of Teacher Education*, *53*(4), 286–302. doi:10.1177/0022487102053004002

Daves, D. P., & Roberts, J. G. (2010). Online teacher education programs: Social connectedness and the learning experience. *Journal of Instructional Pedagogies*, *4*. Retrieved from http://www.aabri.com/jip.html

DiGiulio, R. C. (2004). *Great teaching: What matters most in helping students succeed*. Thousand Oaks, CA: Corwin Press.

Eastman, J. K., & Swift, C. (2001). New horizons in distance education: The online learner-centered marketing class. *Journal of Marketing Education*, *23*(1), 25–34. doi:10.1177/0273475301231004

Edwards, B., Flowers, C., & Stephenson-Green, E. (2005). *Comprehensive assessment system*. UNC Charlotte College of Education. Retrieved from http://education.uncc.edu/assessment/

Emihovich, C. (2008). Preparing global educators: New challenges for teacher education. *Teacher Education and Practice*, *21*(4), 446–448.

Ernst. J. V. (2008). A comparison of traditional and hybrid online instructional presentations in communication technology. *Journal of Technology Education*, *19*(2). Retrieved from http://scholar.lib.vt.edu/ejournals/JTE/v19n2/pdf/ernst.pdf

Feistritzer, C. (2005). *Profile of alternative route teachers*. National Center for Education Information. Retrieved from http://www.ncei.com/PART.pdf

Fenstermacher, G. D., & Richardson, V. (2005). On making determinations of quality in teaching. *Teachers College Record*, *107*(1), 186–213. doi:10.1111/j.1467-9620.2005.00462.x

Gallo, R., & Little, E. (2003). Classroom behavior problems: The relationship between preparedness, classroom experiences, and self-efficacy in graduate and student teachers. *Australian Journal of Educational & Developmental Psychology* (3), 21-34.

Gliem, J. A., & Gliem, J. R. (2003). *Calculating, interpreting, and reporting Cronbach's alpha reliability coefficient for Likert-type scales*. Midwest Research to Practice Conference in Adult, Continuing, and Community Education. Retrieved from https://scholarworks.iupui.edu/bitstream/handle/1805/344/Gliem+&+Gliem.pdf? sequence=1

Goddard, R. D. (2003). The impact of schools on teacher beliefs, influence, and student achievement: The role of collective efficacy beliefs. In Raths, J. D., & McAninch, A. R. (Eds.), *Teacher beliefs and classroom performance: The impact of teacher education. Advances in teacher education* (*Vol. 6*, pp. 183–202). Greenwich, CT: Information Age Pub.

Goddard, R. D., Hoy, W. K., & Hoy, A. W. (2000). Collective teacher efficacy: Its meaning, measure, and impact on student achievement. *American Educational Research Journal*, *37*(2), 479–507.

Goe, L. (2007). *The link between teacher quality and student outcomes: A research synthesis*. National Comprehensive Center for Teacher Quality. Retrieved from http://www.tqsource.org/publications/LinkBetweenTQandStudentOutcomes.pdf

Goe, L. (2007). Linking teacher quality and student outcomes. In C. A. Dwyer (Ed.), *America's challenge: Effective teachers for at-risk schools and students.* National Comprehensive Center for Teacher Quality. Retrieved from http://www.tqsource.org/publications/NCCTQBiennialReport.pdf

Goe, L., & Stickler, L. M. (2008). *Teacher quality and student achievement: Making the most of recent research.* National Comprehensive Center for Teacher Quality. Retrieved from http://www.tqsource.org/publications/March2008Brief.pdf

Harrell, P. E., & Harris, M. (2006). Teacher preparation without boundaries: A two-year study of an online teacher certification program. *Journal of Technology and Teacher Education*, *14*(4), 755–774.

Hart, P. D. (2010). *Career changes in the classroom: A national portrait.* The Woodrow Wilson National Fellowship Foundation. Retrieved from http://www.woodrow.org/images/pdf/policy/CareerChangersClassroom_0210.pdf

Hess, F. M., Rotherham, A. J., & Walsh, K. (2004). *A qualified teacher in every classroom? Appraising old answers and new ideas*. Cambridge, MA: Harvard Education Press.

Hollister, C. D., McGahey, L., & Mehrotra, C. (2001). *Distance learning: Principles for effective design, delivery, and evaluation*. Thousand Oaks, CA: Sage Publications.

Hoy, W. K., & Woolfolk, A. E. (1990). Socialization of student teachers. *American Educational Research Journal*, *27*(2), 279–300.

Hraslinski, S. (2008). Asynchronous and synchronous e-learing. *EDUCAUSE Quarterly*, *31*(4).

Institute for Higher Education Policy. (1999). *What's the difference? A review of contemporary research on the effectiveness of distance learning in higher education*. Washington, DC: National Education Association. Retrieved from http://www.ihep.org/assets/files/publications/s-z/WhatDifference.pdf

Interstate New Teacher Assessment and Support Consortium. (1992). *Model standards for beginning teacher licensing, assessment and development: A resource for state dialogue*. Washington, DC: Council of Chief State School Officers.

Interstate Teacher Assessment and Support Consortium. (2010). *Official website*. Retrieved from http://www.ccsso.org/Resources/Programs/ Interstate_Teacher_Assessment_Consortium_(InTASC).html

Johnson, J. L. (2003). *Distance education: The complete guide to design, delivery, and improvement*. New York, NY: Teachers College Press.

Labaree, D. F. (2004). *The trouble with ed schools*. New Haven, CT: Yale University Press.

Lang, W., & Wilkinson, J. (2008, March). *Measuring teacher dispositions with different item structures: An application of the Rasch model to a complex accreditation requirement*. Paper presented at Annual Meeting of the American Educational Research Association, New York, NY.

Ludvig, M., Kirshstein, R., Sidana, A., Ardila-Rey, A., & Bae, Y. (2010). *An emerging picture of the teacher preparation pipeline: A report by the American Association of Colleges for Teacher Education and the American Institutes for Research*. Retrieved from http://aacte.org/pdf/Publications/Resources/PEDS%20Report%20-%20An%20Emerging%20Picture%20of%20the%20Teacher%20Preparation%20Pipeline.pdf

Maeroff, G. I. (2003). *A classroom of one: How online learning is changing our schools and colleges*. New York, NY: Palgrave Macmillan.

Mehrotra, C. M., Hollister, C. D., & McGahey, L. (2001). *Distance learning: Principles for effective design, delivery, and evaluation*. Thousand Oaks, CA: Sage Publications.

Merton, A. G. (2002). *Improving education outcomes: In colleges, universities, and beyond*. Panel Discussion, Conference Series 47, Education in the 21st Century: Meeting the Challenges of a Changing World. Retrieved from http://www.bos.frb.org/economic/conf/conf47/conf47u.pdf

MetLife. (2010). *Survey of the American teacher: Collaborating for student success*. Retrieved from http://www.metlife.com/assets/cao/contributions/foundation/american-teacher/MetLife_Teacher_Survey_2009.pdf

Muijs, D. (2004). *Doing quantitative research in education with SPSS*. London, UK: Sage Publications.

National Center for Alternative Licensure. (2010). *A state by state analysis: Introduction*. Retrieved from http://www.teach-now.org/intro.cfm

National Commission on Teaching and America's Future. (2003). *No dream denied: A pledge to America's children: A summary*. Retrieved from http://www.ecs.org/html/Document.asp?chouseid=4269

Olson, S. J., & Werhan, C. (2005). Teacher preparation via online learning: A growing alternative for many. *Action in Teacher Education*, *27*(3), 76–84. doi:10.1080/01626620.2005.10463392

Paquette, G. (2004). *Instructional engineering in networked environments. Instructional technology & training series*. San Francisco, CA: Pfeiffer.

Parsad, B., Lewis, L., & Tice, P. (2008). *Distance education at degree-granting postsecondary institutions: 2006-07: First look*. National Center for Education Statistics. Retrieved from http://nces.ed.gov/pubs2009/2009044.pdf

Richardson, V., & Roosevelt, D. (2004). Teacher preparation and the improvement of teacher education. In Smylie, M. A., & Miretzky, D. (Eds.), *Developing the teacher workforce*. Chicago, IL: National Society for the Study of Education. doi:10.1111/j.1744-7984.2004.tb00032.x

Rivkin, S. G., Hanushek, E. A., & Kain, J. F. (2005). Teachers, schools, and academic achievement. *Econometrica: Journal of the Econometric Society*, *73*(2), 417–458. doi:10.1111/j.1468-0262.2005.00584.x

Schrum, L., Burbank, M. D., & Capps, R. (2007). Preparing future teachers for diverse schools in an online learning community: Perceptions and practice. *The Internet and Higher Education*, *10*(3), 204–211. doi:10.1016/j.iheduc.2007.06.002

Singer, N., Catapano, S., & Huisman, S. (2010). The university's role in preparing teachers for urban schools. *Teaching Education*, *21*(2), 119–130. doi:10.1080/10476210903215027

Sirkin, R. M. (2005). *Statistics for the social sciences* (3rd ed.). Thousand Oaks, CA: Sage Publications.

Smith, S., Smith, S., & Boone, R. (2000). Increasing access to teacher preparation: The effectiveness of traditional instructional methods in an online learning environment. *Journal of Special Education Technology, 15*(2), 37–46.

Steinweg, S., Davis, M., & Thomson, W. (2005). A comparison of traditional and online instruction in an introduction to special education course. *Teacher Education and Special Education, 28*(1), 62–73. doi:10.1177/088840640502800107

Tschannen-Moran, M., & Barr, M. (2004). Fostering student learning: the relationship of collective teacher efficacy and student achievement. *Leadership and Policy in Schools, 3*(3), 189–209. doi:10.1080/15700760490503706

Tschannen-Moran, M., & Woolfolk Hoy, A. (2001). Teacher efficacy: Capturing an elusive construct. *Teaching and Teacher Education, 17*, 783–805. doi:10.1016/S0742-051X(01)00036-1

US Department of Education. (2009). *Race to the top application for initial funding*. Retrieved from http://www2.ed.gov/programs/racetothetop/phase1-applications/minnesota.pdf

Wenglinsky, H. (2000). *How teaching matters: Bringing the classroom back into discussions of teacher quality*. Policy Information Center, Mail Stop 04-R, Educational Testing Service. Retrieved from http://www.csa.com

Wenglinsky, H. (2000). *Teaching the teachers: Different settings, different results. policy information report.* Policy Information Center. Retrieved from http://www.csa.com

Wenglinsky, H. (2002). How schools matter: The link between teacher classroom practices and student academic performance. *Education Policy Analysis Archives, 10*(12). Retrieved from http://www.csa.com

Wilson, S. M., Floden, R. E., & Ferrini-Mundy, J. (2002). Teacher preparation research: An insider's view from the outside. *Journal of Teacher Education, 53*(3), 190–204. Retrieved from http://jte.sagepub.com/content/53/3/190.full.pdf+html doi:10.1177/0022487102053003002

Wright, S. P. Horn, S. P., & Sanders, W. L. (1997). Teacher and classroom context effects on student achievement: Implications for teacher evaluation. *Journal of Personnel Evaluation in Education, 11*(1), 57-67. Retrieved from http://www.sas.com/govedu/edu/teacher_eval.pdf

Zhu, E. (2006). Interaction and cognitive engagement: An analysis of four asynchronous online discussions. *Instructional Science, 34*(6), 451–480. doi:10.1007/s11251-006-0004-0

ADDITIONAL READING

Allen, E. I., & Seaman, J. (2009). *Learning on demand: Online education in the United States, 2009*. The Sloan Consortium, Babson Survey Research Group, USA. Retrieved from http://sloanconsortium.org/publications/survey/pdf/learningondemand.pdf

Allen, E. I., & Seaman, J. (2011). *Going the distance: Online education in the United States, 2011*. The Sloan Consortium, Babson Survey Research Group, USA. Retrieved from http://www.onlinelearningsurvey.com/reports/goingthedistance.pdf

Caywood, K., & Duckett, J. (2003). Online vs. on-campus learning in teacher education. *The Journal of Teacher Education Division of the Council for Exceptional Children, 26*(2), 98–105. doi:10.1177/088840640302600203

Darling-Hammond, L. (2010). Teacher education and the American future. *Journal of Teacher Education, 61*(1/2), 35–47. doi:10.1177/0022487109348024

Insung, J., & Latchem, C. (Eds.). (2012). *Quality assurance and accreditation in distance education and e-learning: Models, policies and research*. New York, NY: Routledge.

Paechter, M., & Maier, B. (2010). Online or face-to-face? Students' experiences and preferences in e-learning. *The Internet and Higher Education, 13*(4), 292–297. doi:10.1016/j.iheduc.2010.09.004

Tournaki, N., Lyublinskaya, I., & Carolan, B. V. (2009). Pathways to teacher certification: Does it really matter when it comes to efficacy and effectiveness? *Action in Teacher Education, 30*(4), 96–109.

Ward, M. E., Peters, G., & Shelley, K. (2010). Student and faculty perceptions of the quality of online learning experiences. *International Review of Research in Open and Distance Learning, 11*(3), 57–77.

KEY TERMS AND DEFINITIONS

Alternative Licensure: A non-traditional licensure pathway where someone with a minimum of a bachelor's degree can complete a teacher preparation program and obtain teacher certification.

Asynchronous: Refers to online communication and interaction that does not occur in real time.

Distance Education: Synonymous with online learning or distance learning.

Graduate Certificate Program: This non-degree program leads to a recommendation for licensure and centers on knowledge, skills, and dispositions deemed crucial for a teacher commencing a career in the classroom.

Sense of Preparedness: An assurance in one's pre-service training and subsequent abilities necessary for the classroom.

Synchronous: Refers to real time online communication and interaction.

Teacher Efficacy: A belief in personal abilities to impact student learning.

Section 2
Online Teaching and Learning Initiatives in Online Teacher Education

Chapter 4
Virtually Unprepared:
Examining the Preparation of K-12 Online Teachers

Michael K. Barbour
Wayne State University, USA

Jason Siko
Grand Valley State University, USA

Elizabeth Gross
Wayne State University, USA

Kecia Waddell
Wayne State University, USA

ABSTRACT

At present, there are very few examples of the preparation of teachers for the online environment in teacher education. Even more unfortunate is that less than 40% of all online teachers in the United States reported receiving any professional development before they began teaching online. While some virtual schools provide some training to their own teachers, in most instances, no such training is provided to the school-based personnel. This is unfortunate, as K-12 student success in online learning environments require support from both the online teacher and the local school-based teacher. Clearly, there is a need for teacher education programs to equip all teachers with initial training in how to design, deliver, and – in particular – support K-12 online learning. This chapter begins with an examination of the act of teaching online and how that differs from teaching in a face-to-face environment. Next, the chapter describes existing teacher education initiatives targeted to pre-service teachers (i.e., undergraduate students) and then in-service teachers (i.e., graduate students). This is followed by an evaluation of current state-based initiatives to formalize online teaching as an endorsement area. Finally, a summary of the unique aspects of teaching online and how some initiatives have attempted to address these unique skills, before outlining a course of action that all teacher education programs should consider adopting.

DOI: 10.4018/978-1-4666-1906-7.ch004

INTRODUCTION

In the United States, the first K-12 school to begin using online learning was the private Laurel Springs School in California around 1994. This was followed by the Utah eSchool in 1994-95, which primarily used a correspondence model, but did offer some online courses (Barbour, 2009). In 1996-97, the Florida Virtual School (FLVS) and Virtual High School Global Consortium, which were created using state or federal grants, came into being (Clark, 2007). At the turn of the millennia, Clark (2001) estimated that there were between 40,000 and 50,000 virtual school enrolments. Almost a decade later, Picciano and Seaman (2009) indicated that there were over 1,000,000 students enrolled in online courses, while Watson, Murin, Vashaw, Gemin, & Rapp (2010) reported significant online learning activity in 48 states, and the District of Columbia. In 2006, Michigan became the first state in the US to require that all students complete an online learning experience in order to graduate from high school (a move that has been followed by other states, such as New Mexico, Alabama and Florida). Finally, some have gone so far to predict that the majority of K-12 education will be delivered using online learning by the year 2020 (Christensen, Horn & Johnson, 2008).

Wood (2005) stated there was a "persistent opinion that people who have never taught in this medium [i.e., online] can jump in and teach a class, [however], a good classroom teacher is not necessarily a good online teacher" (p. 36). Roblyer and McKenzie (2000) indicated that many of the factors that make a successful online teacher, such as good communication and classroom organization skills, were similar to those for any successful teacher, yet Davis, Roblyer, Charania, Ferdig, Harms, Compton and Cho (2007) discovered "effective virtual teachers have qualities and skills that often set them apart from traditional teachers" (p. 28). Some of the skills necessary for teaching in an online environment are consistent with those provided by traditional teacher education programs, but there are other necessary skills that are largely absent (Davis & Roblyer, 2005).

At present, there are very few examples of the preparation of teachers for the online environment in teacher education. Even more unfortunate is that Rice and Dawley (2007) found that less than 40% of all online teachers in the United States reported to receiving any professional development before they began teaching online. While some virtual schools provide some training to their own teachers, in most instances no such training is provided to the school-based personnel. This is unfortunate, as Aronson and Timms (2003) indicated that K-12 student success in online learning environment required support from both the online teacher and the local school-based teacher. Clearly there is a need for teacher education programs to equip all teachers with initial training in how to design, deliver, and – in particular – support K-12 online learning.

This chapter begins with an examination of the act of teaching online and how that differs from teaching in a face-to-face environment. Next, we describe existing teacher education initiatives targeted to pre-service teachers (i.e., undergraduate students), and then in-service teachers (i.e., graduate students). This is followed by an evaluation of current state-based initiatives to formalize online teaching as an endorsement area. Finally, we summarize the unique aspects of teaching online and how some initiatives have attempted to address these unique skills, before outlining a course of action that all teacher education programs should consider adopting.

EXAMINING ONLINE TEACHING

Many of us can think of instances where we thought poorly of our professor's ability to teach. Perhaps it was due to poor preparation, a lack of content knowledge, or an inability to explain complex concepts in terms a novice could understand. In other instances, we have all had excellent teachers who motivated us to do our best and helped us get through a class that we never thought we could. Currently, more and more people are able to say the same thing about their online teachers. Some of us have had excellent online instructors and some of us have had horrible online instructors.

Students enrolled in online courses encounter a variety of formats for delivery of instruction (Kaseman & Kaseman, 2000), and thus the skills required of teachers will vary. In the independent model of instruction, students are primarily self-taught, progressing through the content at their own pace and completing much of the work offline or through database-driven online systems. Students will take assessments throughout the course, but there is little in the way of feedback from the instructor. In this delivery model the teacher has little interaction with the student, and therefore does not need much in the way of communication skills. In this case, unless the course is prepackaged, the teacher needs to be skilled in the technical aspects of delivery and the organization of the course.

In an asynchronous course, students work through the content when it is convenient for them. Asynchronous courses have little to no live or real-time interaction with an instructor. However, that does not mean that there is no communication between the teacher, student, and classmates (Zucker & Kozma, 2003). Teachers need to provide feedback on assignments, and students must often interact with one another via discussion boards or group assignments (Friend & Johnston, 2005). However, unless the student is taking an asynchronous course with classmates in a brick-and-mortar school, the potential for student isolation is still present. Therefore, online instructors in an asynchronous environment must be able to provide opportunities for interaction when convenient for the student, provide authentic feedback without ever coming in direct contact with the student, and be able to monitor students who are becoming isolated from the rest of the class.

In a synchronous course, students interact with the teacher and other students in real time; they are separated by distance but come together during regularly scheduled periods. One could consider the courses to simply be traditional courses mediated by technology (Barbour, 2011). Synchronous courses are the most similar to traditional face-to-face courses, although the course may have both synchronous and asynchronous elements. Instructors must have the capabilities to effortlessly work with the new communication technology and be able to integrate synchronous activities with any asynchronous events or discussions that occur when the class is working offline.

On the surface, it would appear that the skills required for teaching online are quite similar to those for teaching in a traditional format. Teachers in both environments must carry out procedural duties (e.g., grading and attendance), provide students with feedback, manage behavior, and cater to the needs of both low-achieving and high-achieving students. Davis and Niederhauser (2007) discussed several similarities between the skill sets of online and face-to-face teachers, among them the ability to stay organized and to communicate effectively with students. In fact, Davis and Rose (2007) found that most online teachers teach in the way that they were once taught, and they transferred their teaching style to the online realm. However, to simply say that the skill sets are exactly the same would be incorrect.

Several problems exist with defining the skill set necessary for successful online teaching. The first problem is obviously identifying those skills. Easton (2003) stated that online instructors needed advanced skills in the management of instructional activities and assessments, as well as stronger engagement skills. In a traditional classroom, all of the students are in one area and can interact with one another based on proximity. In an online environment, the experiences must be engineered so that students separated by both space and time can have engaging interactions with one another. Morris (2003) believed that online instructors needed to be tech-savvy and have a genuine excitement for teaching in the online environment. Instructors also needed to be very familiar with the curriculum. The technical acumen and excitement could be helpful in overcoming technical problems with a content management system and the loss of enthusiasm that could arise when the problems are frequent.

The second problem is validating through research whether such skills are truly unique to online instruction. The aforementioned skills are based primarily on anecdotal evidence, and much of the research that has been done on essential skills has been narrow in scope (Harms, Niederhauser, Davis, Roblyer & Gilbert, 2006). Clearly more research is needed in this area to validate which skills are essential to teaching online. Without strong empirical research backing principles of online instruction, teacher preparation programs may do more harm than good by teaching pre-service teachers faulty methods for teaching courses online.

The third problem is translating this knowledge into training for pre-service and in-service teachers, as it appears that online instruction will be an inevitable part of teachers' duties in the future. Smith, Clark, and Blomeyer (2005) found that only about one percent of K-12 teachers have been trained to teach online. Barbour (2011) stated that most online teacher training is gained through professional development, and this professional development is mainly focused on the technical aspects of a content management system rather than pedagogy.

The training for online teachers is only one aspect of the success of online learners, since more than one person is often responsible for all of the different aspects of delivering online curriculum. The instructor may not have designed the course, and thus online course designers must be able to create quality online courses rather than simply digitize materials from a traditional course. Collis (1999) and Barbour (2007) provide design principles for online courses. In addition to the instructor, another adult is often involved in the monitoring the student (Davis & Niederhauser, 2007). This facilitator is often located at the student's physical school. Research involving the role of the on-site support teachers (Roblyer, Freeman, Stabler & Schneidmiller, 2007) showed that based on surveys of online teachers, the most frequently reported problems concerned the facilitator's ability to monitor student progress. Because of the importance of these facilitators to the success of students, the skills necessary to be a successful facilitator need to be researched and distilled into teacher education programs. In the following two sections, we will look at attempts to prepare teachers on the skills necessary to design online learning, teach online or support students learning online.

EXISTING PRE-SERVICE TEACHER EDUCATION INITIATIVES

Existing pre-service teacher education initiatives for future teachers that attempt to support K-12 online learning are faced with a variety of challenges such as a lack of research and few models to guide their development. Other critical barriers to effective pre-service K-12 online learning teacher education arise from constrictive geographic regulations around the teacher certification process that vary from state to state. Such policies and procedures are more suited to traditional brick and mortar environments and complicate the reach of K-12 online learning's broad development. It is generally agreed that teacher education is currently unprepared for the burgeoning demand for K-12 online learning (Kennedy & Archambault, 2011). Given such consensus, how has pre-service teacher education prepared teachers for K-12 online learning? In this section, we will examine how a small number of universities have attempted to prepare their students for K-12 online learning their pre-service teacher education initiatives. This discussion is not exhaustive, but is fairly representative of the initiatives underway (and is pretty close to an exhaustive listing).

Teacher Education Goes into Virtual Schools (Iowa State University)

Iowa State University, in collaboration with the University of Florida, the University of Virginia, Graceland University and Iowa Learning Online, developed the Teacher Education Goes Into Virtual Schools (TEGIVS) project; the first comprehensive attempt at designing a national model for pre-service teacher education with an emphasis on K-12 online learning. The TEGIVS project sought to identify and develop online teaching competences that would be valuable for all K-12 teachers to support K-12 online learning in the traditional setting, to develop tools that permitted engagement with K-12 online learning practices from multiple perspectives (e.g., the online student, the online teacher, the online course developer, and local school site facilitator), and, ultimately, to build a national community of K-12 online learning practice amongst peers who might constructively critique and challenge the model (Davis et al, 2007).

The Iowa State University model for implementing the actual K-12 online learning training took on various formats within four pre-service teacher education degree programs in four different states (i.e., Iowa, Florida, Virginia and Missouri):

- Secondary lab & lecture (4 hours of training)
- Elementary lab & lecture (4 hours of training)
- Theme within course on distance education (45 hours of training)
- Unit in instructional design course (12 hours of training)
- Theme within a regular methods course (12 hours of training)
- Field experience in a K-12 online learning program (5-24 hours of training) (Davis, 2007)

The variation that occurred in the way each of the pre-service teacher education programs integrated the K-12 online learning curriculum was welcomed by design as researchers sought to gather data on the effectiveness of these varying models.

Field experience in an actual classroom is a foundation in pre-service teacher education programs in North America.

"The field experience in Iowa matched two pre-service teachers with one virtual school teacher. The pre-service teachers were enrolled in a one-credit course that allowed them to work with the virtual school teacher via guided observation and with the online K-12 students via virtual interactions. Pre-service teachers used reflection journals, discussion forums, and interviews to reflect on their practicum experience. Through the study and their involvement in the virtual school field experience, the pre-service teachers experienced a growth of understanding about virtual schooling and formed new personal theories regarding K-12 online learning." (Kennedy & Archambault, 2011)

The Iowa State University model attempted to offer an authentic field experience in a K-12 online learning environment to provide these pre-service teachers the opportunity to be mentored by a teacher comfortable with facilitating learning in this new environment.

While federal funding for the TEGIVS project spanned from 2004-2007, TEGIVS's K-12 online learning lab tools still serve to encourage pre-service and existing teachers to reflect on these topics even as technological tools advance and public policy changes (see http://ctlt.iastate.

edu/~tegivs/TEGIVS/homepage.html). These TEGIVS tools allow pre-service teacher to explore archived scenarios around issues of Internet safety, cheating, and assisting students who cannot take a class due to an illness or even their location are addressed; use a tour tool for observations, and offer a discursive portfolio tool for supervision and mentoring. Likely the greatest impact of TEGIVS is the continued availability of these curriculum materials that can be used as a model for future initiatives. However, as others use these materials, Demiraslan-Cevik (2008) advised them "to help yourself to our resources and adapt them to the ecology of your program, while also forming partnerships with Virtual Schools that parallel those you have with traditional schools" (p. 11).

Student Teaching Partnerships (Florida Virtual School)

Long before K-12 online learning's mass appeal as an educational delivery option, FLVS was providing online opportunities for students in Florida. The success of FLVS's K-12 online learning activities placed the organization in a prominent place to affect change in teacher education programs in that state. Unlike TEGVIS and other universities who struggle with the challenge of identifying K-12 online learning environments to have authentic experiences for their pre-service teachers, this partnership involves pre-service teachers directly with online instruction. In the Fall 2008 the University of Central Florida (UCF) formed a partnership with FLVS to establish a pre-service student teaching internship that aligned to the Florida Educator Accomplished Practices (i.e., state teacher benchmarks for teacher education), National Council for Accreditation of Teacher Education's Unit Standards, and International Association for K-12 Online Learning's National Standards for Quality Online Teaching. UCF's virtual teaching internship – and more recently the University of South Florida since 2010 – provides an option to education majors to complete their student teaching in this innovative environment.

According to Beth Miller, Outreach/Partnerships manager with FLVS, the virtual pre-service interns are paired with lead or cooperating teachers (i.e., certified teachers employed by FLVS) and share in the responsibility of teaching high school students who are enrolled in FLVS courses. The experience parallels the traditional brick-and-mortar internship, with the main difference being that instruction occurs online and not in the traditional school setting. Pre-service interns plan lessons, communicate with students, and assist with assessment of learning, supervised by a university professor and the FLVS lead teacher. Pre-service interns are required to report to a computer lab at their university for observations by their university professors. Additionally, FVLS uses technology to monitor student teachers' work in much the same way that student teachers mentor the work of their FLVS students' work. Finally, feedback is ongoing between the FLVS administration, the lead teacher and pre-service intern in a variety of ways. The program is designed to meet accountability concerns, so at any time the university or FLVS can demonstrate quality assurance through artifacts and data around the activities that the student teacher has undertaken. Given a structured approach to curriculum planning, lead teachers know exactly what to do each week and student teachers are required to create products (e.g., slide presentations, reflective journals, etc.) to demonstrate their time on task.

The focus of the virtual teaching internship experience is for pre-service teachers to develop transferable skills, pedagogical strategies and perspective to their future teaching career – in either online or the traditional classroom environments. Talking with parents, providing feedback on graded assignments, deep content knowledge for effective multi-student differentiation, technology skills, and time management are practical transferable skills essential for either setting. We can assume that the desire to have a separate online student teaching experience is based on Davis et al.'s (2007) premise that "effective virtual teachers have qualities and skills that often set them apart from traditional teachers" (p. 28). This would be further supported by Davis and Roblyer's (2005) assertion that some of the skills necessary for teaching in an online environment are largely absent from traditional teacher education programs.

Online Teaching Course and Practicum (Queen's University)

In addition to the growth K-12 online learning has experienced in the United States, K-12 online learning is also used in similar ways and at comparable levels in Canada (Barbour, 2010). For the past 25 years, Queen's University has had at least one elective course on using computers or information and communications technology (ICT) in teaching and learning. In early 2006, Dr. Geoffrey Roulet – in response to increased web-based instruction by some teachers and school boards as well as the development of online courses by the Ministry of Education – submitted a proposal to create a new elective course entitled "Teaching and Learning Online," described in Table 1.

This course had two goals:

1. Using online tools and resources to enhance classroom based education
2. Teaching online

The purpose of the course was to address the interests and needs of pre-service teachers who desired employment as developers and teachers of online courses and those who aspired to employ online activities in combination with classroom based instruction. The course was approved and taught for the first time during the 2006-07 academic year; however, enrollment was restricted to pre-service teachers in the intermediate-senior (i.e., grades 9-12) program.

According to Roulet's course outline (syllabus) for 2009-10, the course:

"aimed to critically examine present and proposed uses of the Internet/Web in teaching and learning; to collaboratively construct images of what effective online learning could be; and to increase understanding and skills related to the development, presentation, and delivery of online content and learning resources."

Beyond the formal course content, students also participated in an online teaching practicum. During these practicums, many of the participating online teachers were themselves in the initial stages of online teaching careers, and welcomed assistance from these Queen's University students with the design of web-based learning environments and online interaction with their own students.

Teaching and Learning Online was a half credit course, meaning that the course ran throughout the full academic year (i.e., September to April). Additionally, there was a three-week practicum requirement. Dr. Roulet indicated that enrollment in Teaching and Learning Online was generally low, but sufficient to make the course viable from the 2006-07 academic year to the 2009-10 aca-

Table 1. Course description for Teaching and Learning Online

FOCI 291: Teaching and Learning Online
Candidates explore the organization of curriculum and course content for online presentation, construction of learning objects, leading and moderating online discussions and the development of course websites. Course sessions involve classroom meetings and synchronous and asynchronous online interaction. During alternative practicum placements, candidates work with teachers designing and leading online courses or with classroom teachers building learning objects and course websites. http://www.queensu.ca/calendars/education/Program_Focus__FOCI_.html

demic year. However, in the 2010-11 academic year the course was dropped from the schedule due to low enrollment. The course will again not be run in 2011-12, and with Dr. Roulet's impending retirement it is likely that the course may not be offered in the immediate future. When asked why student interest in the Teaching and Learning Online course began to wane, Roulet attributed it to:

"[a] general attitude towards ICT... students have considerable experience with ICT, but largely in the social domain. They have not generally used ICT for intellectual activities other than possibly looking for information on the Web. Thus, students see limited potential for ICT use within education and feel that they have sufficient skills to employ the Web in the ways they imagine using it in a class."

Roulet's assessment is consistent with the sentiments expressed by Davis and Rose (2007), who believed that most online teachers teach in the way that they were once taught, simply transferring their teaching style to the online environment.

Defunct Diploma in Rural and Telelearning (Memorial University of Newfoundland)

Prior to Dr. Roulet's course, the Centre for Tele-Learning and Rural Education acted as a catalyst in the Faculty of Education at Memorial University of Newfoundland for research and development with a special focus on small schools in rural and remote communities in the Canadian province of Newfoundland and Labrador. In the mid- to late-1990s, there was considerable interest in rural schools and solutions to teaching in rural multi-grade classrooms as the majority of schools in the province of Newfoundland and Labrador are rural. Then Chair and Managing Director of the Centre, Dr. Ken Stevens and Mr. Wilbert Boone, initiated the program for Telelearning and Rural Education (Brown, 2000). According to Dr. Jean Brown, a Professor of Education at Memorial involved in the Centre:

"The first thought was that we would develop a graduate program. However, the Associate Dean of Graduate Studies at the time did not support it. Without her support, it was felt we would not be successful in getting the program through Faculty Council and the Academic Council of Graduate Studies within the university. Reluctantly, we decided to develop an after-degree undergraduate diploma. The Associate Dean (Undergraduate Studies), although not a strong supporter, did not oppose it. Mr. Boone had been successful in obtaining external funding for the development of this Diploma, so we moved ahead with it. In hindsight, that was a mistake. Teacher Certification would permit this diploma to count towards a fifth teaching grade, but to obtain a sixth or seventh teaching grade, a Master's Degree was required. Many teachers already had a fifth teaching grade as they held two undergraduate degrees (a B.A. or B.Sc. plus a B.A. (Ed) or B.Ed). That being the case, there was no real incentive for them to do the Diploma. Rather, if continuing their education, they would be wiser to complete a Master's degree."

However, in 1999 the Diploma in Telelearning and Rural School Teaching program was officially launched for teachers already holding a Bachelor of Education degree to better prepare them for teaching in small rural or remote schools in Newfoundland, as well as other jurisdictions.

The program ran from around 2000-01 until at least 2003-04. The diploma comprised of 10 courses (i.e., 6 core courses and 4 electives from a list of 11 possible courses). Of the electives that students could take, there were options for them to participate in a three week, six week, or nine week field-based experience in a rural school environment that may or may not have included as distance education or telelearning component (See Table 2).

Table 2. Diploma in telelearning and rural school teaching program

Core Courses
ED4900: TeleLearning in a Rural School Intranet
ED4901: Effective Teaching Strategies for Multi-grade/Multi-age Classrooms
ED4902: Special Needs in the Context of Rural Schools
ED4903: Leadership Perspectives in Rural Schools
ED4904: Contemporary Educational Issues in Rural Schools
ED4905: Resource-based Learning in the Context of Rural Schools
Elective Courses
ED4906: Career Development in the Context of Rural Schools
ED4907: Curriculum Connections in Multi-grade/Multi-age Classrooms
ED4908: Rural Schools and Community Relationships
ED4909: Rural Schools as Community Learning Centres
ED4910: Curriculum Implementation in All-grade Rural Schools
ED4911: TeleTeaching in a Virtual Classroom
ED4912: Student Assessment in the Context of Rural Schools
ED4916: General Classroom Music
ED4920-4930: Special Topics in TeleLearning and Rural School Teaching
ED4920: Literacy in Small Rural Schools
ED4921: The teaching of Art in Small Rural Schools
Field-Based Experience
ED4913: Field-based Experience in a Rural School (TeleLearning) – 3 weeks
ED4914: Field-based Experience in a Rural School (TeleTeaching) – 6 weeks
ED4915: Field-based Experience in a Rural School (Multi-grade/Multi-age Classroom) – 9 weeks (Memorial University of Newfoundland, 1999)

All of the courses were web-based and supported through CD-ROM for those with limited Internet access.

Regrettably, very little promotion of the program occurred within the province, and no attempt to promote the program outside the province started the demise of the diploma program. Initially 21 potential teachers expressed interest, but only 11 of them actually registered. Further, after the Dean at the time moved to another university, support for the diploma was limited among senior administration. Even within the faculty, some believed the program was not needed because there was a lack of research to support it, while others argued that K-12 distance learning should be integrated in the existing courses currently offered. Upon reflection, this example served to underscore the belief that existing pre-service teacher education initiatives for future teachers and educational leaders require wide-spread support within a faculty if implementation is to be successful – particularly when such programs are ahead of their times, such as this Diploma in Rural and Telelearning was in 1999.

Summary of Pre-Service Teacher Education Initiatives

Partnerships between K-12 online learning programs and universities are essential to the development of effective pre-service teacher education programming. Driven by public demand – and necessity in some instances – pre-service teacher education initiatives that support K-12 online learning are being given increased attention by universities and state Departments of Education. While the duration, quality and availability of these programs vary, the pioneering efforts of the universities discussed in this section have succeeded in beginning an ongoing process of informing and reforming pre-service teacher education initiatives for the demands of this relatively new method of educational delivery.

We believe that K-12 online learning must not simply be an instructional add-on to existing pre-service teacher education programs. The time has come for pre-service teacher education programs to ensure that K-12 online learning is pervasive throughout the undergraduate experience to allow for each teacher to be prepared to fill the roles of online course designer, online teacher and local site facilitator. Regrettably, advances or developments in pre-service teacher education emphasizing K-12 online learning as a course or field experience, such as those described in this section, have been largely reactionary. Clearly more work is needed

EXISTING IN-SERVICE TEACHER EDUCATION INITIATIVES

In much the same way that there are few examples of pre-service teacher education initiatives related to K-12 online learning, the number of examples of in-service teacher education programs are also quite small. The existing initiatives that are targeted to in-service teachers tend to focus on universities that offer graduate level certificates in online teaching with some kind of K-12 focus and/or universities that offer in-service teachers the opportunity to gain an endorsement to their existing teacher certification. The graduate certificates that are offered to educators who would like to learn more about how to teach in an online environment range from certificates that are part of a graduate curriculum and, in some instances, can be used towards a Master's degree to certificates offered by continuing education divisions to certificates offered by K-12 online learning programs that have partnered with universities. We begin this section with a brief look at each of these kinds of certificates, along with what classes and experiences are included in each.

Graduate Certificates in Online Teaching

There are a number of universities that offer certificates to educators for online learning. Generally, the certificates are not limited to K-12 educators, rather these certificates are offered to trainers in industry and higher education instructors who find themselves in a situation that requires online teaching. To date, those universities that offer training that is part of a graduate curriculum include: Arizona State University, Boise State University, University of Central Florida, University of Florida, University of Wisconsin-Stout, and Wayne State University. The certificates generally follow a similar pattern: the in-service teacher must take three to five courses, generally the courses must be taken in sequence, there may or may not be elective courses, and the certificate may or may not have some form of field experience (see Table 3 and Table 4 for the variations in the different programs).

Most of the courses in these graduate certificates can be used towards a Master's degree in educational or instructional technology. The exception is the University of Central Florida, where students in the Master's Degree in Instructional Design and Technology can choose between an educational technology track, an instructional systems track or an e-learning track.

In addition to the variety in the length and nature of these certificates, there is also a great deal of variety in their course offerings. For example, almost all the aforementioned graduate certificate programs offer a course in online teaching methodology and most also offer a course in online course

Table 3. Summary of graduate certificate programs

University	Number of Courses	Nature of Program	Nature of Courses	Field Experience	Other
ASU	5	3 core courses 1 of 4 electives practicum	K-12 focus	Yes	
BSU	3	3 core courses 1 of 4 electives	K-12 track	No	
UCF	9	5 common courses 4 specialized courses	K-12 content	Optional	Leads to M.A.
UF	3	3 of 4 courses	K-12 content	Optional	Currently on hold
UWS	5	4 core courses practicum	K-12 content	Yes	Meets state's online PD requirement
WSU	5	2 core courses 2 of 6 electives practicum	K-12 track	Yes	

Table 4. Summary of course offerings in graduate certificate programs

University	Required	Elective	Field Experience
ASU	1. Principles & Issues in K-12 Online Learning 2. Methods of Online Teaching 3. Online Course Design	One of: 1. Technology Integration Methods 2. Using the Internet in Education 3. Emerging Technologies 4. Technologies as Mindtools	Practicum
BSU	1. Online Teaching in the K-12 Environment 2. **Advanced Online Teaching Methods**	One of: 1. The Internet for Educators 2. Online Course Design 3. Teaching & Learning In Virtual Worlds 4. Educational Games & Simulations	
UCF	1. Current Trends in Instructional Technology 2. Research in Instructional Technology 3. Measurement & Evaluation OR Statistics for Educational Data 4. Fundamentals of Graduate Research in Education 5. Instructional System Design	All of: 1. Multimedia for Education & Training 2. Distance Education 3. Interactive Online & Virtual Teaching Environments 4. Virtual Teaching & the Digital Educator	
UF	1. Instructional Design 2. Distance Online Teaching & Learning	1. Design & Development of Online Content 2. Virtual Schools Philosophy & Pedagogy	
UWS	1. E-Learning for Educators 2. Assessment in E-Learning 3. Instructional Design for E-Learning 4. Creating Collaborative Communities in E-Learning		E-Learning practicum
WSU	1. Facilitation of Online & Face-To-Face Learning 2. **Foundations of Distance Education**	Two of: 1. Designing Web Tools for the Classroom 2. Internet in the Classroom 3. Web-Based Courseware Development 4. Multimedia for Instruction 5. Advanced Multimedia for Instruction 6. Learning Management Systems	Practicum in Instructional Technology

design. The majority of these graduate certificates include a course in either the foundations or trends in distance education and/or in online learning. Several of these certificates also include a course in instructional design.

While there is some consistency in the nature of core courses offered in these graduate certificates, there are few similarities in the elective courses. Those certificates that offer students some choice range from courses in multimedia to emerging tools such as gaming and virtual worlds to learning management systems to courses that are part of the core requirements of some of the certificates (e.g., online teaching and online course design) to a wide range of courses typically found in a graduate program in educational or instructional technology.

Beyond a certificate approved by the School of Graduate Studies at each of these institutions and the ability to apply some or all of the credits towards a Master's or Educational Specialists, in some instances these programs also lead to additional credentials. For example, the three courses in the certificate at Boise State University can be used towards the seven-course online teaching endorsement program that was only announced by the university in August 2011. Similarly, the three of the five courses in the certificate at Wayne State University can be used towards the six-course endorsement in Educational Technology offered by the State of Michigan. In Wisconsin, as of July 1, 2010, "no person may teach an online course in a public school, including charter school, unless he or she has completed at least 30 hours of professional development designed to prepare a teacher for online learning." (State of Wisconsin, 2010). The graduate certificate at the University of Wisconsin-Stout allows teachers to meet this requirement.

Continuing Education Certificates in Online Teaching

In addition to graduate certificates offered by academic departments, there are a couple of examples of graduate certificate programs that are offered by Continuing Education or Extension divisions within the university environment. There are two examples that have a K-12 focus: California State University, East Bay and University of California-Irvine. While both of these universities have a K-12 focus, they also invite instructors from many different backgrounds (e.g., K-12 teachers, military and corporate trainers, community college faculty, continuing education or in-service facilitators, and educators interested in educational technology).

Both certificates include four courses, although these are not identical (see Table 5).

Similar to the graduate certificates offered by academic departments, both of these certificates offer courses in the foundations or trends in online or virtual learning and in online teaching, and one of them offer a course in online course design. Interestingly, the University of California-Irvine

Table 5. Summary of continued education graduate certificates

University	Courses	Field Experience
CSU	1. Introduction to Online Teaching and Learning 2. Teaching Models for Online Instruction 3. Technology Tools for Online Instruction 4. Designing Curriculum for Online Instruction	No
UCI	1. Foundations of Virtual Instruction 2. Advanced Instructional Strategies 3. Performance Assessment in the Virtual Classroom 4. Virtual Teacher Practicum	No

includes a course in assessment (and the University of Wisconsin-Stout is the only other university to include such a course).

The courses included in these certificates are not traditional fifteen week, semester long courses that one would expect to find at a typical university. The courses offered by California State University, East Bay are six weeks in length, while the courses offered by the University of California-Irvine are twelve weeks in length. The certificate courses at the University of California-Irvine do not naturally lead to graduate credits at the university or other universities. This is not to say that other universities will not accept these courses, but that the University of California-Irvine's Extension Division has yet to establish any articulation agreements with other institutions. It is up to the individual university to which the student may be interested in transferring as to whether they will accept the credits. However, the certificate courses from California State University, East Bay are designed to provide a grounding for students who wish to pursue a Master of Science in Teaching program with an option in online teaching (MS-OTL).

Virtual High School Global Consortium Certificate

In much the same way that the FLVS has partnered with universities to better prepare teachers for K-12 online learning, the Virtual High School Global Consortium (VHS) also has long standing relationships with several universities. Since its inception, VHS has offered its own six-week professional development courses as a part of its 21st Century Teaching Best Practices series:

- 21st Century Teaching and Learning explores the tools to teach using technology
- Web-enhanced Classroom explores ways traditional teachers can enhance their practice using web-based tools
- Online Extended Teaching shows teachers how to promote independent study using web tools
- Web 2.0 Collaborative Instruction shows teachers how to use Web 2.0 tools to enhance the learning experience for students
- Becoming an Online Teacher is a practicum experience for teachers to partner with and experienced online teacher

For most of that history, VHS has had partnerships with various universities to allow teachers who complete these professional development courses to obtain graduate credit. These universities include Endicott College (Beverly, MA), Plymouth State University (Plymouth, NH), Framingham State College (Framingham, MA), Northwest Nazarene University (Nampa, ID), Salem State College (Salem, MA), and North Dakota State University (Fargo, ND). Essentially, teachers can pay an additional fee to the participating institution and receive two to four graduate level credit hours depending on the course. Additionally, the VHS has partnered with the Van Loan School of Graduate and Professional Studies at Endicott College and the College of Graduate Studies at Plymouth State University to allow teachers who have completed all five of the VHS courses to achieve a Graduate Certificate in Online Teaching and Learning.

Summary of In-Service Teacher Education Initiatives

At present there are many opportunities for K-12 educators to increase their knowledge, skills, and practice when it comes to classroom instruction. However, this does not hold for their opportunity to increase their ability to design, deliver and support online instruction. Certificates for online teaching often encompass not only K-12, but also the larger field of online learning (including higher education, corporate, and military environments). While some of these certificate programs may be applied for graduate credit towards a Master's of Educational technology degree, in most states there is no standard or endorsement for online teaching.

This lack of standardization has led to a great deal of inconsistency between programs. Of the existing programs, some are three courses, some are four courses and some are five courses. Some provide a field experience in a K-12 online learning program, others simply provide a field experience in any online learning environment, while some have no field experience at all. Most certificate programs do offer courses to provide insights on methodology and trends in the field of online learning, along with courses in online pedagogy and course design. Most also offer instructional design principles as part of the course offerings.

Beyond the graduate certificate programs, there are also certificates that have been created by the extension departments of some universities. In states where there is no teacher certification endorsement for online teaching, these extension programs offer little more than a glorified professional development experience. In fact, at least one K-12 online learning program has taken it upon itself to elevate its own professional development offerings by partnering with several universities. These professional development courses are similar in nature to the professional development provided by numerous other K-12 online learning programs, and the partnerships to receive graduate credit hours from a variety of universities and even a graduate certificate from some university does not equate to the rigor one would expect to find in a traditional semester-long graduate level university course. Simply put, the opportunities for in-service teachers to become better acquainted with the design, delivery and support of K-12 online learning may be greater than they are for pre-service teachers. However, those opportunities vary considerably in the nature of the experience an in-service teacher will receive.

ONLINE TEACHING ENDORSEMENT INITIATIVES

A number of states have brought the practice of online teaching and learning in the K-12 arena to the attention of the legislature. Beyond the 2010 Wisconsin bill in 2010 that required teachers to have a minimum amount of professional development in order to teach online, several states have introduced some form of endorsement to their teaching certification for online teaching. The earliest adopter of this endorsement was Michigan, followed by Georgia and then Idaho. The purpose of these endorsement initiatives has been to ensure that teachers who teach via distance have prepared for and understand the online environment (Michigan Department of Education, 2008). There are a few universities around the country that offer programs that lead to these endorsements, which we will examine in this section, followed by a discussion as to whether there is specific a need to have endorsements for online teaching.

Michigan: Educational Technology (NP) Endorsement

Michigan's educational technology endorsement initiative began in 2000 by a group of professional educators. After review by various groups and school districts, the State Board of Education (SBE) made the recommendation that "all educators and administrators will be prepared to use information-age tools and learning techniques and processes" (Michigan Department of Education, 2008, p. 5). In addition, it was believed that, in order to denote those teachers who had greater skill and study in the area of the use of technological expertise should be given some sort of recognition that they indeed are highly qualified in this area. The original educational technology endorsement was based on 15 standards in three thematic areas that were measured by 93 performance indicators. However, by 2006 the legislature passed a bill that required students to have an online learning experience in order to graduate from high school. This necessitated an update to the educational technology standards.

The revised educational technology standards included 19 new standards measured by 84 performance indicators under an additional three thematic areas.

- **Online Technology Experience and Skills:** Program will prepare Candidates to participate in an online learning experience and demonstrate knowledge and use of an online learning management system(s), adapt online tools to support effective online instruction, understand internet safety issues as well as knowledge of social, ethical, legal, and human issues surrounding the use of educational technology in online teaching and learning, and be able to apply to principles and practice as they relate to technology experiences and skills.
- **Online Course Design:** Professional studies in online course design prepare Candidates to demonstrate knowledge and understanding of pedagogical issues related to teaching and learning in an online environment, and develop and implement curriculum plans aligned with State content standards that include methods and strategies for applying educational technology to maximize learning in an online environment. Professional studies in online course design prepare Candidates who are certified experts in the content subject area being taught, to demonstrate their knowledge and understanding of how to develop, design, and implement strategies that encourage active learning, interaction, participation, and collaboration in the online environment. Professional studies in online course design prepare candidates to demonstrate knowledge about effective online course design with knowledge and understanding of issues related to accessibility and adaptive technologies. Finally, professional studies in online course design prepare candidates to demonstrate knowledge of social, ethical, legal, and human issues surrounding the use of educational technology teaching and learning as it applies to online course design.
- **Online Course Delivery:** Professional studies culminating in the educational technology endorsement prepare candidates to demonstrate knowledge and understanding of: best practices for online delivery of instruction, effective online course technology management, appropriate online assessment and measurement techniques and tools, modeling, moderation and facilitation skills for appropriate online communication with timely feedback, and thoughtful accommodation of student's special needs in an online environment. Professional studies culminating in the educational technology endorsement prepare candidates to facilitate collaboration and incorporate teaming activities in the online environment informed by knowledge of social, ethical, legal, and human issues surrounding the use of educational technology and can apply to principles and practices in teaching and learning as it relates to online course delivery. (Michigan Department of Education, 2008, pp. 21, 24, 30)

At present there are 12 universities throughout the state that offer programs leading to the educational technology endorsement, and in most instances students can use those courses towards a Master's degree or Educational Specialists Certificate in educational or instructional technology.

Georgia: Online Teaching Endorsement

Georgia was the first state in the United States to have a specific endorsement in online teaching. The criteria for this endorsement are that the teacher must already hold certification in qualifying areas, and that the teacher must have an online practicum before being awarded the endorsement. In addition, the awarding institution must follow guidelines that include the following three thematic areas:

1. Content Knowledge, Skills, and Concepts for Instructional Technology
2. Online Teaching and Learning Methodology, Management, Knowledge, Skills, and Dispositions
3. Effective Online Assessment of Teachers, Students and Course Content

Within these three thematic areas there are 10 different standards and 57 competencies.

At present, three universities in Georgia offer programs that lead to this endorsement: Georgia Southern University, Georgia State University and Valdosta State University (see Table 6 for program descriptions)

Note that the Georgia State program is one course longer than the Georgia Southern and Valdosta programs. Also, based on the course descriptions it is not apparent where the required field experience is contained in the Georgia State program, where it occurs in the Field Experience in Online Teaching and Learning course at Georgia Southern and the Design and Delivery of Instruction for E-Learning course at Valdosta. Because the participants in these online only certification processes are already certified teachers, the emphasis is not as much on educational strategies and pedagogy as it is on the incorporation e-learning strategies into the teacher's own teaching style.

Idaho: Online Teacher Endorsement

In Idaho, an initiative to create an endorsement for teaching certification to reflect the study of online teaching and learning has recently passed the state legislature (i.e., November 2011). The endorsement is comprised of ten standards, which are further divided into 53 different knowledge, disposition and performance indicators.

1. **Knowledge of Online Education:** The online teacher understands the central concepts, tools of inquiry, and structures in online instruction and creates learning experiences that take advantage of the transformative potential in online learning environments.
2. **Knowledge of Human Development and Learning:** The online teacher understands

Table 6. Summary of Georgia endorsement programs

University	Courses
GA Southern	1. Theories and Models of Instructional Design 2. Pedagogy of Online Learning 3. Field Experience in Online Teaching and Learning
GA State	1. Integrating Technology into School-Based Environments 2. Evaluation and Assessment for Online Learning 3. The Internet for Educators 4. E-Learning Environments
Valdosta	1. Course Management Systems for E-Learning 2. Resources and Strategies for E-Learning 3. Design and Delivery of Instruction for E-Learning

how students learn and develop, and provides opportunities that support their intellectual, social, and personal development.

3. **Modifying Instruction for Individual Needs:** The online teacher understands how students differ in their approaches to learning and creates instructional opportunities that are adapted to learners with diverse needs.
4. **Multiple Instructional Strategies:** The online teacher understands and uses a variety of instructional strategies to develop students' critical thinking, problem solving, and performance skills.
5. **Classroom Motivation and Management Skills:** The online teacher understands individual and group motivation and behavior and creates a learning environment that encourages positive social interaction, active engagement in learning, and self-motivation.
6. **Communication Skills, Networking, and Community Building:** The online teacher uses a variety of communication techniques including verbal, nonverbal, and media to foster inquiry, collaboration, and supportive interaction in and beyond the classroom.
7. **Instructional Planning Skills:** The online teacher plans and prepares instruction based upon knowledge of subject matter, students, the community, and curriculum goals.
8. **Assessment of Student Learning:** The online teacher understands, uses, and interprets formal and informal assessment strategies to evaluate and advance student performance and to determine program effectiveness.
9. **Professional Commitment and Responsibility:** The online teacher is a reflective practitioner who demonstrates a commitment to professional standards and is continuously engaged in purposeful mastery of the art and science of online teaching.
10. **Partnerships:** The online teacher interacts in a professional, effective manner with colleagues, parents, and other members of the community to support students' learning and well-being.

The requirements for this endorsement state that the teacher must already be certified in his or her field of study. The teacher must take 20 credit hours of courses in the study of online teaching and learning. As well, the teacher must either take an eight-week online teaching internship or have at least one year of experience as an online teacher and be able to document that experience.

With an announcement in August 2010, Boise State University indicated that it was building upon its Graduate Certificate in Online Teaching to provide in-service teachers three options to take advantage of this new online teaching endorsement (see Table 7).

Under the competency-based options, students have to illustrate that they have met all of the performance indicators in the Idaho K-12 Online Teaching Endorsement Matrix (see https://sites.google.com/a/boisestate.edu/idaho-online-endorsement/idaho-k-12-online-teaching-endorsement-matrix).

Ontario: Qualification for Teaching and Learning through e-Learning

Outside of the United States, the Canadian province of Ontario is the only jurisdiction where there is any kind of recognition for online teaching. The Ontario College of Teachers is responsible for the accreditation of teacher education programs in the province. Recently, the Ontario College of Teachers created an "Additional Qualification Course Guideline Teaching and Learning through E-Learning," a thirteen page document that is much more extensive than the online teaching endorsements in the United States and is intended to provide a comprehensive capture of the important aspects of the professional development

Table 7. Boise State endorsement options

Population	Paths
BSU Students	Complete the following courses: 1. Internet for Educators (3 credits) 2. Theoretical Foundations of Educational Technology (3 credits) 3. Online Course Design (3 credits) 4. Teaching Online in the K-12 Environment (3 credits) 5. Advanced Online Teaching (3 credits) 6. Social Network Learning (3 credits) 7. Internship (2 credits or evidence of one year of online teaching experience)
BSU Students	The endorsement is intended to be competency-based, which means that students can demonstrate their competency in meeting the recommended proficiencies through a combination of course completion and other PD experiences. In this instance, the submission of an e-Portfolio (i.e., EDTECH597: Endorsement Portfolio) is required to demonstrate that the proficiencies have been met.
In-service Teachers	This option would apply to teachers who have been teaching in the K-12 online environment for several years, who have participated in PD training through their employer or in PD workshops, and who have perhaps completed some graduate courses related to online teaching and learning. The submission of an e-Portfolio **(i.e., EDTECH597: Endorsement Portfolio)** is required to demonstrate that the proficiencies have been met.

for teachers interested in teaching in an online environment. The qualifications cover ethics, multiculturalism, pedagogy, instructional design components, assessment, and also has references for further reading. Upon successful completion of the program, candidates receive a certificate of completion. It is not clear from the document whether certification is mandatory to teach online or which Ontario universities offer programs that lead to this qualification.

Are Online Teaching Endorsements Necessary?

With all the push toward online teaching and learning, perhaps teachers who wish to provide these services need to be trained to utilize the unique environment of the web. It is a changing presence and becoming ubiquitous in education today. Teachers who teach online must be able to create engaging online learning in an environment where the student is physically (and, in some instances, psychologically) distant from him or her. In terms of communication alone, teaching and learning online is very different than traditional classrooms. But is this really the domain of teachers who teach exclusively online?

It is predicted that by 2019 half of all high school classes will be taught online (Christensen, Horn, & Johnson, 2008). Recently, the market analyst Ambient Insight (2011) estimated that the current level of participation in K-12 online and blended learning was four million students. There is a need to be prepared for the online environment, but not just for a few self-selected teachers. Since it has been shown that most online teachers have previously been teaching in traditional classrooms for many years (Archambeault & Crippen, 2009), the question should be how will all teachers be prepared to utilize these pedagogies in their practice?

Online students can find themselves in a variety of course models. In these models the teacher has differing levels of engagement and responsibility. In the independent model, the course is primarily self-taught and the teacher mainly provides technical skills so that the student can complete the course. In the asynchronous model, the teacher must provide feedback on assignments, moderate student discussion boards, and generally support and guide the student. In these instances, the teacher must be able to communicate well in situations where communication may never be direct (i.e., by phone, chat, or in person). The teacher must monitor the students to make sure they do not feel or become isolated, and if this does occur the teacher must have the tools to help the student overcome this learning challenge. Finally, in the synchronous model the teacher and students interact directly in a real-time setting; most often during a prearranged online meeting time. In these instances, the teacher must be able to implement many of the same skills they would use in a real-time classroom environment, only with those interactions being mediated by technology.

In some ways, the online environment has its own issues—the teacher has to be both tech-savvy and be able to guide students. The teacher or facilitator at the school may not have access to the student's progress or success, so the online teacher must be vigilant in keeping the student engaged. Online teachers may not have designed the course they teach, even though they understand and teach the content. They need advanced skills in the management of instructional activities and strong engagement skills. They need a genuine excitement for the course content, and familiarity with the curriculum. These teachers will also need to be able to select engaging content, rich multimedia for instruction, nontraditional content delivery methods, sound teaching philosophy, an understanding of the use of the Internet to teach and learn, and innovative teaching strategies.

However, it can be argued that these skills are not exclusively the domain of the teacher who delivers instruction via distance. Is it only the domain of online teaching and learning to understand how to use the Internet or select rich multimedia for instruction? These skills provide a richer experience for students regardless of the instructional delivery method. Teacher should be genuinely engaged with the course content and enthusiastically encourages students will bring energy and success to the course, whether the course is delivered via distance or not. As well, teachers need technology skills for instruction, whether it is delivered face-to-face or online. Successful teachers use non-traditional teaching methods. Communication skills are a necessary part of the instructional process. These skills are necessary for every teacher in service today. The profession needs to have these aspects of what is now considered the domain of online instructional practice incorporated into the traditional teacher preparation curriculum of all teacher education programs. We are doing a disservice to all our students and teachers if we do not demand this. In much the same way that all teachers should be able to integrate technology into their teaching (which means that endorsements to technology integration are redundant at best, demeaning to a professional at worst), all teachers should be able to design, deliver and support instruction in an online as well as a face-to-face environment.

REFERENCES

Ambient Insight. (2011). *2011 learning technology research taxonomy: Research methodology, buyer segmentation, product definitions, and licensing model*. Monroe, WA: Author. Retrieved from http://www.ambientinsight.com/Resources/Documents/AmbientInsight_Learning_Technology_Taxonomy.pdf

Archambault, L., & Crippen, K. (2009). K-12 distance educators at work: who's teaching online across the United States. *Journal of Research on Technology in Education*, *41*(4), 363–376.

Aronson, J. Z., & Timms, M. J. (2003). *Net choices, net gains: Supplementing the high school curriculum with online courses*. San Francisco, CA: WestEd. Retrieved from www.wested.org/online_pubs/KN-03-02.pdf

Barbour, M. K. (2007). Principles of effective web-based content for secondary school students: Teacher and developer perceptions. *Journal of Distance Education, 21*(3), 93–114.

Barbour, M. K. (2009). Today's student and virtual schooling: The reality, the challenges, the promise… . *Journal of Distance Learning, 13*(1), 5–25.

Barbour, M. K. (2011). Training teachers for a virtual school system: A call to action. In Polly, D., Mims, C., & Persichitte, K. (Eds.), *Creating technology-rich teacher education programs: Key issues* (pp. 499–517). Hershey, PA: IGI Global.

Barbour, M. K., & Reeves, T. C. (2009). The reality of virtual schools: A review of the literature. *Computers & Education, 52*(2), 402–416. doi:10.1016/j.compedu.2008.09.009

Brown, K. (2000). *Diploma in teleLearning and rural school teaching.* A presentation at Hook Line & Net, Clarenville, NL. Retrieved from http://www.snn-rdr.ca/snn/hln2000/telelearning.html

Christensen, C. M., Horn, M. B., & Johnson, C. W. (2008). *Disrupting class: How disruptive innovation will change the way the world learns.* New York, NY: McGraw-Hill.

Clark, T. (2001). *Virtual schools: Trends and issues—A study of virtual schools in the United States.* San Francisco, CA: Western Regional Educational Laboratories. Retrieved from http://www.wested.org/online_pubs/virtualschools.pdf

Clark, T. (2007). Virtual and distance education in North American schools. In Moore, M. G. (Ed.), *Handbook of distance education* (2nd ed., pp. 473–490). Mahwah, NJ: Lawrence Erlbaum Associates.

Collis, B. (1999). Designing for differences: Cultural issues in the design of WWW-based course-support sites. *British Journal of Educational Technology, 30*(3), 201–215. doi:10.1111/1467-8535.00110

Cyrs, T. E. (1997). Competence in teaching at a distance. In Cyrs, T. E. (Ed.), *Teaching and learning at a distance: What it takes to effectively design, deliver, and evaluate programs* (pp. 15–18). San Francisco, CA: Jossey-Bass Publishers.

Davis, N., Demiraslan, Y., & Wortmann, K. (2007, October). *Preparing to support online learning in K-12.* A presentation at the Iowa Educational Technology Conference, Des Moines, IA. Retrieved from http://ctlt.iastate.edu/~tegivirtualschool/TEGIVIRTUAL SCHOOL/publications/ITEC2007-presentations.pdf

Davis, N. E., & Niederhauser, D. S. (2007). Virtual schooling. *Learning and Leading with Technology, 34*(7), 10–15.

Davis, N. E., & Roblyer, M. D. (2005). Preparing teachers for the "schools that technology built": Evaluation of a program to train teachers for virtual schooling. *Journal of Research on Technology in Education, 37*(4), 399–409.

Davis, N. E., Roblyer, M. D., Charania, A., Ferdig, R., Harms, C., Compton, L. K. L., & Cho, M. O. (2007). Illustrating the "virtual" in virtual schooling: Challenges and strategies for creating real tools to prepare virtual teachers. *The Internet and Higher Education, 10*(1), 27–39. Retrieved from http://ctlt.iastate.edu/~tegivs/TEGIVS/publications/JP2007%20davis&roblyer.pdf doi:10.1016/j.iheduc.2006.11.001

Dawley, L., Rice, K., & Hinck, G. (2010). *Going Virtual! 2010: The status of professional development and unique needs of K-12 online teachers.* Boise, ID: Boise State University. Retrieved from http://edtech.boisestate.edu/goingvirtual/goingvirtual3.pdf

Demiraslan-Cevik, Y. (2008). *Final report to FIPSE for P116B040216 – TEGIVS: Teacher education goes into virtual schooling.* Ames, IA: Iowa State University. Retrieved from http://yunus.hacettepe.edu.tr/~yasemind/HCIPortfolio/TEGIVSPerformanceNarrative.pdf

DiPietro, M. (2010). Virtual school pedagogy: The instructional practices of K-12 virtual school teachers. *Journal of Educational Computing Research, 42*(3), 327–354. doi:10.2190/EC.42.3.e

Easton, S. (2003). Clarifying the instructor's role in online distance learning. *Communication Education, 52*(2), 87–105. doi:10.1080/03634520302470

Forcheri, P. (2011). Editorial: Reimagining schools: The potential of virtual education. *British Journal of Educational Technology, 42*(3), 363–372. doi:10.1111/j.1467-8535.2011.01178.x

Friend, B., & Johnston, S. (2005). Florida virtual school: A choice for all students. In Berge, Z. L., & Clark, T. (Eds.), *Virtual schools: Planning for success* (pp. 97–117). New York, NY: Teachers College Press.

Harms, C. M., Niederhauser, D. S., Davis, N. E., Roblyer, M. D., & Gilbert, S. B. (2006). Educating educators for virtual schooling: Communicating roles and responsibilities. *The Electronic Journal of Communication, 16*(1-2). Retrieved from http://ctlt.iastate.edu/~tegivs/TEGIVS/publications/JP2007%20harms&niederhauser.pdf

Kaseman, L., & Kaseman, S. (2000). How will virtual schools effect homeschooling? *Home Education Magazine* (November-December), 16-19. Retrieved from http://homeedmag.com/HEM/176/ndtch.html

Kennedy, K., & Archambault, L. (2011). The current state of field experiences in K-12 online learning programs in the U.S. In M. Koehler & P. Mishra (Eds.), *Proceedings of Society for Information Technology & Teacher Education International Conference 2011* (pp. 3454-3461). Chesapeake, VA: AACE.

Memorial University of Newfoundland. (1999). *Courses in telelearning and rural school teaching*. St. John's, NL: Author. Retrieved from http://www.mun.ca/regoff/cal99_00/EducationTeleLearningandRuralSchoolTeachingCourses.htm

Michigan Department of Education. (2008). *Standards for the preparation of teachers: Educational technology*. Lansing, MI: Author. Retrieved from http://www.michigan.gov/documents/mde/EducTech_NP_SBEApprvl.5-13-08.A_236954_7.doc

Morris, S. (2002). *Teaching and learning online: A step-by-step guide for designing an online K-12 school program*. Lanham, MD: Scarecrow Press Inc.

Picciano, A. G., & Seaman, J. (2009). *K-12 online learning: A 2008 follow-up of the survey of U.S. school district administrators*. Needham, MA: Alfred P. Sloan Foundation.

Rice, K., & Dawley, L. (2007). *Going virtual! The status of professional development for K-12 online teachers*. Boise, ID: Boise State University. Retrieved from http://edtech.boisestate.edu/goingvirtual/goingvirtual1.pdf

Roblyer, M. D., Freeman, J., Stabler, M., & Schniedmiller, J. (2007). *External evaluation of the Alabama ACCESS initiative phase 3 report*. Eugene, OR: International Society for Technology in Education.

Roblyer, M. D., & McKenzie, B. (2000). Distant but not out-of-touch: What makes an effective distance learning instructor? *Learning and Leading with Technology, 27*(6), 50–53.

Smith, R., Clark, T., & Blomeyer, R. L. (2005). *A synthesis of new research on K-12 online learning*. Naperville, IL: Learning Point Associates. Retrieved from http://www.ncrel.org/tech/synthesis/synthesis.pdf

State of Wisconsin. (2010). *Guidance on the 30 hours of professional development for teaching online courses*. Madison, WI: Author. Retrieved from http://dpi.wi.gov/imt/pdf/online_course_pd.pdf

Watson, J., Murin, A., Vashaw, L., Gemin, B., & Rapp, C. (2010). *Keeping pace with K–12 online learning: An annual review of policy and practice*. Evergreen, CO: Evergreen Education Group. Retrieved from http://www.kpk12.com/wp-content/uploads/KeepingPaceK12_2010.pdf

Wood, C. (2005). Highschool.com: The virtual classroom redefines education. *Edutopia, 1*(4), 31-44. Retrieved from http://www.edutopia.org/high-school-dot-com

Zucker, A., & Kozma, R. (2003). *The virtual high school: Teaching generation V*. New York, NY: Teachers College Press.

ADDITIONAL READING

Cavanaugh, C., & Blomeyer, R. (2007). *What works in K-12 online learning*. Eugene, OR: International Society for Technology in Education.

Rice, K. (2011). *Making the move to K-12 online teaching: Research-based strategies and practices*. Columbus, OH: Allyn & Bacon.

KEY TERMS AND DEFINITIONS

Asynchronous: Not in real time. For example, a discussion forum is an asynchronous technology where one student posts a message and at a later time another student can read and respond to that message. A non-technical example would be like a community bulletin board where one person posts a for sale poster and at a later time another person may walk by and see that sign.

Cyber School: A full-time K-12 online learning program where students do not attend a traditional or brick-and-mortar school.

K-12 Online Learning: A generic term to encompass all forms of distance education at the K-12 level delivered over the Internet. This includes full-time cyber schooling and supplemental virtual schooling.

Synchronous: In real time. For example, a telephone conversation occurs in real time or is said to be synchronous.

Virtual School: A supplemental K-12 online learning program where students attend a traditional or brick-and-mortar school, but may also be enrolled in one or more online courses.

Chapter 5
Teaching and Assessing Problem Solving in Online Collaborative Environment

Yigal Rosen
University of Haifa, Israel

Rikki Rimor
Open University of Israel, Israel

ABSTRACT

Online teacher programs are diverse in their models, expressing a variety of learning objectives, pedagogies, technological platforms, and evaluation methods. Promoting and assessing collaborative learning of an online teacher programs is one of the major challenges, in part because collaboration includes complex cognitive and social-emotional dimensions. This chapter focuses on teachers' academic program in online learning environment and examines the conditions for effective teaching and assessing collaborative problem solving. The chapter provides readers a look at the rationale, implementation, and assessment of collaborative learning in online teacher program and presents the conditions for effective design of collaborative learning for pre- and in-service teachers. Examples from two empirical studies will be provided on how the collaborative learning environment leverages teachers' constructivist teaching, ongoing feedback, and evaluation to prepare teachers for instruction in technology-rich environments.

INTRODUCTION

According to the Organization for Economic Co-operation and Development (OECD) Teaching and Learning International Survey (TALIS), a significant proportion of teachers think that professional development does not meet their needs (OECD, 2009a). Many teachers emphasize lack of suitable development opportunities, conflict with their work schedule and the need for professional development on Information and Communication Technologies (ICT) teaching skills. This suggests a need not just for better support for teachers' preparation and professional development, but

DOI: 10.4018/978-1-4666-1906-7.ch005

for policy makers and school leaders to ensure that the development opportunities available are effective and meet teachers' needs. Online teachers program can be potentially a key solution for these needs. Using online technologies prepare effective educators and increase their competencies throughout their careers while building the capacity to deliver effective teaching.

Standards and resources within International Society for Technology in Education (ISTE, 2008) and UNESCO's project "ICT Competency Standards for Teachers" (UNESCO, 2008), provide guidelines for planning teacher education programs and training offerings that will prepare them to play an essential role in producing technology-rich learning environments for technology capable students. Being prepared to use technology effectively to support student learning have become integral skills in every teacher's professional program. According to the ISTE ICT professional growth standards (ISTE, 2008), teachers will: (a) participate in local and global learning communities to explore creative applications of technology to improve student learning; (b) exhibit leadership by demonstrating a vision of technology infusion, participating in shared decision making and community building, and developing the leadership and technology skills of others; (c) evaluate and reflect on current research and professional practice on a regular basis to make effective use of existing and emerging digital tools and resources in support of student learning; (d) contribute to the effectiveness, vitality, and self-renewal of the teaching profession and of their school and community. UNESCO ICT competency standards for teachers (UNESCO, 2008), emphasize that teacher training should focus on the development of digital literacy, use of ICT for professional improvement, use technology to guide students through complex problems and manage dynamic learning environments. One of the key ICT components in teacher programs is modeling collaborative knowledge construction by engaging with colleagues and students in face-to-face and online environments.

Online teacher programs are diverse in their models, expressing a variety of learning objectives, pedagogies, technological platforms and evaluation methods (e.g. Barnett, 2002; Dede, 2006; Hawkes, & Romiszowski, 2001; Yang, & Liu, 2004). This chapter focuses on teachers' academic online program and examines the conditions for effective construction of knowledge and skills in online collaborative learning environment. The chapter provides readers a look at the rationale, implementation and results of teachers' academic online course in online collaborative learning environment and presents the conditions for effective construction of knowledge and skills.

BACKGROUND

Online teacher preparation and professional development programs can be broadly divided into two types: blended learning or distance learning. Blended learning is that which involves face-to-face contact between the facilitator and teacher alongside internet based input delivery and interaction. Distance learning consists of input delivered entirely via the internet, and interaction taking place via similar technology environment. Online learning communities break through educators' traditional isolation, enabling them to collaborate with their peers (Fishman, 2007). Educators are no longer limited by where they teach or where and when they are involved in a professional development. Promoting and assessing collaborative learning of an online teacher programs is one of the major challenges, in part because collaboration includes complex cognitive and social-emotional dimensions. What process evidences and outcomes should be collected to show success of particular model for collaborative learning? What pedagogical and technological conditions can support collecting these evidences? This chapter focuses on teachers' academic program in online collaborative learning environment and examines the conditions for effective construction of knowledge and skills. Two studies described in this chapter examined cognitive and social patterns of interaction in a collaborative database learning environment (Google Docs) among pre- and in-service teachers in an Open University course in Israel.

THEORETICAL FRAMEWORK

Collaborative Learning

Collaboration is "coordinated, synchronous activity that is the result of a continued attempt to construct and maintain a shared conception of a problem" (Roschelle, & Teasley, 1995, p. 70). According to Dillenbourg (1999), collaborative learning is a situation in which two or more students learn together. One major goal of collaborative learning is to support social interaction and promote the learner's cognitive processes. The learners express their knowledge to the partners and work to co-construct knowledge collaboratively. Learners must externalize their knowledge, that is, they must elaborate on and comprehensibly explain their knowledge to the learning partner. Slavin (1997) associates cooperative learning with well-structured knowledge domains, and collaborative learning with ill-structured knowledge domains. However, collaborative context may vary with respect to the degree to which the actions that can be performed to achieve the goals are specified.

The theoretical framework for learning as a social process was developed by Vygotsky (1978), emphasized the social context of the learning process. He claimed that the personal potential could be realized through a process of interaction with and support from the human environment and from various tools. Interpersonal activity when appropriately implemented could lead to intrapersonal mental development. When trying to solve a problem together through the exchange of ideas, a team of learners constructs shared meanings that the individual would not have attained alone. Furthermore, collaboration is linked to a number of important educational skills, including critical thinking and metacognition (e.g. Heyman, 2008; Kramarski, & Mevarech, 2003; Kuhn, & Dean, 2004; Schraw, Crippen, & Hartley, 2006). Researchers found that online groups, compared to face-to-face groups, engaged in more complex and cognitively challenging discussions develop higher order thinking skills, as well as complex patterns of interactions (Benbunan-Fich, Hiltz, & Turoff, 2003; Wegerif, 2006; Rimor, Rosen, & Naser, 2010).

According to Dillenbourg (1999), effective collaboration is characterized by a relatively symmetrical structure. *Symmetry of knowledge* occurs when all participants have roughly the same level of knowledge, although they may have difference perspectives. *Symmetry of status* involves collaboration among peers rather than interactions involving facilitator relationships. Finally, *symmetry of goals* involves common group goals rather than individual goals that may conflict. The degree of interactivity and negotiability is an additional indicator of collaboration (Dillenbourg, 1999). For example, trivial, obvious, and unambiguous tasks provide few opportunities to observe negotiation because there is nothing about which to disagree. According to Kreijns et al. (2003) a successful collaboration in online environment is a function of interdependence, interaction, individual accountability, interpersonal and small group skills, and active group processing. Specifically, collaborative problem solving in online database environment requires the learner to be active by defining groups of data, classifying them in a database, and presenting arguments for justifying the classification (Jonassen, 1999; Rimor, 2002; Rosen, & Rimor, 2009). Learners interact with and influence each other in the process of problem-solving; these interactions form important units of analyses for research and reflection. The reflective practitioner is able to respond to problematic contexts through reflection, effectively solving the particular problem while at the same time learning from the experience (Schon, 1987, 1991). Reflection-in-practice involves taking time during the activity to gauge one's performance and evaluate how the activity is progressing, while reflection-on-practice involves examining a particular experience after the fact in use for making sense of the experience. Meaningful reflection, as describes by Dewey (1933), occurs through the cycle of disequilibrium and restoration of equilibrium that entails translating abstract concepts into action.

Learning To Teach with Technology Studio (LTTS) is one of possible examples for intertwining online environment in teacher professional development (Duffy et al., 2006). The LTTS is an online professional development system created in 1999 by the Indiana University's Center for Research on Learning Technology to help K-12 teachers learn investigate technology into their content-focused teaching (See http://ltts.indiana.edu for more information). The LLTS was designed to aid teachers in designing teaching units or lesson plans in which technology is used to support student inquiry. The system consists of 60 courses delivered through web-based learning management system. Approximately 600 students, 70% are in-service teachers, 23% are pre-service teachers, and 7% are college faculty and school administrators. The constructivist theoretical framework guided the design of LTTS, emphasizing three principals (Savery, & Duffy, 1996): (a) understanding comes from interactions with the environment, (b) cognitive conflict or puzzlement is the stimulus for learning and is a major factor in determining the organization and nature of what is learned, (c) knowledge evolves through social negotiation and through the evaluation of the viability of individual understanding. The core pedagogical strategy in LTTS is creating a problem-centered environment in which the learner is guided through reflective cycles of inquiry.

Reflective thinking involves rigorously examining the contexts of teaching, framing and reframing problems, generating a range of possible solutions, and evaluating those solutions on the basis of their likely consequences at the personal, academic, social, and ethical levels. Many teacher educators believe in the importance of encouraging reflective thinking in teachers through communities of peers engaged in dialogue about educational issues (Boler, 2004; Grossman, Wineburg, & Woolworth, 2001). Rodgers (2002) described a reflective teacher as one who "does not merely seek solutions, nor does he or she do things the same way every day without an awareness of both the source and the impact of his or her actions" (p. 849). According to Putnam and Borko (2000), when different groups of teachers with different types of knowledge and experience involved in collaborative discourse, the teachers can draw upon and incorporate each other's expertise to create rich discussions and new insights into teaching and learning. Kelly (2011) suggests that "teachers with more reflective and discursive orientation may adopt attitudes which respond to their students' difficulties, seek to collaborate with students and peers in resolving these, look for ways forward in professional guidance, and adopt complex measures of success. Reflective thinking emerges while solving practical problems, in particular in learning situations that are not structured and whose solution is not clear. The importance of reflective processes in learning is even more prominent in interactive computerized environments that serve as the outer representation of knowledge-based tools and discussion-based tools. The combination of interaction and asynchronism encourages students to reflect on their own learning and that of their peers (Wadmany, Rimor, & Rosner, 2011). Because asynchronous forums give participants time, space, and the freedom to express themselves whenever they wish, participants are more likely to express in-depth an individual, even if it challenges others' views (Brookfield, & Preskill, 2005). Major advantages for online teacher learning include the greater flexibility it offers, and the opportunities it provides to utilize resources and reflection (Dede, Ketelhut, Whitehouse, Breit, & McCloskey, 2009).

Assessing Collaborative Problem Solving

Problem solving in technology-rich environments encompasses using ICT to acquire and evaluate information, communicate with others and accomplish practical tasks (OECD, 2009b). The individual is engaged in cognitive processing to understand and resolve problem situations where a method of solution is not immediately obvious. It includes representing and manipulating different types of knowledge in the problem solver's cognitive system (Mayer, & Wittrock, 2006). The problem situation may change during the solving process as a result of interaction with the individual.

Online collaborative problem solving offers new assessment opportunities, while the implications of the social factor and pedagogical structure of the tasks serve as the major challenges. Assessment in online collaborative problem solving can be focused on assessing the individual about the individual, assessing the individual about the group, and assessing the group as a whole. In order to analyze collaborative activity it is necessary to collect information about the collaboration process, recording information about the participants, actions performed, messages sent and received and time of each action. Because of the very complex interactions that occur in online collaborative tasks, where learning occurs through interaction among group members, understanding and analyzing the collaborative learning process requires analysis of group interaction in the context of learning goals. A number of different theoretical and methodological approaches have been taken to deal with these challenges. Collazos et al (2007) suggested five system-based indicators of the success of the collaborative learning process:

1. **Use of strategies:** the ability of the group members to generate, communicate and consistently use a strategy to jointly solve the problem
2. **Intra-group cooperation:** application of collaborative strategies during the process of group work
3. **Reviewing success criteria:** the degree of involvement of the group members in reviewing boundaries, guidelines and roles during the group activity
4. **Monitoring:** the extent to which the group maintains the chosen strategies to solve the problem, keeping focused on the goals and the success criteria
5. **The performance of the group:** how good is the result of collaborative work, total elapsed time while working, and total amount of work done

Another model suggested by the Center for Research on Evaluation, Standards, and Student Testing (CRESST) consists of six measures (O'Neil, Chung, & Brown, 1997):

1. **Adaptability:** refers to the group's ability to monitor the source and nature of problems
2. **Coordination:** a group's process by which group resources, activities, and responses are organized to ensure success
3. **Decision making:** ability to integrate information, use judgment, identify possible alternatives, select the optimal solution, and evaluate the consequences
4. **Interpersonal:** the ability to improve the quality of team member interactions
5. **Leadership:** the ability to direct and coordinate the activities of the team, assess performance, assign tasks, plan and organize, and establish a positive atmosphere

6. **Communication:** information exchange between team members in the agreed manner and by using proper terms, and the ability of clarification and acknowledgement

One of the leading fields of research in the context of online collaborative research is the Computer-Supported Collaborative Learning (CSCL). CSCL is emerging as a dynamic, interdisciplinary, and international field of research focused on how technology can facilitate the sharing and creation of knowledge and expertise through peer interaction and group learning processes. The main objective of CSCL is to provide an environment that supports collaboration between students to enhance their collective learning and group cognition (Kreijns, Kirschner, & Jochems, 2003; Stahl, 2006). In order to examine social interaction in CSCL, the different aspects of the process must be conceptualized. Weinberger and Fischer (2006) propose a multi-dimensional approach to analyze multiple process dimensions of knowledge construction in online learning environment:

1. **The participation dimension:** the quantity of participation and the heterogeneity of participation
2. **The epistemic dimension:** refers to the content of learners' contributions
3. **The argument dimension:** the construction of arguments and the construction of sequences of arguments
4. **The dimension of social modes of co-construction:** On the argument dimension, the construction of single arguments and the construction of sequences of arguments can be differentiated.

   Argumentation comprises the elements of claim, ground with warrant, and qualifier with regard to the construction of single arguments. Sequences of arguments consist of at least one argument in favor of a specific point. New warranted or qualified claim is coded as an argument that has not been preceded by a convicting argument. Counterarguments consequently attack the existing arguments by putting up contrary or alternative claims. Claim supporting points of more than one line of preceding arguments is thus regarded as a reply. The five dimensions of social modes of co-construction in online collaborative learning include:

   a. **Externalization:** discussion usually begins with externalization, when the learner contributes to the discussion without referring to the contributions of the other learners
   b. **Initiative:** group members reciprocally serve as sources of knowledge by asking questions and by obtaining knowledge from their learning partners
   c. **Rapid consensus:** in order to generate collaboration among group members, the learners accept the opinions of their peers, not necessarily because they agree with them or have been persuaded, but because it is a way to quickly advance with the discussion
   d. **Integrative consensus:** the learners reach a consensus through an integration of their various opinions and points of view
   e. **Consensus through conflict:** when building a consensus, the learners must either present their objections clearly and persuasively or present an alternative.

The Weinberger and Fischer (2006) framework has been applied to analyze dialogic argumentation in online learning environments that incorporate asynchronous discussion boards. The automated TagHelper tool was used to code dialogue on all dimensions of the analytic framework for comparison with human coders (Dönmez et al., 2005). Along the dimensions of the framework, the interaction of learners was structured to facilitate specific participation patterns, sequences of epistemic activities, the construction of single arguments, the construction of argumentation sequences, and social modes of co-construction (e.g. Weinberger et al., 2007).

In another study, Sorensen and Takle (1998) examined the quality of electronic dialogue, which involves the following indicators:

- **Brainstorming:** introduces new ideas concerned with the topic or task and provides a perspective not previously considered
- **Articulating:** explains complex or difficult concepts
- **Reacting:** provides an alternative or amplified perspective of a concept previously introduced by a student
- **Organizing:** assembles existing thoughts or perspectives such that a new perspective emerges
- **Analysis:** compares or contrasts previously articulated perspectives or derives new understandings from existing data
- **Generalization:** takes comments or data that are already available and extracts new information or knowledge that applies to a broader set of conditions than that previously extracted

A record of activity as well as product can be kept, replayed, and even modified in online collaborative learning. In the collaborative computing settings, the interaction sequences can be captured automatically and systematically analyzed. Design for improved meaning making requires to be informed by analysis; however analysis also depends on design in its orientation. Collaboration scripts focus on the learners exchanging task information, replicating important information, and reflecting on the relevance of the information to their collaborative task solution.

Methodology

Findings from two empirical studies (Rosen, & Rimor, 2009; Rimor, Rosen, & Naser, 2010) on teachers' collaborative problem solving processes and outcomes in online database environment, shed a light on pedagogical and assessment possibilities and challenges. The studies are based on data collected among fifty eight pre- and in-service teachers participated in a combined academic online course that referred to knowledge construction processes in technological based environments. The teachers were required to take part in the forum of the course and in an online collaborative task based on constructing a shared database on Google Docs. At the beginning of the task, the participants were asked to sort individually different statements from the forum and re-locating them in the proper fields of the database, relating to the variety of knowledge types: declarative, procedural, structural, meta-cognitive. This task was accompanied by individual argumentation, group feedback, and instructor's feedback. At the end a final group decision on the classification of the statements was made. All teachers' activities were documented in the online database. The task was conducted in groups of 3-4 participants, over the course of six weeks, and assessed by two main dimensions: (a) the participant's process of individual knowledge construction and (b) the group's process of collaborative knowledge construction. Each participant was examined according to his/her individual contribution to the database and according to the extent of his/her previous activity in the shared forum. Individual knowledge was measured according to the participant's success in correctly classifying the forum's statements according to the four types of knowledge mentioned above. Classification was based on terminology and articles that were studied during the course. Collective knowledge was measured operationally by the index of accuracy of the group assignment, which verifies the individual classifications of various types of knowledge. Completion of this assignment was a result of the teams' discussion and argumentation process to gain an agreement among the participants.

Additionally, social interactions were analyzed. In the first phase, participants' responses were qualitatively classified according to the Weinberger and Fischer (2006) model. This model examines the degree of collaboration in an online learning environment using five dimensions: Externalization, initiation, rapid consensus, integrative consensus, consensus through conflict. Frequency of each dimension was calculated based on 238 single items. The second phase focused on qualitative characterization of strings which were represented in the database by the protocols of the discussions for each team of students. Identifying different strings allowed classifying various types of social interactions. The final phase referred to the qualitative examination of the complexity of the social interactions and their frequency. The most complex interaction was composed of the dimensions of "integrative consensus" and "consensus through conflict," whereas the least complex interaction was represented by the dimension of externalization.

Results

Findings from these two combined studies indicated differential achievements among participants with different learning orientations. The participants with collaborative learning orientation succeeded more in the collective criteria of knowledge construction compared to the less collaborative ones ($M=17.0$ versus. $M=15.2$; $t=5.1$, $p<0.1$). On the other hand, the less collaborative participants gained higher scores in the personal criteria of knowledge construction compared to the collaborative ones ($M=14.9$ versus $M=13.9$; $t=2.1$, $p<0.5$). While more collaborative participants contributed more to a collective knowledge, the individual-oriented participants focused on constructing their own personal knowledge.

Analysis of social interactions examined the degree of collaboration during the online collaborative task. The findings showed that the interaction among participants was composed of various dimensions of interaction and that most of these dimensions represent a relatively low level of complexity. This includes participant contribution to the discussion without referring to the contributions of other participants, initiation of a group member to reciprocally serve as source of knowledge by asking questions and by obtaining knowledge, and accepting the opinions of their peers not necessarily because they agree with them or have been persuaded, but rather because it is a way to quickly advance with the discussion. Participants tended to initiate investigations regarding perspectives and attitudes of others at a relatively higher frequency than merely externalizing their own personal knowledge. Externalization-Rapid consensus (21%), and Externalization-Initiative (15%) were found as the most frequent low level (simple) interactions. More complex patterns characterized by multiplicity and variety of dimensions (at least three dimensions). Two main patterns were found in this category: Externalization-Rapid consensus- Integrative consensus (23%), and Externalization-Initiative-Rapid consensus-Integrative consensus-Consensus through conflict (24%). In addition, the findings showed that the frequency of complex interactive sequences of discussions that evolved into integrative and conflict agreements consisted of close to 50% of all discussions developing amongst participants. In other cases, the discussions reached the stage of rapid agreement which indicates no change in participants' perceptions.

FUTURE RESEARCH DIRECTIONS

Further work is needed to study the influence of online collaborative learning on teachers' professional development. Research on different types of online learning environments, subject of the collaborative task, homogeneous vs. heterogeneous groups can be highly promising directions. Clarifying the nature of online collaborative learning in the context of teachers' professional development and verifying the impact depends upon further rounds of design research. This research requires illumination of treatment variables, of processes and direct outcomes of professional development, and of desired indirect outcomes. Finally, development and refinements of the assessment valid and reliable instruments to gauge online collaborative problem solving processes and outcomes are vital.

CONCLUSION

Online collaborative environment was discussed here as a constructivist mean's for building knowledge and empowering the learner through goal driven group work. Other researchers are concerned with pedagogical-technological variables of implementing innovative technologies in school in general, as well as teachers' personal motivations and attitudes towards teaching in online learning environments (Avidor-Ungar, & Eshet-Alkalay, 2011). Conclusions of the research evidence emphasize the importance of teachers' perceiving their school as a learning organization. Recent studies show that the extent of teachers' perceiving their school as a learning organization affects their consent to implement new learning environments. Teachers' readiness to become an active partner in technology-integration projects was found to affect their success of various projects (Levin, & Fullan, 2008; Ogobonna, & Harris, 2003; Zimmerman, 2006; Avidor-Ungar, & Eshet-Alkalay, 2011). Consequently, teachers' training strategies of integrating online collaborative environments into the school agenda should gain special attention. These strategies prompt teachers to share their own class material and activities among their school community, peers as well as their students, through new online communication platforms (e.g. Blogs, Wiki, Google docs, Facebook). This enlarges the teacher's role as an active participant and a support provider during online class-community sessions. On occasion, the need to adhere to the group's learning rate or to the group's level of discussion and consensus can harm aspirations and cause some participants to refrain from stating views or expressing opinions (Rimor, Rosen, & Naser, 2010). Peer pressure may not be optimal for collaboration and might not allow for original independent thought. This may diminish the value of cooperative learning and collaboration work based on equality and mutual contributions (Rovai, 2002; Oliver, & Herrington, 2003; Alavi, & Gallupe, 2003). In order to succeed in online group work, teachers need to invest in developing shared norms and work procedures in their class.

The new role of the teacher as an active participant and pedagogical coacher in these environments elicits the need to adopt new ways of online teaching literacy as well as new methodologies of evaluating students' progress as co constructors of knowledge. A qualitative approach to evaluate the process of collaboration becomes more prevalent and urgent. Peer assessment is considered a strategy of evaluating students' collaborative work in class and was found to be effective in working with graduate students in Wiki learning environment (Meishar-Tal, & Schencks, 2010). In order to foster teaching in online collaborative environments we must rethink the role of the teacher, by providing this new pedagogical challenge with methodologies to assess the group's work, its impact on collaborative knowledge construction, as well as individual learning. Automated classification and coding may offer potential affordances for teachers (e.g. Erkens et al., 2006). Teachers face significant challenges when attempting to support argumentation practices within their classrooms. Online teaching and learning environments that integrate automated analytic frameworks could model argumentation practices for the teachers themselves by facilitating the teachers interpret the argumentation processes of their students.

Collaborative problem-solving skills are considered necessary skills for success in today's world of work and school. Online collaborative learning is enabling teachers to work in partnership with their peers. Teachers are no longer limited in time and place of being involved in professional development initiatives. It is necessary for teachers not only to monitor the activities of a particular student but also the activities of his peers to encourage interaction that could influence the individual learning and the development of collaborative skills. Encouraging the articulation and exchange of knowledge, developing a common language, and supporting a culture of professional learning are among the desired educational improvements that online teacher programs are intended to support (Barab, Barnett, & Squire, 2002; Riel, & Polin, 2004). Many teacher education programs choose to prepare teachers who fit into the patterns of current practices of teaching and learning, instead of improve educational practice (Raths, 2001, Ertmer, & Ottenbreit-Leftwich, 2010). To achieve a change in teaching and learning practices we need to help teachers understand how to use technology to facilitate meaningful learning which can be applied to real-life situations. Teachers must first experience the collaboration, reflection, and group meaning discourse in order to create such learning processes in their own classrooms. ICT teacher competency standards (ISTE, 2008; UNESCO, 2008), emphasize modeling collaborative knowledge construction by engaging with colleagues and students in online environments.

Researchers, practitioners, and policymakers are engaged in vigorous debate about the effectiveness and future promise of online professional development for teachers. Professional development programs typically are designed to target short-term needs through traditional educational strategies, incorporating familiar and widely available technologies. Transforming instructional and learning patterns is one of the main challenges of online teachers' professional development in collaborative environment. Transformation is taught with difficulty because it targets societal changes. Teachers and educational institutions often assimilate into their existing patterns what they learn in professional development rather than accommodating their practice to new principles (Johnson et al., 2009; Wiske, Perkins, & Spicer, 2006). The forces of assimilation and accommodations are at work in teachers' professional development, just as they are with all learners. An experienced teacher established beliefs about good practice, the way he or she interpret curriculum and pedagogy. Accommodation may involve a variety of components (Wiske, Perkins, & Spicer, 2006), such as: (a) change of old practices and beliefs with new ones, (b) compromise through gradual or partial integration of new practices, and (c) synthesis by emphasizing and extending practices that are consistent with the new framework. One aspect of teacher support looks teachers' individual strengths and difficulties. Another aspect of support involves helping teachers identify colleagues and administrators with whom they can collaborate to build a professional community.

Collaborative problem solving suggests active co-construction of knowledge, emphasizing intrapersonal interactions between learners. Countervailing forces at both the individual and the institutional levels tend to maintain the status quo focusing in individual learning and achievement. Online professional development programs are becoming more prevalent and have been used for active participation in collaborative construction of knowledge. However, in many cases, the technology is implemented for traditional practices, while paradigmatic change in teaching, learning, and assessment in technology-rich environments is rare. To achieve this change, a school system must go through major processes. It requires setting new educational objectives, preparing new curricula, developing digital instructional material aligned with ICT standards, designing a new teaching and learning environments, training teachers, creating a school climate that is conducive to collaborative educational technology, and so on. Innovative approaches in learning science, technology, and assessment combined with professional development for teachers can provide a foundation for new and better ways to enhance students' knowledge and skills.

REFERENCES

Alavi, M., & Gallupe, R. (2003). Using information technology in learning: Case studies in business and management education programs. *Academy of Management Learning & Education*, *2*, 139–153. doi:10.5465/AMLE.2003.9901667

Avidov-Ungar, O., & Eshet-Alkakay, Y. (2011). Teachers in a world of change: Teachers' knowledge and attitudes towards the implementation of innovative technologies in schools. *Interdisciplinary Journal of E-Learning and Learning Objects*, *7*, 291–303.

Barab, S., Barnett, M., & Squire, K. (2002). Developing an empirical account of a community of practice: Characterizing the essential tensions. *Journal of the Learning Sciences*, *11*(4), 489–542. doi:10.1207/S15327809JLS1104_3

Barnett, M. (2002). *Issues and trends concerning electronic networking technologies for teacher professional development: A critical review of the literature*. Paper presented at the Annual Meeting of the American Educational Research Association, New Orleans, LA.

Barron, B. (2003). When smart groups fail. *Journal of the Learning Sciences*, *12*(3), 307–359. doi:10.1207/S15327809JLS1203_1

Benbunan-Fich, R., Hiltz, S. R., & Turoff, M. (2003). A comparative content analysis of face-to-face vs. asynchronous group decision making. *Decision Support Systems*, *34*(4), 457–469. doi:10.1016/S0167-9236(02)00072-6

Boler, M. (Ed.). (2004). *Democratic dialogue in education: Troubling speech, disturbing silence*. New York, NY: Peter Lang.

Brookfield, S. D., & Preskill, S. (2005). *Discussion as a way of teaching: Tools and techniques for democratic classrooms* (2nd ed.). San Francisco, CA: Jossey-Bass.

Cohen, E. G., Lotan, R. A., Abram, P. L., Scarloss, B. A., & Schultz, S. E. (2002). Can groups learn? *Teachers College Record*, *104*(6), 1045–1068. doi:10.1111/1467-9620.00196

Collazos, C. A., Guerrero, L. A., Pino, J. A., Renzi, S., Klobas, J., & Ortega, M. (2007). Evaluating collaborative learning processes using system-based measurement. *Journal of Educational Technology & Society*, *10*(3), 257–274.

Dede, C. (2006). *Online professional development for teachers: Emerging models and methods*. Cambridge, MA: Harvard Education Press.

Dede, C., Ketelhut, D., Whitehouse, P., Breit, L., & McCloskey, E. (2009). A research agenda for online teacher professional development. *Journal of Teacher Education*, *60*, 8–19. doi:10.1177/0022487108327554

Dewey, J. (1933). *How we think*. New York, NY: D.C. Heath.

Dillenbourg, P. (1999). What do you mean by 'collaborative learning? In Dillenbourg, P. (Ed.), *Collaborative-learning: Cognitive and computational approaches* (pp. 1–19). Oxford, UK: Elsevier.

Dönmez, P., Rosé, C. P., Stegmann, K., Weinberger, A., & Fischer, F. (2005). Supporting CSCL with automatic corpus analysis technology. In T. Koschmann, D. Suthers & T. W. Chan (Eds.), *Proceedings of the International Conference on Computer Supported Collaborative Learning – CSCL 2005* (pp. 125–134). Taipei, Taiwan: Erlbaum.

Duffy, T., Kirkley, J., del Valle, R., Malopinsky, L., Scholten, C., & Neely, G. (2006). Online teacher professional development: A learning architecture. In Dede, C. (Ed.), *Online professional development for teachers* (pp. 175–197). Cambridge, MA: Harvard Education Press.

Erkens, G., Janssen, J., Jaspers, J., & Kanselaar, G. (2006). Visualizing participation to facilitate argumentation. In S. A. Barab, K. E. Hay, & D. T. Hickey (Eds.), *Proceedings of the 7th International Conference of the Learning Sciences (ICLS)* (Vol. 2, pp. 1095–1096). Mahwah, NJ: Lawrence Erlbaum Associates.

Ertmer, P., & Ottenbreit-Leftwich, A. (2010). Teacher technology change: How knowledge, confidence, beliefs, and culture intersect. *Journal of Research on Technology in Education, 42*(3), 255–284.

Fishman, B. (2007). Fostering community knowledge sharing using ubiquitous records of practice. In Goldman, R., Pea, R. D., Barron, B., & Derry, S. J. (Eds.), *Video research in the learning sciences* (pp. 495–506). Mahwah, NJ: Erlbaum.

Grossman, P., Wineburg, S., & Woolworth, S. (2001). Toward a theory of teacher community. *Teachers College Record, 103*, 942–1012. doi:10.1111/0161-4681.00140

Hawkes, M., & Romiszowski, A. (2001). Examining the reflective outcomes of asynchronous computer-mediated communication on inservice teacher development. *Journal of Technology and Teacher Education, 9*(2), 283–306.

Heyman, G. D. (2008). Children's critical thinking when learning from others. *Current Directions in Psychological Science, 17*(5), 344–347. doi:10.1111/j.1467-8721.2008.00603.x

ISTE. (2008). *The ISTE NETS and performance indicators for teachers*. International Society for Technology in Education.

Johnson, L., Levine, A., Scott, C., Smith, R., & Stone, S. (2009). *The Horizon Report: 2009 economic development edition*. Austin, TX: The New Media Consortium.

Jonassen, D. H. (1999). *Computers in the classroom: Mindtools for critical thinking*. Englewood Cliffs, NJ: Prentice Hall.

Kelly, P. (2011). What is teacher learning? A sociocultural perspective. *Oxford Review of Education, 32*(4), 505–519. doi:10.1080/03054980600884227

Kramarski, B., & Mevarech, Z. R. (2003). Enhancing mathematical reasoning in the classroom: The effects of cooperative learning and metacognitive training. *American Educational Research Journal, 40*(1), 281–310. doi:10.3102/00028312040001281

Kreijns, K., Kirschner, P. A., & Jochems, W. (2003). Identifying the pitfalls for social interaction in computer-supported collaborative learning environments: A review of the research. *Computers in Human Behavior, 19*, 335–353. doi:10.1016/S0747-5632(02)00057-2

Kuhn, D., & Dean, D. (2004). A bridge between cognitive psychology and educational practice. *Theory into Practice, 43*(4), 268–273. doi:10.1207/s15430421tip4304_4

Levin, B., & Fullan, M. (2008). Learning about system renewal. *Educational Management Administration & Leadership, 36*(2), 289–303. doi:10.1177/1741143207087778

Mayer, R. E., & Wittrock, M. C. (2006). Problem solving. In Alexander, P. A., & Winne, P. H. (Eds.), *Handbook of educational psychology* (2nd ed.). Mahwah, NJ: Lawrence Erlbaum Associates.

OECD. (2009a). *Creating effective teaching and learning environments: First results from TALIS*. Paris, France: OECD.

OECD. (2009b). *PIAAC problem solving in technology rich environments: Conceptual framework*. Paris, France: OECD.

Ogobonna, E., & Harris, L. C. (2003). Innovation organizational structure and performance. *Journal of Organizational Change Management*, *16*(5), 512–533. doi:10.1108/09534810310494919

Oliver, R., & Herrington, J. (2003). Exploring technology-mediated learning from a pedagogical perspective. *Interactive Learning Environments*, *11*(2), 111–126. doi:10.1076/ilee.11.2.111.14136

O'Neil, H. F. Jr, Chung, G. K. W. K., & Brown, R. (1997). Use of networked simulations as a context to measure team competencies. In O'Neil, H. F. Jr., (Ed.), *Workforce readiness: Competencies and assessment* (pp. 411–452). Mahwah, NJ: Erlbaum.

Putnam, R. T., & Borko, H. (2000). What do new views of knowledge and thinking have to say about research on teacher learning? *Educational Researcher*, *29*, 4–15.

Raths, J. (2001). Teachers' beliefs and teaching beliefs. *Early Childhood Research & Practice, 3*(1). Retrieved from http://ecrp.uiuc.edu/v3n1/raths.html

Riel, M., & Polin, L. (2004). Online learning communities: Common ground and critical differences in designing technical environments. In Barab, S. (Eds.), *Designing for virtual communities* (pp. 16–50). Cambridge, UK: Cambridge University Press.

Rimor, R. (2002). *From search for information to construction of knowledge: Organization and construction of knowledge in database environment*. Unpublished Doctoral Dissertation, Ben-Gurion University of the Negev, Israel.

Rimor, R., Rosen, Y., & Naser, K. (2010). Complexity of social interactions in collaborative learning: The case of online database environment. *Interdisciplinary Journal of E-Learning and Learning Objects*, *6*, 355–365.

Rodgers, C. (2002). Defining reflection: Another look at John Dewey and reflective thinking. *Teachers College Record*, *104*, 842–866. doi:10.1111/1467-9620.00181

Roschelle, J., & Teasley, S. D. (1995). The construction of shared knowledge in collaborative problem-solving. In O'Malley, C. E. (Ed.), *Computer-supported collaborative learning* (pp. 69–97). Berlin, Germany: Springer-Verlag. doi:10.1007/978-3-642-85098-1_5

Rosen, Y., & Rimor, R. (2009). Using collaborative database to enhance students' knowledge construction. *Interdisciplinary Journal of E-Learning and Learning Objects*, *5*, 187–195.

Rovai, A. P. (2002). Building sense of community at a distance. *International Review of Research in Open and Distance Learning*, *3*(1), 1–16.

Savery, J. R., & Duffy, T. M. (1996). Problem-based learning: An instructional model and its constructivist design. In Wilson, B. G. (Ed.), *Constructivist learning environments: Case study in instructional design* (pp. 135–148). Englewood Cliffs, NJ: Educational Tech Pubs.

Schon, D. A. (1987). *Educating for reflective practitioner: Toward a new design for teaching and learning in professions*. San Francisco, CA: Jossey-Bass.

Schon, D. A. (1991). *The reflective turn: Case studies in and on educational practice*. New York, NY: Teachers College.

Schraw, G., Crippen, K. J., & Hartley, K. (2006). Promoting self-regulation in science education: Metacognition as part of a broader perspective on learning. *Research in Science Education*, *36*, 111–139. doi:10.1007/s11165-005-3917-8

Slavin, R. E. (1997). *Educational psychology: Theory and practice* (5th ed.). Needham Heights, MA: Allyn & Bacon.

Sorensen, E. K., & Takle, E. S. (1998). *Collaborative knowledge building in web-based learning: assessing the quality of dialogue*. Paper presented at the Annual Meeting of ED-MEDIA 1998 – World Conference on Educational Multimedia, Hypermedia and Telecommunications Finland: Tampere.

Stahl, G. (2006). *Group cognition: Computer support for building collaborative knowledge*. Cambridge, MA: MIT Press.

UNESCO. (2008). *ICT competency standards for teachers*. Paris, France: UNESCO.

Vygotsky, L. (1978). *Mind and society: The development of higher mental processes*. Cambridge, MA: Harvard University Press.

Wadmany, R., Rimor, R., & Rosner, E. (2011). The relationship between attitude, thinking and activity of students in an e-learning course. *Research on Education and Media, 3*(1).

Wegerif, R. (2006). Towards a dialogic understanding of the relationship between teaching thinking and CSCL. *International Journal of Computer-Supported Collaborative Learning*, *1*(1), 143–157. doi:10.1007/s11412-006-6840-8

Weinberger, A., & Fischer, F. (2006). A framework to analyze argumentative knowledge construction in computer-supported collaborative learning. *Computers & Education*, *46*, 71–95. doi:10.1016/j.compedu.2005.04.003

Weinberger, A., Stegmann, K., Fischer, F., & Mandl, H. (2007). Scripting argumentative knowledge construction in computer-supported learning environments. In Fischer, F., Kollar, I., Mandl, H., & Haake, J. (Eds.), *Scripting computer-supported communication of knowledge—Cognitive, computational and educational perspectives* (pp. 191–211). New York, NY: Springer. doi:10.1007/978-0-387-36949-5_12

Wiske, M. S., Perkins, D., & Spicer, D. E. (2006). Piaget goes digital: Negotiating accommodation of practice to principles. In Dede, C. (Ed.), *Online professional development for teachers* (pp. 49–67). Cambridge, MA: Harvard Education Press.

Yang, S. C., & Liu, S. F. (2004). Case study of online workshop for the professional development of teachers. *Computers in Human Behavior*, *20*, 733–761. doi:10.1016/j.chb.2004.02.005

Zimmerman, J. (2006). Why some teachers resist change and what principals can do about it. *NASSP Bulletin*, *90*(3), 238–249. doi:10.1177/0192636506291521

ADDITIONAL READING

Andriessen, J., Baker, M., & Suthers, D. (Eds.). (2003). *Arguing to learn: Confronting cognitions in computer-supported collaborative learning environments*. Dordrecht, The Netherlands: Kluwer Academic Publishers.

Baker, M., & Lund, K. (1997). Promoting reflective interactions in a CSCL environment. *Journal of Computer Assisted Learning*, *13*, 175–193. doi:10.1046/j.1365-2729.1997.00019.x

Bromme, R., Hesse, F. W., & Spada, H. (Eds.). (2005). *Barriers and biases in computer mediated knowledge communication, and how they may be overcome*. New York, NY: Springer. doi:10.1007/b105100

Bruffee, K. (1993). *Collaborative learning*. Baltimore, MD: Johns Hopkins University Press.

Chung, G., O'Neil, H., & Herl, H. (1999). The use of computer based collaborative knowledge mapping to measure team processes and team outcomes. *Computers in Human Behavior*, *15*, 463–494. doi:10.1016/S0747-5632(99)00032-1

Dillenbourg, P. (Ed.). (1999). *Collaborative learning: Cognitive and computational approaches*. Amsterdam, The Netherlands: Pergamon, Elsevier Science.

Fischer, F., Bruhn, J., Grasel, C., & Mandl, H. (2002). Fostering collaborative knowledge construction with visualization tools. *Learning and Instruction*, *12*, 213–232. doi:10.1016/S0959-4752(01)00005-6

Fischer, F., Kollar, I., Mandl, H., & Haake, J. M. (2007). *Scripting computer supported collaborative learning - Cognitive, computational and educational perspectives*. Berlin, Germany: Springer. doi:10.1007/978-0-387-36949-5

Fung, Y. Y. H. (2004). Collaborative online learning: interaction patterns and limiting factors. *Open Learning*, *19*(2), 135–149. doi:10.1080/0268051042000224743

Gottman, J. M., & Roy, A. K. (1990). *Sequential analyses: A guide for behavioral researchers*. New York, NY: Cambridge University Press. doi:10.1017/CBO9780511529696

Hathorn, L. G., & Ingram, A. L. (2002). Cooperation and collaboration using computer-mediated communication. *Journal of Educational Computing Research*, *26*, 325–347. doi:10.2190/7MKH-QVVN-G4CQ-XRDU

Hmelo-Silver, C. E. (2003). Analyzing collaborative knowledge construction: Multiple methods for integrated understanding. *Computers & Education*, *41*, 397–420. doi:10.1016/j.compedu.2003.07.001

Hsieh, I., & O'Neil, H. (2002). Types of feedback in a computer-based collaborative problem solving group task. *Computers in Human Behavior*, *18*, 699–715. doi:10.1016/S0747-5632(02)00025-0

Janetzko, D., & Fischer, F. (2003). Analyzing sequential data in computer-supported collaborative learning. *Journal of Educational Computing Research*, *28*(4), 341–353. doi:10.2190/805X-VG4A-DNND-9NTC

Kollar, I., Fischer, F., & Hesse, F. W. (2006). Collaboration scripts: A conceptual analysis. *Educational Psychology Review*, *18*(2), 159–185. doi:10.1007/s10648-006-9007-2

Koschmann, T., Hall, R., & Miyake, N. (Eds.). (2002). *CSCL2: Carrying forward the conversation*. Mahwah, NJ: Lawrence Erlbaum Associates.

Kreijns, C. J., Kirschner, P. A., & Jochems, W. M. G. (2002). The sociability of computer-supported collaborative learning environments. *Journal of Educational Technology & Society*, *5*(1), 8–22.

Mäkitalo, K., Weinberger, A., Häkkinen, P., Järvelä, S., & Fischer, F. (2005). Epistemic cooperation scripts in online learning environments: Fostering learning by reducing uncertainty in discourse? *Computers in Human Behavior*, *21*, 603–622. doi:10.1016/j.chb.2004.10.033

Nicolopoulou, A., & Cole, M. (1993). Generation and transmission of shared knowledge in the culture of collaborative learning: The fifth dimension, its play world and its institutional contexts. In Forman, E., Minnick, N., & Stone, C. A. (Eds.), *Contexts for learning: Sociocultural dynamics in children's development*. New York, NY: Oxford University Press.

O'Malley, C. (1995). *Computer supported collaborative learning*. Berlin, Germany: Springer Verlag. doi:10.1007/978-3-642-85098-1

Rummel, N., & Spada, H. (2005). Sustainable support for computer-mediated collaboration: How to achieve and how to assess it. In Bromme, R., Hesse, F., & Spada, H. (Eds.), *Barriers and biases in computer-mediated knowledge communication and how they may be overcome*. Dordrecht, The Netherlands: Kluwer Academic Publisher.

Rummel, N., & Spada, H. (2005a). Learning to collaborate: An instructional approach to promoting collaborative problem solving in computer-mediated settings. *Journal of the Learning Sciences*, *14*, 201–241. doi:10.1207/s15327809jls1402_2

Schwarz, B. B., Neuman, Y., Gil, J., & Ilya, M. (2003). Construction of collective and individual knowledge in argumentative activity. *Journal of the Learning Sciences*, *12*(2), 221–258. doi:10.1207/S15327809JLS1202_3

Stahl, G. (2006). *Group cognition: Computer support for building collaborative knowledge*. Cambridge, MA: MIT Press.

Stahl, G., Koschmann, T., & Suthers, D. (2006). Computer-supported collaborative learning: An historical perspective. In Sawyer, R. K. (Ed.), *Cambridge handbook of the learning sciences* (pp. 409–426). Cambridge, UK: Cambridge University Press.

Suthers, D., & Hundhausen, C. (2003). An empirical study of the effects of representational guidance on collaborative learning. *Journal of the Learning Sciences*, *12*(2), 183–219. doi:10.1207/S15327809JLS1202_2

Weinberger, A., Ertl, B., Fischer, F., & Mandl, H. (2005). Epistemic and social scripts in computer-supported collaborative learning. *Instructional Science*, *33*(1), 1–30. doi:10.1007/s11251-004-2322-4

Weinberger, A., Stegmann, K., & Fischer, F. (2010). Learning to argue online: Scripted groups surpass individuals (unscripted groups do not). *Computers in Human Behavior*, *26*, 506–515. doi:10.1016/j.chb.2009.08.007

KEY TERMS AND DEFINITIONS

Collaborative Learning: Coordinated joint activity in which two or more learners attempt to construct and maintain a shared conception of a construct, process or a problem.

Computer-Supported Collaborative Learning (CSCL): Is emerging as a dynamic, interdisciplinary, and international field of research focused on how technology can facilitate the sharing and creation of knowledge and expertise through peer interaction and group learning processes.

Information And Communication Technologies (ICT) Teaching Skills: Teacher's competency to use technology effectively for retrieving, analyzing and creating data, content, resources, expertise, and learning experiences that can provide better teaching for all learners.

Chapter 6
Online Teacher Education:
A Case Study from New Zealand

Anne Yates
Victoria University of Wellington, New Zealand

ABSTRACT

This chapter reports a study which examined experiences of nine beginning teachers who completed their initial teacher education in the online mode. The study investigated reported perceptions during their first six months teaching. Participants found the content of the online program comprehensive, prepared them well to begin teaching, and provided an opportunity for Māori (New Zealand's indigenous people) to become high school teachers. Main advantages of studying online were: flexibility; saving time and money; developing skills and personal attributes such as independence; and for some, the only way to become teachers. The major disadvantage was the difficulty of studying alone despite an interactive delivery platform. Also, participants were concerned learning online did not allow modelling of teaching skills and this impacted on participants' classroom practice. Recommendations include creating connections for online learners; using skilled staff; creating culturally appropriate online environments; and incorporating opportunities for face-to-face interaction in online initial teacher education.

INTRODUCTION

A beginning high school teacher enters a complex system and is charged with the education of up to 100 teenagers annually. Teachers have substantial impact on student success; as Hattie (2009) claimed, the effect size of the teacher on student learning is 30 percent. Therefore, all initial teacher education (ITE), including that delivered online, must produce effective, high-quality teachers. To ensure high quality ITE in all modes research is needed to investigate their effectiveness.

Research into distance ITE in New Zealand is quite limited due to distance ITE being a relatively recent phenomenon. It began in the 1990s after a major restructure of the education system which allowed tertiary education institutions to set their own directions (Simpson, 2003). Major (2005) noted some ITE providers moved to develop distance programs to reach an untapped market of

DOI: 10.4018/978-1-4666-1906-7.ch006

students who were unable, or unwilling, to move to access teacher education. The program in this study had its origins in this deregulated market. Kane (2005) reports offering teacher qualifications by distance presents exciting opportunities, but it brings challenges, and more in-depth research is needed to shed light on the benefits and challenges of the different modes of delivery. The international literature on distance ITE reports a general consensus that distance ITE is effective if the underlying program is sound and based on the principles of effective teacher education. However, few have studied the graduates into teaching, citing inherent difficulties in doing this.

This chapter presents a case study from New Zealand which took up some of these challenges and followed nine beginning teachers, who completed ITE in the online mode, into their first six months of teaching. The main purpose of the study was to discover the ways an online program prepared student teachers for the reality of face-to-face teaching. The aims were to find out if beginning high school teachers who completed their ITE in the online medium perceived they were suitably prepared to teach in a face-to-face classroom and to make recommendations to assist the learning of online student teachers. Therefore the study investigated the participants' reported perceptions or constructions of the online program and their perceived readiness to begin teaching. It was the participants' journey to becoming a teacher through distance study that inspired this research.

The chapter begins with a short background to New Zealand and its education system, and then presents views from the existing literature on successful online ITE. Also, the literature focuses on the ability of online learning to meet the needs of indigenous populations. As with other former colonies, New Zealand has a minority indigenous population (the New Zealand Māori). New Zealand strives to be a bicultural nation in recognition of Māori as the original people (tangatawhenua). While biculturism pervades all aspects of New Zealand society, it is particularly relevant in education. There is strong demand for Māori teachers, so much so that in 2011 the New Zealand government offered 115 scholarships to Māori to train as high school teachers.

This chapter will provide evidence that distance learning can be successful for ITE and that an online program can provide for Māori student teachers in a way campus-based courses cannot. It will describe the ways in which the beginning teachers felt prepared and discusses the perceived advantages and disadvantages of studying to be a teacher online. It concludes by providing suggestions to make online ITE more effective, because, as Moore (2007) states, the question is no longer should we teach online, but *how*.

BACKGROUND

New Zealand

New Zealand is a small South Pacific country comprising two main islands and approximately the size of Great Britain or Colorado. The majority of New Zealand's 4.4 million population is of European descent (67%); Māori are the largest minority (14.6%), followed by Asians (9.2%) and non-Māori Polynesians (6.9%) (Statistics New Zealand, 2011). New Zealand has a modern, prosperous and developed market economy with an estimated gross domestic product at purchasing power parity per capita of about US$28,250. According to the Human Development Index, New Zealand was the 3rd "most developed" country in the world (United Nations Development Programme, 2010) and ranked 4th in the 2011 Index of Economic Freedom (Heritage Foundation, 2011).

International tests show the New Zealand education system is highly ranked. Schooling is compulsory for children aged 6 to 16, although the majority of children start primary school at age five. At 12.5 mean years of schooling, New Zealand is the second highest in the world after Norway. New Zealand rates highly in the Programme for International Student Assessment (PISA) with only two out of 30 OECD countries achieving significantly higher mean scores than New Zealand in reading literacy (Ministry of Education, 2010a). ITE in New Zealand is conducted in 26 institutions, although the majority of high school teacher graduates (over 96%) attend one of the six universities offering ITE (Kane, 2006).

New Zealand has a long tradition of offering distance education to students from preschool to tertiary and was a foundation member of the International Council for Open and Distance learning. Institutions such as the New Zealand Correspondence School and the Open Polytechnic of New Zealand have been solely devoted to providing distance education since 1922 and 1946 respectively.

The Program in the Study

In 2002, Wellington College of Education (now the Faculty of Education at Victoria University of Wellington) established an online program for the one year Graduate Diploma of Teaching (Secondary) "in order to meet the needs of 21st Century learners" (Wellington College of Education, 2001, p. 2). As the program offers a Graduate Diploma, all entrants to the program have completed at least an undergraduate degree, and complete their teacher education in one year. The online ITE program was developed with the same aims and structure as the campus-based course. This structure was a program consisting of 14 independent courses including curriculum-specific courses, general pedagogy courses and electives. In the curriculum courses the students study the pedagogical content knowledge specific to their teaching speciality, for example teaching second languages, mathematics, English. The general pedagogy courses include learning theories, knowledge of general principles of instruction, models and strategies of teaching, classroom management, assessment and teaching diverse learners. The program includes 14 weeks in schools on teaching experience, where student teachers work in classrooms under the guidance of the regular classroom teacher, and are observed and assessed by faculty staff.

The main difference between the campus and online program is that the online program delivers the theoretical components via a purposely designed website and a short residency. The website is designed not only to deliver content, but enables collaborative learning and social interaction through the use of discussion boards and blogs. The content is delivered through the website, course readings, textbooks, video and voice files. Students complete the work asynchronously, but have mandatory requirements to contribute to weekly blogs and discussion boards. Also, they attend a four-day residency where they have the opportunity to meet the staff and each other, and to participate in curriculum-specific workshops. Access to lecturers and fellow students is via email, Skype and a toll-free telephone number.

In New Zealand, all ITE programs, including that in this study, must ensure that graduates meet the Graduating Teacher Standards (New Zealand Teachers Council, 2011) and programs are designed to enable student teachers meet the criteria in the standards. The seven standards are that graduating teachers: know what to teach; know about learners and how they learn; understand how contextual factors influence teaching and learning; use professional knowledge to plan for a safe, high quality teaching and learning environment; use evidence to promote learning; develop positive relationships with learners and members of learning communities; and are committed members of the profession.

LITERATURE REVIEW

Online ITE

There is no doubt people can and do learn when physically separated from a teacher. Holmberg (2005) simply stated "Distance education works" (p. 37) and that although the pioneers of distance education had a meagre theoretical background, they proceeded on the hypothesis that learning could be possible and effective without the teacher and the learner meeting face-to-face. Since the inception of distance learning research this early hypothesis has been supported by others. The milestone work by Dubin and Taveggia (1968) analysed data from seven million academic records and concluded that the mode of instruction made no difference to student performance. Russell (2001) summarised 355 different research studies dating back to 1928 that support the conclusion that no significant difference exists between the effectiveness of classroom education and distance learning.

However, the perceived disadvantages of distance education, such as lack of interaction and communication with other students and teachers, and therefore the lack of development of interpersonal skills, would seem to preclude distance education as being appropriate for ITE. The image of a distance learner as an isolated individual working alone is at odds with the image of a teacher in constant contact with students and/or colleagues.

Despite these perceived shortcomings, there is a body of literature that has reported teacher education as able to be successfully delivered at a distance. Most agree this success depends upon the convergence of effective ITE and effective online learning.

Effective ITE

As Simpson and Kehrwald (2010) claim, every country is "unique and teacher education programmes must reflect local societal values and cultural differences, and there must be content within each programme that acknowledges and celebrates those differences" (p. 23). To recognise this uniqueness, all ITE programs in New Zealand must be approved and accredited by the New Zealand Teachers Council once every three years to ensure graduates will meet the Graduating Teacher Standards.

Also, it is generally accepted in New Zealand that Hammerness's et al. (2005) seven common features of highly successful teacher education programs should be used as guiding principles. These principles include:

- A common, clear vision of good teaching permeates all course work and clinical experiences
- Well defined standards of practice and performance are used to guide and evaluate course work and clinical work
- Curriculum is grounded in knowledge of child and adolescent development, learning, social contexts, and subject matter pedagogy, taught in the context of practice
- Extended clinical experiences are carefully developed to support the ideas and practices presented in simultaneous, closely interwoven course work
- Explicit strategies help student teachers confront their own deep-seated beliefs and assumptions about learning and students and learn about the experiences of people different from themselves
- Strong relationships, common knowledge, and shared beliefs link school and university-based faculty
- Case study methods, teacher research, performance assessments, and portfolio evaluation apply learning to real problems of practice.(p. 41)

However, Kane (2005) expressed concern that while the conceptual frameworks of New Zealand

ITE are grounded in research, they lacked coherence. She found there was considerable variation in the way conceptual frameworks were understood and a lack of clarity on the purpose of a conceptual framework. Kane argued providers need to critically examine the conceptual coherence and integration of their qualifications to ensure they are built on a strong vision of good teaching practice and supported by sound theoretical information. In addition, she questioned the coherence of some qualifications as they were fragmented into several separate papers, claiming this raised issues in terms of staff and student workload. This could leave student teachers having to make sense of the connections among various papers and parts of the program if this is not made explicit by teacher educators.

The program in this study bases its conceptual framework on Hammerness's et al. (2005) principles and is accredited every three years by the New Zealand Teachers Council, but as Kane (2005) claimed providers need to critically examine their use of conceptual frameworks and the coherence of their qualification. It may be that the student experience somewhat differs from the institution's perception.

Effective Online Learning

In addition to being based on sound ITE principles, successful online ITE programs must meet the requirements of effective online learning. As a relatively new area of research, theories of online learning are still emerging.

In keeping with social constructivist views of teaching and learning, successful online learning needs to be learner-centred and community-based, rather than focused on how the material will be delivered (Paloff & Pratt, 2003). According to Garrison, Anderson and Archer (2000), effective online learning is based on the assumption that learning is embedded in a community of inquiry composed of teachers and students. There is no doubt online learning communities can exist (e.g. Johnson & Johnson, 1996; Paloff & Pratt, 1999; Rovai, 2001). Communities are no longer place-based; the internet is a global village square where people can, and do, come together to trade, meet, talk, and be educated. In the 21st century, communities exist in cyberspace and computer networks provide education with powerful opportunities for collaborative learning (Paloff & Pratt, 1999).

This is a departure from theories of distance learning that promote student independence, and the image of a distance learner studying alone. However, there can be the assumption that because technology allows collaboration this naturally will occur. Online learning is a unique blend of technology and pedagogy (Herie, 2005) and online teachers provide "the container, pedagogically speaking, through which students can explore the territory of the course" (Paloff & Pratt, 2003, p. xv).

For the 'container' to be effective, Paloff and Pratt (2003) provided a framework for courses, institutions, instructors and students. Courses should be kept simple and avoid information overload, encourage communication among students, and make expectations of the students clear. The institution should provide efficient pastoral and technical support, ensure reliable technology, provide an orientation to online learning and counsel students on their suitability for this type of learning. Online instructors first and foremost need to remember there are real people at the end of the line who need support and to be kept on track. They should develop a sense of community, provide prompt and unambiguous feedback, and create a reasonable workload. Finally, an effective online environment needs effective students. Paloff and Pratt suggested a portrait of a successful online student where students: need technical skills self-motivation and self-discipline, are willing and able to commit the time, are willing to contribute to the learning of others, and realise this is not the easy option.

Effective Online ITE

Simpson (2003) analysed 21 programmes of distance delivery pre-service teacher education and found these were not homogenous. Those that were successful converged an understanding of the two disciplines: distance education and ITE. This understanding included attention to student support, student selection, skilled tutors, quality materials, a cohesive programme, field experience, and institutional support. Simpson found a developed/developing country divide where developed countries like New Zealand and the United Kingdom built their distance ITE on decades of face-to-face teacher education, whereas in some developing countries this was not the case and distance ITE had been designed to alleviate teacher shortages.

Oliveira and Orivel (2003) agreed that successful distance ITE programs are based on effective teacher education. They commented that there is "nothing intrinsically good or bad about distance learning and technology to make them more or less effective than face-to-face instruction" (p. 232) but the effectiveness of such programs depends on the quality of the teaching and the materials. Distance ITE is part of ITE and cannot be isolated from the "complex web of issues, policies and resource decisions that in any country affect teaching and teacher education" (Robinson & Latchem, 2003, p. 2). Perraton et al. (2002) agreed that distance teacher education is not an isolated phenomenon but teacher education using distance methods can be effective if the ITE program addresses key issues of knowledge and understanding of the subjects, pedagogy and understanding of students and learning, and the development of practical skills through teaching practice.

In New Zealand the research field of online or distance ITE is relatively small, but tends to concur with the international literature that this form of ITE can be effective. In their reflections on a distance ITE program, Hall, Yates and Campbell (1998) noted students were generally positive about their experiences and valued the block courses. Almost a decade later, Anderson and Simpson (2005) sought to discover how an online program prepared student teachers for their role as beginning teachers. The overall conclusion was, although beginning teachers study faced challenges, they felt confident about the ability of on online program to prepare them to teach.

Not all of the literature is favourable towards distance teacher education. Peterson and Bond (2004) carried out a quasi-experimental investigation into two pairs of asynchronous online and face-to-face courses that prepared pre-service teachers for instructional planning. They found that both groups made significant gains in their learning and assessment results did not reveal any significant difference between the two groups. Participant interviews, however, raised some reservations about online preparation for the transition to classrooms. Face-to-face modelling demonstrations and feedback were preferred over that given online.

In contrast, there is an emerging body of evidence that online ITE may actually help new teachers to deal with the challenges of teaching in the technological age (Robinson & Latchem, 2003). They contended teachers need to become independent learners and to deal with information technology, and that new models and realities of teaching require new modes of teacher training. Anderson and Simpson (2005) similarly found that distance delivered ITE had positive effects, such as teaching self-management, organisation and ICT skills. In contrast with reports of beginning teachers having difficulty communicating with parents and caregivers, the participants in Anderson and Simpson's study already felt a sense of belonging to, and knowing, the community:

"Of course teacher education students who stay in the community are important for small country localities where recruiting and retaining skilled and qualified teachers is difficult. Thus community embeddedness both helped prepare participants

for their teaching role, and helped ensure they won such a role" (p. 6).

However, when examining the success of teacher education, the ability to transfer the theory and knowledge gained to the classroom must be considered. Nielsen (1997) agreed that any teacher education program should be judged on the ability of its graduates to teach well. Simpson (2003) stated that the ability of graduates to teach well should be one measure of the quality of the program; however, she indicated the difficulty of using this measure due to the links between the graduates and the pre-service institutions generally being broken on graduation.

Providing for Indigenous Learners

Unique to New Zealand is ensuring online learning provides for, and respects, Māori students. Online learning has the potential to provide Māori with opportunities to participate in tertiary education and to offer a Māori pedagogy within an e-learning framework. While Tiakiwai and Tiakiwai (2010) lament a dearth of research on e-learning and indigenous peoples, they did conclude e-learning was able to overcome barriers such as work or family commitments and allowed indigenous students to remain in familiar cultural and social environments while pursuing tertiary study. Ham and Wenmoth (2007) specifically stated that "e-learning within the Māori community is seen as a 'leveller' and, in particular, a way to enable learning to take place for Māori communities in remote areas" (p. 60).

The possibilities of e-learning to meet the needs of indigenous populations has been acknowledged by Wall (2008) whereby e-learning allowed Canadian Aboriginal students to continue studying in the familiar context of home while at the same time communities did not lose students to larger centres. Grant (1996) described a mixed-mode programme developed in 1989, in part, to overcome the shortage of Aboriginal teachers in Australia. The mixed-mode course was an outcome of negotiations with Aboriginal communities and was intended to provide maximum opportunity for the students to sustain their family, work, and other responsibilities. This theme is noted by Porima (2005) where Māori participants in her study cited the flexibility of studying from home rather than attending prescribed classes was a major reason for taking online classes.

There is some concern that e-learning may have a negative impact on the preservation of indigenous cultures (Bowers, Vasquez & Raof, 2000; Carr-Chellman, 2005; Hodson, 2004). However, in New Zealand the e-Learning Advisory Group (2002) suggested e-learning can not only improve Māori participation, but also offer a Māori pedagogy within an e-learning framework. This should include Māori principles, such as ako (reciprocal teaching and learning), whanaungatanga (building relationships), aroha (love or sharing), and manaakitanga (supporting others) (Porima, 2005). Porima adds that effective e-learning for Māori should have adequate support to sustain motivation, user friendly software and the opportunity to meet face-to-face with other learners and tutors. Selby (2006) found there is potential for e-learning to make a significant contribution to the teaching of the Māori language (te reo Māori), but cautioned the "e-learning space must be identifiably Māori in its appearance and behaviour" (p. 85). She claimed using Māori pedagogy to develop the e-learning space could actually enhance the preservation of Māori culture.

There is a general consensus in the literature that distance ITE can be effective if the underlying program is sound, and the principles of effective teacher education and conditions for effective online learning are in place. This includes having skilled online lecturers who recognise the importance of building a community of learners. There are some who argue that in the technological age it may have advantages over traditional methods. However, trying to make generalisations about online learning is problematic. It must be considered that online learning is not a single entity, that there is such a variety of online courses that it is impossible to make comparisons, and "that to try and characterise online learning is about as possible as trying to characterise animals in the zoo" (Rovai & Barnum, 2003, p. 59). Factors such as the course design, student support, pedagogy, student characteristics and technology need to be considered, and comparisons and generalisations can only be made with online courses with similar characteristics.

There is a growing body of literature that online learning is providing access to tertiary education for indigenous people, and in New Zealand helping to meet the Government's goals to increase the number of Māori teachers. E-learning in New Zealand has a unique opportunity to improve not only access to tertiary education for Māori students, but also to develop e-learning within a culturally appropriate pedagogy that provides for, and respects, Māori students.

Therefore, there is a continuing need to investigate the outcomes of different modes of teacher education. As Kane (2005) noted, "while offering qualifications by distance, flexibly and/or through web-based delivery brings with it exciting opportunities for student teachers (and staff), it also brings some challenges with respect to quality" (p. 211). Further research is needed to shed light on both the benefits and challenges of different modes of delivery and to provide evidence that student teachers experience quality training.

METHODOLOGY

The study reported in this chapter examined experiences of nine beginning high school teachers who completed their ITE in the online mode. A qualitative research design within a constructivist paradigm was used and a multiple case study was chosen as the methodology. A constructivist paradigm was appropriate as this study hoped to gain an understanding of the constructions held by participants about beginning a face-to-face career, after having studied for much of it in the online medium, and multiple realities of the beginning teachers' experiences may exist. The research sought the participants' reported perceptions and beliefs. It did not intend to explain, predict or critique the situation. Patton (2002) stated that a constructivist perspective is appropriate when the research aims to locate reported perceptions, truths, reality, explanations, and beliefs.

The participants ranged in age from 24 to 45, with five males and four females, four participants identifying as Māori. They were all employed in state co-educational schools in a range of social settings and geographically located around New Zealand. The study investigated the participants' reported perceptions of the online program and their perceived readiness to begin teaching. The research focused on nine beginning teachers during their first six months of teaching. Qualitative data were collected through two surveys and one face-to-face interview. The first survey was sent to the participants between one and two months of their beginning teaching, and the second was administered two months later; the interviews occurred at the end of the six months. The questions in Surveys 1 and 2 were identical with the intention to ascertain if there was a shift in the participants' perceptions over the first few months of their teaching career.

The surveys included reporting of both self-assessment and self-reflection against the Graduating Teacher Standards, specifically those used to assess teaching experiences during the ITE programme. During teaching experiences the student teachers were assessed by associate teachers and visiting lecturers on a range of teaching skills that are expected of beginning teachers and taught within the program. Thus, the participants were familiar with the skills and measurement nomenclature. Furthermore, these standards were thought to guide their thinking to the pedagogy required to begin teaching rather than administrative tasks that will occupy much of their time.

The surveys contained both closed and open-ended questions. The closed questions asked the participants to rate themselves against a four point Likert scale on their perceived competence in a range of teaching skills and their perception of how well their ITE prepared them for these skills. The skills included: planning and preparation; curriculum and subject knowledge; teaching strategies and routines; management of, and communication with, the students; assessment practice and knowledge; questioning techniques; awareness of student needs and abilities; and professional behaviour. The open-ended questions invited comments on each of the closed questions, and there were also several open-ended questions where the participants shared their major successes and challenges as well as any other comments they wanted to make.

The interviews focused on similar topics asking how they perceived ITE had prepared them for teaching, and the advantages and disadvantages of studying online. The researcher conducted one face-to-face, semi-structured interview with each of the participants who agreed to be interviewed, which took place near the end of the first six months of the beginning teacher's career. Interviews allowed flexibility to suit individual participants and for the creation of an empathetic environment where participants might be willing to share negative experiences. Semi-structured interviews allowed the researcher some control over the interview (Creswell, 2003) and enabled some comparison among the participants, but they gave the participants the opportunity to add comments and raise issues to give breadth to the discussion.

The data were analysed using content analysis, in which inductive coding and sorting allowed themes to emerge. Berg (2004) suggests that in content analysis researchers "examine the artefacts of social communication" (p. 267). These artefacts are typically written documents and transcriptions of verbal communications. Holsti (cited in Berg, 2004) stated that content analysis is "any technique for making inferences by systematically and objectively identifying special characteristics of messages" (p. 267).

Following the inductive approach suggested by Abrahamson (cited in Berg, 2004), the researcher analysed the transcripts in order to identify dimensions or themes, and followed Berg's (2004) suggestion that transcripts be coded into content units and categorised according to similar themes and patterns. Researchers have to decide at what level they will sample (e.g., words, phrases, sentences) and what to count (e.g., themes, items or concepts). As this study was identifying themes, the content could be any of single words, phrases and sentences. All content was coded, as the researcher did not want to judge what was worthwhile and what was not; however, content such as social talk, introductions and interruptions were coded as irrelevant. As Charmaz (2006) recommended, careful coding helps the researcher to refrain from putting one's own motives, fears and issues into the data.

DISCUSSION AND FINDINGS

Two major themes emerged from the analysis of the data: how the program prepared the participants to begin teaching and the experience of learning online.

Preparation for Teaching

All data sources confirmed the content delivered in the program was comprehensive and participants agreed they were prepared to begin teaching. They reported competence in teaching skills taught by the program, however felt more prepared for some skills than others. They reported most prepared and most confident in the skills of planning and preparation, teaching strategies and routines, managing student behaviour, and acting professionally. As Table 1 shows, all participants perceived they were either strong or competent in these skills, and agreed the ITE provided good preparation for these. For example, in terms of planning all participants agreed the content was very useful; this included both lesson and unit planning. Comments supported these ratings:

"In this area of teaching (planning) I felt well prepared by the preparation VUW provided. When I arrived in school I had many of the templates, concepts and resources to begin planning lessons that were relevant, interesting and used a range of learning activities and strategies." (P1)

"Doing the unit planning in the course was really useful." (P7)

"I felt well prepared in terms of how to deliver a lesson." (P2)

While success in managing student behaviour was not universal, the ratings and comments support the usefulness of the content in relation to this.

"My teacher training gave me a range of class management strategies and learning activities to put in place. This has helped my professional practice (attempting different approaches rather than giving up and crying)." (P6)

"I felt well equipped with classroom management ideas and theory, but even though it might not have gone that well with one class I certainly felt I knew where I stood in terms of what I needed to accept and what I didn't from the kids." (P2)

"I felt well prepared to teach – the teaching experiences were extremely valuable plus putting into practice theories learned in the online programme, such as behaviour management." (P2)

Interestingly, over the period of the surveys, three participants changed their rating for behaviour management skills from competent to needing further development. The interviews took the opportunity to probe more deeply into this fall in efficacy. The participants had entered a range of teaching contexts and the beginning teacher support they received varied. Those with good support reported managing student behaviour as a success and even with difficulties the strategies learnt in ITE helped considerably. Those who struggled most had the least support and in some cases felt blamed for not coping. The varying levels of support would not be known to the participants as they were unaware of each other's situations which may have exacerbated the feelings of blame and inadequacy. Those receiving little support assumed this was the norm and therefore their responsibility that they were not coping, not realising that their colleagues in other schools were much better supported.

Bartell (2005) and Antony and Kane (2008) discuss how the needs of beginning teachers vary depending on the context in which they are teaching. The generic knowledge learnt in ITE needs to be transferred to the specific context and effective transfer of learning from ITE into the new context needs to occur for teachers to be successful. Baldwin and Ford's (1988) seminal transfer of learning model states workplace support, in terms of feedback and reinforcement is essential for effective transfer. This appears to be the case in this study as the two participants who reported the most difficulty managing student behaviour had the least induction support.

As Table 1 shows, several participants perceived that assessment, questioning and awareness of student needs were the areas in which they felt least competent and least prepared. With regards to questioning comments indicated that despite having learnt this skill, the difficulty was in the execution. One participant (P2) commented that he knew to use wait time, yet felt uncomfortable doing so. Comments included:

Table 1. Summary of survey closed questions

	Survey 1				Survey 2			
	Skill level		Teacher Ed		Skill level		Teacher Ed	
Planning & Preparation	S	3	SA	5	S	3	SA	2
	C	6	A	4	C	3	A	4
	FDR	0	D	0	FDR	0	D	0
Curriculum & Subject Knowledge	S	3	SA	2	S	2	SA	1
	C	3	A	5	C	2	A	2
	FDR	2	D	1	FDR	2	D	3
Strategies & Routines	S	1	SA	1	S	0	SA	0
	C	7	A	7	C	6	A	6
	FDR	0	D	0	FDR	0	D	0
Management & Communication	S	2	SA	2	S	0	SA	0
	C	6	A	5	C	3	A	5
	FDR	0	D	1	FDR	3	D	1
Assessment	S	1	SA	1	S	0	SA	0
	C	3	A	6	C	4	A	4
	FDR	5	D	2	FDR	2	D	2
Questioning	S	1	SA	1	S	2	SA	1
	C	3	A	6	C	3	A	4
	FDR	5	D	2	FDR	1	D	1
Awareness of Student Needs	S	4	SA	5	S	1	SA	0
	C	3	A	3	C	4	A	6
	FDR	2	D	1	FDR	1	D	0
Professionalism	S	6	SA	3	S	4	SA	3
	C	3	A	6	C	2	A	3
	FDR	0	D	0	FDR	0	D	0

Key to Abbreviations
Aspect of Teaching, e.g., Planning and Preparation:
(S): Strong
(C): Competent
(FDR): Further Development Required
(UAN): Urgent Attention Needed
My initial teacher education prepared me well for this aspect of teaching:
(SA): Strongly Agree
(A): Agree
(D): Disagree
(SD): Strongly Disagree

"I don't think it was necessarily the teacher education that didn't prepare me (for questioning). I think it would have been better for me to learn these techniques in a classroom/face-to face-teaching." (P7)

"I still tend to provide answers when questioning so I need to keep developing these skills more." (P9)

In terms of being aware of needs of individual students, the participants were generally aware of the expectations in terms of meeting student needs, but time or other constraints prevented this from occurring. As one participant offered:

"I am aware of the different ability levels that students in my classes have; however, catering for the different levels is a challenge! I'm teaching a one size fits all at the moment." (P1)

When comparing the skills in which participants felt most confident (e.g., planning, and implementing teaching strategies) with those in which they felt least competent, it appears they more easily grasped the skills that could be considered more concrete. The skills reported as more challenging (e.g., being aware of student needs) are often not easily learnt in ITE but develop with experience, implementation, and reflection. This is the complexity of teaching which Schön (1984) refers to as the "swampy lowland," to which there are no simple solutions (p. 42).

These findings concur with the body of literature which reported ITE is able to be delivered successfully at a distance (Perraton et al., 2002; Peterson & Bond, 2004; Robinson & Latchem, 2003; Simpson, 2003). New Zealand studies on ITE, in general, have found beginning teachers to be favourable towards their ITE, feeling well prepared for the classroom and confident in their abilities (Anthony & Kane, 2008; Brocklesby & Sandford, 2006; Cameron & Baker, 2004).

LEARNING ONLINE

Interestingly the mode of study was not the major theme from the data analysis. This supports Robinson and Latchem's (2003) argument that a beginning teacher's readiness to teach cannot be wholly attributed to the mode in which the ITE was undertaken. However, the mode of study was *a* feature and this section discusses the disadvantages and advantages of undertaking ITE online.

Disadvantages

Isolation

Although the delivery platform contained tools by which communication with other students was possible, and most courses included contributions to a weekly forum, the surveys revealed that difficulties of learning online focused on the isolation of the mode of delivery which resulted in a lack of contact with, and support of, other learners. This is exemplified by this comment from Survey 1:

"I really felt the online program lacked the team spirit and collegiality that comes with being part of a physical class. This was highlighted to me when we had our residency – I really think it should be longer or there should be two of them. Working in isolation can really lack emotional support." (P1)

It appears that even though structures were in place to enable student-to-student interaction these did not engender a sense of community. Garrison and Anderson (2003) have argued that because technology allows collaboration, it does not necessarily occur, and that effective facilitation is crucial in establishing online learning communities.

Consequently, the interviews probed more deeply into the lack of community feeling. These revealed that although some students lived nearby each other, they did not, or seldom, contacted one another, citing lack of confidence or time as barriers, and instead preferred lecturer assistance in setting up contacts. Garrison and Anderson (2003) agreed that the role of the lecturer is crucial for successful online learning. The one part of the program which involved face-to-face contact, the residency, was singled out by one participant as being the 'best part of the program' and by another two as being enjoyable because of the contact with other students. So, while the participants valued and wanted to belong to a learning community they did not seem to think it was their responsibility to create one. Successful online learning communities can be created but require active participation from both the teachers and the students. As Paloff and Pratt (2003) pointed out, online learners need to be willing to contribute to the learning of others and have some onus for creating the community. Lecturers, too, need to play their role in developing the learning community and to remember that online learners need support and encouragement to participate. Lecturers have a responsibility to develop a sense of community by creating connections between the students and technological spaces in which the students can communicate socially and academically, and lecturers need to consider their responses and interactions to ensure the students feel neither stifled nor abandoned (Garrison & Anderson, 2003).

While face-to-face interaction is not essential for successful online learning, it is beneficial. People naturally make connections with each other in a face-to-face environment, more easily identifying compatibility. For example, one participant noted that while they found studying online isolating and felt reluctant to contact 'a face in the forum', while at the residency they readily made friends and wished this face-to-face contact had occurred earlier.

Lack of Modelling

The lack of modelling of teaching skills was cited by some participants as a disadvantage of studying online and they considered teaching experiences particularly valuable for learning these. For example, one cited that it was difficult to gain experience of implementing group work from studying about it online and presumed experiencing it on campus would be more beneficial. As they explained:

"It is difficult to gain as much experience of group work online as compared to on campus." (P6)

"Being online it's harder to absorb strategies and routines as in comparison to a 'real' class - there is no teacher modelling these strategies and routines. TEs [Teaching Experiences] were valuable for this, as well as online discussions." (P4)

"I think it is more difficult without having a teacher modelling these skills – but there was a huge amount of consideration given in the online course on this and I think ideas are half the solution. Again, it was great for this, but nothing prepares you for the full-time classroom like being there as a teacher rather than a student teacher." (P.2)

Perraton et al. (2002) discussed similar concerns and provided suggestions on how to provide practical experience for online student teachers. The participants in this study presumed the challenges to implementing successful co-operative learning stemmed from learning about it online but as Gillies and Boyle (2010) have found, many teachers have difficulty implementing co-operative learning and need intensive professional development to use it successfully.

Lack of Consistency

A final disadvantage reported by these students was a lack of consistency among the courses which led to confusion and additional workload. As mentioned previously, distance learning can be an isolated and lonely experience which is exacerbated if students are confused and unsure of how to proceed. This confusion and time spent finding their way through a plethora of material exacerbates an already intense workload.

"I found it difficult to follow what work was due when, and how long to allocate to tasks. Just when I thought I was on top of things I would discover something (unknown to me) was due, or what I thought was going to a 'normal' forum or journal actually required a lot of planning or organisation." (P1)

"Something that would have made life easier would have been uniformity of layout of assignment, forum, journal, deadline templates." (P6)

Unlike campus-based students, online students cannot easily seek clarification from fellow students in the cafe or corridors, so it is all the more important that distance courses are kept simple (Paloff & Pratt, 2003) and consistent (Theissen & Ambrock, 2008). A lack of consistency and coherence is a challenge that ITE programmes, both online and on campus, face in many faculties (Kane, 2005). According to Harasim, Hiltz, Teles, and Turoff (1995), attention to instructional design is critical for successful learning in the online mode.

Advantages

The benefits of studying online included learning new skills, the flexibility of the study option, financial advantages and providing for Māori learners.

Skill Development

In the interviews the participants revealed a wide range of skill development which was attributed to learning online, including improved ICT, time management and research skills. Online learning also enhanced personal qualities such as self-discipline, organisation, resourcefulness and independence, and prepared the beginning teachers well for these expectations in schools. Some comments from the participants include:

"Having to self manage the online course I think was good preparation for multi tasking." (P4)

"I learnt skill of using my computer to plan all my lessons and also as a research tool. At my school there is no library, I mean there is a library for the students, but there is no teacher resource part of the library – so often all I've got is my computer and my Internet connection – which is exactly what I had for the Diploma. If I was given an assignment I searched for resources and found templates and all that and now I can easily take all that to my new school." (P2)

"It definitely improved my time management." (P8)

This supports other studies which are tentatively theorising that while distance study requires independence, organisation and time management, it further develops these skills (Anderson & Simpson, 2005; Holmberg, 2005). Learning online may actually help new teachers to deal with the challenges of teaching in the technological age in a way traditional learning does not (Robinson & Latchem, 2003).

Flexibility and Financial Advantages

Distance education was originally developed to allow those who were unable to attend face-to-face instruction access to education. The word 'distance' would suggest a geographical separation from the education source, yet none of the participants in this study were geographically isolated. They all lived in locations where access to face-to-face teacher education was possible, but chose online learning for its flexibility and financial advantages. This flexibility in terms of location was important for two participants, one of whom started the program while still living overseas and another who moved from a city to a small town during the course of the program, neither of which is possible for those enrolled in a face-to-face program.

Flexibility in terms of time use and having study autonomy was a further advantage of studying online. As Lund and Volet (1998) noted, the most frequently mentioned positive aspects of studying online were flexibility in learning times and pace, and this is consistent with the present study. This aligns with the basic tenet of distance education which permits students to study at times and in ways suitable to them (Garrison, 1989).

Another advantage identified in this study and supported by Verduin and Clark (1991) related to monetary issues. For some participants their family and financial responsibilities were too great to consider attending a campus full time, so studying online created the financial advantage of being able to work part time as well as study. Those who lived in the cities also noted saving the cost of commuting to a campus. These advantages are summed up by this participant:

"Another was I had the freedom to live out in a beautiful area, a beautiful spot by the beach and when I had time to take a break – which wasn't often – I could go down and have a swim – I couldn't do that at a College. And because my wife was studying (online) as well we could spend time together, and we could save money because cooking at home and no travel cost. So we saved money and we had freedom to plan our own schedule. We could decide on Monday for example that we would go and visit friends – that was ok we could just go and do and then on Saturday do the work – that was a great benefit." (P1).

Providing for New Zealand's Indigenous People (Māori)

The program in this study was perceived to be providing an opportunity for Māori learners. As previously noted, four participants identified ethnically as Māori, and all of them successfully completed an intense program of online study and gained employment immediately upon graduation. Three graduated to teach the Māori language and indicated they would not have been able to become a teacher without an online option. Since 2006, the program has graduated a further 31 Māori teachers. A key goal of the Ministry of Education (2010b) is to increase the number of Māori teachers proficient in the Māori language, and the e-Learning Advisory Group (2002) claims that a focus on the potential of e-learning will provide Māori with flexible learning options to enable participation in tertiary education. The findings from this study support these policies.

The possibilities of e-learning to meet the needs of indigenous populations have been mooted by others (Grant, 1996; Porima 2005; Tiakiwai & Tiakiwai 2010; Wall, 2008) whereby e-learning was able to overcome barriers such as work or family commitments that hindered indigenous people from accessing higher education. E-learning in New Zealand can offer similar opportunities and may be an important mode of education provision to further Māori educational aspirations. Potentially e-learning could provide improved access to tertiary education opportunities for Māori learners, particularly in light of the demographic profile of Māori learners involved in tertiary education. According to the New Zealand Council for Educational Research (2004) 60% of Māori students are studying part time, and women make up 68% of the Māori tertiary student population, which is greater than the comparable non-Māori figures. There is a difference in the age profile for Māori students compared with non-Māori students. Māori are younger in the general population but are older in the student population and tend to participate in tertiary education at a later age. Accordingly, Māori may have their educational needs more appropriately met through e-learning given its flexibility and financial advantages.

To ensure participation in e-learning by Māori learners, New Zealand should take the opportunity to develop e-learning environments based on Māori pedagogy. The Institutes of Technology and Polytechnics of New Zealand (2004), Porima (2005) and Ferguson (2008) suggest this would include a model which honours the students, people and the subject. These learning environments would be based on Māori values such as building relationships (whanaungatanga), caring (manaakitanga) and ako (reciprocal teaching and learning), and would incorporate opportunities for face-to-face interaction.

The participants in this study identified successful aspects of the online ITE programme that are congruent with these studies. These are that Māori students value opportunities for face-to-face contact, the ability to form learning communities and the importance of supportive staff to develop and maintain motivation. As Māori one participant stated:

"A strength was the block course at the university as it gave us a face-to-face opportunity to work with those who were lecturing us and studying beside us." (P3)

Participants also noted a lack of Māori protocol indicating e-learning would be more appropriate to them as Māori if this was incorporated, as exemplified by this comment:

"One other thing from talking to our other friends like xxx (a former student in the Diploma) and our other Māori friends is they felt there isn't enough Māori kaupapa (protocol)." (P9)

E-learning in New Zealand has a unique opportunity to not only improve access to tertiary education for Māori students, but also to develop e-learning within a culturally appropriate pedagogy that provides for, and respects, Māori students.

CONCLUSION

The findings from this study report general agreement that the online ITE program was comprehensive and prepared the participants to begin teaching. They felt well prepared for the skills of planning and preparation, and professionalism, and a particular strength was the diversity and quality of the content offered.

Studying online had advantages such as being flexible, so that study could fit around other commitments, saving time and money, and for some it was the only way they could become teachers. Additional benefits were developing ICT skills, time management, research skills, self-discipline, organisational skills, resourcefulness, and independence.

The major disadvantage was the difficulty of studying alone. Although the delivery platform contained tools by which communication with other students was possible, the participants still felt isolated and lacking contact with, and support of, other learners. In addition, they were concerned that learning online deprived them of seeing certain teaching skills and feared that this lack of modelling had an impact on their using them.

The program was perceived to be providing an opportunity for Māori learners. Four participants identified ethnically as Māori and all of them successfully completed an intense program of online study and gained employment immediately upon graduation. Three of these students graduated to teach the Māori language and indicated they would not have been able to become teachers without the online option.

RECOMMENDATIONS

Although this was a qualitative study and as such had limited transferability, the findings concur with the themes evident in the literature. Hence, the recommendations are particularly important in the context of New Zealand but also have relevance for online teacher education in other settings.

Online learners favour connections rather than mere content delivery and therefore it is essential that lecturers to create these. It seems students look to lecturers to develop online learning communities, therefore it is vital to have skilled staff. Lecturers need to be aware of their role in establishing and supporting a learning community, and to develop tasks and activities that encourage collaboration and sharing. Despite current technological advances and the unpredictability of what will be available in the future, it is the pedagogy that should be foremost. All online teachers must acknowledge there is a real person at the end of the line with hopes and aspirations. In the New Zealand context, Ferguson (2008) emphasises the importance of a community for Māori learners to encourage a sense of belonging to the e-learning environment. She also notes the importance of lecturer visibility in the e-learning environment and says students need to know that "their pouako (tutors) are with them" (p. 3). Academic staff teaching online need professional development in both technology and pedagogy so they are effective online teachers.

The participants in this study signalled that information overload, lack of clarity and inconsistencies hindered their learning. Online courses need to heed Paloff and Pratt's (2003) advice that online course material should be kept simple and not overload the student. As Harasim et al., (1995) advocate, attention to instructional design is critical for successful learning in the online mode.

The lack of modeling of teaching skills was cited by some participants as a disadvantage of studying online. Recommendations to bridge this gap include: student teachers observing and practicing modelling while on teaching experiences, under the supervision of the teacher or visiting lecturers; lecturers can ensure a variety of teaching skills are modeled and practiced at residential courses; and in addition, lecturers can implement the use of video case studies so student teachers can observe teaching skills in practice.

Joyce and Showers (1980) suggested a five component model which has been effective for teacher professional development and could be useful for ITE. The model consists of the presentation of theory, demonstration or modelling of the new skill, initial practice in a protected or simulated setting, feedback about performance of the practice, and coaching in a real setting. The five components of this model could be used with online students through website materials, videos of modelling, feedback to student teacher practice via the internet (e.g., Skype), coaching in the authentic setting of teaching experience, and at residencies.

While face-to-face interaction is not essential for success in online learning there are benefits to incorporating residencies in online ITE. In terms of building a learning community, all participants in this study appreciated the opportunity to make face-to-face connections with the lecturers and fellow students. They noted the quality of online discussions improved after the residency and they felt more comfortable approaching and supporting each other. Residencies are also a useful opportunity for student teachers to experience modelling of teaching skills, to practice these and receive feedback about their performance.

In New Zealand online programs must meet the needs of Māori learners and encourage their participation in tertiary education; therefore, Māori education values such as ako, manaakitanga, and whanaungatanga need to be incorporated into courses within the program. All nations can strive to ensure online ITE is delivered in culturally appropriate ways.

REFERENCES

Anderson, B., & Simpson, M. (2005). *From distance teacher education to beginning teaching: What impacts on practice?* Paper presented at the Open and Distance Learning Association of Australia Conference, Adelaide. Retrieved from http://www.odlaa.org/events/2005conf/ref/ODLAA2005Anderson-Simpson.pdf

Anthony, G., & Kane, R. (2008). *Making a difference: The role of initial teacher education and induction in the preparation of secondary teachers.* New Zealand: Crown.

Baldwin, T. T., & Ford, J. K. (1988). Transfer of training: A review and directions for future research. *Personnel Psychology*, *41*, 63–105. doi:10.1111/j.1744-6570.1988.tb00632.x

Bartell, C. (2005). *Cultivating high-quality teaching through induction and mentoring*. Thousand Oaks, CA: Corwin Press.

Berg, B. L. (2004). *Qualitative research methods for the social sciences* (5th ed.). Boston, MA: Pearson.

Bowers, C., Vasquez, M., & Roaf, M. (2000). Native people and the challenge of computers: Reservation schools, individualism and consumerism. *American Indian Quarterly*, *24*(2), 182–199.

Brocklesby, J., & Sandford, M. (2006). *Evaluation of the graduate teacher training programme*. Prepared for Victoria University of Wellington College of Education. Wellington, New Zealand: Research New Zealand. A Powerpoint presentation.

Cameron, M., & Baker, R. (2004). *Research on initial teacher education in New Zealand: 1993-2004 literature review and annotated bibliography*. Wellington, New Zealand: New Zealand Council for Educational Research.

Carr-Chellman, A. (2005). Stealing our smarts: Indigenous knowledge in on-line learning. *International Journal of Media. Technology and Lifelong Learning*, *1*(2), 1–10.

Charmaz, K. (2006). *Constructing grounded theory. A practical guide through qualitative analysis*. London, UK: Sage.

Creswell, J. (2003). *Research design: Qualitative, quantitative, and mixed methods approaches* (2nd ed.). Thousand Oaks, CA: Sage.

E-Learning Advisory Group. (2002). *Highways and pathways: Exploring New Zealand's e-learning opportunities*. The Report of the E-Learning Advisory Group, March 2002.

Ferguson, S. (2008). Key elements for a Māori e-learning framework. *MAI Review*, *3*(3), 1–7.

Garrison, D. R. (1989). *Understanding distance education: A framework for the future*. London, UK: Routledge.

Garrison, D. R., & Anderson, T. (2003). *E-learning in the 21st century: A framework for research and practice*. London, UK: Routledge/Falmer. doi:10.4324/9780203166093

Garrison, D. R., Anderson, T., & Archer, W. (2000). Critical inquiry in a text-based environment: Computer conferencing in higher education. *The Internet and Higher Education*, *2*(23), 87–105.

Gillies, R., & Boyle, M. (2010). Teachers' reflections on cooperative learning: Issues of implementation. *Teaching and Teacher Education*, *26*, 933–940. doi:10.1016/j.tate.2009.10.034

Grant, M. (1996). Development of a model using information technology for support of rural Aboriginal students off-campus learning. *Australian Journal of Educational Technology*, *12*(2), 94–108.

Hall, A., Yates, R., & Campbell, N. (1998). *Teacher education at a distance: Reflections on the first year of a mixed media teacher education programme*. Unpublished paper presented to VITAL day at the University of Waikato, 17 February, 1998.

Ham, V., & Wenmoth, D. (2007). *Evaluation of the e-Learning collaborative development fund*. Final Report to Tertiary Education Commission, Wellington. Retrieved from www.tec.govt.nz/templates/standard.aspx?id=755

Hammerness, K., Darling-Hammond, L., Grossman, P., Rust, F., & Shulman, L. (2005). The design of teacher education programs. In Darling-Hammond, L., & Bransford, J. (Eds.), *Preparing teachers for a changing world: What teachers should learn and be able to do* (pp. 390–441). San Francisco, CA: Jossey-Bass.

Harasim, L., Hiltz, S., Teles, L., & Turoff, M. (1995). *Learning networks: A field guide to teaching and learning online*. Cambridge, MA: MIT Press.

Hattie, J. (2009). *Visible learning*. London, UK: Routledge.

Herie, M. (2005). Theoretical perspectives in online pedagogy. *Journal of Technology in Human Services*, *2*(1), 29–52. doi:10.1300/J017v23n01_03

Heritage Foundation. (2011). *2011 index of economic freedom*. Retrieved from http://www.heritage.org/index/Ranking

Hodson, J. (2004). Aboriginal learning and healing in a virtual world. *Canadian Journal of Native Education*, *28*(1/2), 111–122.

Holmberg, B. (2005). *The evolution, principles and practices of distance education*. Oldenburg, Germany: BIS-Verlag Carl von Ossietzky Universitat.

Institutes of Technology and Polytechnics of New Zealand. (2004). *Critical success factors for effective use of e-learning with Māori learners*. Retrieved from http://elearning.itpnz.ac.nz/index.htm

Johnson, D. W., & Johnson, R. T. (1996). Cooperation and the use of technology. In Jonassen, D. H. (Ed.), *Handbook of research for educational communications and technology* (pp. 1017–1044). New York, NY: Simon & Schuster Macmillan.

Joyce, B., & Showers, B. (1980). Improving inservice training: The messages of research. *Educational Leadership*, *37*(5), 379–385.

Kane, R. (2005). *Initial teacher education policy and practice. (A research report to the Ministry of Education and the New Zealand Teachers Council)*. Wellington, New Zealand: Ministry of Education.

Lund, C. P., & Volet, S. (1998). Barriers to studying online for the first time: Students' perceptions. In C. McBeath & R. Atkinson (Eds.), *Proceedings EdTech'98: Planning for Progress, Partnership and Profit*: Perth, Australia: Australian Society for Educational Technology. Retrieved from http://www.aset.org.au/confs/edtech98/pubs/articles/lund.html

Major, J. (2005). Teacher education for cultural diversity: Online and at a distance. *Journal of Distance Learning*, *9*(1), 15–26.

Ministry of Education. (2010a). *OECD review on evaluation and assessment frameworks for improving school outcomes: New Zealand country background report*. Retrieved from http://www.educationcounts.govt.nz/__data/assets/pdf_file/0009/90729/966_OECD-report.pdf

Ministry of Education. (2010b). *Ngā haeata mātauranga: The annual report on Māori education, 2008/09*. Wellington, New Zealand: Author. Retrieved from http://www.educationcounts.govt.nz/publications/series/5851/75954/4#discussion

Moore, M. (2007). The theory of transactional distance. In Moore, M. (Ed.), *The handbook of distance education* (2nd ed., pp. 89–108). Mahwah, NJ: Lawrence Erlbaum.

New Zealand Council for Educational Research. (2004). *Statistical profile of Māori in tertiary education and engagement in e-learning*. Retrieved from http://www.itpnz.ac.nz/

New Zealand Teachers Council. (2011). *Graduating teacher standards*. Retrieved from http://www.teacherscouncil.govt.nz/te/gts/index.stm

Nielsen, H. D. (1997). Quality assessment and quality assurance in distance teacher education. *Distance Education*, *18*(2), 284–317. doi:10.1080/0158791970180207

Oliveira, J., & Orivel, F. (2003). The cost of distance education for training teachers. In Robinson, B., & Latchem, C. (Eds.), *Teacher education through open and distance learning*. London, UK: Routledge-Falmer.

Paloff, R., & Pratt, K. (2003). *The virtual student: A profile and guide to working with online learners*. San Francisco, CA: Jossey-Bass.

Patton, M. Q. (2002). *Qualitative research and evaluation methods* (3rd ed.). Thousand Oaks, CA: Sage.

Perraton, H., Creed, C., & Robinson, B. (2002). *Teacher education guidelines: Using open and distance learning*. Paris, France: UNESCO.

Perraton, H., & Potashnik, M. (1997). Teacher education at a distance. *Education and Technology Series, 2*(2).

Peterson, C. L., & Bond, N. (2004). Online compared to face-to-face teacher preparation for learning standards-based planning skills. *Journal of Research on Technology in Education, 36*(4), 345–360.

Porima, L. (2005). *Understanding the needs of Māori learners for the effective use of elearning*. Retrieved from http://www.itpnz.ac.nz/

Robinson, B., & Latchem, C. (Eds.). (2003). *Teacher education through open and distance learning*. London, UK: Routledge-Falmer.

Rovai, A. P. (2001). Building classroom community at a distance: A case study. *Educational Technology Research and Development, 49*(4), 33–48. doi:10.1007/BF02504946

Rovai, A. P., & Barnum, K. T. (2003). On-line course effectiveness: An analysis of student interactions and perceptions of learning. *Journal of Distance Education, 18*(1), 57–73. Retrieved from http://proquest.umi.com.helicon.vuw.ac.nz/pqdweb?RQT=318&pmid=57387&cfc=1

Russell, T. (2001). *The no significant difference phenomenon: A comparative research annotated bibliography on technology for distance education* (5th ed.). Littleton, CO: IDECC.

Schön, D. A. (1984). *The reflective practitioner: How professionals think in action*. Aldershot, UK: Arena.

Selby, M. (2006). Language, matauranga Māori and technology? *He Puna Korero: Journal of Māori and Pacific Development, 7*(2), 79–86.

Simpson, M. (2003). *Distance delivery of pre-service teacher education: Lessons for good practice from twenty-one international programs*. Unpublished Doctoral dissertation, The Pennsylvania State University, State College, PA.

Simpson, M., & Kehrwald, B. (2010). Educational principles and policies framing teacher education through open and distance learning. In P. Danaher & A. Umar (Eds.), *Teacher education through open and distance learning* (pp. 23-34). Vancouver, Canada: Commonwealth of Learning.

Statistics New Zealand. (2011). *National population projections: 2009 to 2061*. Retrieved from http://www.stats.govt.nz/tools_and_services/tools/TableBuilder/population-projections-tables.aspx

Theissen, J., & Ambrock, V. (2008). Value added: The editor in design and development of online courses. In T. Anderson (Ed.), *The theory and practice of online learning* (2nd ed.) (pp. 265-276). Edmonton, Canada: Athabasca Press.

Tiakiwai, S.-J., & Tiakiwai, H. (2010). *A literature review focused on virtual learning environments (VLEs) and e-learning in the context of te reo Māori and kaupapa Māori education: Report to the Ministry of Education*. Wellington, New Zealand: Ministry of Education. Retrieved from http://www.educationcounts.govt.nz/__data/assets/pdf_file/0004/72670/936_LitRev-VLEs-FINALv2.pdf

United Nations Development Programme. (2010). *Human development report: 20th anniversary edition: The real wealth of nations: Pathways to human development*. New York, NY: Palgrave. Retrieved from http://www.beta.undp.org/content/dam/undp/library/corporate/HDR/HDR_2010_EN_Complete_reprint-1.pdf

Verduin, J. R., & Clark, T. A. (1991). *Distance education: The foundations of effective practice*. San Francisco, CA: Jossey-Bass.

Wall, K. (2008). Reinventing the wheel? Designing an Aboriginal recreation and community development program. *Canadian Journal of Native Education*, *31*(2), 70–93.

Wellington College of Education. (2001). *Approval documentation, secondary online*. Wellington, New Zealand: Author.

ADDITIONAL READING

Bates, A. (2000). *Managing technological change*. San Francisco, CA: Jossey-Bass.

Danaher, P., & Umar, A. (Eds.). (2010). *Teacher education through open and distance learning*. Vancouver, Canada: Commonwealth of Learning.

Darling-Hammond, L. (2006). *Powerful teacher education: Lessons from exemplary programs*. San Francisco, CA: Jossey-Bass.

Dixon, J. S., & Crooks, H. (2006). Breaking the ice – Supporting collaboration and the development of community online. *Canadian Journal of Learning and Technology*, *32*(2). Retrieved from http://www.cjlt.ca/index.php/cjlt/article/view/51/48

Dolan, S. (2008). Forming communities of learners online: The culture and presence of asynchronous discussion groups. *He Kupu*, *1*(5), 32-47. Retrieved from http://www.nztertiarycollege.ac.nz/HeKupu

Gillies, D. (2008). Student perspectives on videoconferencing in teacher education at a distance. *Distance Education*, *29*(1), 107–118. doi:10.1080/01587910802004878

Hargis, J. (2005). Collaboration, community and project-based learning – Does it still work online? *International Journal of Instructional Media*, *32*(2), 157–161.

Hiltz, S. R., & Goldman, R. (Eds.). (2005). *Learning together online: Research on asynchronous learning networks*. Mahwah, NJ: Lawrence Erlbaum.

Kehrwald, B. (2008). Understanding social presence in text-based online learning environments. *Distance Education*, *29*(1), 89–106. doi:10.1080/01587910802004860

Kelly, H. F., Ponton, M. K., & Rovai, A. P. (2007). A comparison of student evaluations of teaching between online and face-to-face courses. *The Internet and Higher Education*, *10*, 89–101. doi:10.1016/j.iheduc.2007.02.001

Laurillard, D. (2002). *Rethinking university teaching: A conversational framework for the effective use of learning technologies*. New York, NY: Routledge-Falmer. doi:10.4324/9780203304846

Laurillard, D. (2006). E-learning in higher education. In Ashwin, P. (Ed.), *Changing higher education: The development of learning and teaching* (pp. 1–12). London, UK: Routledge.

Lock, J. V. (2002). Laying the groundwork for the development of learning communities within online courses. *The Quarterly Review of Distance Education*, *3*, 395–408.

OFSTED. (Office for Standards in Education, Children's Services and Skills). (2008). *The Open University: A secondary initial teacher training inspection report*. Retrieved from www.ofsted.gov.uk/oxedu_reports/download/(id)/97592/(as)/70096_302793.pdf

Perraton, H. (2010). *Teacher education: The role of open and distance learning*. Vancouver, Canada: Commonwealth of Learning.

Siemens, G. (2005). Connectivism: A learning theory for the digital age. *Instructional Technology and Distance Learning, 2*(1). Retrieved from http://www.itdl.org/Journal/Jan_05/article01.htm

Simpson, M., & Anderson, B. (2009). Redesigning initial teacher education. In Stacey, E., & Gerbic, P. (Eds.), *Effective blended learning practices: Evidence-based perspectives in ICT facilitated education* (pp. 62–78). Hershey, PA: IGI Global. doi:10.4018/978-1-60566-296-1.ch004

Tallent-Runnels, M., Thomas, J., Lan, W., Cooper, S., Ahern, T., Shaw, S., & Liu, X. (2006). Teaching courses online: A review of the literature. *Review of Educational Research, 76*(1), 93–135. doi:10.3102/00346543076001093

UNESCO. (2002). *Teacher education guidelines: Using open and distance learning*. Paris, France: Author.

U.S. Department of Education, Office of Planning, Evaluation, and Policy Development. (2009). *Evaluation of evidence-based practices in online learning: A meta-analysis and review of online learning studies.* Washington, DC.

KEY TERMS AND DEFINITIONS

Blogs: An electronic journal where students post their ideas about teaching and learning.

Collaborative Learning: A learning situation where students work together to learn.

Discussion Board: Students post comments and questions related to specific topics. Other students and lecturers respond.

Graduate Diploma of Teaching: A one year program of study towards a teaching qualification for students already holding an undergraduate degree.

Initial Teacher Education: A program of study towards a teaching qualification.

Instructional Design: The design and development of learning experiences to ensure logical presentation of material, student participation and ease of use.

Learning Community: Students actively learn from each other and support each others' learning.

Modelling: Student teachers have the opportunity to see best practice.

Online Learning: Distance computer-based learning.

Social Constructivist: A learning situation that is social and collaborative.

Chapter 7
Training Teachers for Virtual Classrooms:
A Description of an Experimental Course in Online Pedagogy

Wayne Journell
University of North Carolina at Greensboro, USA

Miguel Gomez
University of North Carolina at Greensboro, USA

Melissa Walker Beeson
University of North Carolina at Greensboro, USA

Jayme Nixon Linton
University of North Carolina at Greensboro, USA

Jerad J. Crave
University of North Carolina at Greensboro, USA

Mary O. Taylor
University of North Carolina at Greensboro, USA

ABSTRACT

The increased demand for online instruction within higher and K-12 education has created a need for teacher education programs to provide pre-service and practicing teachers with training in online pedagogy; however, research has shown that such courses are rare within most teacher training programs. This chapter describes "Theory and Practice in Online Education," an experimental course designed to train teachers for virtual instruction that was offered by the first author in Spring 2011. In this course, students explored the history of online education, online learning theories, the creation of online communities, online assessments, and ways to differentiate online courses for learners with special needs. Students were then able to put this theoretical knowledge into practice by experiencing various forms of synchronous and asynchronous communication and designing their own online course. The authors provide this description in hopes that others may use it as a starting point to create their own courses in online pedagogy.

DOI: 10.4018/978-1-4666-1906-7.ch007

INTRODUCTION

The impetus for this chapter can be traced back to a transformational experience in the first author's professional career. At the end of his first year of teaching high school social studies, the first author was asked by his district technology supervisor to design and teach an online United States Government course for the district's new e-learning program. After accepting the task, he quickly realized that he knew very little about developing online curricula. Without much guidance, he created a course that *seemed* like a good online experience—his students visited websites, submitted work electronically, and discussed content-related topics on the course discussion board. Yet, in the years that followed, he taught that course several times, often growing frustrated at his students' lack of engagement with both the course content and the other students in the course, especially when compared to the academic performances of students in his face-to-face classes. It was only years later, as a doctoral student exposed to literature on online education, that he realized how fundamentally flawed his approach to designing and teaching that course had been.

Sadly, this type of experience with online teaching is not unique. Universities and school districts around the world are increasing their use of online delivery systems, but the preparation of educators specializing in online instruction is not keeping up with the demand for quality online education. Research suggests that distance education programs have the potential to offer instruction that is equivalent to and may even exceed that which is found in face-to-face classrooms (Bernard et al., 2004), but only if online education is subject to the same types of quality-control measures that are currently being used to monitor face-to-face instruction. One such measure is effective teacher training, and we believe, as do others (e.g., Davis & Roblyer, 2005; National Education Association, 2006), that teacher education programs need to take greater responsibility in preparing pre-service and practicing teachers for virtual classrooms.

This chapter describes an experimental course the first author offered in Spring 2011 entitled *Theory and Practice in Online Education* that we believe can serve as a model for the preparation of online teachers. In this course, students experienced multiple forms of synchronous and asynchronous communication and explored the history of online education, online learning theories, the creation of online communities, online assessments, and ways of differentiating online courses for learners with special needs. The culminating project in the course required students to create their own online course using Blackboard technology that implemented the best practices discussed and modeled in class.

BACKGROUND

Current State of Virtual Teacher Preparation

Over the past two decades, online education has become a staple of higher education and, increasingly, K-12 education (Tallent-Runnels et al., 2006). Although research on the perceived cost efficiency of online education is mixed (Guri-Rosenblit, 2005; Njenga & Fourie, 2010), increased budget concerns and advancements in technology have forced universities and school districts to consider online education as a way to reduce costs and offer students the flexibility to take courses off campus (Burbules, 2004; Journell, in press). Many universities are increasingly offering certain programs exclusively online (e.g., Haythorthwaite & Kazmer, 2004), and open universities that operate completely online are becoming more popular throughout the world for both students and faculty (Ng, 2006). Within K-12 education, research has found that many nations have begun to offer "virtual high schools" for students who cannot attend regular schools or whose regular schools cannot offer the necessary courses for graduation or admission to college (Conceicao & Drummond, 2005; Kapitzke & Pendergast, 2005; Schrum, 2004; Tunison & Noonan, 2001). Specifically within the United States, Picciano and Seaman (2009) estimated over one million K-12 students took online courses during the 2007-2008 school year, and the state of Michigan recently took the unprecedented step of requiring all students to take at least one online course by the time they graduate high school (Michigan Department of Education, 2006).

However, one of the challenges in creating viable online education programs is the preparation of online course instructors. If virtual teacher preparation fails to keep up with the demand for online education, then the courses students take will lack engagement and rigor, turning them into the digital diploma mills predicted by skeptics (e.g., Noble, 2001). Too often, instructors are asked to teach online simply because they have been recognized by administrators as exceptional classroom teachers or particularly adept at technology, neither of which automatically translates into effective online instruction (Garrison & Anderson, 2003; Journell, 2008). Although the same basic learning theories often apply to both online and face-to-face instruction, it is a common mistake among online teachers to assume that the same instructional strategies and habits that work in their face-to-face classrooms will transfer to their online courses. Teaching online requires a unique set of skills and dispositions that are often different from face-to-face contexts (Quinlan, 2011).

Much has been written within the literature on teacher education about the importance of preparing teachers for a variety of educational contexts, ranging from ensuring that teachers are equipped to teach in demographically diverse environments to preparing teachers for classrooms containing students with special needs (e.g., Ladson-Billings, 2000; Lucas, Villegas, & Freedson-Gonzalez, 2008; Van Laarhoven, Munk, Lynch, Bosma, & Rouse, 2007). However, the literature has been noticeably silent on how to prepare teachers for virtual classrooms. The irony in this lack of research, of course, is that teacher education programs across the United States and throughout the world are turning to online technology with increased frequency, using online tools to deliver coursework, engage in professional development and teacher networking, and even aid in the supervision of student teachers (e.g., Dede, Ketelhut, Whitehouse, Breit, & McCloskey, 2009; Heafner, 2011; Heafner & Petty, 2010; Levin, He, & Robbins, 2006; O'Connor, Good, & Greene, 2006; Rock et al., 2009; Schlager, Farooq, Fusco, Schank, & Dwyer, 2009).

Of the research that exists on the training of teachers in online pedagogy, most describe professional development in-service programs for university faculty. These types of programs are typically either "crash courses" delivered in a short amount of time and focus on how to use existing course management systems, such as Blackboard or Moodle, without much emphasis on online learning theories (e.g., Gold, 2001; Pankowski, 2003; Wolf, 2006), or a series of short professional development sessions over an extended period of time (e.g., Maor, 2006). Despite the increased demand for online education in both higher and K-12 education, there exist few examples within the literature of in-depth courses that provide teachers with the opportunity to blend online learning theory with the practice of designing and implementing their own online curriculum.

One notable exception can be found in Duncan and Barnett's (2010) study of a Canadian university course designed to train pre-service K-12 teachers in online instruction. The semester-long course was offered in a blended format with students attending face-to-face sessions the first and last two weeks of class and participating in asynchronous threaded discussions during the remainder of the course. During the online sessions, students discussed various aspects of online communication and networking, ethical practices of online learning, online privacy, access and equity online, and online assessment. For the final project in the course, students created an online teaching module in their respective content area. Overall, the authors found that students seemed to better understand best practices for teaching online as a result of both the cognitive presence developed through the online discussions and the experiential nature of developing their own online courses.

The Need for Rich Training in Online Pedagogy

According to the National Educational Association (2006), "all new educators should be equipped to instruct online" (p. 22), and we would amend that statement to include practicing and experienced teachers as well, given the current economic climate in the United States and throughout the world. In order to truly prepare teachers for virtual environments, however, training in online instruction must move beyond technological literacy and basic understanding of online course management systems to include engagement with online learning theories and hands-on practice designing and implementing online curricula, similar to the aforementioned process described by Duncan and Barnett (2010). The reason for this type of approach is simple; online learning, although guided by the same basic principles as face-to-face instruction, often appears quite different than classroom learning, and virtual learners often exhibit considerably different dispositions and approaches to learning than their face-to-face counterparts.

From an instructional standpoint, research has shown that most online courses, even those that offer students the opportunity to interact with each other, rarely move beyond basic learning objectives (Garrison & Cleveland-Innis, 2005). Creating truly constructivist learning environments online is difficult and requires an understanding of online course design, facilitation, and direction (Berge, 2002; Garrison & Anderson, 2003). Building classroom community and social presence online where students cannot see each other and may have never met each other in person requires a different set of strategies and practices than what is typically used in a face-to-face context (Haythornthwaite, Kazmer, Robins, & Shoemaker, 2004; Rovai, 2000a; Swan & Shih, 2005; Tu & McIsaac, 2002). Similarly, assessing student knowledge in online environments where assessments are not proctored and students have unfettered access to learning materials may require teachers to alter their traditional conceptions of student assessment (Gaytan & McEwen, 2007; Rovai, 2000b). Even the process of providing students with feedback on their assignments is different online; since teachers may not have the luxury of being able to communicate non-verbally, written feedback becomes more important and, often, more time consuming (Quinlan, 2011).

Research also suggests that online instructors would benefit from a better understanding of the learning styles and attitudes of virtual learners. There is a growing body of literature that has shown virtual learners to be less academically motivated than their face-to-face counterparts (e.g., Kickul & Kickul, 2006; Weiner, 2003), although this lack of motivation may also be attributed to the negative perceptions about online learning that many students, and some teachers, take into their online courses (Journell, 2008, 2010). Regardless, multiple studies have shown that the percentage of students who drop out of virtual courses exceeds that of face-to-face classes (Jun, 2005), prompting Roblyer and colleagues (Roblyer, 1999; Roblyer, Davis, Mills, Marshall, & Pape, 2008) to argue that online learning is not for everyone. Finally, researchers are only beginning to develop strategies for making online environments more inclusive, especially for English Language Learners and students with learning disabilities (e.g., Coryell & Chulp, 2007; Keeler & Horney, 2007).

If online teachers are not provided with training that delves into this level of depth, one can easily see how teachers can grow frustrated and cynical towards online education and their virtual learners (Journell, 2008). Perhaps the most apt comparison we can make is to compare teachers who are thrust into teaching online without any training in online pedagogy with alternatively-licensed teachers who enter public school classrooms without any formal teacher training. Although some alternatively-licensed teachers flourish to a level that exceeds that of their colleagues who went through traditional teacher training programs (Zeichner & Schulte, 2001), many alternatively-licensed teachers struggle because they do not have the theoretical knowledge and practical experience needed to succeed in the classroom (Darling-Hammond, 2003; Stevens & Dial, 1993). As a result, many of these alternatively-licensed teachers become so disenchanted with teaching that they leave the profession within a few years (Corbell, Booth, & Reiman, 2010; Cleveland, 2003).

Even for technology-proficient teachers who have been educated about learning theories and have spent considerable time in face-to-face classrooms, their first attempt at teaching online will most likely be daunting if they have not received proper training. For teachers to be adequately trained to teach online, we believe they need quality educational experiences in which they are able to apply learning theories to virtual instruction and be exposed to current research pertaining to online instruction and virtual learners. We also believe, as do others (Duncan, 2005; Wilson & Stacey, 2003), that in their virtual training, online teachers should have the opportunity to first experience aspects of online instruction as students and then subsequently be able to put their knowledge and experiences into practice by creating their own online curriculum. The remainder of this chapter describes one such attempt at providing a systematic approach to training teachers for online instruction that we hope can serve as a starting point for others who may be interested in incorporating a course in online pedagogy into their teacher education programs.

CONTEXT AND METHODS

Students and Instructor

The context for this chapter is the first author's graduate-level experimental course, *Theory and Practice in Online Education*, which he offered during the Spring 2011 semester. As the vignette at the beginning of the chapter shows, the first author came into this course having taught online courses as part of his K-12 teaching experience. In addition, he had taken courses in e-learning theory as a doctoral student, had conducted and published research on K-12 online education, and had taught online courses at the university level.

Eight students registered for the course, five of whom were female and three of whom were male. The students' ages ranged between mid-twenties to early forties. Five of the students, who are the co-authors of this chapter, were doctoral students. One of the doctoral students was a practicing K-12 classroom teacher and another was a technology specialist for a local school district. The other three doctoral students were former K-12 educators who held teaching positions in the university's teacher education program. The remaining three students were master's degree candidates, two of whom were practicing K-12 educators. The other master's student was a post-baccalaureate student who was taking courses to receive her initial teaching license.

All of the students had taken at least one online course during their academic careers. However, only one of the students had actually taught online prior to enrolling in the course (a one-credit hour course on educational technology within the School of Education). In a survey completed at the end of the course, all of the students stated that they had enrolled in the course because they believed online education was the future of higher and public education in the United States, and they wanted to be properly trained in online education as a way of making themselves more marketable to future employers as well as ensuring that they would be prepared should they be asked to teach online during their professional careers.

Course Design

The course was structured using a blended approach in which half of the class meetings occurred face-to-face on campus and half occurred online with students participating from their homes. Moreover, half of the weeks in which the class met online were asynchronous and half were synchronous. The purpose of this structure was to give students practice with various forms of online communication while maintaining opportunities to discuss their online experiences in a face-to-face seminar format. The asynchronous class meetings took place on the threaded discussion boards that were part of the Blackboard course space shared by all members of the class. The synchronous class meetings used Elluminate technology that was included as part of the university's Blackboard package.

Throughout the course, various aspects of online pedagogy and learning theory were discussed. Each week, students read a collection of research-based articles that acted as the foundation for that week's discussion. At the face-to-face meetings, one of the five doctoral students had the responsibility to lead the discussion of the readings for that day. The first author led the discussions at the synchronous and asynchronous course meetings so that the students could experience both types of online delivery systems. Table 1 provides an overview of how the course was structured.

Besides class participation and facilitating discussions, students completed two major assignments in the course, both of which will be explained in greater detail later in the chapter. The first assignment was designed to ensure basic technological competency and required students to make a video using a webcam, a podcast, and a screencast. The final project required students to design their own online course based on their respective areas of expertise. Each of the students received their own Blackboard course space to create their courses, and students were expected to incorporate the theory and best practices discussed throughout the semester into the online

Table 1. Course structure

Week	Topic	Format
Week 1	History of Online Education	Face-to-Face
Week 2	Synchronous versus Asynchronous Modes of Online Communication	Face-to-Face
Week 3	Learning Theory in Online Education	Whole Class Asynchronous Discussion
Week 4	Structure and Assessment in Online Education	Small Group Asynchronous Discussions
Week 5	Student Perceptions of Online Learning	Face-to-Face
Week 6	Student Motivation in Online Learning	Synchronous Discussion Using Just Chat/Text Features
Week 7	Creating Community in Online Learning	Synchronous Discussion Using Microphones and Chat/Text Features
Week 8	Role of Teachers in Online Learning	Face-to-Face
Week 9	Creating Inclusive Online Environments	Synchronous Discussion Using Microphones and Chat/Text Features
Week 10	The Future of Online Education	Face-to-Face

courses they designed. Prior to submitting their courses to the first author, each student had the opportunity to enroll a classmate into his or her course and have the classmate navigate the course as a student and provide feedback to the course designer.

Data Collection and Analysis

The data presented in this chapter are taken from the first author's self-study of his experimental course in online pedagogy. Self-studies of teacher practices constitute a growing body of literature (e.g., Dinkelman, 2003; Feldman, 2003; Freese, 2006) that Zeichner (2007) argues can be situated into broader research on substantive issues in teacher education. Throughout the course, the first author wrote researcher memos after each class meeting detailing what had occurred during that session. In addition, the online portions of the course were archived, either through the asynchronous discussion boards on Blackboard or through the Elluminate recording feature, which provided an accurate record of both the comments that students typed into the chat windows as well as the audio from students' comments made via microphone. Similarly, students' technological competency and final projects were posted on their individual Blackboard spaces and archived for analysis. Finally, students' perspectives about the course were gained through an anonymous survey given at the end of the semester which asked students to evaluate the quality of the course as it pertained to preparing them for online instruction.

Data analysis involved wading through the available data and looking for patterns that could then be compared to the interpretations the first author had of the course and his instruction. However, reflective self-study often requires the researcher to seek interpretations from other actors associated with the study (Ellis & Bochner, 2000). Since the course ended, the first author has formally and informally discussed aspects of the course with many of his students, and in preparation to write this chapter, the first author and the five doctoral students participated in a recorded focus group discussion (Patton, 2002) on Elluminate in which the first author was able to pose questions and challenge his own assumptions. Further, the collaborative process of writing this chapter has served as a method of inquiry unto itself by allowing the first author to posit assumptions and receive continued feedback from his students (Richardson, 2000).

Self-studies are innately personal and often specific to the researcher. Although we realize that the description of this course is specific to the time and place in which it occurred, we believe that others can use our experiences to develop naturalistic generalizations (Lincoln & Guba, 1985) about the preparation of teachers for virtual instruction that could transfer to other contexts. Finally, the results of self-studies are reported so that those who read them will reflect upon their own practices (McClam & Sevier, 2010). We offer this course description in the hopes that others will critically evaluate the effectiveness of their own teacher education programs, at least in terms of preparing students to meet the challenges of online instruction in the 21st century.

LEARNING OUTCOMES: THEORY INTO PRACTICE

We feel the most effective way to describe how the course influenced students' pedagogical approaches to online learning is to explain how the course merged theory and practice. According to students' comments to the first author and on the anonymous survey given at the end of the semester, being able to not only learn about the theory behind online learning, but also experience it as students and implement it as course designers made the course instruction richer and more practical. As one student wrote on her end-of-course survey, "I love that we were able to experience it all, [and] not just talk about it. I think experiencing the strategies you are going to use to teach is an important thing for a teacher to do." In the remainder of this section, we will discuss various aspects of the course and explain how they helped students make the connection between theory and practice.

Exploring Online Learning Theory

Identifying Myths and Preconceived Notions

On the first day of class, the first author had students participate in a "four corners" exercise in which he read value statements about online learning, and students moved to certain parts of the classroom that corresponded with their level of agreement or disagreement with each statement.[1] The nine statements asked students to compare online learning with face-to-face instruction as well as assess their perceptions of the technological competence required of students and teachers in order to create successful online experiences. What began as an icebreaker activity quickly evolved into a forum for students to share their prior, and often unsatisfactory, experiences with online instruction. The vast majority of the class appeared skeptical about the equivalency of online and face-to-face instruction, and nearly every student was able to share a story about an ineffective online learning experience in their past that was caused by seemingly lazy instructors, feelings of social alienation, courses built entirely upon completion of worksheets and other "busy work," an array of technical difficulties, or an excess amount of reading and posting to threaded discussion boards.

This initial conversation set the stage for the rest of the course by allowing the class to frame subsequent discussions around the theme of improving online education to the point that it is equivalent to or exceeds the quality of instruction found in face-to-face courses. Since all of the students had spent their lives in face-to-face classrooms and all but one had received traditional teacher training prior to their careers as K-12 classroom teachers, these students were already familiar with what constituted effective teaching. The focus of the course from that point forward was to better understand, through both theory and practice, how teachers can shape online environments to appeal to different learning styles and incorporate opportunities for the type of social and collaborative learning that the students had come to expect from their face-to-face classes.

Supporting Diverse Learning Styles with Technology

One of the themes that ran throughout the course was the notion that "good teaching is good teaching" regardless of the medium being used. Based on their classroom teaching experience and prior educational training, all of the students came to class knowing that effective teaching required catering one's instruction to a variety of learning styles. To further our understanding of this concept, we took an online quiz that had students assess their own learning styles, and even among graduate students who were all adept at using technology, their learning styles varied significantly.

However, many of the students were stymied about how to create online instruction that was not text-based due to the previous online courses they had taken that relied exclusively on reading texts and posting reactions on a threaded discussion board. Again, making comparisons to face-to-face instruction was useful for stimulating ideas about how to make online content acquisition friendlier to all types of learners. The students surmised that in a good ninety-minute face-to-face class, teachers might give a brief lecture that contains audiovisual components as well as narration from the teacher, show a short video that reinforces the main points of the lecture, provide an opportunity for students to work in small groups on an aspect of the topic, have students complete a worksheet or other type of activity that demonstrates individual understanding of the topic, and finish with an individual or group-based project that requires students to either construct something related to the topic or solve an issue that emanated from the prior discussion of the topic. Everyone agreed that this type of lesson is constructivist in nature and would offer something for visual, auditory, sensory, intuitive, reflective, and, depending on the nature of the final project, active/kinesthetic learners.

When asked how they could create the same type of lesson online, students quickly pointed to Web 2.0 technologies, such as YouTube, wikis, and blogs. This initial brainstorming session led to a variety of different interactive technologies being introduced throughout the semester as students began finding resources for their final projects. For example, two of the doctoral students introduced the class to Edmodo (http://www.edmodo.com), a free social networking site designed specifically for education, and Bubbl (https://bubbl.us), a free online graphic organizing tool, when they led the class discussion on the readings for their respective week. Also, once students were able to break away from the norms they had learned in their previous online courses, they began to think outside the box and realize that online learning did not necessarily mean that every learning experience had to take place in front of a computer screen. For example, teachers could easily have their students create projects and conduct experiments offline and then report their findings to their teachers and classmates after signing back online.

The technological competency assignment also aided in this development by showing students that they could create visual and audio resources even when none already exist online. The first step in this assignment was to create a podcast using Audacity (http://audacity.sourceforge.net/), a free, open-source digital recording software, and a free digital encoder (http://download.cnet.com/WAV-to-MP3-Encoder/3000-2140_4-10060500.html) that converts .wav files into .mp3 files that could be downloaded and listened to on an iPod or other portable audio device. Students used these podcasts in their final projects to provide an alternative to text-based lectures and to explain complex materials for struggling readers.

The other two steps in the technological competency assignment required students make videos using a webcam and an open-source video editing software program, such as Debut Video Recorder (http://www.nchsoftware.com/capture/index.html?ref=cj), QuickTime Broadcaster (http://support.apple.com/kb/DL764), or Jing (http://www.techsmith.com/jing/) depending on the operating system being used.[2] For the first video assignment, students made an introductory welcome video for their course, which was also designed to increase feelings of belonging and social presence for students (Tu & McIssac, 2002). In subsequent class discussions, the class expanded on ways in which teachers could use these types of self-made videos, which ranged from a science teacher modeling an experiment for his or her students to a mathematics teacher outlining the steps to a complicated problem. In this instance, the class argued that online learning actually exceeded the type of instruction that typically occurs in a face-to-face class because students could re-watch and pause the videos as necessary to increase understanding.

The second video that students made was an audio-dubbed screencast in which they provided step-by-step instruction on how to navigate an aspect of their course. In addition to creating helpful videos for technology-challenged students, the class quickly began applying screencasts to other aspects of online instruction, such as a way to provide detailed feedback to students on graded work. Typically, feedback in online instruction takes the form of using various functions of Microsoft Word, such as track changes or different colored fonts, which can be difficult for struggling readers and English Language Learners to process. Screencasts allow teachers to verbally explain comments, show students where to find information related to their assignments, or even provide students with interactive feedback on how to improve their work.

Using screencasts to assist struggling readers and English Language Learners was one of the few concrete examples developed during the span of the course on how to adapt online instruction for non-traditional learners. Yet, the class remained aware of the notion that if online learning is to have widespread success in the United States, educators need to develop ways for all types of learners to be successful in virtual environments. The lack of research in this aspect of online learning often left students with more questions than answers; however, the fact that the course addressed the challenges of adapting virtual instruction was a lesson unto itself about the viability and future of online education. As one of the co-authors of this chapter noted in one of our reflection sessions,

"Sometimes in classes, all the way from [Kindergarten] to grad school, you can avoid difficult topics or difficult things that don't have easy answers and just kind of move on. I feel with our class we could have easily avoided things like diversity in the sense of how do we meet the needs of diverse learners, especially learners who may have certain labeled disadvantages. I thought the fact that we spent an entire class session talking about that was important, like how would you deal with someone who is visually impaired or hearing impaired or whatever the case may be. I came out of the class thinking, wow, online learning has some work to do here. It's got some things it can do, but it's not a solution for all the needs out there, and it's not that silver bullet that some people think it can be or that I sometimes think it can be."

Experiencing and Learning from Various Types of Online Communication

As noted in our initial face-to-face session, one of the students' main sources of skepticism related to online learning was the ability to create quality social environments online. Most of their prior experiences had dealt exclusively with asynchronous modes of communication, particularly threaded discussion boards, wikis, and blogs, which corresponds with research on teachers' instructional choices in online courses (Murphy, Rodriguez-Manzanares, & Barbour, 2011). In one of the first classes, the class discussed synchronous and asynchronous communication at length and developed a basic understanding of the advantages and disadvantages of each. Their responses were similar to what is commonly found in the literature on online communication (Murphy et al., 2011). They viewed the primary advantage of synchronous communication as being that it occurs in real time, like a typical face-to-face class, which allows students to receive immediate feedback on their ideas. They believed the basic disadvantages of synchronous communication are that it requires quick thinking on the students' part, often resulting in superficial responses and that synchronous class meetings remove the "anytime, anywhere" element of online learning, which can be problematic if learners span different time zones or take courses around their work and family schedules. Conversely, they viewed asynchronous communication as allowing for both flexible scheduling for students and lengthy, thoughtful replies to questions. However, the students cited the greatest deficiency of asynchronous communication as the lack of immediate feedback on one's own posts and general difficulty with engaging in substantive conversations because of the uncertainty surrounding when to check for new posts.

Even after reading and discussing articles detailing these qualities, it was clear that the students could not truly compare and contrast the different forms of communication without experiencing them. Similar to the class described by Duncan and Barnett (2010), the secondary learning objective during the weeks that were spent online was to have students actually experience forms of online communication so that they could determine how they would structure the modes of communication in the courses they were designing. However, unlike Duncan and Barnett's class, which only used asynchronous communication, this course allowed students to experience both synchronous and asynchronous communication with variations within each format. Also, the first author structured the course so that after students participated in one form of online communication, the class would then meet face-to-face to debrief about what they had learned from their experience. In the remainder of this section, we will discuss students' reactions to the synchronous and asynchronous sections of the course separately.

Asynchronous

Since most of the students were familiar with asynchronous discussion, the first author decided to start the online portion of the course by requiring students to post on the Blackboard threaded discussion board. For the first week of asynchronous discussion, the first author intentionally did not provide students with any instruction other than to reply to the first author's initial post and then to respond to each other's posts and engage in a conversation about the readings throughout the remainder of the week. For the second asynchronous week, the first author divided the class into small groups of three students each.[3] He also required that students make their posts to the initial prompt within the first two days of the assignment in order to give the rest of the group time to respond.

When the class met face-to-face after the two weeks of meeting asynchronously, the students had recognized a difference between the whole group and small group discussions. Due to the overwhelming number of posts the first week, several students admitted to only reading a few posts and selectively choosing posts that they deemed important, such as posts made by the first author, a finding that is consistent with other research on students' habits online (Journell, 2010). Moreover, those students who had posted early expressed frustration at not being able to engage in discussion, and those who posted late admitted to succumbing to a lack of time management, a revelation that prompted a discussion about the motivational issues of many undergraduates and K-12 learners.

While many of these issues remained in the small group week, they were minimized by the additional restrictions placed upon the students. This observation led to discussions about how to best assess asynchronous discussions, appropriate levels of instructor participation, and whether teachers should require minimum word requirements for posts among less intrinsically motivated groups of students. Despite moderate success during this second week, there seemed to be a general agreement among the students that asynchronous discussion on its own, while useful, did not provide enough opportunity for sustained discussion and community development.

Synchronous

For the synchronous portion of the course, we used Elluminate conferencing software, which allows students to communicate through both a text-based chat window and streaming audio from microphones used by students. To speak, participants press the microphone button and their voices can be heard through the rest of their classmates' speakers. Elluminate also features a large whiteboard that is shared by the entire class; anything that is typed on the whiteboard can be seen by the rest of the class members. Figure 1 shows a screenshot of the default layout of Elluminate during one of our synchronous discussions.

Figure 1. Screenshot of the Elluminate main screen

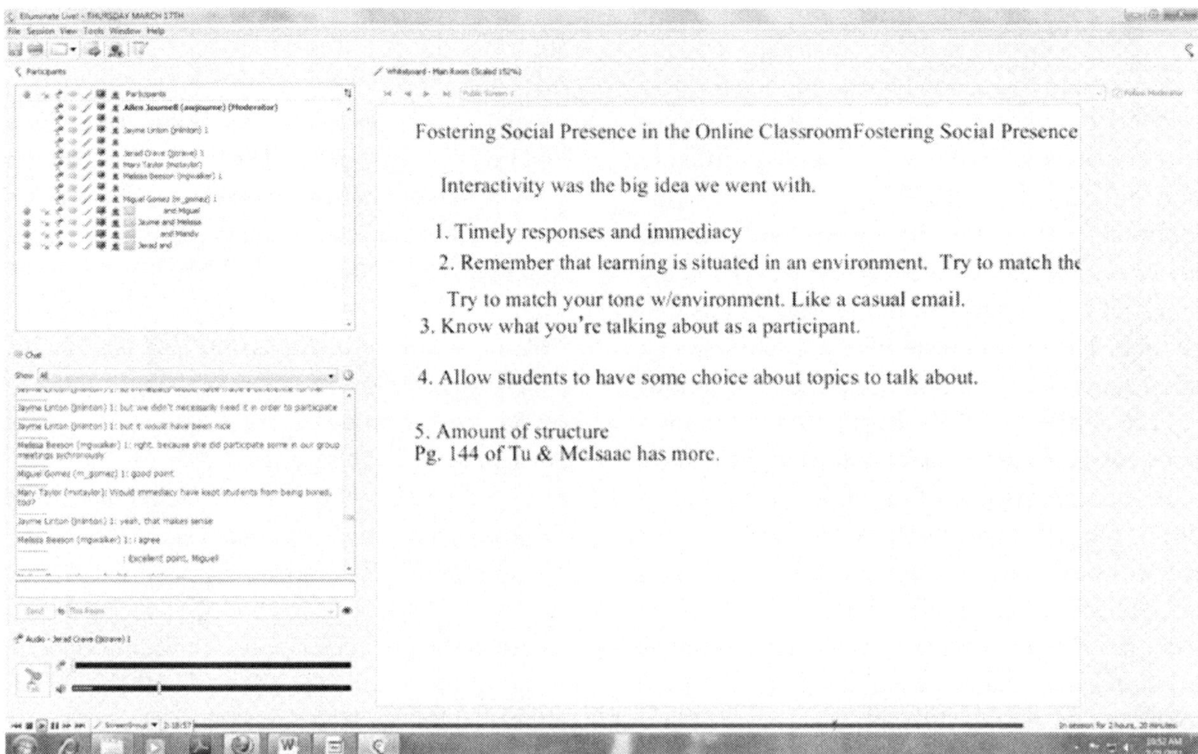

From an instructor's perspective, Elluminate offers as much, if not more, control than what instructors typically have in a face-to-face class. For example, students can be put into "breakout rooms" to work on projects in small groups and can subsequently be brought back into a whole class setting. Instructors can also control students' web browsers through a feature that forces the entire class to look at the same website, and they can regulate student participation by turning off students' chat and speaking abilities or by requiring that students use a tool that simulates raising one's hand before they are allowed to speak. In addition, if a lesson requires that the instructor or a member of the class be seen, Elluminate offers the ability to stream visual images through the use of a webcam.[4]

Despite the availability of these features, the first author decided to only use the chat function of Elluminate during the first synchronous week. The reason for this decision was practical; many universities and school districts may not have adequate bandwidth requirements or the monetary resources to purchase a program like Elluminate (Chen, Ko, Kinshuk, & Lin, 2005; Duncan, 2005), but internet chat rooms and instant messaging programs have made the ability to conduct synchronous, text-based chats almost ubiquitous. Unfortunately, only using the chat window did not provide a very engaging format for discussion. Although the students were able to discuss the readings for that week in depth, they also experienced many of the pitfalls that are associated with text-based chats (Hou & Wu, 2011). The conversation moved quickly, and students often found themselves repeating comments made by others or replying to old comments even after the conversation had shifted to a new topic. The chat was especially frustrating for those students who considered themselves slow typists, and they complained that they were often unable to interject their comments into the discussion because their typing skills did not match up to the speed of the conversation. Even in the breakout rooms where students only had to communicate with one or two classmates, they reported having difficulty moving conversations along. In addition, the informal nature of the text-based chat seemed to invite comments that were off topic, both in the breakout rooms and within the whole class discussion. The overall result was a discussion that often appeared disjointed and difficult to follow.

The addition of the microphones the following week provided a stark contrast in the quality of discussion. As one of the co-authors noted in one of our reflection sessions, "It was night and day different for me. I thought it was a lot better with the microphones. I also thought there were moments we were getting into the [types of] discussions that we were having in the classroom face-to-face." Although Elluminate allows multiple microphones to be used simultaneously, the first author only allowed one person to speak at a time due to bandwidth concerns. Students would "raise their hands" when they wanted to speak, and once students got used to pressing the microphone button to speak as soon as the previous speaker had finished, the conversation flowed smoothly, as if the discussion was being held face-to-face. When students were put into breakout rooms, they were still able to use their microphones, and when they came back to report the results of their discussions to the whole group, the use of audio created a more fluid presentation than what had occurred the previous week in the text-based discussion.

One interesting development that occurred during this second synchronous session was that even with the use of the audio, students were still typing comments into the chat window. These comments displayed approval or disapproval of the audio comments being made, posed thoughtful questions based on the information being discussed via the microphones, and allowed space for informal bantering among students. As one student noted, "I did enjoy having the chat box. If there were two or three people in line waiting to speak [via microphone], I could type something that I was going to say and keep the conversation going." Another student described the relationship between the audio commentary and the conversation occurring in the chat window this way:

"We were having a discussion and then in the chat room there was about an 80/20 split between an 80% focus on what was being talked about through the microphones and 20% of that social presence type stuff, just small banter back and forth that was going on while we were talking about whatever the question was."

According to recent research on multimedia learning, the simultaneous use of both the microphones and chat window may have even provided a more complex, and ultimately richer, environment for discussion than what typically occurs in a face-to-face setting (Debuse, Hede, & Lawley, 2009; Moreno & Mayer, 2002).

At the end of the second synchronous session, the first author used the anonymous survey function of Elluminate to ask his students which of the discussion formats they found most useful. Of the eight students in the class, six stated that they preferred the synchronous discussions over the asynchronous discussion boards, a result that is consistent with previous research (Mabrito, 2006). Of the six students who preferred synchronous discussions, all but one stated that they preferred the combination of the microphones and chat window to just using the text-based chat. Yet, all of the students seemed to recognize that both synchronous and asynchronous communication had certain advantages that could be incorporated into online learning environments. In the end-of-course survey, students were asked their opinion on which form of communication was most effective for online learning, and every student stated that they believed having a mixture of synchronous and asynchronous communication would be beneficial for students. As one student noted, "students need to be able to reflect on discussions and be thoughtful about their replies (asynchronous) as well as be immediate in the moment with a synchronous discussion." Another student explained her answer by saying, "by using both types of mediums for your class, you have a better chance of connecting with your students and allowing them to succeed in the method that they are better suited for."

Students' responses on the end-of-course survey also suggest that they viewed the development of social interaction in online environments as a valuable pedagogical goal. When asked to list the most important factors to teaching online, all but two students listed some variation of community development and social presence. It was also clear that students believed that being able to experience several modes of online communication aided their understanding of how to encourage community development in their own online courses. As one student wrote, "I love the way we explored both synchronous and asynchronous discussions. I feel like we really identified positive aspects and negative aspects of each." The results of these experiences were evident in the students' final projects, the majority of which contained a blend of synchronous and asynchronous opportunities for student interaction.

Merging Theory and Practice

The final project in the course required students to design their own online course based on theory they had learned throughout the semester. On the end-of-course survey, all of the students stated that being able to apply the information learned in the course was "a valuable and useful" learning experience. As one student wrote,

"Creating an online course was a terrific activity that asked us to essentially take the role of an instructor and front-load our course with content, activities, etc. It was very valuable because it allowed us to think carefully about the content, interactions, and activities/assessments that would help students to engage in the material and with one another in meaningful ways."

Reflecting back several months after the course had ended, one of the co-authors added that

"This [course] was theory and practice. I would have not gotten as much out of the class had we not had our own Blackboard space and did it. What you envision in your mind is different than what you can get on the computer screen."

For the final project, students were given their own Blackboard course space and granted instructor privileges. Students were then told to design a course around a topic of their choice with the only real requirement being that their pedagogical decisions should be grounded within the online learning theories presented in class. To help facilitate the latter part of the assignment, students were required to write a short paper that used relevant research and items discussed in class to justify the instructional decisions they made in their courses. In addition, each of the students was enrolled into one of their classmates' courses so that they could provide feedback to the course designer on the functionality of the course.

At the final face-to-face meeting of the course, students gave a brief presentation of the course they had designed to the rest of the class. Overall, student comments indicated that these courses had exceeded their initial expectations about online learning and were considerably more engaging than the previous online courses they had taken. We do not have space to describe all of the courses here; however, we will briefly describe the course created by the second author, which is representa-

Figure 2. The second author's Blackboard home page

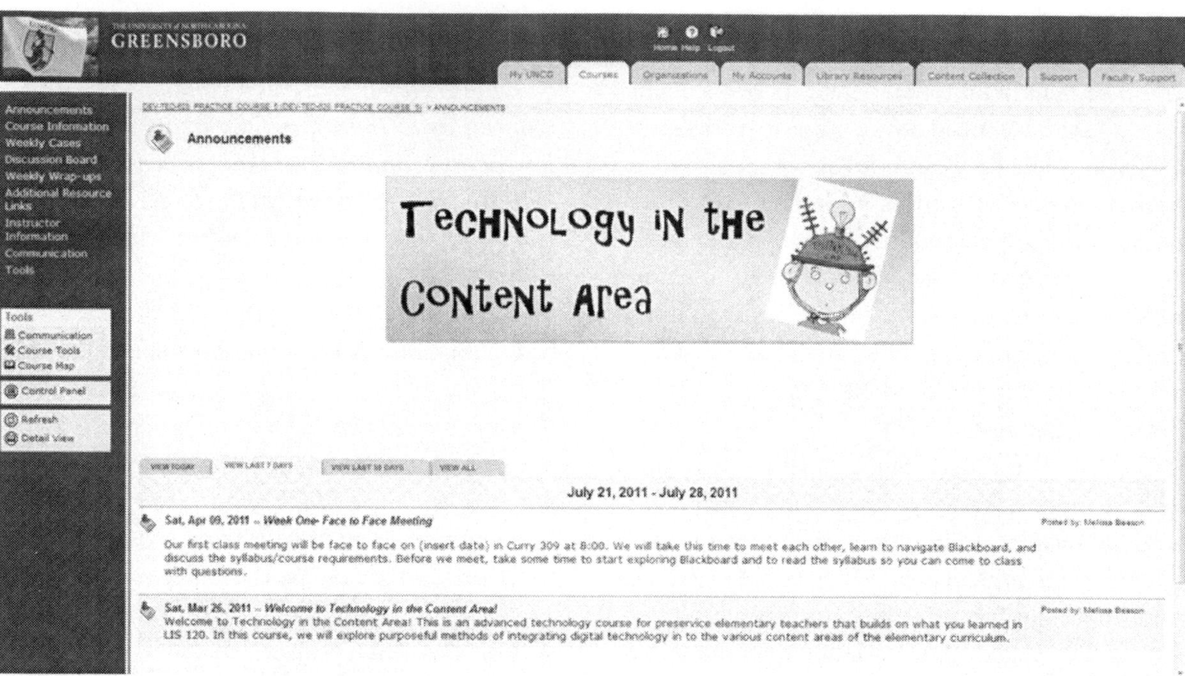

tive of the quality of courses submitted. As Figure 2 shows, the purpose of this course was to prepare pre-service teachers to integrate technology into their respective content areas. The course was designed for an entire semester and used a case-based approach, which is a common instructional design in teacher education that allows students to critically examine theoretical models of teacher practice (e.g., Cherubini, 2009). Other than an initial face-to-face meeting, which research has shown to be beneficial for establishing classroom community (Haythornthwaite et al., 2004), the rest of the course was designed to be delivered exclusively online.

The second author designed the course so that each week students would read a case that described an issue surrounding technology use in the classroom as well as scholarly articles that provided the theoretical framework behind the technology being featured in the case. Then, the second author had students complete activities that catered to a variety of learning styles. Figure 3 shows a typical week in the second author's class. Throughout the span of the course, the second author had students watching videos of successful technology practices in classrooms, creating graphic organizers using free software, listening to and creating podcasts and screencasts, writing e-books using VoiceThread technology, interacting using Web 2.0 technologies, and planning virtual fieldtrips. In order to foster student collaboration, students had to post and respond to weekly reflections on the course discussion board and participate in synchronous Elluminate sessions on a regular basis.

The papers students wrote justifying their pedagogical decisions also played an essential role in this process of merging theory into practice because it allowed them to reflect upon their instruction, which is an important part of teacher development (Chant, 2002; Clandinin, 1986). In her paper, the second author stated that she believed a constructivist approach was important to learning how to use technology, which is why she chose

Figure 3. A typical week in the second author's class

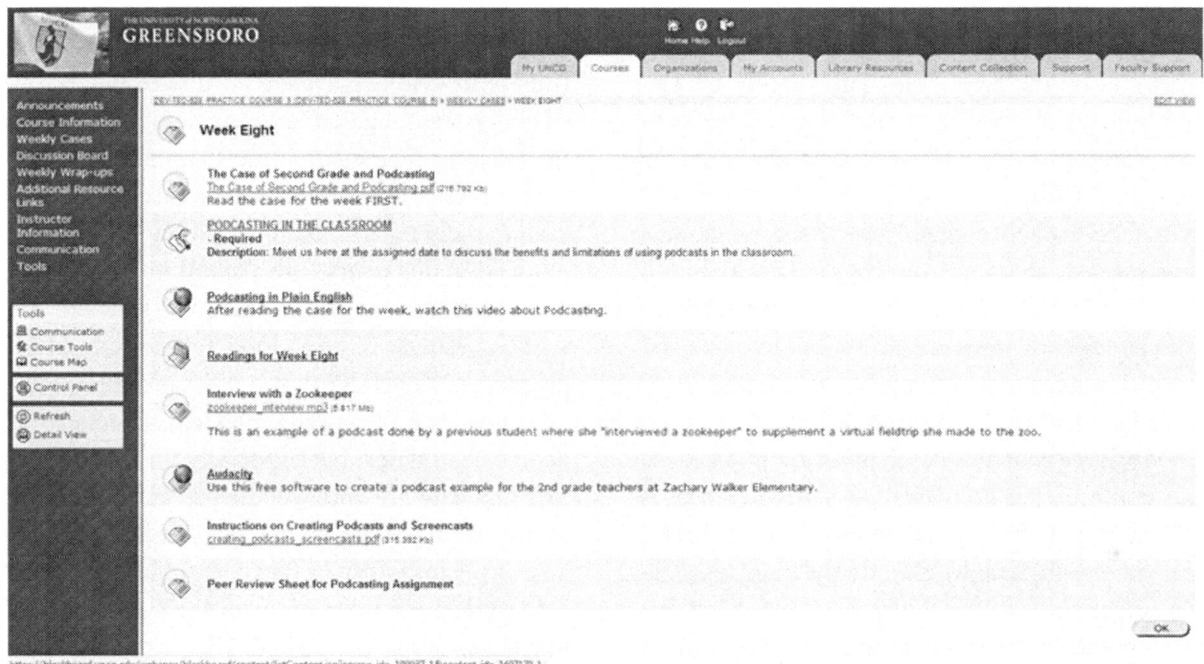

to organize her course using a case-based format. She also stated that effective online courses should provide activities that cater to all learning styles, which is why she arranged her course to include both synchronous and asynchronous communication as well as assignments that moved beyond a text-based understanding of course content. Overall, the combination of the paper and process of creating a course seemed to facilitate a rich understanding of online instruction. As one student wrote about the final project, "it was extremely tedious and time consuming and I spent hours and hours looking at my computer, but it gave me the chance to really think about how I would structure an online environment to provide the type of interaction and community that I would want a face-to-face environment to have."

CONCLUSION AND FUTURE RESEARCH DIRECTIONS

From the perspective of both the instructor and his students, this experimental course was successful in achieving the goal of better preparing students for online instruction. This conclusion is perhaps best exemplified by student comments made on the end-of-course survey. One student wrote,

"Just because you find success in the face-to-face classroom doesn't mean you will find the same success in the online classroom without putting effort into it. Instructional strategies do translate, but not without some tweaking/adjustment. Also, thinking about the choices that are made with technology based on content and pedagogy is so important. We can't just choose to teach synchronously or asynchronously, etc. It has to be a thoughtful decision with content and pedagogy in mind."

Another student noted that the course debunked the familiar perception that "all you have to do is upload content, then sit back and let students work" in online education. The student continued by saying, "we've learned to consider various learning styles and student needs in course design and implementation, just as you would in a face-to-face course." As those comments illustrate, students left the course being able to think not just as teachers, but as *virtual* teachers who understood the needs of virtual learners and the differences that distinguish online pedagogy from face-to-face instruction.

We hope that the description of this course can continue the conversation started by Davis and Roblyer (2005), Duncan and Barnett (2010), and others on how to effectively train teachers for online instruction. We realize, however, that this course had certain limitations and cannot be considered a perfect model for others to emulate. For example, the fact that the class met face-to-face on several occasions affected the sense of community that developed in both the synchronous and asynchronous environments. It is certainly possible that specific aspects of the online portions of the class, such as the use of the microphones in the Elluminate sessions, may have occurred differently had the class never met in person. However, we believe that others who wish to develop similar types of courses at their universities can use our description as a starting point even if they may not have exactly the same resources at their disposal.

Similarly, a few of the limitations that we have discussed in our reflections about the course have led us to think about the potential for future research in this area. For example, many of the students expressed regret at not being able to teach the courses they created. As one student noted in her end-of-course survey, "I think to better understand this, I would need to teach an online course and kind of see what happens and then go from there." Within teacher education, there is no substitute for practice; therefore, we believe that the next step in the evolution of virtual teacher training is to implement a type of "student teaching" for online instructors. Only then will students truly be able to experience the day-to-day rigors of online teaching that are considerably different than face-to-face instruction (Quinlan, 2011).

Finally, we believe that this type of course needs wider student participation. The eight students enrolled in this course are a small fraction of the number of pre-service and practicing teachers that will matriculate through most teacher education programs. We believe that courses in online pedagogy need to be required for all teacher education degree candidates, but especially for pre-service teachers who are entering the profession. Determining what pre-service teachers need to know is an ever-changing and often debatable issue (e.g., Darling-Hammond & Bransford, 2005), but given that the future of education will, most likely, rely heavily on online instruction, we would argue that teacher education programs have a responsibility to prepare their students accordingly.

REFERENCES

Berge, Z. L. (2002). Active, interactive, and reflective elearning. *Quarterly Review of Distance Education*, *3*, 181–190.

Bernard, R. M., Abrami, P. C., Lou, Y., Borokhovski, E., Wade, A., & Wozney, L. (2004). How does distance education compare with classroom instruction? A meta-analysis of the empirical literature. *Review of Educational Research*, *74*(3), 379–439. doi:10.3102/00346543074003379

Burbules, N. C. (2004). Navigating the advantages and disadvantages of online pedagogy. In Haythornthwaite, C., & Kazmer, M. M. (Eds.), *Learning, culture and community in online education: Research and practice* (pp. 3–17). New York, NY: Peter Lang.

Chant, R. H. (2002). The impact of personal theorizing on beginning teaching: Experiences of three social studies teachers. *Theory and Research in Social Education, 30*, 516–540. doi:10.1080/00933104.2002.10473209

Chen, N., & Ko, H., Kinshuk, & Lin, T. (2005). A model for synchronous learning using the Internet. *Innovations in Education and Teaching International, 42*, 181–194. doi:10.1080/14703290500062599

Cherubini, L. (2009). Exploring prospective teachers' critical thinking: Case-based pedagogy and the standards of professional practice. *Teaching and Teacher Education, 25*, 228–234. doi:10.1016/j.tate.2008.10.007

Clandinin, D. J. (1986). *Classroom practice*. London, UK: Falmer.

Cleveland, D. (2003). A semester in the life of alternatively certified teachers: Implications for alternative routes to teaching. *High School Journal, 86*, 17–34. doi:10.1353/hsj.2003.0002

Conceicao, S., & Drummond, S. B. (2005). Online learning in secondary education: A new frontier. *Educational Considerations, 33*, 31–37.

Corbell, K., Booth, S., & Reiman, A. J. (2010). The commitment and retention intentions of traditionally and alternatively licensed math and science beginning teachers. *Journal of Curriculum and Instruction, 4*, 50–69. doi:10.3776/joci.2010.v4n1p50-69

Coryell, J. E., & Chulp, D. T. (2007). Implementing e-learning components with adult English language learners: Vital factors and lessons learned. *Computer Assisted Language Learning, 20*, 263–278. doi:10.1080/09588220701489333

Darling-Hammond, L. (2003). Keeping good teachers: Why it matters, what leaders can do. *Educational Leadership, 60*(8), 6–13.

Darling-Hammond, L., & Bransford, J. (Eds.). (2005). *Preparing teachers for a changing world: What teachers should learn and be able to do*. San Francisco, CA: Jossey-Bass.

Davis, N. E., & Roblyer, M. D. (2005). Preparing teachers for the "schools that technology built": Evaluation of a program to train teachers for virtual schooling. *Journal of Research on Technology in Education, 37*, 399–409.

Debuse, J. C. W., Hede, A., & Lawley, M. (2009). Learning efficacy of simultaneous audio and on-screen text in online lectures. *Australasian Journal of Educational Technology, 25*, 748–762.

Dede, C., Ketelhut, D. J., Whitehouse, P., Breit, L., & McCloskey, E. M. (2009). A research agenda for online teacher professional development. *Journal of Teacher Education, 60*, 8–19. doi:10.1177/0022487108327554

Dinkelman, T. (2003). Self-study in teacher education: A means and end tool for promoting reflective teaching. *Journal of Teacher Education, 54*, 6–18. doi:10.1177/0022487102238654

Duncan, H. E. (2005). Online education for practicing professionals: A case study. *Canadian Journal of Education, 28*, 874–896. doi:10.2307/4126459

Duncan, H. E., & Barnett, J. (2010). Experiencing online pedagogy: A Canadian case study. *Teaching Education, 21*, 247–262. doi:10.1080/10476210903480340

Ellis, C., & Bochner, A. (2000). Autoethnography, personal narrative, reflexivity. In Denzin, N. K., & Lincoln, Y. S. (Eds.), *Sage handbook of qualitative research* (2nd ed., pp. 733–768). Thousand Oaks, CA: Sage.

Feldman, A. (2003). Validity and quality in self-study. *Educational Researcher, 32*(4), 26–28. doi:10.3102/0013189X032003026

Freese, A. (2006). Reframing one's teaching: Discovering our teacher selves through reflection and inquiry. *Teaching and Teacher Education*, *22*, 100–119. doi:10.1016/j.tate.2005.07.003

Garrison, D. R., & Anderson, T. (2003). *E-learning in the 21st century: A framework for research and practice*. London, UK: Routledge Falmer.

Garrison, D. R., & Cleveland-Innis, M. (2005). Facilitating cognitive presence in online learning: Interaction is not enough. *American Journal of Distance Education*, *19*, 133–148. doi:10.1207/s15389286ajde1903_2

Gaytan, J., & McEwen, B. C. (2007). Effective online instructional and assessment strategies. *American Journal of Distance Education*, *21*, 117–132. doi:10.1080/08923640701341653

Gold, S. (2001). A constructivist approach to online training for online teachers. *Journal of Asynchronous Learning Networks*, *5*, 35–57.

Guri-Rosenblit, S. (2005). Eight paradoxes in the implementation process of e-learning in higher education. *Higher Education Policy*, *18*, 5–29. doi:10.1057/palgrave.hep.8300069

Haythornthwaite, C., & Kazmer, M. M. (Eds.). (2004). *Learning, culture and community in online education: Research and practice*. New York, NY: Peter Lang.

Haythornthwaite, C., Kazmer, M. M., Robins, J., & Shoemaker, S. (2004). Community development among distance learners: Temporal and technological dimensions. In Haythornthwaite, C., & Kazmer, M. M. (Eds.), *Learning, culture and community in online education: Research and practice* (pp. 35–57). New York, NY: Peter Lang. doi:10.1111/j.1083-6101.2000.tb00114.x

Heafner, T. L. (2011). *Windows into teaching and learning [WiTL]: Exploring online clinicals for a distance education social studies methods course*. Paper presented at the Annual Meeting of the College and University Faculty Assembly of the National Council for the Social Studies, Washington, DC.

Heafner, T. L., & Petty, T. (2010). Observing graduate interns remotely. *Kappa Delta Pi Record*, *47*, 39–43.

Hou, H., & Wu, S. (2011). Analyzing the social knowledge construction behavioral patterns of an online synchronous collaborative discussion instructional activity using an instant messaging tool: A case study. *Computers & Education*, *57*, 1459–1468. doi:10.1016/j.compedu.2011.02.012

Journell, W. (2008). Facilitating historical discussions using asynchronous communication: The role of the teacher. *Theory and Research in Social Education*, *36*, 317–355. doi:10.1080/00933104.2008.10473379

Journell, W. (2010). Perceptions of e-learning in secondary education: A viable alternative to classroom instruction or a way to bypass engaged learning? *Educational Media International*, *47*, 69–81. doi:10.1080/09523981003654985

Journell, W. (2012). Walk, don't run—to online learning. *Phi Delta Kappan*, *93*(7), 46-50.

Jun, J. (2005). Understanding e-dropout. *International Journal on E-Learning*, *4*, 229–240.

Kapitzke, C., & Pendergast, D. (2005). Virtual schooling service: Productive pedagogies or pedagogical possibilities? *Teachers College Record*, *107*, 1626–1651. doi:10.1111/j.1467-9620.2005.00536.x

Keeler, C. G., & Horney, M. (2007). Online course designs: Are special needs being met? *American Journal of Distance Education*, *21*, 61–75. doi:10.1080/08923640701298985

Kickul, G., & Kickul, J. (2006). Closing the gap: Impact of student productivity and learning goal orientation on e-learning outcomes. *International Journal on E-Learning*, *5*, 361–372.

Ladson-Billings, G. (2000). Fighting for our lives: Preparing teacher to teach African-American students. *Journal of Teacher Education*, *51*, 206–213. doi:10.1177/002248710005100300
8

Levin, B. B., He, Y., & Robbins, H. H. (2006). Comparative analysis of preservice teachers' reflective thinking in synchronous versus asynchronous online case discussions. *Journal of Technology and Teacher Education*, *14*, 439–460.

Lincoln, Y. S., & Guba, E. G. (1985). *Naturalistic inquiry*. Newbury Park, CA: Sage.

Lucas, T., Villegas, A. M., & Freedson-Gonzalez, M. (2008). Linguistically responsive teacher education: Preparing classroom teachers to teach English language learners. *Journal of Teacher Education*, *59*, 361–373. doi:10.1177/0022487108322110

Mabrito, M. (2006). A study of synchronous versus asynchronous collaboration in an online business writing class. *American Journal of Distance Education*, *20*, 93–107. doi:10.1207/s15389286ajde2002_4

Maor, D. (2006). Using reflective diagrams in professional development with university lecturers: A developmental tool in online teaching. *The Internet and Higher Education*, *9*, 133–145. doi:10.1016/j.iheduc.2006.03.005

McClam, S., & Sevier, B. (2010). Troubles with grades, grading, and change: Learning from adventures in alternative assessment practices in teacher education. *Teaching and Teacher Education*, *26*, 1460–1470. doi:10.1016/j.tate.2010.06.002

Michigan Department of Education. (2006). *Michigan merit curriculum guidelines: Online experience*. Retrieved from http://www.michigan.gov/documents/mde/Online10.06_final_175750_7.pdf

Moreno, R., & Mayer, R. E. (2002). Verbal redundancy in multimedia learning: When reading helps listening. *Journal of Educational Psychology*, *94*, 156–163. doi:10.1037/0022-0663.94.1.156

Murphy, E., Rodriguez-Manzanares, M. A., & Barbour, M. (2011). Asynchronous and synchronous online teaching: Perspectives of Canadian high school distance education teachers. *British Journal of Educational Technology*, *42*, 583–591. doi:10.1111/j.1467-8535.2010.01112.x

National Education Association. (2006). *Guide to teaching online courses*. Retrieved from http://www.nea.org/assets/docs/onlineteachguide.pdf

Ng, C. F. (2006). Academics telecommuting in open and distance education universities: Issues, challenges, and opportunities. *International Review of Research in Open and Distance Learning*, *7*(2), 1–16.

Njenga, J. K., & Fourie, L. C. H. (2010). The myths about e-learning in higher education. *British Journal of Educational Technology*, *41*, 199–212. doi:10.1111/j.1467-8535.2008.00910.x

Noble, D. F. (2001). *Digital diploma mills: The automation of higher education*. New York, NY: Monthly Review Press.

O'Connor, K. A., Good, A. J., & Greene, H. C. (2006). Lead by example: The impact of teleobservation on social studies methods courses. *Social Studies Research and Practice*, *1*, 165–178.

Pankowski, M. M. (2003). *How do undergraduate mathematics faculty learn to teach online?* Unpublished Doctoral dissertation, Duquesne University.

Patton, M. Q. (2002). *Qualitative research and evaluation methods*. Thousand Oaks, CA: Sage.

Picciano, A. G., & Seaman, J. (2009). *K-12 online learning survey: A survey of U.S. school district administrators*. Retrieved August 2, 2011, from http://sloanconsortium.org/publications/survey/K-12_06

Quinlan, A. M. (2011). 12 tips for the online teacher. *Phi Delta Kappan, 92*(4), 28–31.

Richardson, L. (2000). Writing: A method of inquiry. In Denzin, N. K., & Lincoln, Y. S. (Eds.), *Sage handbook of qualitative research* (2nd ed., pp. 923–948). Thousand Oaks, CA: Sage.

Roblyer, M. D., Davis, L., Mills, S. C., Marshall, J., & Pape, L. (2008). Toward practical procedures for predicting and promoting success in virtual school students. *American Journal of Distance Education, 22*, 90–109. doi:10.1080/08923640802039040

Robyler, M. D. (1999). Is choice important in distance learning? A study of student motives for taking internet-based courses at the high school and community college levels. *Journal of Research on Computing in Education, 32*, 157–171.

Rock, M. L., Gregg, M., Thead, B. K., Acker, S. E., Gable, R. A., & Zigmond, N. P. (2009). Can you hear me now? Evaluation of an online wireless technology to provide real-time feedback to special education teachers in-training. *Teacher Education and Special Education, 32*, 64–82. doi:10.1177/0888406408330872

Rovai, A. P. (2000a). Building and sustaining community in asynchronous learning networks. *The Internet and Higher Education, 3*, 285–297. doi:10.1016/S1096-7516(01)00037-9

Rovai, A. P. (2000b). Online and traditional assessments: What is the difference? *The Internet and Higher Education, 3*, 141–151. doi:10.1016/S1096-7516(01)00028-8

Schlager, M. S., Farooq, U., Fusco, J., Schank, P., & Dwyer, N. (2009). Analyzing online teacher networks: Cyber networks require cyber research tools. *Journal of Teacher Education, 60*, 86–100. doi:10.1177/0022487108328487

Schrum, L. (2004). The web and virtual schools. *Computers in the Schools, 21*, 81–89. doi:10.1300/J025v21n03_09

Stevens, C. J., & Dial, M. (1993). A qualitative study of alternatively certified teachers. *Education and Urban Society, 26*, 63–77. doi:10.1177/0013124593026001006

Swan, K., & Shih, L. F. (2005). On the nature and development of social presence in online course discussions. *Journal of Asynchronous Learning Networks, 9*, 115–136.

Tallent-Runnels, M. K., Thomas, J. A., Lan, W. Y., Cooper, S., Ahern, T. C., Shaw, S. A., & Liu, X. (2006). Teaching courses online: A review of the research. *Review of Educational Research, 76*, 93–135. doi:10.3102/00346543076001093

Tu, C., & McIsaac, M. (2002). The relationship of social presence and interaction in online classes. *American Journal of Distance Education, 16*, 131–150. doi:10.1207/S15389286AJDE1603_2

Tunison, S., & Noonan, B. (2001). On-line learning: Secondary students' first experience. *Canadian Journal of Education, 26*, 495–514. doi:10.2307/1602179

Van Laarhoven, T. R., Munk, D. D., Lynch, K., Bosma, J., & Rouse, J. (2007). A model for preparing special and general education preservice teachers for inclusive education. *Journal of Teacher Education, 58*, 440–455. doi:10.1177/0022487107306803

Weiner, C. (2003). Key ingredients to online learning: Adolescent students study in cyberspace—The nature of the study. *International Journal on E-Learning, 2*, 44–50.

Wilson, G., & Stacey, E. (2003). Online interaction impacts on learning: Teaching the teachers to teach online. In G. Crisp, D. Thiele, I. Scholten, S. Baker, & J. Baron (Eds.), *Interact, integrate, impact: Proceedings of the 20th Annual Conference of the Australasian Society for Computers in Learning in Tertiary Education.* Adelaide, Australia: ASCILITE.

Wolf, P. D. (2006). Best practices in the training of faculty to teach online. *Journal of Computing in Higher Education*, *17*, 47–78. doi:10.1007/BF03032698

Zeichner, K. M. (2007). Accumulating knowledge across self-studies in teacher education. *Journal of Teacher Education*, *58*, 36–46. doi:10.1177/0022487106296219

Zeichner, K. M., & Schulte, A. K. (2001). What we know and don't know from peer-reviewed research about alternative teacher certification programs. *Journal of Teacher Education*, *52*, 266–282. doi:10.1177/0022487101052004002

ADDITIONAL READING

Anderson, D. L., Standerford, N. S., & Imdieke, S. (2010). A self-study on building community in the online classroom. *Networks*, *12*(2), 1–10.

Archambault, L., & Crippen, K. (2009). K-12 distance educators at work: Who's teaching online across the United States. *Journal of Research on Technology in Education*, *41*, 363–391.

Beebe, R., Vonderwall, S., & Boboc, M. (2010). Emerging patterns in transferring assessment practices from f2f to online environments. *Electronic Journal of E-Learning*, *8*, 1–12.

Bewane, J., & Spector, J. (2009). Prioritization of online instructor roles: Implications for competency-based teacher education programs. *Distance Education*, *30*, 383–397. doi:10.1080/01587910903236536

Bullock, S. M. (2011). Teaching 2.0: (Re) learning to teach online. *Interactive Technology and Smart Education*, *8*, 94–105. doi:10.1108/17415651111141812

Burd, B. A., & Buchanan, L. E. (2004). Teaching and teachers: Teaching and learning online. *RSR. Reference Services Review*, *32*, 404–412. doi:10.1108/00907320410569761

Chan, S. (2010). Designing an online class using a constructivist approach. *Journal of Adult Education*, *39*, 26–39.

Chen, L. W., & Beasley, W. (2005). Type II technology applications in teacher education: Using instant messenger to implement structured online class discussions. *Computers in the Schools*, *22*, 71–84.

Clark, R. C., & Mayer, R. E. (2008). *E-Learning and the science of instruction: Proven guidelines for consumers and designers of multimedia learning* (2nd ed.). San Francisco, CA: Pfeiffer.

Collopy, R. M. B., & Arnold, J. M. (2009). To blend or not to blend: Online learning environments in undergraduate teacher education. *Issues in Teacher Education*, *18*, 85–101.

Condie, R., & Livingston, K. (2007). Blending online learning with traditional approaches: Changing practices. *British Journal of Educational Technology*, *38*, 337–348. doi:10.1111/j.1467-8535.2006.00630.x

Cowie, B., & Khoo, E. (2010). A framework for developing and implementing an online learning community. *Journal of Open. Flexible and Distance Learning*, *15*, 47–59.

Edwards, M., Perry, B., & Janzen, K. (2011). The making of an exemplary online educator. *Distance Education*, *32*, 101–118. doi:10.1080/01587919.2011.565499

Falloon, G. (2011). Making the connection: Moore's theory of transactional distance and its relevance to the use of a virtual classroom in postgraduate online teacher education. *Journal of Research on Technology in Education*, *43*, 187–209.

Ferguson, J. M., & DeFelice, A. E. (2010). Length of online course and student satisfaction, perceived learning, and academic performance. *International Review of Research in Open and Distance Learning*, *11*, 73–84.

Green, C., & Tanner, R. (2005). Multiple intelligences and online teacher education. *English Language Teachers Journal*, *59*, 312–321. doi:10.1093/elt/cci060

Gulati, S. (2008). Compulsory participation in online discussions: Is this constructivism or normalisation of learning? *Innovations in Education and Teaching International*, *45*, 183–192. doi:10.1080/14703290801950427

Horton, W. (2006). *E-learning by design*. San Francisco, CA: Pfeiffer.

Kerr, S. (2011). Tips, tools, and techniques for teaching in the online high school classroom. *TechTrends*, *55*, 28–30. doi:10.1007/s11528-011-0466-z

Meyers, S. A. (2008). Using transformative pedagogy when teaching online. *College Teaching*, *56*, 219–224. doi:10.3200/CTCH.56.4.219-224

Miller, K. W. (2008). Teaching science methods online: Myths about inquiry-based online learning. *Science Educator*, *17*, 80–86.

Oliver, K., Osbourne, J., & Brady, K. (2009). What are secondary students' expectations for teachers in virtual school environments? *Distance Education*, *30*, 23–45. doi:10.1080/01587910902845923

Rosenthal, I. (2010). On line construction: An opportunity to re-examine and re-invent pedagogy. *Contemporary Issues in Education Research*, *3*, 21–26.

Thompson, D. L. (2010). Beyond the classroom walls: Teachers' and students' perspectives on how online learning can meet the needs of gifted students. *Journal of Advanced Academics*, *21*, 662–712. doi:10.1177/1932202X1002100405

Ward, M. E., Peters, G., & Shelley, K. (2010). Student and faculty perceptions of the quality of online learning experiences. *International Review of Research in Open and Distance Learning*, *11*, 57–77.

KEY TERMS AND DEFINITIONS

Asynchronous: Communication that does not occur in real time, such as threaded discussion boards, email, written correspondence, and blogs.

Blackboard: A type of commercial educational software that supports online learning. Similar types of programs include WebCT and Moodle.

Elluminate: A type of commercial software that supports synchronous conferences among multiple participants.

Freeware: Any type of software that can be legally downloaded for free on the Internet.

Podcast: A type of audio recording that can be downloaded from the Internet and played on an iPod or other portable audio device.

Screencast: A type of video that records the actions that occur on a computer screen. The video can then be watched by other users.

Synchronous: Communication that occurs in real time, such as face-to-face conversations, phone calls, internet chat rooms, and instant messaging programs.

ENDNOTES

1 For a copy of the nine statements, email the first author at awjourne@uncg.edu

2 Students also used Any Video Converter (http://www.any-video-converter.com/products/for_video_free/) to easily convert videos into a desired format, or to include multiple video formats in their courses in order to minimize potential technical difficulties for students.

3 Another faculty member within the School of Education began the semester by auditing the class because she wanted to learn more about online instruction, which is why there were nine students in the asynchronous discussion weeks. However, this faculty member stopped attending class soon thereafter due to personal reasons.

4 We do not have space to list all of the features of Elluminate, but interested readers can access an online instructor's manual at http://www.elluminate.com/downloads/support/docs/8.0/Elluminate_Live_V8_Moderator_Guide.pdf

Section 3
Supporting Online Learning in Teacher Education through Collaboration

Chapter 8
Promoting Collaborative Learning in Online Teacher Education

Vassiliki I. Zygouris-Coe
University of Central Florida, USA

ABSTRACT

Online learning continues to grow as a learning option for millions of students in US colleges and universities. Collaboration plays an important role in student learning. This chapter presents information on how collaborative learning was designed and implemented in a comprehensive online course in reading for pre-service and in-service educators in grades P-12. The author presents details on course design issues, instructional practices, benefits, and challenges associated with collaborative learning in this online course, and implications for further development and evaluation of collaborative learning in teacher preparation programs. The author also provides recommendations for promoting collaboration in online teacher education courses.

INTRODUCTION

In this chapter, the author describes how collaborative learning was designed and incorporated in a graduate level online course in reading for preservice and inservice educators. The purpose of this chapter is not to formally assess or evaluate collaborative learning; instead, the author will present her rationale for incorporating collaborative learning experiences in an online education course, the ways in which collaborative learning was incorporated, assessed, and monitored, and she will also share overall observations about benefits and challenges associated with collaborative

DOI: 10.4018/978-1-4666-1906-7.ch008

learning in this situated context. The author will also reflect on the role of collaborative learning in teacher preparation courses. Collaborative learning can support online and teacher preparation learning goals and objectives by promoting critical thinking skills, perspective taking, shared knowledge and decision-making, content knowledge, and reflection.

BACKGROUND

Online Learning and Teacher Education

The 2010 *Class Differences: Online Education in the US* report by the Alfred P. Sloan Foundation (Allen & Seaman, 2010) revealed that US student enrollment rose by almost one million students from a year earlier. According to the survey results from over 2,500 colleges and universities nationwide, approximately 5.6 million students were enrolled in at least one online course in fall 2009. Online learning is a significant choice of learning in US higher education institutions. According to this report, there has been a12-14% annual increase, on average, in enrolment for fully online learning over the five years 2004-2009 in the post-secondary system, compared with an average of approximately two percent per year in enrolments overall.

The convenient accessibility of knowledge, ongoing participation, dialogue, feedback from peers and instructor, availability of formats for presenting materials (Li & Irby, 2008), plethora of readily available tools and resources, learner self-regulation (Li & Irby, 2008; Thomson, 2010), and opportunities for differentiated online instruction (Thomson, 2010), have made online learning a very attractive and relevant learning choice for postsecondary students (Dede, Ketelhut, Whitehouse, Breit, & McCloskey, 2009; Rourke & Kanuka, 2009). Almost 30% of all college and university students now take at least one course online (Allen & Seaman, 2010). As adoption of online learning continues to increase, issues related to quality of online learning become vital. John Sener (2010) proposes that soon online education will become an integral part of the educational experience.

Although many obstacles still remain to full-scale adoption of online higher education, all higher education students will experience online education at some point of their academic career. It is predicted that college students will be able to take online or blended programs in almost any discipline (Sener, 2010). The goal of online learning is to improve the quality of the learning experience for students, offer alternative means of learning, and allow them to experiment, become independent learners, and drivers of change. In spite of its rapid growth and availability, much online learning is still designed using standard educational practices (e.g., lectures, discussions, quizzes, etc.) (Norton & Hathaway, 2008).

Concerns with traditional pedagogy facilitated by course management systems raise questions about the quality of the learner's experiences. Factors such as learner self-monitoring, the social, teaching, and cognitive presence of the online instructor or facilitator, instructional design factors, relevancy and quality of content, collaborative learning opportunities, participants' perceptions of the instruction, collaboration, and online learning, all influence the online learning experience. In the context of steady increase in online courses by US postsecondary students, the types of learning students experience as part of their online learning experiences become critical to program development, delivery, instruction, student satisfaction, and quality assurance.

Preparing teachers who can effectively meet the needs of all students is a major concern of policy-makers, teacher educators, and the public. In recent years, teacher education has been under the political spotlight due to the lack of performance of students in P-12 grades in national and international assessments, the need for technological advancement, and the need to prepare students who will be responsible citizens and effective participants in the global marketplace of the 21st century.

Expectations for teachers are very high in today's era of educational reform. Teachers are expected to be experts in more than one subject. They also need to be prepared to handle the challenges of a growing diverse population of students. Colleges of education need to be preparing teachers for the interconnected world. Teacher preparation should be filled with high quality learning experiences based on sound theoretical principles. Teacher preparation programs should allot significant time for applying theory into practice and reflecting on one's learning (Young, Grant, Montbriand, & Therriault, 2001). According to the American Association of Colleges for Teacher Education (AACTE) (2001), all teachers must be prepared to implement effective instruction designed to meet educational objectives for all learners.

In 2010, the AACTE and The Partnership for 21st Century Skills (P21) released a report titled, *21st Century Knowledge and Skills in Educator Preparation.* The report calls for establishing a shared vision for preparing teachers who will in turn prepare all students with 21st century knowledge and skills. In order to meet the demands of global economy, teachers will need to exemplify, and embed the following in their instruction:

"...the mastery of 21st century skills such as critical thinking, problem-solving, communication, collaboration and creativity and innovation. This includes the application of technology to support more robust instructional methods and understanding the relationship between content, pedagogy and technology through dissemination of Technological Pedagogical Content Knowledge (TPCK) theory and research." (American Association of Colleges of Teacher Education 2008, US Department of Education 2010) (p. 5)

Although there is evidence about the rapid growth in online course offerings, many in higher education have raised concerns about the ability of distance learning to produce teachers who will be problem-solvers and will have the knowledge, skills, and dispositions necessary to handle the real world challenges of classroom life (Duffy, Webb, & Davis, 2009). Some researchers argue that the methods instructors employ to engage the learner in learning are more important than the delivery method; student interactions with instructor and peers effect learning outcomes (Dooley, Lindner, Dooley, & Wilson, 2005).

What role will online teaching and learning teacher education programs play in preparing teachers who will meet current and future challenges in P-12 schools? Since teachers will always need to collaborate with other educators in planning effective instruction, assessing students' needs, and improving curricula and other programs, what role would collaborative learning play in the knowledge, skills, and experiences of preservice and inservice teachers? In this chapter the author will share how she promoted collaborative learning in an online teacher education course by designing authentic, reflective, and connected real-world tasks(Darling-Hammond, Hammerness, Grossman, Rust, & Shulman, 2005), assignments, experiences, and providing scaffolded feedback and support throughout the course.

Andragogy and Online Learning

Michael Knowles (1980, 1984, 1986; Knowles, Swanson, & Holton, 2005) has helped us better understand the ways in which adults learn through his work on andragogy, the science of adult learning. Adults learn differently than children or adolescents. Adult learners have specific learning goals and outcomes, are self-reliant and self-motivated, bring a variety of life experiences, require relevant purpose in learning, have time constraints, seek solutions to problems that affect their daily activities and responsibilities, and tend to be independent thinkers and learners. Adult learners also desire immediacy in feedback and relevancy in tasks. Their "need to know" requires meaningful learning contexts.

Educators of adult learners need to consider andragogy principles when developing learning environment, programs, and other professional experiences. Adults like to explore possibilities and act as mentors. They enjoy organized settings, opportunities to build productive relationships, and choice over tasks and solutions (Knowles, Swanson, & Holton, 2005). Adult learners also enjoy discussions and collaborations with other learners and facilitators in the learning process. It is important the adult learning environments allow for sharing and usage of past experiences, setting of flexible goals, provision of feedback to the learning process, provision for adult learning needs (i.e., vision, hearing, physical and other disabilities), sequencing of tasks, active participation and freedom of expression, creative tasks, and open-ended-type questions. Active participation for adult learners can include a variety of collaborative activities and tasks (e.g., relevant, games, role-playing, simulations, case studies, debates, question and answer sessions, discussions, demonstrations, problem-solving tasks) (Ota, DiCarlo, Burts, Laird, & Gioe, 2006).

Collaborative Learning

According to a recent Horizon Report (2008) a "renewed emphasis on collaborative learning is pushing the educational community to develop new forms of interaction" (p. 5). The Horizon Report (2008) identified six emerging technologies that will likely enter mainstream use for teaching, learning, or other creative applications. One of those six emerging technologies that relates to the focus of this chapter is collaboration webs. Collaboration is no longer difficult or expensive; there are several free and flexible tools people can use to collaborate with others. In addition, the way we work, collaborate, and communicate is ever evolving. This emerging trend of available communication and collaboration tools (e.g., online collaborative spaces, Skype, social networking tools, mobile devices) is also transforming online higher education and student demographics—many international students are able to enroll in online courses and use such tools to collaborate and learn from a distance (The Horizon Report, 2008).

The etymology of the word "collaborate" comes from the Latin word *collaboratus,* past participle of *collaborare* which means "to labor together", from Latin *com-* + *laborare*: to labor (Merriam-Webster Dictionary). The term "collaborative learning" refers to a "co-labor type of learning" in which all members of the group must participate actively in working together toward objectives and goals.

The effectiveness of collaborative learning as a pedagogical approach (in face-to-face learning environments) is well supported (Barkley, Cross & Major, 2005; Johnson, Johnson & Smith, 1998; Schmuck & Schmuck, 2004). Cognitive science tells us that transference of knowledge from teacher to learner cannot happen. Learning is a complex, active, and socially constructed process. Learners must engage meaningfully with learning, establish connections between old and new information, and assimilate new information into existing schemas. Pascarella and Terenzini (2005) stated that as a result of their mammoth review of studies about how college affects students, "the weight of evidence from this research is reasonably consistent in suggesting that collaborative learning approaches can significantly enhance learning" (p. 103).

Dewey (1916/1997), Piaget (1969), Vygotsky (1930/1978), and Lewin (1935), have all highlighted the importance of student-centered learning. Collaborative learning is mainly based on constructivism theory that views knowledge as socially constructed by communities of people (MacGregor, 1990).

Constructivists believe that the way we learn is by interpreting our experiences based on our prior knowledge, making connections between old and new information, constructing meaning, and revising our understanding as a result of new experiences. In a constructivist-learning environment students should be engaged in activities and assignments that include authentic and reflective tasks embedded in real world contexts. The relationship between the instructor and the students is one of cognitive apprenticeship, in which the instructor models problem-solving, creates learning experiences, provides scaffolding as students attempt tasks, and encourages ongoing reflection. Students collaborate with others and confront multiple perspectives on the content being learned. Students are encouraged to reflect on their experiences and learning at all times.

In the online learning environment, students benefit from group work by listening to, and examining multiple perspectives on the same topic, defending propositions by providing supporting evidence, and practicing critical thinking skills by making connections among viewpoints and pieces of information. They learn how to become a productive member of a learning community. Collaborative learning in online environments promotes student participation, socialization (even in the absence of face-to-face interaction), reflection, self-development, and learning.

Collaborative learning can help students to appreciate the importance of collaborative inquiry in learning; students will experience knowledge not as a mere transmittance from instructor to student (Sheridan, 1989) or book to student but instead, as a co-constructed process. Higher education focuses on helping young adults to take responsibility for their own learning and become effective problem solvers and team members in their disciplines and fields of study (Weimer, 2002; Saroyan &Amundsen, 2004). Accountability for learning, decision-making, collaboration, and reflection are basic tenets of teacher education training. Pre-service teachers need to develop necessary knowledge, skills, and dispositions that will help them to become effective teachers in P-12 grades.

Collaborative learning is a wonderful way to introduce active, student-center learning (Conrad & Donaldson, 2004; Palloff & Pratt, 2005). Collaboration and community building depend on one another (Palloff & Pratt, 2005). Community building is a prerequisite to collaboration (Lowell, 2006). In order for collaborative learning to take place in an online environment, the course facilitator has to design and structure activities that will promote it (Lowell, 2006). For collaborative learning to be successful, much attention must be given to the structuring of the learning task(s), the alignment of such tasks with overall learning goals, group placement and size, the conditions set by the instructor for dialogue, the role of the facilitator, student orientation to collaborative learning expectations and individual accountability, and assessment and evaluation of collaborative work (Barkley, Cross, & Major, 2005).

Over ten years ago, Garrison, Anderson, & Archer (2000) introduced the Community of Inquiry (CoI) model in an effort to deal with challenges associated with their online graduate program. Garrison and Anderson (2003) believe that an effective online learning community has to include social, teaching, and cognitive presence. They define defined social presence in an online environment, as "the ability of participants in a community of inquiry to project themselves socially and emotionally as 'real' people, through the medium of communication being used" (2003, p. 29). Social presence refers to the degree to which a person feels "socially present" and is actively taking part in community interactions (Gunawardena, 2005; Wise, Chang, Duffy, & del Valle, 2004). Teaching presence is defined as the design, facilitation, and direction of cognitive and social processes for the purpose of achieving meaningful learning outcomes (Swan, Garrison, & Richardson, 2009). Cognitive presence is defined "as the extent to which learners are able to construct and confirm meaning through sustained reflection, and discourse in a critical community of inquiry" (Garrison & Anderson, 2003, p. 28).

The CoI model has made several contributions to our understanding of online and blended learning (Swan & Ice, 2010). Social, teaching, and cognitive presence are closely related to collaborative learning in any online course. In a long distance medium, the facilitator must gauge the social presence of students in the learning community—e.g., frequency of logins, expression of emotion and humor, personalization, sharing of life details (self-disclosure), replying to others' discussions and continuing on a thread, reflecting to others' messages, asking questions, and sharing resources and ideas. What participants do in an online medium matters for community-building, social presence (Wise, Chang, Duffy, & del Valle, 2004), collaboration, and learning.

Student interaction is essential to successful online learning. Collaboration and group work are important teaching and learning strategies but often leave students dissatisfied by the process. In order to increase student satisfaction and learning effectiveness, students should be supported in both the development of collaborative technology tool use and the development of collaborative group work skills. One successful model for developing online collaborative teams is the Phases of Engagement Model (Conrad & Donaldson, 2004). This model was designed to "transform" a student from being a newcomer to an online course to an actively participating community member. The model includes introductory community-building activities that help build confidence, trust, and a risk-free environment. Here are the phases and shifting roles of engagement:

- **Co-exist:** learner: newcomer; instructor: social negotiator; weeks 1-2; Instructor activities that are interactive and help students to get to know one another (e.g., icebreakers, personal introductions, community Netiquette and other rules). Instructor communicates course expectations for engagement and provides orientation to the course
- **Communicate:** learner: cooperator; facilitator: structural engineer; weeks 3-4; Instructor forms student dyads and provides activities that require critical thinking, reflection, and sharing of ideas
- **Cooperate:** learner: collaborator; instructor: facilitator; weeks 5-6; Instructor provides activities that require small groups to collaborate, problem-solve, solve, and reflect on experiences
- **Collaborate:** learner: initiator/partner; instructor: community member/challenger; weeks 7-end of semester; Activities are learner-led. Discussions reflect group's learning and decisions. Group decides how they will present/share learning

Another influential model for online interaction is the 5-Stage Model by Gilly Salmon (2004). This model presents the "interaction effects" of e-moderating, the role of student use of technology, and student learning; it focuses on how the instructor can moderate, facilitate, or support student technology use. Please see the following summative description of the 5-Stage model.

- **Stage 1:** Access and motivation
  - **E-moderating:** Welcoming and encouraging
  - **Technology:** Setting up system and accessing
- **Stage 2:** Online socialization
  - **E-moderating:** Familiarizing and bridge-building (cultural, social, learning)
  - **Technology:** Sending and receiving messages

- **Stage 3:** Information exchange
    - **E-moderating:** Facilitating tasks and supporting use of learning materials
    - **Technology:** Searching, personalizing software
- **Stage 4:** Knowledge construction
    - **E-moderating:** Facilitating process
    - **Technology:** Conferencing
- **Stage 5:** Development
    - **E-moderating:** Supporting, responding
    - **Technology:** Proving links outside close conferences

The 21st century workplace calls for adults who can work well in collaborative problem-solving teams. In addition, the Millennium generation has been experiencing the power of collective knowledge (McLoughlin & Lee, 2008). Creating successful online teams is based on thoughtful and intentional planning instead of just the availability of emergent tools integrated into more traditional instructional approaches. Although much supporting evidence exists in the role of collaborative learning for student learning in face-to-face environments, research in examining collaborative learning as an instructional practice in online settings, is limited.

Challenges Associated with Collaborative Learning

There are several challenges associated with collaborative learning in an online environment. Aside from standard learner, instructor, and task factors, technology in itself, as well as the inherent restrictions of some LMS can become added barriers to the collaborative learning experience. Online learning requires certain learner characteristics that can compete with collaboration, depending on the characteristics and skills of the learner. For some online learners, collaboration can be challenging. Online learning trends show that the profile of the isolated, independent, place-bound learner which largely characterized the "classic distance education learner," is now being altered by the new generation's online learner and by socially mediated online learning activities that de-emphasize independent learning and stress social interaction and collaboration (Dabbagh, 2007).

Students often enroll in online courses because they are independent learners and thinkers, they have a certain level of knowledge and experience with technology, they have their own time management and learning styles, and they are goal-oriented. Many adult learners value self-reliance and individualism and often do not know how to collaborate with others. Others have difficulty providing or receiving feedback from peers. Collaborative learning calls for a shift in the learner's role. In a collaborative learning setting, the student will be an active participant, a listener, a problem-solver, will need to be socially present (Garrison & Anderson, 2003), and a co-developer of common group goals, rules, and tasks; sometimes, he or she will play the role of the group encourager, rule enforcer, note-taker, problem-solver, etc.

Research indicates that interpersonal and communication skills (which include writing skills) and fluency in the use of collaborative online learning technologies are critical competencies for the online learner (Dabbagh & Bannan-Ritland, 2005; Dabbagh, 2007). On the other hand, competency in communication and collaborative technologies does not ensure meaningful interaction, collaboration, and knowledge building in online learning environments (Lindblom-Ylanne & Pihlajamaki, 2003). It is important that online learners develop collaborative learning skills (e.g., conflict resolution, communication, perspective-taking, self-reflection, group reflection, self and group monitoring, and evaluation) independent of these technologies (Orvis & Lassiter, 2008; Graham & Misanchuk, 2004; Roberts, 2004).Other challenging issues with collaborative learning in online courses include, some students' aversion toward group work, the way groups are selected and formed, the lack of essential group skills by some, the free-rider, possible inequalities of student abilities, the withdrawal of certain group members, and the assessment of individuals within the group (Roberts & McInnerney, 2007).

Instructors will need to prepare learners for these role shifts, including the expectation to assume greater responsibility for their own learning. Collaboration implies that the course instructor will practice a model of gradual release of responsibility (Pearson & Gallagher, 1983) in the learning environment, will relinquish control, and will create relevant and meaningful tasks. Instructors have the responsibility of establishing clear objectives, explaining collaborative learning purposes and processes and preparing learners for success by clearly communicating clear expectations about group work, roles, outcomes of group work, the value assigned to group work, support systems, assessment, and monitoring of learning. Adult learners value challenging tasks that allow them to apply their knowledge, experiences, and learning style; it is important that learning tasks meet course learning goals and objectives. Collaboration requires planning, trust, security, community, support, reflection, and monitoring. It also requires intensity, directionality, and intentionality.

MAIN FOCUS OF THE CHAPTER

Course Description

This fully online course is offered each semester at the University of Central Florida (UCF), College of Education, for preservice teachers in initial certification programs (P-12) and inservice elementary and secondary grade educators pursuing graduate studies. It meets state policy for teacher preparation in eight graduate programs (i.e., elementary education, reading education, secondary education (English language arts, etc.), exceptional education, school psychology, school counselor, communicative disorders, and media education), plus initial certification (preservice education). First, the author will share information about the instructional design elements, content development guidelines, delivery, and pedagogy involved with this comprehensive course teacher education course. Secondly, the author will present evidence through student work samples and assessments designed to promote collaborative learning in teacher education programs. Lastly, the author will discuss the role of collaborative learning in building and maintaining a professional learning community and challenges associated with collaborative learning in teacher education online courses.

The course is 14 weeks long and it is offered each semester; the author transformed the face-to-face course into a fully online course a few years ago. The author introduced collaborative learning elements to the course three years ago and since then has been modifying and adjusting collaborative learning activities using student feedback. Major collaborative learning adjustments include designing tasks that will have (a) a "real-classroom" feel for prospective P-12 teachers, tasks that are authentic, job-embedded and relevant; (b) opportunities for preservice teachers to develop their instructional decision-making skills; (c) flexible and allow for student choice; (d) promote self-regulation of learning; and lastly, (e) encourage ongoing collaboration with other preservice teachers throughout the duration of the course. On average, 30 preservice and inservice teachers enroll in the course on a semester basis; 80% of the students are preservice teachers and 20% are inservice teachers and other future educators (i.e., students in school psychology, counselor education, and communicative disorders, exceptional education, and educational programs). Its content focuses on reading development, research, and instruction, for preservice or inservice teachers in grades P-12 and other educators who will be working in elementary and secondary school settings. As part of the *No Child Left Behind Public Act of 2011*, *Teacher Quality* federal initiative, preparation of educators in reading has been an integral part of teacher education preparation programs. This accounts for the diverse student demographics for this course. As result, the wide student diversity, disciplines, and needs represented in this course present a unique challenge to its development and instruction. The course is designed in such a way that course objectives, readings, assignments, activities, and assessments, meet the program and student needs from each discipline.

The learning management system for the course (LMS) was Blackboard, the official LMS for UCF. The author has developed all course content for this course. The course included many multi-media elements and the author utilized Adobe Connect for synchronous course meetings and discussions. The author also used a wide array of other Web 2.0 presentation tools (e.g., Prezi, Slideshare, Glogster, Wordle), video tools (e.g., Animoto, XtraNormal), and community tools (e.g., Google Docs, Wikispaces). Students in the course used additional tools (e.g., social networking tools, exchanged phone numbers, used AIM/ichat, etc.).

Collaborative Learning: A Situated Perspective

Because collaboration is important to teacher and student learning, the author decided to create various structured collaborative learning experiences for students in this course for the following purposes: a) to maintain the learning community; b) to stimulate further interactions among students of various backgrounds and disciplines; c) to promote deeper knowledge development beyond the standard online discussions and synchronous and asynchronous communications; d) to facilitate reflection and critical thinking skills; and, e) experience first-hand the benefits and challenges of collaboration. In planning for collaborative learning in this online teacher education course, the author implemented elements of the CoI model (Garrison, Anderson, & Archer, 2000), the Phases of Engagement Model (Conrad & Donaldson, 2004), and the 5-Stage Model by Salmon (2004). The author created a culture of collaboration in her course by creating authentic tasks, designing learning experiences that involved collaborative work and reflective thinking, guiding student learning, modeling task performance, and providing scaffolded support throughout the learning process.

Collaborative learning took place in the following ways in this course:

- Course discussions boards—coffee shop, weekly online discussion posts to readings (asynchronous communication and interactions), and replies to each other's postings, literacy logs (i.e., students' responses to effective reading strategies), questions, and ideas.
- Online chats using course chat feature (chats were used by instructor and students; students used the chat feature to work in their groups, set-up book group meetings, review course assignment expectations, etc.).
- Synchronous communication/interactions via Adobe Connect (e.g., instructor presentations, discussions with students on course assignments, discussions on course issues as they arose). Skype was also used as needed by instructor and students.

Collaborative learning course tasks included the following.

A Professional Book Group

In addition to the online course content, the course instructor (author) selected four teacher professional books for students to choose to study and reflect upon during the course. The author chose the books based on student characteristics (i.e., program area of study) and on current reading research and instruction issues for grades P-12. Aside from the introductory, community-building tasks of the first week of classes, by the end of that first week of classes, students had to select a book they would like to study during the semester. The instructor asked students to rank their choice from one to four, post their choices in a specific area in the discussion board. The instructor explained that depending on how many people signed up per group she would place students in groups of four and create as many groups as needed per book choice.

The group size was kept small intentionally for the purpose of maximizing student-to-student interactions. Choice of the book was left up to each student but the requirement was mandatory. The instructor (author) presented information on each book, offered suggestions on why books were selected by her, how they would benefit the students, and how she included specific descriptions about the task, the process of collaboration, the value added to this experience, rubrics, examples, and step-by-step instruction on the entire process. Explanations were included in the course (Module 1—Introduction, in the course syllabus, and they were also discussed periodically during instructor meetings and through instructor communications and reminders throughout the course.) By the first day of the second week of classes, the instructor had posted group placement and had emailed each individual group about group composition and book they would be working with throughout the semester.

Upon receiving group placement, the next task students had to complete was to contact one another in their respective groups, introduce themselves again (each student had already done a formal introduction during week one of the course), and decide on group roles. The instructor designed specific student-led group roles for this collaborative task. Each group decided how to communicate with group members. The roles were as follows: a) Group facilitator: Kept group on track and facilitated discussion based on provided questions; b) Group organizer: Sent reminders to group about upcoming chat; c) Chat note-taker: Took any notes to discuss at a later date or in the discussion room. Notes were posted in the discussion room; and, d) Chat cheerleader/time manager: Every group needs an encourager! This group member helped to create a positive atmosphere among group members and also helped to keep everyone on task and meeting deadlines.

Three Online Chats (Synchronous or Asynchronous) Set and Offered by Each Group

According to the course schedule, each group was expected to conduct three formal, public online meetings/chats on their book. Chat one was to take place during week four, chat two during week six, and chat three during week eight. It was important that most of these collaborative activities took place at the beginning through the middle of the course with the hope that relationships that were formed between students remained throughout the course. The instructor had broken each book into manageable sections and had also written guiding questions to be answered by each group during each meeting. Group members had the freedom to go beyond the assigned questions and discuss other topic or course-relevant issues. These meetings were designed for student-student collaborations. The instructor had explained (and also stated in syllabus and course) that she would visit various group meetings but did not guarantee attendance. Groups knew that these online meetings were to be held by them, for them, and for interactions with other course members who could attend their meeting. Students used their own ways of communicating and making group decisions; they met several times to negotiate on schedules, deadlines, and discuss their book—some groups used Wikispaces, other Google Docs, Skype, and others, even met face-to-face for coffee at various locations and at the university library.

Group attendance, monitoring, participation, and development of a report on the content of the chat and the group's thoughts, discussions, and questions, was to be sent to the course instructor within 24 hours from the completion of the chat by one student per group (usually the note-taker compiled notes for the report and submitted it online). As part of the report, students had to also state who attended the chat, who submitted group responses and ideas, questions about the material, questions and thoughts from the group and others to the instructor, and which member played which role (i.e., note-taker, chat monitor, etc.). All of these tasks were left entirely up to the students.

The instructor provided a template for the report students could use to summarize their thoughts and ideas. Students would be evaluated on the content of their report (how thoroughly they examined assigned questions from the book) and their attendance to the meeting. The instructor provided a rubric for assessing and evaluating their work; particular emphasis was placed on evidence-based and reflective thinking, critical evaluation of material, Netiquette, respect and collaboration with one another and on following a communication protocol.

A Group-Generated Book Project

As a final group project, each group was expected to develop and share with the class a group presentation on the book they studied, discussed, and reflected upon throughout the course. Groups got together numerous times for the production of their final project. The instructor provided students with a rubric for this task and with examples and tools they could use; each group decided on the choice of presentation tool and content of presentation. Various groups used Publisher to create flyers on their book, others developed a newsletter, and others used Prezi, Glogster, Wordle, and digital storytelling. Students used their creativity to develop, present, and promote a relevant book they had worked on collaboratively during the semester. Groups shared project development responsibilities and were accountable for meeting deadlines, sharing ideas, and developing a final project each group member agreed upon.

Issues, Controversies, and Problems

In order to promote collaborative learning in this online course, the author designed a situated model of collaboration. Literature shows that collaboration is not an automatic result of placing people in groups. There are certain conditions that need to be met in order for collaboration to be developed and maintained in an online setting. In this case, the author used progressive scaffolds and provided guided supports for collaborative learning. First of all, collaboration cannot be "caught"; it has to be planned with intentionality (purpose), directionality, and intensity. Students need to have clear expectations about goals and outcomes, about their roles, and about what is allowed and what is not. In this content, the instructor provided students with "guided" collaborative tasks, assumed the role of facilitator throughout the process, monitored student progress and outcomes, provided feedback as needed on what worked and what did not work with each group's collaborative efforts, and placed value on collaborative tasks beyond the regular, expected types of interactions that "naturally" occur in an online course. To have collaboration, one must plan for collaboration (and be also ready to deal with challenges that will arise).

Collaboration could not have happened without the existence of explicit expectations, support systems, and flexibility (on the part of the instructor and the students). Throughout the course, the instructor explicitly modeled, supported, and guided collaborative learning. Rubrics, examples of tasks/outcomes, step-by step instructions, time to reflect and adjust tasks and processes as needed, flexibility, reflection, and social and teaching presence are important instructor requirements for student collaboration.

Collaboration did not work well for all groups; some became frustrated when others did not fulfill their responsibilities, or came to a meeting unprepared, or did not submit their work on time, which in turn, caused the entire group report to be submitted to the instructor late. Some students disappeared for a while and their group members became frustrated with spending so much time trying to reach them, etc. Group dynamics and conflict resolution were left to the group to work with. The instructor intervened with suggestions only in extreme cases (this happened at least once per semester with a group when two out of the four group members were not participating and the instructor had to work with the group to redesign the group roles). The instructor also intervened and provided support to students when a group member did not display essential group-work social skills.

Information on how to make group work successful was available to students throughout the course. The instructor gave class presentations on collaboration, provided students with information on communication tools and strategies, shared tips from lessons learned, and addressed issues through course e-mail and chats. Close monitoring of the collaborative learning process with each group helped the instructor to provide students with relevant and specific feedback.

Another factor that contributed to the positive collaborative learning experiences was the group size and formation. Choice is important to adult learners; students had choice over which book to choose to study and discuss in a group and they also had choice over which role to fulfill in each group, when to set-up their meetings, and how to present their final project outcome. Keeping the group size small helped the students get to know each other easier and better (as they stayed in the same group for 13 weeks), spend less time turning to meet schedule restrictions, and spending more time with one another—it made collaborative learning tasks more manageable. Once students chose the role they'd like to play in the group, they had clear expectations of what was expected of them. Group members helped keep each other accountable in the process. Another advantage of the group was that its members had shared interests in a topic. In addition, the small group size helped the instructor to monitor and evaluate each group's activities and collaborative learning outcomes.

By the end of the course, students were commenting on how much they learned from one another and from the collaborative learning tasks (overall, 90% of the students each semester report that they enjoy and learn from the collaborations with one another). Students were very vocal about what is working and what is not working with collaboration; they also shared across groups and received tips and suggestions from others on various group-related tasks. In order for collaboration to happen, the instructor has to establish the environment, model desired outcomes, be present to respond to student needs, provide relevant feedback and guidance, and evaluate the process. It is important that students know the outcomes of their activities and efforts. Students will not generate knowledge in the absence of community, support, choice, flexibility, relevance, cooperation, and collaboration.

As part of this course, students had a number of assignments and requirements. The professional book group meetings and project were just two of the overall course assignments. But, what happened as a result of these collaborations, students started to respond more to their group members' postings in discussion boards, offer more feedback, interact more with one another, share more resources relating to different course topics, talking more openly about their own experiences with teaching or preparation to teach. In addition, they developed a better understanding of the roles and responsibilities each student would have to play in P-12 students' learning. Students better understood the role a school psychologist would play in a Response to Intervention (RtI) school team and how important it is that all educators collaborate, share data, ideas, and come together to solve instructional and other problems that would help to improve student learning. Learning to learn collaboratively often involves a shift in one's views of teaching and learning (Dirkx & Smith, 2004).

Solutions and Recommendations

There are many lessons to be shared from the integration of collaborative learning activities in this online teacher education course. Interactions can be rich and sustained over time when students feel safe, when their voices are valued, when they are given choices, and are encouraged (and are expected) to become in charge of their own learning. Rigor, relevance, and relationships are important prerequisites to effective online collaborative learning. The course design facilitated interactions between the course instructor and students and also among students through multiple communication methods, mandatory instructor and student-led synchronous (and asynchronous) meetings, provision of specific deadlines, rubrics, examples, benchmarks, and weekly reminders. All of these design elements are particularly useful to future teachers who will be handling deadlines, collaboration and communication with multiple audiences, and planning effective instruction using a variety of benchmarks and objectives.

Teacher educators who teach online courses need to provide the functions preservice teachers need to collaborate and learn through collaboration (Kirschner, Strijbos, Kreijns, & Beers, 2004). In this course, the process of collaboration was supported through advanced planning, the group formation process (i.e., guidelines, group placement, group size, group roles, helping match students with books), allowing time for group members to get to know each other before they have to produce an artifact, introduce students to communication tools and group work products, monitoring of group collaborations, providing rubrics for the different activities, and also providing students with means for self and group evaluation. The author purposely modeled collaborative learning, scaffolded its elements, provided feedback, and monitored throughout the process. Preservice teachers have to "feel" collaborative learning and its advantages and disadvantages for them as learners and also as future teachers.

Heterogeneous small collaborative groups can be productive (Clark & Mayer, 2008) and they can engage in more critical discourse (Lee, 2008). Group diversity produces different perspectives that can either support of inhibit the collaborative learning process (Posey, 2007). Asynchronous communications between instructor-students and among students facilitate reflection and synchronous communications help to enhance social presence, community building, and even conflict resolution (Harvard, Du, & Xu, 2008). Regular interactions among group members contribute to trust, community building, and group performance (Orvis & Lassiter, 2008). Providing group members with group roles can contribute to individual accountability (Kirschner et al., 2004).

Collaborative assignments should be assessed collaboratively (i.e., self-assessment, peer assessment, and instructor assessment) (Palloff &Pratt, 2007). Student ownership of tasks will happen over time if student learning is scaffolded, individual perspectives and creativity are valued, support systems are readily available, and the course instructor is socially present throughout the learning process.

Building a positive, active, and collaborative online learning community requires time, advanced planning, attention, and monitoring. Collaborative learning is not necessarily easier learning—some have indicated that online group work may be perceived as more challenging than group work in face-to-face settings (Kim, Liu, & Bonk, 2005).Collaborative learning should be relevant, achievable, engaging, and motivating. Building positive relationships among students, using groups as a platform for problem-solving complex learning tasks, designing relevant quality learning tasks, helping students with communication skills and motivation, and assessing and evaluating collaborative learning, when properly modeled and supported can advance the students' online learning experiences.

Because collaborative learning in online courses has its own challenges (e.g., takes time from instructor to plan, model, support and monitor; students have to negotiate with other students for completion of assignment; not everyone knows how to or wants to collaborate, etc.), instructors should build time and a system for promoting and maintaining student socialization in the collaboration process; they should also provide support to new online learners. Students will benefit from examples and feedback on how to resolve conflict resolution and interpersonal teamwork challenges such as lack of participation, the free-rider, inequalities in ability, and other personality clashes. Collaborative learning can be used as a teaching and a learning activity. The likelihood of successful achievement of learning objectives and achieving course competencies increases through collaborative engagement (Palloff, & Pratt, 2005).

FUTURE RESEARCH DIRECTIONS

The number of students enrolling in higher education distance programs is steadily increasing in colleges and universities in the US and abroad. As a result, many institutions have been developing plans to either implement or improve online education. Technology systems to support online education, the training and compensation of online instructors, the needs of today's online students, and the evaluation and assessment of online learning, all present ongoing challenges to online learning. In an era of e-books, social networking, video streaming, online testing, mobile devices and learning, budgetary constraints, wide range of online technologies to incorporate into online instruction, the rise of blended learning, and other emerging technologies, instructors and institutions have to rethink online learning. The future of online learning will be effected by learner needs, technology, pedagogy, and learning paradigms (Bonk, 2004; Kim, Bonk, & Zeng, 2005).

Peer-to-peer collaboration, especially as it related to knowledge-transmission, plays a critical role in student learning and satisfaction. There is a need for online environments that facilitate deeper student learning, reflection, real-world problem solving, and engagement (Bonk, Wisher, & Lee, 2003). Online courses should promote learner-centered experiences that involve students in critical thinking, peer learning, and interdisciplinary experiences (Weigel, 2005). In addition, online teacher education courses that involve preservice and inservice teachers in using technology will help teachers value the positive effects of technology on student engagement with learning and its connection to 21st century skills. Online collaborative learning is not only beneficial to preservice teachers as learners, but it will also prove to be useful for their future students. The US Secretary of Education, Arne Duncan stated the following in a recent speech (March 3, 2010) about educators, technology, and 21st century skills:

In the 21st century, students must be fully engaged. This requires the use of technology tools and resources, involvement with interesting and relevant projects, and learning environments—including online environments—that are supportive and safe... In the 21st century, educators must be given and be prepared to use technology tools; they must be collaborators in learning—constantly seeking knowledge and acquiring new skills along with their students.

In order for teachers to incorporate technology in their classroom instruction and foster 21st century skills, they have to experience technology use as part of their own learning. Preservice teachers need to learn in a culture of collaboration, discussion, and reflection in their face-to-face and online courses. Online instructors, researchers, and administrators, must continue to examine pedagogical issues in teacher education online learning. More research is needed in how online learning can develop preservice teachers' content, pedagogical, and 21st century skills such as collaboration, critical thinking, personal expression and negotiation skills, and evaluation skills. Technology and 21st century skills will help teacher education programs to achieve critical educational outcomes. Teacher education programs have a responsibility to make technology a core aspect of teacher training and education so teacher educators can model preservice and inservice teachers can use technology in their classrooms to meet their student needs.

A process that involves inquiry confronts the unknown and relies on personal or collective resources to resolve questions. The online environment in which inquiry can flourish is gradually built by collaborative and collective contributions. Such collaboration efforts are likely to result in better outcomes, designs, practices, or products (Collison, Elbaum, Haavind, & Tinker, 2000, p. 30).

CONCLUSION

In the context of global educational collaborations, online learning has the potential to revolutionize higher education. Although results from many studies indicate that online courses can be as effective, if not more so, than face-to-face courses (Means, Toyama, Murphy, Bakia, & Jones, 2009) and although students generally have positive attitudes toward online learning, its use is not without challenges. Some teacher educators and policy makers object to online instruction in teacher preparation programs (Midobuche & Benavides, 2006); they call for the exclusive use of face-to-face experiences for teacher candidates. There is a need for research that will examine the effectiveness of online learning specifically on the preparation of teachers. What types of online instruction and experiences will have the most impact in helping develop preservice teachers' knowledge, skills, and dispositions? Would preservice teachers who participate in authentic collaborative learning experiences develop certain decision-making skills that will serve them well throughout their career? On the other hand, as more and more two year and four year institutions offer more and more online routes for teacher certification, the question remains about what types of teachers such programs will produce.

Other online learning and teaching implications for teacher educators and administrators include learning how to adapt to new learning environments by knowing how to turn face-to-face courses into online ones, creating appropriate adjustments and providing resources for students (Tallent-Runnels, Thomas, Lan, Cooper, Ahern, Shaw, & Liu, 2006), maximizing communication in online learning, designing meaningful collaborative learning experiences, and providing reflective practice to meet learners' and institutions' needs (Palloff & Pratt, 1999). Students in online courses must also shift their perspectives about how learning "happens" and how knowledge is acquired and developed in a 21st century online course. Higher education institutions, and especially colleges of education, must broaden their vision about how they can utilize online learning to "produce" teachers who will be well equipped (in content, pedagogy, and dispositions) to meet the needs of their students in grades P-12. Teacher educators must resist the pressure to just develop quick online courses that are just placing a textbook online. Effective online learning, and collaborative learning, in particular, requires rigor, planning, resources, instructor expertise, technology infrastructure, and ongoing monitoring of student learning. Learning about effective online pedagogical practices and innovative uses of technology is important for teacher educators who teach online courses.

For meaningful, sustained, relevant, and thought-provoking interactions to take place in online courses, student to student and instructor to student interactions in an online environment have to move beyond the discussion boards (Al-Bataineh, Brooks, & Bassoppo-Moyo, 2005). Assignments that require group participation can be enjoyable and effective in online courses (Palloff & Pratt, 2005). Communication between group members working on a significant course project can help promote sustained collaboration (Palloff & Pratt, 2009) as students work together to develop the project ideas, schedule chat/meeting times, assign roles and responsibilities, and negotiate on meaning.

In this chapter, the author described how collaborative learning was intentionally (and intensively) designed and incorporated in a graduate level online course in reading for preservice and inservice educators. Observations from several semesters of implementation reflected benefits on student learning, class community, and teacher preparation. Challenges associated with this type of collaborative learning still remain in the course but ongoing monitoring and student input have helped the instructor to streamline many of the related processes and make the collaborative experiences smoother, clearer, and more student-centered.

For collaborative learning to be productive, the instructor must, first and foremost, work throughout the duration of the course (and not only at the beginning) to develop and sustain a positive, safe, and professional learning community. Engaging students in some type of socialization process that meets their academic goals and personal interests contributes toward a community of learners. The author believes that teacher educators who plan for online collaborative learning experiences must begin with the student in mind. If the goal is student involvement, problem-solving, synthesizing evidence from multiple sources, and deep comprehension of texts/content, then, student activity, interactivity, and engagement with learning throughout the course become prototypical course design and instructional elements. Observations from teaching this collaborative course over the past few years have shown that advanced planning, resources, clear and ongoing communication, community-building, student choice, relevant and authentic assignments and materials, rubrics, support and guidance, instructor social presence, flexibility, and ongoing monitoring and evaluation contribute to the success of collaboration among students and between instructor and students in this course. Collaboration is a core daily activity for every educator, and although it has its own challenges and demands when used as learning tool in an online learning environment (Zygouris-Coe & Swan, 2010), it still carries value for the training of teachers.

While flexibility and convenience still remain top reasons for students' enrollment in online courses (Palloff & Pratt, 2009), time spent in active and meaningful learning can be another time saver in the long run. Teacher education (and other) online courses presented in interactive, collaborative, authentic and relevant project-based formats can be both enjoyable and productive experiences in students' academic lives. Participating in collaborative learning experiences online will help teacher candidates and other educators to not only use collaboration as a learning tool for them as learners but to also use it as a learning tool for developing their students' 21st century skills. Building a positive, active, and collaborative online learning community requires time, advanced planning, attention, and monitoring but it can also shape preservice and inservice teachers' instruction. There is a need for monitoring effective student-to student interactions in online collaborative groups and assessing the impact of collaborative learning on student learning (Calvani, Fini, Molino, & Ranieri, 2010). Lastly, colleges of education need to provide teacher educators with the training, resources, and support necessary for them to continue to explore new ways to make online learning more meaningful, relevant, enjoyable, and productive for preservice and inservice teachers.

REFERENCES

Al-Bataineh, A., Brooks, S. L., & Bassoppo-Moyo, T. C. (2005). Implications of online teaching and learning. *International Journal of Instructional Media, 32*(3), 285–294.

Allen, I. E., & Seaman, J. (2010). *Class differences: Online education in the United States, 2010.* The Alfred P. Sloan Foundation: Author. Retrieved from http://sloanconsortium.org/publications/survey/class_differences

American Association of Colleges for Teacher Education and The Partnership for 21st Century Skills. (2010). *21st century knowledge and skills in educator preparation.* Upper Saddle River, NJ: Pearson.

Barkley, E. F., Cross, P. K., & Major, C. H. (2005). *Collaborative learning techniques. A handbook for college faculty*. San Francisco, CA: Jossey-Bass.

Bonk, C. J. (2004). *The perfect E-storm: Emerging technologies, enhanced pedagogy, enormous learner demand, and erased budgets*. London, UK: The Observatory on Borderless Higher Education.

Bonk, C. J., Wisher, R. A., & Lee, J. Y. (2003). Moderating learner-centered e-learning: Problems and solutions, benefits and implications. In Roberts, T. S. (Ed.), *Online collaborative learning: Theory and practice* (pp. 54–85). Hershey, PA: Information Science Publishing. doi:10.4018/978-1-59140-174-2.ch003

Calvani, A., Fini, A., Molino, M., & Ranieri, M. (2010). Visualizing and monitoring effective interactions in online collaborative groups. *British Journal of Educational Technology, 41*(2), 213–226. doi:10.1111/j.1467-8535.2008.00911.x

Clark, R. C., & Mayer, R. E. (2008). *E-learning and the science of instruction: Proven guidelines for consumers and designers of multimedia learning* (2nd ed.). San Francisco, CA: Wiley.

Collison, G., Elbaum, B., Haavind, S., & Tinker, R. (2000). *Facilitating online learning*. Madison, WI: Atwood Publishing.

Conrad, R. M., & Donaldson, J. A. (2004). *Engaging the online learner: Activities and resources for creative instruction*. San Francisco, CA: Jossey-Bass.

Dabbagh, N. (2007). The online learner: Characteristics and pedagogical implications. *Contemporary Issues in Technology & Teacher Education, 7*(3), 217–226.

Dabbagh, N., & Bannan-Ritland, B. (2005). *Online learning: Concepts, strategies, and application*. Upper Saddle River, NJ: Prentice Hall.

Darling-Hammond, L., Hammerness, K., Grossman, P., Rust, F., & Shulman, L. (2005). The design of teacher education programs. In Darling-Hammond, L., & Bransford, K. (Eds.), *Preparing teachers for a changing world: What teachers should learn and be able to do* (pp. 390–440). San Francisco, CA: Jossey-Bass.

Dede, C., Ketelhut, D. J., Whitehouse, P., Breit, L., & McCloskey, E. M. (2009). A research agenda for online teacher professional development. *Journal of Teacher Education, 60*, 8–19. doi:10.1177/0022487108327554

Dewey, J. (1997). *Democracy and education: An introduction to the philosophy of education*. New York, NY: The Free Press. (Original work published 1916)

Dirkx, J., & Smith, R. (2004). Thinking out of a bowl of spaghetti: Learning to learn in online collaborative groups. In Roberts, T. S. (Ed.), *Online collaborative learning: Theory and practice* (pp. 132–159). Hersey, PA: Information Science Publishing.

Dooley, K., Lindner, J., Dooley, L., & Wilson, S. (2005). Adult learning principles and learner differences. In Dooley, K., Lindner, J., & Dooley, L. (Eds.), *Advanced methods in distance education: Applications and practices* (pp. 56–75). Hershey, PA: Information Science Publishing. doi:10.4018/978-1-59140-485-9.ch004

Duffy, G. G., Webb, S., & Davis, S. G. (2009). Literacy education at a crossroad: A strategy for countering the trend to marginalize quality teacher education. In Hoffman, J. V., & Goodman, Y. (Eds.), *Changing literacies for changing times* (pp. 189–197). New York, NY: Routledge, Taylor and Francis.

Duncan, A. (March 3, 2010). *Using technology to transform schools: Remarks by Secretary Arne Duncan at the Association of American Publishers Annual meeting*. Retrieved from http://www.ed.gov/news/speeches/using-technology-transform-schools%E2%80%94remarks-secretary-arne-duncan-association-american-

Garrison, D. R., & Anderson, T. (2003). *E-learning in the 21st century: A framework for research and practice*. London, UK: Routledge. doi:10.4324/9780203166093

Garrison, D. R., Anderson, T., & Archer, W. (2000). Critical inquiry in a text-based environment: Computer conferencing in higher education. *The Internet and Higher Education*, *2*(2–3), 1–19.

Graham, C. R., & Misanchuk, M. (2004). Computer-mediated learning groups: Benefits and challenges to using group work in online learning environments. In Roberts, T. S. (Ed.), *Online collaborative learning: Theory and practice* (pp. 181–214). Hershey, PA: Information Science Publishing.

Gunawardena, C. N. (2005). *Social presence and implications for designing online learning communities*. Paper presented at the Fourth International Conference on Educational Technology, Nanchang, China. Retrieved from http://www.edu.cn/include/new_jiaoyuxxh/xiazai/gunawardena.ppt

Harvard, B., Du, J., & Xu, J. (2008). Online collaborative learning and communication media. *Journal of Interactive Learning Research*, *19*(1), 37–50.

Johnson, D. W., Johnson, R. T., & Smith, K. A. (1998). *Active learning: Cooperation in the college classroom*. Edina, MN: Interaction Book Company.

Kim, K.-J., Bonk, C. J., & Zeng, T. (2005, June). Surveying the future of workplace e-learning: The rise of blending, interactivity, and authentic learning. *E-Learn Magazine, 6*. Retrieved from http://elearnmag.acm.org/archive.cfm?aid=1073202

Kim, K.-J., Liu, S., & Bonk, C. J. (2005). Online MBA students' perceptions of online learning: Benefits, challenges and suggestions. *The Internet and Higher Education*, *8*(4), 335–344. doi:10.1016/j.iheduc.2005.09.005

Kirschner, P. A., Strijbos, J. W., Kreijns, K., & Beers, P. J. (2004). Designing electronic collaborative learning environments. *Educational Technology Research and Development*, *52*(3), 47–66. doi:10.1007/BF02504675

Knowles, M. (1980). *The modern practice of adult education*. Chicago, IL: Association Press.

Knowles, M. (1984). *The adult learner: A neglected species* (3rd ed.). Houston, TX: Gulf Publishing Co.

Knowles, M. (1986). *Using learning contracts: Practical approaches to individualizing and structuring learning*. San Francisco, CA: Jossey-Bass.

Knowles, M. S., Swanson, R. A., & Holton, E. F. III. (2005). *The adult learner: The definitive classic in adult education and human resource development* (6th ed.). CA: Elsevier Science and Technology Books.

Lewin, K. (1935). *A dynamic theory of personality*. New York, NY: McGraw Hill.

Li, C., & Irby, B. (2008). An overview of online education: Attractiveness, benefits, challenges, concerns and recommendations. *College Student Journal, 42*, 449–458.

Lindblom-Ylanne, S., & Pihlajamaki, H. (2003). Can a collaborative network environment enhance essay-writing processes? *British Journal of Educational Technology, 34*(1), 17–30. doi:10.1111/1467-8535.00301

Lowell, N. (2006). Collaborating online: Learning together in community. *Quarterly Review of Distance Education, 7*(2), 211–214.

MacGregor, J. (1990). Collaborative learning: Shared inquiry as a process of reform. In Svinicki, M. (Ed.), *The changing face of college teaching, New Directions for Teaching and Learning No. 42*. San Francisco, CA: Jossey-Bass. doi:10.1002/tl.37219904204

McLoughlin, C., & Lee, M. J. W. (2008). Future learning landscapes: Transforming pedagogy through social software. *Innovate: Journal of Online Education, 4*(5). Retrieved from http://innovateonline.info/?view=article&id=539

Means, B., Toyama, Y., Murphy, R., Bakia, M., & Jones, K. (2009). *Evaluation of evidence-based practices in online learning: A meta-analysis and review of online learning studies*. Washington, DC: U.S. Department of Education. Retrieved from http://www.ed.gov/rschstat/eval/tech/evidence-basedpractices/finalreport.pdf

Midobuche, E., & Benavides, A. H. (2006). Preparing teachers to teach English language learners: Best practices for school and after-school programs. In Cowart, M., & Dam, P. (Eds.), *Cultural and linguistic issues for English language learners* (pp. 83–107). Texas Woman's University: Cahn Nam Publishers, Inc.

Norton, P., & Hathaway, D. (2008). On its way to classrooms, Web 2.0 goes to graduate school. *Computers in the Schools, 25*(3), 163–180. doi:10.1080/07380560802368116

Orvis, K. L., & Lassiter, A. L. R. (2008). *Computer-supported collaborative learning: Best practices and principles for instructors*. Hersey, PA: Information Science Publishing. doi:10.4018/978-1-59904-753-9

Ota, C. DiCarlo, C., Burts, D., Laird, R., & Gioe, C. (2006, December). Training and the needs of adult learners. *Journal of Extension, 44*(6). Article Number: 6TOT5. Retrieved from www.joe.org/joe/2006december/tt5.shtml

Palloff, R., & Pratt, K. (2007). *Building online learning communities: Effective strategies for the virtual classroom*. San Francisco, CA: Jossey-Bass.

Palloff, R. M., & Pratt, K. (2005). *Collaborating online: Learning together in community*. San Francisco, CA: Jossey-Bass.

Palloff, R. M., & Pratt, K. (2009). *Assessing the online learner: Resources and strategies for faculty*. San Francisco, CA: Jossey-Bass.

Pascarella, E. T., & Terenzini, P. T. (2005). *How college affects students: A third decade of research* (*Vol. 2*). San Francisco, CA: Jossey-Bass.

Pearson, P. D., & Gallagher, M. C. (1983). The instruction of reading comprehension. *Contemporary Educational Psychology, 8*, 317–344. doi:10.1016/0361-476X(83)90019-X

Piaget, J. (1969). *The mechanisms of perception*. London, UK: Routledge and Kegan Paul.

Posey, L. (2007). Critical thinking and collaboration in post-secondary online education. In T. Bastiaens & S. Carliner (Eds.), *Proceedings of World Conference on E-Learning in Corporate, Government, Healthcare, and Higher Education 2007* (pp. 1770-1774). Chesapeake, VA: AACE.

Report, H. (2008). *The Horizon Report: 2008 edition*. A collaboration between The New Media Consortium and the EDUCAUSE Learning Initiative (ELI), an EDUCAUSE program. Retrieved from http://www.nmc.org/pdf/2008-Horizon-Report.pdf

Roberts, T. S., & McInnerney, J. M. (2007). Seven problems of online group learning (and their solutions). *Journal of Educational Technology & Society*, *10*(4), 257–268.

Rourke, L., & Kanuka, H. (2009). Learning in communities of inquiry: A review of the literature. *Journal of Distance Education*, *23*(1), 19–48.

Salmon, G. (2004). *E-moderating: The key to teaching and learning online* (2nd ed.). London, UK: Routledge.

Saroyan, A., & Amundsen, C. (Eds.). (2004). *Rethinking teaching in higher education*. Sterling, VA: Stylus Press.

Schmuck, R. A., & Schmuck, P. A. (2004). *Group processes in the classroom* (8th ed.). Madison, WI: Brown & Benchmark.

Sener, J. (2010). Why online education will attain full scale. *Journal of Asynchronous Learning Networks*, *14*(4), 3–16.

Sheridan, J. (1989). Rethinking andragogy: The case for collaborative learning in continuing higher education. *Journal of Continuing Higher Education*, *37*(2), 2–6. doi:10.1080/07377366.1989.10401167

Swan, K., Garrison, D. R., & Richardson, J. C. (2009). A constructivist approach to online learning: The Community of Inquiry framework. In Payne, C. R. (Ed.), *Information technology and constructivism in higher education: Progressive learning frameworks* (pp. 43–57). Hershey, PA: Information Science Publishing. doi:10.4018/978-1-60566-654-9.ch004

Swan, K., & Ice, P. (2010). The community of inquiry framework ten years later: Introduction to the special issue. *The Internet and Higher Education*, *13*(1-2), 1–4. doi:10.1016/j.iheduc.2009.11.003

Tallent-Runnels, M. K., Thomas, J. A., Lan, W. Y., Cooper, S., Ahern, T. C., Shaw, S. M., & Liu, X. (2006). Teaching courses online: A review of the research. *Review of Educational Research*, *76*(1), 93–135. doi:10.3102/00346543076001093

Thomson, D. L. (2010). Beyond the classroom walls: teachers' and students' perspectives on how online learning can meet the needs of gifted students. *Journal of Advanced Academics*, *21*, 662–712. doi:10.1177/1932202X1002100405

Vygotsky, L. (1930/1978). *Mind in society*. Cambridge, MA: Harvard University Press.

Weigel, V. (2005). From course management to curricular capabilities: A capabilities approach for the next-Generation CMS. *EDUCAUSE Review*, *40*(3), 54–67.

Weimer, M. (2002). *Learner-centered teaching*. San Francisco, CA: Jossey-Bass.

Wise, A., Chang, J., Duffy, T., & del Valle, R. (2004). The effects of teacher social presence on student satisfaction, engagement, and learning. *Journal of Educational Computing Research*, *31*(3), 247–271. doi:10.2190/V0LB-1M37-RNR8-Y2U1

Young, E. E., Grant, P. A., Montbriand, C., & Therriault, D. J. (2001). *Educating preservice teachers: The state of affairs*. Naperville, IL: North Central Regional Educational Laboratory.

Zygouris-Coe, V., & Swan, B. (2010). Challenges of online teacher professional development communities: A statewide case study in the United States. In Lindberg, J. O., & Oloffson, A. D. (Eds.), *Online learning communities and teacher professional development: Methods for improved education delivery* (pp. 114–133). Hershey, PA: Information Science Publishing.

ADDITIONAL READING

Barkley, E. F., Cross, P. K., & Major, C. H. (2005). *Collaborative learning techniques. A handbook for college faculty*. San Francisco, CA: Jossey-Bass.

Bates, A., & Sangra, A. (2011). *Managing technology in higher education: Strategies for transforming teaching and learning*. San Francisco: Jossey-Bass.

Bonk, C. J. (2009). *The world is open: How web technology is revolutionizing education*. San Francisco, CA: Jossey-Bass.

Bonk, C. J., & Graham, C. R. (Eds.). (2006). *Handbook of blended learning: Global perspectives, local designs*. San Francisco, CA: Pfeiffer Publishing.

Bonk, C. J., & Zhang, K. (2008). *Empowering online learning: 100+ activities for reading, reflecting, displaying, and doing*. San Francisco, CA: Jossey-Bass.

Chih-Hsiung, T. (2004). *Collaborative learning communities: Twenty-one designs to building an online collaborative learning community*. Westport, CT: Library Unlimited.

Conrad, R., & Donaldson, J. A. (2004). *Engaging the online learner: Activities and resources for creative learning*. San Francisco: Jossey-Bass.

Doyle, T., & Tagg, J. (2008). *Helping students learn in a learner-centered environment: A guide to facilitating learning in higher education*. Sterling, VA: Stylus Publishing.

Friedman, T. L. (2005). *The world is flat: A brief history of the twenty-first century*. New York, NY: Farrar, Straus, and Giroux.

Garrison, D. R., & Anderson, T. (2003). *E-learning in the 21st century: A framework for research and practice*. London, UK: Routledge. doi:10.4324/9780203166093

Gay, G., & Hembrooke, H. (2004). *Activity centered design*. Cambridge, MA: MIT Press.

Kear, K. (2010). *Online and social networking communities: A best practice guide for educators*. London, UK: Routledge.

Lehman, R. M., & Conceicao, S. C. O. (2010). *Creating a sense of presence in online teaching: How to be "there" for distance learners*. San Francisco, CA: Jossey-Bass.

Lindberg, J. O., & Olofsson, A. D. (2010). *Online learning communities and teacher professional development: Methods for improved education delivery*. Hersey, PA: IGI Global.

Moore, M. (Ed.). (2007). *Handbook of distance education* (2nd ed.). Mahwah, NJ: Lawrence Erlbaum.

Orvis, K. L., & Lassiter, A. L. R. (2008). *Computer-supported collaborative learning: Best practices and principles for instructors*. Hershey, PA: IGI Global. doi:10.4018/978-1-59904-753-9

Palloff, R. M., & Pratt, K. (2005). *Collaborating online: Learning together in community*. San Francisco, CA: Jossey-Bass.

Palloff, R. M., & Pratt, K. (2009). *Building online learning communities: Effective strategies for the virtual classroom*. San Francisco, CA: Wiley.

Palloff, R. M., & Pratt, K. (2011). *The excellent online instructor: Strategies for professional development*. San Francisco, CA: Wiley.

Ragusa, A. T. (Ed.). (2010). *Interaction in communication technologies and virtual learning environments: Human factors*. Hershey, PA: IGI Global. doi:10.4018/978-1-60566-874-1

Roberts, T. S. (2004). *Online collaborative learning in higher education*. Hershey, PA: IGI Global. doi:10.4018/978-1-59140-408-8

Russell, D. (2009). *Cases on collaboration in virtual learning environments: Processes and interactions*. Hershey, PA: IGI Global. doi:10.4018/978-1-60566-878-9

Salmon, G. (2011). *E-moderating: The key to teaching and learning online* (3rd ed.). London, UK: Routledge.

Stahl, G. (2006). *Group cognition: Computer support for building collaborative knowledge*. Cambridge, MA: MIT Press.

Stavredes, T. (2011). *Effective online teaching: Foundations and strategies for student success*. San Francisco, CA: Wiley.

West, J. A., & West, M. L. (2009). *Using wikis for online collaboration: The power of the read-write web*. San Francisco, CA: Jossey-Bass.

KEY TERMS AND DEFINITIONS

Andragogy: The study and science of adult learning.

Collaboration Or Collaborative Learning: Refers to the process of co-laboring; a group of people working together toward a common goal(s).

Constructivism: A type of learning theory that views human learning as an active effort to construct meaning in the world around us. Constructivists believe that actual learning takes place through accommodation, which occurs when students allow new information to change their existing ideas or knowledge.

Engagement: The process of active participation in a learning task. An engaged learner is actively participating, thinking, and questioning, in the learning process; he or she makes connections to existing knowledge and experiences and is reflective about learning.

Instructional Practices: Teaching and learning techniques and activities.

Learning Community: A group of people who share common interests and are actively engaged in learning from one another.

Online Learning: All forms of electronically supported learning and teaching. The delivery of a training, or education program by electronic means. Online earning involves the use of a computer or other electronic devices and means to provide training, educational or learning material.

Pedagogy: The art, science, and profession of teaching. Also refers to philosophy, beliefs, and strategies of instruction.

Social Presence: Refers to the ability of participants in a learning community to project themselves socially and emotionally as 'real' people, through communication.

Student-Centered Learning: Learning that places the student in the center of the learning process.

Chapter 9
Creating Virtual Collaborative Learning Experiences for Aspiring Teachers

David M. Dunaway
University of North Carolina at Charlotte

ABSTRACT

Finding time for reflection and collaboration presents challenges for teachers. Combined with this are feelings of isolation from colleagues. Web 2.0 tools can assist in alleviating these difficulties for teachers. This chapter discusses the potential of Web 2.0 tools, the development and uses of these tools, and considerations to make when using Web 2.0 tools. The chapter also presents ways colleges of education can support reflection and collaboration while diminishing feelings of isolation. The experiences of one instructor in implementing Web 2.0 strategies in Master's of School Administration classes are shared throughout the chapter to support the rationale for utilization of Web 2.0 tools.

INTRODUCTION

There is a changing definition of school leadership. School leadership is no longer the domain of principals, assistant principals, or department heads. School leadership today is much broader and includes the entire professional staff (Harris, Leithwood, Day, Sammons, & Hopkins, 2007).

As examples, the recently implemented North Carolina Teacher Evaluation Process (2009) includes teacher leadership as the first of five major evaluated areas, and in part says: "Teachers demonstrate leadership in the school. Teachers work collaboratively with school personnel to create a professional learning community"(p.20). The North Carolina School Executive Evalua-

DOI: 10.4018/978-1-4666-1906-7.ch009

tion Process (2008) under Standard I, Strategic Leadership, describes the following expectation: "Distributive Leadership: The principal creates and utilizes processes to distribute leadership and decision-making throughout the school" (n.p.). "School leaders" in this chapter refers to both teachers and principals.

BACKGROUND

Much is unchanged about education in the last 90 years–especially the processes that generally organize, frequently direct, often confine, and ultimately determine student learning. What have changed dramatically are the expectations of both students and teachers. And, the old processes cannot produce what they were never intended to produce. An aphorism of organizations is that when good people are caught in bad (or ineffective) processes, the bad processes overcome the best of intentions. This juxtaposition of the old processes and the new expectations creates the critical question for this chapter: *If the processes that underpin the profession in the K-12 environment have little changed, but the expectations have greatly changed, how can the use of technology create new processes?*

ISOLATION FROM COLLEAGUES: A SIGNIFICANT PROBLEM FOR TEACHERS

A key practice or process which separates education as a profession from most others is the isolation of the professionals from other professionals. This practice is found throughout the profession and in schools particularly. Teachers teach classes of students normally with the classroom door closed and even more archetypically without professional contact and collaboration with other teachers. There is little dispute that collegiality and collaboration with associates plays in high stressed professional cultures. Where this practice confounds high expectations for learning is when children in a classroom experience problems in learning, the solution to problems in the traditional/current process lies with the expertise of the individual teacher not with a group of professional colleagues. With the increased demands to teach more and at a higher intellectual level, Richard Elmore (2000) wrote that isolation is the enemy of innovation and the development of unique solutions to learning problems.

A NOT-SO-OBVIOUS SOLUTION FOR ISOLATION, COLLABORATION, AND REFLECTION

Like Elmore (2000), others, both in and outside of education have echoed the issues associated with professionals working in isolation. According to Bridgstock, Dawson, and Hearn (2011),

"[M]ost theorists now agree that while individual skills and knowledge, and traits like personality and intelligence, are important foundations for innovation, in actuality innovation thrives on social interaction and collaborative efforts. It involves the active combination of people, knowledge, and resources" (p. 105).

Schmoker (2006) wrote that the typical school functions as a group of private freelancers united by the school parking lot. Michael Fullan (2001), in *Leading in a Culture of Change,* writes of the importance of social learning or learning in context. "Learning in the setting where you work, or learning in context, is the learning with the greatest payoff because it is more specific (customized to the situation) and because it is social (involves a group)" (p.136).

WEB 2.0 DEFINED

So what is Web 2.0, exactly? How is it different from Web 1.0? Can it be used within the typical policies of school districts? Web 1.0 is the information web. It is what happens when anyone accesses the Internet to retrieve any kind of information. When you think of Web 1.0, think of *Internet Explorer, Safari,* or Google *Chrome.* They allow users to access all kinds of content. School districts have hesitatingly embraced Web 1.0 use – usually with significant site restrictions on both teachers and students when using school owned technology.

Web 2.0 is the social web. When you think of Web 2.0, think of *Facebook* or *Twitter*, but for our purposes think more of *Google docs.* Web 2.0 allows users not only to access information, but to share and to collaborate around information in almost any form online with other users. Like Web 1.0, Web 2.0 has not been universally accepted by school districts. Students and teachers are not routinely allowed to use *Facebook* or *Twitter* or personal email while using district technology resources. However, based on observations from my work in schools, schools are increasingly allowing teachers to use *Google docs* to collaborate around school routines such as lesson planning.

WEB 2.0 POTENTIAL

Lesson planning is an example of a potential Web 2.0 operation, but not really a collaborative process focused on problem solving. When Web 2.0 is expanded beyond routine school tasks, its technologies can provide the time and the means for virtual professional collaboration around problems of practice allowing teachers to engage with each other on not just writing lesson plans, but improving planning so that it engages students in learning. The key problem of practice for any school is how to provide for the child who fails to master expectations. Professional learning communities – teachers meeting face-to-face – are beginning to show teachers how to engage with each other to solve common or uncommon problems of practice. But there is a seldom discussed drawback – time needed to meet. In my 43 years of involvement with K-12 education, I have seldom found teachers willing to give up time. There is no more precious resource, and today's high pressures of standard-driven assessments have only increased the reluctance of teachers to meet. Thus, isolation again is the norm.

Web 2.0 can provide schools today with at least a partial solution to the very large problem of working in isolation. One major aspect of Web 2.0 is that it does not require synchronous (at the same time) participation by all parties. Teachers (and administrators or even parents) can offer solutions or feedback asynchronously (when it is convenient to them), thus largely eliminating the significant pressure of meetings during or after school. If a group wants to set up a time to meet that is common to everyone, they can easily do that. At the same time, once a framework is established for Web 2.0 collaboration, most contributions, reflections, and edits can be done at the convenience of each group member. In either case virtual cooperative learning is occurring through Web 2.0 collaboration.

Toward achieving that goal, this chapter is written. The beginning of this school culture-altering solution must begin in pre-service education for teachers and school leaders. The remainder of this chapter will be devoted to how colleges of education can implement this change and draws on my experiences of implementing Web 2.0 strategies in my Master's of School Administration classes.

THE DEVELOPMENT AND USE OF WEB 2.0 TOOLS

In my third year as an assistant professor, two events occurred which significantly changed my teaching strategies. First, I received a survey from a colleague asking several questions about how or if I used Web 2.0 tools in my master's classrooms. I completed the survey, but honestly, I did not even know what was meant by Web 2.0 tools. Second, I polled my master's classes to ask how many of my students routinely used *Facebook* or *MySpace* – it was over 90%.

From these two events I concluded that, though I considered myself significantly above the mean in my understanding and use of technology in teaching, there was a segment of technology about which I was almost totally ignorant, and my ignorance was potentially harming my students' current learning and future successes. I applied for and received a grant from UNCC to develop teaching strategies using Web 2.0 to create virtual cooperative learning. The remainder of this chapter is based on my experiences and findings from a student survey at the conclusion of the grant period.

With the above description in mind, this section describes the Web 2.0 tools I chose and helped design to meet my goal of providing virtual cooperative learning experiences as part of my regular teaching strategies. The relative value of each tool will be described later in the chapter. Here I describe the tools and the purposes for each.

Social Network Creation with NING

One of my goals that was congruent with collaboration within my classes was also to provide an avenue for professional collaboration between students, former students, former students now practitioners, department professors, and a select number of people not directly connected with my university. I chose *NING* as the platform. It was free of cost at the time, decently easy to design, and provided a safe environment for networking. The safety issue was a concern as several local teachers had recently been terminated for inappropriate *Facebook* postings, and I wanted to stay completely away from potential problems by association. As I set up the network, I controlled membership through invitations and approvals, and now three years into it, continue with this safety measure.

NING was purposefully designed with a webpage-like environment. The network is called the **IN*SITE** *network*© with the tagline: *Engaging Colleagues Around Problems of Practice.* I set up the network around the mainstay of *Effective Schools* research and the Correlates of Effective Schools. The first groups to be created for membership included a group for each of the seven correlates. As the **IN*SITE** *network*© grew, and as I grew in familiarity with it, groups were added for each class I was teaching each semester as well as some other interest groups.

The **IN*SITE** *network*© has grown from an initial membership of ten to a current count of more than 300 members. Membership has naturally included my students but it has also been extended to many department members and practitioners whom I often use as mentors for new students, thus enhancing the use of collaboration.

The **IN*SITE** *network*© is designed to be a place where members can find resources, upload resources, include professional reflections through blogs and forum discussions, and ask for help from other members. These opportunities are open to all members, but students participate in most or all of the options. Whereas commenting on a blog is optional but encouraged, posting blogs on professor designed topics related to the course or student presentation handouts, for instance are expected.

There are a couple of other things to note about the **IN*SITE** *network*©. As can be seen in the screen shot above I use the **IN*SITE** name and combine it with other descriptors such as **IN*SITE** *links* where members, and especially students, can find helpful links to everything from a class-supportive URL to the department's webpage to directions on how to access the library remotely, to state education statues.

NING provides a secure environment in which members can provide opinions from comments to entire blogs or videos. Given the openness of

Facebook and MySpace and recent pronouncements from *Facebook* founder, Steven Zukerburg, about his view of privacy, the security features of *NING* at the control ofthe designer has been critical to the growth and use of the **IN*SITE** *network*©. The **IN*SITE** *network*© is promoted as a by-invitation-only, professional, social network.

As my experience with Web 2.0 as an instructional strategy has increased, *NING* has taken on a greater and greater role as an instructional tool. Each semester, for each class I create a group specifically for the members of the class. In this group, they can find classroom materials (previously these were accessed through *PBWorks*). On their group page, they respond to discussions started by me as their professor or by group members. Additionally, on their group page, students are expected to post helpful comments by their mentors so that all students can have the benefit of the advice of all mentors. As well, when students make individual or group presentations, those presentations are uploaded to the group page as a current and future resource for all class members.

A major emphasis of the **IN*SITE** *network*© experience is the use of blogs to create opportunities and meet expectations for out-of-class reflections. The interface for the blogs is straight forward as is the space for comments on posted blogs. I initially made this an optional expectation for students, but have since required a certain number of blogs from each class, but still do not determine or suggest their content. As a side note, as a former K-12 teacher and administrator, I have 36 years of experiences from which to draw to add depth to the understanding of my students. Unlike peer reviewed writing, which really is not written for a K-12 audience, I use the blogs as a means to relate my experiences and/or to comment on current topics of interest. My contributions act as an incubator for student blogs which are very often as insightful as my own.

The *NING* **IN*SITE** *network*© has a familiar web-page and Facebook feel to it. With location tabs across the top of the page, navigation has only a very short implementation dip, and the individual profile page allows for a certain amount of individual creativity by the member. In an effort to build a sense of community, I ask members to include a photo of themselves. This simple act helps me and my students get to know each other more quickly.

So, does *NING* have any drawbacks? Without a doubt, it has its limitations. While it provides email and chat functions, it does provide the functionality for collaboration of *Google docs* or even *PBworks*. *NING*, as well, now has a cost associated with it. At the time of this writing, the yearly cost for the version which allows up to 10,000 members with most of the network amenities approaches $240 a year. Not an unreasonable cost but one that is significant in times of budget crunching for every dollar.

If you are your own *NING* webmaster, it does take some time. Initially, the interface as an instructional tool must be set up to meet your instructional goals. Since I set security as a major goal, I approve each member personally for general membership and as members of each group, and this takes time but is quickly done during the first week of class. I have also found that I must "prime the pump" occasionally to increase contributions, and again this takes time. Typically, this involves posting a video or news story of interest and then sending an email blast to all members to invite them to take note of the new addition. Another way of generating visits by my students is that I periodically change the logo just a bit (think *Google* logos) to reflect monthly or holiday themes, and I award a small gift such as a coffee shop gift card or a jar of my wife's fig preserves to the first student who notices the change. The drawback to being your own webmaster is simply time constraints.

Google Docs

Google Docs was chosen to be the toolset where student virtual interaction around projects would take place. The various tools within the toolset were much like *Microsoft Office* so there was not a significant learning curve, which was important. Students who are experiencing a new paradigm in terms of course content and an expanded view of the profession benefited significantly from the familiarity of the *Google Docs* interfaces. Important to the support of the overall project was that *Google Docs* could be used synchronously with several people editing the same document at the same time and, just as importantly, asynchronously which provided each person the freedom to choose the time which worked best for them.

Google Docs proved to be the almost perfect tool for student virtual collaboration on class projects. Its similarity to *Microsoft Office* provided a familiar and comfortable connection between computer-based document creation and cloud-collaboration. Compared to *PBWorks, Google Docs* word processing and presentation tools were far superior. *PBworks* provided only the most unsophisticated word processing functions whereas *Google Docs* provided *Word*-like functionality including the ability to upload *Word* documents and to download *Google Docs* created documents in Word or RTF formats.

The synchronous and asynchronous abilities of *Google Docs* provided for scheduled collaboration or collaboration and editing when the time was convenient to the students. In the real world of schools, finding time to collaborate is a major obstacle to successful collaboration. Cloud-collaboration eliminates, for all practical purposes, this detriment to teachers and administrators collaborating on a common document.

Finally, *Google Docs* provided the perfect way for me as the professor to keep up with the pace and extent of virtual collaboration and to participate. I simply required each group to include me as a member with whom all work is automatically shared. So each time a person adds to the project, I receive an email notice of the addition.

Are there drawbacks? Certainly. Most students have not engaged, at least to their knowledge, in cloud computing, and getting comfortable with the *Google Docs* interface takes a little getting used to. Compared to CPU-based document processing, it is a little clunky. But its similarity to *Word* quickly relieves this initial anxiety by students. I typically introduce them to the world of *Google Docs* by creating a spreadsheet where I have entered their names, and they are asked to add additional contact information such as alternate emails or work phone numbers. It is not uncommon for students to complete this process during the first class even as I have the *Google Docs* spreadsheet live on the overhead. They use a familiar process, and I am able to collaborate with my students – a win-win.

PBWorks

PBWorks was chosen initially as the depository for professor-uploaded class materials such as syllabi, PowerPoint presentations, and class resource materials. It was also chosen as the depository for student-uploaded materials such as presentations, advice from class mentors, and as a place for comments and concerns about the course, and occasional comments from students about the contribution of other students.

As a Web 2.0 tool *PBworks* did what it has famously done – provide a location that is easily accessible, and where navigation is rather intuitive. A link from *PBworks* to the **IN*SITE** *network*© was provided as was a link from the **IN*SITE** *network*© to *PBworks.*

After the pilot and one semester's classes, *PBworks* succumbed to student suggestions that going back and forth between the **IN*SITE** *network*© was confusing, and *Google docs* provided a more user friendly interface for virtual collaboration.

PBWorks was not an ineffective tool. It did what it was designed to do. However, *PBWorks* when combined with the **IN*SITE** *network*© (*NING*), students quickly recognized a significant duplication of many aspects and expectations of *PBWorks* with the **IN*SITE** *network*©. Whereas

I had envisioned and intended *PBWorks* as a collaboration tool, Google docs simply provided a more familiar interface for collaboration. As a result, after a full semester of using *PBWorks*, it was put on a backburner and no longer was used as an instructional tool. I have not completely abandoned its usefulness. Currently, I am using it as a warehouse for resources, ideas and solutions generated in each class and will continue to explore this use in the future.

Gmail

As I worked with my grant focus group to design/select the tools and design their use, the group early on suggested that each student be required to obtain a *Gmail* account so that they could have full access to the *Google Docs* toolset. Many students already had a *Gmail* account which was connected to a smart phone and/or to Microsoft *Outlook.* I found early on in my university teaching that students are notoriously unreliable when it comes to checking their university provided email accounts, so the use of *Gmail* significantly improved my communication with my students. But there was also a final and very significant use of *Gmail*. A value-added plus was that as graduate students, most of my students were graduated in two or three years and contact was lost. Having a *Gmail* account that was directly connected to the **IN*SITE***network*© (it was their login) meant that contact was easily maintained with them after graduation through the **IN*SITE** *network*© as they moved from student to practitioner and mentor for future students, again supporting the goals of building a Web 2.0 learning community.

The initial value of these tools resided (1) in their availability, (2) that no initial costs were involved, and (3) that they were easily malleable to goals of using Web 2.0 instructionally. A clear benefit to me personally was that they were user friendly to such a level that after the grant expired and the funds for a technology design specialist no longer existed, I was able to continue to develop and expand their uses.

PROJECTS AND ASSIGNMENTS APPROPRIATE FOR VIRTUAL COOPERATIVE LEARNING

Prior to the first night of class, I contact the students and asked them to obtain a *Gmail* account and send me their *Gmail* address. As I noted earlier, the first introduction to *Google docs* is my creation of a spreadsheet containing the names of the students with cells for additional contact information such as work email and mobile phones. Using the students' *Gmail* addresses, I share the spreadsheet with them. I introduce *Google docs* and the spreadsheet on the first night of class, and if students have brought laptops, they often complete their information as I am showing the rest of the class how to do it. This, in itself, is great reinforcement for the intro to cloud-based documents.

Weblogs

Weblogs provide an exceptional avenue for student reflections on topics introduced in class. Early on as a professor I found that although *I* had reflected on my lesson and was always looking for practical examples of sometimes seemingly esoteric topics, my students were not. Research clearly and consistently shows that reflection on what has been learned is a significant means of increasing the depth of understanding of and the retaining of new knowledge and skills (Johnson & Aragon,2003).

When I initially introduced the concept of weblogs in my classes, I made students' contributions to the **IN*SITE** *network*© voluntary, or as I told the students, "contribute when the spirit moves you." I found out rather quickly that the spirit of reflection was largely absent in my first classes' experience with Web 2.0 teaching techniques. Later, I made the contribution of a set number of weblogs an assessed requirement. This helped, but students being students, many waited until the final two weeks of the semester to make contributions, which defeated the purpose of reflections on issues in close proximity to their introduction in class. My final strategy tweak to the use of blogs was to require that the blogs be posted within a week of the conclusion of the class topic on which they were to be based. I found that this was received well with students and contributed to other students adding comments to the weblogs of other students and to actually adding blogs on non-required issues.

I mentioned earlier in the chapter of the need to "prime the pump" as related to student use of the **IN*SITE** *network*©. Such is also the case with student blogs. So, I write a blog about every two weeks. This gives the students an idea of the content and style that I would like to see, and they usually draw comments and, not infrequently, start virtual discussions about the topic. My general rule for weblogs is that they should be significantly longer than a *Tweet* and significantly shorter than an essay. At the time of this writing (eight semesters of my use of Web 2.0 tools) 285 weblogs have been submitted to the **IN*SITE** *network*©.

Virtual Collaboration

Virtual Collaboration (or virtual cooperative learning) is a skillset for school leaders and teachers that should be regarded as required pre-service learning. Time is a precious resource in schools. Especially today with emphasis on high-stakes testing every minute counts, so it is understandable that teachers may resist giving up "their time" to help solve the problems found in every school. Making the time ultimately flexible by the use of virtual collaboration is a key stress reducer, and will very likely produce better solutions.

As a stress reducer, the members of the virtual group can meet synchronously at a time agreed upon by the group not determined by the principal, or often even better, they can participate in the discussion asynchronously at a time of each person's choosing. So if Sandra is a night owl, she might prefer to make her contribution at 11:00 p.m., but if John is an early riser, he might make a contribution at 4:00 a.m. William who is single without kids, may find that 7:00 p.m. is the perfect time for him. The flexibility to contribute, to comment, to edit on your own schedule is a powerful option. What's more, the principal can "observe" the working of each group and choose to participate to help them stay on the right path, or simply be an interested observer.

Likely the group will eventually need to meet face-to-face, and when this occurs, this meeting is significantly more productive since much of the minutia has been solved virtually. Better solutions come about largely because each person in the virtual collaborative group can "speak" without fear of being cut off or dismissed. Young educators with fresh ideas are typically reticent about speaking up in large group faculty meetings of seasoned veterans (Feiman-Nemser, 2003). Virtual meetings allow each staff voice to be heard and even provides them the opportunity to edit the contributions of others – something they would not do in a large group meeting.

Group Projects

Group projects that would have traditionally been included as an assessed course product lend themselves equally well, if not better, to virtual collaboration. I found many times in my classes that when I assigned a group project with a presentation at the end, that members of the group simply divided the assignment up so that each individual was responsible for a separate piece that was added to the other separate pieces at the time of the presentation of the project's results to the class. The presentations almost always were disjointed, lacking transitions from one part to another – rather resembling the old cartoon of the horse designed by a committee. Virtual collaboration gives the professor much more control and opportunity for guidance over the group process. By using *Google Docs*, I am able to monitor the progress of the group from brainstorming ideas to sorting and prioritizing those ideas to fleshing them out and creating a final presentation.

A typical project that I use in intro leadership classes is a scenario that requires the group to design a plan for engaging parents in the learning of their children beyond typical parental involvement. One group where this was recently assigned began by creating an "Idea Board" where all members could brainstorm ideas, and then moved forward from there.

Here I share a couple of suggestions from experience. First, I appoint a member to lead the group in getting started by creating an initial document (such as the "Idea Board"), including other members of the group and the professor as editors of the document, and staying on task. Once other members see this process they can assume the leader role for other collaborative projects. Second, always require that you, as professor, are included as an editor so that any changes are automatically forwarded to you as this is the key to monitoring. I have the group create their presentation using the *Google Docs* presentation program which can be downloaded to PowerPoint, or the classroom presentation can be run from *Google Docs* if internet connections are available. Finally, I have the students upload their presentations to **IN*SITE** *network*© where other students can access it and download them as resources for future use. The final product of this activity is the presentation, but the value added virtual collaboration experiences, which can be used with colleagues back in schools, are likely more lasting.

Paired Learning

In courses such as School Law, where there is a steep learning curve in mastering how to read, summarize, and report court opinions in the *case brief* format, I began by pairing students up virtually to create their first briefs. I have found that this significantly shortens the learning time necessary to master the process of creating case briefs for both the students and for me. The process is the same as that described above except all work is completed in the *Google Docs* word processing program.

Mentors

As mentioned earlier in the chapter I use former students as mentors to enhance the collaboration between students and practitioners, but mentors also provide a vital role in the virtual collaboration model. I encourage students to communicate with their mentors for any shortcuts the mentors may have created or learned as they mastered the art of virtual collaboration. The use of mentors has created an ongoing bond to the university program, but more importantly, their participation communicates the heart of cooperative learning as a process necessary in this day of significantly increased expectations and changing processes and so is as much a critical aspect of virtual cooperation as is the technology itself.

Virtual Collaboration between Professors and Classes

Prior to my first semester of fully implementing Web 2.0 as a teaching strategy, I had a visit from a colleague in another department with the college of education. She had been assigned to teach a course–*Advanced Practicum in Teaching, Learning, and Leadership*–which was typically known as *Teacher Leadership*. My colleague approached me about leadership resources since I taught aspiring school administrators. As we explored potential resources, we had an epiphany – why not use the virtual collaboration model to bring her aspiring teacher leaders "together" with my aspiring school leaders.

She and I created a series of three projects where teachers' and leaders' interest would intersect and that could and would be completed virtually. We then set up the groups so that each group had equal numbers of students from each class. Members from both classes became members of the **IN*SITE***network*© and used *Google Docs* as the virtual collaboration tool. Finally, the products of the virtual collaboration were posted on the **IN*SITE** *network*© where members of both classes could have future access to them as resources. Purely by luck of the draw our classes were on the same night at the same time, and our final collaboration was a combined class where faces and names could finally be matched.

One of the major outcomes of this experience with my colleague was an understanding that much of what we do in colleges of education is done in isolation. Colleagues work in well-defined areas even when those areas could frequently be connected virtually to other related programs with very little work involved providing enrichment to both programs and increased depth of understanding. Perhaps more importantly we found that the students in our virtual exploration saw real merit in working with others not in their individual programs but directly connected to them.

CLASS PREPARATION AND OTHER CONSIDERATIONS

Preparing the Course for Web 2.0

There are five key elements to consider in adding Web 2.0 strategies to an existing course. First, decide on the purpose and goals that Web 2.0 strategies will play if introduced to the course. For instance, in my courses, collaboration is a power-skill – a practical and pragmatic skill – that aspiring leaders need to master to be successful as practitioners. Additionally, it cannot be mastered in the abstract. It must be practiced. Adding the element of virtual collaboration met the goal of using the kinds of technology tools my students were already using and showing them how to adapt it to the school environment.

Second, identify the activities that can successfully be adapted or designed for virtual environment of cloud-computing, social networks, or wikis. Just as people who are involved in face-to-face collaboration produce better solutions when they are engaged by the problem facing them as a group, so, too, should virtual collaboration be engaging both in the content and the technological experiences.

Third, assess how much out-of-class time should be expected of students and whether or not this virtual learning time should be subtracted from assigned in-class time whether that in-class time is online or face-to-face. When I first introduced the Web 2.0 strategies, I subtracted 30 minutes from each class to accommodate for the virtual learning time. In total this amounted to about two classes over the semester. I quickly concluded that I was not willing to give up two classes worth of content, and after my second semester, I no longer allocated class time for virtual learning.

Fourth, you must assess your skill level with Web 2.0 tools as you prepare for a course where they will be part of your teaching strategies. As with all new learning, there is an implementation dip as you become familiar with the tools. Even if you are proficient in their use, adapting them as classroom strategies will likely produce some implementation dip when compared to other semesters where your pre-course preparation may have been limited to an ongoing revision of a relatively standardized syllabus. If you, as I did, have to learn to use the tools before trying to teach their use to students, then I would suggest a mini-pilot of a couple of weeks in a course where you are really comfortable with the students and the content, and then even consider a full-pilot of the strategies the next semester in only one course. My mini-pilot took place in the spring semester, and then I did a full-pilot in a short summer session.

Finally, if you decide to use Web 2.0 tools, the design or adaptation of the wiki, social network, or clouddocuments for your particular coursewill add significant time to course preparation. Each course will require its own Web 2.0 identity which actually is the first step in engaging the students in the virtual learning world.

Considerations when Including Web 2.0 Strategies

When considering the inclusion of Web 2.0 strategies, there is a long list of logistical considerations that should be considered. The list below was developed from my experience. If it existed before I left into this new teaching dimension, I did not find it in my preparations.

- Where will class resources be stored for student access?
- How will the students be able to access the different class web 2.0 sites?
- What level of security will you employ for access to resources?
- If you prefer to use a non-education organization email address, how will you communicate this to students? Beware that your university or college may require that you use official communication channels for all class communication.
- How will you orient students to the use of the tools?
- How will you monitor student progress?
- How will you assess students' products of virtual collaboration?
- How will you accommodate special needs students?
- How will you accommodate students who are not active Web 2.0 users?
- How will you design "prime the pump" during the course to assure ongoing student engagement with the Web 2.0 tools?
- How engaged in your strategies will you personally be outside class time?

A Comparison of Blended and Face-to-Face Instructional Strategies

The reader needs to know that I am primarily a face-to-face teacher. Never taught an online course. Never took a course online. My foray into blended instructional strategies using face-to-face and Web 2.0 strategies was undertaken as an effort to stay connected with a generation of graduate students for whom Web 2.0 experiences were the norm on a daily basis in their personal lives and, as I could begin to see, at the cusp of being important in their professional and academic lives as well.

That being said, I would not go back to a purely face-to-face classroom experience. Whereas I hoped for and occasionally could see evidence of in-depth reflections on class content, usually this was most evident on take-home exams (the only kind I employ). Here students were expected to employ higher level cognitive skills of application, analysis, and evaluation. I cared not at all for their ability to recall facts, dates, or names, and, in fact, observed that they often would actually add depth to their knowledge and understanding by completing some research as needed to better understand the nature of the question. My grading techniques for these take-home exams always centered around Microsoft Word where I would use the "comment" feature to note issues or exemplary thinking. I still tell my students today when they get an exam returned to them filled with comment boxes that they should think of my comments as my side of a conversation with them that they had started with their exam responses. The exams were submitted to me and returned to the student by email.

As I look back on those first Web 2.0-like "conversations" taking place with my exams, perhaps this was the conception-moment where I realized the power of virtual conversations to extend classroom learning and that it need not occur only a few times during a semester with exams, but on a routine basis – both professor-planned and spontaneous student eureka-moments.

AN UNEXPECTED PARADIGM SHIFT

I expect there is no one reading this chapter who hasn't yet concluded from this chapter that I am, as they say, "old school." The paradigm of old school teaching is that teachers teach and students take notes and participate in discussion by raising hands or being called on. This is direct teaching in a nutshell and is still a powerful technique. I had always invited my students to bring laptops to class so that they could take notes as the class discussion merited, but about a fourth of the way through my second semester of integrating Web 2.0 tools and techniques, a student, Jake, was diligently typing away at his computer at a time when I just asked a question of the class. Clearly, he was not paying attention to me, and my old school radar immediately activated and honed in on Jake. "Even though this is a master's class, Jake, I expect you to pay attention to me and participate when a question is asked," I quietly said to him as I bent down over his shoulder so no one else would hear. Just as quietly, Jake said to me, "But I am, Dr. D.; I was just *Googling* your question to see if I could find some additional information."

That unexpected but epiphanous encounter with Web 2.0 led me to shift my teaching paradigm regarding Web 2.0 strategies, again. I had been focusing on Web 2.0 strategies *outside* the classroom. Clearly there was a place for using the Web 2.0 strategy of going, getting, and sharing of information during a class discussion. It simply required one simple thing: *trust in my students.* The precursor to trusting that my students were using the omnipresent wireless technology of laptops and smartphones was to insure that they were engaged in the lesson, and although engagement is a function of each student, it starts with me, the professor.

CONCLUSION

It is not just a worthwhile endeavor of colleges of education to expose their undergraduates to Web 2.0 as a tool of virtual collaboration; it is truly a game-changer. Virtual collaboration with Web 2.0 can give to teachers and to the profession that which legislatures cannot – Time. The reorganization, reuse, and rethinking of the use of time as a resource of educational reform must begin with a simple idea: not all problems need to be solved face-to-face. Asynchronous problem solving using Web 2.0 tools can change the very culture of the profession. Consultation with colleagues, as in other professions, can become the norm instead of the exception.

RECOMMENDATIONS

There are three primary recommendations if this vision of the future is to be recognized. First, pre-service teachers and administrators must be trained in its use. What better way to do that than to experience its power as students. Second, school districts must adopt policies that recognize, permit, and emphasize the critical role of professional social networking in solving the problems of practice of schools. Finally, there is simply a dearth of research into the use of Web 2.0 as an instructional tool in K-12 or the university setting. This must be remedied so that we move forward, not on limited experience of a few, but on research-proven best practices.

A FINAL CAVEAT

Who of us has not sat face-to-face in the presence of master-teachers enthralled by their voice inflections, their stories, their knowledge, the challenges they give to us, and the understanding they show toward us as novices? Web 2.0 strategies in the blended classroom environment need not replace the master-teacher. Rather, by bringing back to the classroom insights gained from Web 2.0 interactions the richness found in the classrooms of great teachers can be significantly enhanced by increased depth of student cooperation and understanding.

REFERENCES

Bridgstock, R., Dawson, S., & Hearn, G. (2011). Innovation through social relationships: A qualitative study of outstanding Australian innovators in science, technology, and the creative industries. In Mesquita, A. (Ed.), *Technology for creativity and innovation: Tools, techniques and applications*. Hershey, PA: IGI Global. doi:10.4018/978-1-60960-519-3.ch005

Elmore, R. (2000). *Building a new structure for school leadership*. Washington, DC: Alberta Shanker Institute.

Feiman-Nemser, S. (2003). Keeping good teachers: What new teachers need to learn. *Educational Leadership*, *60*(8), 25–29.

Fullan, M. (2001). *Leading in a culture of change*. San Francisco, CA: Jossey Bass.

Harris, A., Leithwood, K., Day, C., Sammons, P., & Hopkins, D. (2007). Distributed leadership and organizational change: Reviewing the evidence. *Journal of Educational Change*, *8*(4), 337–347. doi:10.1007/s10833-007-9048-4

Johnson, S., & Aragon, S. (2003). An instructional strategy framework for online learning environments. *New Directions for Adult and Continuing Education*, (100): 31–43. doi:10.1002/ace.117

North Carolina school executive principal and assistant principal evaluation process. (2008). *North Carolina Department of Public Instruction*. Retrieved from http://www.dpi.state.nc.us/docs/profdev/training/principal/evaluationprocess.pdf

North Carolina teacher evaluation process. (2009). *North Carolina Department of Public Instruction*. Retrieved from http://www.dpi.state.nc.us/docs/profdev/training/teacher/teacher-eval.pdf

Schmoker, M. (2006). *Results now*. Alexandria, VA: Association for Supervision and Curriculum Development.

ADDITIONAL READING

Cheng, G., & Chow, J. (2011). A comparative study of using blogs and wikis for collaborative knowledge construction. *International Journal of Instructional Media*, *38*(1), 71–78.

Cifuentes, L., Xochihua, O. A., & Edwards, J. C. (2011). Learning in web 2.0 environments: Surface learning and chaos or deep learning and self-regulation? *Quarterly Review of Distance Education*, *12*(1), 1–21.

Conole, G. (2010). Facilitating new forms of discourse for learning and teaching: Harnessing the power of Web 2.0 practices. *Open Learning*, *25*(2), 141–151. doi:10.1080/02680511003787438

Grant, M. M., & Mims, C. (2009). Web 2.0 in teacher education: Characteristics, implications, and limitations. In Kidd, T. T., & Chan, I. (Eds.), *Wired for learning: An educator's guide to Web 2.0* (pp. 343–360). Charlotte, NC: Information Age Publishing.

Kear, K., Woodthorpe, J., Robertson, S., & Hutchinson, M. (2010). From forums to wikis: Perspectives on tools for collaboration. *The Internet and Higher Education*, *13*(4), 218–225. doi:10.1016/j.iheduc.2010.05.004

Nicholas, H., & Ng, W. (2009). Fostering online social construction of science knowledge with primary pre-service teachers working in virtual teams. *Asia-Pacific Journal of Teacher Education*, *37*(4), 379–398. doi:10.1080/13598660903050336

KEY TERMS AND DEFINITIONS

Facebook: A free social networking service.

Google Docs: A free web-based word processing and data storage offered by Google.

NING: Online platform where people can create their own social networks.

PBWorks: A wiki site commonly used by educators.

School Leaders: Those who demonstrate leadership in a school environment, including teachers and administrators.

Chapter 10
Cyber-Place Learning in an Online Teacher Preparation Program:
Engaging Learning Opportunities through Collaborations and Facilitation of Learning

Victoria M. Cardullo
University of Central Florida, USA

ABSTRACT

Learning and technology skills required for the 21st century can be developed through online pre-service teaching preparation programs. This chapter is an exploratory look at the implications of learner-centered and place-based approaches. These approaches to teaching and learning are collaborative and distributed through online learning. In this chapter, it is the author's intent to offer guidelines for transference of classroom best practices to a cyber-place learning environment that will align with teacher preparation programs. The main objective is to improve access to advanced educational experiences by allowing students and instructors to participate in remote learning communities that foster skills needed for the 21st century. Online learning communities provide collaboration that is flexible and convenient and opportunities for individuals who may not otherwise have their voices heard.

INTRODUCTION

Over 5.5 million students, or nearly 1 in 4 students, took at least one online course during the fall of 2009, and approximately 84 percent were undergraduates (Allen & Seaman, 2010). As universities prepare undergraduates, are instructors of education programs producing quality online instruction? What role will education programs have in developing and training educators in the technological future of education? In Florida there are currently 63,675 students attending Florida Virtual Schools. Are educators ready to educate these tech-savvy students? "The Department of

DOI: 10.4018/978-1-4666-1906-7.ch010

Education views virtual schools as a powerful technology innovation expanding opportunities for learning anytime, anyplace in support of the No Child Left Behind Act" (Hassel, 2004, p. 4). Learning is a social activity as well as an individual one. Online learning offers multiple opportunities for a high social learning environment (Kearsley, 2000). Interactivity can vary drastically from course to course. Swan (2001) states course design is critical when determining the amount of interaction and the depth of discussion. Shaw (2006) states that skills gained through collaborative experiences are highly transferable to team-based work environments, which are elements of preparation for the 21st century. Educator preparation programs need to prepare pre-service teachers for the preparation of the students entering the work force. Chapman, Ramondt, and Smiley (2005) confirm that the workplace environment requires the learner to apply, analyze, synthesize, and evaluate information bringing the learner into the higher order thinking skills needed in the 21st century.

The driving force of this chapter is to explore the primary role of the learner in a teacher preparation program. Teacher preparation programs must mirror or align with goals and objectives that will not only propel pre-service teachers but also potential students they may influence. This chapter explores online learning and technology skills that are required for the 21st century as well as the need for collaboration and critical thinking skills. Problem solving and communication skills that are necessary for further development of learners are also discussed. Implications of learner-centered and place-based approaches that are collaborative and distributed through some form of online learning program are investigated. It is the intent of this chapter to offer guidelines for transference of classroom best practices to a cyber-place learning environment that will align with teacher preparation programs.

BACKGROUND

Distance learning environments or virtual classrooms are defined as modes of instructional delivery in which either place or time separates the instructor and the student. The virtual classroom is a teaching and learning environment located within a computer-mediated communication system. The full objective is to improve access to advanced educational experiences by allowing students to participate in remote learning environments, promoting learning that is time and place independent (Deal, 2002).

Distance learning provides opportunities for individuals who may not otherwise be able to participate due to time or place constraints. It offers flexibility and convenience at any time. According to Dillenbourg and Schneider (1995), the distinction between cooperative and collaborative learning is imperious. In a distance learning environment, collaboration becomes imperative for learning. The protocols for cooperative learning (classroom based) are as follows: dissemination takes place prior to the project, the subtask separated amongst group members, and finally the partners solve their portions independently. Collaborative learning is the opportunity to build synchronously two or more subjects and design a solution to the problem jointly. In a collaborative environment, the teacher becomes a facilitator in a learning-centered environment. According to Schiro, (2008) learning-centered ideology takes place when the teacher becomes the facilitator and observer. Teachers and students develop intellectually, socially and emotionally in accordance with their own innate nature and that of their culture. Their development of knowledge intensifies as the curriculum enhances the ability to make real world connections.

Virtual schools have the ability to extend equitable access to high quality education to urban and rural areas, low-income students, and students with learning challenges (Hassel, 2004, p. 5). Online learning can promote an environment that develops the reserves of knowledge instead of disseminating limited thinking and inquiry.

Mann states we live in troubled times: "Here is a society that manifests the most extraordinary contradictions: dire poverty walks hand in hand with the most extravagant living ever known; an abundance of goods of all kinds is coupled with privation, misery, starvation" (as cited in Slattery, 2006, p. 231). School, like the nation, is in need of a central balance, a unification of intellectual plans through the critical bridge between brick and mortar schools and online learning. As a nation, we must integrate the need to reevaluate the quality of online teaching. The problem has many facets and needs unification to solidify a foundation for our future. The digital imprint of our schools is historic by nature. Education is undergoing major changes that are multiple and complex. Educational changes are a constant condition rather than event (Goodson, 2003). The revolution of digital based learning has shifted to a constant state of renewal and development as new technologies and resources emerge. We can combine the previous concept of educational change with the concept of New Literacies as defined by Coiro (2009). Literacies are defined by constant change; change that is defining teaching and learning, change that is constant as technologies emerge, and change that redefines education. The main objective is to improve access to advanced educational experiences by allowing students and instructors to participate in remote learning communities.

Learning communities provide opportunities for individuals who may not otherwise be able to let their voice be heard. Online learning communities provide collaboration that is flexible and convenient. Marquardt and Kearsley (1998) found online collaboration to be just as effective as face-to-face collaboration. They posited that online learning communities that have established clear instruction or guidelines can lead to the development of a sense of individual accountability, team support, positive group interaction, and consensus building. This methodology moves away from the teacher-centered model into a student-centered model. When student collaboration is encouraged, students develop a stronger metacognitive knowledge, moving away from cognitive function of procedural knowledge.

The invention of the World Wide Web in 1992 made online learning accessible and permitted new forms of education to emerge. The 21st century shed light to the shift in attitudes toward online education. In the early 1980s, innovation and expansion imploded the networking levels for online communities of learning. In 1981, the first fully online course emerged for adult education, and soon networked classrooms followed. In the brief twenty-five years since the inception of the World Wide Web, online education has tackled many questions and developed new methods for learning and teaching. As access to computers and networks grew, educators realized that this new technology could be used for a wide variety of tools within the classrooms. Development of new modes of delivery and learning began to advance opportunities for improved quality of learning. Three distinct delivery methods emerged in online teaching: online courses, blended or hybrid courses, and web facilitated courses.

"Online course" refers to a course where most, if not all of the content is delivered online; typically, there are no face-to-face meetings. Course activities all take place within the realm of the online class environment. Collaboration occurs through group projects, discussion boards, or chat tools. Online interaction is not hindered by technology but, rather expanded or enhanced due to technology.

A blended or hybrid model is developed to deliver a significant portion of the teaching either in a face-to-face classroom or distance course. The time frame fluctuates based on teacher preferences. Networking tools are fully integrated into the online environment and become a regular part of the course curriculum. Typically, they reduce the number of face-to-face meetings.

A web facilitated model utilizes the network to enhance the traditional face-to-face teaching course; it is usually not a required element or graded element of the course. It can be used as enrichment to a face-to-face course where students have the opportunity to tap into the rich resources developed through the World Wide Web. Web enhanced is used to support face-to-face teaching; it can house assignments, grades, emails, and other classroom management tools.

In any format discourse is essential, for virtual space is social space. Collaboration is the most important concept when developing online pedagogy. Learning is enhanced by articulation of newly acquired knowledge. Articulation is a cognitive act in which the student defends, develops, and refines ideas.

The ease of asynchronous and synchronous communication has influenced many aspects of society. The ease of communication has developed e-governments, e- learning, e-commerce, and e-business, to name a few. Skills needed for effective communication to fortify efficacy for the 21st century can be procured through learning communities as students engender and dismantle content through discussions and activities generated online. Hoadley and Kilner (2005) call use of communication the C4P process. This process generates a greater element of communication through content, conversation, connections, and (informational) content. This form of learning and communication is a nonlinear process: each key stroke becomes a differentiation of learning for all learners based on schema and cognition. Hoadley and Kilner (2005) propose that "the greater these elements are present within a community the more likely and effective the knowledge generation and transfer will be" (p. 33).

Situated learning can be defined as an environment in which learning takes place. Aspects of the learning environment can support or hinder the learning. Shulman and Shulman (2004) state motivation is key to teacher preparation as they extend energy and willingness to sustain teaching that matches their vision. If pre-service teachers are prepared to teach within the confines of digital literacies, the barriers will be limited and the motivation to transfer previous learning exemplified. Willingness and desire to engage in innovative technologies must be closely related to one's vision, knowledge, and cognition. These networked technologies can be supported through online teacher preparation programs or programs that are enhanced by technology. Programs that are blended, fully online, or mediated in a face-to-face web enhanced environment can promote place of learning that will propel learners into the 21st century using programs that enhance teaching through the use of technology not bound by place and time.

LEARNER-CENTERED IDEOLOGY AND ONLINE LEARNING

The exploration of online collaboration parallels Schiro's (2008) vision of learner-centered ideology. The aim of learner-centered ideology focuses on nurturing and stimulating growth through engaged curriculum. Teaching online requires educators to create a democratic environment that encourages student collaboration (Muirhead, 2002). Brown, Collins, and Duguid (1989) affirmed that students must work together and listen actively to others in order to develop skills related to complex problem solving. Knowledge simply cannot be dispersed from teacher to student or student to student; knowledge must be developed through collaborative learning that instills deep knowledge of content.

Collaborative learning takes place when students are purposely placed in groups to work together toward a common goal or outcome. This aspect of collaboration is used to enhance the learning experiences. Slattery (2006) discusses democratic education as a society that is in need of constant reform and change that must involve both structural changes in education and the use of reconstructing society.

Many employers are already demanding not only core requirements but also creativity, critical thinking, idea generation, presentation skills, team building, and confidence. In other words, skills that will prepare students for the 21st century workforce are necessary. Are educators fortifying this type of learning in classrooms throughout the nation? Team collaboration, self-programmable workers, and critical thinking skills need to be incorporated into the restructuring process, endowing individuals with need competencies and tools that are vital to their future.

The postmodern paradigm shift characterized by fast-changing recurring concepts known as "global information revolution" is changing the vision of learning modalities (Slattery, 2006). Acculturation in the academic community is no longer education for the discipline but rather education for life (Schiro, 2008). Online learning, virtual schools, and hybrid classrooms are moving in the direction of learner-centered ideology. Constructivism modifies the role of the teacher. Teachers become facilitators instead of dispensers of knowledge. Teacher preparation programs must train potential teachers on how to become a facilitator. This can be modeled effectively through a learner-centered classroom online or face-to-face that focuses on collaboration and reflection of practice. In turn, this process transforms students from a passive recipient of knowledge to an active participant in the learning process.

Constructivist teachers pose questions and problems through online discussions that guide students' inquiries by helping them find their own answers. Constructivism encourages inquiry, which can lead to multiple interpretations through collaborative group work. Piaget, Dewey, and Vygostsky all developed theories of childhood development and education. Their theories led to progressive education, which in turn evolved into constructivism. There are numerous benefits related to constructivism: active involvement rather than passive learning, development of deep thinking and learning, learning that is transferable, and learning that develops ownership and promotes social and communicative skills as students learn to negotiate and evaluate their own contributions to the work.

LEARNER-CENTERED VIEW AND THE CONSTRUCTIVIST THEORY A COMBINED THOUGHT

Learner-centered ideologies are formulated on the constructivist view. The constructivist view encourages the learner to gain deeper understanding of content through reflective practice. Jonassen (1993) posits that constructivism proposes that the learning environment should support multiple perspectives or interpretations of reality, knowledge, and rich experienced-based activities. The online learning environment offers an enormous amount of information, tools for creativity, and development embedded with multiple environments or 'places.' These new technology tools provide multiple opportunities for the student-centered learner to develop inquiry and engagement in complex content. This process can lead to expert learners. No longer are students passive learners within a learner-centered classroom. As students reflect through online discussions, ideas that are generated gain complexity as they begin to integrate new concepts. Educators are no longer looking at best practices in the classroom. They are now beginning to look at best practices in cyber-place.

Constructivism is a theory of knowledge and not a theory of pedagogy. To bridge the pedagogical aspect, constructivist pedagogy bridges the gap between how knowledge is acquired and the link between theory and practice (Doolittle, 1999). To develop strong pedagogy, learning should take place in authentic, real world environments, be it brick and mortar or an online environment. No Child Left Behind hopes to intensify learning by fostering independent learning and leveling the playing field for minorities, low-income students, and students with disabilities through unbiased teaching in a virtual setting.

McCombs (2001) states all major changes have required some form of transformation: transformation of thinking, transformation of seeing, or transformation of interpreting reality. In an era of educational reform, shifts in thinking are taking place. Virtual schools are developing quality online instruction. Online teaching is becoming even more popular, and learning is being shared by all learners. Several structural components are utilized throughout virtual environments to create that central balance of online teaching. Table 1 below offers helpful suggestions for course development at a glance that can support a central balance:

PLACE-BASED LEARNING A NEW PLACE: "CYBER-PLACE"

The digital revolution and the inception of the World Wide Web have effectively created an alternate space for place-based learning. Cyber-place has created an alternate space, an extension or substitute for 'physical' place (Gibson, 1984). This space supports many daily activities, which are supported within a physical environment. A place according to Chastain and Elliot (1998) is a territory defined by sense of being somewhere as opposed to just anywhere. Moore (2001) defines the sense of place as inter-subjective construction of conditions and experiences through realities that give place a character or quality of life. Place is defined as a setting that transforms space and activity into a socio-cultural event, coming together at the same time and same location for the purpose of an authentic activity (Chastain & Elliott, 1998). Remote learning rich cultural and social phenomenon of 'cyber-place' can be designed to represent a 3-dimensional physical

Table 1. Structural components to create a central balance

	Clear Guidelines	Students	Faculty	Structure
Collaboration and communication	Multiple opportunities for student, teacher, and peer interaction	Cooperation and collaboration among students and faculty	Prompt feedback that is critical and informative	Well designed, discussions, small groups, learner engagement
Goals and expectations	Establish policy, goals, expectations and procedures	Active listening, clear communication	Informational feedback to increase and encourage collaboration and communication	Challenging task, purposeful assignments, pacing of assignments and work load
Objectives and activities	Objectives and standards related to instruction	Purposeful feedback, feedback that is specific and suggestions for growth	Acknowledgement and feedback	Sample cases, simulations and assignments that are problem based
Time line and feedback	Timelines that support and facilitate learning	Communication that is positive, critical and developmental	Praise for quality work Resources for struggling students	Choice of assignments directly related to goal and objectives
Outcomes	Grading – up front guidelines and expectations	Instill high expectations, promote self-efficacy	High expectation, Publication of quality work	Time on task, purposeful assignments and deadline flexibility

place using common web tools such as Java 3D. This human interaction will foster a sense of place. Equipping a cyber-place environment with learning tools such as a white board will further enhance the sense of place, affording different levels of student interaction. Development of outside resources and environments in a virtual world (i.e., library, museum, organizations) will foster a deeper level of collaboration through place-based cyber-place learning environments.

Sobel (2005) states, "Place-based learning is a process that requires hands-on learning, real world application, and integrated curriculum. Any genuine teaching will result, if successful, in someone's knowing how to bring about a better condition of things than existed earlier" (Dewey, 1980). Genuine teaching as Dewey stated will result in "knowing." To understand the notion of "knowing," one must stand within the world, interacting with the elements, rather than standing outside, looking in (Sobel, 2004). Genuine teaching emerges from pedagogy of place, rather than application of skills in a digital environment. Place-based education integrates curriculum that emphasis project based learning. Teachers collaborate and create an environment that is conducive to real world application. John Dewey (1890, 1980) saw this need almost one hundred years ago when he wrote *School and Society;* he stated from the standpoint of the child, "isolation from the school-is isolation from life" (p. 18). Dewey felt that the child was unable to apply what he was learning in school to real world. Real world projects cultivate the students, creating a sense of community within their environment. Authentic learning needs to take place in an authentic environment that propels students into the new literacies of the 21st century. Educators should no longer request electronic devices be left outside the classroom. Devices such as the iPad, iPod, smart phones, and eReaders are an important part of the 21st century learner's environment. They will propel student learning into a deeper understanding of content, preparing them for the skills needed to succeed in the future.

Place-based education supports and enhances "Closing the Achievement Gap." Lieberman and Hoody (1998) discussed the latest research involving place-based education and Environment as an Integrating Context (EIC). It showed that students, teachers, and administration all developed problem-solving skills, critical thinking skills, and increased engagement in learning. In the study, they compared attitude, attendance, and behavior. The study showed a decrease in behavioral referrals, lower attendance rates, and overall increased engagement in learning.

Smith states, "the primary strength of place-based education is that it can adapt to the unique characteristics of any environment, even digital, and it can help overcome the disjuncture between school and children's lives" (as cited in Sobel, 2005). Woodhouse states, "knowledge of the nearest things should be acquired first" (as cited in Sobel, 2005, p. 4). What are the nearest things in regards to digital technologies? Text message, IM, email. Sobel (2005) shows a paradigm of place-based education from the concept of near to far through his question. He asks, "Why are educators using textbooks that focus on land formations in Arizona when we have such amazing resources right in our own back yard" (p. 4). When teaching geography using the textbook, pictures depict the concept, instead of real world application. Often students fail to see the real world connections. Technology can and will enhance the visual learning through videos, photos, and site based organizations.

Sobel (2005) talks about "speciation," the ability to let the curriculum transform into a unique and different learning guideline. He states that what emerges is pedagogy of place, a desire to cultivate the child's strengths.

Teachers and students formulate and solve real world problems; they foster a community of collaboration and commitment; they develop an authentication to learning. In his pedagogical creed Dewey (1897) states that education is a process of living and not a preparation for future living. As educators it is our job to instill this in the youth of the country. Sobel's theoretical frameworks echo the words of nineteenth century educator John Dewey. Dewey affirms

"from the standpoint of the child, the great waste in the school comes from his inability to utilize the experiences he gets outside the school in any complete and free way within the school itself; while, on the other hand, he is unable to apply in daily life what he is learning in school. That is the isolation of the school, its isolation from life" (Loveland, 2002).

Johnson & Johnson (1996) have provided a theoretical basis for collaborative learning; they state a Vygotskian approach would enhance behavioral learning; social interdependence and positive social interdependence would enhance the competitive work environment. Instructors can enhance social interdependence, behavioral learning, and critical thinking by moving from cooperation in a place-based brick and mortar classroom to a collaborative cyber-place environment. Can it effectively incorporate pedagogy of learning (technology, content, and critical thinking) online, or do we need to stay in the era of brick and mortar learning?

In an age of digital immigrants, lecture style learning would have sufficed, but as Reich states, new immigrants are "symbolic-analytic workers" (as cited in Williams, 1991), workers who have the ability to solve problems. They have the capacity to create new knowledge products through analysis and synthesis of existing information. As educators explore possible alternatives in the delivery method of curriculum, it is imperative to view the pedagogy in conjunction with the new medium of instruction. Either delivery method must have specific guidelines and protocols in place, clarity, expressiveness, feedback, methodological quality, and completeness. Promotion of positive interdependence, which is at the heart of collaboration (Johnson & Johnson, 2004), will breed success within the group. Development of trustworthiness by members through the acknowledgement and challenge of each other's ideas and perspectives developed through online social skills will foster the development of collaboration. Activities that reinforce clear communication, quality work, social interaction, and reciprocal interaction will enhance the positive interaction required for deep collaboration in an online environment. Hiltz, Coppola, Rotter, and Turoff (2002) suggest students must possess or be trained to have enough computer literacies so that technology does not impede communication and collaboration. As instructors develop online modules, learning needs to move away from the cognitive process, which is task-oriented into a more complex discussion that allows cognition to lead to the development of metacognitive knowledge. This process will be nurtured through collaborative trust activities, sense of belonging in the online community and consensus building opportunities.

As educators continue to explore diverse modes of delivery, we have to remember Vygotsky (1978) and Piaget (1972) had a strong influence on the formation of learner-centered development. Vygotsky conceived that knowledge constructed through relevant activities developed an immersion that students contextualized, constructed, collaborated, and communicated, that same knowledge which can be constructed within online courses. Vygotsky's "Zone of Proximal Development" allows for social interaction between the novice learner and the expert. Piaget (1972) describes four stages of cognitive development: sensorimotor, preoperational, concrete, and formal operations. Development of these four stages relates to a person's ability to understand and assimilate new information.

As educators, we must become facilitators in a technological classroom, be it brick and mortar or cyber place. Whiteboards need to be more than just an extension of the chalkboard; schools need to embrace the creativity that technology offers. Teachers need to function outside of institutionalized settings (Williams, 2008). The social dynamic of the learning environment has changed; we can promote students' critical thinking skills through well-designed online discussions. We can foster a deeper understanding through communication and respect for diversity. Educators can give pre-service teachers the tools employers seek by developing collaborative problem solving skills, but to do this educators must move from scholar academic to learner-centered ideologies where the professors and the students are a community of learners.

If educators are interested in changing the current status of education then educators must focus on teacher education programs. Educators must focus on teacher knowledge, delivery, and the learning experience. Knowledge building through reflective learning experiences for pre-service teachers that are powerful, relevant to practice, and meaningful enough to transfer the application to their students when they begin teaching is required. Teacher learning is an active process in which knowledge is enacted, constructed, or revised. This process needs to be applicable to the 21st century and the needs of digital learners.

Digital activities within online courses can lead to engagement of activity. Hargreaves (2001) states that teachers are teaching and learning in ways they have never been taught; teacher knowledge is moving beyond procedural and conditional knowledge. Teachers need to move beyond knowing the "how" and "that." They are moving deeper in complex cognition that is leading teachers to reflect and think more critically about their practice. Teachers learn and develop their professional pedagogy best when they have an objective or a goal. The activity becomes relevant to learning. When they can use the tools to express their desired outcome, the transference of learning develops. Networked technologies can support teacher preparation programs through collaborative modules that reinforce knowledge building activities such as video conferencing, email, discussion posting, case study scenarios, simulations, and interactive animations. Networked technologies can support teacher preparation programs by reducing isolation and fostering reflective practice.

Digital technologies developed early in teacher preparation programs allow the learner to interact and engage within an activity that prepares students for the skills necessary for 21st century digital literacies. This interaction may extend or enhance the users objectives based on constraints or affordances within the technology. When pre-service teachers are interested in changing the dynamics of education related to digital technology then teacher preparation programs will need to prepare future educators to move beyond the constraints of technology.

SOLUTIONS AND RECOMMENDATIONS

Collaboration in an online environment is critical to depth of understanding, moving the student into a deeper understanding of pedagogy through real world experiences; thus, the outcome of the reflective process leads to self-examination of knowledge. The constructivist framework emphasizes the growth of pre-service teachers. Students who received specific guidelines and feedback on their work develop stronger ties to their learning community and foster a deeper understanding of the content. This in turn may nurture patience and empathy for their fellow learners in the community. Reflective teaching is a necessity to develop metacognitive skills and strategies to propel extended depth of understanding of concept. Merely putting students together in a group setting does not equate to student engagement through the collaborative learning community process in place-based learning. Faculty must expand clear and concise guidelines and expectations as they develop and foster a sense of community.

Constructivism, which is a learner-centered approach, helps to fortify the main goal of online teaching and learning, creating meaningful environments that are inclusive of communication and collaboration (Gold, 2001). These environments are no longer confined to brick and mortar physical spaces. Gold (2001) indicates that online facilitators can enhance learning for their students in three ways. First and foremost learning can be enhanced though organization. Online course content must be clear, concise, and organized, showing strong correlation to goals and objectives. Students are the facilitators, developing the strategies and schema within the learner-centered community. As the community of learning progresses and develops, communication is the next step in development. Communication can take many forms in both online synchronous or asynchronous environments. Announcements via email, chat rooms, and online discussions that offer specific feedback develop a reflective stance. Discussion board postings become an active process of assessment of factual knowledge and serve as artifacts that show preparation of students as facilitators for 21st century learning. Finally, an online presence is crucial to the success of students as facilitators and active learners. Instructors become a role model in the facilitation process; the process that leads to learner-centered learning as students become active in their learning process. Teacher modeling of best practices within the digital environment will support student growth. As instructors guide student inquiry into the exploration of new cognition and schema collaboration, reflection develops a deeper understanding of student-centered education.

Teacher preparation programs need to provide opportunities for learners to incorporate elements of the classroom that bridge both the traditional brick and mortar and the online environment. As we prepare students for the world, we [educators and facilitators] must develop real world application within their environment. Environments are no longer bound by place and physical components. Place-based learning has evolved to "cyber place" learning in which learning is constructed through collaboration and facilitation. The constructivist theory and place based learning are necessary components to online learning; it develops learning that fosters collaboration and technology cognition. As Doolittle (1999) stated, development of strong pedagogy should take place in authentic real world environments. The 21st century is propelling change, change that is driving teacher preparation programs to revisit learning and delivery of material.

FUTURE RESEARCH

A common characteristic of online teaching is student-centered instruction, where instructors define parameters and goals needed for student success. The intent is to establish a classroom-learning environment that is not bound by place or time. One of the largest changes brought about within the online environment is the movement from Computer-Mediated Instruction (CMI) in which the interaction between students and content has moved beyond the drill and skill tutorials of earlier CMI programs. Today, the paradigm shift has propelled learners into Computer Mediated Communication (CMC), where the focus is on the interaction between student, instructor, and peers. This interaction is mediated by means of the computer (Kearsley, 2000). Online learning and teaching encourages the facilitator to take the problem-based approach to learning, which is very compatible to online learning. This is made possible because of the wealth of resources available. As educators develop online content, it is important to develop a few parameters: learners must possess initiative, self-discipline, and follow through. Students who lack these skills usually do poorly in an online environment. As one develops the online content, it is imperative to develop scaffolding opportunities to support students who may not have these characteristics.

The nature of online education provides a great deal of autonomy to the learner. Choice must be considered as the educator develops timelines and activities within the course. Choice of when, where, and how enhances the learner-centered approach, further developing the constructivist pedagogy. Lastly, the learner-centered approaches must move away from the teacher being the dispenser of knowledge. Student problem solving and collaboration should become the pivotal point of learning and so social skills are extremely important especially when collaboration is involved. Interaction with peers and outside experts in the field depend on one's ability to communicate, whether it is face-to-face or virtually. The benefits of this collaboration illustrate the development of metacognitive skills in a supportive environment. Kearsley (2000) states that the milieu of online learning activities requires a different set of skills and behaviors when compared to face-to-face interaction. Online interaction is more complex due to the nuances of digital literacy.

Formal education has traditionally relied upon the objectivist view of knowledge (Gulati, 2004). Learning is a social experience and assumes that flexibility offered by online environments can support the diverse learner. Student centered learning is rooted in the constructivist view and online learning emphasizes the student-centered, problem-solving approaches. Constructivism is a learning process in which students create long-term relationships with knowledge, that relationship will be transferable to other environments of learning.

CONCLUSION

In conclusion, collaborative learning that emphasizes the constructivist views in a cyber-place environment will enhance the individual learners' potential to achieve the objectives and goals. Learner- centered environments require collaboration and discussion between all participants within the environment. Online discussions and chats will lead to reflective practices, resulting in a higher quality of pedagogical knowledge. We cannot ask pre-service teachers to prepare 21st century students if they are not prepared. Learning opportunities must be embedded within the environment in which it takes place. This does not just happen; it must be created and nurtured throughout the place-based experience, be it brick and mortar or cyber-place. What we learn and how we learn go hand-in-hand; cognition is not just within the individual but also within the content. Course design in which content is directly related to student outcomes and objectives will instill a deeper conversation and extension of knowledge. Once students see value in assignments as they relate to specific learning objectives and goals, they begin to take ownership of tasks. Tasks must be authentic and present cognitive challenges where critical thinking and discussion are involved. The complexity of the task and the learning environment should complement each other, bringing the learner closer to preparedness in the 21st century framework. This framework will support and encourage the learner through the negotiation of knowledge.

Students must have multiple opportunities to manipulate content with extensive feedback and comments. Piaget (1972) stated that learning occurs through the cognitive process of environmental interactions and the corresponding construction of mental structures to make sense of the process. Cognitive learning is often labeled constructivism, emphasizing that learning is taking place as the learners construct their own understanding. This mental process is schema; if we wish to prepare our pre-service teachers for the instruction of skills needed for the preparation of the 21st century and beyond, learning should take place within those structures. Many of the theories presented in this chapter overlap and complement online teaching and to some extent deepen the understanding of content present within web courses. The needs and challenges that both K-12 and teacher preparation programs face with respect to technology show that collaboration through learner-centered learning can provide a foundation for skills required of our students. Technology can enhance collective intelligence through collaboration and discussion. It is said that teacher education programs that support and develop technological opportunities for pre-service teachers through authentic learning opportunities that emphasis-using technology have an impact on student learning. These skills will prepare students for the workforce of the 21st century.

REFERENCES

Allen, I. E., & Seaman, J. (2010). *Class differences: Online education in the United States, 2010*. The Alfred P. Sloan Foundation: Author. Retrieved from http://sloanconsortium.org/publications/survey/class_differences

Brown, J. S., Collins, A., & Duguid, P. (1989). Situated cognition and the culture of learning. *Educational Researcher*, *18*(1), 32–42.

Chapman, C., Ramondt, L., & Smiley, G. (2005). Strong community, deep learning: Exploring the link. *Innovations in Education and Teaching International*, *47*(3), 217–230. doi:10.1080/01587910500167910

Chastain, T., & Elliot, A. (2001). Cultivating design competence: Online support for beginning design studio. In *Proceedings of Association of Computer Aided Design in Architecture* (pp. 130-139). Quebec City, Canada.

Coiro, J. (2009). Rethinking online reading assessment. *Educational Leadership*, *66*(6), 59–63.

Dewey, J. (1897). My pedagogic creed. *The School Journal*, *55*(3), 77–80.

Dewey, J. (1916). *How we think*. Boston, MA: Houghton Mifflin Company.

Dewey, J. (1891, 1980). *School and society*. Carbondale, IL: Southern Illinois University Press.

Doolittle, P. (1999). Constructivist pedagogy. *Educational Psychology*.

Gibson, W. (1984). *Neuromancer*. New York, NY: Ace Books.

Gold, S. (2001). A constructivist approach to online training for online teachers. *Journal of Asynchronous Learning Networks*, *5*(1), 35–57.

Goodson, I. F. (2003). *Professional knowledge, professional lives: Studies in educational change*. Philadelphia, PA: Open University Press.

Gulati, S. (2004). *Constructivism and emerging online learning pedagogy: A discussion for formal to acknowledge and promote the informal*. London, UK: University of Glamorgan.

Hargreaves, A., Earl, L., Moore, S., & Manning, S. (2001). *Learning to change: Teaching beyond subject and standards*. San Francisco, CA: Jossey Bass.

Harrison, S., & Dourish, P. (1996). Re-place-ing space: The roles of place and space in collaborative systems. *Proceedings of the 1996 ACM Conference on Computer Supported Collaborative Work*.

Hassel, B., & Terrell, M. (2004). *How can virtual schools be a vibrant part of meeting the choice provisions of the No Child Left Behind Act?* [Virtual School Report, White Paper], Summer 2004. Retrieved from http://www.connectionsacademy.com/virtualreport.asp

Hiltz, S. R., Coppola, N., Rotter, N., & Turoff, M. (2001). Measuring the importance of learning for the effectiveness of ALN: A multi-measure, multi-method. *Journal of Asynchronous Learning Networks*, *4*(2), 103–125.

Hoadley, C., & Kilner, P. G. (2005). Using technology to transform communities of practice into knowledge-building communities. *SIGGROUP Bulletin*, *25*(1), 31–40.

Johnson, D. W., & Johnson, R. T. (1996). Cooperation and the use of technology. In Jonassen, D. H. (Ed.), *Handbook of research for educational communications and technology* (pp. 1017–1044). New York, NY: Simon and Shuster Macmillan.

Kearsley, G. (2000). *Online education cyberspace learning and teaching*. Belmont, CA: Wadsworth.

Lieberman, A., & Hoody, L. L. (1998). *Closing the achievement gap: Using the environment as an integrating context for learning*. San Diego, CA: State Environment and Education Roundtable.

Loveland, E. (2002). Connecting communities and classrooms. *Rural Roots Newsletter*, *3*(4). Retrieved from http://www.seer.org/pages/rsct.html

McCombs, B. L. (2001). What do we know about learner and learning? The learner-centered framework: Bringing the educational system into balance. *Educational Horizons*, *53*, 82–191.

Moore, S. A. (2001). Technology, place, and nonmodern thesis. *Journal of Architectural Education*, *54*(3), 130-139.

Muirhead, B. (2002). *Salmon's e-tivities: The key to active online learning*. Sterling, VA: Stylus.

Piaget, J. (1972). *The psychology of the child*. New York, NY: Basic Books.

Schiro, M. (2008). *Curriculum theory: Conflicting visions and enduring concerns*. Los Angeles, CA: Sage.

Shulman, L. S., & Shulman, J. (2004). How and what teachers learn: A shifting perspective. *Journal of Curriculum Studies*, *36*(2), 257–271. doi:10.1080/0022027032000148298

Slattery, P. (2006). *Curriculum development in the postmodern era* (2nd ed.). New York, NY: Rutledge.

Sobel, D. (2004). *Place based education: Connecting classrooms and communities*. Great Barrington, MA: Orion Society.

Sobel, D. (2005). *Place-based education: Connecting classrooms &communities*. Great Barrington, MA: Orion Society.

Vygotsky, L. (1978). *Mind in society*. Cambridge, MA: Harvard University Press.

Williams, P. (2008). Leading schools in the digital age: a clash of cultures. *School Leadership & Management*, *28*(3), 213–228. doi:10.1080/13632430802145779

ADDITIONAL READING

Bartolome, I. I. (1994). Beyond the methods fetish: Toward a humanizing pedagogy. *Harvard Educational Review*, *64*(2), 173–194.

Brown, A. L. (1990). Domain-specific principles affect learning and transfer in children. *Cognitive Science*, *14*(1), 107–133.

Bruffee, K. A. (1999). *Collaborative learning: Higher education, interdependance and the authority of knowledge* (2nd ed.). Baltimore, MD: John's Hopkins University Press.

Bruner, J. (1990). *Acts of meaning*. Cambridge, MA: Harvard University Press.

Burch, C. B. (1999). When students (who are preservice teachers) don't want to engage. *Journal of Teacher Education*, *50*(3), 165–172. doi:10.1177/002248719905000302

Chester, A., & Gwyne, G. (1998). Online teaching: Encouraging collaboration through anonnymity. *Journal of Computer-Mediated Communication*, *4*(2).

Chou, C. C. (2001). Formative evaluation of syncronous cmc systems for learner-centered online course. *Journal of Interactive Learning Research, 12*, 169–188.

Coppola, N. W., Hiltz, S. R., & Rotter, N. (2001). Becoming a virtual professor: Pedagogical roles and ALN. *Proceedings of the 34th Annual Hawaii International Conference on System Science*. IEEE Press.

Curtis, D. D., & Lawson, M. J. (2001). Exploring collaborative online learning. *Journal of Asynchronous Learning Networks, 5*(1), 21–34.

Dillenbourg, P., & Schneider, D. (1995). *Collaborative learning and the Internet*. Retrieved from http://cfa.unige.ch/tecfa/research/CMC/olla/iccai95_1.html

Duffy, T. M., & Cunningham, D. J. (1996). Constructivism: Implications for the design and delivery of instruction. In Jonassen, D. H. (Ed.), *Handbook of research for educational communication and technology*. New York, NY: Simon & Schuster Macmillian.

Espinoza, S., & McKinzie, L. (1994). *Internet activities open new worlds for educators. Technology and Teacher Education Annual* (pp. 666–670). Charlottesville, VA: Association for the Advancement of Computing in Education.

Forrester, J. W. (1992). System dynamics and learner-centered-learning in kindergarten through 12th grade education. *Lecture at Sloan School of Management, 43*(3), 11–21.

Garrison, D. R. (2003). Cognitive presence for effective asynchronous online learning: The role of reflective inquiry, self-direction, and metacognition. In Bourne, J., & Moore, J. C. (Eds.), *Elements of quality online education: Practice and direction* (pp. 47–58). Needham, MA: Sloan Center for Online Education.

Gillette, D. H. (1996, Fall). Using electronic tools to promote active learning. *New Directions for Teaching and Learning, 67*, 59–70. doi:10.1002/tl.37219966708

Harrison, S., & Dourish, P. (1996). Re-place-ing space: The roles of place and space in collaborative systems. *Proceedings of the 1996 ACM Conference on Computer Supported Cooperative Work.*

Hickman, L. A. (1990). *John Dewey's pragmatic technology*. Bloomington, IN: Indiana University.

Hiltz, S. R. (1986). The virtual classroom: Using computer mediated communication for university teaching. *The Journal of Communication, 36*(4), 95–104. doi:10.1111/j.1460-2466.1986.tb01427.x

Hiltz, S. R., Coppola, N., Rotter, N., & Turoff, M. (2001). Measuring the importance of learning for the effectiveness of ALN: A mult-measure, multi-method. *Journal of Asynchronous Learning Networks, 4*(2).

Hung, D., & Nichani, M. (2001). Constructivism and e-learning: Balancing between the individual and social levels of cognition. *Educational Technology, 41*(2), 40–44.

Jonassen, D. H., & Reeves, T. C. (1996). Learning with technology: Using computers as cognitive tools. In Jonassen, D. H. (Ed.), *Handbook of research for educational communication and technology* (pp. 693–720). New York, NY: Simon & Schuster Macmillan.

Kalay, Y. (2004). Virtual learning environment. *ITcon, 9*, 195–207.

Knapp, C. E. (1996). Just *beyond the classroom: Community adventures for interdisciplinary learning*. Charleston, WV: ERIC Clearinghouse on Rural Education and Small Schools. (ERIC Document Reproduction Service No. ED388485).

Koohang, A. (2009). A learner-centered model for blended learning design. *International Journal of Innovation and Learning*, *6*(1), 76–91. doi:10.1504/IJIL.2009.021685

Kosnick, C., & Beck, C. (1999) *Looking back: Six teachers reflect on the action research experience in their education program*. Paper presented at Educational Research Association, Montreal, Canada

LaMaster, K. L., Martin, D. B., & Vinge, S. (1999). Does electronic communication inhance guided reflection among graduate students? *SITE- Association for the Advancement of Computers in Education* (pp. 1380-1386).

Lambert, N., & McCombs, B. I. (1998). *How students learn: Reforming schools through learner centered education*. Washington, DC: APA Books. doi:10.1037/10258-000

Marquardt, M. J., & Kearsley, G. (1998). *Technology based learning: Maximizing human performance and corperate success*. Washington, DC: CRC Press.

McCombs, B. L. (2001). What do we know about learner and learning? The learner centered framework: Bringing the education sytem into balance. *Educational Horizons*, *79*(4), 182–193.

Meyer, K. A. (2004). Evaluating online dicussions. Four different frames of analysis. *Journal of Asynchronous Learning Networks*, *8*(2), 101–114.

Palloff, R., & Pratt, K. (1999). *Building learning communities in cyberspace: Effective strategies for online classrooms*. San Francisco, CA: Jossey-Bass.

Parker, D., & Gemino, A. (2001). Inside online learning: Comparing conceptual and technique learning performance in place-based and ALN formats. *Journal of Asynchronous Learning Networks*, *5*(2), 64–74.

Reil, M. (1994). Educational change in technology-rich environment. *Journal of Research on Computing in Education*, *26*(4), 452–474.

Renninger, K. A., & Shumar, W. (2002). *Building virtual communities: Learning and change in cyperspace*. New York, NY: Camridge University Press. doi:10.1017/CBO9780511606373

Richardson, W., & Mancabelli, R. (2011). *Personal learning networks: Using the power of connections to transform education*. Bloomington, IN: Solution Tree.

Salomon, G. (1997). *Distributed cognition: Psychological and educational considerations*. New York, NY: Cambridge University Press.

Schworm, P. (2008, November 13). College students flocking to online classes. *The Boston Globe*, p. 1.

Sosniak, L. A. (1999). Professional and subject matter knowldgefor teacher education. In Griffin, G. A. (Ed.), *The education of teachers* (pp. 185–204). Chicago, IL: University of Chicago Press.

Stone, E. (1994). Reform in teacher education the power of pedagogy. *Journal of Teacher Education*, *45*(4), 310–319. doi:10.1177/0022487194045004012

Teles, L., & Duxbury, N. (1992). *The networked classroom: Creating an online environment for k-12 education*. Burnaby, Canada: Faculty of Education, Simon Fraser University.

Woodhouse, J. L., & Knapp, C. E. (2000). *Place-based curriculum and instruction: Outdoor and environmental education approaches*. Charleston, WV: ERIC Clearinghouse on Rural Education and Small Schools. (ERIC Document Reproduction Service No. ED 448012).

Zuga, K. F. (1992). Social reconstruction curriculum and technology education. *Journal of Technology Education*, *3*(2), 48–58.

KEY TERMS AND DEFINITIONS

Collaborative Learning Communities: Consist of students who work together in small groups to enhance the learning of all members of the group.

Cyber-Place Based Learning: Teaching and learning in an alternate environment, one that is not bound by time or place, integrating curriculum that emphasizes project based learning within the confines of the students' learning environment.

Cognition: The active mental process of receiving, processing, storing, and using information.

Constructivist View: A theory of knowledge. Knowing and interacting within your experiences and environment to construct meaning.

Distance Learning or Virtual Classroom: Learning that takes place anytime, anywhere, without the constraint of time or place within a computer-mediated communication system.

Learner-Centered: Learning that utilizes a problem solving approach and construction to develop meaning and understanding. Often learner-centered is developed under constructivism theory.

Metacognition: The ability to monitor one's thinking, it relies upon the students' ability to adjust strategies when meaning breaks down.

Online Learning Communities: Cyber-place communities that encourage opportunities for individuals to collaborate and communicate. Online learning communities provide collaboration that is flexible and convenient.

Place-Based Learning: Teacher created environment that is conducive to real world application. Place-based education integrates curriculum that emphasis project based learning within the confines of the students environment.

Positive Interdependence: Accomplishments of a group as a whole, the amount of knowledge achieved by all members of the group that lead to the success of all members.

Section 4
Literacy Education in Online Learning

Chapter 11
Differentiating Instruction for Adult Learners in an Online Environment

Dixie Massey
University of Washington, USA

ABSTRACT

Teacher education courses offered online are becoming increasingly common. Unfortunately, few instructors of online teacher education courses have specific preparation for teaching adult learners or in teaching online courses, resulting in faltering attempts to transfer traditional methodology such as lectures to online platforms. This chapter considers the background of distance education and examines relevant literature on adult learners. Differentiated instruction is proffered as a means of meeting the needs of adult learners in online teacher education courses. Specific examples of differentiating content, process, and product are suggested.

INTRODUCTION

Distance education has received increased attention since 1981 when totally online courses were first offered to adult education students (Harasim, 2006; Li & Irby, 2008). Increased offerings of online courses within higher education have raised concerns within the academic community (Duffy, Webb, & Davis, 2009). While many argue about the advantages of differing delivery format, many researchers document that the actual instruction provided within any delivery mode is what determines the effectiveness of any learning experience (Parsons, Massey, Vaughn, Scales, Faircloth, Howerton, Griffith, & Atkinson, 2011). Asking "Is online learning as effective as face to face learning?" is the wrong question. Instead, teacher educators need to explore how online learning can be an effective vehicle for bettering the education offered in teacher education courses.

One area that deserves further attention and research at the higher education level is differen-

DOI: 10.4018/978-1-4666-1906-7.ch011

tiated instruction. While there is little published about differentiated instruction beyond the K-12 settings, online courses are rich environments with many supportive elements for differentiated instruction. Further, typical teacher education distance courses are full of a wide array of adult learners who range in age from teenagers to senior citizens. Online teacher education students often do not share similar geographical places, socio-economic status, family situations, cultural, or linguistic backgrounds. They vary widely in life and teaching experience, making it an ideal environment for differentiation.

The objectives for this chapter are to further the nascent discussion of differentiated instruction at both the college level and in online courses. After providing foundational background from the literature regarding distance education, adult learners, and differentiated instruction, this chapter concludes with an example of differentiation in an online learning environment.

BACKGROUND

When considering differentiated learning in an online teacher-education course, three distinct areas provide important frameworks for consideration: online learning, adult learners, and differentiation.

Online Learning

The advancements of the Internet have naturally led to increased opportunities for learners. Universities have begun to capitalize on the increased student population available world-wide for online courses. Research on online learning has suggested multiple benefits from online courses. First, there is the opportunity for increasingly individualized instruction (Thomson, 2010). Thomson's work with gifted high school students learning online reported that students and teachers believed the format of online learning seemed more personalized with much of the instruction taking place through extended email conversations between the instructor and a single student. Also, online instructors often make use of multiple formats for presenting the material (Li & Irby, 2008), targeting specific learners' strengths and weaknesses (Blake, 2009). Additionally, the learner can self-regulate the learning that occurs (Li & Irby, 2008; Thomson, 2010). Further, some research has suggested that enhanced communication between instructor and student is one frequent outcome of online interaction (Dykman & Davis, 2008; Li & Irby, 2008; Thomson, 2010) and that students may feel more comfortable sharing because they are given time to compose thoughtful responses rather than being called on to respond immediately in class (Li & Irby, 2008). Finally, the terms "convenience" and "flexibility" are used repeatedly in the online learning literature, emphasizing the variety of learning options that make online courses ultimately more inviting and available for many students. As Li and Irby (2008) noted, "Online education provides a time-independent and place-independent learning environment" (p. 453). Thomson (2010) concluded:

> *"There are number of benefits that the online environment can offer students above and beyond what the traditional classroom environment offers. As a result, instead of trying to recreate the face-to-face environment to whatever degree possible, online teachers and program administrators should instead, try to capitalize on the unique benefits that the online environment can offer students." (p. 702)*

Increased offerings of online courses have raised concerns within the academic community. Cited as a major barrier to more extensive adoption of online learning within these settings, many faculty members continue to devalue such course experiences (Allen & Seaman, 2007). Particular to teacher education, Duffy and colleagues (Duffy, Webb, & Davis, 2009) questioned if distance learning was an effective mode of teaching when trying to develop teachers' judgment in dealing with the complexities of classrooms.

Dooley, Lindner, Dooley, and Wilson (2005) made a strong argument that it is not the delivery method that impacts learner achievement as much as the methods used to engage the learner. However, "How learners interact among themselves and with the instructor does have an effect on learner satisfaction" (Dooley, Lindner, Dooley, & Wilson, 2005, p. 57). Learning online may actually help teacher education candidates become more adept at using technology (Barone & Morrell, 2007). Barone and Morrell (2007) considered skills particular to literacy content and wrote, "These new technologies require teachers who are competent in using them themselves and guiding students in their use….Computers become more than a once-a-week visit in the computer lab. They become essential tools in literacy and in sharing reading and writing" (p. 178). These researchers concluded, "Teacher candidates will need to be prepared to use computers and other new media technologies as tools to facilitate the development of academic and new media literacies" (p. 178).

In order for online learning to be effective in teacher education programs, online instructors need to move beyond trying to transfer face-to-face content to an online platform. Technology allows and even enhances the ease with which teacher educators can communicate content with students and individualize instruction. Special consideration must also be given to the learners taking the courses. These adult learners are an increasingly diverse population (Blake, 2009).

Adult Learners

Recognition of the university student as a unique learner is not new. Parr (1929) described special programs for those who struggled to read college material. The GI Bill of Rights: Servicemen's Readjustment Act following World War II brought an increased need for a focus beyond the K-12 student. This Bill offered free tuition to returning soldiers. The GIs took advantage of free tuition, though they were not always prepared academically to succeed in the existing college programs and specialized remedial programs began to emerge across multiple universities (Stahl & Smith-Burke, 1999). Even though attention to adult learners is not new in post-secondary education, the specific needs of a changing population continue to prove challenging for universities. Merriam and Cafferella (1999) pointed out that change from industrial to informational-based society, based on ever-increasing technology capabilities and uses, demands ongoing and evolving education for an older and more diverse population.

Researchers acknowledge that adults have distinct characteristics that make their approach to learning unique (Knowles, 1990). Adult learners rely less on the teacher for direction and content than do younger counterparts. They have more experiences to draw from, which may influence their learning positively or negatively. They are often highly motivated to learn information that has utility for them. The adult learner enters a new learning situation in order to generate change—a change in environment, skills, or knowledge (Russell, 2006). This sense of change carries distinct expectations that the learning will be immediately relevant to their own goals (Knowles, Holten, & Swanson, 1998).

What does instruction that furthers adult learners' learning look like? Rather than talk about the pedagogy appropriate for adult learners, researchers use the term "andragogy" to refer to learner-focused (as opposed to teacher-focused) education for adults (Conner, 2004). Andragogy is based on several assumptions about the learner: (1) learner's need to know, (2) self-concept of the learner, (3) prior experience of the learner, (4) readiness to learn, (5) orientation to learning, and (6) motivation to learn (Knowles, Holton, & Swanson, 1998). Instruction that involves the adult learner in the process of setting goals and outcomes is often a hallmark of androgogical approaches (Dooley, Lindner, Dooley, & Wilson,

2005). This participation in the learning process can only be successful if the learner is self-directed, a principle that is also a critical component of success for distance education students. Further, Dooley, Lindner, Dooley, & Wilson (2005) noted that activities that require students to use and share their own prior experiences are often valued by adult learners. Dollisso and Martin (1999) summarized adult learning by writing,

"A common thread in the literature concerning adult learning is the premise that adult educators or program planners should respond to the needs, interests, and real-life problems of adult learners. Customers frequent a business that satisfies their needs. The same is true with adult learners." (p. 38)

Such a description clearly aligns with the principles of differentiated learning, discussed in the following section.

Differentiated Instruction

Rooted in cognitive psychology (Rock, Gregg, Ellis, & Gable, 2008), differentiated learning has most frequently been linked to the type of instruction necessary for students in special education or gifted education; however, it is more recently being repositioned as the type of instruction that is beneficial to multiple special populations, such as culturally and linguistically diverse students and even to all learners (Lawrence-Brown, 2004; Santamaria, 2009; Waldron & McClesky, 1998). Differentiated instruction is typically defined as providing varied instruction to meet the individual needs of all students (Landrum & McDuffie, 2010; Tomlinson, 1999). There is some debate in the literature about individualized vs. differentiated instruction. While some researchers viewed individualized instruction as occurring exclusively in a one-on-one setting, other researchers labeled this a misperception. Landrum & McDuffie (2010) emphasized that differentiated instruction is "the teacher's response to students' individual needs" (p. 14). Teachers might respond to student needs might occur one-on-one, small group, or even whole-class (Kauffman, Mock, Tankersley, & Landrum, 2008). Landrum & McDuffie (2010) posited,

"Instruction is individualized when (a) it is planned in a way that builds on what individual students currently know and can do and targets meaningful goals regarding what they need to learn next; and (b) accommodations and modifications to teaching and testing routines are made in order to provide students with full and meaningful access to the content they need to learn." (p. 9)

For the purposes of this chapter, I will not make a distinction between individualized or differentiated instruction. Instead, any instruction that includes clearly stated goals and ongoing adjustments of teaching routines within content, process, or product will be labeled as differentiated instruction.

Differentiation is framed by four principles, according to Rock, Gregg, Ellis, and Gable's (2008) extensive review of the literature: (a) focus on goals or ideas of the learning, (b) responsiveness to student differences, (c) assessment that is viewed as part of instruction, and (d) continual adjustment of content, process, and product. These tenets echo many of the benefits listed previously with online learning, highlighting the theoretical link between differentiated instruction and distance education. In order to enact the supporting principles of differentiation, Tomlinson (1999) conceptualized three instructional areas of impact. These three instructional areas are the content, the process, and the product.

Differentiated learning is not a focus on student learning styles. Though previously emphasized in the literature, current research has noted that while students may have different learning preferences, those preferences are not the determining factor in their achievement (Landrum & McDuffie, 2010; Stahl, 1999). Stahl (1999) bluntly stated, "The reason researchers roll their eyes at learning styles is the utter failure to find that assessing children's learning styles and matching to instructional methods has any effect o their learning" (p. 1). Instead of catering to student preferences, Stahl (1999), Landrum & McDuffie (2010) and others (Chamberlin & Powers, 2010) continually emphasized the importance of articulating a clear goal for instruction and matching different methods to the goal and to the different stages of development, not to the individual students. Additionally, open-ended activities have been positively received by a variety of learners (Hertzog, 1998; Parsons, et al., 2011).

While differentiated learning is emphasized in the K-12 classrooms, less attention is given to differentiation at the university level and in the online environment One possible explanation for the lack of research about differentiated instruction at the collegiate level may be the use of terminology. Most often, differentiated instruction has been used to refer to instruction in the K-12 classroom, while adaptive teaching has been used to describe the kind of teaching that teacher education programs seek to promote from their teachers. For example, Tomlinson (2000), when targeting instruction for the K-12 learner, suggested that differentiated instruction occurs when teachers adjust instruction to accommodate student differences. The same adjustment of instruction is referred to as adaptation or in-the-moment decisions in teacher education literature (Duffy, Miller, Kear, Parsons, Davis, & Williams, 2008; Parsons, 2008; Parsons & Burrowbridge, 2011). However, many of these pieces remain theoretical and do not report the systematic study of the impact of differentiated or adaptive teaching on post-secondary learners... Regardless of the term used—adaptive, reflective, individualized, or differentiated--the underlying foundation is that effective teachers are those who are constantly revising their own methodology to better meet student needs (Anders, Hoffman, & Duffy, 2000; Hammerness, Darling-Hammond, Bransford, Berliner, Cochran-Smith, McDonald, & Zeichner, 2005; Snow, Griffin, & Burns, 2005).

One notable exception to the dearth of research on differentiated teaching at the university level is Chamberlin and Powers (2010) work. The researchers taught undergraduate mathematics courses. In each of four units, an undifferentiated lesson was taught initially, followed by formative assessment. The following class then presented two separate activities differentiated for those who were not yet comfortable with the topic and for those who were comfortable with the topic and had demonstrated their understanding in the formative assessment. Following the differentiated activity, a whole-class discussion was used to summarize and extend the learning. Individual, whole group, and small group instruction were utilized as a means of differentiating as needed, though small groups were particularly referenced as a common way to differentiate. When working on particular problems, students were grouped by similar personal interests, similar learning profiles, or self-selection. Assessment played a key role in ongoing modifications, including pre-assessments, formative assessments, interest inventories, learning style inventories, and summative assessments that were evaluated as Developing or Proficient against the seven to ten learning objectives identified for each unit. All students completed the same quizzes and tests. Students all completed homework problems, with some of the problems being given to everyone and some problems being differentiated for particular individuals or students. Students were also assigned projects and were allowed to choose the product format from a list of five to six options. The authors' results suggested that all students advanced in their mathematical learning, regardless of their initial pre-test scores. The authors emphasized that "it was not necessary to differentiate every class or every assignment. When done purposely in response to students' needs, most instructors differentiate their instruction one-third to one-half of the time" (Chamberlin & Powers, 2010, p. 130).

The authors noted particular differences in differentiated instruction for K-12 and undergraduate levels. K-12 teachers have more classroom contact time with students and are therefore better able to extend the amount of time addressing particular goals. Further, college professors rarely have their own classrooms and may be limited in how much they can vary the classroom environment. The authors noted what they viewed as a positive aspect of technology, stating that websites, email, and online materials can be used to add additional differentiation, though they did not record instances of that in their study.

ENACTING DIFFERENTIATION

The review of the literature establishes clear support for the usefulness of differentiated instruction in classes where multiple levels exist. These multiple levels are certainly evident in online teacher education courses (Blake, 2009; Inyega & Ratliff, 2007). What are the components, then, to help frame an online teacher education course and serve as a model for future teachers' own differentiated instruction? As stated previously, differentiation is generally outlined in three components: differentiation of content, differentiation of process, and differentiation of product. The following sections offer suggestions for differentiating these components in the online environment.

Content

Content components refer to the goals and concepts of the course. Content can also refer to adjustment of complexity for learners. In both online and face to face classes, goals should be clearly stated and established at multiple points. In a face to face class, the teacher might list objectives on the board and revisit those objectives during the course of class. Because information communicated online is often viewed sequentially, objectives need to be presented to students frequently. In an online course, not only should the syllabus state the overall objectives of the course, those objectives should be matched to specific content of the course. For each unit presented, the specific objectives of the unit should be restated. Assignments should be reviewed with attention to understanding of these objectives and clarification should be offered via feedback to individuals, small groups, or the whole class. Individual feedback on assignments should re-emphasize the overarching objectives and identify where students are in relationship to their proficiency with the objectives. For example, a ChecBric (Lewin, 2006) can be used as a way to clearly break down the assignment into specific components and link each piece to the course objectives (see Table 1). Students complete the ChecBric and submit it with the assignment to verify their familiarity with the assignment components and the objectives covered. It also serves as a way for the instructor to communicate with the student about progress on certain objectives. In certain cases, the ChecBric can be altered to further differentiate the objectives that need to be met for each student.

Process

The process components of differentiated instruction change the way in which students complete their work. Typically, a major part of differentiating process includes allowing students to work individually and in groups. Additionally, a variety of grouping techniques are used instead of just a single grouping technique in order to emphasize critical and creative thinking, engaging all learners, and balancing teacher- and student-selected tasks (Tomlinson, 2000).

One of the challenges of online learning is creating community within the class and between the students. Grouping students becomes imperative as a way to differentiate and as a way to create community between students (Karber, 2001). On-

Table 1. ChecBric for final case study project (Atkinson, 2009)

Directions: Complete the ChecBric and Student Comments in order to receive full credit.			
Items & Objectives	**Related Tasks**	**Student Verification of Completion**	**Instructor Rating**
Item 1 Objectives 1, 2	**Edited Initial Student Profile and Plan:** Include the entire Initial Student Profile and Plan from Module 4. All corrections and amendments should be evident (different color ink or highlighted). This is a polished version of what you submitted in Module 4.	✓	✓
Item 2 Objectives 2, 3	**Photo Journal:** A copy of your student's entire Photo Journal should be scanned and included in this assignment.	✓	✓
Item 3 Objectives 2, 3	**Photo Journal Narrative:** Add a narrative explaining the value of the photo journal experience and what it revealed about your student.	✓	✓
Item 4 Objectives 2, 4, 5, 6	**Lesson Plans and Reflections:** Number each lesson and lesson reflection. Submit a plan for each lesson, including detailed instructional plans and goals, followed by a reflection of each session. The reflections should detail student progress and what is planned for "next steps." Include short-term goals for the students and how you plan to meet those goals. (Refer to the texts, the word study sequence in WTW, and the Important Behaviors to Notice and Support handout.) Citations from texts should be clearly evident in these reflective paragraphs indicating the incorporation of knowledge gained during this semester.	✓	8/8
Item 5 Objectives 2, 4, 6, 7	**Thoughtfully Adaptive Teaching Reflections:** Reflect on times when you were thoughtfully adaptive in your teaching. Describe those adaptations and the rationale for making these adaptations.	✓	4/4
Item 6 Objectives 2, 3, 7	**Running Records (optional but recommended):** Running records are valuable tools for documenting student's reading progress. Accuracy rates, self correction rates, and analysis of miscues provide a window into the child's reading process. Scanned copies of running records or narratives detailing the results from the running records may be included as well as a reflection on how the running record results influenced your lesson plans (think text selection, strategy support, and prompting here).	✓	✓
Item 7 Objectives 2, 3, 4, 5, 6, 7, 8	**Final Assessments and Results:** results from ALL assessments administered (pre and post) should be included in this section. Final QRI-4 calculation sheets were included, as well as a Developmental Spelling Assessment feature guide. You do not need to include ALL QRI-4 passages and words lists administered unless you want to	✓	5/5
Item 8 Objectives 2, 4, 5, 6	**Assessment Findings/Discussion:** Within several paragraphs, describe the trends and patterns that were noted in your student's reading across the entire time that you met. Address how this student is "literate" and whether or not the assessments that were used accurately documented his/her understandings. Discuss whether or not the assessments documented your student's gains in the area or areas you focused upon in his/her instruction.	v	3/5
Item 9 Objectives 9, 10	**Home School Partnership Part A:** A narrative describing the recommendations that were made for 1) your tutee's teacher (or another designee) and 2) your tutee's parent/guardian so that they can help him/her maintain the progress that has been made. (see Module 5 Commentary for more details)	✓	✓
Item 10 Objectives 9, 10	**Home School Partnership Part B:** Include a schedule of your completed meetings and the agendas of these meetings. There should be evidence that you met with each of these parties at least 3 times across the tutoring experience and that you followed through with your plan. Evidence should be in the form of the other parties' reactions and responses and how your partnership influenced your work with the student.	✓	4/4

continued on following page

Table 1. Continued

Item 11 Objectives 2, 4, 5, 6	**Final Reflection:** What did you learn from this tutoring experience about a) supporting your student so he/she could succeed, b) using formative assessment to plan instruction and c) teaching struggling readers in general? Citations from texts (those used in this course and other sources) must be clearly evident in this reflection. THIS PIECE IS CRITICAL. IT SHOULD PROVIDE CONVINCING EVIDENCE OF YORU EXPERTISE AS A MASTER TEACHER OF READING.	✓	8/10
Item 12 Objective 1, 7	Correct Grammar, Usage, and Mechanics (no more than 2 errors)	✓	1/1
Item 13 Objective 7	ChecBric items checked off and Student Comments reflect professional knowledge and thoughtful composition	✓	1/1
Total:		38	34
Student Comments:			
Instructor Comments:			

line platforms allow for clearly defining multiple groups within the same class. Some topics may utilize self-selected groups while other topics may include instructor-assigned groups. For example, teacher education courses that include students with different concentrations may first establish groups based on discipline. These groups may explore the use of a particular technique within a discipline or work together to create lesson plans appropriate for discipline. At the same time, the instructor may offer a select number of texts that support a topic or objective and allow students to self-select the book of their choice. Students may then become a part of a book club group that is self-selected, supports the objectives of the course, and enhances particular areas of interest for the student. For example, a series of texts might be offered on how to teach comprehension, but each text would offer a different level of focus (elementary, middle, high school, or special education). Table 2 offers some possible grouping strategies, including suggested objectives and assignments. Ideally, students would be members of at least two different kinds of groups. Separate groups might be linked through the jigsaw grouping method, where the home group becomes an expert group in one area. This group is broken up into separate groups and the information that was learned in the expert group is then shared with the new group. This allows everyone to contribute new information to a group. Group variation would allow further opportunity for specific differentiation of content, process, and product.

Product

The product components of instruction include the ways in which students demonstrate they have met the established goals. Some assignments may require the same end-product, while other assignments have differentiated end-products based on student choice or teacher assignment. Most objectives can be met in multiple ways that sometimes capitalize on student interest or preference. For example, demonstrating understanding of content might take the final form of a WebQuest, a video, a written paper, or a presentation.

In the online classroom, students are often asked to show their competence through some written work (DeCandido, 2006; Gilman, 2010). True differentiation values other forms of expression beyond just writing, such as speaking, performance, or demonstration. Thus final products should include more possibilities than just written products, such as observing students via posted video, live Webcam, or listening to discussion via Skype or other media.

Table 2. Grouping strategies

	Description	Objectives	Assignments
Disciplinary Group	Disciplinary Groups are assigned by the instructor based on the student's major or area of concentration.	• Students will gain in-depth knowledge of a particular area, either discipline-specific or how another subject relates to discipline-specific content. • Students will provide feedback to one another about the strengths and areas for improvement in a content-area lesson.	• Threaded discussions about the content of the reading • Demonstration of the content of the reading, including a PowerPoint, video, Glog, or other media that can be uploaded or viewed online. • Discussion responses • Feedback forms submitted to peers • Use of track changes or Jing video responses.
Book Club Group	Book Club Groups are student-selected based on a set of texts posed by the instructor that support the objectives of the course.	• Students will identify and explore the pros and cons of particular methodologies. • Students will analyze and critique a particular theory.	• Threaded discussions • Blogs created to summarize and categorize resources for other group members/readers. • Demonstration of the theory or methodology presented in the reading, including a PowerPoint, video, Glog, or other media that can be uploaded or viewed online.
Media Groups	Media groups are groups that view a particular video, set of videos, or websites. They may be student-chosen or assigned by the instructor.	• Students will explore the resources available for teaching a particular content. • Students will view and analyze a teaching method in action. • Students will create their own video and share it with the group members.	• Threaded discussions • Blogs created to summarize and categorize resources for other group members/readers. • Video editing to show particular methods.
Instructor Discussion Group	Instructor Discussion Groups are chosen by the students and are centered on topics of interest that are posed both by the instructor and by the students.	• Students will gain in-depth knowledge of the course content. • Students will consider alternate theories of education.	• Case studies of particular educational issues. • Wikis that build on the group's prior knowledge and include new knowledge gained through course content.
Technology Support Group	May include the whole class and be used as an informal site to pose and answer questions about how to use particular technologies or solve technological issues.	• Students will demonstrate the use of a range of technologies available to support teaching.	• Try one new technology unfamiliar to you at the beginning of the course and document a) the steps you took to learn and use this technology and b) your personal reaction to this process.

Assessments are critical to a differentiated classroom. Tomlinson (1999) emphasized that assessment is "ongoing and diagnostic" as a means to understand the needed modifications on a daily basis (p. 10). Assessment must be ongoing and frequent enough to provide further learning opportunities. Assessments should include a variety of formative and summative assessments. Examples of possible formative and summative assessments are listed in Table 3.

Feedback: Pulling it All Together

While feedback is not formally one of the official components of differentiated instruction, feedback forms the backbone of all three components. The feedback should flow between instructor and student and from student to instructor. Students in non-differentiated classrooms often receive infrequent feedback and feedback only on the summative assessments. In a differentiated online class, feedback should happen frequently,

Table 3. Possible formative and summative online assessments

Assessment Type	Feedback
Formative • Threaded discussions • Wikis • Reports submitted in parts rather than as a whole • Group discussions via Skype, posted video, live Webcams	• Instructor poses questions, adds new information to a discussion or a wiki. • Instructor suggests new resource for further learning. • Instructor "conferences" with individual students via email, phone, or Skype to update student about progress. • Instructor completes rubric on ongoing drafts for assignment goals that students will review before submitting assignment for summative assessment.
Summative • Presentations • Glogs • Papers • Videos of demonstrations or performances • Observations via Webcam	• Rubrics and/or Checbrics that show performance against established goals. • Individual conferences to document progress.

generally at least weekly, so that students have a clear picture of their proficiency with the course objectives and to support the community created within the class.

Feedback on formative assessments might take the form of instructor comments on a threaded discussion where the instructor can pose questions or problems for further discussion. Ongoing individual emails help the instructor form individualized plans with each student. The use of rubrics at the formative stage of feedback can be particularly helpful to aid students in understanding their proficiency against the course objectives. For example, a ChecBric (See Table 1) can be emailed back to individual students at several different points in the semester so that they can continue to monitor their progress on specific components of an ongoing assignment. Feedback for summative assessments will be based on similar feedback systems, including the use of rubrics, conferencing, individual emails or live conversations. Additionally, differentiated feedback will often suggest next steps for the individual, whether future ideas to consider, additional resources to reference, or future steps to try in a classroom.

Students should provide feedback to the instructor in several ways. Feedback might take the form of an "exit slip" or email at the end of each topic. The ChecBric in Table One presents a section for student comments where students might add their comments. Students should be invited to offer summaries of what they learned, ongoing questions or confusions, and suggestions. Keeping ongoing individual conversations open allows students to feel more comfortable emailing for further clarification throughout the course. Formal course evaluations allow for a final round of feedback between student and instructor, though this type of feedback is only good for adjusting upcoming courses and not for fine-tuning the existing class time.

As with any methodology, it is important to make the online students very aware of the differentiation--how differentiation is being achieved, and why differentiation is being utilized. Ultimately, the differentiation of online learning should serve to not only meet the learner needs more efficiently and effectively, but also serve as a model for the future teachers' own classroom instruction.

CHALLENGES AND SUGGESTIONS FOR DIFFERENTIATED LEARNING IN ONLINE COURSES

Though there is a strong theoretical background in support of differentiated online learning, there are many challenges to creating differentiated learning situations for online courses. First, online instructors may be very familiar with K-12 teaching but ill-prepared to teach online and/or to teach adult learners. Many, if not most, online instructors in teacher education programs are trying to adapt traditional formats, such as lecturing, and recreate them as online formats instead of creating a new method matched to online learning (Dooley, Lindner, Dooley, & Wilson, 2005). Zuckerbrod (2011) noted that many teachers have a "shallow repertoire of instructional tools" (p. 36) and this has been noticeably true for online instructors.

Another barrier for differentiation comes from one of the unique idiosyncrasies of online learning. Typically the entire course is uploaded at the beginning of the course or semester, as opposed to face-to-face courses, which are only outlined at the beginning of the course. Many researchers suggested that online courses require more upfront planning than traditional face-to-face courses (Dykman & Davis, 2008; Karber, 2001; Li and Irby, 2008). What often results in an inflexible framework for online courses and thus, formative assessment is nearly non-existent. In face-to-face situations, much of the differentiation can be created during the instruction, occurring as frequent, even daily adaptations based on observed needs. This is not possible in an online course where students must know the layout of the course at the beginning and groups should be established quickly to establish community (Dykman & Davis, 2008). Thus, differentiation has to be a planned-for event and is best implemented in a course that the instructor has taught before in an online environment. That previous experience can help identify probable components that would benefit from differentiated learning.

Additionally, because of the intense time commitments of online instruction, it is unrealistic to think about differentiating all of the learning modules or assignments. Instead, identifying one or two assignments or modules for differentiated instruction allows the instructor to create a schedule that shows particular assignments will be given but notes that specific details will be given or the module will be opened later in the course. That gives students knowledge of what is coming and which objectives will be met while also giving the instructor time to get to know students and understand more about the varying needs of the students.

WHAT DOES IT LOOK LIKE? DIFFERENTIATION IN AN ONLINE COURSE

What could all of this really look like in action? I considered this question as part of a larger research project on the impact of teacher researcher and online instruction (Massey, 2009). The course, Teachers as Researchers, was taught entirely online to a group of 25 graduate students in teacher education. I conducted my own teacher research to better understand online instruction. One of the questions I considered was how I could plan an online teacher research course to better meet the needs of my students.

Teacher research is a methodology useful as a way of studying one's own practice and is both a way of researching and a way of revising one's own instruction (Chiseri-Strater & Sunstein, 2006; Hubbard & Power, 1999). I collected data throughout the year-long course and during the year after the course was completed. Data included student assignments, student feedback, interviews with five students after the course was complete, and a researcher notebook. I analyzed the data using qualitative content analysis (Patton, 1990). Based on the results from this research, I planned successive iterations of the course to purposefully include differentiated elements. I used content, process, and product as an outline for differentiation throughout the course (Tomlinson, 1999).

The first step for differentiating was to understand all that I could about each student. I reviewed the student data two weeks prior to class beginning. I also used this time to email individuals and ask for more information if they were from a major that was unfamiliar to me. Right before class started, I emailed a questionnaire to everyone that asked general information. During the first days of class, I asked them to make personal introductions to one another on discussion pages. Finally, within the first weeks, I emailed each student individually, asked some follow-up questions about the initial information they provided, and tried to create a sense of communication so that they would feel comfortable communicating with me. In instances when they expressed concern over learning online, I invited them to an individual phone conversation so that we could talk through the online platform and some students participated in this with me. I kept detailed electronic notes for each student and added to these throughout the semester.

As the instructor, I determined to implement differentiated learning through a limited number of assignments. The goals for these assignments were:

- Read, analyze, and synthesize literature related to teacher research.
- Examine issues of data collection, data analysis, research design, and writing for publication in relation to teacher research.
- Explore various research strategies used in qualitative and ethnographic research with a specific focus on the issues and adaptations involved when teachers in their own classrooms use these strategies.
- Write a research proposal.
- Develop small research communities that allow ongoing support in researching one's own classroom.
- Acquire skills to plan and implement a teacher research project in a professional educational setting with appropriate methodology that will answer specific questions about teaching and learning.
- Consider issues related to the ethical implications of teacher research, assumptions about personal teaching practices, and the purpose of democratic, action research.

Differentiation of Content

To meet the course objectives, I organized a grouping structure at the beginning of the course. Students selected a Book Club group based upon their choice of texts. The texts supported the overall course objectives and dealt with different elements of practice that were of interest to the teachers based on the initial information they provided to me. These topics included writing instruction, reading instruction, and grouping students. As a group, they established a reading schedule for the book they selected and then submitted their agreed-on schedule to me. Students participated in discussion group postings about their Book Club readings through the first half of the course and they established their own discussion norms.

Additionally, each student selected a research question of their choice, completed a literature review about that particular topic, and conducted individual research. If more than one student chose a similar topic, I referred them to one another to share ideas and references.

Differentiation of Process

Midway through the course, students were grouped into a second group—their Research Group. The Research Group was based on exposing students to questions and research problems that others were pursuing. This was a teacher-selected group studying different research topics so that they would be able to learn more about the research of others who were researching a different question and/or a different level (elementary, middle school, high school, administrative). Students shared their data collection instruments (surveys, interviews, notebook entries), as well as data analysis and rough drafts in this group.

Discussion posts were completed weekly and groups established a variety of structures—some with a discussion leader and some without. Towards the end of the semester, students were asked to share a rough draft of their proposal with their original Book Club group for an additional round of feedback. Throughout all group postings and responses, I added additional comments within the group posts and sent individual feedback via email to redirect, encourage, and offer new ideas. Grades were based on completion of the discussions and quality of the posted discussions. If a post showed lack of understanding, I emailed the individual separately and suggested ways to strengthen their understanding, in some cases asking them to post additional information instead of merely assigning a grade value.

Finally, the process of collecting data was differentiated when students were given the objectives and guidelines, as well as multiple (and differing) models of data collection, but they were allowed to determine the number of data sources, the number of participants, and their method for analysis. Each of these things was guided through ongoing and individualized emailed feedback from me.

Differentiation of Product

Not every product that students completed was differentiated. Everyone completed an assignment that collected survey data and an assignment that collected interview data. This gave the students a similar experience and practice with data analysis. However, after they gained experience and received feedback from peers and from me, students were asked to choose their third data source from multiple options: research notebook, observations, artifact collection, additional informant interviews, archival records, or other relevant data to their research question.

Because the process of data collection and analysis had been differentiated, the product that resulted was different for each student. Though the overall product was a research paper, the paper was divided up into multiple drafts with drafts due throughout the semester. After reading and reflecting on the draft, I returned feedback to individual students via email with an evaluation of progress towards the final objectives. Each component was marked with a Complete or Developing rating. Where the work was Developing, detailed feedback was provided about how to move the rating to a Complete. I added all the records of my feedback to my earlier electronic notes about each student and referred to these frequently when creating my next round of feedback.

FUTURE RESEARCH DIRECTIONS

As previously reviewed, we do not have a robust body of research that informs our research in the nexus of adult learning, distance education, teacher education, and differentiated learning. While work has reported results in these areas separately, little work has been done to understand the impact of these areas in combination. Parsons et al. (2011) found that it was the type of task, not the format of the course (distance courses vs. face to face) that influenced the learning of the teacher education students. Massey (2009) followed a small number of students into the first year past coursework and found that students still demonstrated habits of thinking about students and data in a way that was consistent with the objectives for the differentiated online course. However, we need many more of these studies that offer comparisons and examine effectiveness of teacher education online courses. Research should address the practices of the online instructors implementing differentiated learning, the experiences of the teacher education candidates, their satisfaction with the differentiation components of online courses and programs.

Further, research supports the effectiveness of differentiated instruction at the K-12 level, but we have limited supporting evidence to show substantial results or provide outlines for providing differentiated instruction simply because very little research is being published in this area. Research should address student and teacher perceptions of differentiated instruction, as well as the teacher education candidates' abilities to differentiate learning for their own students

CONCLUSION

Teacher education courses offered online are here to stay. They benefit universities from an economic standpoint. They benefit students from a convenience standpoint. Given this new reality, it is up to teacher educators to continue to refine the methods of online instruction. This methodology must not be simply transferring face-to-face methods of teaching into online teaching methods. Instead, teacher educators must become learners again, considering what makes online learners unique and what methods are most effective when utilized online. Differentiated learning is one component that has been part of the special education and gifted education K-12 literature for decades but has not been applied with any regularity to distance education. This gap continues in spite of numerous researchers noting the unique benefits of online learning that include being able to individualize learning because of increased communication and the access to a wide variety of media.

In this chapter, I have attempted to link the differentiated learning literature with distance education and adult learner research as a way of highlighting the possibilities and continuing the conversation about what makes online learning effective for teacher education students and for the K-12 students they will teach. To date, much of the teacher education literature has been suspicious, if not hostile, towards teacher education courses offered online. The time for that discussion has ended and we as teacher educators must now move towards refining our own teaching and becoming the learners we wish our students to be.

REFERENCES

Allen, E., & Seaman, J. (2007). *Online nation: Five years of growth in online learning*. Needham, MA: Sloan Consortium.

Anders, P., Hoffman, J., & Duffy, G. (2000). Teaching teachers to teach reading: Paradigm shifts, persistent problems, and challenges. In Kamil, M. L., Mosenthal, P. B., Pearson, P. D., & Barr, R. (Eds.), *Handbook of reading research* (*Vol. III*, pp. 719–742). Mahwah, NJ: Lawrence Erlbaum Associates.

Atkinson, T. A. (2009). *Syllabus for READ 6422 clinical procedures in the identification and evaluation of reading disabilities*. East Carolina University.

Barone, D., & Morrell, E. (2007). Multiple perspectives on preparing teachers to teach reading. *Reading Research Quarterly*, *42*, 167–180. doi:10.1598/RRQ.42.1.10

Blake, D. (2009, December). What I learned from teaching adult learners online. *Elearn Magazine*. Retrieved from http://elearningmag.acm.org/archive.cfm?aid+1692866

Chamberlin, M., & Powers, R. (2010). The promise of differentiated instruction for enhancing the mathematical understandings of college students. *Teaching Mathematics and Its Applications*, *29*, 113–139. doi:10.1093/teamat/hrq006

Chiseri-Strater, E., & Sunstein, B. (2006). *What works? A practical guide for teacher research*. Portsmouth, NH: Heinemann.

Conner, M. (2004). *Pedagogy + andragogy*. Ageless Learner. Retrieved from http://agelesslearner.com/intros/andragogy.html

DeCandido, G. A. (2006). On my mind. *American Libraries*, *37*, 23.

Dollisso, A., & Martin, R. (1999). Perceptions regarding adult learners' motivation to participate in educational programs. *Journal of Agricultural Education*, *40*, 38–46. doi:10.5032/jae.1999.04038

Dooley, K., Lindner, J., & Dooley, L. (2005). Engaging learners and fostering self-directedness. In Dooley, K., Lindner, J., & Dooley, L. (Eds.), *Advanced methods in distance education: Applications and practices* (pp. 76–97). Hershey, PA: Information Science Publishing. doi:10.4018/978-1-59140-485-9.ch005

Dooley, K., Lindner, J., Dooley, L., & Wilson, S. (2005). Adult learning principles and learner differences. In Dooley, K., Lindner, J., & Dooley, L. (Eds.), *Advanced methods in distance education: Applications and practices* (pp. 56–75). Hershey, PA: Information Science Publishing. doi:10.4018/978-1-59140-485-9.ch004

Duffy, G. G., Miller, S. D., Kear, K., Parsons, S., Davis, S., & Williams, J. (2008). Teachers' instructional adaptations during literacy instruction. In Y. Kim, V. Risko, D. Compton, D. Dickinson, M. Hundley, R. Jimenez, K. Leander, & D. Rowe (Eds.), *57th yearbook of the National Reading Conference* (pp. 160-171). Oak Creek, WI: National Reading Conference.

Duffy, G. G., Webb, S., & Davis, S. G. (2009). Literacy education at a crossroad: A strategy for countering the trend to marginalize quality teacher education. In Hoffman, J. V., & Goodman, Y. (Eds.), *Changing literacies for changing times* (pp. 189–197). New York, NY: Routledge, Taylor and Francis.

Dykman, C., & Davis, C. (2008). Online education forum: Part two--Teaching online versus teaching conventionally. *Journal of Information Systems Education*, *19*, 157–164.

Gilman, T. (2010). Designing effective online assignments. *The Chronicle of Higher Education*, *56*, 44–45.

Hammerness, K., Darling-Hammond, L., Bransford, J., Berliner, D., Cochran-Smith, M., McDonald, M., & Zeichner, K. (2005). How teacher learn and develop. In Darling-Hammond, L., & Bransford, J. (Eds.), *Preparing teachers for a changing world: What teachers should learn and be able to do* (pp. 358–389). San Francisco, CA: Jossey-Bass.

Harasim, L. (2006). A history of e-learning: shift happened. In Weiss, J., Nolan, J., & Trifonas, P. (Eds.), *International handbook of virtual learning environments* (pp. 25–60). Dordrecht, The Netherlands: Kluwer. doi:10.1007/978-1-4020-3803-7_2

Hertzog, N. (1998). Open-ended activities: Differentiation through learner responses. *Gifted Child Quarterly*, *42*, 212–227. doi:10.1177/001698629804200405

Hubbard, R. S., & Power, B. M. (1999). *Living the questions: A guide for teacher-researchers*. New York, NY: Stenhouse.

Huebner, T. A. (2010). Differentiated instruction. *Educational Leadership*, *67*, 79–81.

Inyega, H., & Ratliff, J. (2007). Teaching online course: Lessons learned. In Sampson, M. B., Szabo, S., Falk-Ross, F., Foote, M., & Linder, P. (Eds.), *Multiple literacies in the 21st century* (pp. 344–363). Commerce, TX: Texas A & M University Commerce.

Karber, D. (2001). Comparisons and contrasts in traditional versus on-line teaching in management. *Higher Education in Europe*, *16*, 533–536. doi:10.1080/03797720220141852

Kauffman, J., Mock, D., Tankersley, M., & Landrum, T. (2008). Effective service delivery models. In Morris, R. J., & Mather, N. (Eds.), *Evidence-based interventions for students with learning and behavioral challenges* (pp. 359–378). Mahwah, NJ: Lawrence Erlbaum Associates.

Knowles, M. S. (1990). *The adult learner: A neglected species*. Houston, TX: Gulf.

Knowles, M. S., Holton, E. F., & Swanson, R. A. (1998). *The adult learner: The definitive classic in adult education and human resource development*. Woburn, MA: Butterworth-Heinemann.

Landrum, T. J., & McDuffie, K. A. (2010). Learning styles in the age of differentiated instruction. *Exceptionality*, *18*, 6–17. doi:10.1080/09362830903462441

Lawrence-Brown, D. (2004). Differentiated instruction. *American Secondary Education*, *32*, 34–62.

Lewin, L. (2006). *Teaching resources from Larry Lewin.* Retrieved from http://www.larrylewin.com/teachingresources/checbrics.html

Li, C., & Irby, B. (2008). An overview of online education: Attractiveness, benefits, challenges, concerns and recommendations. *College Student Journal*, *42*, 449–458.

Lindner, J. R., Dooley, K. E., & Williams, J. R. (2003). Teaching, coaching, mentoring, facilitating, motivating, directing … What is a teacher to do? *The Agricultural Education Magazine*, *76*, 26–27.

Massey, D. D. (2009). Teacher research: What's the point and who's it for? *Journal of Curriculum and Instruction, 3.* Retrieved from http://www.joci.ecu.edu/index.php/JoCI/issue/view/20

Merriam, S. B., & Caffarella, R. (1999). *Learning in adulthood: A comprehensive guide*. San Francisco, CA: Jossey-Bass Publishers.

Parr, F. W. (1929). *A remedial program for the inefficient silent reader*. An unpublished Doctoral dissertation, State University of Iowa.

Parsons, S. (2008). Providing all students ACCESS to self-regulated literacy learning. *The Reading Teacher*, *61*, 628–635. doi:10.1598/RT.61.8.4

Parsons, S., & Burrowbridge, S. (2011). *Thoughtfully adaptive teaching: An overlooked form of differentiated instruction*. Manuscript submitted for publication.

Parsons, S., Massey, D., Vaughn, M., Scales, R., Faircloth, B., & Howerton, W. (2011). Developing teachers' reflective thinking and adaptability in graduate courses. *Journal of School Connections*, *3*, 91–111.

Patton, M. Q. (1990). *Qualitative evaluation and research methods* (2nd ed.). Newbury Park, CA: Sage.

Perna, L. (2010). Understanding the working college student. *Academe Online, 96.* Retrieved from http://www.aaup.org/AAUP/pubsres/academe/2010/JA/

Rock, M., Gregg, M., Ellis, E., & Gable, R. (2008). REACH: A framework for differentiating classroom instruction. *Preventing School Failure*, *52*, 31–47. doi:10.3200/PSFL.52.2.31-47

Russell, S. (2006). *An overview of adult learning processes: Adult learning principles*. Retrieved from http://www.medscape.com/viewarticle/547417_2

Santamaria, L. (2009). Culturally responsive differentiated instruction: Narrowing gaps between best pedagogical practices benefiting all learners. *Teachers College Record*, *111*, 214–247.

Snow, C. E., Griffin, P., & Burns, M. (Eds.). (2005). *Knowledge to support the teaching of reading: Preparing teachers for a changing word.* San Francisco, CA: Jossey-Bass.

Stahl, N., & Smith-Burke, M. T. (1999). National Reading Conference: The college and adult reading years. *Journal of Literacy Research*, *31*(1), 47–66. Retrieved from http://findarticles.com/p/articles/mi_qa3785/is_199903/ai_n8848968/?tag=content;col1 doi:10.1080/10862969909548036

Stahl, S. (1999). Different strokes for different folks? *American Educator*, *23*, 27–31.

Thomson, D. L. (2010). Beyond the classroom walls: Teachers' and students' perspectives on how online learning can meet the needs of gifted students. *Journal of Advanced Academics*, *21*, 662–712. doi:10.1177/1932202X1002100405

Tomlinson, C. (1999). *The differentiated classroom: Responding to the needs of all learners*. Alexandria, VA: Association for Supervision and Curriculum Development.

Tomlinson, C. (2000). Reconcilable differences: Standards-based teaching and differentiation. *Educational Leadership*, *58*, 6–11.

Waldron, N., & McClesky, J. (1998). The effects of an inclusive school program on students with mild and severe learning disabilities. *Exceptional Children*, *64*, 395–405.

Zuckerbrod, N. (2011). From readers' theater to math dances: Bright ideas to make differentiation happen. *Instructor*, *120*, 33–38.

ADDITIONAL READING

Chamberlin, M., & Powers, R. (2010). The promise of differentiated instruction for enhancing the mathematical understandings of college students. *Teaching Mathematics and Its Applications*, *29*, 113–139. doi:10.1093/teamat/hrq006

Dooley, K., Lindner, J., & Dooley, L. (2005). Engaging learners and fostering self-directedness. In Dooley, K., Linder, J., & Dooley, L. (Eds.), *Advanced methods in distance education: Applications and practices for educators, administrators, and learners* (pp. 76–98). Hershey, PA: IGI Global. doi:10.4018/978-1-59140-485-9.ch005

Dooley, K., Lindner, J., Dooley, L., & Wilson, S. (2005). Adult learning principles and learner differences. In Dooley, K., Lindner, J., & Dooley, L. (Eds.), *Advanced methods in distance education: Applications and practices* (pp. 56–75). Hershey, PA: Information Science Publishing. doi:10.4018/978-1-59140-485-9.ch004

Dykman, C., & Davis, C. (2008). Online education forum: Part one – The shift toward online education. *Journal of Information Systems Education*, *19*, 11–16.

Dykman, C., & Davis, C. (2008). Online education forum: Part two – Teaching online versus teaching conventionally. *Journal of Information Systems Education*, *19*, 157–164.

Harasim, L. (2006). A history of e-learning: Shift happened. In Weiss, J., Nolan, J., & Trifonas, P. (Eds.), *International handbook of virtual learning environments* (pp. 25–60). Dordrecht, The Netherlands: Kluwer. doi:10.1007/978-1-4020-3803-7_2

Knowles, M. S., Holton, E. F., & Swanson, R. A. (1998). *The adult learner: The definitive classic in adult education and human resource development*. Woburn, MA: Butterworth-Heinemann.

Landrum, T. J., & McDuffie, K. A. (2010). Learning styles in the age of differentiated instruction. *Exceptionality*, *18*, 6–17. doi:10.1080/09362830903462441

Oblinger, D. (2003). Boomers, gen-xers, and millennials: Understanding the "new students.". *EDUCAUSE Review*, *38*, 36–40.

Parsons, S., Massey, D. D., Vaughn, M., Scales, R., Faircloth, B., & Howerton, S. (2011). Developing teachers' reflective thinking and adaptability in graduate courses. *Journal of School Connections*, *3*, 91–111.

Rock, M., Gregg, M., Ellis, E., & Gable, R. (2008). REACH: A framework for differentiating classroom instruction. *Preventing School Failure*, *52*, 31–47. doi:10.3200/PSFL.52.2.31-47

Stahl, S. (1999). Different strokes for different folks? *American Educator*, *23*, 27–31.

Thomas, J. (2005). Underprepared adult learners: Their passage through higher education. *New Horizons in Adult Education, 19*, 4 -11. Retrieved from http://education.fiu.edu/newhorizons

KEY TERMS AND DEFINITIONS

Andragogy: Used to describe a particular belief system and method for teaching adult learners. This term is used to distinguish the teaching of adults from the teaching of younger students and the methodology (or pedagogy) of such instruction.

Adaptive Instruction: Teachers making both planned and "on-the-fly" adjustments to instruction in order to better meet student needs.

Adult Learner: In the context of this chapter, an adult learner is anyone beyond the PreK-12 spectrum of education.

Differentiated Instruction: Changes made in content, process, and product in order to help students of diverse levels and needs reach a common set of goals.

Individualized Instruction: Individualized instruction most often refers to instruction tailored to a single student's needs; however, individualized instruction may also be used to refer to meeting the shared needs of a small group of students.

Learning Preferences: In spite of much research, little evidence has been found to support learning styles that are deterministic of a student's approach to learning. Instead, each student manifests a particular set of learning preferences that often evolve over time and depending on the particular area to be learned.

Chapter 12
Teacher Education in Online Contexts:
Course Design and Learning Experiences to Facilitate Literacy Instruction for Teacher Candidates

Salika A. Lawrence
William Paterson University, USA

ABSTRACT

Teacher candidates in online courses engage in authentic learning to foster 21st-century practices similar to those of their K–12 students, namely information and technology literacy and media production. This chapter describes instructional practices used in six online literacy courses for pre-service and in-service teacher candidates. The instructor assumed multiple roles during online instruction, including pedagogue, technologist, and evaluator. Although the course designs were highly structured, the instructor incorporated multiple resources to support diverse learners, to foster independent learning, to promote critical thinking and reflection on how instructional strategies can be used in K–12 classrooms, and to facilitate small group collaboration through authentic problem-solving tasks. Online courses for teacher education programs can serve as a vehicle for supporting candidates' information and technology skills. Online instructors can assume the primary role of pedagogue to help candidates connect their content area with best practices in literacy and technology.

INTRODUCTION

Today's learners "are increasingly exposed to an array of sophisticated learning resources and technology tools such as hypertexts, streaming video, and visualization tools" (Jeong & Hmelo-Silver, 2010, p. 84). Consequently college students today are more technologically savvy than their predecessors (Fisher & Wright, 2010; Prensky, 2001). Despite these capabilities, some online learners in teacher education programs are not able to draw upon their out-of-school practices

DOI: 10.4018/978-1-4666-1906-7.ch012

with technology to complete some academic tasks, such as conducting online research and evaluating information (O'Hanlon & Diaz, 2010). This limitation is problematic because in order to support K–12 students, contemporary teachers are expected to use technology to support learning (International Society for Technology in Education, 2008; Partnership for 21st Century Skills, 2004).

Another literacy challenge we face as educators is the inundation of information and the abundance of media; educators need guidance on how to prepare students to critically evaluate information retrieved from media sources. Research shows that some people tend to make "snap judgments about relevance of information content based on surface observations of [media and technology] sources" (O'Hanlon & Diaz, 2010, p. 43), rather than evaluating information and its source for credibility, reliability, authenticity, purpose, and validity. Online teaching and learning are particularly important areas in a globalized, information-rich society where people are bombarded with information from multiple sources, and where many people use technology and media as a vehicle for reading, for access to information, and for communication.

Allen and Seaman (2010) reported that enrollment in online courses has increased annually by 11.4% since 2006. In fall 2009, 29.3% of students enrolled in postsecondary institutions were taking online courses; an average annual increase of 2.8% since 2002 (Allen & Seaman, 2010). The trend of higher education institutions offering more online courses indicates the need for more research on how online instructors facilitate and design courses. In addition to understanding the pedagogy of online instruction, research in online teacher education courses is important because what teacher candidates experience in these courses may have implications for K–12 instruction. There is also a need to "pay more attention to the role of resources in learning" (Jeong & Hmelo-Silver, 2010, p. 84) and to the types of strategies online instructors use to facilitate learning. Most educators will agree that the research reports in this area need more rich descriptions of pedagogical methods, assignments, and assessments (Dell, Low, & Wilker, 2010).

With a focus on pedagogy, this chapter explores the effectiveness of online teacher education and how educators can use technology to prepare teacher candidates, particularly in online literacy courses offered to pre-service and in-service teachers. Although the chapter will focus on fully online literacy courses, a replicable model for structuring hybrid (blended) or fully online teacher education courses is offered. The instructional model presented here fostered opportunities for teacher candidates to practice using technology, to explore ways to incorporate it into K–12 practice, and to use online resources to develop a better understanding of effective literacy instruction. Novice online instructors and others seeking to develop a more structured course format might benefit from this type of course design model.

The chapter will (a) provide a brief overview of teaching and learning trends in online courses; (b) describe course experiences and instructional strategies used in six online literacy courses; (c) describe how course content and design can foster critical thinking, technology, and information literacy skills of teacher candidates; (d) identify the roles online teacher educators assume to promote learning in online contexts; (e) explain how highly structured course designs support different learning styles and provide access to varied resources; and (f) highlight implications for teachers' use of technology-based learning in K–12 contexts.

LITERATURE REVIEW

Organizational Structure of Online Courses

The literature on online instruction shows that effective courses use various assignments and assessments, multimedia and video, peer-to-peer interaction, and problem-based learning. Students enrolled in effective courses can also engage in mediated communication with the instructor. Ke and Xie (2009) characterized online course design models in three ways: (a) *course and support*—highly structured course content around a textbook; (b) *wrap around*—partially structured content around a textbook with readings and references periodically updated; and (c) *integrated*—fluid and unstructured content with no textbook, but with readings and reference links updated weekly. In the online literacy courses described in this chapter, the instructor used a combination of the course and support model and the wrap around model to design highly structured courses without textbooks. Although no textbook was used in the courses consistently, research studies and articles were identified and frontloaded into the course.

Researchers have found that instructors assume different roles when teaching online courses: instructors ranked pedagogical, professional, and evaluator roles as high, and they ranked the researcher role as low (Baran & Correia, 2009; Bewane & Spector, 2009). As an evaluator, instructors assess student performance (Bewane & Spector, 2009). Some online instructors use corrective feedback to assess student performance; motivational feedback to encourage learners to meet goals and complete tasks; and technological feedback to provide support for hardware and software issues that students encountered in online contexts (Pyke & Sherlock, 2010). Pyke and Sherlock (2010) showed that in asynchronous discussion posts, corrective feedback was more evident than motivational and technological feedback; corrective feedback accounted for two-thirds of all feedback during student-teacher interactions; and all instructor feedback decreased as the course progressed. In the technologist role, instructors select and develop appropriate materials as learning resources (Bawane & Spector, 2009). In this capacity, effective online instructors "scaffold the search process by providing students with pre-selected resources and/or technical tools to assist the handling of information" (Jeong & Hmelo-Silver, 2010, p. 85) by designing tasks and environments to facilitate students' active use of the resources.

For most online programs, instructors structure the course around discussion tasks to reinforce content through weekly units, modules, and chapters (Ke & Xie, 2009) that are incorporated as part of the course assessments. In general, instructors choose online discussions as a pedagogical strategy to foster student response to assigned readings, to pose questions, and to promote opportunities for students to reflect on how to apply what they have learned to real-world scenarios (Baran & Correia, 2009) and long-term tasks (Baran & Correia, 2009; Wise, Padmanabhan, & Duffy, 2009); to foster leadership where students assume roles as discussion leaders (Slagter van Tryon & Bishop, 2009; Dell, Low, & Wilker, 2010); and "to promote meaningful dialogue and to produce high levels of participation and quality conversation around the weekly topics" (Baran & Correia, 2009, p. 356). Other assessments in online courses include virtual labs, computer simulation models, case studies, quizzes, multiple choice questions, interactive tutorials, and games (Sewell, Frith, & Colvin, 2010). Sewell, Frith, and Colvin (2010) found exemplary learning assessments are authentic, challenging, coherent, engaging, respectful, responsive, rigorous, and valid.

Educators also apply structure to online courses by using the approach of problem-based learning: online instructors can design problem-based learning as independent or group tasks. In these instances, online learners are not told what to learn. Instead instructors allow independent learners to "identify their own learning goals in the context of a given problem and then research these learning issues more or less in a self-directed manner" (Jeong & Hmelo-Silver, 2010, p. 86). For group tasks, learners can work together and solve a problem as a group or co-construct knowledge through a shared inquiry experience.

In highly structured online courses that employ problem-based learning, the instructor answers questions via e-mail or phone, lectures through pre-recorded video and PowerPoint presentations, and, in integrated courses with no structured references to textbooks but with connections to weekly readings, instructors actively facilitate weekly discussions. In the literature, researchers suggest that to teach effectively online an instructor must expend a significant amount of time frontloading and identifying resources and structuring learning experiences for candidates to complete when they begin the course. To appropriately apply these research-based online teaching practices to teacher education, instructors should develop online courses for teacher candidates that structure the course content with opportunities for candidates to reflect on the implications of what they are asked to do and to consider the applications to K–12 contexts. Online courses can provide an authentic opportunity for teacher candidates to engage in learning experiences they might use in their own classrooms, such as effective uses of technology and multimedia resources to support students' literacy development.

Learning in Online Contexts

Participating in Discussions

Students in online courses learn and participate in part through "asynchronous discussions [using] text-based information, which lacks immediate instructors' verbal feedback used in face-to-face classrooms as well as nonverbal and contextual cues for communication" (Baran & Correia, 2009, p. 339).

Consequently, online learners are largely independent because they can draw upon resources to facilitate their learning, and they can work online to complete readings before participating in online discussions (Ke & Xie, 2009). In fact "when [online learners] are charged with handling their own learning process, support materials become the most important contributor to success in the online classroom" (Miner, 2009, p. 29). Online discussions among students can facilitate knowledge construction and shared meaning because the experiences in the course serve as a shared context (Wise, Padmanabhan, & Duffy, 2009) that provides a common experience that they may not share outside of the course.

Accessing and Using Resources

The resources that online instructors use to promote learning through active engagement, inquiry, reflection, and collaboration are particularly important. Jeong and Hmelo-Silver (2010) point out that in online contexts, learners interact with and use resources, technology, and information tools such as hypertexts and streaming video. By referring to resources accessible through their coursework, online learners can construct new understanding and can gain diverse perspectives about textbooks (Jeong & Hmelo-Silver, 2010). Adult learners, seem to prefer structured learning experiences in online contexts that include "major areas of reading, lectures (comprising PowerPoint slides, instructional videos, and instructor's notes), and assignments (including handouts, evaluation rubrics, and sample answers or exemplar student works) … [and] the syllabi generally had a clear timeline on what learning tasks were to be and when students should complete them" (Ke & Xie, 2009, p. 144).

Within the environment of online courses, many online learners develop proficiencies for finding, evaluating, and using information from various sources to make meaning and to build upon their prior knowledge. It is important to note that when online learners complete online tasks, they can be overconfident, and consequently they make errors in assignments where they must apply knowledge (O'Hanlon & Diaz, 2010). In fact, some researchers suggest that time on task is not necessarily an indicator of success in online learning (O'Hanlon & Diaz, 2010). This evidence suggests that to effectively evaluate online learning, course instructors should incorporate some element of self-assessment or reflection for the online learner to complete. O'Hanlon and Diaz (2010) used a self-regulated learning survey to encourage students to reflect on their learning strategies and to become more aware of their own learning processes. Then the researchers evaluated student responses to determine how students perceived themselves as learners and how students identified bias in online text.

By knowing how online learners use resources, instructors can better determine how to design online courses, in particular how to organize and present information (e.g., formatting and extended narratives). Fisher and Wright (2010) found that when most online learners do not complete assignments it is because (a) they have difficulty locating information, (b) course tools were not visible or clearly identified, or (c) they may have only scanned the materials instead of reading through the narratives posted. Therefore the resources identified for the course should be meaningful, useful, and accessible for teacher candidates to navigate independently.

Using Multiple Literacy Skills

Educators recognize that, in order to work productively with current technology, students need to demonstrate critical thinking, as well as information, media, and technology literacy, and teachers need to help students develop these skills. From Paul and Elder's (2008) perspective, an effective critical thinker (a) demonstrates the ability to think and examine issues from different points of view and to consider alternatives; (b) uses processes for conceptualizing, gathering, generalizing, examining, reflecting, reasoning, communicating information, and inferring; (c) applies intellectual practices that cross disciplines, namely evidence, fairness, consistency, and precision; and (d) emphasizes process, rather than merely retaining information. Critical thinkers ask questions, conduct research, and evaluate information—especially by interpreting information from different sources—and subsequently draw conclusions; critical thinkers are open-minded and willing to look at issues from different perspectives while evaluating their own assumptions, and their ways of communicating, sharing ideas, and solving problems with others (Paul & Elder, 2008).

Today, students need to use critical thinking, problem solving, decision making in authentic problems, and digital citizenship to demonstrate creativity and innovation. They use digital media to communicate and collaborate, to conduct research, and to specifically locate, analyze, and evaluate information through digital tools and technology skills across core subject areas (International Society for Technology in Education, 2007). Sewell, Frith, and Colvin (2010) found that teacher candidates in online courses demonstrate critical thinking through narrative posts and reflections, where candidates summarize what they learn and reflect on questions about their understanding.

Researchers found that online learners must engage in various problem-solving activities including learning how to work in online contexts, and in the case of teacher education, using online video resources to examine practices in authentic classrooms while conducting online research about educational theory (Jeong & Hmelo-Silver, 2010). Completing these kinds of analyses promotes critical thinking and reflection. Wise, Padmanabhan, and Duffy (2009) learned that when students in online courses analyze classroom practice by reviewing videos, the students can develop a shared context and common experience that frames their discussions.

AN INSTRUCTIONAL MODEL FOR ONLINE LITERACY COURSES IN TEACHER EDUCATION

The Context

The author of this chapter teaches at the College of Education in a mid-sized state university in New Jersey. Teacher candidates in the pre-service, initial certification programs are required to enroll in and complete a subject-specific program (e.g., Math, Science, English, Social Studies) and take 30 credits in education coursework. In this education program, candidates are required to take one literacy course based on the certification they seek (Table 1).

Another program available for graduate, in-service teacher candidates is a 33-credit Masters in Literacy program, which leads to reading specialist certification. As shown in Table 1, some of the online literacy courses available to these candidates range in length and are frequently modified to accommodate candidates. The largest enrollment is in Literacy, Instruction, and Technology (LIT), which typically runs four sections in the fall and another four sections in the spring with about 20 teacher candidates enrolled in each section. Different instructors teach each section of LIT and one of the four sections offered each semester is taught face-to-face.

Although this author has taught these six graduate and undergraduate courses in classroom and online formats, this chapter focuses on the online versions of these courses. Blackboard is the online course management system used for the courses. It provides a platform for course development where courses can include digital media and text-based resource materials. The Blackboard platform also includes templates, which instructors can use to organize the course material. By choosing certain templates, instructors can upload assignments, pictures, video, and word-processed documents and files. With other templates, the instructor can supply links to websites and other resources on the Internet.

After students register for an online course, they can access the course using their student identification. Although Blackboard can deliver

Table 1. Online teacher education course offerings

Course	Session	Timeframe	Candidates	Certification Area
Reading and Writing Across the Grades (RWG) 5–8 Curriculum*	fall spring summer	15-week 3-week	graduate and undergraduate	Middle school certification (grades 5–8)
Reading and Writing Across the Curriculum (RWC)*	spring	8-week	graduate, alternative route	Middle and high school certification (grades 6–12)
Literacy, Instruction, and Technology (LIT)*	winter summer	3-week 6-week	undergraduate	K–12 initial certification
Literacy and Technology Across the Curriculum (LTC)*	winter summer	3-week, 6-week	graduate	K–12 initial certification
Advanced Inquiry into Literature for Children and Youth (AIL)**	winter summer	3-week, 6-week	graduate	Reading Specialist
Administration and Supervision of Reading Programs (ASRP)*	summer fall spring	6-week, 15-week	graduate	Reading Specialist

*Required course
**Elective

courses asynchronously or synchronously, all courses described in this chapter used asynchronous course delivery. In an asynchronous course, class members are not required to be online at the same time and the instructor does not give immediate feedback. In contrast, in synchronous courses students participate in specific real-time discussions and complete assignments online, and they can interact with the course instructor immediately (e.g., via chats).

Designing asynchronous courses requires thoughtful frontloading. Before the candidates begin the course, this instructor identifies readings, develops or locates multimedia learning aids, and develops course assignments and assessments so the entire course content is available on the first day. The course design depends on the timeframe for course delivery. For all courses described in this chapter, the instructor defines a *day* and a *week* so that students understand when assignments are due. This instructor defined the course timeframe and rules in the following manner:

- One day typically runs from 7:00 a.m. to 11:00 p.m. (candidates must complete tasks between 7:00 a.m. and 11:00 p.m.).
- A week may be Monday to Sunday or Thursday to the following Wednesday.
- Three-week courses are organized by days, while 6-week, 8-week, and 15-week courses are organized around week-long units.
- Quantity of work depends on course type (e.g., 3-credit courses require 15 hours more work and time than 2-credit courses).
- All work assigned for the week (or day) must be completed before the week (or day) is over.
- Candidates must participate in discussions by responding to peers. (Make an initial post and then select at least one or two posts from peers to respond to. If working in groups, respond to everyone in the group. Each week, initial responses must be made by the third day so that peers have time to respond to each other.)

Data Collection and Analysis

Course assessments and assignments such as projects, online discussion posts, syllabi, the instructor's notes, surveys and course evaluations completed by the candidates were collected and analyzed from six online literacy courses taught by the instructor in a teacher education program. During initial analysis this author used the data from each course to identify patterns between and among the data sources (Hubbard & Power, 1999) specifically to highlight similarities in pedagogical practice.

All of the courses help teacher candidates develop a better understanding of literacy and the reading process. The courses also show candidates how to use technology and multimedia to support literacy development and engagement with diverse learners. In these courses, teacher candidates learn how to incorporate literacy into a content area to support student learning through meaningful planning.

In the LIT course, for example, pre-service candidates are introduced to strategies for selecting supplemental texts and technology in K–12 contexts; these strategies build reading comprehension, critical thinking, and critical literacy across the curriculum. Candidates then work in cross-curricular groups (English, Math, History, etc.) of four or five to develop an integrated unit plan that incorporates the texts, technologies, and strategies introduced in the course. Similarly, in the ASRP course, in-service candidates learn about various reading intervention programs, school-wide evaluation, and teacher leadership in school contexts. They must then evaluate the literacy program in their own schools, using the evaluation tools and methods from the course, and make recommendations for improvement. Candidates present their data, findings, and a description of the processes to the class in a PowerPoint presentation online.

Through course evaluations, feedback, and surveys, this author learned that teacher candidates prefer having access to all content when they begin the course so they can plan accordingly. Most teacher candidates enrolled in the six online literacy courses hold full-time or part-time jobs and take a demanding credit load at the university. In previous semesters, the courses were taught using a roll-out method (gradually releasing course content periodically, such as weekly or bi-weekly, throughout the course), as well as all-access (providing all of the course materials at the onset of the course). Now, all online literacy courses taught by the instructor are highly structured and use an asynchronous format that allows flexibility throughout the week. Teacher candidates can work at their own pace, they can access resources to increase their understanding of topics, and they can collaborate while they work.

Recursive document and content analysis (Henn, Weinstein, & Foard, 2006) was used to revisit the data to extract examples that supported the themes that emerged, to conceptualize the instructional model and develop an understanding of practices in the online literacy courses, and to provide recommendations for online instructors seeking to design online teacher education courses. For example, data analysis helped to conceptualize decision-making, particularly how and why changes in the courses were made, how resources were used, and how learning was facilitated in the online courses. Although the strategies presented below were gleaned from fully online courses, many of the solutions presented can be generalized to blended formats.

The proposed course design model, which is presented graphically in Figure 1, consists of three components: Design a Structured Course, Facilitate Learning, and Use Resources and Tasks to Foster Multiple Literacy Skills and Learning. These three components are described in further detail below and illustrative examples, from the data obtained from the courses, is also included.

Solutions and Recommendations

Design a Structured Course

Structured content in literacy courses promotes interactive learning, active engagement and participation, and better understanding of course material, specifically when candidates experience how to apply strategies in the course to K–12 contexts with students. Before starting a course, teacher candidates received an email announcing that the course was available and explaining how to access Blackboard. When candidates logged in, they saw the course shell with the following tabs: *Syllabus*, *Announcements*, *Readings*, *Lectures*, *Examples & Resources*, *Discussion Board*, *Groups*, *Contact Instructor*, *External Links*, and *Tools*. Teacher candidates then referred to specific required lectures, readings, and resources during their initial posts and responses.

The following descriptions show how this author structured online courses according to the tabs available through Blackboard and provides a rationale for the decision-making while designing the online courses. As previously mentioned, the strategies used can be adapted for use with other platforms for online and hybrid courses.

- **Syllabus:** The course content was presented in the syllabus as well as on Blackboard. A tentative schedule allowed for slight modifications in the course. For example, as the course progressed and teacher candidates demonstrated their knowledge base for the material, course reading(s) might be adjusted later in the course to address a knowledge gap or to provide additional information on a particular topic. Course readings and resources were embedded in the tentative schedule, which was format-

Figure 1. Schematic representation of course design model for online teacher education

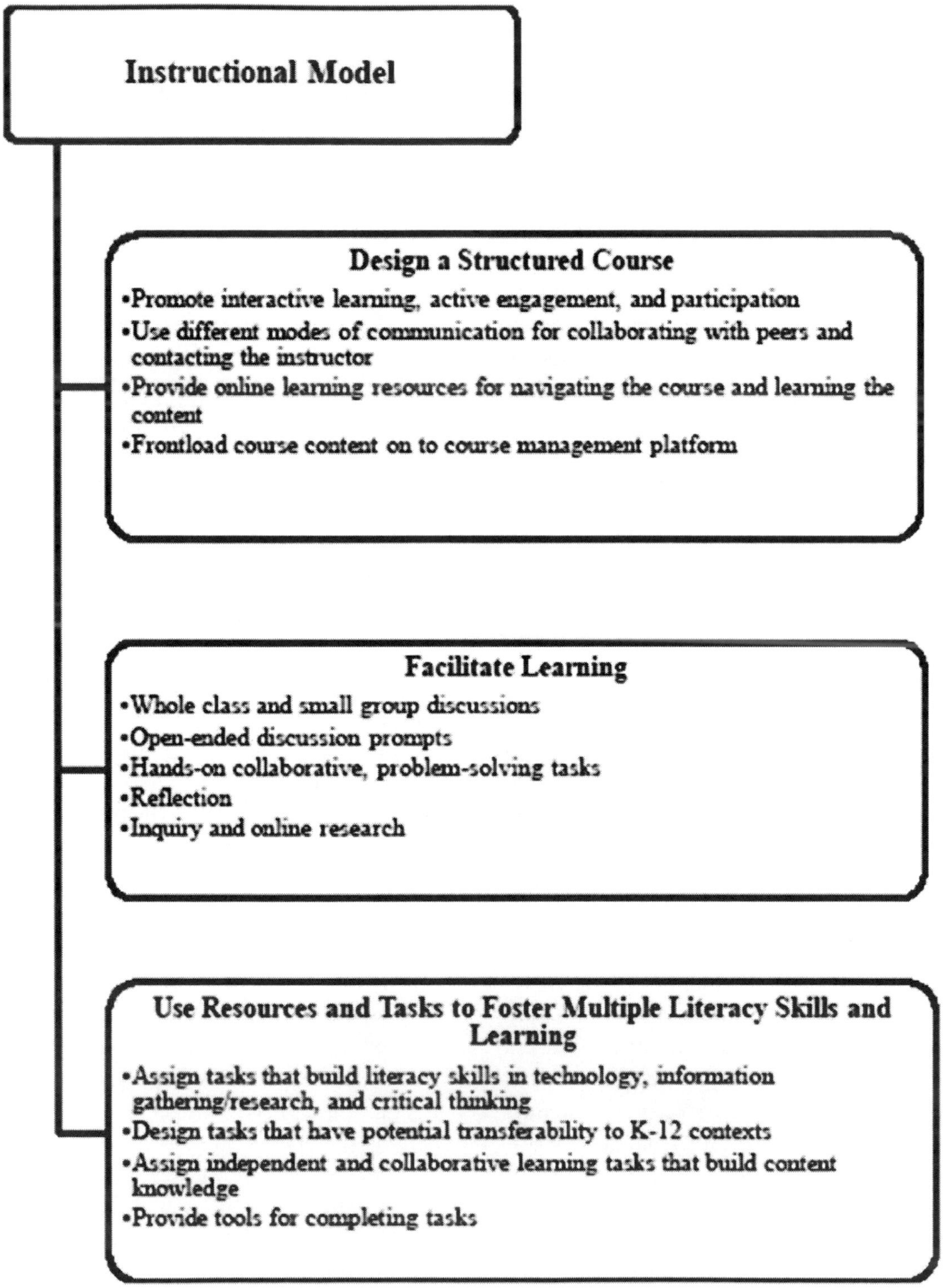

ted on a table. The syllabus included the following elements:

- course description
- course objectives
- learning outcomes
- class procedures
- expectations for participation and a rubric for online learning
- tentative schedule for all online tasks, with course assessments and rubrics for each

- **Announcements:** Updates, such as revised readings and resources or technology issues, appeared in Announcements. Announcements were also used to summarize discussions and clarify information shared during the discussions. Announcement posts let the instructor scaffold candidates' understanding of course materials, based on observations of the tasks completed in the course. Announcements were posted as PowerPoint summaries, audio summaries, or documents.
- **Readings:** Reading and resources were identified in the syllabus and on Blackboard. Some materials had hyperlinks that accessed research articles through the university's library, video clips, and other online resources.
- **Lectures:** The lectures were in PowerPoint presentations or Microsoft Word documents. Some lectures were required for a particular assignment, while other lectures were posted as additional resources to help clarify concepts.
- **Examples & Resources:** Examples and resources for successfully completing all expected outcomes in the course (e.g., discussions, assessments, assignment sheets describing the task, rubrics, and examples of previously completed projects) were posted.
- **Discussion Board:** Teacher candidates participated in the course through the Discussion Board. Everyone in the course had access to all forums in the discussion board. To create the discussion board this instructor opened a *forum*, and identified the due date for initial posts (e.g., Tuesday 11 p.m.). Each discussion board had a specific prompt to facilitate discussion and active participation. Candidates were provided with a rubric and instructions for making online discussion posts, which was modified from Nielsen (2003) http://www2.uwstout.edu/content/profdev/rubrics/discussionrubric.html. For some discussions, teacher candidates were assigned to specific groups. This instructor periodically participated in discussions to clarify posted comments or to reply to a specific post to commend someone for a comment. In some discussion forums the instructor would post a summary to clarify and scaffold understanding of a topic.

Although all teacher candidates were instructed to frequently access the Announcements page, some did not. Therefore, this instructor updated and posted some course modifications in the respective Discussion Board window. For example, during the spring 2010 semester, a pre-service math teacher in the LIT course inquired about creating a WebQuest for the assignment, and asked for other resources to identify WebQuests in her subject area. Figure 2 shows how the Discussion Board was modified to identify and include additional resources and to clarify expectations for completing the assignment.

- **Groups:** Groups was a modified Discussion Board. Only assigned teacher candidates in a group could participate. To complete collaborative assignments, the team worked on the planning within the Group page and then presented the completed project to the class on the Discussion Board. The instructor created and set up the groups manually

Figure 2. Example of modified page to clarify assignment

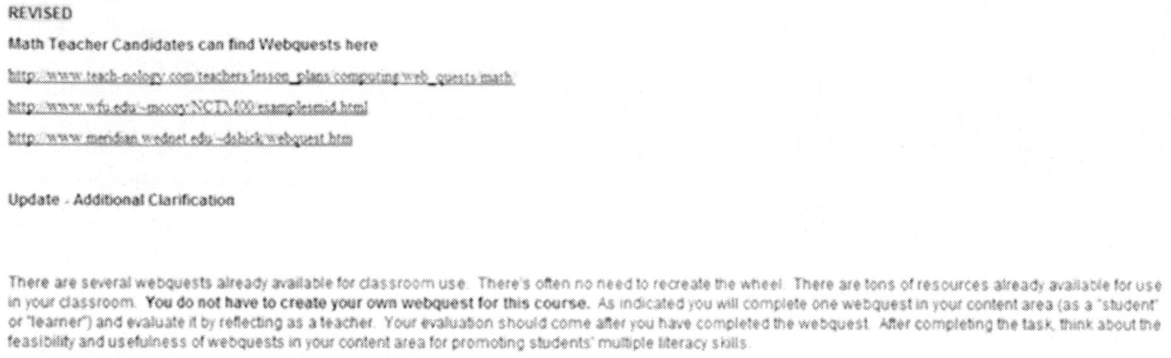

REVISED

Math Teacher Candidates can find Webquests here

http://www.teach-nology.com/teachers/lesson_plans/computing/web_quests/math/

http://www.wfu.edu/~mccoy/NCTM00/examplesmid.html

http://www.meridian.wednet.edu/~dshick/webquest.htm

Update - Additional Clarification

There are several webquests already available for classroom use. There's often no need to recreate the wheel. There are tons of resources already available for use in your classroom. **You do not have to create your own webquest for this course.** As indicated you will complete one webquest in your content area (as a "student" or "learner") and evaluate it by reflecting as a teacher. Your evaluation should come after you have completed the webquest. After completing the task, think about the feasibility and usefulness of webquests in your content area for promoting students' multiple literacy skills.

and was included as a group member to observe how teacher candidates' participated in each group.

- **Contact Instructor:** For questions about the course content or personal issues that might impede participation, teacher candidates could contact the instructor via email or phone. The instructor also set up an "Ask the Instructor," forum, where teacher candidates could ask questions about course material that might be relevant to others. If a candidate sent a question via email that was pertinent to the whole class, that query (without the teacher candidate's name) and the response were posted in Announcements and/or on the Ask the Instructor forum in Discussion Board.
- **External Links:** Additional materials for teacher candidates appeared under External Links. This area contained hyperlinked information that supplemented the course content. Through External Links, candidates could expand their understanding of a particular topic; for example, with videos that show models of teaching and theoretical concepts introduced in the course. These resources compensated for the lack of face-to-face time candidates would have encountered in a typical class, where the instructor might model a particular strategy.
- **Tools:** Through the Tools tab, teacher candidates could access a wide range of media links. They could also access other Blackboard tabs, send a direct email to others in the course and view the grade center, where they could monitor their progress in the course—grades for different assignments were updated in each candidate's grade dashboard.

The content in the courses was organized around assigned readings, and the readings generated discussions among the online learners. The articles provided a common reference for discussions and connections throughout the course. Teacher candidates were expected to refer to course readings when they made their initial post on the Discussion Board or when they responded to peers. The instructor then updated or modified course readings and resources based on information gleaned from discussions and coursework. In other words, the initial discussions and assignments were formative assessments because they guided instructional decisions: to make the course experience more meaningful, this instructor posted additional outside resources, supplemented the required readings or links to additional video examples, or placed candidates in discussion groups or partnered them for projects based on their work.

It is important to note that illustrative examples were particularly important in an online teacher education course because teacher candidates were presented with theories and information about best practices without the benefit of hands-on and real-time opportunities to enact the practice. Candidates could also use the External Links when they made connections to course readings and when they completed course assignments such as designing and planning activities for K–12 students. External Links might include information about literature circles, information about guided reading, and using comic books and graphic novels. Each teacher candidate brought a different knowledge base to the course. Therefore information and resources were added to External Links as this instructor noticed gaps in knowledge during discussions and when reviewing assignments (e.g., how to use APA format or how K–12 teachers can use authentic assessments in content classrooms).

Facilitate Learning through Discussions

An integral part of online instruction was structuring and guiding discussions. Through weekly open-ended discussion prompts (see Figure 3) candidates could reflect on the resources (readings, videos, etc.) that related to the week's topic. Although traditional assessments such as research papers are useful in all teacher education courses (both online and face-to-face), for this instructor, the responses to open-ended discussion prompts became meaningful and useful formative assessments, which also served as on-going assessments throughout the course.

The teacher candidates brought different backgrounds and experiences to the course, experiences that influenced their perspectives and their responses to the discussion prompts. Some can-

Figure 3. Examples of prompts used for discussions about online readings and videos

- 3-2-1
- I was surprised by…
- I noticed that …
- I wondered why …
- What **specific** literacy practices (behaviors) are evident in the classrooms?
- Do you see evidence of the learning and literacy theories introduced in this course, being enacted in these classrooms?
- What do you think are the advantages and limitations/challenges with using these kinds of texts?
- In today's classrooms we encounter a wide range of students with different cultural and academic backgrounds. Several strategies were identified in the readings for supporting the literacy development of the diverse students including ESL, and struggling readers. Identify 2 strategies and how they can be implemented in the classroom (either by the classroom teacher or in collaboration with the librarian).
- Curriculum and instruction are intricately connected. Building on the discussion from … what do we expect to see during classroom observation? When we look at a curriculum does it tell us what we would expect to see when we walk into a classroom (how the curriculum is enacted)? Watch the Video "Learning is Hard." This is a Math lesson but focus on the **instructional strategies used by the teacher**.

didates, for instance, were seeking certification for new teaching contexts (e.g., an elementary substitute teacher now seeking a high school teaching license). By using open-ended prompts, this instructor could gain insight into the prior knowledge candidates brought to a particular topic and determine how candidates interpreted information from the course readings. The instructor could then use course resources and assigned readings to give candidates a similar context for their discussions and thus provide candidates with a shared background from which to explore a topic.

For example, in one weekly discussion task, candidates in the LIT course viewed a PowerPoint lecture on assessment and then completed one online tutorial about assessment. Either

- Concept to Classroom – Assessment, Evaluation, and Curriculum Redesign http://www.thirteen.org/edonline/concept2class/assessment/index.html, or
- Rubrics, informal reading inventories, portfolios, and anecdotal records http://jonathan.mueller.faculty.noctrl.edu/toolbox/index.htm.

Candidates then worked in small jigsaw groups (where they interacted with someone who completed a different task from their own) to share what they learned through an open-ended prompt. The following prompt facilitated this kind of discussion:

3-2-1 is a strategy teachers can use to quickly assess students' understanding of a topic or concept. This technique can be used at the beginning of the period as part of a Do Now as a formative assessment to find out what students already know about a topic before you teach it (or as a way for students to connect to a previous lesson); or at the end of the period to see what students learned about the topic before they leave class (as an exit slip). For this discussion, you will use 3-2-1 in the following way. Discuss **three (3)** *new insights you have gained about assessment. Identify* **two (2)** *practical benefits of using assessments to monitor students' progress. Discuss* **one (1)** *benefit of including students in the assessment process. Remember to cite your sources.*

Through these open-ended discussion prompts (e.g., 3-2-1), candidates could share what they learned in course readings and connect to their prior knowledge; this instructor used these prompts as examples of a quick alternative assessment that candidates might use in their own classroom to facilitate reflection and to assess student understanding. For another open-ended prompt used in the LIT, RWC, and LTC courses pre-service candidates selected two articles—one on the literacy needs of struggling adolescent readers and one on elementary students' literacy needs—to learn the distinctions of working with adolescents as compared to elementary students. Candidates then posted an initial response to the following Discussion Board prompt and then collaborated with peers to construct, through their discussion, an understanding of the unique literacy needs of students in different age groups.

Compare the literacy needs and practices of young students to those of adolescents. Use the following prompts to reflect on what you have learned about the literacy needs and practices of students in these age groups.

- I was surprised by…
- I noticed that …
- I wondered why …

Use Resources and Tasks to Foster Multiple Literacy Skills and Learning

The skills fostered through these online courses include critical thinking, media and video production and analysis, online research and information literacy skills (e.g. retrieving information and evaluating online resources), problem-solving through use of simulation, and collaborative learning. Open-ended course assignments helped teacher candidates develop their media and technology skills when candidates developed their own multimedia product using online resources.

Multimedia Products

Recently, this instructor shifted the approach of online literacy courses for pre-service candidates to incorporate more multimedia resources. In prior semesters, pre-service candidates used only websites like puzzlemaker.com, where they generated a crossword puzzle as a sample classroom product. While in prior literacy courses (Lawrence & Mongillo, 2010), only in-service candidates gained experience with multimedia resources and media production.

With the new shift to multimedia resources like a Jeopardy template, or video production software such as Windows Movie Maker or iMovie, and Microsoft Photostory 3, pre-service teacher candidates could create learning tools for their students that were more interactive and engaging, tools that were similar to the multimedia and technology tools that their students might encounter outside of school. For example, pre-service teacher candidates in the RWC, LIT, LTC, and RWG courses were trained on multimedia resources and created interactive and engaging vocabulary activities and study aides for K–12 students (Figures 4 and 5). The task required candidates to select 10 vocabulary words for a unit they might teach in their content area and use a multimedia program—namely Jeopardy, iMovie, or Photostory 3—to create a visual artifact K–12 students can use as a learning object. After completing the multimedia study aide, teacher candidates posted their product on their designated class Discussion Board, selected one or two study aides created by peers in the class, and provided each other with feedback using the 2+2 protocol: give two commendations and two recommendations on a peer's work.

Some teacher candidates used Jeopardy to create a review of vocabulary in a specific text; for instance, to introduce vocabulary words encountered in chapters 1-5 of the Great Gatsby (e.g. permeate, clad, innuendo, and vehement). Other candidates used the Jeopardy software to review a specific topic such as cell biology (Figure 4). Similarly, a candidate used software to create a movie that reviews vocabulary concepts pertinent to a specific art technique as shown in Figure 4.

The Jeopardy game and the movies created by the teacher candidates can serve as study aides for K–12 students. The teacher candidates were able to use technology as a vehicle for presenting information about concepts across subject areas. Rachel (all names were changed to pseudonyms), a pre-service Spanish teacher who created a Jeopardy game using the categories feminine, masculine, singular, and plural, to introduce students to phonological patterns in Spanish words, received the following 2+2 feedback from Mary, a peer in the class.

I like [the] categorization of your vocabulary. I know that while I was learning a foreign language one of the things that always confused me was which were masculine and which ones were feminine. I also like the vocabulary that is used in the jeopardy game. With words that concern the family and school it is something that students are more likely to relate to.

A recommendation is maybe making the word the question and the definition of the word the answer. For example, the jeopardy clue would read "This means brother in Spanish." and the answer slide would read "What is hermano?" With that format I believe the students would learn the meaning of the word as well as practicing how to say it as they answer. Another recommendation is maybe

Figure 4. Vocabulary learning object created by pre-service science teacher using Jeopardy template

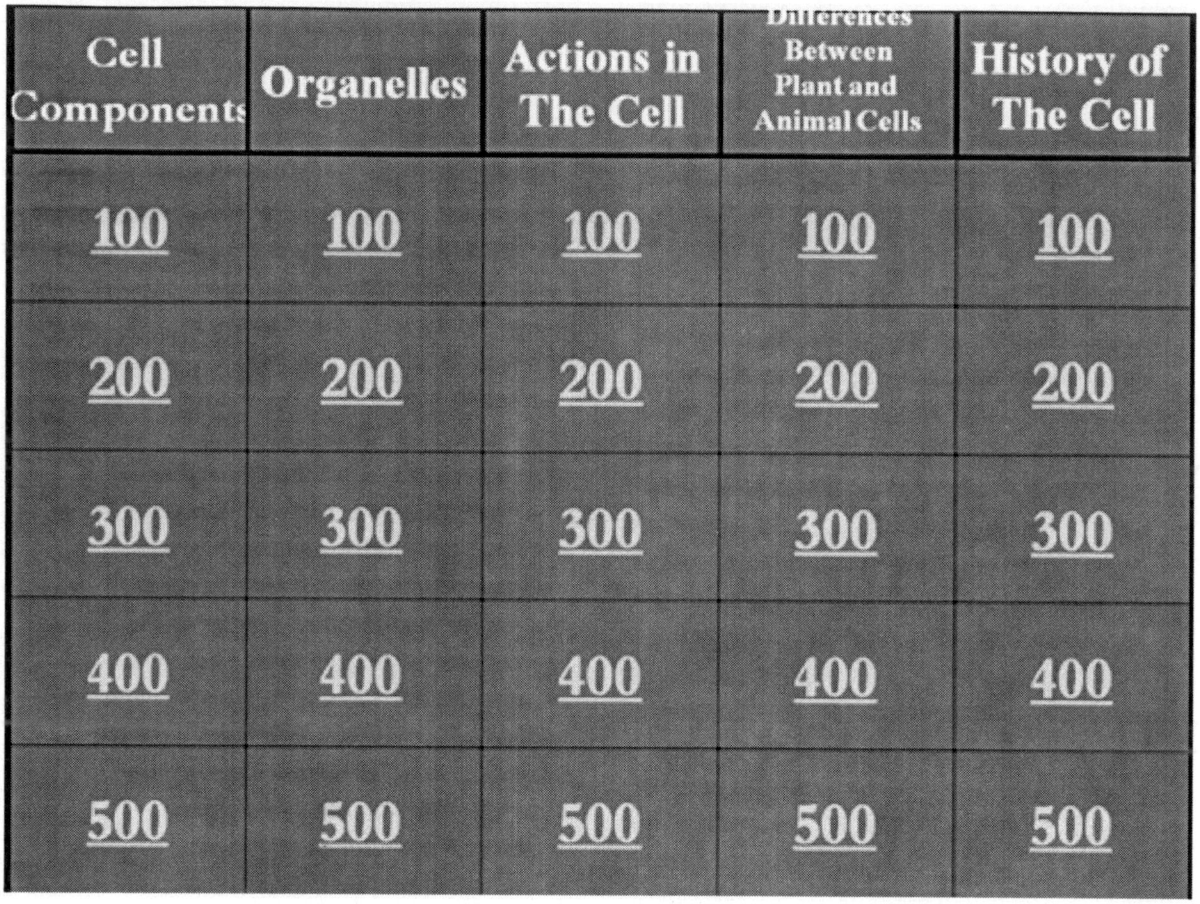

Cell Components	Organelles	Actions in The Cell	Differences Between Plant and Animal Cells	History of The Cell
100	100	100	100	100
200	200	200	200	200
300	300	300	300	300
400	400	400	400	400
500	500	500	500	500

adding images to some of the slides. The images may help students in the classroom who are visual spatial learners.

WebQuest and Evaluation of Online Resources

Two assignments—the WebQuest review and the Online Resources & Multimedia Evaluation project—also cultivated teacher candidates' experience with technology and online resources. In both tasks pre-service teacher candidates interacted with online materials and then reflected on how to incorporate resources into their content area classroom. In contrast to the WebQuest review assignment—where teacher candidates chose an available WebQuest provided by the instructor, reviewed it, reflected, posted, and discussed with peers on the Discussion Board—candidates submitted the Online Resources project (Appendix A) directly to the instructor for evaluation. For the Online Resources project, teacher candidates developed their information literacy and research skills by locating their own online and multimedia resources, resources that they might use to teach a particular topic in their content area. The teacher candidates then evaluated their selected online resources, preparing an annotated bibliography in which they explained why they selected the resource, how they might have modified it, and how the resource addressed the needs of diverse learners.

Figure 5. Selected clips captured from a 2:55 min. engraving movie created by a pre-service art teacher

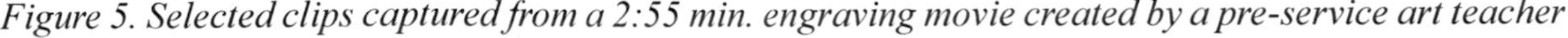

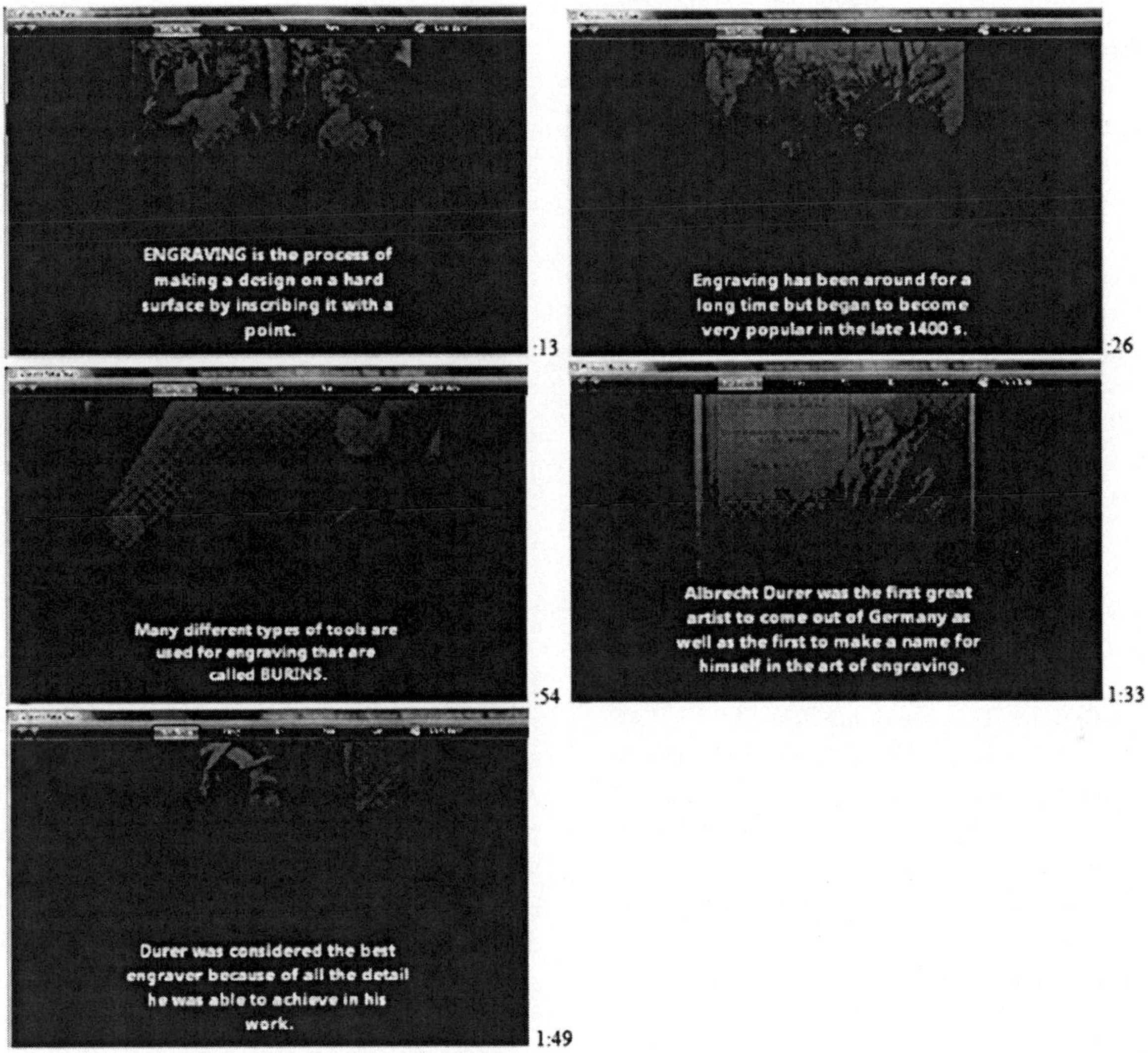

For the WebQuest assignment (Appendix B), teacher candidates evaluated a WebQuest they might ask their own students to complete in their content area (e.g., Math, Science, or History). Teacher candidates completed the WebQuest as a "student," reflected on the experience as the student (a hypothetical learner), and then reflected on the experience as a teacher. They identified the skills the WebQuest fostered and they indicated whether or not they perceived the WebQuest as a beneficial learning tool.

This assignment allowed teacher candidates to use the WebQuest as an inquiry tool, where they could increase their knowledge of a topic through online research. Teacher candidates used their insights as learners to consider and explore potential implications of technology-based learning tools found in K-12 classrooms. Teacher candidates completed WebQuests on such topics as The Middle Ages, Nutrition, and The Crucible, and then posted their analysis in the Discussion Forum. Isabella, a Social Studies/History teacher candidate, completed a WebQuest on the Middle Ages for students in grades 6–9. Isabella noted that as a learner the WebQuest increased her understanding of culture during the Middle Ages. She said that the "websites were all informative, but at the same time they did not give confusing, drawn out explanations of certain topics [and learners are able to] gain further comprehension skills by reading text and relating it to their own lives as if they actually lived during the Medieval period." She believed that the WebQuest blended learning by merging classroom instruction with online research. She said the teacher involvement facilitated through the WebQuest she completed was important because "by keeping the project in steps it allows the teacher to make sure that the students are on the right track and focusing on the correct main ideas of each task. The student is not allowed to advance without the teacher's consent. This is also beneficial to the students because they can see where they need improvement and what positive things they are doing as well."

Most teacher candidates expressed that the WebQuest was a good learning tool for K–12 students. All of the candidates expressed that the WebQuest they completed was a good tool to support K–12 students' understanding of specific topics in their respective content areas while fostering their inquiry skills. They noted that the WebQuest facilitated "hands-on learning," "independent learning through real life application and research," fostered "cross-content connections," provided a structured learning opportunity to build students' vocabulary, technology, writing, and collaborative skills.

Online Video Analysis

Online video resources facilitate learning by allowing teacher candidates to connect theory and practice. Videos can be illustrative examples of specific K–12 strategies introduced in the online literacy courses. In the online literacy courses, teacher candidates examined classroom practice by watching videos that focused on different aspects of literacy instruction. Candidates in all courses watched videos and used guided viewing format to structure their analysis and focus. All pre-service teacher candidates were required to use a video analysis form (Appendix C) to examine literacy instruction in three ways: teacher questioning and discussion techniques; strategies used to engage students in literacy and learning; and the ways the teacher assessed students.

Several videos were posted on Blackboard and in the syllabus; these videos allowed the teacher candidates to acquire greater understanding of specific areas. The videos were assigned to the whole class, or to specific candidates, or as additional resources. This instructor also modified the video analysis task by assigning one group to watch the same video but asking each member of the group to watch the video through a different lens, such as how the teacher asked questions or how the teacher assessed students. Depending on the size of the group, two teacher candidates sometimes might be assigned the same lens. After viewing the videos, the teacher candidates participated in a small group jigsaw discussion to report their insights on the video from their assigned perspective. Through these videos candidates gained a shared experience, which served as a context for small, weekly group discussions about classroom practice.

Some of the video resources used in the online literacy courses included web resources such as Edutopia.org, Annenberg Learner, PBS, and videos from YouTube or TeacherTube. During one class discussion, pre-service teacher candidates were each assigned to read one article listed in the syllabus and to watch one of the following videos from Annenberg:

- Strategy

http://www.learner.org/workshops/teachreading35/classrooms/cv5.html

- Metacognition

http://www.learner.org/courses/learningclassroom/session_overviews/metacog_home9.html

- Content Area

http://www.learner.org/courses/learningclassroom/session_overviews/arch_knowledge_home10.html

- Context Clues

http://www.learner.org/workshops/teachreading35/classrooms/cv1.html

- Building Comprehension

http://www.learner.org/workshops/teachreading35/session3/index.html

- Teaching Cause and Effect

http://www.learner.org/workshops/teachreading35/classrooms/cv6.html

In small jigsaw groups, candidates analyzed their specific video using the analysis form and determined how practices in the video related to the article they were assigned to read (each member of the group read a different article and watched a different video on the same topic). Then in small groups on Discussion Board (each candidate's name was listed on a specific forum), the candidates used their analyses (Table 3) to construct an understanding of classroom practice that is conducive to literacy instruction. All of the forums could be viewed by anyone in the course, but each group responded only to their group members.

Because each group member is assigned to watch a different video, they must provide an overview for the group as well as their own interpretation so that other group members can respond. Group members also collaborate in order to co-construct knowledge of best practices in literacy instruction. By completing these kinds of video analyses, teacher candidates used critical thinking to analyze the videos and determined how course readings on literacy instruction were enacted in the teachers' classrooms in relation to literacy practices such as vocabulary instruction, teacher read aloud, teacher-student conference, reading and writing instruction. In their analyses the teacher candidates closely examined the teacher's use of questioning, scaffolding, and modeling, as well as other instructional strategies introduced in their online literacy course.

The candidates accessed problem-solving skills when they reviewed videos to examine classroom practice, determining effective and potentially ineffective techniques. For instance, while describing the lesson she watched, Marin determined the teacher, "...could have encouraged [students] to think aloud as they consider the questions he was going to ask them, as I learned about the benefits of doing so from the Boulware-Gooden study. This supports the students as they learn how to think and learn." In her analysis she connects back to one of the assigned readings to explain how the lesson could have been enhanced. Carmen's comments reveal her perceptions of students' literacy capabilities and expectations for what she believes the writing skills of fifth grade students should be: "an essay with proper sentence structure and paragraph formation." She also draws upon her prior knowledge to critique the video by pointing out what literacy practices are missing and should be incorporated to support students' writing development—peer editing.

By responding to open-ended discussion prompts, candidates could role-play, or simulate the learning experiences of K–12 students. Assignments such as discussions, evaluating online resources including videos of classroom literacy instruction, unit planning, using multimedia software, required candidates to solve problems and apply their knowledge to authentic scenarios.

Table 3. Excerpted examples of online collaborative video analysis

Video	Group Analysis
Context Clues Lesson	Camille: [The teacher] uses 'guided reading' by the end of the video, which assists all of the students individually, and on their own level, which is very helpful to the student. The teacher is asking very open-ended questions, and listening to multiple responses to scaffold off each student. She then moves on to the 'mind map', which is group brainstorming to come up with answers for the questions, which is also very helpful. With the information on the board, the children can look at it and absorb the information better, and from the other students, not just themselves. She then breaks the students into groups of two in order for them to work together as word detectives, which is also an interesting concept. [The students] take 'clues' from the text and their 'reader voices' to determine what the new word means. This is a great way for the students to learn new vocabulary, while not feeling bad about not knowing right away. April: I love how the teacher was sharing her own experiences with the class. This is great because it allows the students to be able to know someone who went through the same steps they are going through. You mentioned in your response how the teacher used guided reading groups. My video also showed the teacher using guided reading groups. I think this is a great strategy and a great way for the teacher to be one-on-one with each student
Strategy Lesson	Marin: [The teacher] taught lessons to the entire class while also including conferencing when he helped students with questions and unfamiliar words.... He also models how to react when there is an unfamiliar word (locket), and how to mark it but move along in the text. He then goes back and follows through steps for the students to decode this new word... encouraging the students to access their prior knowledge.... His lessons concentrated heavily on discussion, so he would easily be aware of a student who was lost or frustrated. These are all major strengths because these strategies encourage better vocabulary retention and reading comprehension.... While it may have happened and was not shown in the video, I did not see any summative or peer assessments. This class could have benefited from having both assessment types as [he] could have instructed the teams he created do peer assessments, and the summative assessment would show him which students still needed more literacy practice. Carmen: I think that it is important that you point out that [the teacher] did not allow on camera for the individual process that writing in the 5th grade really should be. By this point in their academic career, a student should be able to compile an essay with proper sentence structure and paragraph formation. It is during this individual process that a teacher would be able to determine which basic skills need to be worked on, either as a class or on a one on one basis. I also think it is very important to give the students as much practice through this procedure to prepare them for the upcoming skill requirements, and practice makes perfect.

These online learning experiences also allowed candidates to collaborate with peers to co-construct knowledge of teaching and learning practice, to hypothesize and reflect on their own practice, and to identify uses of technology-based resources such as WebQuests and multimedia resources they could incorporate into the 21st-century classroom.

FUTURE RESEARCH

Future researchers who seek to investigate how to design effective online courses in teacher education programs should explore the long-term impact on K–12 students who are taught by teachers who completed most of their certification work through online courses. Through these studies, researchers can examine how teacher candidates apply knowledge gained from online learning experiences when they transition to a K–12 context; in particular, investigators can explore how these teacher candidates use technology and multimedia to support teaching and to facilitate student learning.

Longitudinal research or case studies in this area can follow teacher candidates' progress through a teacher preparation program that incorporates online coursework and then into their first few years of teaching to see how they apply techniques introduced by online teacher educators. Longitudinal research can also examine how the teacher educator develops and implements online courses. As suggested by Pyke and Sherlock (2010), online instructors tend to gradually release responsibility to online learners by initially scaffolding and providing significant support through feedback at the onset so that learners develop independence and efficacy by the end of the course.

A case study can follow instructors through the stages of course concept, design, and implementation to determine how the instructor made and modified instructional decisions throughout the course and upon course completion. A comparative framework that captures and analyzes data from instructors across disciplines, online platforms, formats (e.g. structured and unstructured), and grade levels will also help to characterize the specific processes and factors that might help to determine instructional effectiveness in online contexts.

Another potential research direction is to investigate and measure how effectively teacher candidates transfer knowledge from an online course and how they gain knowledge of their content area and pedagogical approaches from within an online context. Through this kind of research, educators can examine such questions as: (a) to what extent do teacher candidates conduct outside research to develop a better understanding of concepts discussed in teacher education courses and (b) what qualities or attributes characterize teacher candidates who demonstrate pedagogical proficiency and content expertise in online contexts? One hypothesis that emerges is that teacher candidates who demonstrate leadership and assertiveness in online courses tend to seek additional information from outside resources and are more likely to apply strategies introduced in the course more readily when they transition to K–12 contexts. During these inquiries, researchers can evaluate teacher candidates' reflections and responses to discussion prompts using rubrics to characterize their understanding of teaching and learning. This kind of close examination of teacher candidates' understanding of course materials, which includes how they interpret research and video cases on literacy lessons, can be used as pre- and post-assessments as well as indicators or precursors to teacher efficacy in K–12 contexts.

CONCLUSION

Empowering teachers to use technology in their K–12 classrooms begins with their teacher education training (Summerville & Reid-Griffin, 2008). Online literacy courses foster learning experiences that promote critical thinking, reflection, and the opportunity for teacher candidates to explore how to apply literacy-based activities in K–12 contexts. To help their K–12 students meet the demands of current technology standards, teacher candidates can draw on their experiences as learners in online courses to support student learning and to design digital-age learning experiences (International Society for Technology in Education, 2008). The online assignments teacher candidates complete—such as conducting Internet searches for resources, viewing video clips of classroom practice, and accessing predetermined web-based resources assembled in a course database—introduce the candidates to best practices and ideas they can use to expand their own instructional repertoire and to nurture their students' literacy practices. Online courses can be an ideal way to provide pre-service and in-service teachers with opportunities to practice using the technology-based learning tools, such as WebQuests and movie production software that they might incorporate into their own classrooms.

By participating in the online literacy courses discussed in this chapter, teacher candidates completed simulations where they experienced how multiple resources, examples of expected learner outcomes, collaborative learning experiences, open-ended questions and problem-solving tasks, and hands-on learning opportunities with technology can engage learners, support knowledge acquisition, and encourage active student participation. Through the learning experiences candidates encounter in these literacy courses—experiences that connect K–12 students' prior knowledge and real life—teacher candidates face authentic situations where they learn how to cultivate their own students' critical-thinking and technology skills, such as media production and information literacy, by working with the same resources in these literacy courses. The teacher candidates in these online literacy courses used a wide range of technology tools during their learning experiences; through these multimedia experiences, they could reflect on classroom practice and enhance their technological, information-gathering, evaluation, and critical-thinking skills; these are all skills educators expect K–12 students to learn and apply in the classroom.

The online courses discussed in this chapter promoted opportunities for interactive learning. The coursework generated an array of assessment materials that this instructor used to evaluate teacher candidates' understanding of literacy, teaching, and learning. Through simulations facilitated in the online courses, teacher candidates developed a better understanding of the significance of pedagogical practices such as frontloading, planning, setting clear expectations, and facilitating student-centered learning. Additionally, the resources in these literacy courses were specifically organized to support the needs of diverse learners, so that learners can gain understanding about best practices from videos, readings, PowerPoint lectures, and peers during group discussions.

Dell, Low, and Wilker (2010) believe online instructors should focus on developing quality courses. Courses should be highly structured and designed to promote usability (Fisher & Wright, 2010)—which includes meaningful coursework; varied interactions with peers and instructor; instructor feedback; and opportunities for independent learning that support different learning styles and give access to many varied resources—and to support learning through purposeful examples and clear expectations. Therefore, the primary role of the online teacher educator is pedagogy and providing good examples of instructional approaches online. The goal of focusing on pedagogy and incorporating these strategies into teacher education courses is to model for teacher candidates how to use instructional design techniques and strategies introduced in the course to incorporate technology and to promote critical thinking in K–12 contexts.

REFERENCES

Allen, I. E., & Seaman, J. (2010). *Class differences: Online education in the United States, 2010*. Babson Survey Research Group. Retrieved from http://sloanconsortium.org/publications/survey/pdf/class_differences.pdf

Baran, E., & Correia, A. (2009). Student-led facilitation strategies in online discussions. *Distance Education*, *30*(3), 339–361. doi:10.1080/01587910903236510

Bawane, J., & Spector, M. (2009). Prioritization of online instructor roles: Implications for competency-based teacher education programs. *Distance Education*, *30*(3), 383–397. doi:10.1080/01587910903236536

Dell, C. A., Low, C., & Wilker, J. F. (2010). Comparing student achievement in online and face-to-face class formats. *MERLOT Journal of Online Learning and Teaching*, *6*(1), 30–42.

Fisher, E. A., & Wright, V. H. (2010). Improving online course design through usability testing. *MERLOT Journal of Online Learning and Teaching*, *6*(1), 228–245.

Henn, M., Weinstein, M., & Foard, N. (2006). *A short introduction to social research*. London, UK: Sage Publications.

Hubbard, R. S., & Power, B. M. (1999). *Living the questions: A guide for teacher-researchers*. York, ME: Stenhouse.

International Society for Technology in Education. (2007). *NETS for students*. Retrieved from http://www.iste.org/Content/NavigationMenu/NETS/ForStudents/NETS_for_Students.htm

International Society for Technology in Education. (2008). *NETS for teachers 2008*. Retrieved from http://www.iste.org/Content/NavigationMenu/NETS/ForTeachers/2008Standards/NETS_for_Teachers_2008.htm

Jeong, H., & Hmelo-Silver, C. E. (2010). Productive use of learning resources in an online problem-based learning environment. *Computers in Human Behavior*, *26*, 84–99. doi:10.1016/j.chb.2009.08.001

Ke, F., & Xie, K. (2009). Toward deep learning for adult students in online courses. *The Internet and Higher Education*, *12*(3/4), 136–145. doi:10.1016/j.iheduc.2009.08.001

Lawrence, S. A., & Mongillo, G. (2010). Multimedia literacy projects: Strategies for using technology with K to 20 learners. *The Language and Literacy Spectrum. Journal of New York State Reading Association*, *20*, 39–49.

Nielsen, L. E. (2003). *Online discussion rubric*. Retrieved from http://www2.uwstout.edu/content/profdev/rubrics/discussionrubric.html

O'Hanlon, N., & Diaz, K. R. (2010). Techniques for enhancing reflection and learning in an online course. *MERLOT Journal of Online Learning and Teaching*, *6*(1), 43–54.

Partnership for 21st Century Skills. (2004). *Framework for 21st century learning*. Retrieved from http://www.p21.org/index.php?option=com_content&task=view&id=254&Itemid=120

Paul, R., & Elder, L. (2008). *The miniature guide to critical thinking: Concepts and tools*. Dillon Beach, CA: The Foundation for Critical Thinking Press.

Prensky, M. (2001). Digital natives, digital immigrants. *Horizon*, *9*(5), 1–6. Retrieved from http://www.marcprensky.com/writing/Prensky%20%20Digital%20Natives,%20Digital%20Immigrants%20-%20Part1.pdf doi:10.1108/10748120110424816

Sewell, J. P., Frith, K. H., & Colvin, M. M. (2010). Online assessment strategies: A primer. *MERLOT Journal of Online Learning and Teaching*, *6*(1), 297–305.

Slagter van Tryon, P., & Bishop, M. J. (2009). Theoretical foundations for enhancing social connectedness in online learning environments. *Distance Education*, *30*(3), 291–315. doi:10.1080/01587910903236312

Summerville, J., & Reid-Griffin, A. (2008). Technology integration and instructional design. *TechTrends*, *52*(5), 45–51. doi:10.1007/s11528-008-0196-z

Wise, A. F., Padmanabhan, P., & Duffy, T. M. (2009). Connecting online learners with diverse local practices: The design of effective common reference points for conversation. *Distance Education*, *30*(3), 317–338. doi:10.1080/01587910903236320

ADDITIONAL READING

Abbitt, J. T. (2011). Measuring technological pedagogical content knowledge in preservice teacher education: A review of current methods and instruments. *Journal of Research on Technology in Education*, *43*(4), 281–300.

Barseghian, T. (May, 2011). Hybrid learning comes to life at Rocketship. *Mind shift: How we learn, KQED*. Retrieved from http://mindshift.kqed.org/2011/05/hybrid-learning-comes-to-life-at-rocketship/

Barton, P. E. (2006). The dropout problem: Losing ground. *Educational Leadership, 63*(5), 14–18.

Brookhart, S. M., & Freeman, D. J. (1992). Characteristics of entering teacher candidates. *Review of Educational Research, 62*(1), 37–60.

Brooks-Young, S. (2010). *Teaching with the tools kids really use: Learning with web and mobile technologies*. Thousand Oaks, CA: Corwin Publishers.

Corbett, D., Wilson, B., & Williams, B. (2002). *Effort and excellence in urban classrooms: Expecting—and getting—success with all students*. New York, NY: Teachers College Press.

Dantas, M. L. (2007). Building teacher competency to work with diverse learners in the context of international education. *Teacher Education Quarterly, 37*(1), 75–94.

Darling-Hammond, L., & Bransford, J. (Eds.). (2005). *Preparing teachers for a changing world: What teachers should learn and be able to do*. San Francisco, CA: Jossey-Bass.

Fisher, D., & Ivey, G. (2007). Farewell to *A Farewell to Arms*: Deemphasizing the whole-class novel. *Phi Delta Kappan, 88*(7), 494–497.

Garcia, E., Arias, M. B., Murri, N. J. H., & Serna, C. (2010). Developing responsive teachers: A challenge for a demographic reality. *Journal of Teacher Education, 61*(1-2), 132–142. doi:10.1177/0022487109347878

Gay, G. (2000). *Culturally responsive teaching: Theory, research, & practice*. New York, NY: Teachers College Press.

Hinchman, K., Alvermann, D., Boyd, F., Brozo, W., & Vacca, R. (2003/2004). Supporting older students' in- and out-of-school literacies. *Journal of Adolescent & Adult Literacy, 47*(4), 304–310.

Honigsfeld, A., & Schiering, M. (2004). Diverse approaches to the diversity of learning styles in teacher education. *Educational Psychology, 24*(4), 487–507. doi:10.1080/0144341042000228861

International Reading Association. (2004). *Standards for reading professionals*. Retrieved from http://www.reading.org/downloads/resources/545standards2003/index.html

International Reading Association. (2009). *Standards for reading professionals—draft*. Retrieved from http://www.reading.org/General/Current-Research/Standards/ProfessionalStandards.aspx

Kea, C., Campbell-Whatley, G. D., & Richards, H. V. (2006). *Becoming culturally responsive educators: Rethinking teacher education pedagogy*. Retrieved from http://www.nccrest.org/Briefs/Teacher_Ed_Brief.pdf

Korthagen, F. A. J., & Kessels, J. P. A. M. (1999). Linking theory and practice: Changing the pedagogy of teacher education. *Educational Researcher, 28*(4), 4–17.

Kupermintz, H. (2003). Teacher effects and teacher effectiveness: A validity investigation of the Tennessee Value Added Assessment System. *Educational Evaluation and Policy Analysis, 25*(3), 287–298. doi:10.3102/01623737025003287

L'Allier, S. K., & Elish-Piper, L. (2007). "Walking the walk" with teacher education candidates: Strategies for promoting active engagement with assigned readings. *Journal of Adolescent & Adult Literacy, 50*(5), 338–353. doi:10.1598/JAAL.50.5.2

Lei, J. (2009). Digital natives as preservice teachers: What technology preparation is needed? *Journal of Computing in Teacher Education, 25*(3), 87–97.

Malu, K. F., Lawrence, S. A., & Mongillo, G. (2008). The reimagination of a graduate reading program: Roles and responsibilities, themes, reflections, and implications. In Craig, C. J., & Deretchin, L. F. (Eds.), *Imagining a renaissance in teacher education: Teacher education yearbook, XVI* (pp. 157–169). Lanham, MD: Association of Teacher Educators and Rowman & Littlefield Education.

Marcos, J. J. M., Sanchez, E., & Tillema, H. (2008). Teachers reflecting on their work: Articulating what is said about what is done. *Teachers and Teaching: Theory and Practice*, *14*(2), 95–114. doi:10.1080/13540600801965887

Mintu-Wimsatt, A., Kernek, C., & Lozada, H. R. (2010). Netiquette: Make it part of your syllabus. *MERLOT Journal of Online Learning and Teaching*, *6*(1), 264–267.

O'Hara, S., & Pritchard, R. H. (2008). Meeting the challenge of diversity: Professional development for teacher-educators. *Teacher Education Quarterly*, *35*(1), 43–61.

Pyke, J. G., & Sherlock, J. J. (2010). A closer look at instructor-student feedback online: A case study analysis of the types and frequency. *MERLOT Journal of Online Learning and Teaching*, *6*(1), 110–121.

Quatroche, D. J., & Wepner, S. B. (2008). Developing reading specialists as leaders: New directions for program development. *Literacy Research and Instruction*, *47*, 99–115. doi:10.1080/19388070701878816

Reagan, T. G., Case, C. W., & Brubacher, J. W. (2000). *Becoming a reflective educator: How to build a culture of inquiry in the schools*. Thousand Oaks, CA: Corwin Press.

Santoro, N. (2007). 'Outsiders' and 'others': 'Different' teachers teaching in culturally diverse classrooms. *Teachers and Teaching: Theory and Practice*, *13*(1), 81–97. doi:10.1080/13540600601106104

Staker, H. (2011). *The rise of K–12 blended learning: Profiles of emerging models*. Innosight Institute, Inc. Retrieved from http://www.innosightinstitute.org/innosight/wp-content/uploads/2011/05/The-Rise-of-K-12-Blended-Learning.pdf

Swinglehurst, D., Russell, J., & Greenhalgh, T. (2008). Peer observation of teaching in the online environment: An action research approach. *Journal of Computer Assisted Learning*, *24*, 383–393. doi:10.1111/j.1365-2729.2007.00274.x

Tomlinson, C. A. (2008). Reconcilable differences? Standards-based teaching and differentiation. *Educational Leadership*, *58*(1), 6–11.

Tschannen-Moran, M., Hoy, A. W., & Hoy, W. K. (1998). Teacher efficacy: Its meaning and measure. *Review of Educational Research*, *68*(2), 202–248.

Wayne, A. J., & Youngs, P. (2003). Teacher characteristics and student achievement gains: A review. *Review of Educational Research*, *73*(1), 89–122. doi:10.3102/00346543073001089

Williams, B. (Ed.). (2003). *Closing the achievement gap: A vision for changing beliefs and practices* (2nd ed.). Alexandra, VA: Association for Supervision and Curriculum Development.

Willingham, D. T. (2007). Critical thinking: Why is it so hard to teach? *American Educator*, *31*(2), 8–19.

KEY TERMS AND DEFINITIONS

Course Design: The structure and format used to organize online coursework and learning experiences

Engagement: active participation in a learning experience.

Online Instructor/Teacher: Facilitator of online learning experiences and courses.

Online Course: Interactive learning experience that uses technology and multimedia and is facilitated by an instructor; the course may be completely online, blended to allow some face-to-face time, synchronous (with immediate feedback from the instructor and real-time interaction with peers) or asynchronous (no immediate feedback from instructor and real-time interaction with peers).

Online Resources: Course content and materials provided to (and accessible to) learners throughout the course.

Open-Ended Assignments: Authentic tasks used by the course instructor as formative and ongoing assessments, which also provide opportunities for online learners to apply their knowledge, think critically, and reflect on what they already know.

Teacher Candidates: Undergraduate and graduate students enrolled in teacher education courses.

APPENDIX A

Online Resources and Multimedia Evaluation Project Assignment Sheet

Description of Assignment

As a group you will work with peers who have similar content area background to evaluate current online and multimedia resources you can use to support students' literacy skills: reading (comprehension, fluency, vocabulary), writing, technology, speaking, listening, etc. Your evaluation of these resources will take the form of an **Annotated Bibliography.**

1. Identify a grade level your group will target for this assignment.
2. Find **10 technology-based (multimedia) or online resources** you can use to support students' literacy skills in your content area. Your resources can include online lesson plans already developed in your subject area, media and technology, educational software, and professional resources on the web that will support you in teaching this topic. Be sure to include both print (e.g. lesson plan, newspaper article) and non print media (e.g. YouTube video clip, PBS video). You should not create these materials. Instead you should look for and identify these resources. Then evaluate them to determine how you might use them to support students' literacy skills.
3. Organization/ Example of Template: In your Annotated Bibliography you should Summarize (Describe), Evaluate (Assess), and Reflect (How you might use it) on the resource. Therefore your evaluation can include the following sections: identify why you selected it (e.g. its potential use as an instructional resource), and evaluate its potential (to what extent to you think it will be effective for supporting students' literacy skills, will you have to make any modifications, and why).
4. Attach all resources or include active web links to the sources you have identified. Consider the diverse needs of students when assembling the resources. Be sure to clearly identify your grade level, subject, and topic.
5. Create an annotated bibliography of the resources you have found. Use APA style for your annotations. More instructions on Annotated Bibliographies http://owl.english.purdue.edu/owl/resource/614/01/

Examples completed by previous students are posted on Blackboard.

APPENDIX B

WebQuest Assignment

1. Complete one content area WebQuest – "an inquiry-oriented lesson format in which most or all the information that learners work with comes from the web" (www.webquest.org).
   a. Go to www.webquest.org
   b. Click " Find WebQuest" on the left side
   c. Scroll down to " Curriculum X Grade Level Matrix"
   d. Select a subject and grade level. **Select a topic/ activity based on the content and grade level you want to teach.**
   e. Click " Search Matrix"
   f. Browse through the WebQuest options
   g. Select one WebQuest
   h. Review the components of the WebQuest
   i. Complete the WebQuest. You are the learner; complete the Webquest as if you were the student.
2. Is the WebQuest an effective learning tool? To what extent did you learn **content** by completing the WebQuest (remember: you completed the activity as an adolescent learner)?
3. What **other skills** do you believe were developed by the WebQuest? What **other knowledge** did you gain?
4. What kinds of issues might you and/or students encounter when using WebQuest as a teaching and learning tool?

Additional Information

Websites for PE/Health

http://www.gecdsb.on.ca/d&g/DP/locatorc.asp

http://www.safehealthyschools.org/webquests/physical_education/index.htm - click on "student instructions".

http://www.safehealthyschools.org/webquests/choose_webquest.htm

Music Educators: WebQuests

http://oben.powertolearn.com/Music/wq.html

http://www.nashua-plainfield.k12.ia.us/projects/chad/

In case some of the links are no longer active on the webquest.org website provided above, you can also access Webquests across content areas using the link below – Scroll down to "Content WebQuests" …some of the links do not work, but it provides an additional resource

http://www.mapacourse.com/webquest%20project/webquestlinks.htm

APPENDIX C

Video Analysis Form for Classroom Observation: Literacy in the Content Area

Instructions: Watch the assigned video. Complete the chart below based on the information requested for each focus area. (This is also the prompt each of you will use for your Discussion post)

Table 4.

Focus of Observation/ Perspective (Lens)	What to Look For in the Video
Using Questioning and Discussion Techniques	• Describe the quality/level of questions, discussion techniques, and student participation. Identify examples if they are evident.
Engaging Students in Literacy & Learning	• Describe how literacy is being taught and connect with content to share information about the subject to students (reading, writing, listening, speaking, technology, etc.), activities and assignments, students are grouped, instructional materials and resources, etc. Identify examples if they are evident.
Assessing Literacy & Learning	• How is the teacher checking for understanding? How is the teacher determining what students' know and don't know? • What other kinds of assessment strategies could the teacher incorporate into this classroom?

1. What do you think are the strengths and weaknesses in the lesson?
2. Based on the insights you have gained from the course readings/resources what other kinds of strategies could the teacher have used to support students' literacy development?

Chapter 13
Putting Multiliteracies into Practice in Teacher Education:
Tools for Teaching and Learning in a Flat World

D. Bruce Taylor
University of North Carolina at Charlotte, USA

Lindsay Sheronick Yearta
University of North Carolina at Charlotte, USA

ABSTRACT

While technology has always played a role in teaching and learning, with the advent of Information Communication Technologies (ICTs), schools have struggled to keep pace with Web 2.0 tools available for teaching and learning. Multiliteracies, a term coined by scholars who published under the name The New London Group in 1996, has helped provide a theoretical foundation for applying new texts and tools to teaching and learning; however, much of the scholarship around Multiliteracies remains in the academic and theoretical domain. The authors suggest a pedagogic framework or metastructure for applying Multiliteracies to teacher education and by extension to P-12 classrooms. They document Web 2.0 tools and discuss how they have used them in undergraduate and graduate teacher education courses.

INTRODUCTION

Technology has always been integral to teaching and learning. Incremental changes over time from clay tablets to papyrus scrolls to paper and ink have, through human history, shaped the work of teachers and students in their work together. The rate of advancement and impact on teaching and learning took a leap forward with the invention of the printing press. The invention of mass production printing technologies was one factor among others that in the 19th Century increased access of education from the wealthy and elite to the children of middle and lower socioeconomic families in Europe and the United States (Heath, 1991; Resnick & Resnick, 1977). The rise of Information Communications Technologies (ICTs) over the past three decades has further enriched

DOI: 10.4018/978-1-4666-1906-7.ch013

and complicated teaching and learning in P-12 classrooms and, at the same time, accelerated the rate of change. One challenge that remains is how to prepare teachers in the 21st Century to use technology in their classrooms in meaningful ways.

Multiliteracies, a term coined by the New London Group (1996), captures this shifting notion of literacy to include the "multiplicity of communications channels and increasing cultural and linguistic diversity in the world today call for a much broader view of literacy than portrayed by traditional language-based approaches." In short, these scholars sought to overcome the limitations of traditional notions of literacy to include social and cultural changes in society and the emergence of new technologies that enable students to negotiate "the evolving language of work, power, and community, and fostering the critical engagement necessary for them to design their social futures and achieve success through fulfilling employment."

While Multiliteracies has been a fruitful area for scholarship and received attention in academic circles, it has been less visible as a concept in P-12 classrooms and teacher education programs as a construct for linking post-modern thinking about teaching and learning. The New London Group addresses a pedagogy of Multiliteracies by introducing a new metalanguage based on the concept of design and the Designs of Meaning which according to the authors includes Available Designs, Designing, and the Redesigned. Available Designs, they write, include "the 'grammars' of various semiotic systems: the grammars of languages, and the grammars of other semiotic systems such as film, photography, or gesture." We see this work as groundbreaking but in working with teachers and students have struggled to translate the academic discourse into meaningful classroom practices.

While the scholarship of literacy has in many ways attempted to keep up with the rapid changes of an increasingly "flat world" (Friedman, 2007), schools and other educational institutions including teacher education programs have not (Collins & Halverson, 2009). States have attempted to enact policy to help bridge this gap between technologies used beyond the school those used in classrooms. For example, The North Carolina State Board of Education established the *Strategic Plan for Reading Literacy* in April 2007, which states:

Reading is the fundamental skill needed for success in life, especially in the 21st century. While students must be at proficiency or above in basic literacy (reading, writing, listening, speaking, using conventional or technology-based media), these skills are no longer sufficient for college- and work-ready high school graduates. As the world continues to change rapidly, schools must evolve to meet future needs.

Learning and communicating requires an ability to read, write and locate information. Today's technology driven society also requires digital literacy, which means that an individual can read and write digitally in order to access the Internet; find, manage and edit digital information; join in communications; and otherwise engage with an online information and communications network.

The *Strategic Plan for Reading Literacy* as well as national organizations including the State Educational Technology Directors Association (SETDA), an association representing technology leadership in all 50 states, the District of Columbia, the U.S. Virgin Islands, American Samoa, and the Bureau of Indian Affairs, calls for teachers and schools to prepare students to be globally competitive 21st Century citizens and professionals able to provide leadership for innovation using 21st Century technologies (Jones, Fox, & Levin, 2011).

Preparing teachers to teach in an increasingly connected and digital world is part of the challenge of teacher education. Scholarship documents the challenge of preparing teachers to effectively use ICTs in the classroom (Kumar & Vigil, 2011; Means, 1994; Brooks & Kopp, 1989). This chapter seeks to bridge the gap between the academic discourse of a theory of Multiliteracies by providing pedagogic metastructure for teachers and students that we believe can help them identify Web 2.0 tools useful for teaching and learning. We discuss this pedagogic structure, identify key tools for teaching and learning, and provide practical classroom applications of Web 2.0 technologies to teacher education and P-12 teaching and learning including ways these tools were integrated into teacher education assignments.

A BRIEF REVIEW OF THE LITERATURE

Multiliteracies, Rethinking Literacy, and Learning in the 21st Century

The term Multiliteracies was coined by The New London Group in the mid 1990s. These scholars, who met in New London, Connecticut, published their ideas in a seminal piece in the *Harvard Educational Review* (New London Group, 1996). Since the publication of this article, scholars from several disciplines have explored the meaning of Multiliteracies (Cope & Kalantzis, 2000; Gee, 2003; Luke & Elkins, 1998). New ideas sprang from these discussions and notions of being literate expanded beyond the traditions of print media. Terms such as "new literacies" emerged as a result of new information communication technologies (Lankshear & Knobel, 2003; Leu, 2002; O'Brien & Bauer, 2005). Luke and Elkins (1998) captured this period ushered in by new technologies as "New Times," a period in which literacy is viewed as multi-modal processes that include the reading of print-based and electronic texts, use of visual, spatial, gestural, and aural representations. The ideas behind the term Multiliteracies provide an expanded and expanding notion of text and what it means to be literate to include a multitude of texts including but not limited to web pages, e-readers, social networking sites, video, audio and performance texts.

As our society has moved from an industrial era to an information era, changes in thinking about literacy are seen by some as a radical shift away from long-held traditions of what it means to read and write and have given rise to new literacy communities not imagined two decades ago (Luke & Elkins, 1998; Leu, 2002). Author and *New York Times* columnist Thomas Friedman documents this change in his book *The World is Flat* (2006). Trilling and Fadel (2009) place switch from an industrial to knowledge-based economy in 1991 when, for the first time, knowledge-based expenditures like computers, servers, software and phones exceeded industrial expenditures (for example, engines and machines for agriculture, mining, transportation, manufacturing and energy production). Collins and Halverson (2009) acknowledge the tensions these changes cause teachers and students but see this shift as a "digital revolution" that schools must embrace.

As literacy educators we embrace the term Multiliteracies because it captures in a single word the rapidly evolving nature of literacy. As teacher educators who work with pre-service and in-service teachers across many disciplines, we see it more broadly as a frame for teaching and learning in the 21st Century. It is this second meaning that we seek to address in this chapter and hope that educators across academic disciplines find useful the ideas and pedagogies we explore.

Technology at the P-12 Education Level

Research shows that despite the rise of technology over the past two to three decades, schools have been slow to respond (Cuban, 2001). Scholarship that examines school change suggests that P-12 schools have long been resistant to adopting new technologies (Cuban, 1986; Collins & Halverson, 2009) as well as new teaching methods (Evans, 1996; Tyack & Cuban, 1995). This scholarship suggests that teachers, immersed in a nearly constant flow of change, often resist calls to adopt new practices (Evans, 1996). Tyack and Cuban (1995) note that schools as institutions are also entrenched in their practices and argue that the "grammar of schooling"—long-held institutional traditions of teaching and learning—work against change.

Technology itself presents barriers especially regarding cost and access. The ratio of students to computers in the U.S. is around five to one, but in urban schools it is a 9:1 ratio (Collins & Halverson, 2009). As this research suggests, socioeconomic factors are part of the challenge. In 2005, 89% of all households owned a personal computer, and 81% of all households had Internet access, but only 68% percent of households with income less than $25,000 had computers (Carroll, et al., 2005).

Despite the challenges, calls for the use of technology in education are widespread. Friedman (2006) points to the pervasive use of ICTs and other technologies in the workplace and their impact on the global economy. Others including Leander and Boldt (2008) point to the multiple social uses of technology within and beyond schools and offices. Collins and Halverson (2009) present research suggesting that technology offers learners greater customization in meeting their own learning needs and greater control of the process while Gee (2003) suggests that some technologies offer greater interaction including more immediate feedback. There is some evidence that technology may have a positive impact on learning. A review of research commissioned by Cisco Systems (2006) concludes that while the research is still emergent, "technology does provide a small, but significant, increase in learning" (p. 15).

Teacher Education and ICTs

Teacher education programs offer an opportunity to help P-12 teachers learn about and integrate technology into their teaching. The National Council for Accreditation of Teacher Education (NCATE), the accrediting body for many teacher education programs in the United States, Standard 1: Candidate Knowledge, Skills, and Professional Dispositions includes the following:

Teacher candidates reflect a thorough understanding of the relationship of content and content-specific pedagogy delineated in professional, state, and institutional standards. They have in-depth understanding of the content that they plan to teach and are able to provide multiple explanations and instructional strategies so that all students learn. They present the content to students in challenging, clear, and compelling ways, using real-world contexts and integrating technology appropriately. (National Council for Accreditation of Teacher Education, 2008).

Despite standards such as those imposed by NCATE and other accrediting bodies to integrate technology into teacher education programs, research suggests that there is a gap between the goals for preparing teachers to use technology and their use of technology in their teaching. Research by Stobaugh and Tassell (2011) shows that while universities have sought to infuse technology use into teacher education, it competes with other program needs and goals and that even when university programs meet NCATE goals, students do not necessarily transfer those skills to their practice as teachers (Teclehaimanot, Mentzer, & Hickman, 2011).

Although the research is not conclusive, one line of thought suggests that teacher education has focused too much on how to use specific technologies is not likely to improve the practice of teaching and learning (Brand, 1997; Moursund & Bielefeldt, 1999; Koehler & Mishra, 2005; So & Kim, 2009). Studies on technology in teaching suggest a shift from an emphasis on technology skills in isolation to integrating pedagogy and content with technology—what Koehler and Mishra (2005) call technological pedagogical content knowledge (TPACK).

The challenges lie not only with the programs but also with pre-service and in-service teacher education students. Hammond, Reynolds and Ingram (2011) found that pre-service teachers were receptive to using ICTs but that their levels of use varied with most utilizing interactive white boards (IWB) but fewer utilizing Web 2.0 technologies citing limitations of access to ICT, a lack of self-efficacy regarding technology, and their beliefs regarding ICT's limited impact on student learning. Even pre-service teachers who were deemed proficient in the use of technology faced obstacles to implementing it in their teaching. Kumar and Vigil (2011) found a significant gap between pre-service teachers' personal uses of Web 2.0 technologies such as social media and their use of technology in the classroom.

What seems clear is that while teacher education is promoting the use of technology in its programs, there is a lack of transfer of those skills to P-12 classrooms. While there are many reasons for this disconnect, we believe that the more work needs to be done to bridge the divide between theory and practice in technology use in teacher education. To that end we offer a Multiliteracies framework that our students have helped us develop to identify digital technologies and their application to teaching and learning in P-12 classrooms.

DEVELOPING A PEDAGOGIC FRAMEWORK OF MULTILITERACIES

We agree with the scholarship on Multiliteracies that notions of reading and writing are expanding and believe that ICTs have increased the rate of change to a pace rarely seen since the development of the printing press. Further, we acknowledge the role that language and cultural diversity play in an increasingly global world. But, we have wondered what this means for education, and, in our case, for teacher education.

In 2006, one of the authors began thinking about Multiliteracies and its implications for teacher education. He began work to develop a course for his university's graduate literacy and reading programs and in 2007 piloted a course that came to be titled *Multiliteracies in a Global World: Reading and Writing in New Times*. The description for this course includes:

Multiliteracies is a way of rethinking about what it means to be literate today in a globally connected world. It takes literacy beyond a focus on traditional print-based literacy to multiple-forms of knowing, including print, images, video, and combinations of forms in digital contexts, which are represented in inter-related and complex ways. A theory of Mulitliteracies complicates what we know about teaching and learning because it is constantly evolving as new information and communications technologies (ICTs) emerge. This course immerses students in both the theory and practice of Multiliteracies and considers how globalization has created a more complex environment for teachers and students.

The creation of the Multiliteracies course at our institution along with the development of online teacher education courses and programs had an impact on our thinking. This is documented in more detail in another publication (Taylor, 2012), but suffice it to say that the digital texts and tools we envisioned in the Multiliteracies course influenced our teaching in other courses.

Each semester that the Multilitearcies course was taught, we have read and discussed the New London Group's 1996 article and explored the theoretical foundations of the concept and then turned to apply that theory to our teaching. That move from theory to practice has given rise to a pedagogic metastructure, an overarching set of categories or themes, that both faculty and students in the course have helped to develop. We do not go into a lengthy description here of how these categories were developed but mention a seminal moment in what was an ongoing discussion that arose through an assignment in the course. One of the authors, who was teaching the course, worked with students to create a Mulitliteracies wiki (Figure 1), an online portfolio of tools and ideas for teaching using technology (http://multiliteraciesatuncc.pbworks.com). Its development required a scheme or categories in which to organize the various digital tools, websites, programs and texts we found. The categories came as a result of those discussions.

These categories focus on activities that students and teachers engage in when using technology (traditional or digital) which fit with our beliefs of learning as a social and active process (Leontiev, 1978; Vygotsky, 1978). Because of this focus on action, we have used gerunds—nouns formed from verbs with *–ing* endings—to name the elements. The categories or actions include tools for:

- Researching
- Communicating
- Collaborating
- Exploring
- Creating
- Sharing

These categories allow us to organize practical tools for teaching and learning by how they are used. Figure 2 describes each Action (e.g., *Researching*) in terms of its Content or Text (e.g., search engines) as well as the digital Tools (e.g., Google.com and Bing.com) associated with it. We acknowledge that these are not discrete but an overlapping set of categories. For example, blogs are texts and tools that foster collaboration but also serve as a vehicle for sharing. Dipity.com,

Figure 1. The Multiliteracies Wiki

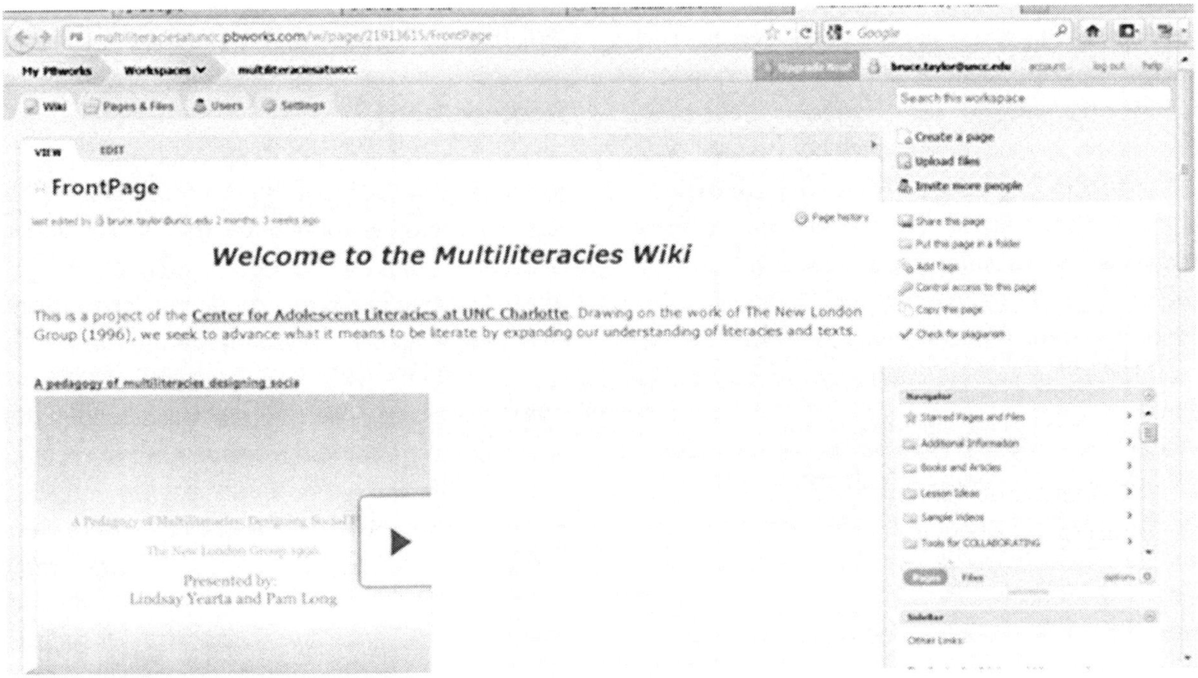

Figure 2. A pedagogic metastructure for teaching multiliteacries

Action	Content or Text	Tools
Researching	Search engines Video content sites Events Libraries & Encyclopedias	Google, Bing, Yahoo, DogPile YouTube, Teacher Tube Dipity.com Wikipedia.com, LibrarySpot.com, and IPL.com.
Communicating	Video conferencing Social Media Chat and email	Skype, FaceTime Twitter, Facebook ePals, email programs, Gaggle.net
Collaborating	Wikis Blogs Shared Documents	PBWorks, Wikispaces Edublogs.org, Blogger, Tumblr Google Documents, Adobe Acrobat (Buzzword)
Exploring	Geography WebQuests	Google Maps, Google Earth WebQuest.org
Creating	Photographs Video Audio Multimedia	iPhoto, Picture Manager, Photoshop.com iMovie, MovieMaker, Jing Audacity, Garageband VoiceThread, Dipity
Sharing	Blogs Photos and Images Video	Edublogs.org, Blogger, Tumblr Picasa, YouTube, Vimeo, TeacherTube, Jing

an online timeline tool and website, can be used for conducting research on current events or to create and share content.

Over the past four years, students in the Multiliteracies course have contributed many Web 2.0 and technological tools for teaching and learning to the wiki. They have posted lesson plans and units, links to websites, and reviews of books and articles. The Multiliteracies Wiki is a dynamic resource that continues to be used in our teaching at our university and by our students in their schools.

Next, we discuss each of the six categories of actions and offer examples of their use in our teacher education programs. Some of the examples include work between our pre-service and in-service teachers and P-12 students that came as a result of required clinical and/or service learning requirements in some of our teacher education courses.

PUTTING MULTILITERACIES TO WORK IN TEACHER EDUCATION

Researching

Research is vital to learning in any context from the Kindergarten classroom to the university teacher education course. While the term conjures images of students taking notes on 3-by-5 cards from articles and books found in a school, university or public library with the advent of ICTs that image has changed. The Internet has expanded the repertoire of traditional research tools to include a diverse range of online media that now includes websites and hypertext, digital images and video, and online access to books and articles. Less formal research tools also include blogs, news feeds, and email, to name just a few. Search engines such as Google, Bing, Yahoo, and metasearch engines such as Dogpile allow students to "surf" for information.

The Internet also provides access to digital libraries (LibrarySpot.com and IPL.com, for example) and online encyclopedias (Wikipedia.com and Britannica.com). Less obvious but useful tools for researching also include YouTube.com, Vimeo.com and TeacherTube.com video content sites.

Students have also found the following websites useful in their own research or in working with P-12 students:

- *Go2web20* (http://www.go2web20.net) offers a frequently updated resource of Web 2.0 tools for use by teachers and students.
- *Fact Monster* (http://www.factmonster.com) is an online encyclopedia, dictionary, thesaurus and almanac for children. Our students often include this resource in lessons they create for P-12 students.
- *Quiki* (http://www.qwiki.com) is a research source that presents short multimedia presentations about topics.
- *CIA World Factbook* (https://www.cia.gov/library/publications/the-world-factbook) is an excellent resource for studying the governments, culture, and economies of countries around the world.

While our students draw on many of these resources for assignments in their teacher education coursework, two forms of research have benefitted from this expansion of research tools offered online. The I-search paper (Macrorie, 1988) and multigenre research (Romano, 1990; Romano, 1995) are both forms of writing that allow for greater choice in genre and less formality in style than traditional research papers, and both allow for a greater variety of sources. For example, students in our undergraduate reading methods courses have emailed scholars in the fields of literacy to ask and get answers to specific questions arising from their I-search questions and research. Other students found useful information in YouTube and TeacherTube videos.

Researching in a digital environment fits with our understanding of learning as a social and active process. Not only do students have access to a broad range of resources, they can view presentations and videos posted in sites such as YouTube by experts on a topic. As our students have shown us, they can also use Skype and email to conduct email or live interviews with sources. This kind of interaction with scholars and informants makes researching more engaging but also provides access to the most current information.

Communicating

Communication is at the heart of what it means to teach. Lecture and discussion are deeply rooted in the "grammar of schooling" (Tyack & Cuban, 1995), but ICTs and other new technologies have created opportunities for teachers and students to move away teacher-centered formats. They also hold the potential to expand communication beyond the walls of the school.

We have used video conferencing tools like Skype to bring teachers and scholars into our undergraduate and graduate teacher education courses to join our discussions and share their experiences. Our students have also used Skype for group meetings and planning for collaborative projects and, on more than one occasion, to join our class when they could not attend in person.

Screen capture software is another tool that has become invaluable to us over the past three years. Jing and Camtasia allow us to narrate feedback for students particularly of projects developed using online collaborative tools such as wikis and shared documents such as Google Documents. As we describe in the next section on collaborating, students create unit plans in shared document spaces and portfolios of instructional tools in wikis. We use Jing (http://www.techsmith.com/jing.html), a free download from Techsmith which also makes Camtasia, to record feedback of these assignments for our students at points in time during the semester. Jing allows for recordings of up to five minutes in length. These can be uploaded to Screencast.com, a server site provided by TechSmith. We email students a link to the recorded screen capture in which they hear our comments over a video of their unit or wiki. In online sections of our courses, we ask students to create their own Jing videos of their unit plans and post these in a threaded discussion so that others in the course can provide feedback before the final due date.

More recently we've begun recording how-to videos using Jing for short videos and Camtasia for videos requiring more than five minutes in length. We have created videos showing students how to set up their wikis and shared documents, use video and photo editing tools, and publishing video to YouTube and Vimeo. We've also created video overviews of our syllabi and assignments for online sections of the courses we teach. Our goal is to better communicate with our students but also to model the use of these tools and, in some cases, have them use them as part of their assignments. Students have asked us to make these videos available to their colleagues, so we have posted some to a blog we host, Teaching & Learning in the 21st Century (http://teach-learn21.blogspot.com).

Just as Web 2.0 tools make researching more active and engaging, they also allow for a higher level of interaction among students, between teachers and students, and between students and those outside the classroom and school. Skype, email, video, and screen capture software such as Jing help to minimize distance and break down traditional barriers that exist between schools and communities. These tools literally minimize classroom walls and school fences and to echo the phrase coined by Thomas Friedman, help to flatten our world.

Collaborating

Collaboration is one of the hallmarks of Web 2.0 technologies. Wikis, blogs and shared document spaces such as Google Documents and Adobe Acrobat Buzzword allow groups to work together in a single environment.

Blogs are among the first Web 2.0 tools we began using a few years ago. Initially, we used blogs to foster communication among students when we could not access threaded discussion tools such as those provided in Blackboard and Moodle course software. At the time, access to this course software was limited to online courses only at our university, so we created blogs in Edublogs (www.edublogs.org) and Blogger, Google application. Figure 3 is an example of one of our class blogs used in an undergraduate literacy methods course.

Blogs allowed us to move some discussions outside of our class so that we could use class time for other activities. Over time we learned that blogs have other benefits. In particular, blogs allowed us to invite others into our discussions. Teachers in schools in which we were conducting clinicals in P-12 classrooms joined our discussions of topics or readings. In 2008, students in a graduate literacy course were discussing an article written by a noted scholar who lived in Asia. To our delight, the author of the article joined our discussion in the blog. He began his comments by saying that someone had emailed him about our blog and discussion.

In several of our literacy methods courses, wikis have taken the place of traditional portfolios and collections of teaching tools. In our content-area literacy courses, a toolkit of teaching strategies that was compiled in a lengthy word processing document has for the past three or four years been assembled in a wiki. These toolkit wikis are course assignments in which students identify strategies to support comprehension, vocabulary and concept learning, writing-to-learn strategies and texts useful for learning in their subject area.

Figure 3. Reading methods class blog

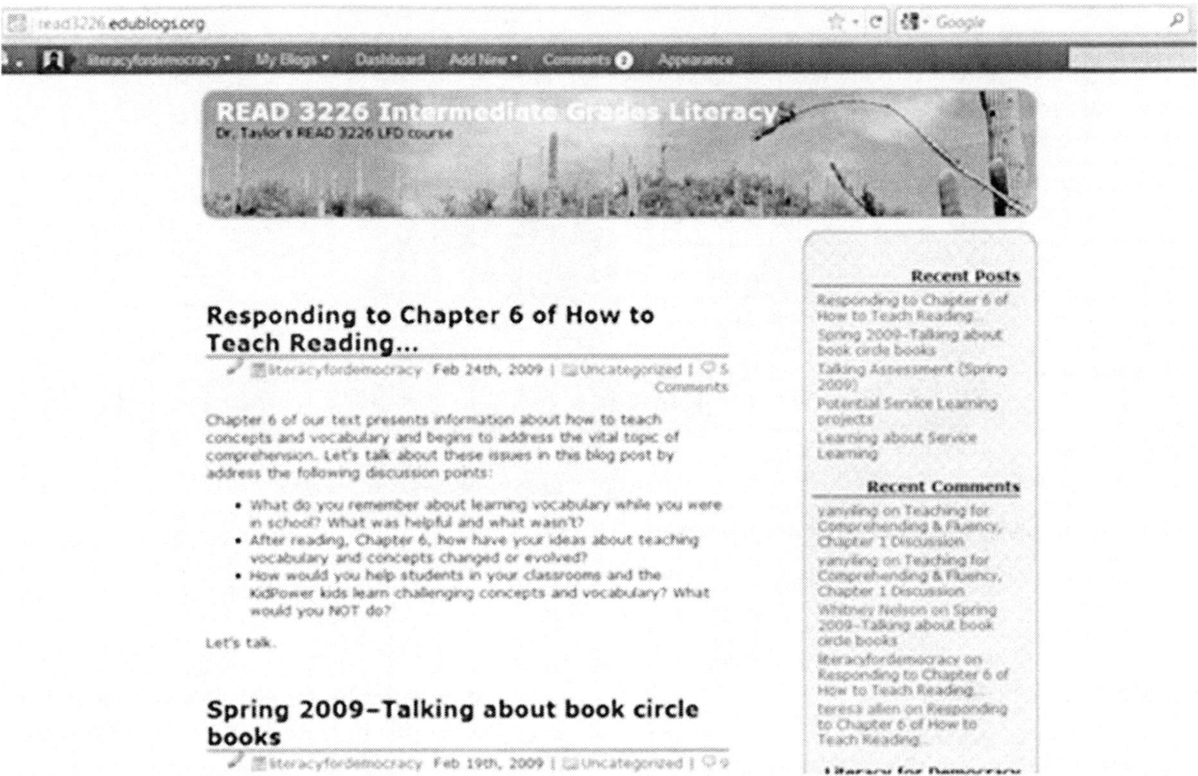

Groups of students who share a common teaching focus (social studies, mathematics, English language arts, science, fine arts, etc.) collaborate on the development of their wiki. Figure 4 shows an English language arts wiki developed by students.

The wikis—created in PBWorks (pbworks.com) or WikiSpaces (www.wikispaces.com) – allow for greater flexibility and access. Students can upload word processing documents, PDF files, videos, audio files, digital photographs and links to other websites. Many of these wikis are used solely for credit in our course; however, some students have continued to update and use their wikis long after the course has ended.

Shared document spaces such as Google Documents and Adobe Acrobat Buzzword are another tool we find invaluable for collaboration among our students. In the past, students would create unit plans in a word processing document. If they worked in a group, this document was often email among the students and at times content was lost among the different versions. The shared documents spaces provide a web-based solution that allows groups of students to make contributions without losing content. They also provide access to us as instructors so that we can provide feedback at any time without asking for a current copy to be printed off and turned into us or emailed as an attachment.

We began using these tools—blogs, wikis and shared document spaces—primarily in online courses but have found over time that we use them face-to-face and hybrid courses as well. Like other Web 2.0 tools, they minimize traditional barriers between students and the world in which they live and, at the same time, make learning a more active and social process.

Figure 4. Content-area literacy course toolkit wiki

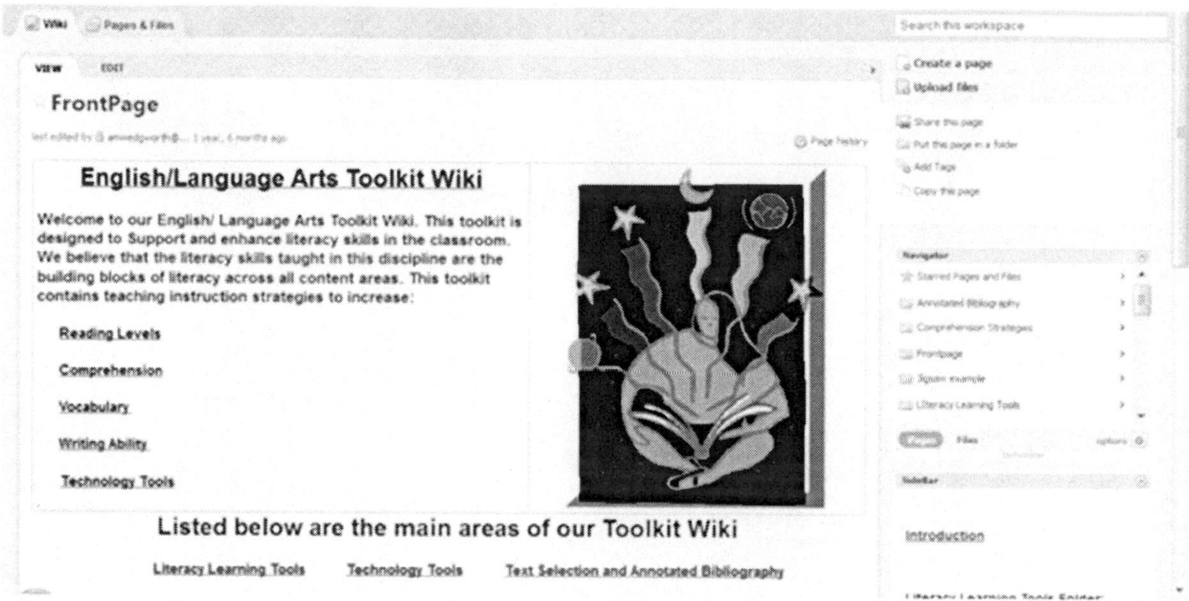

Exploring

The Internet offers us a vehicle for exploring the world without having to leave the classroom. There are innumerable websites that allow students and teachers to tour museums online, visit historic and scenic places, watch video, or see the world. Websites such as Google Lit Trips (http://www.googlelittrips.com/GoogleLit/Home.html) allow students to trace the routes of characters from literary works. Google Earth and Google Maps provide a virtual tour of the globe that includes satellite and digital images, facts, and information.

Figure 5 shows an image of a Google Map our students in an undergraduate reading methods course created with third and fourth grade students at an elementary school at which we were partnering. Our pre-service teachers met twice a week with these third and fourth graders to research, read and write around the topic of *Family Stories: Past and Present.* Both groups of students used Google Maps to document where their ancestors had come from by marking the map of the world with virtual "push pins." The students took this information as well as family stories and published it in a 'Zine that they created in a word processing document and published on the school's and university's websites as a PDF.

WebQuests (Dodge, 1995) provide another tool for using the Internet to "travel" outside the classroom. WebQuests are inquiry-based lessons that primarily utilize content found on the Internet. While there are several websites that offer WebQuests and help teachers create their own, WebQuest.org, which was founded by Bernie Dodge and Tom March in the mid 1990s. Their site provides robust information about WebQuests as well as a database of WebQuests that is searchable by subject area and topic.

Creating

The first term that came to mind when we were developing a category that captured the notion of authorship was *writing*. However, we soon decided that *creating* was more encompassing. We use the term *creating* to capture the authoring of not only written texts but also digital video,

Figure 5. Google map showing family origins of elementary and college-age students

audio and photography. We examine all three in the Multiliteracies course but also use these tools in our literacy methods and content-area literacy courses that we teach. There are many web-based tools for creating digital content and we share just a few here from the many that we have used in our teacher education programs. We focus on four types of content or texts: photographs, video, audio and multimedia.

There are many tools for editing digital photographs including software that comes on many Windows and Macintosh computers. Both iPhoto (Macintosh) and Windows Picture Manager offer basic tools for photo editing. Software such as Adobe Photoshop and Photoshop Elements provide a greater range of tools for photo editing. A rarity a few years ago, digital photography has become a valuable tool in our courses. Students use digital cameras to document clinical work in schools, to capture examples of P-12 student work, and to illustrate projects. In our undergraduate and graduate writing courses, photographs have become a text of their own or to illustrate poems, essays and other creative multigenre writing. Students publish their photographs in various formats including word processing documents, wikis, and blogs.

Digital video has become a tool that our pre-service and in-service teachers use for documenting work with P-12 students. Most computers feature video editing software—iMovie on Macintosh and Movie Maker for Windows-based computers. Below are a few of the projects our students have created in collaboration with P-12 students:

- **Public Service Announcements:** As part of a service learning project in a literacy reading methods course, undergraduate education majors and students at an elementary school conducted a food drive. The two groups collaborated to create short video public service announcements aired on the school's morning announcements.

- **Video Expose:** Students in a content-area literacy course collaborated with students in an urban high school American history course. The high school students were studying the Gilded Age and came across the work of photojournalist Jacob Riis, author of *How the Other Half Lives* (1890). The students created a video expose with interviews of teachers, students and community members that explored the question: are we the other half?
- **Documentary Reporting:** Students in our Multiliteracies course, which is often taught during the summer, have worked with elementary and middle school students in a summer program. Video projects were developed documenting healthy living and nutrition, the environment, and community building. These videos were posted to YouTube.

We have also used podcasts in our teaching as a tool for sharing content. Our students have found existing podcasts published in iTunes and Podcast Alley that are excellent resources for teaching and learning. Podcasts from these sources explore language learning, history, current events, religion, politics, economics, and technology to name a few subjects. For example, the website Howstuffworks.com has podcasts on history, current events and technology. Radio Lingua offers language learning podcasts such as CoffeeBreak Spanish and CoffeeBreak French.

We use Audacity, a free, open-source program available from http://audacity.sourceforge.net/ to create our own podcasts. Our students have used Audacity to create assignment updates and how-to podcasts, and prior to using Jing to create screen casts, we used Audacity to create assignment overview podcasts for online courses.

There are many Web 2.0 tools that allow users to create multimedia presentations or "mash ups" of different content including photos, images, video, audio and text files. VoiceThread (www.voicethread.com) is such tool that allows for group conversations around a set of texts that are uploaded to the site. A VoiceThread may contain video, word processing files, photos and image files, audio files, Power Point or other texts that are strung together into a movie-like presentation. Participants and viewers can discuss the VoiceThread by posting audio, video or text comments.

Figure 6 shows a VoiceThread one of our students created to discuss chapters from a book used in the Multilitearcies course we teach. As this figure shows, other students were able to post audio and text comments in response.

The tools that we and our students have identified for creating epitomize sociocultural understandings of teaching and learning. They are transactional and discursive in their nature. Moreover, they foster collaboration and creativity which in our experience make learning more engaging than traditional classroom assignments.

Sharing

As literacy educators, we know the value of having an authentic audience that goes beyond the classroom or university. We stress this in the writing courses we teach and have come to see its value in other courses. The Internet provides multiple venues for sharing or publishing content including blogs, wikis, video content sites (YouTube, Vimeo, and Teacher Tube, for example), and social media sites (Facebook, Google+, and Twitter). Multimedia sites such as VoiceThread and Dipity also provide viable ways to share student created content.

As mentioned above in the section on Creating, we have used YouTube and Vimeo video content sites to share student created video content. Figure 7 a video created by a middle school student titled *Planning for the Future.* The video was one of several developed around a theme that arose through our work with these students.

Not only are these videos available to our students and the P-12 students with whom we

Figure 6. A VoiceThread discussion of chapters from a book

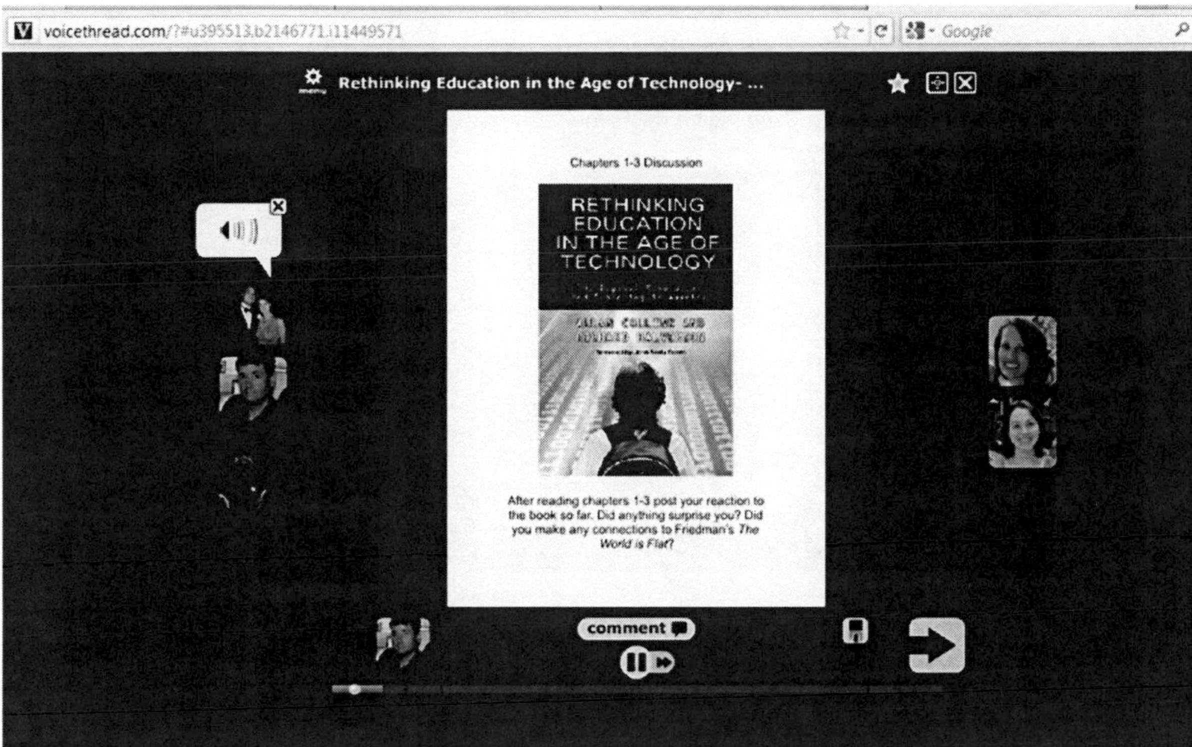

work, they reach an audience outside the classroom that includes family members, other teachers and students. Over time, our videos have attracted a small group of followers in YouTube who subscribe to our uploaded videos. As faculty, we also use YouTube and Vimeo to upload and archive how-to videos for our students (see http://teachlearn21.blogspot.com).

We described the Dipity timeline website in the section about Researching, and it is a good source for studying current events. Dipity also allows users to create their own timelines. Students in our undergraduate literacy methods and content-area literacy courses have created reading autobiographies in Dipity. Figure 8 shows a timeline created by students in our graduate Multiliteracies course that charts the many events that author Tom Friedman lists as "flatteners" in his book, *The World is Flat* (2006).

We use other tools as well for sharing including wikis and blogs. The key as we see it is finding a way to enlarge our audience beyond the classroom walls. When we do this, we find that our students and the P-12 students with whom they work are more motivated and engaged in learning.

CONCLUSIONS...FOR THE MOMENT

It's tempting to name this section "Intermission" rather than "Conclusions" due to the ever changing nature of technology. We recognize that we share no more than a sliver of the amount of digital tools and information that can be used by teachers as they learn to teach in their own classrooms. We know that the tools—the technologies, software, and websites— will change and be replaced by

Figure 7. YouTube video created by pre-service teachers and middle school students

Figure 8. The world is flat Dipity timeline created by students

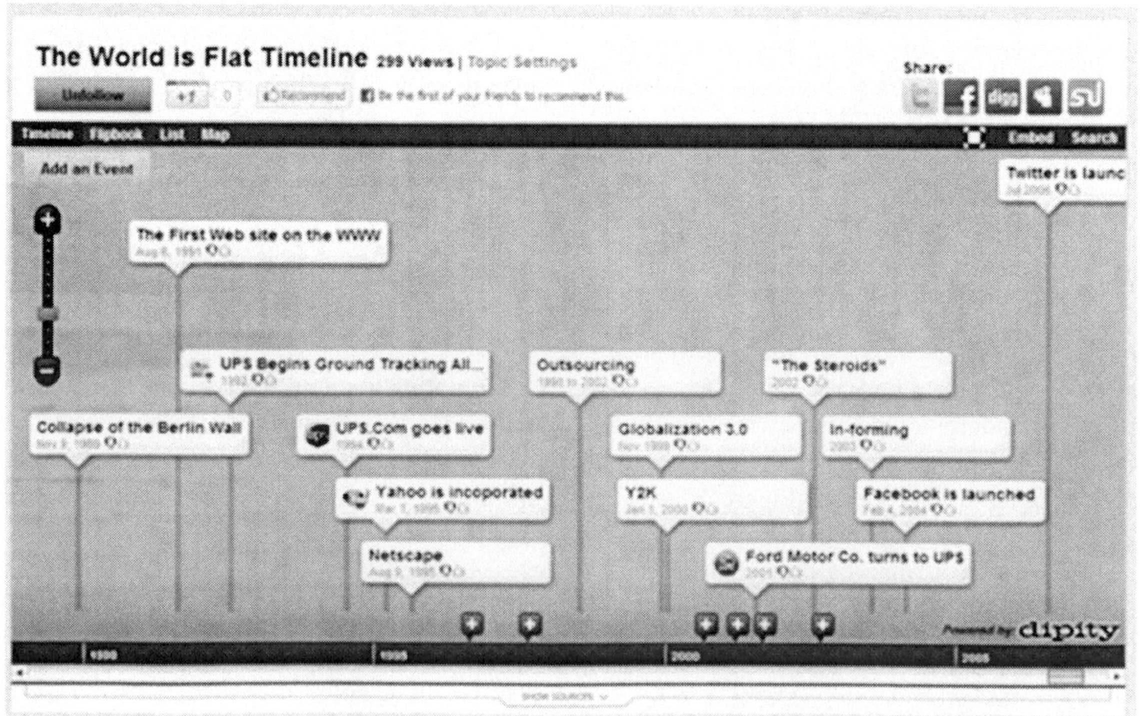

others. The tools we share here may soon be obsolete. The ways we access that content are also evolving. Two or three years ago computing tablets were thought to be a failed line in the evolution of the computer. The debut of the Apple iPad and the other tablet computing devices that have followed have changed that. While we do not include in this chapter our uses of tablets and "apps" many of the tools we share in this chapter are finding their way to tablet computing.

What we hope we've done is to suggest a durable set of categories that helps teachers and students identify and use technologies to do the work of teaching and learning in their classrooms. We purposefully chose actions to denote our pedagogic categories because we see teaching and learning as an active process. This development has helped us in identifying and using a multitude of tools and texts and to better communicate them to our students. Our goal in this work has been, in part, to create a pedagogic discourse that is helpful for enacting a pedagogy envisioned by The New London Group. We also see a pedagogy of Multiliteracies using Web 2.0 tools as one better suited to teaching and learning in a globally connected "flat" world than our traditional modes of teaching which typically bound learning within the classroom and school walls. By reaching outside the school into the community and world, we believe that students have more motivation to learn and avenues for applying what they learn in real-world contexts. Moreover, we see this Multiliteracies metastructure (Researching, Communicating, Collaborating, Exploring, Creating, and Sharing) as a tool to enhance teacher professional development by engaging teachers with their students in learning. As teacher educators, we conduct professional development for inservice teachers and have used this framework in workshops for teachers. We find the categories make clearer ways of integrating technology into teaching and learning for our teacher participants. The adoption of the Common Core curriculum, which pushes the integration of technology in teaching and learning, provides further impetus to make technology integration manageable for teachers and students.

This process of teaching and learning is transactional. We have learned with and from our students but also sought to model uses of Web 2.0 technologies in our teaching and through our assignments so that our students see them in use in classrooms. They, in turn, have introduced us to many digital teaching and learning tools. We hope that by being engaged in their use in our classrooms that all of us—P-12 and university educators—will be more likely to use these tools in effective ways.

REFERENCES

Brand, G. (1997). What research says: Training teachers for using technology. *Journal of Staff Development*, *19*(1), 10–13.

Brooks, D. M., & Kopp, T. W. (1989). Technology in teacher education. *Journal of Teacher Education*, *40*, 2–8. doi:10.1177/002248718904000402

Carroll, A. E., Rivera, F. P., Ebel, B., Zimmerman, F. J., & Christakis, D. A. (2005). Household computer and Internet access: The digital divide in a pediatric clinical population. Proceedings AMIA Annual Symposium, (pp. 111-115).

Cisco Systems. (2006). *Technology in schools: What the research says*. Retrieved from http://www.cisco.com/web/strategy/docs/education/TechnologyinSchoolsReport.pdf

Collins, A., & Halverson, R. (2009). *Rethinking education in the age of technology: The digital revolution and schooling in America*. New York, NY: Teachers College Press.

Cope, B., & Kalantzis, M. (2000). *Multiliteracies: Literacy learning and the design of social futures*. New York, NY: Routledge.

Cuban, L. (1986). *Teachers and machines*. New York, NY: Teachers College Press.

Cuban, L. (2001). *Oversold and underused: Computers in the classroom*. Cambridge, MA: Harvard University Press.

Dodge, B. (1995). WebQuests: A technique for internet-based learning. *Distance Education, 1*(2), 10–13.

Evans, R. (1996). *The human side of school change: Reform, resistance, and the real-life problems of innovation*. San Francisco, CA: Jossey-Bass Publishers.

Friedman, T. L. (2006). *The world is flat [updated and expanded]: A brief history of the twenty-first century*. New York, NY: Farrar, Straus and Giroux.

Gee, J. P. (2003). *What video games have to teach us about literacy and learning*. New York, NY: Palgrave MacMillan Press.

Hammond, M., Reynolds, L., & Ingram, J. (2011). How and why do student teachers use ICT? *Journal of Computer Assisted Learning, 27*(3), 191–203. doi:10.1111/j.1365-2729.2010.00389.x

Heath, S. B. (1991). The sense of being literate: Historical and cross-cultural features. In Barr, R., Kamil, M. L., Mosenthal, P., & Pearson, P. D. (Eds.), *Handbook of reading research* (*Vol. II*, pp. 3–25). New York, NY: Longman Publishing Group.

Jones, R., Fox, C., & Levin, D. (2011). *National educational technology trends: 2011. Transforming education to ensure all students are successful in the 21st century.* State Educational Technology Directors Association. (ERIC: ED522777).

Koehler, M. J., & Mishra, P. (2005). Teachers learning technology by design. *Journal of Computing in Teacher Education, 21*(3), 94–102.

Kumar, S., & Vigil, K. (2011). The Net generation as preservice teachers: Transferring familiarity with new technologies to educational environments. *Journal of Digital Learning in Teacher Education, 27*(4), 144–153.

Lankshear, C., & Knobel, M. (2003). *New literacies: Changing knowledge and classroom teaching*. Philadelphia, PA: Open University Press.

Leander, K., & Boldt, G. (2008). *New literacies in old literacy skins*. Paper presented at the Annual Meeting of the American Educational Research Association, New York.

Leontiev, A. N. (1978). *Activity, consciousness, and personality*. Hillsdale, NJ: Prentice-Hall.

Leu, D. J. Jr. (2002). The new literacies: Research on reading instruction with the Internet and other digital technologies. In Samuels, J., & Farstrup, A. E. (Eds.), *What research has to say about reading instruction* (pp. 310–336). Newark, DE: International Reading Association.

Luke, A., & Elkins, J. (1998). Reinventing literacy in "New Times". *Journal of Adolescent & Adult Literacy, 42*(1), 4–7.

Macrorie, K. (1988). *The I-search paper: Revised edition of searching writing*. Portsmouth, NH: Boynton/Cook Publishers.

Means, B. (Ed.). (1994). *Technology and education reform*. Sand Francisco, CA: Jossey-Bass Publishers.

Moursund, D., & Bielefeldt, T. (1999). *Will new teachers be prepared to teach in a digital age? A national survey on information technology in teacher education*. Santa Monica, CA: Milken Exchange on Education Technology.

National Council for Accreditation of Teacher Education. (2008). *Unit standards*. Retrieved from http://www.ncate.org/standards/ncateunit-standards/unitstandardsineffect2008/tabid/476/default.aspx

New London Group. (1996). A pedagogy of multiliteracies: Designing social futures. *Harvard Educational Review*, *66*, 60–92.

North Carolina State Board of Education, Department of Public Instruction. (1997). *A strategic plan for reading literacy*. Retrieved from http://www.ncpublicschools.org/docs/curriculum/languagearts/elementary/strategicplanforreadingliteracy.pdf

O'Brien, D. G., & Bauer, E. (2005). New literacies and the institution of old learning. *Reading Research Quarterly. Essay Book Review*, *40*, 120–131.

Resnick, D. P., & Resnick, L. B. (1977). The nature of literacy: An historical exploration. *Harvard Educational Review*, *47*(3), 370385.

Riis, J. (1890). *How the other half lives*. New York, NY: Charles Scribner's Sons. doi:10.1037/12986-000

Romano, T. (1990). The multigenre research paper: Melding fact, interpretation, and imagination. In Daiker, D., & Morenberg, M. (Eds.), *The writing teacher as researcher: Essays in the theory and practice of class-based research* (pp. 123–141). New Hampshire: Boyton/Cook Publishers.

Romano, T. (1995). *Writing with passion: Life stories, multiple genres*. Portsmouth, NH: Boyton/Cook.

So, H.-J., & Kim, B. (2009). Learning about problem based learning: Student teachers integrating technology, pedagogy and content knowledge. *Australasian Journal of Educational Technology*, *25*(1), 101–116.

Stobaugh, R. R., & Tassell, J. L. (2011). Analyzing the degree of technology use occurring in pre-service teacher education. *Educational Assessment, Evaluation and Accountability*, *23*(2), 143–157. doi:10.1007/s11092-011-9118-2

Taylor, D. B. (2012). Multiliteracies: Moving from theory to practice in teacher education. In Polly, A. B., Mims, C., & Persichitte, K. (Eds.), *Creating technology-rich teacher education programs: Key issues* (pp. 266–287). Hershey, PA: IGI Global. doi:10.4018/978-1-4666-0014-0.ch018

Teclehaimanot, B., Mentzer, G., & Hickman, T. (2011). A mixed methods comparison of teacher education faculty perceptions of the integration of technology into their courses and student feedback on technology proficiency. *Journal of Technology and Teacher Education*, *19*(1), 5–21.

Trilling, B., & Fadel, C. (2009). *21st century skills: Learning for life in our times*. San Francisco, CA: Jossey-Bass.

Tyack, D., & Cuban, L. (1995). *Tinkering toward utopia: A century of public school reform*. Cambridge, MA: Harvard University Press.

Vygotsky, L. S. (1978). *Mind in society: The development of higher psychological processes*. Cambridge, MA: Harvard University Press.

ADDITIONAL READING

Alvermann, D. E., Hinchman, K. A., Moore, D. W., Phelps, S. F., & Waff, D. R. (2006). *Reconceptualization the literacies in adolescents' lives* (2nd ed.). Mahwah, NJ: Erlbaum.

Anstey, M., & Bull, G. (2006). *Teaching and learning multiliteracies: Changing times, changing literacies*. Newark, DE: International Reading Association.

Bonk, C. J. (2009). *The world is open: How web technology is revolutionizing education*. San Francisco, CA: Jossey-Bass.

Chandler-Olcott, K., & Mahar, D. (2003). Tech-savviness meets multiliteracies: An exploration of adolescent girls' technology-mediated literacy practices. *Reading Research Quarterly*, *38*, 356–385. doi:10.1598/RRQ.38.3.3

Collins, A., & Halverson, R. (2009). *Rethinking education in the age of technology: The digital revolution and schooling in America*. New York, NY: Teachers College Press.

Cope, B., & Kalantzis, M. (2000). *Multiliteracies: Literacy learning and the design of social futures*. New York, NY: Routledge.

Hendron, J. G. (2008). *RSS for educators: Blogs, newsfeeds, podcasts, and wikis for the classroom*. Washington, DC: International Society for Technology in Education.

Karchmer, R. A., Mallette, M. H., Kara-Soteriou, J., & Leu, D. J. (Eds.). (2005). *Innovative approaches to literacy education: Using the internet to support new literacies*. Newark, DE: International Reading Association.

Lankshear, C., & Knobel, M. (2006). *New literacies: Everyday practices and classroom learning* (2nd ed.). Berkshire, UK: McGraw-Hill/Open University Press.

Lankshear, C., & Knobel, M. (2007). Sampling the "new" in new literacies. In Knobel, M., & Lankshear, C. (Eds.), *A new literacies sampler* (pp. 1–24). New York, NY: Peter Lang.

Solomon, G., & Schrum, L. (2007). *Web 2.0: New tools, new schools*. Washington, DC: International Society for Technology in Education.

Taylor, D. B. (2012). Multiliteracies: Moving from theory to practice in teacher education. In Polly, A. B., Mims, C., & Persichitte, K. (Eds.), *Creating technology-rich teacher education programs: Key issues* (pp. 266–287). Hershey, PA: IGI Global. doi:10.4018/978-1-4666-0014-0.ch018

Taylor, D. B., Hartshorne, R., Eneman, S., Wilkins, P., & Polly, A. B. (2012). Lessons learned from the implementation of a technology-focused professional learning community. In Polly, A. B., Mims, C., & Persichitte, K. (Eds.), *Creating technology-rich teacher education programs: Key issues* (pp. 535–550). Hershey, PA: IGI Global. doi:10.4018/978-1-4666-0014-0.ch034

KEY TERMS AND DEFINITIONS

Blog: A contraction of the term "web log"; a blog is a website maintained by an individual and may include regular posts, pictures and other media, RSS feeds, and commentary from guests or visitors to the blog. Popular blogging tools include WordPress, EduBlogs, Blogger, and LiveJournal.

Information Communication Technologies (ICTs): Computer-based technologies that facilitate communication. These include radio, television, cellular phones, computer and network hardware and software, and satellite systems. Computer applications including Skype, email, chat, and texting would forms computer software applications that are ICTs.

Multiliteracies: A term coined by the New London Group in 1996 to describe the rapidly evolving forms of literacy beyond traditional print-based forms used to communicate within diverse cultural and social settings.

Web 1.0: A retrofit term that arose after the term Web 2.0 came into popular usage. The primary concept the term Web 1.0 attempts to capture is the static nature of some web content and applications. Web pages that can only be read most often fall into this category. Some applications such as word processing programs and spreadsheets that are computer specific (i.e., the software exists on each computer on which it is used) are sometimes considered to be Web 1.0 applications.

Web 2.0: A term coined by Dale Dougherty of O'Reilly Media to capture changes in the Internet and web-based technologies that allow for my dynamic use of the web. Instead of passively consuming text on the web, users use tools such as blogs, wikis, comments, RSS feeds, and social networking sites (to name just a few) to read, write, and edit to the web.

Wiki: A web-based application that allows multiple users to create and edit content, which can include text, hypertext, audio, video, and more. Popular wiki tools and applications include Wikipedia, PBWorks, and WetPaint.

Section 5
Comparing Delivery Options in Teacher Education

Chapter 14
(Re)Assessing Student Thinking in Online Threaded Discussions

Felicia Saffold
University of Wisconsin- Milwaukee, USA

ABSTRACT

A teacher educator examines the level of critical thinking of her preservice teachers participating in an urban education course through online discussions. The objective was to see if online discussions, which were the heart of the learning process, could be an effective strategy to promote critical thinking skills. Using the revised version of Bloom's Taxonomy (Anderson & Krathwohl, 2001) as a guide, participants' posts and responses were assessed to determine the quality of thinking that occurred in the online discussion forum. Results show that utilizing online discussion forums can be an effective pedagogy for classes where complex, often controversial issues such as social justice, equity, and white privilege are discussed.

INTRODUCTION

This study took place at a large Midwestern university in the United States where I teach. Our mission is to prepare future teachers to teach children of racially and linguistically different backgrounds than their own and to promote meaningful, engaged learning for all students, regardless of their race, gender, ethnic heritage, or cultural background. Like most city-based teacher preparation programs, many students attending our university are not familiar with the unique assets children in the city bring to the classroom nor have they experienced the structural inequities around race, class, culture, abilities, and language that permeate urban schools. Our students match the typical profile of white, female, middle class, and rural or suburban teacher candidates (Banks

DOI: 10.4018/978-1-4666-1906-7.ch014

2000; Sleeter 1994). Yet, we know their future classrooms are not likely to have such homogeneous or affluent student populations.

Like many teacher educators, I have struggled with what my students take away from my urban education course. The course was originally designed to challenge prospective teachers' beliefs and transform their ideas about teaching, particularly in urban schools. My intention was to engage my students' existing beliefs about teaching and learning and to encourage their consideration of alternative beliefs presented in educational literature. I require my students to reflect on their field experience and discuss their learning. I plan lessons to arouse curiosity and to push my students to a higher level of knowledge. Students are encouraged to learn by doing and to interact with one another. Typically, I have students participate through discussions, group projects, case studies, and presentations. I strongly believe that through these types of activities, students will have richer experiences in the course and will retain more information long after the course is over. Unfortunately, often times the field experience reinforced or reproduced more stereotypes rather than changing beliefs (Tiezzi & Cross, 1997; Haberman & Post, 1992; Wiggins & Follo, 1999). With that in mind, I have been making changes in my course hoping to determine the best ways to scaffold high levels of reflection about field experiences and urban issues. In this chapter, I discuss my investigation into using an online discussion forum as a tool for promoting quality student participation and increasing the level of critical thinking amongst preservice teachers who are asked to discuss complex urban issues.

BACKGROUND

Students begin their professional coursework with an introductory teaching class. A capstone course for the School of Education, Introduction to Teaching, offers a thorough introduction and examination of urban education issues to approximately 200 teacher education students each semester who have an interest in pursuing a teaching career. Students are primarily white females from backgrounds very different from the students they will teach in urban areas. Students participate in weekly discussions designed to help them explore and understand the dynamics of teaching and learning in urban schools and complete 50 hours of field experience in a local school setting. Typically, students are uncomfortable addressing issues related to urban schooling, fearing awkwardness or conflict. For this reason, I use a variety of techniques to try and help overcome these barriers. For example, I begin with less controversial topics like parental involvement equity in urban schools before tackling more sensitive issues like white privilege. I share my expectations for class participation, based on an agreement to honor each other's differences and experiences; and I use role-playing or debates to help students see how others might perceive an issue differently. These activities require students to do some thinking beyond what they can recall from the textbook.

According to a recent NCEI report, 84% of teachers are white and students of color are growing. Therefore, it is imperative that we prepare future teachers who can effectively teach students whose cultural backgrounds are different from their own (Banks, 2000; Howard, 2006). Unless preservice teachers become aware of their own preconceptions and how those preconceptions affect their notion of teaching and learning they are unlikely to deconstruct their preconceptions and construct a new vision that includes culturally responsive practices that are fair and equitable for all students.

During one urban education class, a teacher educator cannot possibly place special emphasis on every dimension of the complex issues facing urban schools. However, she can teach students to reason well through any issue, and, through this emphasis, help students develop the habits of mind they need to deal with issues facing urban schools. In the Introduction to Teaching class, I prepare students to deal with complex issues in urban schools by learning how to think critically through such issues. Therefore, in addition to being competent in what it means to be a teacher, I prepare my students to confront race and racism in themselves, their classrooms, schools, and society. Specifically, I give students the opportunity to examine their own cultural frame of reference and try to get them to understand how social inequality is reflected and reproduced in schools.

Although the traditional face-to-face course was redesigned into an online format, I did not receive the formal training on how to facilitate online discussions as Kyong-Jee and Bonk (2006) suggested will be essential in the coming years, but, like many such online courses, it was designed to build community and interactivity, with the discussion forum being the "heart" of the course, the "safe place" where students would meet to discuss readings and field experiences and to exchange ideas with each other. Whereas the field experience provides students with real life experiences in urban schools, the online discussion forum allows for effective processing of the field experience (Bennett & Green, 2001).

I used the Desire to Learn (D2L) platform to create what I considered "safe places" for critical conversations. These forums were designed so that students had the opportunity and support to grapple with the complexities of their own learning to teach histories. I did this by creating discussion forums that asked them to interrogate the relationships between and among the way they were taught, how they learn, and how they envisioned their classroom practice in an urban school. Prospective teachers need considerable guidance and support to examine their own cultural frame of reference and to think critically about their experiences in schools and, especially, how social inequality is reflected and reproduced in schools (Cochran-Smith, 1999, 2004; Weiner, 1999). Each week students were required to make original postings to the discussion forum and make comments and replies to others' post, so that overall the students were expected to post several times a week. My intention for the redesign of the online course was to provide a similar experience as the face-to-face course; however, I was hoping to see more depth and reflection in the students' postings by providing a "safe place" for them to address the more uncomfortable issues of power and equity--namely, racism (Banks & McGee-Banks, 2006; Nieto & Bode, 2008).

My ongoing research in this area is aligned with recent research in teacher education that involves teacher educators using web-based discussions to explore teacher candidates' thinking. Although my goal in Introduction to Teaching is to get preservice teachers to think critically and analytically about urban issues, I often struggle with the most effective methods to encourage critical thinking. Researchers suggest that threaded discussions are useful in promoting critical thinking because they allow for more reflective and less spontaneous discourse (Garrison & Anderson, 2003; Meyer, 2003). Therefore, this was one of the reasons that I decided to transform my face-to-face course to the online format. I thought the online course would allow more time for my students to engage in their field experience and it would offer them the flexibility to post responses after they had time to reflect critically on the course material. This chapter explores the quality of participation and contributions of online threaded discussions and assesses the levels of critical thinking that occur when preservice teachers discuss complex urban issues.

The Emergence of Online Discussions

One of the most common means of asynchronous communication is the "threaded" discussion. It is viewed as an essential part of online courses because it helps to build a community of learners. According to Swan (2005), threaded discussions are asynchronous online conversations that take the form of a series of linked messages organized around a common topic. It is a form of computer-mediated communication where conversations are organized around a theme and every one in a class is able to read, reflect and respond to the ideas posted. Rovai (2000) suggested that threaded discussions are fundamental to the learning process because learners share and develop alternative viewpoints and help each other negotiate higher levels of understanding. Some researchers report that this type of web-based communication has a higher level of teacher-student and student-student interaction than in most traditional classroom settings (Kearsley & Lynch, 1991).

Some researchers report that online discussions offer greater opportunity for reflection and comments made by other students (Givens, Generett & Hicks, 2004; Meyers, 2003; Salmon 2005). Boswell (2003) asserted that it is easier to encourage critical thinking in the discussion forum because it forces students to participate, whereas more outgoing students tend to dominate face-to-face discussions. It is the additional time for reflection that may encourage those who may be hesitant to participate in a face-to face environment to participate and produce deeper and more reflective contributions (Bender, 2003). Drops (2003) supported the strength of the online discussion and stated that in the face-to-face classroom students do not give all their time and attention to things because they know they can just come and listen. Previous research has shown that while participating in online discussions people were able to express themselves without having the sneers, rejection, and embarrassment that sometimes occurs when in a face-to-face classroom environment (Hammond, 2000). According to Greenlaw and DeLoch (2003), "electronic discussions can provide a natural framework for teaching critical thinking to a group, capturing the best features of traditional writing assignments and in-class discussions" (p. 41). Yet, these claims have not been confirmed by other studies which have reported unbalanced participation patterns similar to that found in face-to-face environments (Poole, 2000).

While many researchers support the use of online discussions, there are differing opinions about whether or not to assess student participation in threaded discussions. Yeh (2005) noted that student participation increases as instructors place an importance on posting by assigning grades to forum use. Edelstein and Edwards (2002) reported that by requiring participants to post their work and respond to others' work the learning process is enhanced and more engaging. Ho (2002) and McKenzie and Murphy (2000) asserted that assessment criteria can be used as a guide for students and the expected quality of thinking. On the other hand, some researchers believe that assessing online discussions is counterproductive to facilitating good learning outcomes through discussions. These critiques argue that assessing participation restricts free expression and new thinking, hinders students from gaining deeper understanding of course outcomes, and results in the regurgitation of content, discussing merely for the sake of meeting course requirements rather than engaging in meaningful dialogue (Davis, 1993; Lacross, Chylack, 1998). Brescia, Swartz, Pearman, Williams, & Balkin (2004) concurred that there is a clear risk of students posting for the sake of meeting requirements rather than learning from the material when instructors attach online participation with assessment.

Research on the analysis and assessment of the content of online discussions has been prevalent for many years. Several frameworks have been used to assess the learning present in the discussion forums. For example, Henri (1992), a pioneer in the development of criteria for content analysis, developed a useful tool for online discussion analysis to code postings using five dimensions. Zhu (1998) created a model for the patterns of knowledge construction in online discussion forums, which shed light on how new knowledge results from instructional scaffolding. Garrison, Anderson & Archer (2001) developed a four stage critical thinking model to assess the nature and quality of critical discourse and reflection in online discussions. Hara, Bonk & Angeli (2002) built on Henri's (1992) work with a purpose of constructing better guidelines on how computer conferencing could be analyzed. Meyer (2005) used Bloom's Taxonomy as a tool for her analysis of online discussions and reviewed them as a group endeavor as opposed to individual progress. Her research showed that patterns occur in discussion threads based on the initial discussion prompt.

Bloom's Taxonomy has stood the test of time, is very familiar to educators, and has been used as a method of classification for thinking behaviors (Bloom & Krathwohl, 1956). I am very familiar with this framework. However, in this study, I used the revised Bloom's Taxonomy developed by Anderson and Krathwohl (2001), based on Bloom's Taxonomy, to assess the level of critical thinking in my students' online postings. The Revised Bloom's Taxonomy is two dimensional instead of one dimensional and identifies both the Cognitive Dimension (remember, understand, apply, analyze, evaluate and create) and the Knowledge Dimension (factual, conceptual, procedural, metacognitive). It has been reported as a "more authentic tool for curriculum planning, instructional delivery and assessment" (oz-TeacherNet, 2001). Therefore, in the current context, the Anderson and Krathwol (2001) framework was chosen because it focuses the analysis on the students' level of thinking. Furthermore, the thinking dimensions were utilized in this study to create a rubric for evaluating both the online discussions and the discussion prompts used to initiate the discussions.

METHODOLOGY

This study took place within an introductory teaching, undergraduate level course, at a Midwestern university in the United States. The class was held for 15 weeks and was conducted fully online. Twenty-four students were in the class, 7 males and 17 females. Each week students corresponded to a distinct topic pertaining to specific urban education issues. Four major discussion topics were analyzed in this study: inequity in urban schools, parental involvement, white privilege and why urban education.

Desire 2 Learn (D2L), a web-based online course management system, allowed me to set up and manage threaded discussions as structured online conversations where my students were required to post and comment on others' posts in an asynchronous environment. This type of format for the online discussions allowed students time to reflect on the subject matter and respond at a time most convenient for them.

After the selected discussion topics were chosen, I printed the set of prompts and responses for that week to simplify analysis. A rubric, based on Bloom's Revised Taxonomy (Anderson and Krathwhol, 2001), was helpful in determining classifications of postings. The rubric defined three types of contributions to online discussions. Low (knowledge and comprehension) contributions are those which recall information or simply explain a concept. Medium (application and analysis) contributions are those which consider the problem situation from a new perspective. Finally, High (evaluate or create) contributions are those where judging the value of ideas takes place or ideas are put together to create a plan.

I had a research assistant help with the coding of the discussion responses. We both had a description of Bloom's Revised Taxonomy that included all the categories and a short summary that defined each category. It was determined that we would use only one code per online post; however, we broke apart an entire response to determine the highest possible category that would be used. Each of us independently coded the responses. We came together to negotiate 100% agreement. After this analysis, we used a second rubric adapted from Bloom's Revised Taxonomy to determine the complexity level in the actual discussion prompts.

RESULTS

The study sought to explore the quality of participation and contributions of online threaded discussions and assess the levels of thinking that occur when preservice teachers discuss complex urban issues. The majority of the postings were categorized at the lower level of thinking across all six question types in Bloom's Revised Taxonomy. In this category, it was common that students shared a story from their field experience or personal background that demonstrated their understanding of the topic. For example, when the prompt asked students to give an example from their field experience that demonstrated their understanding of tracking, we considered the response below to be Low level. Notice the student did not formulate any opinions about the situation or critique it any meaningful way. She simply recalled the incident.

In a classroom observation I did last semester, the 1st graders were grouped together and told to do their readings. The highest level group (2 black children and the only 3 white children in the class) was given worksheets, new books, and lots of attention. The lowest group, which was obviously the children with behavior problems and who needed the most attention, were left alone in the corner with alphabet books in which they would look at the pictures and pick up another one. I couldn't believe that a teacher would actually teach like that, but it is done. Those kids will probably always be treated like that, and like you said it will carry through with them.

Similarly, a post that basically reported facts from the course readings was considered Low level. Notice the way in which the student recalled the information and listed her responses.

Tracking is evident in my field site too... BIG time. All the exceptional education kids are located on the ground level, are put in a corner for instruction, and only white kids are in the advanced classes. I think that is what tracking was telling us about.

In another discussion forum, when students were asked to explain the role of teachers and parents as coworkers in helping students have academic success, lower level thinking was also prevalent in the responses. Notice the way this student used the course readings to explain her thoughts on the parent /teacher connection but did not elaborate beyond the text reference.

I think that many teachers assume that because a parent might not show up to volunteer at a school activity or come to conferences that they don't care about their children's education. According to the Finders and Lewis article, this is not the case. Most of the time there are other obstacles and barriers that keep parents from being involved.

Then again, in a discussion about white privilege, we considered this student's response about the chance to take a pill to no longer be white, as Medium level. She admitted, "I could never imagine what it would be like to not have white privilege and what that life would be like, so I wouldn't take the pill." At this level of thinking one is able to apply what they know to new and related situations. Notice how she compared and contrasted as she related her experience outside of the United States with her course knowledge when she discussed the issue in more detail:

Unfortunately white privilege is in more societies than just America's where there is a lot of diversity. While in Thailand, a girl I taught with and I, both white and blonde were offered everything. They loved our hair, didn't know why we wanted to be tan, and men constantly brought us gifts because we were white. They have creams to bleach your skin and the little girls will wear white powder on their face to look white. I think that in American society it demonstrates white power because, as Tim Wise points out, it's as if white privilege is some unwritten rule in our society that nobody really speaks about. We hide it because we are ashamed and know it's wrong and don't want to address it on a whole society level. Whereas places like Thailand, their society accepts that the whiter someone looks the better. I don't know which is sadder a society that knows it's wrong and decides not to do anything or a society who doesn't believe it is.

Following the same prompt, the student below analyzed the notion of white privilege and really questioned it by using a scenario from the text as a way to offer alternative perspectives to a given situation. However, he fails to respond to the actual prompt. Note we would have rated this as a High level post if he would have used all of the information in his post as a way to formulate or discuss his final decision about taking the pill.

The sad, complex, web of truth that I think we might need to face is that now these attitudes, decided rather arbitrarily within a larger fight over who gets what, are etched into us now; and they are not easy to ascribe to simplistic dichotomies in many cases. There's a great example in Because Teaching Matters of the way these patterns reinforce themselves almost automatically, and to my mind how they're not clear cut: when Erika Thompson recounts her story of witnessing a White teacher ushering out poor, drug-addicted Black parents and welcoming White ones into the classroom (Pugach, PP. 214-215), I found myself asking a question. Were the Black parents ushered out because they were Black, or because the teacher was intimidated by the drug issue? Were the White parents welcomed because they were White, or because they were affluent? Or was it a little of both? How can we say? Maybe we can't, no matter how long we try to differentiate notions of privilege. What "White Privilege" ultimately doesn't seem to address is how to solve the problem.

There really was not a consistent pattern of thinking evidenced in the responses. The same students did not always post Low level or High level responses within a discussion thread or across the discussion topics. However, it was rare to see posts at the High level range. The two highest levels of cognitive thought, according to Bloom's Revised Taxonomy, are evaluate or create. This type of thinking is more complex and involves evaluating various perspectives and making choices based on reasoned argument. For instance, we rated as High level the posting of a student who challenged one of her classmate's posts and generated a series of reflective questions to create a more in-depth discussion. She acknowledged, "Yes, this is a controversial topic and not one that most like to talk about, but it needs to be talked about." She was not afraid to pass judgment on another student's comment:

In my opinion, you are misunderstanding how we need to treat everyone equal. Your post seems to suggest that we should ignore differences and go back to embracing the colorblind ideology. I'm kind of surprised to see that especially after the digital story our professor provided. I'm curious as to what you took away from that. We should all be given equal opportunities, but we should still be noticing that everyone is different and comes from different cultures/backgrounds.

Then she questioned:

Do you really believe that there is no white privilege or that there has never been? Clearly, whites have been given more opportunities than blacks or minorities. This curriculum needs to be around or how will we change the way our society functions? How effective would you be working in an urban school if you didn't have to wrestle with these issues before hand? Have you gone to fieldwork yet? Do you think that these students in these urban schools are given the same educational opportunities as a white suburban community school?

The level of this questioning demonstrated critical thinking. The next example also confirmed higher level thinking. This student clearly demonstrated his ability to reflect on course content, generate ideas and create a plan for working effectively with culturally diverse students. When the discussion prompt asked students to provide a rationale for taking an introductory urban education course, this student had an extensive response. Notice this example of how a multi-level post was ultimately coded as Medium level as shown in Table 1 after the full quote:

After spending the last few days thinking about this class, here is what I have come away with:

1. *When making lesson plans, I know and realize the importance of taking into account the diversity of the students' backgrounds in order to ensure that I make the content relevant to as many students as possible. I will not rely on textbooks only as my sole resource. I will make sure that my students see themselves in the curriculum and understand that they are not always portrayed in a negative light. As a social studies teacher, I will also need to examine my lessons to make sure I'm checking whose perspective I am telling stories from and check that there is balance.*
2. *When students identify differences between themselves, I will use this as a learning opportunity so the students can begin to understand the different perspectives of their peers. Through the discussion forum, I have learned to appreciate the various perspectives in class.*
3. *I also need to realize that even if my classroom is not very diverse, several of my students will likely move outside of their current community. I need to prepare them for interactions with a diverse set of people. I used to ask why an all white class would need to be exposed to multicultural literature or why would a black class need to have black literature. I now see how both populations can benefit from having materials from their own culture and various cultures. Culturally relevant teaching makes sense to me now.*
4. *I have started a "great ideas" journal that I will continue to use while I'm in field experiences. Urban teachers face soooo many challenges. Having a reference to how I saw urban teachers work in difficult situations will be helpful.*

   I learned so much in the field. That was the best part of the class, but, dissecting the issues with peers, who basically came from the same place as me, was very beneficial. I usually "suck up" everything in classes and don't actually participate much, but this time, I think I actually learned a lot more. I'm not sure if urban ed is for me, but I do feel more knowledgeable about the issues and where I stand on them.

After all of the individual posts were analyzed, Table 2 was created to display the overall ratings of the four major discussion topics.

As Table 2 illustrates students in the course demonstrated thinking at all levels of Bloom's Revised Taxonomy. There were a total of 305 responses from the 24 students. More than half of the postings (211) were classified as Low

Table 1. Sample of coding discussion responses (adapted from Anderson & Krathwohl, 2001)

Bloom Level	Quote from Discussion Thread
Low: Remember or Understand Key verbs: recognizing, recalling, summarizing, explaining	"After spending the last few days thinking about this class, here is what I have come away with: 1. When making lesson plans, I know and realize the importance of taking into account the diversity of the students' backgrounds in order to ensure that I make the content relevant to as many students as possible.
Medium: Apply or Analyze Key verbs: relating, applying, identifying, pointing out	I will not rely on textbooks only as my sole resource. I will make sure that my students see themselves in the curriculum and understand that they are not always portrayed in a negative light. As a social studies teacher, I will also need to examine my lessons to make sure I'm checking whose perspective I am telling stories from and is there variety in this. 2. When students identify differences between themselves, I will use this as a learning opportunity so the students can begin to understand the different perspectives of their peers. Through the discussion forum, I have learned to appreciate the various perspectives in class. 3. I also need to realize that even if my classroom is not very diverse, several of my students will likely move outside of their current community. I need to prepare them for interactions with a diverse set of people. I used to ask why an all white class would need to be exposed to multicultural literature or why would a black class need to have black literature. I now see how both populations can benefit from having materials from their own culture and various cultures. Culturally relevant teaching makes sense to me now.
High: Evaluate or Create Key verbs: creating, preparing, assessing, critiquing	4. I have started a "great ideas" journal that I will continue to use while I'm in field experiences. Urban teachers face soooo many challenges. Having a reference to how I saw urban teachers work in difficult situations will be helpful. I learned so much in the field. That was the best part of the class, but dissecting the issues with peers who basically came from the same place as me was very beneficial. I usually "suck up" everything in classes and don't actually participate much, but this time, I think I actually learned a lot more. I'm not sure if urban ed is for me, but I do feel more knowledgeable about the issues and where I stand on them."

Table 2. Levels of thinking in major discussion threads (adapted from Anderson & Krathwohl, 2001)

Bloom Level	Parental Involvement	Why Urban Education	Inequity in Schools	White Privilege
Low: Remember or Understand	65	50	66	30
Medium: Apply or Analyze	11	22	8	16
High: Evaluate or Create	3	11	3	14
Total	**79**	**89**	**77**	**60**

level, revealing that most discussions were likely oriented toward contributing knowledge from the field experience or summarizing the assigned readings. Students used readings and field experiences to show their understanding of the topics, but they were not used to problematize or interpret issues, which would have caused them to extend the discussions or push them to higher levels of thought.

The majority of the High level postings (14), those classified as evaluate or create, occurred in the white privilege discussion forum which required students to justify their decision about taking a pill to make them no longer white. Even though the white privilege discussion revealed the most variety in terms of the level of thinking used in the forum, with 30 responses at the Low level, 16 at the Medium level, and 14 at the High level, it also was the forum that had the lowest number of students engaging in the conversation (60 posts out of the 305 total posting).

This data shows that discussion prompts can be used in the online discussion forum in order to scaffold students' critical thinking and give them the opportunity to think and interact with their peers at high levels; however, students can

Table 3. Analysis of discussion prompt (adapted from Anderson & Krathwohl, 2001)

	Cognitive Dimension					
Knowledge Dimension	1. Remember Retrieve relevant knowledge from long-term memory	2. Understand Construct meaning from instructional messages	3. Apply Carry out or use a procedure in a given situation	4. Analyze Break material into its constituent parts and determine how the parts relate to one another and to an overall structure	5. Evaluate Make judgments based on criteria and standards	6. Create Put elements together to form a coherent or functional whole
Factual knowledge						
Conceptual knowledge		Question 1 Question 2				
Procedural knowledge						
Metacognitive knowledge					Question 3	Question 4

still opt out of the discussion regardless of having guidelines for participation. In other words, uneven patterns of participation can be seen in the online format similar to that in the face-to-face format as Poole (2000) reported. .

The final analysis attempted to use Anderson and Krathwohl's (2001) revised taxonomy table to analyze the discussion prompts to see if the nature of the questions might have influenced the level of response from the students. Mapping the questions, as seen in Table 3 below, gave an indication to the relative complexity of the questions.

As Krathwohl (2002) reported, the structure of the revised taxonomy table provided a clear, concise, visual representation of the alignment between standards and educational goals, objectives, products, and activities. The mapping revealed that the questions used in the discussion prompts were not diverse in that there was not a balance of questions to assess knowledge and cognitive thinking processes. Table 3 illustrates that the discussion prompts could be classified as follows:

Q1. Give an example from your field experience that demonstrates your understanding of tracking. (understand, conceptual knowledge)

Q2. Explain the role of teachers and parents as coworkers in helping students achieve academic success. (understand, conceptual knowledge)

Q3. If you had an opportunity to take a pill that would make you no longer a white person, would you take it? Why or why not? (evaluate, metacognitive)

Q4. Provide a rationale for taking an introduction to urban education class. (create, metacognitive)

As they were written, two of the questions were designed to assess low level cognitive skills (Q1 and Q2) and two of the questions (Q3 and Q4) were designed to assess higher level cognitive skills. In addition, there was little variety in the questions in terms of assessing knowledge skills. In Q1, 66 of the 77 responses were rated as low. Similarly, in Q2, 65 of the 79 responses were rated low. Questions 3 and 4 both had a greater variety in responses and fewer responses were in the lower level, 50 out of 89, and 30 out of 60, respectively. The table results thus indicate that the complexity of the question prompt may be a critical factor in determining the level of thinking in which students will engage.

Solutions and Recommendations

As result of this study, in the next offering of the course, my discussion prompts will be revised to ensure that I am challenging my class to address difficult topics at a high level of abstraction by using questions that utilize multiple levels of thinking. In order to facilitate complex thinking, I have to make sure that there is a strong congruence between objectives and discussion prompts in the forum. The online discussion forum can be a great technology tool that can provide students with the flexibility to participate, reflect on issues, appreciate the various perspectives of their peers, and offer a safe space for students to participate in the dialogue and not just "suck" up the information as illustrated in the sample discussion posts. However, careful attention needs to be made as to how the cognitive interaction will occur or the "safe spaces" for critical conversations can produce an abundance of unproductive threaded posts or communication that does not show students engaging beyond the Low level of response.

From this study, I offer the following recommendations. First, the Revised Bloom's Taxonomy was used to evaluate students' online thinking and may be a useful framework for faculty to use who wish to better understand the dynamics of online discussions and who want to identify the level of thinking that occurs when students are asked to engage in critical conversations such as equity, social justice, and white privilege. Furthermore, the taxonomy may be beneficial for instructors looking to evaluate the complexity of their discussion prompts or the balance that is being assessed in terms of cognitive and knowledge skills. Although I used Bloom's Revised Taxonomy as a framework to analyze the level of thinking in my discussion posts, I believe that other frameworks for assessing the learning in the discussion forum can be utilized. Regardless of the framework used, the online discussion forum should be modified or adjusted based on the results of the analysis. As seen in this study, the type of questions posed in the forum may be a critical factor in the complexity of the discussion posts. Clearly, my discussion prompts must be modified to provide opportunities to create dissonance, debate or controversy if the goal of higher level thinking is to be accomplished.

Second, my intention for the online version of the course was to provide a similar experience as the face-to-face course; however, I did not take the material that I had success with and convert it, instead, I avoided using it. In the face-to-face course, I typically used strategies such as role-playing or debates to help students see how others might perceive an issue differently. These activities resulted in active learning, reflection, and sharing of views. Whether having an online discussion or one that occurs face-to-face, authentic learning within Bloom's framework must be planned for and good pedagogical strategies must be integrated into the online classroom. I never considered the vast possibilities of online discussion forums such as creating debates for students or having small groups of students dissect a scenario and present their solution to the class. These are the type of activities that an online discussion format supports because they take into account the time that students need to reflect on an issue and they are designed in such a way that students can take ownership of content, push each other, and allow students to demonstrate higher level thinking skills.

The nature of the online discussion forum provides a unique opportunity for all students in a classroom to share ideas, challenge them, and be full participants in the dialogue and not silent observers who, "suck up" all the information. However, merely making the technology available will not guarantee that critical thinking will occur, but, if the technology platform is utilized effectively, it can offer opportunities for significant teaching and learning benefits.

FUTURE RESEARCH DIRECTIONS

There are many implications for further research deriving from this study. Research shows that online classes have the potential to promote student engagement and interactivity. While this study focused on the level of thinking skills demonstrated by the students, future research could examine the interactions among students in the forum and; a) provide insight into how collaborative learning communities are formed; b) explore how students move through a discussion thread as a group having a conversation, rather than focusing on individual posts and responses; and, c) determine what types of questions are posed amongst students that give them the opportunity to move through the different levels of critical thinking. Furthermore, interviews could be conducted with students participating in online discussions to unearth students' perceptions about discussing critical issues in the forum.

Moreover, future research could focus on the role of the instructor in facilitating online discussions. Research suggests that facilitation skills of online instructors will be vital in the years to come. Studies that examine the role of the instructor, the type of questions he poses and the frequency of his interactions might also add to the body of research on online teaching and provide guidance for new and veteran instructors on how to facilitate discussions so that students move up to higher levels of thinking as defined by the Revised Bloom's Taxonomy.

CONCLUSION

Change is here. Either teacher educators are going to continue to resist emerging technologies and continue to prepare digital age learners with traditional teaching methods or we will have to attempt to understand these new technologies and find ways to utilize them to transform our classrooms. These tools can expand what teacher educators do when they teach and they can expand the classroom beyond its traditional boundaries. Like I discovered in this study, technology, if used as a means and not an end, has the potential to make learning environments more effective, specifically, in terms of computer-mediated communications, where students can have the time and space to reflect on course content and engage in more meaningful dialogue about the content.

In this technological age, faculty need to consider the possibilities of (re)designing teacher preparation courses that will utilize technology in ways that allow teacher education students, who are primarily white and female, to discuss issues of equity, social justice, and white privilege in a "safe" environment. Using an online discussion forum can be an effective tool to enhance critical conversations about complex issues that require reflection. Using a framework such as the revised version of Bloom's Taxonomy can provide teacher educators with the knowledge that could help them design discussion prompts which require higher level thinking and increase student engagement in the online forum.

REFERENCES

Anderson, L., & Krathwohl, D. (Eds.). (2001). *A taxonomy for learning, teaching, and assessing: A revision of Bloom's taxonomy of educational objectives*. New York, NY: Longman Publishers.

Banks, J. A. (2000). *Cultural diversity and education: Foundations, curriculum, and teaching*. Boston, MA: Allyn & Bacon.

Banks, J. A., & McGee Banks, C. A. (Eds.). (2006). *Multicultural education: Issues and perspectives*. New York, NY: John Wiley.

Bender, T. (2003). *Discussion-based online teaching to enhance student learning*. Sterling, VA: Stylis.

Bennett, G., & Green, F. (2001). Promoting service learning via online instruction. *College Student Journal, 35*(4), 491–497.

Boswell, C. (2003). Well-planned online discussions promote critical thinking. *Online Cl@ssroom*, 1–8.

Bloom, B. S., & Krathwohl, D. R. (1956). *Taxonomy of educational objectives: The classification of educational goals*. New York, NY: Longmans, Green.

Brescia, W., Swartz, J., Pearman, C., Williams, D., & Balkin, R. (2004). Peer teaching in web based threaded discussions. *Journal of Interactive Online Learning, 3*(2). Retrieved from http://www.ncolr.org/jiol/issues/pdf/3.2.1.pdf

Cochran-Smith, M. (1999). Learning to teach for social justice . In Griffin, G. (Ed.), *The education of teachers: Ninety-eighth yearbook of the National Society for the Study of Education* (pp. 114–144). Chicago, IL: University of Chicago Press.

Cochran-Smith, M. (2004). *Walking the road: Race, diversity and social justice in teacher education*. New York, NY: Teacher College Press.

Davis, B. (1993). *Tools for teaching*. San Francisco, CA: Jossey-Bass.

Drops, G. (2003). Assessing online chat sessions. *Online Cl@ssroom*, 1–8.

Edelstein, S., & Edwards, J. (2002). If you build it, they will come: Building learning communities through learning discussions. *Online Journal of Distance Learning Administration, 5*(1). Retrieved from http://www.westga.edu/~distance/ojdla/spring51/edelstein51.html

Garrison, D. R., Anderson, T., & Archer, W. (2001). Critical thinking, cognitive presence, and computer conferencing in distance education. *American Journal of Distance Education, 15*(1), 7–23. doi:10.1080/08923640109527071

Garrison, D. R., & Anderson, T. (2003). *E-learning in the 21st century: A framework for research and practice*. London, UK: Routledge/Falmer. doi:10.4324/9780203166093

Givens Generett, G., & Hicks, M. A. (2004). Beyond reflective competency: Teaching for audacious hope-in-action. *Journal of Transformative Education, 2*, 187–203. doi:10.1177/1541344604265169

Greenlaw, S. A., & DeLoch, S. B. (2003). Teaching critical thinking with electronic discussion. *The Journal of Economic Education, 34*(1), 36–52. doi:10.1080/00220480309595199

Haberman, M., & Post, L. (1992). Does direct experience change education students' perceptions of low-income minority students? *Mid-Western Educational Researcher, 5*(2), 29–31.

Hammond, M. (2000). Communication within online forums: The opportunities, the constraints and the value of communicative approach . *Computers & Education, 35*, 251–262. doi:10.1016/S0360-1315(00)00037-3

Hara, N., Bonk, C. J., & Angeli, C. (2002). Content analysis of online discussion in an applied educational psychology course. *Instructional Science, 28*, 115–152. doi:10.1023/A:1003764722829

Henri, F. (1992). Computer conferencing and content analysis . In Kaye, A. R. (Ed.), *Collaborative learning through computer conferencing: The Najaden Papers* (pp. 116–136). Berlin, Germany: Springer-Verlag. doi:10.1007/978-3-642-77684-7_8

Ho, S. (2002). *Evaluating students' participation in on-line discussions*. Retrieved from http://ausweb.scu.edu.au/aw02/papers/refereed/ho/paper.html

Howard, G. (2006). *We can't teach what we don't know: White teachers, multiracial schools* (2nd ed.). New York, NY: Teachers College Press.

Kearsley, G., & Lynch, W. (1991). Computer networks for teaching and research: Changing the nature of educational practice and theory. *DEOSNEWS – The Distance Education Online Symposium, 1*(18). BITNET.

Krathwohl, D. R. (2002). A revision of Bloom's taxonomy: An overview. *Theory into Practice*, *41*(4), 212–218. doi:10.1207/s15430421tip4104_2

Kyong-Jee, K., & Bonk, C. J. (2006). The future of online teaching and learning in higher education: The survey says…. *EDUCAUSE Quarterly*, *29*(4). Retrieved from http://www.educause.edu/apps/eq/eqm06/eqm0644.asp?bhcp=1

Lacoss, J., & Chylack, J. (1998). What makes a discussion section productive? *Teaching Concerns*, Fall.

McKenzie, W., & Murphy, D. (2000). "I hope this goes somewhere": Evaluation of an online discussion group. *Australian Journal of Educational Technology*, *16*(3), 239–257. Retrieved from http://cleo.murdoch.edu.au/ajet/ajet16/mckenzie.html

Meyer, K. A. (2003). Face-to-face versus threaded discussions: The role of time and higher-order thinking. *Journal of Asynchronous Learning Networks*, *7*(3), 55–65. Retrieved from http://www.sloan-c.org/publications/jaln/v7n3/pdf/v7n3_meyer

Meyer, K. A. (2005). The ebb and flow of online discussions: What Bloom can tell us about our students' conversations. *Journal of Asynchronous Learning Networks*, *9*(1), 53–63.

Nieto, S., & Bode, P. (2008). *Affirming diversity: The sociopolitical context of multicultural education*. New York, NY: Addison-Wesley.

oz-TeacherNet. (2001). *oz-TeacherNet: Teachers helping teachers: Revised Bloom's Taxonomy*. Retrieved from http://rite.ed.qut.edu.au/oz-teachernet/index.php?module=ContentExpress&func=display&ceid=29

Poole, D. M. (2000). Student participation in a discussion-oriented online course: A case study. *Journal of Research on Computing in Education*, *33*(2), 162–177.

Pugach, M. (2009). *Because teaching matters* (2nd ed.). Hoboken, NJ: Wiley and Sons, Inc.

Rovai, A. P. (2000). Online and traditional assessments: What is the difference? *The Internet and Higher Education*, *3*(3), 141–151. doi:10.1016/S1096-7516(01)00028-8

Swan, K. (2005). *Threaded discussion*. Retrieved from www.oln.org/conferences/ODCE2006/papers/Swan_Threaded_Discussion.pdf

Tiezzi, L., & Cross, B. (1997). Utilizing research on prospective teachers' beliefs to inform urban field experiences. *The Urban Review*, *29*(2), 113–125. doi:10.1023/A:1024634623688

Weiner, L. (1999). *Urban teaching: The essentials*. New York, NY: Teachers College Press.

Wiggins, R. A., & Follo, E. J. (1999). Development of knowledge, attitudes, and commitment to teach diverse student populations. *Journal of Teacher Education*, *50*, 94–105. doi:10.1177/002248719905000203

Yeh, H. (2005). *The use of instructor's feedback and grading in enhancing students' participation in asynchronous online discussion*. Advanced Learning Technologies (ICALT 2005).

Zhu, P. (1998). Learning and mentoring: Electronic discussion in a distance learning course . In Bonk, C. J., & King, K. S. (Eds.), *Electronic collaborators: Learner-centered technologies for literacy, apprenticeship, and discourse* (pp. 233–259). Mahwah, NJ: Erlbaum.

ADDITIONAL READING

Anderson, G., & Szabo, S. (2007). The "power" to change multicultural attitudes. *Academic Exchange Quarterly*, *11*(2). Retrieved from http://www.rapidintellect.com/AEQweb/

Bennett, G., & Green, F. (2001). Promoting service learning via online instruction. *College Student Journal*, *35*(4), 491–497.

Bonk, C. J., & Cunningham, D. J. (1998). Searching for learner-centered, constructivist, and sociocultural components of collaborative educational learning tools . In Bonk, C. J., & King, K. S. (Eds.), *Electronic collaborators: Learner-centered technologies for literacy, apprenticeship, and discourse* (pp. 25–50). Mahwah, NJ: Erlbaum.

Bonk, C. J., Daytner, K., Daytner, G., Dennen, V., & Malikowski, S. (2001). Using Web-based cases to enhance, extend, and transform preservice teacher training: Two years in review. *Computers in the Schools*, *18*(1), 189–211. doi:10.1300/J025v18n01_01

Bonk, C. J., & Dennen, V. (2003). Frameworks for research, design, benchmarks, training, and pedagogy in Web-based distance education . In Moore, M. G., & Anderson, B. (Eds.), *Handbook of distance education* (pp. 331–348). Mahwah, NJ: Lawrence Erlbaum Associates.

Bonk, C. J., Malikowski, S., Angeli, C., & East, J. (1998). Web-based case conferencing for preservice teacher education: Electronic discourse from the field. *Journal of Educational Computing Research*, *19*(3), 267–304. doi:10.2190/L298-TGTL-4X38-9FH2

Dennen, V. P. (2005). From message posting to learning dialogues: Factors affecting learner participation in asynchronous discussion. *Distance Education*, *26*(1), 125–146.

Ehman, L., H., Bonk, C. J., & Yamagata-Lynch, E. (2005). A model of teacher professional development to support technology integration. *AACE Journal*, *13*(3), 251–270.

Ellis, R. (2006, March). Synchronous e-learning survey results. *Learning Circuits*. Retrieved from http://www.learningcircuits.org/2006/March/2006synch_poll.htm

Harasim, L. (1987). Teaching and learning on-line: Issues in computer mediated graduate courses. *Canadian Journal of Educational Communication*, *16*(2), 117–135.

Hiltz, S. R. (1990). Evaluating the virtual classroom . In Harasim, L. (Ed.), *Online education* (pp. 134–184). New York, NY: Praeger.

Hofmann, J. (2004). *Live and online!: Tips, techniques, and ready-to-use activities for the virtual classroom*. San Francisco, CA: Pfeiffer Publishing.

James-Deramo, M. (Ed.). (1999). *Best practices in cyber-serve: Integrating technology with service learning instruction*. Blacksburg, VA: Virginia Tech Service Learning Center.

Ko, S., & Rossen, S. (2004). *Teaching online: A practical guide* (2nd ed.). Boston, MA: Houghton Mifflin.

King, A. (1995). Designing the instructional process to enhance critical thinking across the curriculum. *Teaching of Psychology*, *22*(1), 13–17. doi:10.1207/s15328023top2201_5

MacKnight, C. B. (2000). Teaching critical thinking through online discussions. *EDUCAUSE Quarterly*, *4*, 38–41.

Palloff, R. M., & Pratt, K. (2001). *Lessons from the cyberspace classroom: The realities of online teaching*. San Francisco, CA: Jossey-Bass.

Salmon, G. (2000). *E-moderating: The key to teaching and learning online*. London, UK: Kogan Page.

Salmon, G. (2002). *E-tivities: The key to active online learning*. Sterling, VA: Stylus Publishing.

Santo, S. (2006). Relationships between learning styles and online learning: Myth or reality? *Performance Improvement Quarterly*, *19*(3), 73–88. doi:10.1111/j.1937-8327.2006.tb00378.x

Schrage, M. (1990). *Shared minds: The technologies of collaboration*. New York, NY: Random House.

Singh, H. (2006). Blending learning and work: Real-time work flow learning . In Bonk, C. J., & Graham, C. R. (Eds.), *Handbook of blended learning: Global perspectives, local designs* (pp. 474–490). San Francisco, CA: Pfeiffer Publishing.

Wade, R. C., Boyle-Baise, M., & O'Grady, C. (2001). Multicultural service-learning in teacher education . In Anderson, J., Swick, K., & Yff, J. (Eds.), *Service-learning in teacher education enhancing the growth of new teachers, their students, and communities* (pp. 248–259). Washington, DC: AACTE Publications.

Watkins, R. (2005, May/June). Developing interactive e-learning activities. *Performance Improvement*, *44*(5), 5–7. doi:10.1002/pfi.4140440504

Wenger, M. S., & Ferguson, C. (2006). A learning ecology model for blended learning from Sun Microsystems . In Bonk, C. J., & Graham, C. R. (Eds.), *Handbook of blended learning: Global perspectives, local designs* (pp. 76–91). San Francisco, CA: Pfeiffer Publishing.

Warren, K. (1998). Educating students for social justice in service learning. *Journal of Experiential Education*, *21*(3), 1–8.

Zhang, K., & Peck, K. L. (2003). The effects of peer-controlled or moderated online collaboration on group problem solving and related attitudes. *Canadian Journal of Learning and Technology*, *29*(3), 93–112.

KEY TERMS AND DEFINITIONS

Asynchronous Discussion: A form of discussion that allows students and faculty to post and respond to discussion board topics when it is convenient for them.

Computer Mediated Communication: Communication that takes place between people via the computer such as threaded discussions, chats or email communication.

Critical Thinking: Thinking that moves beyond low level tasks like recalling information and restating information and requires one to apply to knowledge in new contexts.

Field Experience: An opportunity for preservice students to get real life experience working in schools.

Online Discussion: A conversation similar to a class discussion but is held on the Web.

Threaded Discussion: A Web based discussion held online that is organized by topics.

Urban Education: Educating students in public school systems that may lack the resources to handle the challenges faced in educating every student given the diversity and poverty they represent.

Chapter 15
Investigation of Blended vs. Fully Web-Based Instruction for Pre-Teacher Candidates in a Large Section Special Education Survey Course

Chris O'Brien
University of North Carolina at Charlotte, USA

Shaqwana M. Freeman
University of North Carolina at Charlotte, USA

John Beattie
University of North Carolina at Charlotte, USA

LuAnn Jordan
University of North Carolina at Charlotte, USA

Richard Hartshorne
University of Central Florida, USA

ABSTRACT

This chapter summarizes the results of a quasi-experiment conducted to determine the relative effectiveness of preparing pre-teacher education university students using a fully web-based course conducted asynchronously versus a blended model of instruction using the same LMS for forty percent of instructional time. The project evaluated two large sections of SPED 2100, "Introduction to Students with Special Needs." Data was collected to evaluate the extent to which pre-teacher education students developed understanding of critical information related to human development factors, psychological, sociological, and policy foundations of teaching students with special needs. Further, data collection examined student preferences in learning and the extent to which students developed comparable perception of preparedness for the future teaching roles. Results indicated no significant differences regarding content knowledge, but varying perspectives on the potential for success in fully web-based courses dependent largely on learner profile and the point of development in university coursework.

DOI: 10.4018/978-1-4666-1906-7.ch015

INTRODUCTION

Increasing external demands on student time, calls for more cost effective programs, and a growing demand for course offerings are just a few factors that have prompted colleges of education to expand and diversify the way they deliver instruction. Online coursework is one method of course delivery that has grown in popularity and provides alternatives to traditional face-to-face instruction (Allen & Seaman, 2011; DEST, 2002). With advancements in the functionality and usability of "distance education" technologies offered via the web, colleges and universities have opportunities for more robust online courses in both synchronous and asynchronous formats (Kim & Bonk, 2006; McGreal & Elliott, 2004;).

Are the learning experiences comparable? To answer this question, ongoing research at the national and university level are critical to assure the credibility of online preparation experiences and the potential for such programs to fully mirror or improve upon existing models (Ludlow, 2006). It appears that attention to quality has resulted in efficient and effective web-based course offerings in various areas of concern. For example, student achievement is critical to any discussion of parity between web-based and face-to-face instruction. Literature on distance education recognizes "no significant difference" in numerous studies that compare student achievement in online courses with achievement in both blended and face-to-face settings (Allen & Seaman, 2011; Beile & Boote, 2002; Caywood & Duckett, 2003; Hartshorne, Heafner, & Petty, 2011; McNamara, Swalm, Stearne, & Covassin, 2008; Scoville & Buskirk, 2007; Steinweg, Davis, & Thomson, 2005). In addition, students' perceptions of their own learning affect their satisfaction with coursework. There is evidence that students perceive instruction in distance education to be of equal quality to coursework offered in the traditional format (Beattie, Spooner, Jordan, Algozzine, & Spooner, 2002; Hartshorne, Heafner, & Hartshorne, 2011; Petty, Heafner, & Hartshorne, 2009; Spooner, Jordan, Algozzine, & Spooner, 1999).

Other research has addressed specific components of instruction, particularly course content knowledge. Students enrolled in graduate special education courses reported that they were able to appropriately and satisfactorily acquire course content knowledge when participating in online coursework (Korir Bore, 2008). Steinweg, Davis, and Thomson (2005) also reported that students were as successful in acquiring the course content knowledge via online course offerings as they were in face-to-face class settings. Further, these teachers-in-training connected their learning to future professional behavior, and believed that their knowledge of students with disabilities would have a positive impact on their work with all students in an inclusion classroom (Steinweg et al., 2005).

Not all the evidence is positive for online learning, however. Students in the Korir Bore study (2008) reported that there was something missing in the academic experience, and that they did not experience a true sense of connection with the online course instructor. Current research does not erase the reservations about web-based learning among many teacher educators in special education.

In hybrid or blended learning environments, instructors can incorporate the best of both web-based and face-to-face instruction by reducing lecture time and supplementing instruction with web-based instruction/assessments and/or learning materials. The emergence of hybrid or blended learning environments has further expanded the instructional offerings at universities. Although there is no standard format for offering blended courses, the most consistent interpretation is a 25-50% reduction in face-to-face meeting times by reducing the time of individual class sessions or reducing the number of class meetings (Dziuban, Moskal, & Hartman, 2005). Because of its inherent flexibility, students and university faculty are drawn to hybrid or blended learning environments. These environments are particularly conducive to special education teacher preparation.

Trends in Instructional Opportunities for Early College Students

In light of expanding instructional options, which incorporate emerging technologies, a natural area of interest for change is the large survey or introductory course. Riffell and Sibley (2005) suggest that most of these courses offered at large colleges and universities are taught in traditional style and in large lecture halls with students passively receiving instruction. The literature is sparse regarding the ability of early college students in large lecture survey courses to experience success in web-based or blended web-supported instructional contexts. However, on a positive note, Tallent-Runnels et al., (2006) reviewed existing literature that indicates that university students see a clear appeal to the flexibility and convenience of web-based instruction. Students are more likely to feel this way if they have previous exposure to learning with computers and feel both web and computer-savvy.

Riffell and Sibley (2005) offered a large lecture undergraduate biology course using a blended or hybrid model and evaluated student perceptions and performance in the course. The authors found that the students in the blended course performed equally or better when compared to a traditional lecture format; they also appreciated the web-based activities that offered more opportunities for active learning and interaction with the instructor. Notably, freshmen in the blended course exhibited smaller gains in class performance when compared to their more mature undergraduate counterparts. Ward (2004) compared an introductory statistics course taught in both blended and face-to-face formats. Students in the blended course had a more positive attitude toward the course and the effectiveness of the instruction and took on more ownership of their learning in the course. This outcome was attributed to the use of online activities. Althaus (1997) demonstrated higher academic performance with 142 undergraduate students who participated in a blended course using online discussion forums to supplement face-to-face class instruction. The author suggested that the online discussions provided more opportunity to thoroughly read and consider the material and post carefully developed written responses. In contrast, when Faux and Black-Hughes (2000) compared the instructional experiences of undergraduate social work majors, they found the most positive results in the face-to-face classroom. The students felt uncomfortable learning the course content online and desired more interaction with the instructor.

Brown and Liedholm (2002) explained potential differences in undergraduate student performance in terms of time commitment. The authors suggested that students, given the liberty of web-based instruction, were spending considerably less time on course content than their face-to-face peers who were bound to allocated in-class instructional time. Ward (2004) found that students in the blended course did not take on as many extra credit tasks as the face-to-face group due to the perception that online activities were time consuming. While blended courses offer the "best of both worlds," they may also be perceived as more work than the typical large face-to-face course.

Purpose of the Study

Questions remain as to whether pre-teacher education majors can effectively learn critical content in a large lecture course offered using fully web-based instruction or blended instruction. The majority of research on distance and web-based instruction measuring positive perceptions and comparability in performance has involved graduate coursework and older students. Tallent-Runnels, et al., (2006) note the average age of the online learner is 29 with a typical range of 30-35. In consideration of the available information regarding the overall effectiveness of online coursework, the purpose of the current evaluation study was to examine the apparent feasibility of both fully web-based instruction and blended web and lecture instruction in a gateway course that is typically required of *all* students majoring in education, specifically the course *Introduction to Students with Special Needs*. Essentially, the gateway special education course in our College of Education emphasizes two areas of importance in introductory-level preparation of undergraduate students:

1. Critical foundations of special education content and disability-related issues including the foundations of public policy impacting schools, the daily role of teachers, foundations of educational psychology, and the disability categorization process (i.e., diagnoses, characteristics, etc.).
2. Shaping of dispositions, attitudes, and confidence toward adequately addressing the needs of diverse populations, particularly related to inclusive belief systems and expectations associated with performance outcomes for students with special needs.

Given the established literature which suggests students will experience no significant differences in content acquisition in these different modes of instruction, the purpose of this study was to evaluate pre-teacher education majors perception of course effectiveness, confidence regarding future application of course content, and knowledge of critical content relative to the effectiveness of their learning experience in fully online versus blended instructional contexts. Specifically, the research questions included the following:

1. What are the effects of fully web-based and blended modes of instruction on students' confidence and perception of preparedness in teaching students with special needs?
2. What are students' perceptions of their effectiveness as learners in fully web-based and blended modes of instruction for an introduction to special education course?
3. What are the effects of fully web-based and blended modes of instruction on students' knowledge of critical content in a course on teaching students with special needs?

METHODS

Participants

A total of 138 undergraduate students preparing for admission to one of several teacher education programs offered in the college of education participated in the study, with the majority of the students majoring in elementary education followed by special education, middle grades education, and child and family development. Of those 138 students, 69 were enrolled in the fully web-based section, and 69 students in the blended section (one late withdrawal from each section). Roughly 80% of students were female and approximately 25% of students were transfers from local community colleges who were practicing teacher assistance or career changers seeking their first degree in education. The demographics of the classes included a predominance of not only female students, but also students of White European descent from small, regional communities. Although not explicitly collected, race/CLD information in the courses reflected a population of less than 10% CLD students.

Setting and Course Description

The study was conducted at a large state university in the southeastern part of the United States, located in a large metropolitan city. Because of the large number of education major programs requiring the course, SPED 2100: Introduction to Students with Special Needs, as a prerequisite for students, multiple modes of instruction were needed as an alternative to the traditionally large face-to-face traditional lecture sections of the course that were previously offered (i.e., a fully web-based section and a blended section).

Consistent between the two sections of this course was a broad foundation of knowledge of psychological and sociological factors that impact special education, and the same textbook was used in both sections of the course in this study, *Exceptional Children* by Heward (2006).

Course Instructor

One junior faculty member with 4 years of web-based teaching experience and extensive training in the use of distance education technology taught both sections of the course. The instructor had multiple advanced professional development trainings for programs in WebCT, Blackboard Vista, podcasting, web design, and multimedia development. Additionally, the instructor held a doctoral degree in special education with advanced knowledge in high-incidence disabilities, predominantly in learning disabilities and communicative disorders.

Research Design

A post-only quasi-experimental design, using a convenience sample and including quantitative and qualitative measures was used to evaluate student perspectives on web-based and blended modes of instruction. The effects of mode of instruction were measured on a) perceptions of preparedness for teaching students with special needs, b) effectiveness of the learning experience in each group's mode of instruction, and c) academic performance on a summative measure of content knowledge.

Because the mode of instruction is self-selected by students, random selection of students and random assignment to treatment group was not feasible. There is however a level of randomization built into the process of university course registration through the availability of classes, times of offerings, etc. There was no difference in the listing of the compared sections in the university catalog or the registration system. The university registration system includes notes for students indicating that sections will be offered once or twice a week and by primarily lecture or fully web-based. The blended course was listed as a lecture course with additional instruction requirements provided via a web-based learning management system. Participation in all aspects of the study (e.g., focus groups), with the exception of the final examination, was voluntary and did not impact student grades in the course.

Procedure

Fully Web-Based Course

The fully web-based section of the course was delivered exclusively through the Blackboard online instructional management system, meaning there were no required face-to-face meetings. Weekly comprehension checks in a repeatable quiz format, interactive discussions conducted in the discussions application of Blackboard, learning modules and notes, and archived video of traditional lectures were all used as a means to provide equivalent learning opportunities to online students that were available in the face-to-face lecture-based classroom. Participants in the fully web-based course could view and listen to the same lectures as students in face-to-face lecture sections of the course. Discussions in Blackboard were also generally equivalent to weekly discussion topics in the face-to-face component of the blended class.

Blended Course

The hybrid, or blended, course combined characteristics of both face-to-face and fully web-based instructional settings. Students in the blended course were provided with opportunities for the inclusion of both asynchronous (a student-centered teaching method in which students learn the same things just at different times) and synchronous (students learning the same things at the same time in the same place) interactions with various stakeholders in the instructional setting (Dennis, El-Gayar, & Zhou, 2002). Similar to the fully web-based course, the online components of the blended course consisted of weekly comprehension checks in a repeatable quiz format, learning modules, and notes which organize the text material. The traditional lecture presentations (blended and full-lecture) included presentation of new concepts (aligned with chapters in the text), in-class notes on key points (consistent with the web-based presentation), PowerPoint presentations, and interactive discussions.

Data Collection

Questionnaire

During the final week of classes, an eighteen-item questionnaire was administered to both sections of SPED 2100 to evaluate the opinions of the course instructor, student outlook, perceptions of preparedness related to special education and perceptions of success learning a vast amount of new special education content in their selected course section. To evaluate student's confidence in preparedness regarding special education topics and experience of learning in the course (web-based versus blended), eight questionnaire items was specifically developed to analyze these two main areas. Questionnaire items required participants to rate their agreement with a statement related to the above areas of evaluation based on a 5-point Likert scale (1 = Strongly Disagree, 2 = Disagree, 3 = Undecided, 4 = Agree, 5 = Strongly Agree).

The College of Education's Conceptual Framework associated with NCATE accreditation and state legislature resolutions was used to develop the language of the questionnaire items associated with confidence toward teaching students with special needs. The Center for Teaching and Learning at the site of the study and the nationally recognized *Center for Distributed Learning* (www.cdl.ucf.edu) at the University of Central Florida was used to develop the questionnaire items associated with perspectives on learning in differing modes of instruction.

To ensure clarity, intent, and content validity of the questions the questionnaire was piloted with volunteer undergraduate student assistants in the department and was reviewed by three peer faculty members. Minor adjustments were made to the questionnaire based on this feedback. To account for both section of the course two formats, a hard copy format and an online format using SurveyShare® was developed. Students in the blended course completed the hardy copy questionnaire, while students in the web-based section completed the identical version of the questionnaire presented via SurveyShare®.

Interviews

In addition to the questionnaires, semi-structured interviews were conducted among focus groups; one representative group was created from each section (web-based and blended) of the course during final exam week. Balanced groups (students for whom this represented a first experience in college and students who were returning to college and balancing school with work and family) were created to investigate the overall effectiveness of the two sections and for their effectiveness with specific subgroups. The groups consisted of traditional freshmen, junior-level transfer students, and non-traditional students. An effort was made to achieve balance in the gender, cultural/linguistic background and the extremes of the groups.

Interviews were conducted by a research assistant who was not involved with student grading or assessment of performance in either of the courses, with the assumption that students would be less apprehensive about sharing their perspectives with another student. Student comments were audio recorded and transcribed by the research assistant for later analysis by the research team. All of the students in the blended section participated in on-campus focus group interviews. The focus group for the five students in the fully web-based course was conducted in two ways; two responded through in person interviews and three responded to questions posed by the interviewer via a telephone conference call.

Focus Group 1

The focus group for the blended course was comprised of four female and one male student. The female students included one traditional undergraduate and three non-traditional undergraduates all of whom were working part-time, raising families, and returning to school later in life. The traditional undergraduate student was an education major and age appropriate for a typical first-semester college student. Additionally, she commuted to campus like her non-traditional counterparts. The one male student was a traditional undergraduate living near the university campus.

Focus Group 2

The focus group for the web-based course included five students—four female and one male. This group consisted of four female students who were traditional undergraduates living on campus or nearby. The other two students (one female and one male) were both non-traditional students returning to school later in life. The female student was raising a family and running an in-home business while the male student worked full-time and attempted to balance a return to school with his work and family responsibilities.

Data Analysis

Using the data analysis software, Statistical Package for the Social Sciences (SPSS Inc.), independent samples t-tests were conducted to determine if statistically significant differences existed between the two groups of students. The alpha level was set at $p < .05$ for all statistical analyses.

Qualitative data obtained from the focus group discussions, after transcription, were read three times by each of the researchers. Findings of commonalities in the transcripts were highlighted by each researcher and then compared for possible themes using the constant comparative method of data analysis (Bogden & Biklen, 1982; Corbin & Strauss, 1990). Constant comparative data analysis is an iterative process of data reduction that helps to identify a higher order structure in data and organize codes into emergent themes. Members of the research team initially conducted iterations independently in order to allow for comparison and validity checks. After the initial independent iterations, the research team convened several times for comparison of themes. At each meeting, data were read and reread to confirm researcher agreement on themes and to explore participant meaning. After patterns were identified and data sorted into domains (Huberman & Miles, 2002; LeCompte & Schensul, 1999), exemplary quotes from the narratives were identified and cited to support each of the emergent themes.

A summative examination was created to measure content knowledge gained in the special education course. The exam was developed using 1) a collection of 100 items were selected from the test bank created by the textbook author, 2) items were organized by topics of the course in an attempt to provide multiple items from each critical content area with topics of broad relevance (e.g., legal issues, foundations of special education) prioritized, 3) 50 items were selected to represent the 15 major topics of the course (disability categories and broad policy and instructional issues in special education), 4) items were modified in language to mirror class discussions and to present a range of challenge across the examination, 5) the complete examination was reviewed by graduate student assistants for clarity, 6) the complete examination was reviewed by a number of content experts in the special education department to determine content accuracy and item clarity.

The examination was created as an online assessment in Blackboard for students in both classes. Students were given 60 minutes to complete the exam, and were informed that use of the textbook would not be allowed, nor would it be practical given the short amount of time allotted to take the assessment. On the other hand students were permitted to use basic review sheets developed in class that simply served as outlines of course content and keys to acronyms and other professional jargon.

RESULTS

Means, standard deviations, and statistical analyses are reported in Table 1. Overall response rate for the questionnaire was approximately 83%; 82% for the blended course, and 85% for the web-based course.

Perception of Preparedness and Confidence in Teaching Students with Special Needs

Three survey items (1, 2, and 3) addressed the first research question (i.e., what are the effects of mode of instruction on students' confidence and perception of preparedness in special education content). Across all three items, mean responses were highest for the blended course when compared to the web-based section. A statistically significant difference was indicated between the blended group (M= 4.34, SD= .67) and the web-based group (M= 3.91, SD= .81) on item 2 ("I believe I will be successful teaching students with disabilities in my classroom."), $t(111)=3.06$, $p = .003$. There were no significant differences between the groups, although item one approached significance, $t(111)=1.98$, $p=.05$. Students in the blended class reported consistently greater levels of confidence in their future work with students with disabilities as compared to students in the web-based class.

Table 1. Means, standard deviations, and t-scores for perceptions across courses

	Course		
Item	**Blended**	**Web**	***t***
1. I am comfortable working with students with disabilities in schools.	4.18 (.77)	3.91 (.66)	1.98
2. I believe I will be successful teaching students with disabilities in my classroom.	4.34 (.67)	3.91 (.81)	3.06*
3. I believe students with disabilities can be successful in my classroom.	4.42 (.65)	4.25 (.64)	1.45
4. The instructional approach used in this course helped me to learn new information about children with disabilities in schools.	4.45 (.80)	4.53 (.63)	-.58
5. I felt successful learning with my chosen mode of instruction.	4.36 (.87)	4.51 (.78)	-.95
6. The instructional approach used in this course was effective for adapting to my schedule and personal life needs.	4.55 (.65)	4.81 (.40)	-2.54*
7. I can effectively learn new information from web-based resources.	4.17 (.90)	4.51 (.60)	-2.36*
8. The workload was reasonable for me to complete throughout the semester.	4.72 (.52)	4.67 (.51)	.60

$*p < .05$

Effectiveness of Mode of Instruction

Items 4-8 evaluated perceptions of the instructional options based on research question 2 (What are students' perceptions of their effectiveness as learners in web-based and blended modes of instruction for an introduction to special education course?). Trends in mean response scores generally favored the web-based instruction group (e.g., Items 4-7). Only item 8 showed a mean higher score for the blended section, but no significant difference was noted. A statistically significant difference was indicated between the blended group (M= 4.55, SD= .65) and the web-based instruction group (M= 4.81, SD= .40) for Item 8 ("The instructional approach used in this course was effective for adapting to my schedule and personal life needs."), t (113) = -2.54, p = .013. A statistically significant difference was also found between the blended group (M= 4.17, SD= .90) and the web-based instruction group (M= 4.51, SD= .60) for Item 9 ("I can effectively learn new information from web-based resources."), t (113) = -2.36, p = .020. Students in the web-based class reported significantly higher perceptions of flexibility for adapting the course to their personal schedules. Further, students in the web-based class confirmed that they had a greater preference for learning new information from web-based resources compared to students in the lecture class.

Academic Performance

Students in the blended and fully web-based instructional modes scored almost identically on the summative examination of special education content. No statistically significant differences were noted between the test scores, t (132) = .275, p=.784, of the blended group (M= 77.32, SD= 9.93) and the web-based instruction group (M= 76.80, SD= 11.93).

Focus Groups

Evaluation of open-ended data presented a richer understanding of student feelings and perceptions of each instructional mode (web-based, blended). Analyses of qualitative data revealed emergent themes regarding differences in the instructional modes related to: *perceptions of preparedness and confidence*, *perceptions of the instructional modes*, and *recommended modifications to courses based on instructional modes*.

Perceptions of Preparedness and Confidence

The analysis of the focus-group transcripts revealed four emergent themes (opinions about the general content, identification and awareness, uncertainty, and personalization) regarding the level of preparedness and confidence focus group participants felt after completion of the course. These data present the best description for how each instructional approach influenced the level of preparedness and the confidence for students.

The data revealed that students from focus groups representing each instructional mode felt that the *course content provided a general overview* of the instructional material but did not provide in depth analysis of course topics. There was a sense that the course provided an introduction to issues that may not have been previously considered by these pre-teacher education students. This was consistent across focus groups for each instructional mode and aligned with the stated course objectives.

Participants in both focus groups indicated that the course provided *a general awareness* for students with special needs that did not previously exist. Again, this was consistent across each focus group and was consistent with the stated course objectives.

Students in the web-based version of the course indicated more *uncertainty* regarding their level of preparedness. While being able to identify students that would need additional assistance and instructional modifications, participants in the web-based section indicated an uncertainty with knowing how to address these issues ("I don't know that I would really know how to deal with it," "I think it just ... opened my eyes, but that is pretty much it," "I guess it is more of an awareness than really knowing how to deal with it"). This was a theme that only emerged in the focus group with online students. Additionally, while there was a significant attempt made to establish equality in the learning experiences of students in the instructional modes, students in the web-based course section seemed to indicate a sense that they were teaching themselves, which led to additional uncertainty regarding their level of preparedness upon completion of the course ("It is not somebody putting it into your head. It is kind of like reading a book, you are not going to remember all the details that you may need, you are just going to remember the overview"). One reason for this might have been the lack of familiarity with web-based learning, as most students in the web-based section focus group had no previous experience with web-based courses.

Notably absent from the commentaries of web-based class participants were the discussions of *personalization* among participants in the blended class. Students in the blended class consistently noted the importance of the enthusiasm and passion of their instructor that created a stronger sense of connection with the course content ("He is so passionate, it is easy to pay attention."). The students emphasized the value of relating the course content to personal experiences and passions of the instructor ("I think that he really personalized it, to hear it from him and not just, reading a book is one thing but hearing about somebody's experiences is another.").

Perceptions of Instructional Modes

Another issue examined in the focus groups included the student perceptions of the instructional modes. Specifically, these data present the best description for the reasons for selecting the instructional modes, as well as the perceptions of the effectiveness of the various learning modes.

Participants enrolled in each of the instructional modes cited differences in *reasons for selecting the instructional modes*, as well as general perceptions of their experiences with the instructional mode of choice. As would be expected, students that enrolled in the web-based section cited convenience and flexibility as the rationale for enrolling.

There was a clear distinction between the *perceptions of effectiveness* of the web-based course from the perspectives of traditional and non-traditional students. Traditional students felt that the web-based course was much harder than expected and took a significant amount of time to acclimate themselves to the learning environment and structure ("...it took me a while." "I really was surprised by that, me remembering..."). They found the environment confusing and had a more difficult time with issues that are common with new online learners, such as organization, persistence, scheduling ("I know he has office hours Monday and Friday, but that is optional. Where there are those optional big lecture classes, you know they don't take attendance, I mean I go, but it is an easier way for me not to go. But if it is required to meet at that time to see where everyone is. I think it is a lot better because it pushes you to have to do it" "As good as it sounds to be online and not have to go to class, I don't think you retain as much, and I don't think you get as much from it, or give as much to it"). They viewed learning as a more passive process (rather than an active process), which added to their feelings of being overwhelmed by the web-based learning environment. On the other hand, the non-traditional students enrolled in the web-based section of the course felt that the learning environment made them more comfortable and had a positive impact on their learning experience. Contrary to the views of the traditional students, the non-traditional students did not feel the web-based course hindered their learning experience, but did cite the need to be organized and self-motivated in order to be successful.

There was no distinction between the views of the traditional and non-traditional students enrolled in the blended courses related to perceptions of effectiveness. Students enrolled in the blended section of the course believed the environment was effective and facilitated a positive learning experience. They believed they were more prepared and that the additional discussion and focus on higher order skills associated with the blended environment resulted in a more interesting, successful, and interactive class. Some participants did note that the blended environment was somewhat confusing at first, but that it "wore off pretty quickly."

Recommended Modifications to Courses Based on Instructional Modes

There was a consensus among the blended students that no modifications needed to be made to the instructional mode or content of the course. This was not the case with the web-based students, as a number of recommended modifications emerged from the focus group discussions. Because students were somewhat unsure of how to approach the learning environment, students suggested a need for more structure in the layout and approach to the course. While there was a significant attempt to align the structure and layout of the web-based course with those of the blended courses, there was clearly a perception among students that this was not the case. A number of students indicated a desire for a blended version of the course, or a face-to-face component of the web-based course.

DISCUSSION

This study sought to evaluate the experience of two groups of pre-teacher education students in large sections of a special education survey course offered as a fully web-based section and a blended web and lecture section. The research design of this project should be viewed as an evaluation study of intact programming at a large state university with all of the research limitations assumed in such conditions. Although, this project was a quasi-experiment, considerable effort was put forth to establish a detailed description of student experiences including both quantitative and qualitative methods.

Overall, results indicate no significant differences in acquisition of critical course content consistent with an extensive line of research on comparisons of web-based and traditional instruction. There are, however, some notable differences observed in perceptions of preparedness or confidence and flexibility of instructional experiences suggesting some trends in favor of perceptions of preparedness (similar to a self-efficacy construct) for the blended class and some trends in favor of the flexibility and convenience of learning for the web-based group. Focus group data give an even more detailed and nuanced picture of the aggregate findings from the questionnaire.

Regarding perceptions of preparedness, student responses suggest that students in the blended course had a greater perception of their potential to be successful teaching students with special needs in their future careers than students who participated in the fully web-based course. Notable in student responses for these items is the idea that students in the web-based course demonstrated some parity with the blended course in perceptions related to the ability to recognize the potential for students with special needs to be successful in their classrooms (item 6) based on the content of the course. This perception of preparedness did not, however, seem to translate to an expectation that those future students would be successful in *their* classrooms (item 4) as compared to student responses in the blended course. The differences between these two items seem to indicate a greater sense of preparedness among those students who met in person with an instructor in the blended section. Focus group data reinforce the questionnaire data regarding perceptions of preparedness or confidence suggesting that students in the web-based class may have felt more distanced from course content feeling uncertain of the connection of their new "awareness" of disabilities in schools to their future role as educator of a diverse population of children. Despite the availability of web-based lectures (e.g., video archives) designed to be comparable to the blended course's live lectures, the students in the web-based course indicated difficulty "putting it all together." Students in the blended class, however, noted the personal connections and passion of their instructor as meaningful in enhancing their feeling of "connection" the critical course concepts.

Questionnaire data for student perceptions of their instructional mode suggest that students had relatively comparable perceptions of the reasonable nature of the workload and ability to be successful learning in their mode of instruction despite some small trends in mean scores favoring the fully web-based instruction group. Not surprisingly, students in the web-based course reported the most positive perceptions of flexibility and the ability to adapt to the course to their personal schedules and needs. Students in the web-based course also demonstrated greater perception of their ability to learn using web-based materials. Although this is hardly uncommon in research in web-based instruction, given a typical comparison to face-to-face instruction, this finding is notable, as the blended section also included a large amount of web-based instructional materials and learning opportunities. Both groups were aware that there should have been aware of web-based learning activities in their course. It could be hypothesized that students in the web-based course developed more savvy and preference for the web-based materials given a semester of practice; this perhaps occurred out of necessity, whereas students in the blended course might have continued to "lean" on the in-person lecture and discussion opportunities.

Focus group data add substantially to the examination of success with the chosen mode of instruction, as we find disparities among subgroups of the web-based students. Students in the web-based course focus group commented in a manner consistent with previous research findings suggesting that the average, successful online student is older and more mature than the traditional undergraduate with a more complicated home life and schedule that requires students to be motivated, organized, and self-regulated. Perhaps building on the notion suggested by Tallent-Runnels et al., (2006) that successful online students are experienced and well versed in the university learning process and familiar with web-based learning, the traditional first year students in this study expressed great anxiety and frustration with their initial experiences in an asynchronous web-based course. In contrast, focus group comments were uniformly positive with regard to the instructional design of the blended course as it impacted their learning and preparedness despite some momentary confusion about the idea of multiple learning access points—lecture and web-based. Possibly echoing the observations of Brown and Liedholm (2002), traditional early college undergraduates in the web-based course expressed apprehension and poor ability to dedicate time in their schedules to the web-based course, instead preferring a situation in which they would meet at a predictable, scheduled, and mandatory time with a professor. The desire for flexibility was not at the forefront of their priorities and the need to be "managed" by university faculty was surprisingly desirable to these traditional university students. Riffell and Sibley (2005) similarly noted that perceptions of success in online learning experiences were less substantial for students in their first year of school.

The data on academic performance are notable despite the lack of difference in performance. Summative testing experiences were completely comparable between the two groups. Instructional opportunities and course content were held consistent between the two sections and the summative assessment attempted to measure mastery of critical special education content knowledge. There were no significant differences between the sections reinforcing the conclusions of an extensive line of research indicating that students are capable of mastering critical content in web-based courses. Despite perceptions of differing experiences, the main measure of accountability was equivalent. The unknown long-term outcome would certainly relate to the extent to which these students can apply the information they learned, but that is beyond the scope of this current study.

Limitations

A relevant, but challenging limitation of this study relates to the nature of studying existing university coursework with limited ability to actualize random selection and random assignment to the two sections of the course. These factors are nearly unavoidable and mirror historical challenges in educational research. Another major limitation of this and many other studies on web-based and blended instructional opportunities pertains to the lack of clarity and consistency in models used in the studies. In this study, we examined an asynchronous web-based course offering using Blackboard and providing extensive resources including, but not limited to web-based video archives of weekly lectures. Other scholars examining topics of distance education might develop a course quite differently or use another emerging LMS such as Moodle. They might even blend a fully web-based course with a web-based conferencing system like Wimba. It could be difficult to replicate many of the aspects of this particular study because of the specific nature of the course instructor's teaching style and the instructional design he employed across the two sections. Blended courses may create even more challenges for replication, as there exist only vague definitions of the construct with the commonality being the use of a mix of traditional instruction and computer-based, web-based instructional opportunities.

Implications for Teacher Education Programs and Large Lecture Courses

Results of this study are relevant to contemporary teacher education. Universities prepare a diverse cross-section of pre-teacher education majors who range from traditional, on-campus students to an increasingly large number of non-traditional students balancing the return to school with work and family responsibilities. Contemporary teacher education is characterized by an increase in web-based course offerings and a diverse population of pre-service teachers in need of flexible educational opportunities (Beattie, Spooner, Jordan, Algozzine, & Spooner, 2002; Hagie, Hughes, & Smith, 2005). A major impetus to this current work was the need to expand course offerings to students traditionally prepared in large lecture courses with minimal interactivity without risking a loss of impact on their knowledge and skills. In this study, students had options in their course delivery method and participants from both sections were relatively satisfied with their choice of course delivery in terms of manageability and effectiveness.

The generally higher perception scores related to preparedness and confidence for students in the blended course, which included live lectures, echoes Korir Bore's (2008) identification of a core element of the academic experience missing from web-based instruction; these students did not perceive a true sense of connection with their instructor when taking web-based courses. This sense of connection is considered a positive aspect of a course according to typical course evaluations. Teacher educators should make note that instructional designers for web-based course offerings may need to "work harder" to create a sense of connection with the instructor and enhance community building. Related to the previous concern about apprehension and anxiety for new university students taking web-based courses, the instructor might connect with students and create a sense of community at the start of the course by providing an extensive orientation to course processes, technology, and communication systems.

Consistent with previous commentaries by Dziuban et al., (2005) and evidence from Althaus (1997) and Ward (2004), the blended course appeared to reflect strengths in terms of the connectedness between instructors and students. For the teacher educator, having the choice of offering a blended or hybrid course allows for a focus on what would best serve student needs without sacrificing the most effective components of either live class sessions or web-based course activities.

Given various asynchronous and synchronous web-based and blended instructional opportunities continuing emerge at universities across the nation, continued attention to research on efficacy and comparability are critical to furthering the core mission of teacher education (Scheetz & Gunter, 2004). Specifically, this study examined whether various web-based and blended educational models are equally effective in terms of preparing pre-teacher education majors for their future critical role as teachers. The results affirm general comparability in terms of knowledge and perceptions of course experiences, but raise questions regarding the assumption of student preferences for web-based courses, particularly for early college students and the need to foster connection with students to overcome feelings of anxiety and isolation.

REFERENCES

Allen, E. I., & Seaman, J. (2011). *Going the distance: Online education in the United States, 2011*. Needham, MA: Sloan Consortium. Retrieved from http://sloanconsortium.org/publications/survey/going_distance_2011

Althaus, S. L. (1997). Computer-mediated communication in the university classroom: An experiment with on-line discussions. *Communication Education*, *46*, 158–174. doi:10.1080/03634529709379088

Beattie, J., Spooner, F., Jordan, L., Algozzine, B., & Spooner, M. (2002). Evaluating instruction in distance learning classes. *Teacher Education and Special Education*, *25*(2), 124–132. doi:10.1177/088840640202500204

Beile, P. M., & Boote, D. N. (2002). Library instruction and graduate professional development: Exploring the effect of learning environments on self-efficacy and learning outcomes. *The Alberta Journal of Educational Research*, *48*, 364–367.

Bogden, R. C., & Biklen, S. K. (1982). *Qualitative research for education: An introduction to theory and methods*. Boston, MA: Allyn and Bacon.

Brown, B. W., & Liedholm, C. E. (2002). Can web courses replace the classroom in principles of microeconomics? *The American Economic Review*, *92*(2), 1–12. doi:10.1257/000282802320191778

Caywood, K., & Duckett, J. (2003). Online vs. on-campus learning in teacher education. *Teacher Education and Special Education*, *26*(2), 98–105. doi:10.1177/088840640302600203

Corbin, J., & Strauss, A. (1990). Grounded theory research: Procedures, cannons, and evaluative criteria. *Qualitative Sociology*, *13*, 3–31. doi:10.1007/BF00988593

Dennis, T., El-Gayar, O. F., & Zhou, Z. (2002). A conceptual framework for hybrid distance delivery for information system programs. *Issues in Information Systems*, *3*, 137–143.

Department of Education, Science and Training. (DEST). (2002). *Universities online: A survey of online education and services in Australia*. Canberra, Australia: Author.

Dziuban, C. D., Moskal, P. D., & Hartman, J. (2005). Higher education, blended learning, and the generations: Knowledge is power: No more. In Bourne, J., & Moore, J. C. (Eds.), *Elements of quality online education: Engaging communities*. Needham, MA: Sloan Center for Online Education.

Faux, T. L., & Black-Hughes, C. (2000). A comparison of using the Internet versus lectures to teach social work history. *Research on Social Work Practice*, *10*, 454–466.

Hagie, C., Hughes, M., & Smith, S. J. (2005). The positive and challenging aspects of learning online and in traditional face-to-face classrooms: A student perspective. *Journal of Special Education Technology*, *20*(2), 52–59.

Hartshorne, R., Heafner, T., & Petty, T. (2011). Examining the effectiveness of the remote observation of graduate interns. *Journal of Technology and Teacher Education*, *19*(4), 395–422.

Heward, W. (2006). *Exceptional children: An introduction to special education* (8th ed.). Upper Saddle River, NJ: Prentice Hall.

Huberman, A. M., & Miles, M. B. (2002). *The qualitative researcher's companion*. Thousand Oaks, CA: Sage Publications.

Kim, K. J., & Bonk, C. J. (2006). The future of online learning: The survey says.... *EDUCAUSE Quarterly*, *4*, 22–30.

Korir Bore, J. C. (2008). Perceptions of graduate students on the use of web-based instruction in special education personnel preparation. *Teacher Education and Special Education*, *31*, 1–11. doi:10.1177/088840640803100101

LeCompte, M., & Schensul, J. (Eds.). (1999). *The ethnographers' toolkit*. Walnut Creek, CA: AltaMira Press.

Ludlow, B. L. (2006). Overview of online instruction. In Ludlow, B., Collins, B. C., & Menlove, R. (Eds.), *Online instruction for distance education: Preparing special educators in and for rural areas. Special monograph of the American council on rural special education* (pp. 2–42). Victoria, BC: Trafford Publishing.

McGreal, R., & Elliott, M. (2004). Technologies of online learning. In Anderson, T., & Elloumi, F. (Eds.), *Theory and practice of online learning* (pp. 115–135). Athabasca, Canada: Athabasca University.

McNamara, J. M., Swalm, R. L., Stearne, D. J., & Covassin, T. M. (2008). Online weight training. *Journal of Strength and Conditioning Research*, *22*, 1164–1168. doi:10.1519/JSC.0b013e31816eb4e0

Petty, T., Heafner, T., & Hartshorne, R. (2009, March). Examining a pilot program for the remote observation of graduate interns. In R. Weber, K. McFerrin, R. Carlsen, & D. A. Willis, (Eds.), *2009 Society for Information Technology and Teacher Education Annual: Proceedings of SITE 2009* (pp. 2658-2660). Norfolk, VA: Association for the Advancement of Computing in Education (AACE).

Riffel, S., & Sibley, D. (2005). Using web-based instruction to improve large undergraduate biology courses: An evaluation of a hybrid course format. *Computers & Education*, *44*, 217–235. doi:10.1016/j.compedu.2004.01.005

Scheetz, N. A., & Gunter, P. L. (2004). Online versus traditional classroom delivery of a course in manual communication. *Exceptional Children*, *71*, 109.

Scoville, S. A., & Buskirk, T. D. (2007). Traditional and virtual microscopy compared experimentally in a classroom setting. *Clinical Anatomy (New York, N.Y.)*, *20*, 565–570. doi:10.1002/ca.20440

Spooner, F., Jordan, L., Algozzine, B., & Spooner, M. (1999). Evaluating instruction in distance learning classes. *The Journal of Educational Research*, *92*, 132–140. doi:10.1080/00220679909597588

Steinweg, S. B., Davis, M. L., & Thomson, W. S. (2005). A comparison of traditional and online instruction in an introduction to special education course. *Teacher Education and Special Education*, *28*, 62–73. doi:10.1177/088840640502800107

Tallent-Runnels, M. K., Thomas, J. A., Lan, W. Y., Cooper, S., Ahern, T. C., Shaw, S. M., & Liu, X. (2006). Teaching courses online: A review of the research. *Review of Educational Research*, *76*, 93–135. doi:10.3102/00346543076001093

Ward, B. (2004). The best of both worlds: A hybrid statistics course. *Journal of Statistics Education*, *12*(3), 74–79.

ADDITIONAL READING

Allen, E. I., & Seaman, J. (2007). *Online nation: Five years of growth in online education,* 2007. Needham, MA: Sloan Consortium. Retrieved from http://www.sloanconsortium.org/publications/survey/pdf/online_nation.pdf

Carlson, E., Lee, H., Schroll, K., & Pei, Y. (2004). Identifying attributes of high quality special education teachers. *Teacher Education and Special Education*, *27*, 350–339. doi:10.1177/088840640402700403

Caywood, K., & Duckett, J. (2003). Online vs. on-campus learning in teacher education. *Teacher Education and Special Education*, *26*(2), 98–105. doi:10.1177/088840640302600203

Charp, S. (2002). Online learning. *T.H.E. Journal*, *29*(8), 8–10.

Collins, B. C., Schuster, J. W., Ludlow, B. L., & Duff, M. (2002). Planning and delivery of online coursework in special education. *Teacher Education and Special Education*, *25*(2), 171–186. doi:10.1177/088840640202500209

Fitch, F. (2003). Inclusion, exclusion, and ideology: Special education students' changing sense of self. *The Urban Review*, *35*(3), 233–252. doi:10.1023/A:1025733719935

Glendinning, M. (2002). Beyond the digital fun factor. *Independent School*, *62*(1), 90–96.

Hartshorne, R., Heafner, T., & Petty, T. (2011). Examining the effectiveness of the remote observation of graduate interns. *Journal of Technology and Teacher Education*, *19*(4), 395–422.

Heward, W. (2006). *Exceptional children: An introduction to special education* (8th ed.). Upper Saddle River, NJ: Prentice Hall.

Hines, R. A., & Pearl, C. E. (2004). Increasing interaction in web-based instruction: Using synchronous chats and asynchronous discussions. *Rural Special Education Quarterly*, *23*(2), 33–36.

Jung, W. S. (2007). Preservice teacher training for successful inclusion. *Education*, *128*, 106–113.

LePage, P., Nielsen, S., & Fearn, E. F. (2008). Charting the dispositional knowledge of beginning teachers in special education. *Teacher Education and Special Education*, *31*, 77–92. doi:10.1177/088840640803100202

Ludlow, B. L. (2001). Technology and teacher education in special education: Disaster or deliverance? *Teacher Education and Special Education*, *24*(2), 143–163. doi:10.1177/088840640102400209

O'Brien, C., Aguinaga, N. J., Hines, R., & Hartshorne, R. (2011). Using contemporary technology tools to improve the effectiveness of teacher educators in special education. *Rural Special Education Quarterly*, *30*(3), 33–40.

O'Brien, C., & Beattie, J. R. (2011). *Teaching students with special needs: A guide for future educators*. Dubuque, IA: Kendall Hunt.

O'Neal, K., Jones, W. P., Miller, S. P., Campbell, P., & Pierce, T. (2007). Comparing web-based to traditional instruction for teaching special education content. *Teacher Education and Special Education*, *30*, 34–41. doi:10.1177/088840640703000104

Scheines, R., Leinhardt, G., Smith, J., & Cho, K. (2005). Replacing lecture with web-based course materials. *Journal of Educational Computing Research*, *32*(1), 1–25. doi:10.2190/F59B-382T-E785-E4J4

Spooner, F., & Lo, Y. (2009). Synchronous and asynchronous distance delivery experiences from two faculty members in special education at UNC Charlotte. *Rural Special Education Quarterly*, *28*(3), 23–29.

Steinweg, S. B., Davis, M. L., & Thomson, W. S. (2005). A comparison of traditional and online instruction in an introduction to special education course. *Teacher Education and Special Education*, *28*, 62–73. doi:10.1177/088840640502800107

Sun, L., Bender, W. N., & Fore, C. III. (2003). Web-based certification courses: The future of teacher preparation in special education? *Teacher Education and Special Education*, *26*, 87–97. doi:10.1177/088840640302600202

KEY TERMS AND DEFINITIONS

Asynchronous: Use of distance learning or web-based instruction in which learning activities are designed to be student-centered and at least somewhat self-paced. Instructors create learning modules or activities that can be completed without direct faculty interaction.

Dispositions: Term used in teacher education referring to a teacher candidate's readiness and internalized attitudes, values, and beliefs related to meeting student needs or holding high expectations for students in their classes.

Hybrid/Blended Course: Considered the "best of both worlds" by combining the most effective elements of both face-to-face meetings and online attributes. Typically, results in a 25-25% reduction in traditional meeting time.

Inclusion: "Special education philosophy that values social integration and access to general curriculum standards for children with disabilities by providing specialized services in regular classes" (O'Brien & Beattie, 2011, p. 23).

LMS: Learning management systems; a software application (such as Moodle or Blackboard) that enables online teaching and learning interaction used in web-based or hybrid/blended courses.

Special Education: A system in U.S. public schools premised on the Individuals with Disabilities Education Act (2004) that offers individualized education programs and specialized instructional supports for school-age children determined to exhibit special educational needs significantly outside the mainstream of the school population.

Synchronous: Use of distance learning or web-based instruction that often mirrors traditional instruction allowing instructors to provide lectures in real time and interact with students online.

Web-Based Instruction: For the purposes of this research the term reflects an orientation toward use of learning management systems like Moodle or Blackboard as the dominant mode of instruction using electronic learning modules, multimedia, and discussion forums.

Chapter 16
Examining Student Behaviors in and Perceptions of Traditional Field-Based and Virtual Models of Early Field Experiences

Hyo-Jeong So
National Institute of Education, Nanyang Technological University, Singapore

Emily Hixon
Purdue University Calumet, USA

ABSTRACT

The purpose of this study was to explore preservice teachers' behaviors in and perceptions of traditional field-based and virtual models of early field experiences. Specifically, this study examined some of the strengths and limitations associated with each model. Fifty undergraduate students participated in either a traditional field-based or a virtual field experience and completed an online questionnaire that examines various behaviors and student perspectives related to each model of early field experiences. The virtual field experiences include activities in the Inquiry Learning Forum (ILF), a web-based environment where students can observe and discuss diverse pedagogical practices and conceptual issues captured in a collection of video-based classrooms. The results of this study suggest that a virtual field experience which utilizes video-based cases may promote reflective practices which could be especially valuable to students early in their teacher education program. In addition, this study suggests that the strengths and limitations of each format need to be considered in relation to the goals and objectives of the early field experience, and discusses the possibility of a hybrid model of field experiences.

INTRODUCTION

A typical component of most teacher education programs are *early field experiences* where preservice teachers are placed in local schools to observe for a set amount of time over the course of a semester. Since there may be multiple classes that include field-based components and all preservice teachers must also be placed for student teaching, many universities and colleges find that the local schools are becoming overwhelmed with the number of preservice teachers in their buildings. To alleviate some of this burden, some universities have begun exploring alternatives to the traditional

DOI: 10.4018/978-1-4666-1906-7.ch016

format of early field experiences. One such option is a technology-enhanced or virtual field experience that utilizes various technological tools such as online discussion forums, video-based cases and virtual simulations.

Before universities adopt such a program, it is important to understand how the experience of the student completing the technology-enhanced field experience compares with that of the student completing the traditional field-based experience. This study is designed to explore students' experiences in and perceptions of the different field experience settings and attempt to understand some of the mechanisms underlying student views.

THEORETICAL BACKGROUND

The Role and Format of Technology-Enhanced Field Experiences

Field experiences have played a prominent role in teacher preparation. The importance of university-school collaborations to provide structured and beneficial experiences has been a focus for Schools of Education. Field experiences in teacher preparation programs can serve a variety of purposes and may take on many different formats. There has been much discussion regarding the purpose of field experiences. Aiken and Day (1999) identify objectives for field experiences including to "decide if teaching is an appropriate career choice, decide upon certification area; understand school and classroom differences; and better understand the process of educating students beyond the scope of a particular subject area or grade level" (p. 9). Another common goal in field experiences is to help students understand various theoretical concepts and issues learned in teacher training courses and link theoretical knowledge to practice (Frey, 2008; Moore, 2003). These are desirable goals that emphasize the importance of field experiences in teacher education.

Field experiences can take many different formats, especially with the application of various Web, video, and communication technologies. Drawing on the classification scheme of field experience formats by Paese (1996), Hixon and So (2009) present three types of technology-enhanced field experiences according to the degree of reality and virtuality: (a) Type I - concrete direct experience in reality, (b) Type II - vicarious indirect experience with reality, and (c) Type III – abstract experience with model of reality. Blended approaches are also possible by using different types of field experiences in conjunction with one another

Type I experiences refer to a field experience where student teachers are placed in real classrooms for observations. Concrete experiences take place at school sites where students observe an actual live classroom and/or actively participate in the instructional process (student teaching is the capstone concrete experience). In Type I, technologies are often used for flexible communication among university supervisors, school mentors and student teachers. Also technologies can be used to provide opportunities for sharing experiences and reflection. For instance, Wu and Kao (2008) discuss a web-based system that allows student teachers placed in different schools to view field-teaching sessions and share constructive feedback for improvement.

Type II experiences utilize various technologies for gaining vicarious experiences with real classrooms. Type II examples include student teachers observing real classrooms by video-conferencing or watching pre-recorded video cases, sometimes with the possibility of also being able to interact with the teacher remotely. In particular, Type II field experiences are useful for accessing teaching and learning situations that are not readily available to student teachers. For example, Lehman and Richardson (2003) report positive outcomes of virtual field experiences where student teachers were able to observe classrooms with diverse types of learners via a two-way video conferencing technology.

Type III experiences are totally virtual, meaning that field experiences happen in simulated environments. In Type III experiences, student teachers are interacting with artificial models of students and teachers in a virtual practicum. While there is not much research available regarding totally virtual field experiences, Type III experiences appear to be a promising format for "bridging apprenticeship experience allowing a practice space for those acquiring skills in designing and implementing a teacher work sample" (Girod & Girod, 2006, p. 483), especially within the early teacher education context. Further, virtual simulations can be a safe environment for student teachers to be focused on observing and learning certain teaching strategies without interruptions of irrelevant information and interaction (Payr, 2005).

In Table 1, we use the classification scheme of professional education teaching experiences proposed by Cruickshank and Armaline (1986) to highlight both the similarities and differences between the two models of field experiences examined in this study on the six characteristics (each format is described in more detail in the Methods section). It is helpful to understand the variety of field experiences and not be limited by a restricted definition.

Why Technology in Field Experiences?

As field experiences become more popular in teacher education programs, it is generally believed that students should be provided as many opportunities in actual classrooms as possible. Although much research has reinforced the value of field experiences as necessary components of teacher education programs, concerns and limitations have also been expressed related to the quality and impact of field experiences. Students sometimes view the field experience as an off-campus activity as opposed to on-the-job training, and believe that the field experiences do not provide *real teaching experiences* (Aiken & Day, 1999; Moore, 2003). Such beliefs indicate that students may not be gaining as much from these experiences as teacher educators had hoped.

Technology's role in field experiences has been examined as a way of possibly addressing these concerns related to what students gain from the traditional format of field experiences. Technology-enhanced/virtual field experiences may offer benefits related to the level and quality of student reflection and observation. Video-based cases are one of the most commonly used formats for technology-supported field experiences with several advantages over traditional field-based approaches for teacher learning (Cannings & Talley, 2002; Knight, Pedersen, & Peters, 2004; So, Lossman, Lim, & Jacobson, 2009).

Table 1. Classification of field experience types

Type of Field Experience	Traditional Field-Based	Technology-Enhanced/Virtual
Concreteness	Concrete	Vicarious
Context/Setting	Field-based	Campus-based/Online
Directness/Reality	Direct; with reality	Indirect; model of reality
Purpose	Professionals (vs. Craftsman)	Professionals (vs. Craftsman)
Duration	Part-time (vs. Full-time)	Part-time (vs. Full-time)
Sequence in Program	Pre-professional	Pre-professional

It has been reported that video-based cases can help students cognitively prepare for field experiences and train their eyes to be critical about classroom events. The use of video technologies, such as video-based cases and videoconferencing, in connection with field experiences may encourage students to actively process and reflect upon their experiences due to the representational power of video and the affordance of reviewing a scene as needed. Students observing in a live classroom obviously cannot ask the teacher and students to repeat the event that just took place. But this is exactly what students using video-based cases are able to do when they come to a particularly interesting, challenging, or meaningful scene. Santagata and colleagues (2007) found that after student teachers had participated in observation-based activities using a video-based program, they were able to significantly improve their ability to be more analytical and descriptive in the analysis of lesson events in terms of elaboration, math content, student learning, critical approaches, and alternative strategies. Similarly, Green (2009) compared traditional face-to-face observations to CD-ROM based and video-conferencing based observations. He found that while most students preferred traditional face-to-face observations over technology-mediated approaches, technology-supported forms of observations impacted higher levels of learning outcomes over the traditional format.

Another benefit of technology-enhanced/virtual field experiences is that they provide a common experience for students and the instructor to discuss. Having a common experience to reference is noted by faculty and student teachers as a key benefit of video-based cases and may result in more meaningful discussions and deeper reflection (Rhine & Bryant, 2007; Rosaen, Lundeberg, Cooper, Fritzen, & Terpstra, 2008). Being able to select and identify especially relevant and thought-provoking cases also provides some level of quality control that is not possible when students are placed in different classrooms for traditional field-based experiences. An instructor can select clips that address a particular concept or theory and provide students with structured activities that will help them make the relevant connections, thereby enhancing the connection between theory and practice. Many video-based cases also contain teacher commentary or other notes/materials provided by the classroom teacher. These materials can provide insights that will help students, especially those early in their education program, begin to see events from the teacher's perspective (Lambdin, Duffy, & Moore, 1996; Santagata et al., 2007). Although some students participating in live observations will have the opportunity to debrief a lesson with the classroom teacher, this may not be a common occurrence.

THE CURRENT STUDY

Research Purpose and Questions

To date, there has been much discussion on the benefit of technology in complementing, enhancing or even replacing a traditional field-based model of field experiences. However, it is important to understand that while technology may offer some advantages (as suggested above), it may also present new challenges. Hence, the purpose of this pilot study is to provide a balanced view by identifying both benefits and limitations of traditional field-based and virtual models. The contribution of this study is twofold: first, unlike many previous studies that focus on a single model of field experiences, this study examines both traditional field-based and virtual models collectively for better understanding of benefits and limitations that each model offers; second, this study has a particular interest in the role of technology within the context of *early* field experiences that are commonly associated with foundations and methods courses offered in an early stage of teacher education, prior to student teaching (e.g., final year teaching practicum).

While there is a handful of research studies on the role of various technologies for field experiences in general, not much attention has been given to the potential of technology within the context of early field experiences where the main emphasis is on helping students better understand theoretical and conceptual issues of teaching and learning and connect them to actual classroom situations. Therefore, beyond the general claim about the affordances of technology, specifically this study aims to examine what technology can offer to students who are early in their teacher training and how their early field experiences can be enhanced with the affordances of technology tools.

Toward this end, this study investigates whether preservice teachers recognize and take advantage of opportunities that each field experience format offers by exploring their perceptions of and experiences in different models. Special attention is also paid to students' reflective practices and perceptions of authenticity and satisfaction with each format. Research questions are as follows:

1. Do students recognize and take advantage of opportunities that each field experience format offers (e.g., interacting with teacher, observation of a variety of teaching styles and student characteristics, opportunities to see class concepts *in action*)?
2. How do students perceive the authenticity of the field experiences?
3. How do students completing each field experience format differ in terms of their reflective practices, satisfaction with and value gained from their field experience?
4. What strengths and weaknesses do students perceive for their specific field experience format?

METHOD

Participants

One hundred six preservice teachers from four sections of an Introduction to Educational Psychology course at a large Midwestern university in the US were invited to participate in this study. As shown in Table 2, two of these sections were traditional *live* offerings of the course, while the other two sections were web-based classes offered completely online. In terms of an early field experience component, students in one of the web-based sections (n=20) participated in the virtual field experience using video-based classrooms; students in the other web based class (n=22) and in the live sections (n=64) completed a traditional field-based format. Since the field experience format is the focus of this pilot study, all students participating in a field-based experience (whether their course was live or online) were grouped together.

Inquiry Learning Forum (ILF)

The core of the virtual field experience is the Inquiry Learning Forum (ILF), a web-based environment designed to "support teachers with diverse experiences and expertise coming together in a virtual space to observe, discuss, and reflect on pedagogical theory and practice anchored to

Table 2. Sections of introduction to educational psychology

	Sections 1 & 2	Section 3	Section 4
Course	Live	Web-based	Web-based
Format of field experience	Traditional Field-based	Traditional Field-based	Virtual
Number of students	64	22	20

actual teaching vignettes" (Barab, MaKinster, Moore, & Cunningham, 2001, pp. 72-73). The ILF (http://ilf.crlt.indiana.edu) has many components including a library of lesson plans and web resources, a collection of inquiry-based professional development labs, a collaboratory with a variety of public and private asynchronous discussion forums, and a collection of video-based classrooms. ILF members, who are primarily inservice and preservice teachers, use the various resources in the ILF environment to create and maintain a community of professionals interested in inquiry-based pedagogy.

The video-based classrooms are the heart of the Inquiry Learning Forum, and are the foundation for the virtual field experience. The video vignettes are intended to capture the "everyday practices" of teachers "who have a variety of strengths and weaknesses" (Barab et al., 2001, p. 78). Each video classroom shows a lesson which

Figure 1. The ILF video-based classroom page

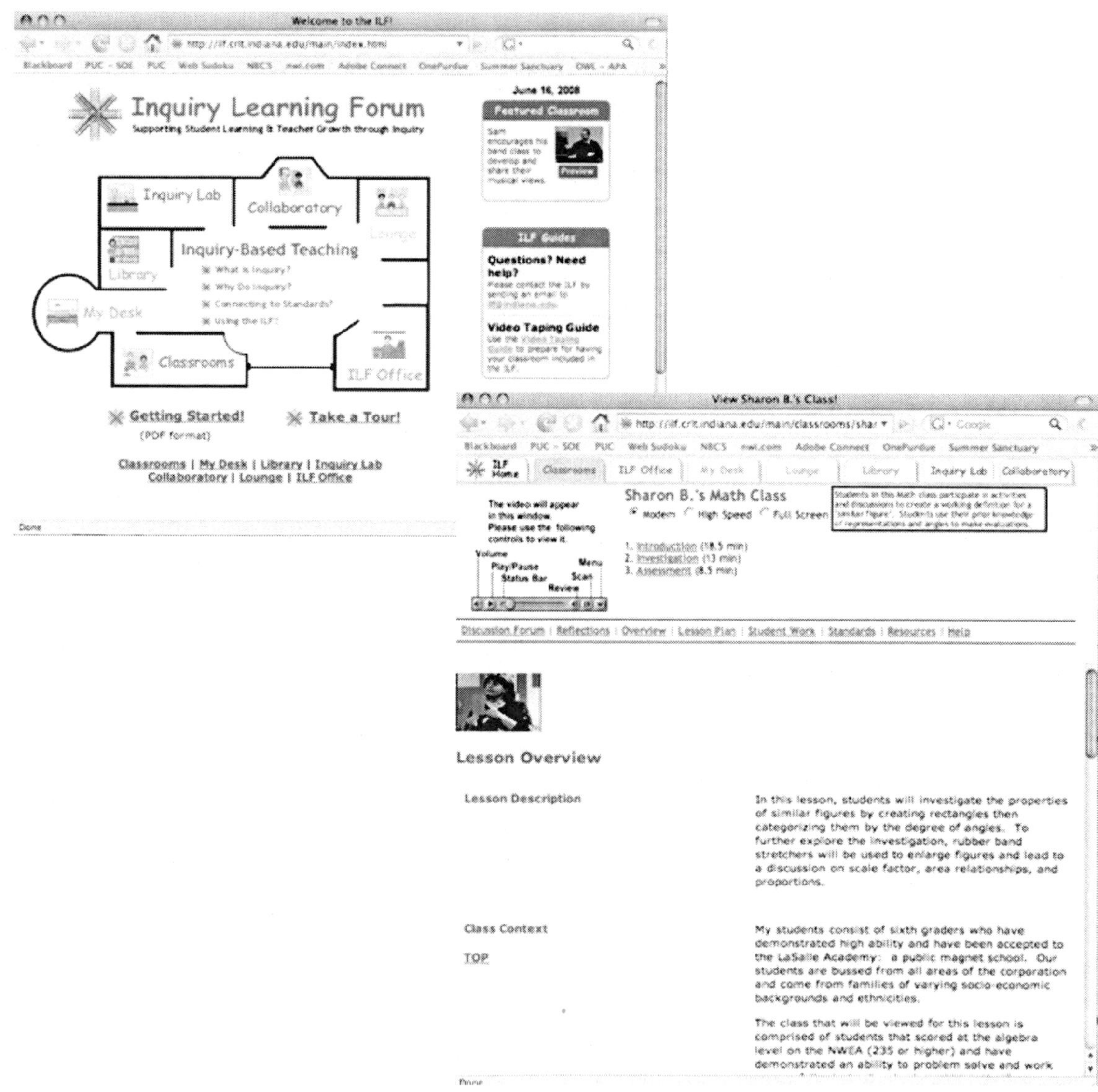

has been edited and broken into several 3-10 minute video clips. Each classroom also has several additional resources supplied by the participating teacher including: a lesson overview (description, class context, lesson context, lesson goals, assessment, explanation of how lesson is inquiry-based/student-centered, description of what happened the next day, and brief descriptions of each video clip), a detailed lesson plan, the teacher's reflections for each video clip, examples of student work, relation to relevant state and national standards, and links to accompanying resources.

Field Experience Activities: Traditional Field-Based vs. Virtual

In this study, a traditional field-based experience involves preservice teachers being placed in real classrooms where they observe and interact with real students and teachers. The virtual field experience refers to a format where preservice teachers gain vicarious experiences by observing teaching and learning in real classrooms through video-based cases and supplemental materials. Below, we describe the detailed activities that each group of students engaged in during their early field experiences.

Traditional Group

Students completing the traditional field-based model were placed in a local classroom where they were expected to observe for at least 20 hours over the course of the semester. Over a six- to eight-week period, students visited their assigned classroom one day a week for approximately 2-4 hours each visit. While at their field placement, students reportedly spent most of their time observing, but also frequently engaged in other activities including: one-on-one tutoring, assistance with instructional planning, teacher support, and supervision of students. Although it is not common, some students do also teach a lesson and/or attend staff meetings. Beyond the required number of observation hours, individual instructors have autonomy over the assignments and activities related to the field experience. For the two live sections of the course where the class meets face-to-face, the traditional field experience included a weekly one-hour meeting dedicated to discussion and activities related to the field experience. Both instructors teaching the face-to-face sections of the course used the one-hour lab meeting in a similar fashion where they used every other weekly meeting to process and debrief the field experience. These debriefing sessions typically took the form of large group discussions often focused on a specific topic that was currently being addressed in the class (e.g., classroom management, constructivism, motivation, etc.).

Each instructor provided her students with specific observation questions to focus on during each field session. The questions focused on concepts and issues that were being addressed in class at the same time. Students were required to write reflection papers about each visit based on these questions. One of the instructors also utilized written case studies that were discussed on alternating weeks (when field experience processing was not occurring) as a way of helping students reflect on and process what occurs in a classroom.

For the web-based section of the course that completed a traditional field-based experience (where they observed for 20 hours at a local school), all discussion/activities related to the field experience took place via online discussions. The instructor often provided students with focus questions for the field observations and required students to post and discuss responses to those questions in an online discussion forum. Students were also required to submit two field experience summaries where students reflected on their field experience as it related to various concepts studied in the course.

Virtual Group

Students completing the technology-enhanced field experience were enrolled in a web-based section of the course and used the video-based classrooms in the Inquiry Learning Forum (ILF) as the foundation for their field experience. Throughout the semester, students were required to view several video cases from the ILF collection and to complete a variety of activities related to the cases they viewed and the concepts they were learning in the course. Students participated in three different types of activities:

- **Video Case Reviews:** Students viewed a video case and read all accompanying materials. They then wrote a detailed and structured analysis including the following sections: Initial Thoughts, Clip-by-Clip Analysis, Comments on Student Work, Overall Evaluation and Alternative Approaches. Students were encouraged to link the concepts they were learning to cases they viewed. Students completed three such reviews over the course of the semester.
- **Multi-Case Activities:** These structured activities required students to *review portions of several different video cases and make comparisons and contrasts.* Each activity focused on an educational concept the students were learning about at the time. Students completed two of these activities during the semester.
- **Design Your Own Multi-Case Activity:** After having completed two multi-case activities provided by the instructor, students were required to design their own multi-case activity using clips from at least 3 different video cases. Students were instructed that their activity must focus around an educational concept they learned about in class.

In addition to these structured activities, students were assigned a constructive friend (a fellow classmate planning to teach a similar grade level and/or subject area) who gave feedback and engaged in virtual discussion related to the above assignments. Each constructive friend pair was required to comment and provide constructive feedback to each other on the above assignments. Students participating in the virtual field experience were also required to keep a journal detailing the time spent on video-related activities, what they did during that time, and their thoughts and reflections related to the activities.

DATA COLLECTION AND ANALYSIS

Procedure

A list of student email addresses was obtained from course instructors. Students received an email message inviting them to participate in the study and providing them with a link to the content form and online questionnaire (see below). Two reminder emails were sent to all invited participants during the following week.

Questionnaire

To address the specific research questions, the researchers developed a questionnaire, including (a) basic demographic questions, (b) reports on various behaviors related to field experiences (e.g., hours spent observing and reflecting on field experiences, interacting with teachers, etc.), and (c) Likert-scale questions about students' perceptions of and experiences in their field experience. The overall Cronbach Alpha coefficient of the questionnaire is quite high ($\alpha = 0.92$).

The majority of questions were close-ended, asking students to indicate their agreement with each statement on a 5-point Likert-type scale (1- Strongly Disagree to 5- Strongly Agree). Questions related to the opportunities available in each type of field experience (Research Question #1) addressed the following topics: reflection time, re-watching of video clips, discussion with classmates, observation of various student and teacher characteristics, and links to class content. A series of questions related to the authenticity of the field experience (Research Question #2) asked about the field experience's impact on students' understanding of teaching, their decision to teach, their view of the teacher as a mentor, their discussions with the classroom teacher, and their involvement in the classroom. Students' satisfaction with the field experiences (Research Question #3) was addressed by questions asking about what they learned from the teacher, whether it added value to the course, if they enjoyed the experience, and whether they got ideas for their future classroom. The final questions on each questionnaire were open-ended, allowing students to comment about the perceived strengths and limitations of their field experience format (Research Question #4). The objective items on the questionnaire were analyzed with ANOVAs as well as descriptive statistics. The two open-ended questionnaire items were analyzed by the researchers to identify emerging themes.

RESULTS

Among 106 students who took the course, 50 students completed the online questionnaire: 42 of the 86 students in the traditional group and 8 of the 20 students in the virtual group. The total response rate is 47%. Participants who completed a traditional field-based experience included 28 women and 14 men, and their mean age was 21.14 years. Participants in the virtual field experience group included 3 women and 5 men, and their mean age was 21.38 years. Most students who completed the questionnaires were sophomores: 83.3% in the traditional group and 62.5% in the virtual group.

Behavioral Patterns

Behavioral data were collected to examine how students across the different groups engaged in various activities related to early field experiences, including observations, discussions, and reflections (Research Question #1). First, students were asked to report their actual and expected observation hours[1]. At the time of questionnaire administration, students in the traditional group reported that they had completed an average of 15.27 hours of observation in the classroom and expected to spend a total of 21.26 hours by the end of the semester. Students completing the virtual field experience reported spending 12.69 hours watching the video cases at the time of questionnaire administration and expected to spend a total of 15.81 hours by the end of the semester. Expected hours for observations were statistically significant, $F(1, 45) = 21.14, p < .001, \eta 2 = .32$ with students completing the traditional field-based experience expecting to spend more time observing in the classroom ($M = 21.26$, $SD = 3.06$) than the virtual field experience students expected to spend watching the video cases ($M = 15.81$, $SD = 5.25$).

One advantage of using video-based cases for field experiences is that students can review specific portions of clips that are of interest for focused observations. All students in the virtual group reported that they did re-watch video clips or portions of video clips when completing the video-based assignments; six students reported re-watching clips occasionally, two reported doing so frequently. Another advantage of the virtual option is an increased flexibility in how time is spent, possibly allowing more time for reflection. To examine this, students were also asked to indicate how much time they spent reflecting on their observations, including completing assigned activities and discussing observations with classmates. The scale used asked students to indicate how much time they spent reflecting on each hour of observation: 1 = no time; 2 = less than .5

hours; 3 = .5 to 1 hour; 4 = 1 to 1.5 hours; 5 = 1.5 to 2 hours; 6 = 2 to 2.5 hours; 7 = 2.5 – 3 hours. Students completing the virtual field experience reported spending more time reflecting on their observations (M = 4.50, SD = 1.85) than students completing the field-based experience (M = 3.26, SD = 1.25), F(1, 45) = 5.87, p = .019, $\eta 2$ = .11. Students completing the virtual field experience on average reported spending 1 – 2 hours reflecting on each hour of observation, whereas those completing the traditional field-based experience generally spent less than 1 hour reflecting on each hour of classroom observation.

Engaging in meaningful discussion with classroom teachers is another opportunity of field experiences. Students were asked to rate the frequency for discussions with teachers on the scale (1) Never, (2) A couple of times, (3) Occasionally, or (4) Frequently. Students participating in the traditional field-based experience reported engaging in more meaningful discussions with the classroom teacher(s) (M = 3.08, SD = .89) than did those completing the virtual field experience (M = 1.88, SD = 1.24), F(1, 46) = 10.60, p = .002, $\eta 2$ = .19. Perhaps not surprisingly, while the majority of the virtual field experience respondents (62%) reported that they never engaged in discussion with the classroom teacher, only one traditional field experience student (2%) reported not having engaged in any such discussions. Meaningful discussion with classmates was also examined as participants in this study engaged in some forms of discussions of their field experiences with classmates through face-to-face or virtual discourses; however, there were no significant differences across the two groups.

Student Perceptions

Regarding overall satisfaction with each field experience format, students completing the traditional field-based experience rated their time spent observing classroom(s) as significantly more enjoyable (M = 4.20, SD = 1.11) than those completing the virtual field experience (M = 3.12, SD = 1.13), F(1, 46) = 6.19, p = .017, $\eta 2$ = .12. Interestingly, 75% of respondents in the traditional field-based group disagreed (over 50% strongly disagreed) that they would have preferred to complete the field experience in the virtual format. Similar patterns emerged when asked if they felt they would have learned more in a virtual field experience.

Open-Ended Responses

Regarding the last research question, students were also asked to respond to open-ended questions regarding what they perceived to be the strengths and limitations of their field experience format. The trends that emerged for each field experience format are identified in Table 3 and discussed below with representative participant responses included.

Strengths and Limitations of Field-Based Experiences

Of the 42 respondents in the traditional group, 35 of them completed the open-ended item asking about the strengths of the traditional field-based model. Based on their responses, a number of perceived strengths were identified: (a) one-on-one interaction with students and teachers, (b) getting a feel for what a classroom is really like, (c) hands-on experience, (d) observing teaching and management strategies, (e) influencing teaching as a future career, and (f) exposure to student diversity. Representative responses regarding the strengths of traditional field-based experiences with the number of similar responses are shown below (f indicates a frequency of each item mentioned in the open-ended question):

- One-on-one interaction with students and teachers (f = 21): "I love that fact that I get to personally interact with a real child." "It

Table 3. Summary of strengths and limitations as identified by respondents

	Strengths	Limitations
Traditional Field-based	• one-on-one interaction with students and teachers • getting a feel for what a classroom is really like • hands-on experience • observing teaching and management strategies • influencing teaching as a future career • exposure to student diversity	• meeting 20-hour requirement • distance and transportation • problems related to the structure of field experience • observing only one classroom and one teacher • limited role in experience • placed in classroom unrelated to subject area • no diversity related to students and schools • placed with poor classroom teacher
Technology-Enhanced/ Virtual	• flexibility in scheduling • no need to drive • quality of video clips	• lack of classroom feel • lack of interaction with teacher and students • limited scope of observation

is nice to actually be there and be able to interact with the teacher."

- Getting a feel for what a classroom is really like (f = 18): "You get a real feel for what could be your classroom in a couple years." "It is like the old saying a picture is like a thousand words, well an actual classroom is the same way. A future teacher cannot get the whole picture without actually being there."
- Hands-on experience (f = 7): "I feel that the strengths of observing in local schools are that I was able to get a firsthand experience of the classroom atmosphere through observing."
- Observing teaching and management strategies (f = 5): "...learning how classroom management works..." "Seeing everything, good and bad and how it is dealt with."
- Influencing teaching as a future career (f = 4): "It reinforced my ambitions to be a teacher." "This experience has given me a positive outlook on teaching."
- Exposure to student diversity (f = 3): "It opened my eyes to different socio-economic backgrounds" "...it gives us the opportunity to realize what children's learning styles are at that age..."

Thirty-one of the 42 respondents in the traditional group responded to the question asking about perceived limitations of the traditional field-based model. Some of the limitations identified include: (a) meeting the 20-hour requirement, (b) distance and transportation, (c) problems related to the structure of field experience, (d) observing only one classroom and one teacher, (e) limited role in experience, (f) placed in classroom unrelated to subject area, (g) no diversity related to students and schools, and (h) placed with poor classroom teacher. Interestingly, 8 students (19% of respondents) reported that there was no limitation in the traditional field experience. Below are examples of student responses related to the limitations of the traditional field-based format with the number of similar responses indicated:

- Meeting the 20-hour requirement (f = 7): "I somewhat feel that twenty hours out in the field is pointless." "Make a minimum number of visits with a shorter time span (say 6 visits for 2 hours)."
- Distance and transportation (f = 6): "A limitation for me personally is trying to have an effective form of transportation that runs smoothly with the time schedules."
- Problems related to the structure of field experience (f = 4): "When you only meet once a week for a short period you only are able to observe a limited amount."

- Observing only one classroom and one teacher (f = 4): "Sitting in the same classroom with the same teacher giving the same lesson plan three times in a row is not helpful."
- Limited role in experience (f = 4): "I felt more like hired help than someone there that wanted to learn more about teaching."
- Placed in classroom unrelated to subject area (f = 3): "The main limitations I saw are the limitations of finding a placement in a specific area of the student's study. I was placed in a Physics class. I will not be teaching Physics."
- No diversity related to students and schools (f = 3): "On the other hand I don't get to observe classrooms all over the US with different cultures."
- Placed with poor classroom teacher (f = 2): "I think that the only limitation would be a teacher that was not using this opportunity to best benefit the student who is observing."

Strengths and Limitations of the Virtual Field Experience

Of the 8 respondents who completed the technology-enhanced field experience, all of them completed the open-ended question asking about the strengths of the video-based virtual field experience. Strengths of the virtual field experience identified by students include: (a) flexibility in scheduling, (b) no need to drive, and (c) quality of video clips. Below are representative responses for each strength with the frequency of similar responses indicated:

- Flexibility in scheduling (f = 4): "They allow you to complete assignments and observation on your own schedule."
- No need to drive (f = 2): "Students do not have to drive to a surrounding school."
- Quality of video clips (f = 2): "Another strength is that we are provided with the teacher reflections, flexible observation clips, and student works. These items are rarely, if ever, available in a classroom observation."

All eight students in the virtual group completed the question about perceived limitations of the virtual field experience. Among the limitations identified were: (a) lack of classroom feel, (b) lack of interaction with teacher and students, and (c) limited scope of observation. Representative responses and frequencies of similar responses are as follows:

- Lack of classroom feel (f = 3): "You don't get to sit in on an actual classroom. In fact I think that if there is one class that shouldn't be web based it is this one."
- Lack of interaction with teacher and students (f = 2): "To actually see a child and how they react to a teacher is much better than just 'trying' to see him/her on a computer screen."
- Limited scope of observation (f = 2): "You don't get to see all aspects all the time. You really only get one viewpoint and its portrayed as the teacher wants you to see it."

DISCUSSION

Traditional Field-Based and Virtual Field Experiences

The purpose of this study was not to determine which field experience format (traditional field-based or virtual) was more effective than the other. Each format has strengths and weaknesses that must be considered when alternative field experience formats are being considered. As is well-documented in the literature, field experiences can serve a number of purposes ranging from exploring teaching as a career option to linking theory with practice (e.g., Aiken & Day, 1999; Hopkins, 1995). The overall goals of each specific field experience should be the cornerstone of all field experience format decisions. The virtual field experience option may be more effective in meeting some field experience objectives than others, and the same is true for the traditional field-based model.

It is interesting to note that some of the most prevalent strengths mentioned by those completing the traditional field-based experiences are the most commonly mentioned weaknesses of the virtual field experience and vice versa. The three most frequently listed strengths of the traditional field-based experience are the opportunity for one-on-one interactions with students and the teacher, the ability to *get a feel* for the classroom and what teaching is really like, and having the opportunity to gain hands-on experience with teaching. Students completing the traditional field-based experience emphasized the importance of being physically present in a classroom and being part of that environment. Virtual field experience students also recognize the value of being in the classroom which is demonstrated by what they identified as the two most prevalent limitations of the virtual field experience –inability to get a feel for the classroom and lack of interactions with the students and teacher.

Similarly, the top strengths of the virtual field experience are the most commonly mentioned weaknesses of the field-based experience. The virtual field experience students appreciated the fact that they were not tied to a schedule allowing them to complete observations whenever it was convenient, and that they did not have to travel to get to their observations. Similar topics were addressed by the traditional field-based students who reported having difficulty meeting the 20-hour observation requirement and traveling to their field placement. Especially as students are being placed at increasingly distant locations, the logistics of getting to that location to complete 20 hours of observation can be problematic.

One of the commonly found limitations of the traditional field-based format is that students often do not spend adequate time processing and reflecting on their observations, thereby not gaining as much as they perhaps could (or should) from the experience (Feiman-Nemser & Buchmann, 1985; Goodman, 1986; Johnston, 1994). The findings of this study highlight this issue and provide some insights into addressing this concern in the future. The virtual field experience students reported spending at least as much time reflecting on the video-based cases as they did watching them, which is contrasted with the students in the traditional field-based group who spent less time reflecting than they did observing. It appears that students in the two field experience formats demonstrate different patterns of activity. While the virtual field experience students spend more time reflecting and less time watching the video cases, the traditional field-based students spend more time observing in the classroom and less time reflecting on those observations. There are many possible explanations for this observed difference in reflective behavior. For example, it is possible that reflection is more integrated into live observations making it difficult for the traditional field experience students to separate out reflective behavior from their observation time. It is also possible that viewing and processing the video-based cases required more reflective thought from the virtual field experience students because they lacked familiarity with the setting. Similarly, it is possible that some characteristics of the video-based cases that were not present in the traditional field placements (e.g., ability to re-watch clips, supplementary materials) provided stimuli which promoted further reflection. The type of activities required for each field experience format may also have impacted the time students were prompted to engage in reflective thought. It is obviously important for preservice teachers to understand the value of reflection and to be encouraged to become reflective practitioners (Schon, 1983); a successful field experience needs to promote such practice (e.g., Frieberg, 1995; Norlander-Case, Reagan, & Case, 1999; Posner, 2005). Identifying the mechanisms underlying the differences in reflective thought between the groups is critical to creating a field experience (of any format) that supports the development of reflective practitioners. This topic is discussed further in the following paragraphs.

As mentioned above, a key quality of the video-based cases is that students can re-watch all or part of the video case. All students participating in the virtual field experience reported that they did indeed take advantage of this feature and re-watched clips as they completed their reflective assignments. This ability to rewind and watch a specific event or interaction again, which is obviously not possible in a live classroom, may help overcome students' *selective memories* and develop more *trained eyes*. It is possible that being able to review video clips at various stages in the thought process promotes more and deeper reflection.

Similarly, the fact that students in the virtual group had observed the same classroom events (as depicted in the video cases) may have enhanced their discussion with classmates and promoted deeper reflection. Knight and colleagues (2004) report that students participating in a virtual classroom observation recognize the value of having a shared experience which allows them to "check their perceptions of the class with other students and the instructor" (Knight et al., 2004, p. 146). The structured activities students in the virtual group were asked to complete in relation to specific commonly viewed clips may give students early in their program the cognitive scaffolding they need to make connections between their observations and the course content, and thereby benefit more fully from their observations.

An important feature of field-based experiences is that students have the opportunity to engage in meaningful discussion with the classroom teacher. While all but one of the traditional field-based students reportedly engaged in at least one meaningful discussion with the classroom teacher, the majority of virtual field experience respondents did not engage in any such discussions. This finding is not surprising given the minimal opportunity for such interactions in the virtual environment (through minimally active discussion forums). While there are discussion forums related to each video classroom open to all Inquiry Learning Forum members (including the classroom teachers in the videos), these discussions are not very active and it is possible (or likely) that the participating teacher may not see and/or respond to posted messages in a timely fashion (if at all). Providing extensive notes about each teacher's reflections on a given lesson may compensate for some of this lack of interaction, but other avenues for communication with the teacher should also be investigated.

Although students in the virtual group did not have the same opportunities to engage in discussions with the classroom teachers they observed, the video-based classroom environment in the Inquiry Learning Forum does include each teacher's reflections on the given lesson (clip-by-clip). The teacher's detailed commentary on each clip of the video case provides a model of the reflective thought process that students in a traditional field-based experience are unlikely to experience. Without having a video of a lesson to re-watch, it is unrealistic to expect classroom teachers to spontaneously reflect on each aspect of their lesson. Also, preservice teachers early in their education program would likely not know what to ask the classroom teacher to get the same level of relevant and meaningful reflection provided in the structured online commentary. So while students completing a virtual field experience may lack the personal connection with the classroom teacher, the extensive reflections provided by the teachers in the video-based cases may actually be more effective at promoting quality reflection in students.

Another key finding from this study is that students who participated in the virtual format did not enjoy their experience as much as those who completed the traditional field-based format. One of the field-based experience students said it best, "It is like the old saying a picture is like a thousand words, well an actual classroom is the same way. A future teacher cannot get the whole picture without actually being there." It is not entirely surprising that students may not find sitting in front of a computer screen as enjoyable as interacting with students and the teacher in an actual classroom. However, this issue is one that should be acknowledged by teacher educators and taken into consideration when discussing field experience formats. Future research should investigate what qualities of the virtual field experience may impact student enjoyment and satisfaction level, so efforts can be made to address those issues and make the experience more enjoyable.

Possibility of Hybrid Model

Each format has its strengths and limitations, and perhaps it is possible to create a *hybrid model* that builds on the strengths and minimizes the weaknesses of each format. Many of the virtual field experience students commented that they did not get a sense of what it was like to be a teacher in a classroom. As stated previously, there can be a variety of objectives for any given field experience, but if *getting a feel* for the classroom and interacting with students and teachers is the primary goal of the field experience, the virtual field experience may be limiting. Various multimedia components can be used to create a rich learning environment, however nothing can replace being in an actual classroom to see, hear, and smell what it is like to be a teacher. On the other hand, there are situations when the virtual field experience may be more promising or even the only option available for students. For instance, when students who need to complete an early field experience during the summer when schools are not in session, the virtual field experience may provide a more flexible option for students who have difficulty scheduling and getting to an off-campus field placement. Students who are enrolled in distance learning programs or online courses are also in need of a flexible format of field experiences (Simpson, 2006). Within such contexts of teacher education, the traditional and virtual formats can be blended instead of relying on a single mode of field experiences. A hybrid model can be structured to use video-based cases to help students cognitively prepare for real field experiences, and then place them in the traditional field-based experience to help transfer knowledge and skills learned in relevant courses and virtual environments to real classrooms.

Limitations and Suggestions for Future Research

Some limitations of this pilot study should be acknowledged. The first limitation of this study is the small number of students in the virtual group (n=8) who completed the questionnaire. While this represents 40% of the total population (n=20) which is an acceptable response rate for a survey, statistical analyses involving a small sample size must be interpreted with caution. Replicating this study with more students from the virtual field experience is necessary to corroborate and perhaps extend the current findings. It is also important to note that the field experiences were facilitated by different instructors, which allowed the researchers little control over the conditions of the field experiences. While efforts were made to document the conditions of and activities included in each field experience, variations among the instructors may have presented confounding variables which were not fully explored in this study. Future research should seek to standardize the field placement activities across formats as much as is possible.

It would also be beneficial to include an additional group of students who are completing a virtual field experience while taking the course face-to-face. That group was not included in this study, but including it may help understand the mechanisms underlying the current findings.

Future research also needs to investigate the quality of reflection done by students in various field experience formats and how that reflection relates to the overall learning and value gained through the field experience. In order to structure experiences to maximize reflection and learning, the specific qualities of the field experiences that impact the level and quality of student reflection and overall learning need to be identified. This pilot study was limited to self-report data, but a more focused study that involves content analysis of student reflections would undoubtedly yield interesting information that would extend the findings reported here.

CONCLUSION

As was stated previously, the purpose of this pilot study was not to identify which field experience was *better*, but rather to better understand the students' experiences in and perceptions of different types of field experiences. It is hoped that the information provided here will be of use to individuals who are considering a virtual field experience as an alternative to the traditional field-based experience. It has become clear that the overall goals and objectives for a specific field experience must be the focus when field experience options are being explored. Additionally, when comparing various formats, it is important to consider both the experience itself and the learning that occurs as a result. The current study addressed only the student experience, and further research is needed to investigate the learning related to each type of field experience.

REFERENCES

Aiken, I. P., & Day, B. D. (1999). Early field experiences in preservice teacher education: Research and student perspectives. *Action in Teacher Education*, *21*(3), 7–12. doi:10.1080/01626620.1999.10462965

Barab, S. A., MaKinster, J. G., Moore, J. A., & Cunningham, D. J. (2001). Designing and building an on line community: The struggle to support sociability in the Inquiry Learning Forum. *Educational Technology Research and Development*, *49*(4), 71–96. doi:10.1007/BF02504948

Cannings, T., & Talley, S. (2002). Multimedia and online video-based studies for preservice teacher preparation. *Education and Information Technologies*, *7*(4), 359–367. doi:10.1023/A:1020969723060

Cruickshank, D. R., & Armaline, W. D. (1986). Field experiences in teacher education: Considerations and recommendations. *Journal of Teacher Education*, *37*(3), 34–40. doi:10.1177/002248718603700307

Feiman-Nemser, S., & Buchmann, M. (1985). Pitfalls of experience in teacher preparation. *Teachers College Record*, *87*(1), 53–65.

Frey, T. (2008). Determining the impact of online practicum facilitation for inservice teachers. *Journal of Technology and Teacher Education*, *16*(2), 181–210.

Frieberg, H. J. (1995). Promoting reflective practices. In Slick, G. A. (Ed.), *Emerging trends in teacher preparation: The future of field experiences* (pp. 25–42). Thousand Oaks, CA: Corwin Press, Inc.

Girod, M., & Girod, G. (2006). Exploring the efficacy of the Cook school district simulation. *Journal of Teacher Education*, *57*(5), 481–497. doi:10.1177/0022487106293742

Goodman, J. (1986). Making early field experience meaningful: a critical approach. *Journal of Education for Teaching*, *12*(2), 109–125. doi:10.1080/0260747860120201

Greene, H. C. (2009). Multimedia observations: Examining the roles and learning outcomes of traditional, CD-ROM based, and videoconference observations in pre-service teacher education. *Current Issues in Education, 11*. Retrieved from http://cie.ed.asu.edu/volume11/number3/

Hixon, E., & So, H. J. (2009). Technology's role in field experiences for preservice teacher training. *Journal of Educational Technology & Society*, *12*(4), 294–304.

Hopkins, S. (1995). Using the past; guiding the future. In Slick, G. A. (Ed.), *Emerging trends in teacher preparation: The future of field experiences* (pp. 1–9). Thousand Oaks, CA: Corwin Press, Inc.

Johnston, S. (1994). Experience is the best teacher; or is it? An analysis of the role of experience in learning to teach. *Journal of Teacher Education*, *45*(3), 199–208. doi:10.1177/0022487194045003006

Knight, S. L., Pedersen, S., & Peters, W. (2004). Connecting the university with a professional development school: Pre-service teachers' attitudes toward the use of compressed video. *Journal of Technology and Teacher Education*, *12*(1), 139–154.

Lambdin, D. V., Duffy, T. M., & Moore, J. A. (1996). *A hypermedia system to aid in preservice teacher education: Instructional design and evaluation.* Paper presented at the National Convention of the Association for Educational Communications and Technology, Indianapolis, IN.

Lehman, J., & Richardson, J. (2003). Virtual field experiences: Helping pre-service teachers learn about diverse classrooms through video conferencing connections with K-12 classrooms. In D. Lassner & C. McNaught (Eds.), *Proceedings of World Conference on Educational Multimedia, Hypermedia and Telecommunications 2003*, Chesapeake, VA: AACE, 1727-1728.

Moore, R. (2003). Reexamining the field experiences of preservice teachers. *Journal of Teacher Education*, *54*(1), 31–42. doi:10.1177/0022487102238656

Norlander-Case, K. A., Reagan, T. G., & Case, C. W. (1999). *The professional teacher: The preparation and nurturance of the reflective practitioner*. San Francisco, CA: Jossey-Bass.

Paese, P. C. (1996). Contexts: Overview and framework. In McIntyre, J., & Byrd, D. M. (Eds.), *Preparing tomorrow's teachers: The field experience* (pp. 1–7). Thousand Oaks, CA: Corwin Press, Inc.

Payr, S. (2005). Not quite an editorial: Educational agents and (e-)learning. *Applied Artificial Intelligence*, *19*(3/4), 199–213. doi:10.1080/08839510590910147

Posner, G. J. (2005). *Field experience: A guide to reflective teaching* (6th ed.). White Plains, NY: Allyn and Bacon.

Rhine, S., & Bryant, J. (2007). Enhancing preservice teachers' reflective practice with digital video-based dialogue. *Reflective Practice*, *8*(3), 345–358. doi:10.1080/14623940701424884

Rosaen, C. L., Lundeberg, M., Cooper, M., Fritzen, A., & Terpstra, M. (2008). Noticing noticing: How does investigation of video-records change how teachers reflect on their experiences? *Journal of Teacher Education*, *59*(4), 347–360. doi:10.1177/0022487108322128

Santagata, R., Zannoni, C., & Stigler, J. W. (2007). The role of lesson analysis in pre-service teacher education: an empirical investigation of teacher learning from a virtual video-based field experience. *Journal of Mathematics Teacher Education*, *10*, 123–140. doi:10.1007/s10857-007-9029-9

Schon, D. A. (1983). *The reflective practitioner: How professionals think in action*. New York, NY: Basic Books.

Simpson, M. (2006). Field experience in distance delivered initial teacher education programmes. *Journal of Technology and Teacher Education*, *14*(2), 241–254.

So, H. J., Lossman, H., Lim, W. Y., & Jacobson, J. M. (2009). Designing an online video-based platform for teacher learning in Singapore. *Australasian Journal of Educational Technology*, *25*(3), 440–457.

Wu, C. C., & Kao, H. C. (2008). Streaming videos in peer assessment to support training pre-service teachers. *Journal of Educational Technology & Society*, *11*(1), 45–55.

ADDITIONAL READING

Armstrong, V., & Curran, S. (2006). Developing a collaborative model of research using digital video. *Computers & Education*, *46*, 336–347. doi:10.1016/j.compedu.2005.11.015

Borko, H., Jacobs, J., Eiteljorg, E., & Pittman, M. E. (2008). Video as a tool for fostering productive discussions in mathematics professional development. *Teaching and Teacher Education*, *24*(2), 417–436. doi:10.1016/j.tate.2006.11.012

Brophy, J. (2004). *Using video in teacher education* (*Vol. 10*). San Diego, CA: Elsevier Advances in Research on Teaching.

Chaney-Cullen, T., & Duffy, T. M. (1999). Strategic teaching framework: Multimedia to support teacher change. *Journal of the Learning Sciences*, *8*(1), 1–40. doi:10.1207/s15327809jls0801_1

Hatch, T., Sun, C., Grossman, P., Neira, P., & Chang, T. (2009). Learning from the practice of veteran and novice teachers: A digital exhibition. *Journal of Teacher Education*, *60*(1), 68–69. doi:10.1177/0022487108328683

Hauge, T. E., & Norenes, S. O. (2009). Changing teamwork practices: Videopaper as a mediating means for teacher professional development. *Technology, Pedagogy and Education*, *18*(3), 279–297. doi:10.1080/14759390903255551

Hennessy, S., & Deaney, R. (2009a). The impact of collaborative video analysis by practitioners and researchers upon pedagogical thinking and practice: A follow-up study. *Teachers and Teaching: Theory and Practice*, *15*(5), 617–638. doi:10.1080/13540600903139621

Hennessy, S., & Deaney, R. (2009b). "Intermediate theory" building: Integrating multiple teacher and researcher perspectives through in-depth video analysis of pedagogic strategies. *Teachers College Record*, *111*(7), 1753–1795.

Jones, L., & McNamara, O. (2004). The possibilities and constraints of multimedia as a basis for critical reflection. *Cambridge Journal of Education*, *34*(3), 279–296. doi:10.1080/03057640420000289929

Koc, Y., Peker, D., & Osmanoglu, A. (2009). Supporting teacher professional development through online video case study discussions: An assemblage of preservice and inservice teachers and the case teacher. *Teaching and Teacher Education*, *25*(8), 1158–1168. doi:10.1016/j.tate.2009.02.020

Lazarus, E., & Olivero, F. (2009). Videopapers as a tool for reflection on practice in initial teacher education. *Technology, Pedagogy and Education*, *18*(3), 255–267. doi:10.1080/14759390903255528

Le Fevre, D. M. (2004). Designing for teacher learning: Video-based curriculum design. In Brophy, J. (Ed.), *Using video in teacher education* (pp. 235–258). New York, NY: Elsevier Science.

Rich, P. J., & Hannafin, M. J. (2008). Decisions and reasons: Examining preservice teacher decision making through video self-analysis. *Journal of Computing in Higher Education*, *20*(1), 62–94. doi:10.1007/BF03033432

Santagata, R. (2009). Designing video-based professional development for mathematics teachers in low performing schools. *Journal of Teacher Education*, *60*(1), 38–51. doi:10.1177/0022487108328485

Seago, N., Mumme, J., & Branca, N. (2004). *Learning and teaching linear functions: Video cases for mathematics professional development* (*Vol. 6-10*). Portmouth.

Seidel, T., Stürmer, K., Blomberg, G., Kobarg, M., & Schwindt, K. (2010). Teacher learning from analysis of videotaped classroom situations: Does it make a difference whether teachers observe their own teaching or that of others? *Teaching and Teacher Education*, *27*(2), 259–267. doi:10.1016/j.tate.2010.08.009

Sherin, M. G., & Han, S. Y. (2004). Teacher learning in the context of a video club. *Teaching and Teacher Education*, *20*(2), 163–183. doi:10.1016/j.tate.2003.08.001

Sherin, M. G., Linsenmeier, K. A., & van Es, E. A. (2009). Selecting video clips to promote mathematics teachers' discussion of student thinking. *Journal of Teacher Education*, *60*(3), 213–230. doi:10.1177/0022487109336967

van Es, E. A., & Sherin, M. G. (2008). Mathematics teachers' "learning to notice" in the context of a video club. *Teaching and Teacher Education*, *24*(2), 244–276. doi:10.1016/j.tate.2006.11.005

Zhang, M., Lundeberg, M., Koehler, M. J., & Eberhardt, J. (2011). Understanding affordances and challenges of three types of video for teacher professional development. *Teaching and Teacher Education*, *27*(2), 454–262. doi:10.1016/j.tate.2010.09.015

KEY TERMS AND DEFINITIONS

Authenticity: A perception of the *realness* of an experience.

Early Field Experience: An intentionally created opportunity for preservice teachers who are in an early stage of their teacher education (e.g., foundational courses) to observe a classroom.

Inquiry Learning Forum (ILF): Web-based environment which contains a collection of video-based teaching vignettes, supplementary commentary and materials, and opportunities for communication.

Preservice Teachers / Student Teachers: Higher education students who are taking coursework leading to their certification as elementary or secondary teachers.

Technology-Enhanced/Virtual Field Experience: A set of activities that utilize technology to allow preservice teachers to observe and reflect on video recordings of classroom lessons.

Traditional Field-Based Field Experience: An intentionally created opportunity for preservice teachers to observe and/or participate in a real classroom in real time.

ENDNOTE

1. Due to time constraints, the questionnaire was administered prior to the end of the semester, so students were asked to report the time they had already spent observing as well as they total amount of time they expected to spend observing by the conclusion of the semester.

Section 6
Innovative Online Teaching and Learning Practices in Teacher Education

Chapter 17
Preparing Pre-Service Secondary English Language Arts Teachers to Support Literacy Learning with Interactive Online Technologies

Luke Rodesiler
University of Florida, USA

Barbara G. Pace
University of Florida, USA

ABSTRACT

In this chapter, the authors present the framework and methods they employ to integrate online learning opportunities into an English teacher education program at a large, public university in the southeastern United States. The authors focus on their efforts to extend pre-service secondary English language arts teachers' understandings of what constitutes literacy and what counts as text in the secondary English language arts classroom in a blended technology- and media literacy-focused methods course, a required component of a three-semester English Education Master's degree program. Specifically, the authors document the ways they nudge pre-service teachers to consider the kinds of literacy events they might design and the types of literacy practices they might promote to support literacy learning with interactive online technologies and popular media in English language arts classrooms.

INTRODUCTION

This chapter highlights our efforts to prompt pre-service secondary English language arts teachers to consider the possibilities for advancing literacy teaching and learning via interactive online technologies in a blended technology- and media literacy-focused methods course. Such work requires encouraging pre-service teachers to expand their understandings of literacy and of texts. Thus, we begin this chapter by establishing our alignment with a sociocultural view of literacy and learning, by describing constructs that help teachers fulfill their roles as designers of classroom events, and

DOI: 10.4018/978-1-4666-1906-7.ch017

by highlighting understandings of literacy made possible by communication technologies and the ubiquity of multimodal texts. The second half of this chapter describes the context of our work, details the integration of interactive online technologies that support our goals, and considers future directions for investigating the integration of online learning opportunities in teacher education programs. Ultimately, by presenting the framework and methods we employ to support the integration of online learning opportunities in English teacher education, we hope to inform teacher educators who share our goal of developing paths toward meaningful technology integration in teacher preparation programs.

BACKGROUND

In its policy brief on 21st-century literacies, the National Council of Teachers of English (2007) acknowledges that evolving technologies, while informing the literacy practices of today's students, provide teachers with opportunities for promoting literacy learning in diverse, participatory contexts (p. 2). Recognizing those opportunities, we have worked to integrate online technologies in an English teacher education program, an effort to prepare pre-service English teachers for designing technology- and media-rich literacy events that may shape the literacy practices of the students they will serve.

This discussion of how we have integrated online technologies in English teacher education is framed by our perspectives on literacy and by how those views have informed our practice. Our goals and the literacy work we assign to pre-service teachers are framed by sociocultural views of literacy. That is, we understand that literacy is part of social experiences and that ways of practicing literacy are variable and evolve as individuals participate in literacy events in multiple settings with various types of texts. This framework provides for consideration of the literacy practices that youth use daily (Goldman, Booker, & McDermott, 2008; Rideout, Foehr, & Roberts, 2010) and for reflection on how those practices can be enhanced and built upon in academic settings.

A Sociocultural View of Literacy and Learning

We advance this view in the teacher education program by focusing pre-service secondary English language arts teachers on the social aspects of literacy and on the uses of literacy in context (Pahl & Rowsell, 2005). We accept that "[l]iteracy does not just reside in people's heads as a set of skills to be learned" but it "is essentially social, and it is located in the interaction between people" (Barton and Hamilton, 1998, p. 3). This broad view shifts us away from the view that literacy is a skill set employed by individuals without regard for social context (Kucer, 2005; Street, 1993/2001). This sociocultural perspective allows us, as Szwed (1981/2001) implores, to consider "the varieties of reading and writing available for choice; the contexts for their performance" (p. 422), and to recognize that the varying contexts in which literacy acts are performed may require different sets of skills. That is, a sociocultural perspective embraces Scribner and Cole's (1981/2001) position on literacy as a social practice, which posits that, beyond merely knowing how to read and write, literacy is the application of such knowledge in particular contexts and for particular purposes. This broad view of literacy aligns with that advanced by the National Council of Teachers of English (1996).

By promoting a sociocultural view of literacy, we aim to broaden pre-service secondary English language arts teachers' understandings of literacy and of texts. In the methods course described here, we encourage students to consider texts as multimodal and literacy as more than alphabetic decoding. We encourage them to acknowledge the literacies that adolescents bring to the classroom and to consider how those literacies can be advanced through the thoughtful design of classroom activities and the meaningful use of interactive online technologies.

Literacy Events and Literacy Practices

Contextualized understandings of literacy open the door for two constructs that have gained traction among New Literacy scholars (e.g., Barton & Hamilton, 1998; Gee, 1996; Heath, 1982/2001; Street, 1993/2001): *literacy events* and *literacy practices*. In the course we report on below, these constructs are used as heuristics that promote ways of thinking about and designing literacy learning activities that draw on interactive online technologies or that use standard print-based texts in a face-to-face setting.

A literacy event is "any occasion in which a piece of writing is integral to the nature of participants' interactions and their interpretive processes" (Heath, 1982/2001, p. 445). To account for multiple forms of texts, Morrell (2004) extends Heath's position and posits that any text, print or otherwise, may be central to a literacy event. Such a broadened notion of the texts central to a literacy event is significant given our advocacy of online technologies, multimodal texts, and popular media in the teaching of English.

As teacher educators, we have found that explicitly using the concept of literacy events with pre-service teachers helps to emphasize the contextual nature of literacy. In particular, it helps when prompting them to design literacy activities, as any consideration of literacy events calls for consideration of how, when, with what texts, and for what purposes students will engage in activities that involve literacy. Moreover, the notion of literacy events is valuable in our own thinking, for it helps us consider how we might design online learning experiences.

The concept of literacy practices, what Barton (2007) describes as "common patterns in using reading and writing in a particular situation" (p. 36), is also key to advancing a sociocultural view of literacy. As Barton suggests, individuals draw upon literacy practices while engaged in literacy events and no one set of practices is essential to all contexts or situations. As pre-service English teachers in the program design various literacy events, we urge them to consider the literacy practices the students they teach will draw upon and the ways in which classroom events may best embellish those practices. As pre-service teachers gain familiarity with the constructs of literacy events and literacy practices, they come to understand that literacy is not a fixed set of skills but, rather, that literacy may vary from one context to another. In English teacher education, ensuring such understandings is essential for moving toward meaningful technology integration and adopting interactive online technologies that may support literacy learning.

New Literacies

The expanded notion of literacy and texts that we promote has also, in part, been informed by the notion of *new literacies*. Advanced by the likes of the New London Group (1996), Cope and Kalantzis (2000), and Lankshear and Knoble (2003), new literacies are the literacies available as a result of technological advancements, economic shifts, and social changes that have altered the nature of texts and the activities carried out with those texts. Such changes are exemplified in Kress' (2003) contention that "[t]he screen more than the page is now the dominant site of representation and communication in general" (p. 65). Moreover, as Anstey and Bull (2006) point out, a "basic, print-dominated literacy toolkit" (p. 2) is not sufficient in today's world. Instead, as those who have studied new literacies contend, "students should be able to both read critically and write functionally, no matter what the medium" (Kist, 2005, p. 11). Such thinking shapes our efforts to prepare pre-service secondary English language arts teachers to teach in continually changing times.

Literacy and Technology in a Methods Course

New understandings of literacy and the evolving purposes of literacy education prompted us to consider how technologies might be infused in meaningful ways in a methods course for secondary English teachers. Like other English educators (e.g., Kajder, 2003; Pope & Golub, 2000; Swenson, Rozema, Young, McGrail, & Whitin, 2005; Young & Bush, 2004), we advocate for technology integration that embellishes subject-area content and underpins processes for building understandings in English language arts. Our thinking has been advanced not only by the aforementioned changes in the field but also by Mishra and Koehler's (2006) model of technology integration.

The TPACK model of technology integration (Mishra & Koehler, 2006) emphasizes the importance of discipline-specific concepts and epistemologies. It includes three core knowledge areas—Technological, Pedagogical, and Content Knowledge (TPACK)—and showcases the complex and dynamic relationship among them. Of particular importance is the role of Shulman's (1986) concept of pedagogical content knowledge in the model. According to Shulman, effective teachers have both a strong foundation in content-area knowledge and an understanding of pedagogy and learning. Shulman's work on how pedagogy and content knowledge comingle in the work of effective teachers was used to frame the secondary education programs at the university where we teach.

Developing an understanding of how technology might be integrated with content knowledge and with theories of literacy learning dovetail with goals that have guided the English education program since its inception. The TPACK model not only advances technology integration but also supports our efforts to conflate technology, pedagogy, and content area understandings. According to Mishra and Koehler (2006), teacher activity is informed by the epistemologies that frame an academic discipline and by the "theories of learning" that teachers hold (p. 1027). As English teacher educators, we agree that technology integration should be "structured for particular subject matter" (Koehler & Mishra, 2009, p. 62). We describe *meaningful technology integration* as instances in which technologies serve authentic purposes that advance subject-area knowledge and that extend the processes and practices that guide how understandings are generated in the academic discipline.

In English language arts, interactive online technologies can support adolescents in developing the sophisticated literacy practices needed to participate in literacy events and with multiple types of texts. Thus, interactive online technologies can support changes in how literacy is practiced—changes increasingly aligned with learning (Bloome, Carter, Christian, Otto, & Shuart-Faris, 2005; Rogoff, 2003). Preparing English language arts teachers to design lessons that strengthen literacy practices is at the core of meaningful technology integration in the discipline. Cervetti, Damico, and Pearson (2006) call for literacy teacher educators to expand how pre-service teachers define literacy and to showcase the possibilities inherent in teaching multimodal texts. Other scholars in the field also link these broader strategies with the affordances of new technologies (e.g., Kinzer & Leander, 2003; Swenson et al., 2005).

Based on new understandings of literacy as a sociocultural process, contemporary views that shifting practices denote learning, and the potential of interactive online technologies to support literacy growth, Pace (2010) developed three overarching concepts that were used to guide the curriculum design of a methods course focused on meaningful technology integration and on media literacy education.

1. Literacy is a social practice that can be embellished and developed through teacher-designed literacy events and through interactive productive processes.
2. Literacy events should include explicit connections to content-area knowledge and to conceptual understandings that unite an academic discipline and the objects of study within the discipline.
3. Both teachers and students should be actively engaged in literacy events.

These principles were taught to pre-service teachers and modeled by faculty in the methods course. In the next section we describe how these principles were operationalized in the course and share three major assignments that showcase their application.

Description of the Technology and Media Literacy Course

Technology and Media Literacy was developed as a graduate-level course to serve pre-service secondary English language arts teachers enrolled in a graduate, NCATE-approved, initial certification program at a large, public university in the southeastern United States. Students enter the program with a bachelor's degree in English and complete a three-semester, master's degree program that leads to state certification and includes ESOL endorsement. During the first two semesters students participate in graduate courses in the afternoon and engage in morning field placements in English language arts classes at local schools. The course described in this chapter is the final methods course in the program.

As a blended course, Technology and Media Literacy requires students to meet face-to-face twice per week during the semester, to participate in discussion forums, and to collaborate online as they complete various assignments. These assignments have been designed to advance understandings of how interactive online technologies may be meaningfully integrated in an English language arts classroom. Though Moodle, an open source learning management system, is used as a central hub for the online portion of the course, many of the student projects incorporate interactive online technologies available across the Web, such as VoiceThread and Xtranormal. Such projects are described in greater detail later in the chapter.

A program requirement for eight years, the course has always included the tenets of media literacy education and has always been aligned with the "pedagogy of multiliteracies" developed by the New London Group (1996), which suggests that discursive differences and multimodality must be recognized in literacy pedagogy. Initially the course focused on how the study of popular culture texts might advance literacy education. The course was popular with students, but the ideas that grounded the course did not make use of interactive online technologies because those technologies were unavailable when the course was first designed.

Three years ago the course was revised to take advantage of the affordances of interactive, Web-based applications and the increasing availability of media texts online. The new version of the course was also anchored by a desire to promote meaningful technology integration that could promote student learning in an English language arts classroom. Changes to the course included the addition of the new conceptual framework described previously. The tenets of media literacy education and the pedagogical principles identified by the New London Group (1996) that had informed the original course were reworked and carried into the new course.

The Role of Media Literacy in the Course

Designing a course that sits at the nexus of media literacy and new interactive online technologies serves multiple purposes in the teacher education program. In addition to fulfilling the technology requirements established by various certification agencies (i.e., Florida Department of Education and the National Council for Accreditation of Teacher Education), the course promotes a working understanding of literacy events and literacy practices and how those constructs are related to learning (Garland & Pace, 2011). It also advances the thoughtful use of both media texts and technology and underscores the collaborative, sociocultural dimensions of literacy (Pace, 2010; Pace, Rodesiler, & Tripp, 2010). The focus on media literacy education was maintained to support one aspect of meaningful technology integration: the preservation of subject-area constructs that are enhanced by rather than eclipsed by technology use.

For the last two decades "media literacy education" has been understood as education about (not just with) media (Buckingham, 2003). The hoped-for outcome of media literacy education is media literacy, which is frequently defined as the ability to access, analyze, evaluate, and construct media texts (Aufderheide, 1997; Hobbs, 1997). This definition of media literacy provides an opportunity for those working with pre-service teachers to identify the literacy practices that are embedded in these abilities. It opens a door early in the course for thinking about literacy events and literacy practices.

A Focus on Literacy Practices and Literacy Events

To help pre-service teachers unpack the idea of literacy practices, a core concept in the course, we also present them with a diagram of Freebody and Luke's (1990) Four Resources Model and its description of a literate learner. Students examine the roles that literate learners play according to Freebody and Luke's model and the actions that they take as "code breakers," "meaning makers," "text users," and "text critics" (Luke & Freebody, 1999). Tying these roles to the outcomes promoted in media literacy education reinforces expanded concepts of literacy. Throughout the semester, we return to this model and to a discussion of literacy practices as we examine assignments. Furthermore, prospective teachers are asked to reflect on their participation in class projects and to identify the literacy practices embedded in their work.

These explicit discussions of what constitutes literacy practices also provide a way of addressing assessment and of connecting learning goals to classroom events. We extend that thinking by emphasizing the teacher's role as a designer and facilitator of literacy events and by modeling the instructional processes that we want these future teachers to adopt. Pre-service teachers are encouraged to examine what specific practices they want to support in classroom literacy events as they design activities and assignments. Thus, literacy practices and literacy events can become heuristics that future teachers may use to bracket classroom experiences and to examine how those experiences support adolescents' developing literacy practices.

Connecting Literacy Practices and Events to Content Area Knowledge and Processes

Anchored by the dimensions of literacy and by the concept of design, the course also promotes the thoughtful use of popular media and an appreciation for the positive role that popular media can play in learning. Classroom use of popular media is anchored by a consideration of overarching concepts in English language arts that can be applied across multiple forms of media. These *transmedia concepts,* or "metalanguages" (New London Group, 1996, p. 77), include ideas, such as narrative form and framing, that provide a language for constructing and deconstructing texts and for examining how the design tools an artist or writer has available in a medium shape the texts that are produced (McLuhan, 1964).

Pre-service teachers also examine how metalanguages can be presented through overt instruction and used to anchor situated, independent practice during classroom literacy events. As they work through course projects focused on various media and apply relevant metalanguages, they also reflect on the ways literacy events that promote increasingly sophisticated literacy practices might be constructed and assessed.

Classroom use of popular media is also described as a way of tapping the "funds of knowledge" (Moje, Ciechanowski, Kramer, Ellis, Carrillo, & Collazo, 2004, p. 41) adolescents access to make sense of school texts. Popular media is also discussed as a way of incorporating culturally relevant pedagogy (Gay, 2002) into the curriculum. Pre-service teachers link assignments and projects to secondary students' life worlds and design literacy events that focus on everyday, common topics and on media that are accessible across economic strata. For example, the teaching television projects described later in this chapter showcase how a media resource that is widely available can be used to promote students' evaluations of how TV narratives shape identities and attitudes (Cortés, 2005; Kellner & Share, 2005).

Active Engagement in Literacy Events

Throughout the course, pre-service teachers are required to consider how the dimensions of a pedagogy of multiliteracies (New London Group, 1996) can be applied in a classroom. Overt instruction is described as instruction that spans the life of a literacy event. Pre-service teachers are asked to consider how to extend their involvement in student work. They are also prompted to engage students in conversations that draw on metalanguages and that provide an organized and academic approach to the analysis and evaluation of media texts. Finally, pre-service teachers are encouraged to design opportunities for students to practice analysis and the creation of new, multimodal texts. These pedagogical principles align with contemporary views of teaching and learning literacy, and they constitute an important aspect of the methods course.

Promoting Meaningful Technology Integration with Interactive Technologies

In keeping with the TPACK model (Mishra & Koehler, 2006) and the work of Shulman (1986), the course includes subject-area concepts and pedagogical knowledge anchored by broad definitions of literacy and by the dimensions of a pedagogy of multiliteracies (New London Group, 1996). These areas are augmented by technologies that support the design of literacy events and that encourage dialogue and analysis of texts.

To facilitate Web–based activities that buttress literacy learning, one class meeting each week was held in a computer lab. During these sessions, prospective teachers first worked independently and then in collaborative groups to complete assigned projects that were both creative and analytical. These groups sometimes met face-to-face, but work was often completed online through Web 2.0 applications or online conferencing tools, such as Skype. Below we describe three of the Web-based projects that pre-service teachers completed in the latest iteration of the course and the literacy practices that each project encouraged.

Advertising Analysis via VoiceThread

The first Web-based project that students completed was facilitated by the use of VoiceThread. Available at <http://voicethread.com>, VoiceThread is an online application that provides a space for individuals to comment on a common text. Spoken comments may be added with a standard desktop microphone, a webcam, a telephone, or a pre-recorded comment. Users may also comment by typing a response. As these options suggest, VoiceThread is capable of housing a wide range of file types, including image and video files. This capability reinforces the emphasis on broadening students' understanding of what constitutes a text.

To begin this project, pre-service teachers were introduced to VoiceThread and given time to examine several VoiceThreads created for or by students in English language arts classes, which are freely available on the VoiceThread website. This exploratory period was followed by a discussion during which pre-service teachers identified how the samples they viewed promoted literacy practices—such as text analysis and dialogue—that are traditionally associated with English language arts and with the activities performed by literate learners.

Pre-service teachers were then introduced to metalanguages commonly associated with print advertising (e.g., framing, vectors, actors, reactors) and to online repositories of vintage advertisements. Working in teams, they identified concepts to teach and a group of advertisements that could be used to showcase those concepts. Pre-service teachers were then tasked with designing interactive, asynchronous online literacy events that promoted secondary students' exploration and application of the metalanguage associated with advertising. They were required to include overt instruction and to model the practices they were trying to teach in the opening screens of their VoiceThread. At the conclusion of the project, pre-service teachers were asked to explore the VoiceThreads created by at least three other teams and to provide feedback based on the objectives each team had established. Not only did the activity showcase the affordances and potential of interactive online technologies, it also provided a way of analyzing how concepts might be taught and how participatory activities may be designed in various ways. Moreover, post-activity debriefing and discussion gave pre-service teachers the opportunity to contemplate the challenges, benefits, and limitations that come with designing interactive, asynchronous online literacy events for use with secondary English language arts students.

Exploring Available Designs: Re-Design via Xtranormal

The second Web-based project in the course drew upon the affordances of Xtranormal, a text-to-movie program whose tagline is "If you can type, you can make movies." Available at <http://www.xtranormal.com>, Xtranormal allows users to manipulate virtual actors and to create a scene in which these actors move and speak. Users select camera angles, character types, backgrounds, and sound effects by simply typing instructions and "playing" the movie they have made.

This lab, like the first one, began with an introduction of the application and with an opportunity for pre-service teachers to explore projects completed by others. They were then asked to select a favorite passage or speech from a work of literature or from history and to "re-design" that passage or speech by maximizing the affordances of Xtranormal. Students chose to re-design monologues, exchanges of dialogue, or other passages from a range of traditional literary texts, including Hurston's (1990) *Their Eyes Were Watching God*, Hemingway's (1997) *A Farewell to Arms*, Shakespeare's (1963) *Hamlet*, Faulkner's (1990) *As I Lay Dying*, and Vonnegut's (2005) *Slaughterhouse-Five*, among others. The activity required students to consider how they might best convey the emotion and meaning of the print-based excerpt by manipulating shots, camera angles, setting, positioning, and other non-verbal cues.

Developed to showcase how "available designs" (New London Group, 1996, p. 74) can limit or enhance the production of texts across media, the activity also emphasizes what Buckingham casts as "translations" (2003, p. 78), the process of transferring a work from one medium to another. Translation activities can promote the literacy practices associated with the role of a "meaning maker" (Luke & Freebody, 1999). They can also prompt pre-service teachers to develop critical understandings of the affordances and limitations of various modes and media and to recognize what is gained and lost in moving from one medium to the next.

Moreover, just as the first project extended pre-service teachers' understandings of the metalanguage of print advertisements, this activity introduced pre-service teachers to the metalanguage of film. They made creative decisions about camera angles, characters, and sound and considered how such decisions influenced meaning-making processes. This project concluded with a reflection on and a discussion of the literacy practices prospective teachers drew upon as they created these translations. The Xtranormal-literature projects not only introduced an interactive online technology but they also reinforced the concept of translations and the idea that all texts are designed both by possibilities and purposes. Again the use of interactive online technology supplemented traditional academic processes rather than supplant them. Finally, this project also prompted pre-service teachers to consider how they might design similar literacy events that could foster both literacy practices and concepts aligned with English language arts.

Designing Narrative and Critical Media Analysis Wiki Resources

A pair of assignments that prompted pre-service teachers to collaboratively construct wikis was also featured in the course. Capable of featuring text, hypertext, images, and embedded videos, wikis are online spaces that any user can edit, whether adding content or removing it (Richardson, 2006). Freely available online by providers such as Wikispaces at <http://wikispaces.com>, a wiki is an ideal online technology for supporting the goals we established for the course described here, particularly as they relate to challenging narrow conceptions of literacy and texts. We have found that the use of wikis has the potential to open up conversations about authorship, text features, collaboration, Web design, and the credibility of online sources, among others.

The wiki assignments were designed literacy events that focused pre-service teachers on collaborative, constructive processes in online spaces. After being introduced to wikis in the computer lab, pre-service teachers were asked to create wikis for two projects. The first wiki assignment included designing a series of lessons that could be used to teach adolescents about television. This assignment was augmented by readings on the structure of television narratives (Pace, 2006) and on the power of television to shape identities and beliefs (Cortés, 2005). With an audience of teachers in mind, pre-service teachers elected a specific genre of television and were charged with developing a short instructional unit that examined television as a medium and that engaged adolescents in the consideration of how television teaches. Groups of pre-service teachers constructed wikis exploring genres that included, among others, mockumentaries (e.g., *The Office* (Silverman et al., 2005) and *Parks and Recreation* (Daniels, Schur, Klein, & Miner, 2009)), teen dramas (e.g., *Degrassi* (Schuyler, Stohn, & Yorke, 2001)), and family sitcoms (e.g., *Everybody Loves Raymond* (Rosenthal et al., 1996) and *Modern Family* (Levitan & Lloyd, 2009)).

As part of the activity, pre-service teachers were asked to prepare materials that could be used to inform other teachers and to provide them with resources for teaching about television. This focus promoted the development of a publishable product, and each wiki had to have specific features: a rationale for teaching television, hand-outs that supported instruction in metalanguages associated with moving images, and day-by-day lessons that could be used in secondary classrooms.

The second wiki assignment was the capstone project at the end of the course. Students worked in teams to prepare a wiki that addressed the teaching of documentary film. This project was not framed by a specific academic goal, such as considering narrative or promoting critical thinking. Rather, students were charged with identifying a purpose and objectives for their lessons.

Pre-service teachers worked collaboratively to develop units for teaching a range of documentary films that included, among others, *American Teen* (Burstein, Huddleston, Gonda, & Roberts, 2008), *Freakonomics: The Movie* (Troutwine et al., 2010), *Grizzly Man* (Nelson et al., 2005), and *Super Size Me* (Spurlock, 2004). In doing so pre-service teachers identified academic tasks that were related to literacy practices in English language arts and they developed activities that drew upon the interactive online technologies introduced in the class. In other words, their projects evinced meaningful technology integration.

FUTURE DIRECTIONS

In the blended course described in the pages of this chapter, we hoped to promote broadened views of literacy among pre-service secondary English language arts teachers. We sought to do this by promoting meaningful technology integration that showcased how interactive online technologies can support and embellish literacy practices. Opportunities to engage others in discussion around various media, to produce multimodal texts, and to write collaboratively with their peers through interactive online technologies provides opportunities for pre-service teachers to examine their own literacy practices and to gain an understanding of how the affordances of online technologies can contribute to learning in their future classrooms.

Understanding those affordances is a beginning point in recognizing how technology and literacy intersect and in seeing how the texts and tools teachers have at their disposal can support student learning and underscore the literacy practices their students bring to the classroom. Finally, experiencing literacy events that incorporate interactive online technologies firsthand helps to challenge notions of literacy that are restricted to the traditional reading and writing of print-based texts. Though such benefits are possible, it is apparent that opportunities for future research remain.

Investigations of how graduates of teacher education programs that feature online learning components apply what they have learned may offer insight into the efficacy of such experiences. For example, as of this writing, we understand the need and the benefit of a formal follow-up that investigates how graduates who have experienced the online learning opportunities described in this chapter make use of media literacy and technology in their classrooms. By following graduates into the field and investigating how they are applying what they have learned about literacy, technology, and popular media, we could learn much about the benefits and limitations of our design. Such research is essential as teacher education programs continue to design online learning opportunities for their students.

CONCLUSION

In the unique context of English teacher education and for our specific purposes, integrating online learning opportunities for pre-service teachers has been aligned with the traditions and practices of English language arts. By adopting a framework for meaningful technology integration, we have developed a curriculum that showcases the connections between developing literacy practices and learning the processes that guide the discipline. Encouraging pre-service secondary English language arts teachers to engage in and reflect on the kinds of literacy events they may one day provide for adolescents is a sensible approach. Moreover, incorporating online tools that invite multiple voices and promote multimodal media production, such as those highlighted in this chapter, is ideal. However, we understand that the possibilities for supporting the development of pre-service teachers with online technologies in varied contexts, in other content areas, and for different purposes are vast. As Koehler and Mishra (2009) contend, "There is no 'one best way' to integrate technology" (p. 62). Rather, productive technology integration is designed for specific subject matter in specific contexts (Mishra & Koehler, 2006). Accordingly, it is only through careful examination of variables, such as subject-area content, pedagogy related to that content, and the affordances of technology, that teacher educators can design meaningful online learning experiences and roadmaps for technology integration that meet the needs of the pre-service teachers they serve.

REFERENCES

Anstey, M., & Bull, G. (2006). *Teaching and learning multiliteracies: Changing times, changing literacies*. Newark, DE: International Reading Association.

Aufderheide, P. (1997). Media literacy: From a report on the National Leadership Conference on media literacy. In R. Kubey (Ed.), *Media literacy in the information age: Current perspectives* (pp. 79-86). New Brunswick, NJ: Transaction Publishers.

Barton, D. (2007). *Literacy: An introduction to the ecology of written language*. Oxford, UK: Blackwell.

Barton, D., & Hamilton, M. (1998). *Local literacies: Reading and writing in one community*. New York, NY: Routledge.

Bloome, D., Carter, S. P., Christian, B. M., Otto, S., & Shuart-Faris, N. (2005). *Discourse analysis and the study of classroom language and literacy events: A mircoethnographic perspective*. Mahwah, NJ: Lawrence Erlbaum Associates.

Buckingham, D. (2003). *Media education: Literacy, learning and contemporary culture*. Malden, MA: Polity Press.

Burstein, N., Huddleston, C., Gonda, E., & Roberts, J. (Producers), & Burstein, N. (Director). (2008). *American teen*. United States: Paramount Vantage.

Cervetti, G., Damico, J., & Pearson, P. D. (2006). Multiple literacies, new literacies, and teacher education. *Theory into Practice*, *45*(4), 378–386. doi:10.1207/s15430421tip4504_12

Cope, B., & Kalantzis, M. (2000). *Multiliteracies: Literacy learning and the design of social futures*. New York, NY: Routledge.

Cortés, C. (2005). How the media teach. In Schwarz, G., & Brown, P. U. (Eds.), *Media literacy: Transforming curriculum and teaching. The 104th yearbook of the National Society for the Study of Education, Part I* (pp. 55–73). Malden, MA: Blackwell Publishing.

Daniels, G., Schur, M., Klein, H., & Miner, D. (Producers). (2009). *Parks and recreation*. Van Nuys, CA: Universal Media Studios.

Faulkner, W. (1990). *As I lay dying: The corrected text*. New York, NY: Vintage Books.

Freebody, P., & Luke, A. (1990). Literacies programs: Debates and demands in cultural context. *Prospect: Australian Journal of TESOL*, *5*(7), 7–16.

Garland, K., & Pace, B. G. (2011, July). *Shifting concepts of literacy: How media literacy education can serve as transformative pedagogy for secondary students and pre-service teachers*. Paper presented at the meeting of the National Association for Media Literacy Education, Philadelphia, PA.

Gay, G. (2002). Preparing for culturally responsive teaching. *Journal of Teacher Education*, *53*(2), 106–116. doi:10.1177/0022487102053002003

Gee, J. P. (1996). *Social linguistics and literacies: Ideology in discourses* (2nd ed.). New York, NY: Routledge.

Goldman, S., Booker, A., & McDermott, M. (2008). Mixing the digital, social, and cultural: Learning, identity and agency in youth participation. In Buckingham, D. (Ed.), *Youth, identity, and digital media* (pp. 185–206). Cambridge, MA: The MIT Press.

Heath, S. B. (2001). Protean shapes in literacy events: Ever-shifting oral and literate traditions. In E. Cushman, et al. (Eds.), *Literacy: A critical sourcebook* (pp. 443-466). Boston, MA: Bedford/ St. Martin's. (Reprinted from *Spoken and written language: Exploring orality and literacy*, pp. 91-117, by D. Tannen, Ed., 1982, Norwood, NJ: Ablex)

Hemingway, E. (1997). *A farewell to arms*. New York, NY: Scribner Classics.

Hobbs, R. (1997). Expanding the concept of literacy. In Kubey, R. (Ed.), *Media literacy in the information age: Current perspectives* (pp. 163–183). New Brunswick, NJ: Transaction Publishers.

Hurston, Z. N. (1990). *Their eyes were watching God: A novel*. New York, NY: Perennial Library.

Kajder, S. (2003). Plugging in: What technology brings to the English/language arts classroom. *Voices from the Middle*, *11*(3), 6–9.

Kellner, D., & Share, J. (2005). Toward critical media literacy: Core concepts, debates, organizations, and policy. *Discourse: Studies in the Cultural Politics of Education*, *26*(3), 369–386. doi:10.1080/01596300500200169

Kinzer, C. K., & Leander, K. M. (2003). Technology and the language arts: Implications of an expanded definition of literacy. In Flood, J., Lapp, D., Squire, J., & Jensen, J. (Eds.), *Handbook of research on teaching the English language arts* (pp. 546–566). Mahwah, NJ: Erlbaum.

Kist, W. (2005). *New literacies in action: Teaching and learning in multiple media*. New York, NY: Teachers College Press.

Koehler, M. J., & Mishra, P. (2009). What is technological pedagogical content knowledge? *Contemporary Issues in Technology & Teacher Education*, *9*(1), 60–70. Retrieved from http://www.citejournal.org/vol9/iss1/general/article1.cfm

Kress, G. (2003). *Literacy in the new media age*. New York, NY: Routledge. doi:10.4324/9780203164754

Kucer, S. B. (2005). *Dimensions of literacy: A conceptual base for teaching reading and writing in school settings* (2nd ed.). Mahwah, NJ: Lawrence Erlbaum Associates.

Lankshear, C., & Knobel, M. (2003). *New literacies: Changing knowledge and classroom learning*. Philadelphia, PA: Open University Press.

Levitan, S., & Lloyd, C. (Producers). (2009). *Modern family*. Los Angeles, CA: 20th Century Fox.

Luke, A., & Freebody, P. (1999, October 26). Further notes on the four resources model. *Reading Online*. Retrieved from http://www.readingonline.org/research/lukefreebody.html#hasan

McLuhan, M. (1964). *Understanding media: The extensions of man*. New York, NY: Mentor.

Mishra, P., & Koehler, M. J. (2006). Technological pedagogical content knowledge: A framework for teacher knowledge. *Teachers College Record*, *108*(6), 1017–1054. doi:10.1111/j.1467-9620.2006.00684.x

Moje, E. B., Ciechanowski, K. M. I., Kramer, K., Ellis, L., Carrillo, R., & Collazo, T. (2004). Working toward third space in content area literacy: An examination of everyday funds of knowledge and discourse. *Reading Research Quarterly*, *39*(1), 38–70. doi:10.1598/RRQ.39.1.4

Morrell, E. (2004). *Linking literacy and popular culture: Finding connections for lifelong learning*. Norwood, MA: Christopher Gordon Publishers.

National Council of Teachers of English, & International Reading Association. (1996). *Standards for the English language arts*. Urbana, IL: NCTE.

National Council of Teachers of English. (2007). *21st-century literacies: A policy research brief*. Retrieved from http://www.ncte.org/library/NCTEFiles/Resources/PolicyResearch/21stCenturyResearchBrief.pdf

Nelson, E., Beggs, K., Campbell, B., Fairclough, P., Meditch, A., Ortenberg, T., & Palovak, J. (Producers), & Herzog, W. (Director). (2005). *Grizzly man*. United States: Lions Gate Films.

New London Group. (1996). A pedagogy of multiliteracies: Designing social futures. *Harvard Educational Review*, *66*(1), 60–92.

Pace, B. G. (2006). *Teaching narrative media online and in other social spaces*. Unpublished manuscript.

Pace, B. G. (2010, November). *Embellishing conceptions of literacy with pre-service English teachers*. Paper presented at the meeting of the National Council of Teachers of English, Orlando, FL.

Pace, B. G., Rodesiler, L. B., & Tripp, L. (2010). Pre-service English teachers and Web 2.0: Teaching and learning literacy with digital applications. In Maddux, C. D., Gibson, D., & Dodge, B. (Eds.), *Research highlights in technology and teacher education 2010* (pp. 177–184). Chesapeake, VA: Society for Information Technology and Teacher Education.

Pahl, K., & Rowsell, J. (2005). *Literacy and education: Understanding the new literacy studies in the classroom*. Thousand Oaks, CA: Sage.

Pope, C., & Golub, J. (2000). Preparing tomorrow's English language arts teachers today: Principles and practices for infusing technology. *Contemporary Issues in Technology & Teacher Education*, *1*(1), 89–97. Retrieved from http://www.citejournal.org/vol1/iss1/currentissues/english/article1.htm

Richardson, W. (2006). *Blogs, wikis, podcasts, and other powerful Web tools for classrooms*. Thousand Oaks, CA: Corwin Press.

Rideout, V. J., Foehr, U. G., & Roberts, D. F. (2010). *Generation M²: Media in the lives of 8- to 18-year olds*. Kaiser Foundation. Retrieved from the http://www.kff.org/entmedia/8010.cfm

Rogoff, B. (2003). *The cultural nature of human development*. Oxford, UK: Oxford University Press.

Rosenthal, P., Romano, R., Smiley, S., Rosegarten, R., Schneider, L., & Cawley, T. … Stevens, J. (Producers). (1996). *Everybody loves Raymond*. Burbank, CA: Warner Bros. Studios.

Schuyler, L., Stohn, S., & Yorke, B. (Producers). (2001). *Degrassi*. Toronto, Canada: Epitome Pictures.

Scribner, S., & Cole, M. (2001). Unpacking literacy. In E. Cushman, et al. (Eds.), *Literacy: A critical sourcebook* (pp. 123-137). Boston, MA: Bedford/St. Martin's. (Reprinted from *Writing: The nature, development, and teaching of written communication*, pp. 71-87, by Marcia Farr Whiteman, Ed., 1981, Mahwah, NJ: Lawrence Erlbaum Associates)

Shakespeare, W. (1963). *The tragedy of Hamlet, prince of Denmark*. New York, NY: Signet Classics.

Shulman, L. S. (1986). Those who understand: Knowledge growth in teaching. *Educational Researcher*, *15*(2), 4–14.

Silverman, B., Daniels, G., Gervais, R., Merchant, S., Klein, H., & Lieberstein, P. (Producers). (2005). *The office*. Van Nuys, CA: Universal Media Studios.

Spurlock, M. (Producer & Director). (2004). *Super size me*. United States: Samuel Goldwyn Films.

Street, B. (2001). The new literacy studies. In E. Cushman, et al. (Eds.), *Literacy: A critical sourcebook* (pp. 430-442). Boston, MA: Bedford/St. Martin's. (Reprinted from *Cross-cultural approaches to literacy*, pp. 1-21, by B. Street, Ed., 1993, London: Cambridge University Press)

Swenson, J., Rozema, R., Young, C. A., McGrail, E., & Whitin, P. (2005). Beliefs about technology and the preparation of English teachers: Beginning the conversation. *Contemporary Issues in Technology & Teacher Education*, *5*(3/4), 210–236. Retrieved from http://www.citejournal.org/vol5/iss3/languagearts/article1.cfm

Szwed, J. F. (2001). The ethnography of literacy. In E. Cushman, et al. (Eds.), *Literacy: A critical sourcebook* (pp. 421-429). Boston, MA: Bedford/St. Martin's. (Reprinted from *Writing: The nature, development, and teaching of written communication*, pp. 13-23, by Marcia Farr Whiteman, Ed., 1981, Mahwah, NJ: Lawrence Erlbaum Associates)

Troutwine, C., Romano, C., & O'Meara, D. (Producers), & Ewing, H., Gibney, A., Gordon, S., Grady, R., Jarecki, E., & Spurlock, M. (Directors). (2010). *Freakonomics*. United States: Magnolia Pictures.

Vonnegut, K. (2005). *Slaughterhouse-five, or, The children's crusade: A duty-dance with death*. New York, NY: Dial Press.

Young, C. A., & Bush, J. (2004). Teaching the English language arts with technology: A critical approach and pedagogical framework. *Contemporary Issues in Technology & Teacher Education*, *4*(1), 1–22. Retrieved from http://www.citejournal.org/vol4/iss1/languagearts/article1.cfm

ADDITIONAL READING

Alvermann, D. E., & Hagood, M. C. (2000). Critical media literacy: Research, theory, and practice in 'new times.'. *The Journal of Educational Research*, *93*(3), 193–205. doi:10.1080/00220670009598707

Barton, D., & Hamilton, M. (2000). Literacy practices. In Barton, D., Hamilton, M., & Ivanic, R. (Eds.), *Situated literacies: Reading and writing in context* (pp. 7–15). New York, NY: Routledge.

Boling, E. C. (2008). Learning from teachers' conceptions of technology integration: What do blogs, instant messages, and 3D chat rooms have to do with it? *Research in the Teaching of English*, *43*(1), 74–100.

Bush, J. (2003). Beyond technical competence: Technologies in English language arts teacher education (A response to Pope and Golub). *Contemporary Issues in Technology & Teacher Education*, *2*(4), 467–471. Retrieved from http://www.citejournal.org/vol2/iss4/english/CITEBushEnglish1commentary.pdf

Coiro, J., Knobel, M., Lankshear, C., & Leu, D. J. (Eds.). (2008). *The handbook of research in new literacies*. New York, NY: Erlbaum.

Doering, A., & Beach, R. (2002). Preservice teachers acquiring literacy practices through technology tools. *Language Learning & Technology*, *6*(3), 135–141.

Doering, A., Beach, R., & O'Brien, D. (2007). Infusing multimodal tools and digital literacies into an English education program. *English Education, 40*, 41–61.

Ferdig, R. E. (2006). Assessing technologies for teaching and learning: Understanding the importance of technological pedagogical content knowledge. *British Journal of Educational Technology, 37*, 749–760. doi:10.1111/j.1467-8535.2006.00559.x

Fulton, H. (with Huisman, R., Murphet, J., & Dunn, A.). (2005). *Narrative and media*. New York, NY: Cambridge University Press.

Grabill, J. T., & Hicks, T. (2005). Multiliteracies meet methods: The case for digital writing in English education. *English Education, 37*(4), 301–311.

Hagood, M. C. (Ed.). (2009). *New literacies practices: Designing literacy learning*. New York, NY: Peter Lang Publishing.

Hagood, M. C., Alvermann, D. E., & Heron-Hruby, A. (2010). *Bring it to class: Unpacking pop culture in literacy learning*. New York, NY: Teachers College Press.

Hicks, T. (2006). Expanding the conversation: A commentary toward revision of Swenson, Rozema, Young, McGrail, and Whitin. *Contemporary Issues in Technology & Teacher Education, 6*(1), 46–55. Retrieved from http://www.citejournal.org/vol6/iss1/languagearts/article3.cfm

Hixon, E., & So, H.-J. (2009). Technology's role in field experiences for preservice teacher training. *Journal of Educational Technology & Society, 12*(4), 294–304.

Hobbs, R. (2007). *Reading the media: Media literacy in high school English*. New York, NY: Teachers College Press.

Jenkins, H. (2008). *Convergence culture: Where old and new media collide*. New York, NY: New York University Press.

Kalantzis, M., & Cope, B. (2000). A multiliteracies pedagogy: A pedagogical supplement. In Cope, B., & Kalantzis, M. (Eds.), *Multiliteracies: Literacy learning and the design of social futures* (pp. 239–248). London: Routledge.

Kellner, D. (2001). New technologies/new literacies: Reconstructing education for the new millennium. *International Journal of Technology and Design Education, 11*, 67–81. doi:10.1023/A:1011270402858

Kist, W. (2000). Beginning to create the new literacy classroom: What does the new literacy look like? *Journal of Adolescent & Adult Literacy, 43*(8), 710–718.

Lacy, N. (2000). *Narrative and genre: Key concepts in media studies*. New York, NY: Palgrave.

Lewis, C., Enciso, P., & Moje, E. B. (Eds.). (2007). *Reframing sociocultural research on literacy: Identity, agency, and power*. Mahwah, NJ: Lawrence Erlbaum Associates.

Luke, A., & Elkins, J. (1998). Reinventing literacy in 'new times'. *Journal of Adolescent & Adult Literacy, 42*(1), 4–7.

Luke, C. (2000). New literacies in teacher education. *Journal of Adolescent & Adult Literacy, 43*(5), 424–435.

Martinson, D. L. (2004). Media literacy education: No longer a curriculum option. *The Educational Forum, 68*(2), 154–160. doi:10.1080/00131720408984622

Masterman, L. (1990). *Teaching the media*. London, UK: Routledge.

Matthew, K. I., Felvegi, E., & Callaway, R. A. (2009). Wiki as a collaborative learning tool in a language arts methods class. *Journal of Research on Technology in Education*, *42*(1), 51–72.

Myers, J., & Beach, R. (2004). Constructing critical literacy practices through technology tools and inquiry. *Contemporary Issues in Technology & Teacher Education*, *4*(3), 257–268. Retrieved from http://www.citejournal.org/vol4/iss3/languagearts/article1.cfm

National Council of Teachers of English. (2006). *Guidelines for the preparation of teachers of English language arts*. Urbana, IL: NCTE. Retrieved from http://www1.ncte.org/library/files/Store/Books/Sample/Guidelines2006Chap1-6.pdf

Schwarz, G. (2001). Literacy expanded: The role of media literacy in teacher education. *Teacher Education Quarterly*, *28*(2), 111–119.

Selber, S. (2004). *Multiliteracies for a digital age*. Carbondale, IL: Southern Illinois University Press.

Selfe, C. L., & Hawisher, G. E. (2004). *Literate lives in the information age: Narratives of literacy from the United States*. Mahwah, NJ: Lawrence Erlbaum Associates.

Smagorinsky, P., & Whiting, M. E. (1995). *How English teachers get taught*. Urbana, IL: National Council of Teachers of English.

Swenson, J. (2006). Guest editorial: On technology and English education. *Contemporary Issues in Technology & Teacher Education*, *6*(2), 163–173. Retrieved from http://www.citejournal.org/vol6/iss2/languagearts/article1.cfm

Swenson, J., Young, C. A., McGrail, E., Rozema, R., & Whitin, P. (2005). Extending the conversation: New technologies, new literacies, and English education. *English Education*, *38*(4), 351–369.

Tyner, K. (1998). *Literacy in a digital world: Teaching and learning in the age of information*. Mahwah, NJ: Lawrence Erlbaum.

Williams, B. T. (2009). *Shimmering literacies: Popular culture and reading and writing online*. New York, NY: Peter Lang Publishing.

Writing in Digital Environments (WIDE) Research Center Collective. (2005). Why teach digital writing? *Kairos, 10*(1). Retrieved from http://english.ttu.edu/kairos/10.1/binder2.html?coverweb/wide/index.html

Wysocki, A. F., Johnson-Eilola, J., Selfe, C. L., & Sirc, G. (2004). *Writing new media: Theory and applications for expanding the teaching of composition*. Logan, UT: Utah State University Press.

KEY TERMS AND DEFINITIONS

Blended Course: A course that affords students the benefit of learning through both routine face-to-face class meetings and regular online class participation. For the course described in this chapter, students met face-to-face twice per week while also participating in discussion forums and collaborating online with their peers to complete numerous assignments, including the activities detailed previously.

Interactive Online Technologies: Those technologies available online via the Web that allow users to construct and manipulate texts and engage others in social interaction. VoiceThread, Xtranormal, and wikis are examples of interactive online technologies featured in this chapter.

Literacy Event: An activity in which people's interactions and interpretations are facilitated by any text, print or otherwise.

Literacy Practices: The ways people carry out literacy in their respective cultural contexts.

Media Literacy: The outcome of media literacy education, it is widely recognized as the ability to access, analyze, evaluate, and construct media texts.

Media Literacy Education: An educational approach that prepares students to access, analyze, evaluate, and create a wide range of media messages.

Metalanguages: Specific concepts used to examine how the tools for design an artist or writer has available in a medium shape the texts that are produced.

New Literacies: Literacies available as a result of technological advancements, economic globalization, and social changes that have altered the nature of texts and the activities carried out with those texts.

New Literacy Studies: A specific approach to literacy research that explores the social, cultural, and historical contexts in which literacy events and practices are carried out. It is noteworthy for challenging approaches that place greater emphasis on basic, de-contextualized skills.

Popular Media: Mass-produced texts, either print-based or non-, that present information through multiple communicative modes.

Pre-Service Teacher: A student who is enrolled in a teacher education program that prepares him or her to work as a teacher in K-12 classrooms.

Transmedia Concepts: Overarching concepts that can be applied across multiple forms of media.

Chapter 18
Designing and Teaching an Online Elementary Mathematics Methods Course:
Promises, Barriers, and Implications

Drew Polly
University of North Carolina at Charlotte, USA

ABSTRACT

This chapter discusses a longitudinal examination of a mathematics methods course for teacher candidates taught in hybrid and a 100% asynchronous online format. Using Guskey's (2000) framework for evaluating learning experiences for teachers, thematic analysis was conducted on teacher candidates' course feedback and two major course assignments. Data analysis indicated that teacher participants valued the amount of support provided by the instructor and communication with classmates, had mixed comments about having to take ownership of their learning, and disliked the amount of work in the course. Participants' work samples reflected the application of emphasized pedagogies in lesson plans and course projects, and participants also positively impacted student learning during their clinical project. Implications for future courses as well as the examination of online methods courses are shared.

INTRODUCTION

Online Learning Opportunities

During the last decade the demand to develop online methods courses in teacher education programs has increased dramatically (Ko & Rosen, 2010). While the benefits of online courses have been established (Tallent-Runnels et al., 2006), online courses in teacher education, especially mathematics education, provide a set of design issues that differ from other online courses outside of teacher education programs or other content areas (Delfino & Persico, 2007; Tallent-Runnels et al., 2006). To address these challenges, researchers have identified specific tools for online teaching that expand learning opportunities. These include: accessibility to the Internet and Web 2.0 technologies, a collaborative framework, and other course structures.

DOI: 10.4018/978-1-4666-1906-7.ch018

Higher education has embraced the possibilities of using the Internet as a medium to teach college courses. Nelson, Christopher, and Mims (2009) suggested that "the Internet and Web 2.0 technologies afford teachers ready access to collaborative, authentic opportunities for students to engage in meaningful experiences related to the curriculum" (p. 85). Oliver (2010) noted that the use of Web 2.0 technologies advances online learning opportunities that were not readily available before. There are numerous technologies to support teaching mathematics online (Hodges & Hunger, 2011). Blogs and wikis have been recognized by faculty as effective tools, especially in mathematics courses (Ajjan & Hartshorne, 2008; Carter, 2009; Peterson, 2009).

These technologies increase opportunities for student interaction and discussion; an essential component of online courses (Levin et al., 2001; Tallent-Runnels et al., 2006). In their comparison study of discussion formats, Im and Lee (2004) found that synchronous discussion was more effective at social interaction, while asynchronous discussion was better suited for task-oriented interaction. Regardless, the presence of peer collaboration and discussions in online courses is empirically-linked to students' perceptions of the course (Schlager & Fusco, 2004).

Further, specific course structures have been found to be effective in online courses. Greene and Land (2000) found that both guiding questions and frequent feedback from instructors helped them understand assignments and led to higher quality work products. Experts have highlighted the importance of relevant and challenging assignments (Levin, Waddoups, Levin, & Buell, 2001), clear expectations and evaluation methods (Moallem, 2003), and opportunities to reflect on assignments (Levin et al., 2001). Furthermore, online courses create new learning spaces for teachers and engender an environment that is learner-centered and align to characteristics of effective learning environments.

Online Learning Opportunities for Teachers

The research base regarding online learning opportunities for teachers includes studies regarding online professional development opportunities for teachers who are currently in classrooms and either working on advanced degrees or pursuing a deeper understanding of either content or pedagogy (e.g., Russell, Carey, Klieman, & Venable, 2009; Signer, 2008). Researchers that have compared online professional development to face-to-face models for mathematics teachers noted that there was no significant difference between the influence of the learning experiences on teachers' beliefs, knowledge or skills (Russell et al., 2009). Online learning experiences have found that the increase in written communication helps to promote more reflective inquiry about course content (Spicer, 2002; Treacy, Kleiman, & Peterson, 2002). In their five-year longitudinal study, Delfino and Persico (2007) found that teachers' written work in the online courses contained frequent instances of critical thinking and in-depth reflection. Further, on the whole teachers enjoy online learning opportunities and are willing to take more in the future (Russell et al., 2009; Signer, 2008).

However, while research on pre-service courses for pre-service teachers with limited or no teaching experience is scant. In a hybrid mathematics education methods course, Schwartz (in press) posed a task to a group of students in an online asynchronous format and compared their strategies and reactions to those in a face-to-face format. While pre-service teachers approached and solved the task in the same way, there was a lack of opportunities for the instructor and pre-service teachers to discuss the pedagogy and in-the-moment teacher decision making. O'Connor (2011) found that while pre-service teachers learned technology and formed effective collaborative relationships with their colleagues, videos of teaching experiences reflected a lack of student-centered pedagogies that were emphasized in the course.

Clearly, there is a need for further research on online learning opportunities for pre-service teachers, specifically in the area of mathematics education. This chapter provides an overview of the development of an online mathematics education methods course for elementary education students, including the theoretical foundation and course activities. The purpose of this study is to present a learner-centered framework for an online mathematics education course for elementary school pre-service teachers. That is followed by a case study that includes data collected and analyzed from course offerings. Lastly, implications are presented that inform current and future work related to this course.

OVERVIEW

Course Overview

The course described in this chapter is the second of two mathematics education courses in our Elementary Education Graduate Certificate in Teaching Program at the University of North Carolina at Charlotte. The course, ELED 5301: Assessing, Modifying, and Integrating Mathematics Instruction, includes numerous projects to prepare teacher candidates to assess students, plan instruction based on data, and then reassess students using formative and summative processes to monitor students' learning. ELED 5301 also includes content and pedagogy tasks related to fractions, algebraic thinking, and geometry.

Students in the Graduate Certificate in Elementary Education Program hold a bachelor's degree in another field, and are working on obtaining their initial teacher license in K-6 education. While some students have volunteered or worked in schools as teacher assistants or tutors, most students have little to no experience in elementary school classrooms. Students complete ELED 5301 when they are one semester away from student teaching, so the course includes a heavy focus on designing mathematics lessons, assessing students, and differentiating instruction.

Motivation to develop an online version of this course started in the 2007-2008 year, through discussions about alternative ways to provide teacher education courses to meet the scheduling needs of our students in the Graduate Certificate in Teaching Program.

While the entire Graduate Certificate in Teaching program is offered in a face-to-face format, there was interest from both College and Department leadership in exploring online courses. One reason for students' overwhelmed feeling is that they work full time during the day and typically take 3 3-hour courses each semester, and every course is offered in the evening. Students often report being overwhelmed with 9 seat hours of course work weekly in addition to 15 hours of clinical assignments per course. An online course provides students with flexibility to complete modules during the week and decrease the number of seat hours that students have, while maintaining the amount of clinical assignments.

The development of online modules in ELED 5301 was supported through a UNC Charlotte Curriculum Improvement and Development grant, which provided resources to purchase video vignettes of elementary school classroom teaching from the Developing Mathematical Ideas professional development materials (Schifter, Bastable & Russell, 2008). While these video vignettes were also used in face-to-face weeks, these videos were effectively used as part of online modules, in which teacher candidates viewed videos and analyzed how the teachers supported students' mathematical learning with pedagogies that were emphasized during the course.

Theoretical Foundation of the Course

The design and activities in this course are driven by two theoretical constructs: learner-centered instruction and mathematical knowledge for teaching. In this section, I provide an overview of each and then describe how each construct aligns to the course activities (Table 1).

Learner-Centered Instruction

ELED 5301 has been designed based on the theoretical framework of learner-centered instruction for teacher learning (Polly & Hannafin, 2010; National Partnership for Educational Accountability and Teaching [NPEAT], 2000).The American Psychological Associations' *Learner-centered Principles* (APA Work Group, 1997), provide the empirical basis for designing learner-centered environments for both students and adults. The NPEAT adopted these *Principles* for teachers' learning. Recently, the *Learner-Centered Principles* were synthesized with empirical research on teacher learning to derive a set of learner-centered principles for the design of learning experiences for teachers and teacher candidates (Polly & Hannafin, 2010). Specifically, these experiences should:

- Prepare teachers to address student learning issues (Heck, Banilower, Weiss, Rosenberg, 2008),
- Ensure ownership of their learning experiences (Garet et al., 2001),
- Promote collaboration among teachers or teacher candidates (Glazer & Hannafin, 2006),
- Emphasize comprehensive change processes through ongoing support (Fishman, Marx, Best, & Tal, 2003; Orrill, 2001)
- Develop knowledge and proficiency related to specific pedagogies, content and the intersection of content and pedagogy (Heck et al., 2008; Garet, et al., 2001); and
- Support reflection on work samples and artifacts from students and classrooms (Cohen, 2005; Loucks-Horsley et al., 2009).

Table 1. Alignment between characteristics of effective teacher learning environments and course activities

Characteristics of Learner-centered environments for teachers	Course Activities Teacher candidates...	Aspects of Mathematical Knowledge for Teaching
Address student learning issues	Complete the Culminating Diagnostic Project, which involves pre-assessing, teaching and post-assessing students.	Knowledge of Content and Students (KCS), Knowledge of Content and Teaching (KCT)
Provide teachers with ownership	Select topics for Culminating Diagnostic Project, Curriculum Evaluation and other activities.	All aspects
Promote collaboration	Collaborate with classmates on blogs and with in-service teachers on all clinical projects.	All aspects
Provide ongoing support	Turn in the Culminating Diagnostic Project at various times during the semester for feedback and support.	All aspects
Develop knowledge of content and pedagogy	Complete cognitively-demanding mathematical tasks in each module, analyze curriculum, and examine video-based and text-based vignettes.	All aspects
Support the reflection process	Write on blogs throughout the semester and reflect on the impact on students during the Culminating Diagnostic Project.	All aspects

Mathematical Knowledge for Teaching

In addition to the *Learner-centered Principles*, which specify how to structure effective learner environments for teachers, ELED 5301 is also grounded in the construct of mathematical knowledge for teaching (MKT). MKT (Figure 1) is a research-based framework that describes the different aspects of knowledge involved with teaching mathematics to students (Ball, Thames & Phelps, 2008; Thames & Ball, 2010). MKT includes both mathematical content knowledge and pedagogical knowledge. Based on MKT, opportunities for teacher learning should address these different knowledge types.

Table 1 shows the alignment between characteristics of learner-centered experiences for teachers and course activities. These activities are described in more detail later in the chapter.

Structure of the Course

Course Format

For every online module in either the online versions or the hybrid versions online content was asynchronous. Students communicated and interacted with each other and the instructor on their own time without any online meetings scheduled.

Learning Management System

All of the course activities were assigned through Moodle (http://moodle.org), an open source Learning Management System (LMS). Within Moodle, students interacted with course content (articlcs, videos, assignments), submitted work, and received feedback and grades on their progress.

Figure 1. Mathematical knowledge for teaching (Thames & Ball, 2010)

Mathematical knowledge for teaching (MKT) consists of distinguishable domains defined in relation to the work of teaching. Some resources that teaching requires resemble math knowledge used in other settings, common content knowledge (CCK). Teachers also need pedagogical content knowledge (PCK). But some math knowledge is specialized.

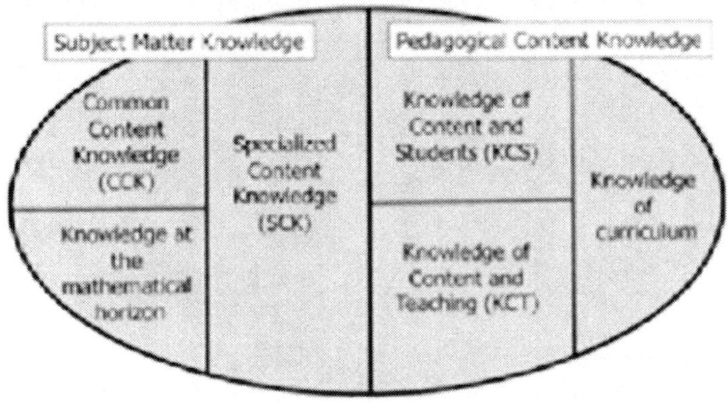

Modules

Assignments were bundled together in a module. Each module was organized as a series of tasks, which included a cognitively-demanding task, an analysis of classroom mathematics teaching, activities to complete based on their work with the state standards, curriculum, or elementary school students, and a reflection that connected the various activities. Modules were designed to last anywhere between 1-3 hours of work. Some modules also served as the foundation for major course projects, which are detailed below.

Collaboration with Classmates

Each teacher candidate maintained their own blog, which was set up through either Word Press (http://wordpress.com) or Posterous (http://posterous.com). Teacher candidates posted to their blog each week, and responded to at least two of their classmates' postings. This communication between classmates provided opportunities for teacher candidates to share their thoughts and experiences with each other, and receive support from their peers. Teacher candidates also had access to a course wiki where they could go and receive support from each other and the instructor about course activities.

Course Activities

Grounded in the principles of learner-centered instruction, the course included various activities. Table 1 provides details about how course activities match these characteristics. These course activities are described in more detail later in the chapter.

Cognitively-Demanding Mathematical Tasks

During each module, teacher candidates completed mathematical tasks that were related to the Kindergarten through sixth grade state mathematics standards. Each of these tasks was a multi-step, complex problem that aligned with the construct of cognitively-demanding mathematical tasks (Stein, Grover, & Henningsen, 1997). These tasks varied in terms of the mathematical concepts that they dealt with, but most of them focused on number sense and algebraic thinking. In the face-to-face and hybrid formats, these tasks were solved during face-to-face course meetings.

Examining Artifacts of Mathematics Teaching

Teacher candidates examined examples of mathematics teaching through vignettes that were freely available on the Internet (http://learner.org) or in the course's text book, *My Kids Can* (Storeygard, 2010). Through viewing these video-based vignettes and focusing their attention on key aspects, such as teacher's questioning, student communication, or the structure of the lesson, teacher candidates developed a deeper sense of knowledge of content as it related to how to teach and how students learn content. These videos were incorporated into online modules during the hybrid versions of the course, and viewed during.

Assessing Students' Mathematical Learning

Teacher candidates conducted mathematical interviews with elementary school students. During these interviews, teacher candidates posed tasks, observed students as they work, and asked students to explain their processes and mathematical thinking. This assignment was a clinical assignment done in schools in all versions of the course.

Writing about Mathematics Teaching and Learning

Students completed a written reflection about activities in the module. These reflections focused on either their reaction to videos that they watched, articles they read, or their own experiences during their time observing and teaching mathematics to elementary school students. These were posted on blogs using either WordPress or Posterous. In all versions of the course, these blogs were used. However, students were required to post more frequently during hybrid and online sections of the course.

Curriculum Evaluation

During the middle of the semester, teacher candidates evaluated and analyzed a unit from a standards-based curriculum. They then read a few articles about the types of mathematics curriculum, the "Math Wars" and then reflected on their ideal math curriculum about being standards-based or more traditional. This assignment was completed in every version of the course. In the face-to-face section of the course, the background information was presented during a course meeting, and it was presented online during hybrid and online versions of the course.

Culminating Diagnostic Project

The culminating experience in the course was a diagnostic project that required teacher candidates to pre-assess, plan and teach five lessons, and post-assess a small group of students to examine students' growth. Teacher candidates were expected to base all of their instructional decisions on data that was collected during the pre-assessment and lessons. Teacher candidates also planned and taught a lesson to an entire class of students, in addition to planning a 10 lesson geometry unit that aligned to the state standards. Through these activities students developed pedagogical content knowledge, including knowledge related to teaching, students and curriculum. This assignment was completed in every version of the course.

METHODS OF EVALUATING THE COURSE

Evaluation Framework

Guskey (2000) provided a multi-level framework (Table 2) for evaluating the impact of learning experiences for teachers. While this framework was intended to evaluate the influence of professional development for practicing teachers, it can also be adapted for teacher candidates. In the remainder of this chapter, data is presented from each of these levels provides insight into teacher candidates' performance as well as the impact of the online course.

Research Questions

For this chapter, data was analyzed to examine:

- What are teacher candidates' reactions to participating in an online mathematics education course? (Level 1)

Table 2. Levels of evaluation (Adopted from Guskey, 2000)

Level	Name of level	Questions examined
Level 1	Teachers' reactions	How do teachers feel about their experiences?
Level 2	Teachers' knowledge and skills	What knowledge and skills do teachers learn during their experiences?
Level 3	Use of knowledge and skills	How do teachers use their new knowledge and skills in their own classroom?
Level 4	Impact on organization and program	How did the experience impact the program or organization responsible for providing the experience?
Level 5	Impact on student learning	How are PK-12 students impacted by teachers' participation in this experience?

- How do teacher candidates' demonstrate their understanding of course content? (Level 3)
- To what extent do teacher candidates' impact student learning on the courses' Culminating Diagnostic Project? (Level 5)
- How has data informed revisions and modifications of the course? (Level 4)

Context

Data in this chapter are shared from teacher candidates in ELED 5301 between 2008 and 2011. Table 3 describes the formats of the various courses. All students in the course were teacher candidates in the Graduate Certificate in Teaching Elementary Education program at UNC Charlotte, and held a bachelor's degree in a different field. Students' bachelor degrees vary, but common fields include Business, English, or Psychology.

Data Sources

Data sources for this study included: post-course feedback, online reflections, curriculum project reflections, and data from the Culminating Project.

Post-Course Feedback

Teacher candidates provided anonymous feedback using both the course evaluation form as well as a more detailed survey that I sent students. The survey asked them to share their opinions about: which assignments were the most beneficial at this stage in the teaching career, which assignments were not very beneficial, and what experiences would they like more of. Lastly, I asked participants to comment about the nature of the course and whether the online or hybrid design should be modified.

Online Reflections

Teacher candidates' online reflections on their blogs were analyzed to examine the topics of the blog posts. In each reflection, participants had freedom to focus their blogs topics of most interest to them. At the beginning of the semester, participants had a structured prompt, such as "what are the elements of an effective mathematics lesson?" or "what are your concerns related to planning a mathematics lesson?" Later in the semester, as participants are more involved with their clinical experiences, they are asked to write about their experiences and things that they are learning from teaching and interviewing students.

Curriculum Project

On their curriculum projects, teacher candidates analyzed standards-based curricula to look at the types of mathematical tasks, the way that content was presented, and the type of teacher support that was provided for them. After their analysis, candidates reflected on the assignment and discussed their perceptions of the curricula

Table 3. Formats of ELED 5301

Semester	Format	Number of modules	Class size
Fall, 2008	Online	10 online modules	28
	Hybrid	3 online modules, 12 face-to-face meetings	28
Fall, 2009	Hybrid	3 online modules, 12 face-to-face meetings	32
Fall, 2010	Hybrid	5 online modules, 10 face-to-face meetings	44
Fall, 2011	Online	10 online modules	39 students

that they examined. They also shared how the curricula aligns or conflicts with what they have learned in their coursework and their beliefs about teaching mathematics.

Culminating Diagnostic Project

Data sources from the diagnostic project included the types of mathematical tasks in the lesson plans as well as student growth between the pre- and post-assessments. Lesson plans were written using an indirect instruction format (http://coedpages.uncc.edu/abpolly/5301/coursedocs/diagnostic/lesson-plan-resources/diagnostic-6pt.doc), which focuses on designing the lesson around a cognitively demanding mathematical task, and supporting students through questions and follow-up tasks. Pre- and post-assessments were identical, and were either created by the teacher candidate, or modified from a curriculum.

DATA ANALYSIS

Course Feedback and Reflections

Course feedback and reflections from teacher candidates were examined using inductive, thematic analysis (Bogden & Biklen, 2003). All of the data was entered into a spreadsheet and labeled. During the data analysis process, data was coded using an open-coding process and then sorted all of the data by code. After coding and sorting, data was reread to confirm that the data matched the codes. The next step was to group the data by code in order to generate themes. Once themes were generated, data was reread again, to make sure that the themes truly reflected the data within that group. The themes were then matched up to the research question. Themes addressed either research questions 1 and 2. The Findings section of this paper describes the most common themes that were found during the data analysis process.

Types of Mathematical Tasks in Lesson Plans

Using the cognitively-demanding tasks framework (Stein et al., 1997), tasks from every lesson plan was coded and analyzed. The analysis was similar to the process detailed above, where tasks were coded, and once organized, reanalyzed to make sure that they were properly categorized. Once each task had been coded and checked, frequencies were determined. Table 4 provides descriptions and examples of the four types of mathematical tasks.

Student Growth on Pre- and Post-Assessments

As part of the Culminating Diagnostic Project, teacher candidates gave a pre- and post-assessment and taught 5 lessons around a concept. Since the assessments varied in length, topic, and grade level, data analysis focused on whether or not students demonstrated growth between the pre- and post-assessments. Since each teacher candidate gave the same assessment as both a pre- and post-assessment, growth was noted anytime a students' score increased on the post-assessment.

FINDINGS

What are Teacher Candidates' Reactions to Participating in an Online Mathematics Education Course?

Data regarding teachers' reactions to participating in an online (or hybrid) course centered on three major themes: responsible for learning, amount of work, and communication with class members and instructors.

Table 4. Types of mathematical tasks

Cognitive Demand	Name of Task Type	Description	Example
High Cognitive Demand	Doing Mathematics	Students explore mathematical tasks that require them to choose an approach, complete the task, and explain their steps and decision-making.	There are 24 yards of fencing for the garden. If you want to make a rectangular garden with side lengths that are whole yards, what are the possible dimensions of the garden? Which garden is the largest?
	Procedures with Mathematical Connections	Students explore tasks that can be solved with an algorithm, but have to generate more than one representation.	There are 9 dozen cookies in the bag. If you eat 6 cookies how many are left? Show your picture using a picture and an equation.
Low Cognitive Demand	Procedures with Mathematical Connections	Students explore tasks that require only an algorithm and only one mathematical representation.	There are 9 dozen cookies in the bag. If you eat 6 cookies how many are left?
	Memorization	Students recall a fact that is expected to be known.	What is the product of 9x6?

Responsible for Learning

Participants across all semesters reported on post-course feedback that they felt more responsible for their own learning during online activities. This was true for the online courses as well as the online modules in hybrid courses. Participants expressed various emotions regarding this theme.

In some cases, students enjoyed the responsibility that they perceived. "It was really hard to concentrate more and learn the material better without having regular meetings with an instructor. I was completely in control of my own learning, which was nice, but I want more regular contact with my instructors." (Student, Fall, 2010 semester).

However, many students reported discontent and dissatisfaction with the amount of responsibility that they had to take on.

"I feel like I am teaching myself how to teach math, and I don't feel confident that I am learning what I need to. The instructor provides modules and helps us if we get stuck, but we basically are on our own to figure things out. I don't think that's fair." (Student, Fall, 2008 semester).

While feedback from each course included both positive and negative reactions about students' responsibility for learning, students in the hybrid courses compared the face-to-face and online learning. Many commented about the benefit of having a blend of both face-to-face and online work.

"I feel like a hybrid model best met my needs. The weeks in class allowed us to see and work with our colleagues, but we were also able to have the flexibility of working on our own during online weeks." (Student, Fall, 2009 semester).

Amount of Work

Data from teacher candidates' reactions also focused on the amount of work in the courses. Nearly every comment compared the amount of work in an online course to a face-to-face course.

"My friend is taking the face-to-face section, and my online section has much more work to do. The concepts are the same, but it seems as if we have a lot more reading and activities to work on. I would rather spend my evenings in a classroom sitting there instead of working on these modules." (Student, Fall, 2008 semester)

Some students, however, reported that they thought there was much less work in a 100% asynchronous online course and that they were able to better differentiate their attention on concepts and activities that they were more interested in and needed to learn more about.

"The amount of work takes up much less time since it is an online class. I think that I still learned a lot, but was able to focus on things that I individually found interesting or needed, instead of sitting in a classroom and having to learn things that are not as relevant to me." (Student, Fall, 2011, semester)

Teacher candidates reported in the online only courses reported mixed reactions about the amount of work. While a face-to-face class is designated as a 3 hour weekly course meeting followed by approximately 3 hours of weekly activities and assignments, students were not comfortable with the idea that an online course may vary in terms of the length of the modules and the amount of work each week. While the amount of work did not change between versions of the online course, more recent teacher candidates reported less frustration with the amount of work involved in the course.

"It was hard to predict how long a module would take. There were some where I knew the math concept and I flew through it, and some like algebra, that took me an entire Saturday. I feel like a face-to-face class would have been better since they are typically more consistent about the amount of time that you have to spend on projects and homework." (Student, Fall, 2008 semester).

Over the past few years, the number of comments about the inconsistent lengths of the online modules has dissipated.

Communication with Class Members and Instructors

Data from participants' reactions also frequently mentioned communication with class members and instructors. Many students commented about the surprisingly high amount of communication that they had with students via their weekly online reflections.

"I did not expect to interact with classmates at all in this course. It was nice to be able to get responses from my weekly reflections, and be able to hear what others thought about the course activities. The interaction helped me feel like there were others in the same situation as myself." (Student, Fall, 2011 semester).

Other teacher candidates shared disappointment with the lack of communication in the asynchronous online course. In the course feedback, they cited the amount of group work that goes on in face-to-face courses in the program.

"I just felt awkward and disappointed in this course. We do group activities in every face-to-face course in our program, including the first math course, but we were very isolated here. Commenting on reflections is definitely not enough communication for a course like this."

One change between the 2008 online and 2011 online course was a more frequent use of the class wiki to share problem solving strategies and ideas. Most students in the Fall, 2011 course commented about the benefit of the wiki and the opportunity to frequently communicate with each other.

"I felt supported, especially when we had to solve those math tasks. Having the wiki where the instructor and classmates put ideas about how to start working on those tasks was a huge help. It lessened the anxiety I had trying to solve hard math problems." (Student, Fall, 2011 semester).

The data showed that participants were fairly satisfied with the amount and frequency of communication between them and the instructor in all versions of the course. However, some students did admit that in the online course, it was difficult to generate questions that they said would most likely be answered during face-to-face course meetings. A majority of the comments related to communication focused on the amount of support they received on drafts of assignments, including the major Culminating Diagnostic Project.

"I felt like this instructor lived on e-mail. He was constantly available to give feedback and answer questions. The fact that he gave detailed feedback on the Diagnostic Project draft was a huge amount of help." (Student, Fall, 2011 semester).

The data indicated that while some participants did not like the perceived large amount of work, and a few felt that they were overly responsibly for their own learning, the communication mechanisms established in the course were beneficial. On the whole, though, participants expressed mixed reactions about their experiences in the course. However, over time student responses seemed more favorable toward online course delivery.

How do Teacher Candidates' Demonstrate their Understanding of Course Content?

In order to explore this question, participants' online reflections and their curriculum projects were analyzed. In both cases, analysis was constrained to topics related to standards-based instruction, specifically pedagogies, such as posing cognitively-demanding mathematical tasks, asking higher-level questions, and the use of manipulatives.

Mathematical Tasks in Lesson Plans

In the Culminating Diagnostic Project, teacher candidates were encouraged to use cognitively-demanding mathematical tasks, specifically either Procedures with Mathematical Connections tasks or Doing Mathematics tasks (Table 4). The analysis of the mathematical tasks found that there were a substantial amount of Procedures with Mathematical Connections tasks in all lesson plans across all courses (Table 5). Further, there was no difference noted between the online and hybrid courses; teacher candidates in both courses used cognitively-demanding tasks regardless of the course format.

Based on the data in Table 5, there was little difference between the number of cognitively-

Table 5. Types of mathematical tasks in lesson plans

			M	P w/out C	P w/ C	DM
Fall, 2008	Online	28 students	0	3 (1.08%)	245 (88.45%)	29 (10.47%)
	Hybrid	28 students	0	5 (1.67%)	262 (87.33%)	33 (11%)
Fall, 2009	Hybrid	32 students	0	6 (1.63%)	325 (88.08%)	38 (10.3%)
Fall, 2010	Hybrid	44 students	0	5 (1.05%)	422 (88.47%)	50 (10.48%)
Fall, 2011	Online	39 students	0	7 (2.21%)	286 (90.22%)	24 (7.57%)
Total		171 students	0	26 (1.49%)	1540 (88.51%)	174 (10.0%)

M: Memorization
P w/ out C: Procedures without Connections
P w/ C: Procedures with Connections
DM: Doing Mathematics

demanding tasks labeled Procedures with Connections or Doing Mathematics. A one-way t-test was conducted to compare the types of mathematical tasks between online and hybrid courses. In SPSS 17.0, tasks were coded in the following way: Memorization (1), Procedures without Connections (2), Procedures with Connections (3), Doing Mathematics (4). The t-test results showed no statistical significant difference between course format and the types of mathematical tasks in lesson plans, $t\,(1738) = -1.159$, $p = 0.247$.

Online Reflections

Data from online reflections focused largely on teacher candidates' responses to the various modules and course activities. Most of these activities included watching a video or reading a classroom vignette and responding about both the teacher's and students' actions.

Data across all courses showed that participants reported favorable impressions of the use of cognitively-demanding mathematical tasks and higher-level questions during the vignettes. Participants shared that the vignettes provided opportunities to see these concepts in action in classrooms, which gave a clearer picture of what they look like.

"The video clips were a huge help. In the first course and this one, we were taught about the value of posing good math tasks and asking high level questions. I haven't seen that though in my clinical classrooms, so I didn't know what it looked like until I saw these videos." (Student, Fall, 2011 semester)

Many students shared how the use of video vignettes coupled with clinical experiences helped increase their buy-in about how the type of tasks and questions influence students' learning.

"Until I saw the videos, I didn't really believe math could be taught this way. The teacher in the first video posed a hard task, let students struggle, and then asked questions to help them. The approach actually worked. I was surprised, but it was nice to see." (Student, Fall, 2011 semester).

In some cases the online reflections provided an opportunity for teacher candidates to spend time on more broad issues, such as classroom management issues.

"In the [vignette] that we read, I was mostly interested in how the teacher managed classroom behavior and her system of disciplining students. It was interesting to see how she dealt with students who acted out during math class." (Student, Fall, 2008 semester).

Another student commented, "I was surprised at how wild the students were on the video. I am not sure I could even start a math lesson if a class were behaving that way." (Student, Fall, 2011 semester).

The online reflections provided an opportunity for teacher candidates to process what they were gleaned from these video and text-based vignettes with topics that they were learning. Still, as seen above, some comments were not focused on the mathematics, but about much more broad issues, such as classroom management.

Curriculum Reflections

The goal of the curriculum reflection was to have students share their views on the standards-based mathematics textbook that they analyzed in light of what they have read and their beliefs about how students learn mathematics. The analysis of teacher candidates' curriculum reflections from the 2008 semester indicated that teacher candidates acquired a value for standards-based curriculum, but their views of curriculum were much more focused on the supplemental materials and teacher support than the types of tasks included in the curriculum.

"I have learned to value this type of mathematics teaching with lots of hands-on activities and engaging activities for students. However, the curriculum that I looked at did not have as many manipulatives or books as the textbook that my school has." (Student, Fall, 2008 semester).

"All of the good tasks in the world don't help, if the textbook doesn't give me a detailed picture of what the lesson should look like. I feel like I need a textbook that has a lot of different resources so I can choose which ones to use with my students." (Student, Fall, 2008 semester).

During the 2011 online course, teacher candidates reported that their preferences were largely influenced by their 15 hours of clinical experiences and by observing their clinical teachers' work with curriculum.

"My teacher uses standards-based curriculum in third grade, and it is amazing to hear how the students are talking about mathematics. At first, I thought that I would favor a traditional book with more problems and just hands-on activities, but I really like the idea of standards-based materials that have rich tasks and a lot of opportunities for students to share their ideas and strategies."

The influence of clinical teachers also negatively influenced teacher candidates' feelings about curriculum. This was especially true in Grades 3 through 5, which have a high-stakes state-wide assessment to take later in the year. "My grade level was forced to use standards-based books and it is too hard for these students. They don't have any freedom and I would hate being in that position."

The reflections from students in the hybrid courses during 2009 and 2010 focused more on the quality of mathematical tasks. For example, "I think that my ideal textbook is one with demanding tasks, and opportunities for students to use a lot of hands-on manipulatives. I think teacher support is important, but not as important as a book with good mathematical tasks." (Student, Fall, 2009 semester).

Another commented, "A textbook is just a resource. If it has high-level tasks which are hard to find, then that is a huge plus for teachers."

In general, reflections from online courses focused much more on the use of manipulatives in 2008, and their observations from clinical experiences in 2011. Responses from teacher candidates in the hybrid sections spent much more time talking about how a curriculum is a set of resources and the quality of mathematical tasks is the top factor in its effectiveness.

To what Extent do Teacher Candidates' Impact Student Learning on the Courses' Culminating Diagnostic Project?

On the Culminating Diagnostic Project, teacher candidates taught a small group of 3-5 students after conducting a pre-assessment. After the lessons, a post-assessment was also given. Table 6 shows the percentage of students showing growth between the pre- and post-assessments. Over 97 percent of students showed growth between assessments in all courses.

As shown in the table nearly every elementary school student (97.84%) made growth between their pre-test and post-test. A t-test comparing online to hybrid models showed no statistically significant difference between course formats, $t(555)= -0.103$, $p = 0.918$.

How Has Data Informed Revisions and Modifications of the Course?

The modifications of ELED 5301, including the development of an online course has been largely based on data collected from teacher candidates during the course, course evaluations, and through conversations with past students after they had completed the course and their student teaching

Table 6. Percent of students showing growth

Semester	Format	Teacher candidates	Total Students	Students Showing Growth	Percent of Students Showing Growth
Fall, 2008	Online	28	111	108	97.30
	Hybrid	28	110	107	97.27
Fall, 2009	Hybrid	32	98	96	97.96
Fall, 2010	Hybrid	44	125	123	98.40
Fall, 2011	Online	39	113	111	98.23
Total		171	557	545	97.84%

semester. The revisions mainly fall into three categories: mathematics content, pedagogies, and preparation to teach.

Mathematics Content

This course was originally designed to address geometry, data and non-number sense concepts. The first course focuses heavily on number sense and the use of manipulatives to facilitate computation. However, as I continued to teach the course, especially in online and hybrid formats, I noticed that students' understanding of mathematical content in many areas, especially fractions and algebra were incredibly weak. This is incredibly troubling, as a large portion of content emphasized in the upper elementary school grades is fractions and algebra. During student teaching, teacher candidates anecdotally report a desire to spend more time working on fractions and algebra. Therefore, the course content since 2008 has focused more on fractions and algebra. Geometry is still taught, but a majority of the content is fractions and algebra.

Pedagogies

One of the goals of the course is to continue to develop teacher candidates' understanding of standards-based mathematics with a focus on cognitively-demanding tasks and high-level questions that address students' mathematical thinking. The online modules in this course have provided a venue to use more video vignettes in the course. Currently, teacher candidates view videos from *My Kids Can* (Storeygard, 2009) and the Annenberg Mathematics Library (http://www.learner.org/resources/series32.html). Prior to the 2010 semester, the Annenburg videos were the only resource used. Teacher candidates responded favorably to the use of videos in online modules, so *My Kids Can* was adopted as the course textbook. It includes video and text-based vignettes.

The focus on mathematical tasks and questions was emphasized in the videos, but also as students observe in classrooms, and plan and teach lessons. As teacher candidates wrote lesson plans and reflections from their experiences, emphasis was heavily placed on the types of mathematical tasks and the questions that were posed.

Teaching Experiences

During the 2008 semester, the number of clinical projects that teacher candidates completed in elementary school classrooms was increased. Based on program data about the lack of time that our teacher candidates spend in classrooms, the current course stresses course experiences with elementary school students (Table 7).

Based on the data collected during the past few years, teacher candidates have expressed a higher level of preparedness to teach mathematics after student teaching. Part of that is likely contributed to the amount of time that they spend in

Table 7. Current clinical experiences in the course

Experience	Estimated Time
Culminating Diagnostic Project	7 hours
Observing Mathematics Lessons	3 hours
Teaching a Whole Group Lesson	1-2 hours
Student Interviews	1-3 hours

mathematics-related activities during their clinical experiences.

The potential downside of the intensive amount of clinical experiences is the quality of the classroom where teacher candidates are placed. With the courses' emphasis on standards-based mathematics instruction, teacher candidates who are placed in classrooms where the teacher does not use cognitively-demanding tasks or high-level questions express frustration and a disconnect between what they are learning in the course and seeing in the field.

DISCUSSION AND IMPLICATIONS

The evaluation data presented above describe the influence of an online mathematics education methods course for elementary school teacher candidates. The findings above extend the knowledge base on online courses and have relevant implications for future teaching of online mathematics education methods courses.

Clearly Stating Course Expectations to Teacher Candidates

Students' feedback of the online course focused heavily on their perceptions that they had to take a lot of responsibility for their own learning and there was more work involved than a face-to-face course. This finding extends the work of prior studies that found that online courses were more successful when they included relevant and challenging assignments (Levin et al., 2001) and opportunities for critical thinking (Delfino & Persico, 2007).

Consistent with Moallem's (2003) work, the data indicated that the gradual refinement of instructions led to less student frustration about the clarity of assignments. At the start of the 2011 semester the instructor e-mailed students to clearly explain the nature of the course, time demands, and the philosophy behind the course's focus on projects and self-paced learning. Providing clarity helped students to have a better understanding of course expectations. Online instructors of methods courses might also lean more towards a face-to-face orientation meeting prior to the beginning of the semester, or incorporating synchronous experiences to increase the social nature of courses (Im & Lee, 2004; Tallent-Runnels et al., 2006). Future studies should examine ways that course instructors can clearly communicate expectations to students, and the influence that those initial and during-course messages have on the quality of students' work.

Ensuring Effective Clinical Activities

All versions of this course described leveraged numerous clinical activities to provide teacher candidates with experiences in elementary school classrooms. The most recent version of the National Council for Accreditation in Teacher Education (NCATE) Standards (2010) heavily emphasizes the need for teacher candidates to participate in worthwhile clinical activities throughout their program. In a 100% online course, clinical activities are more critical since teacher candidates do not have face-to-face course meetings to see the instructor model methods or to discuss these pedagogies. In this study, some teacher candidates completed their clinical requirements with teachers who used standards-based pedagogies that aligned to course objectives. These students reported a high degree of buy-in that these strategies are effective. Schwartz (in press) noted that online courses are limited in that they cannot easily model pedagogies as well as the teacher action of in-the-moment decision making that occurs frequently in classrooms. Instructors of online methods courses for teacher candidates need to ensure that students are placed in classrooms that embody the types of pedagogies emphasized during the course.

Future studies should more closely examine the influence of clinical activities on teacher candidates' knowledge and skills. As O'Connor (2011) noted, the online courses did not effectively prepare pre-service teachers to implement standards-based pedagogies that were emphasized during the course. In an online course, these pedagogies and skills related to in-the-moment decision making must be included in clinical activities and in vignettes that are included in the courses.

Creating a Feeling of Support

Teacher candidates frequently reported comments about the amount of support that they received during the course. In this study, the courses' heavy focus on projects and the application of course content in elementary school classrooms created a high level of stress and the perception that there was a lot of work in the course. For many of the teacher candidates, this was the first time they taught the same group of students multiple lessons. The course instructor provided feedback by helping them to revise their assessments and gave them feedback about their lesson plans. By providing feedback and giving suggestions about lesson plans, teacher candidates reported feeling supported while working on this major assignment. This aligns with Moallem's (2003) findings that students in online courses need to have clear expectations, including information about rubrics and evaluation.

Greene and Land (2000) found that students had clearer expectations of assignments and produced higher quality work products when instructors provided guiding questions and frequent amounts of feedback. This study extends the results of their work. In future studies, there is a need to examine how the different types and manners in which feedback is given influence the quality of students' work products.

Providing Opportunities for Communication

In this study, teacher candidates communicated with each other via their online reflections on either WordPress or Posterous. These required reflections allowed teacher candidates to communicate their thoughts about vignettes that they had examined, their clinical experiences, and other issues related to mathematics teaching and learning. By reading and responding to their classmates' reflections, students were able to interact with their classmates and have shared experiences. Based on the feedback from teacher candidates, this helped reduce the feeling of isolation that is common in online courses (Tallent-Runnels et al., 2006).

The requirement to participate in these online discussions is critical; very few students participated in the optional activities of contributing to and viewing the class wiki. If posting and responding to online reflections was optional, some students would likely not participate at all. This extends the work of Im and Lee (2003/2004), who found that creating asynchronous communication effectively supported discussion of content, but not social conversations. The addition of synchronous communication might have increased participants' willingness to also participate in the optional course activities on the wiki. Future studies should examine types of synchronous and asynchronous communication and compare which types may best support pre-service teachers in a methods course.

CONCLUDING THOUGHTS

Impact of Course

The online mathematics methods course provided an opportunity for teacher candidates to develop the knowledge and skills related to teaching elementary school mathematics. Based on the evaluation data, teacher candidates demonstrated knowledge about mathematics, an understanding of planning cognitively-demanding mathematical tasks, gains in student learning outcomes, and relatively positive perceptions of their experience. While this study was not a direct comparison between the impact of online and face-to-face versions of the same course, the data indicate that the course experience successfully developed aspects of teacher candidates' mathematical knowledge for teaching (MKT) and positively impacted their students.

As program faculty continue to decide whether or not this course is worthwhile to teach in a 100% online format, the evaluation data suggests that it has potential, under the conditions that the course supports communication between students, and the course instructor readily and frequently provides feedback on assignments.

Limitations

While this study provided data about the benefit of an online methods course for pre-service teachers, this study includes some limitations, making it difficult to generalize the findings across various contexts. While the data analysis process included multiple iterations of reading, coding, and reanalyzing the data, the data collected provides a description of the course on a few levels of Guskey's evaluation model. Further, the impact on student learning is limited, as nearly every student showed growth, but it is difficult to acknowledge how much growth was made without a group of control students to compare.

While the findings from this chapter come with limitations, there is enough data regarding the effectiveness to continue to pursue the refinement

Table 8. Research questions for examining online methods courses for teacher candidates

Level		Questions examined	Data Sources
Level 1	Teachers' reactions	How do teachers feel about their experience? How do they feel about course structures (e.g., communication, course feedback)? How do they feel about course activities (e.g., projects, assignments)?	Course evaluation forms Surveys Interviews Focus Groups
Level 2	Teachers' knowledge and skills	What knowledge and skills do teachers learn during their experiences? What knowledge related to content do teachers learn? What knowledge related to pedagogy do teachers learn? What knowledge related to assessment do teachers learn?	Course assignments Lesson plans Exams
Level 3	Use of knowledge and skills	How do teachers use their new knowledge and skills in their own classroom? To what extent and how do teachers apply emphasized pedagogies? To what extent and how do teachers modify instruction while using emphasized pedagogies?	Student work samples Videos of classroom teaching
Level 4	Impact on organization and program	How did the experience impact the program or organization responsible for providing the experience? What revisions are made to course structures between semesters? What revisions are made to course assignments between semesters?	Course documents, including syllabi, assignments, rubrics Interviews Surveys
Level 5	Impact on student learning	How are PK-12 students impacted by the course? How well do PK-12 students perform on assessments compared to those not influenced by the course?	Student work samples Assessments Student interviews

of an online mathematics education course and continue to research the impact of the course. To that end, future studies should continue to evaluate the impact of the course on teacher candidates as well as K-6 students.

Directions for Future Research

Guskey's (2000) framework for evaluating opportunities for teacher learning provided a research-based foundation to examine how the course influenced various aspects of teacher candidates, the institution and students. I close this chapter with a set of possible questions worth examining for those interested in researching online methods courses for teacher candidates (Table 8).

Due to the cyclical nature of teaching, where instructors teach a semester, and then revise their course based on data, including feedback from their students as well as their own reflections, design-based research ([DBR]; Edelson, 2002; Reeves, Herrington, & Oliver, 2005) offers a potential way of framing research studies regarding online teaching. DBR is based off a preliminary theory that is grounded in a combination of empirical and anecdotal information. The theory leads to a proposed intervention, which is then carried out. Based on formative data that is collected and analyzed during the study, researchers make decisions regarding both the implementation of their intervention as well as make conclusions from the data.

In the case of online teaching, course instructors' structures and content is based off a combination of empirical and anecdotal information. As data is collected during and after the semester, the instructor makes empirically-based decisions, which influence the course, as well as provide information about the influence of the course on their students. In the case of teacher candidates, these courses influence not only teacher candidates, but also the students that they work with during their clinical experiences. DBR provides a systematic way for instructors and researchers involved with online teaching to analyze the impact of the course on students and others involved in the teaching and learning process.

REFERENCES

Ajjan, H., & Hartshorne, R. (2008). Investigating faculty decisions to adopt Web 2.0 technologies: Theory and empirical tests. *The Internet and Higher Education*, *11*(2), 71–80. doi:10.1016/j.iheduc.2008.05.002

Ball, D. L., Thames, M. H., & Phelps, G. (2008). Content knowledge for teaching: What makes it special? *Journal of Teacher Education*, *59*(5), 389–407. doi:10.1177/0022487108324554

Bogden, R. R., & Biklen, S. K. (2003). *Qualitative research in education: An introduction to theories and methods* (4th ed.). Boston, MA: Allyn & Bacon.

Carter, J. F. (2009). Lines of communication: Using a wiki in a mathematics course. *Problems, Resources, and Issues in Mathematics Undergraduate Studies*, *19*(1), 1–17.

Cohen, S. (2005). *Teachers' professional development and the elementary mathematics classroom: Bringing understandings to light*. Mahwaw, NJ: Lawrence Erlbaum Associates, Inc.

Delfino, M., & Persico, D. (2007). Online or face-to-face? Experimenting with different techniques in teacher training. *Journal of Computer Assisted Learning*, *23*, 351–365. doi:10.1111/j.1365-2729.2007.00220.x

Edelson, D. C. (2002). Design research: What we learn when we engage in design. *Journal of the Learning Sciences*, *11*(1), 105–121. doi:10.1207/S15327809JLS1101_4

Fishman, B. J., Marx, R. W., Best, S., & Tal, R. T. (2003). Linking teachers and student learning to improve professional development in systemic reform. *Teaching and Teacher Education*, *19*(6), 643–658. doi:10.1016/S0742-051X(03)00059-3

Garet, M., Porter, A., Desimone, L., Briman, B., & Yoon, K. (2001). What makes professional development effective? Analysis of a national sample of teachers. *American Educational Research Journal*, *38*(4), 915–945. doi:10.3102/00028312038004915

Glazer, E. M., & Hannafin, M. J. (2006). The collaborative apprenticeship model: Situated professional development within school settings. *Teaching and Teacher Education*, *22*(2), 179–193. doi:10.1016/j.tate.2005.09.004

Greene, B. A., & Land, S. M. (2000). A qualitative analysis of scaffolding use in a resource-based learning environment involving the World Wide Web. *Journal of Educational Computing Research*, *23*(2), 151–180. doi:10.2190/1GUB-8UE9-NW80-CQAD

Guskey, T. R. (2000). *Evaluating professional development*. Thousand Oaks, CA: Corwin Press.

Heck, D. J., Banilower, E. R., Weiss, I. R., & Rosenberg, S. L. (2008). Studying the effects of professional development: The case of the NSF's local systemic change through teacher enhancement initiative. *Journal for Research in Mathematics Education*, *39*(2), 113–152.

Hodges, C. B., & Hunger, C. M. (2011). Communicating mathematics on the Internet: Synchronous and asynchronous tools. *TechTrends*, *55*(5), 39–44. doi:10.1007/s11528-011-0526-4

Im, Y., & Lee, 0. (2004). Pedagogical implications of online discussion for preservice teacher training. *Journal of Research on Technology in Education*, *36*, 155–170.

Ko, S., & Rosen, T. (2010). *Teaching online: A practical guide*. New York, NY: Routledge.

Levin, S. R., Waddoups, G. L., Levin, J., & Buell, J. (2001). Highly interactive and effective online learning environments for teacher professional development (electronic version). *International Journal of Educational Technology*, *2*. Retrieved from http://smi.curtin.edu.au/ijet/v2n2/slevin/index.html

Loucks-Horsley, S., Love, N., Stiles, K. E., Mundry, S., & Hewson, P. W. (2009). *Designing professional development for teachers of science and mathematics* (3rd ed.). Thousand Oaks, CA: Corwin Press.

Moallem, M. (2003). An interactive online course: a collaborative design model. *Educational Technology Research and Development*, *51*(4), 85–103. doi:10.1007/BF02504545

National Council for Accreditation of Teacher Education. (2010). *NCATE unit standards*. Retrieved from http://www.ncate.org/standards/ncateunitstandards/unitstandardsineffect2008/tabid/476/default.aspx

National Partnership for Excellence and Accountability in Teaching (NPEAT). (2000). *Revisioning professional development: What learner-centered professional development looks like*. Oxford, OH: Author. Retrieved from http://www.nsdc.org/library/policy/npeat213.pdf

Nelson, J., Christopher, A., & Mims, C. (2009). TPACK and web 2.0: Transformation of teaching and learning. *TechTrends*, *53*(5), 80–85. doi:10.1007/s11528-009-0329-z

O'Connor, E. A. (2011). The effect on learning, communication, and assessment when student-centered Youtubes of microteaching were used in an online teacher-education course. *Journal of Educational Technology Systems*, *39*(2), 135–154. doi:10.2190/ET.39.2.d

Oliver, K. (2010). Integrating Web 2.0 across the curriculum. *TechTrends*, *54*(2), 50–60. doi:10.1007/s11528-010-0382-7

Orrill, C. H. (2001). Building learner-centered classrooms: A professional development framework for supporting critical thinking. *Educational Technology Research and Development*, *49*(1), 15–34. doi:10.1007/BF02504504

Peterson, E. (2009). Using a wiki to enhance cooperative learning in a real analysis course. *PRIMUS (Terre Haute, Ind.)*, *19*(1), 18–28. doi:10.1080/10511970802475132

Polly, D., & Hannafin, M. J. (2010). Reexamining technology's role in learner-centered professional development. *Educational Technology Research and Development*, *58*(5), 557–571. doi:10.1007/s11423-009-9146-5

Reeves, T. C., Herrington, J., & Oliver, R. (2005). Design research: A socially responsible approach to instructional technology research in higher education. *Journal of Computing in Higher Education*, *16*(2), 97–116. doi:10.1007/BF02961476

Russell, M., Carey, R., Kleiman, G., & Venable, J. D. (2009). Face-to-face and online professional development for mathematics teachers: A comparative study. *Journal of Asynchronous Learning Networks*, *13*(2), 71–87.

Schifter, D., Bastable, V., & Russell, S. J. (2008). *Developing mathematical ideas*. Parsippany, NJ: Dale Seymour.

Schlager, M. S., & Fusco, J. (2004). Teacher professional development, technology, and communities of practice: Are we putting the cart before the horse? In Barab, S., Kling, R., & Gray, J. (Eds.), *Designing virtual communities in the service of learning* (pp. 120–153). Cambridge, MA: Cambridge University Press. doi:10.1080/01972240309464

Schwartz, C. S. (2012). Counting to 20: Online implementation of a face-to-face, elementary mathematics methods problem-solving activity. *TechTrends*, *56*(1), 34–39. doi:10.1007/s11528-011-0551-3

Signer, B. (2008). Online professional development: Combining best practices from teacher, technology and distance education. *Journal of In-service Education*, *34*(2), 205–218. doi:10.1080/13674580801951079

Spicer, J. (2002). Even better than face-to-face? In Thorson, A. (Ed.), *By your own design: A teacher's professional learning guide* (pp. 32–33). Columbus, OH: Eisenhower National Clearinghouse for Mathematics and Science Education.

Stein, M. K., Grover, B. W., & Henningsen, M. (1996). Building student capacity for mathematical thinking and reasoning: An analysis of mathematical tasks used in reform classrooms. *American Educational Research Journal*, *33*, 455–488.

Storeygard, J. (2009). *My kids can*. Boston, MA: Heinemann.

Tallent-Runnels, M. K., Thomas, J. A., Lan, W. Y., Cooper, S., Ahern, T. C., Shaw, S. M., & Liu, X. (2006). Teaching courses online: A review of the research. *Review of Educational Research*, *76*(1), 93–135. doi:10.3102/00346543076001093

Thames, M. H., & Ball, D. L. (2010). What mathematical knowledge does teaching require? Knowing mathematics in and for teaching. *Teaching Children Mathematics*, *17*(4), 220–225.

Treacy, B., Kleiman, G., & Peterson, K. (2002). Successful online professional development. *Leading & Learning with Technology*, *30*(1), 42–47.

Work, A. P. A. Group of the Board of Educational Affairs. (1997). *Learner-centered psychological principles: A framework for school reform and redesign*. Washington, DC: Author.

ADDITIONAL READING

Ball, D. L., Thames, M. H., & Phelps, G. (2008). Content knowledge for teaching: What makes it special? *Journal of Teacher Education*, *59*(5), 389–407. doi:10.1177/0022487108324554

Ko, S., & Rosen, T. (2010). *Teaching online: A practical guide*. New York, NY: Routledge.

Moallem, M. (2003). An interactive online course: A collaborative design model. *Educational Technology Research and Development*, *51*(4), 85–103. doi:10.1007/BF02504545

Tallent-Runnels, M. K., Thomas, J. A., Lan, W. Y., Cooper, S., Ahern, T. C., Shaw, S. M., & Liu, X. (2006). Teaching courses online: A review of the research. *Review of Educational Research*, *76*(1), 93–135. doi:10.3102/00346543076001093

KEY TERMS AND DEFINITIONS

Asynchronous: A format of an online course in which the students and course instructor communicate electronically but not in real-time.

Blog: A shortened phrase for weblog, an online system in which users can include text, audio, or video, and others can comment on their postings.

Clinical Experience: Activities that are completed by teacher candidates in PK-12 classrooms.

Hybrid: A format of an online course in which students have a combination of face-to-face meetings and online modules.

Mathematical Task: Any mathematical problem or activity completed by a student.

Methods Course: A course in a teacher education program that specifically focuses on content and pedagogy related to a topic (e.g., mathematics, science, social studies).

Synchronous: A format of an online course in which the students and course instructor communicate in real-time, typically in a computer system that has audio and/or a chat room interface.

Chapter 19
Taking Action Research in Teacher Education Online: Exploring the Possibilities

Nancy Fichtman Dana
University of Florida, USA

Desi Krell
University of Florida, USA

Rachel Wolkenhauer
University of Florida, USA

ABSTRACT

The systematic, intentional study by teachers of their own classroom practice is critical for powerful professional development. Action research, or practitioner inquiry, provides teachers with a vehicle to engage in this professional development in order to raise teacher voices in educational reform and capture and share the knowledge generated by teachers within their classrooms. The quality of any piece of action research completed by a practitioner inquirer is directly related to the coaching s/he receives in the process, but the scope and reach of teacher educators' action research coaching is often constrained by limitations of time and space. Extending the coaching of action research to online environments may provide possibilities for negotiating challenges of time and space and enhance both the quantity and quality of the teacher educator's action research coaching opportunities. The purpose of this chapter is to explore online tools that can facilitate distance action research coaching.

INTRODUCTION

Many teacher education programs include action research as a core component of their programs at the pre-service level during the student teaching experience as well as the in-service level as the capstone experience to earn an advanced graduate degree (Cochran-Smith & Lytle, 2009). Simply stated, action research, or practitioner inquiry, is defined as systematic, intentional study by educators of their own professional practice (Cochran-Smith & Lytle, 1993, 2009). Inquiring professionals seek out change by reflecting on their practice. They do this by engaging in a

DOI: 10.4018/978-1-4666-1906-7.ch019

cyclical process of posing questions or "wonderings," collecting data to gain insights into their wonderings, analyzing the data along with reading relevant literature, taking action to make changes in practice based on new understandings developed during inquiry, and sharing findings with others (Dana & Yendol-Hoppey, 2009).

While many educational innovations have come and gone, the systematic study of teachers' own classroom practice is a concept that has proved its staying power, rooted in the work of John Dewey (1933), popularized by Kurt Lewin in the 1940s (Adelman, 1993), and shortly thereafter applied to the field of education by Stephen Corey (1953). Whether we refer to this process as classroom research, teacher research, action research, teacher inquiry, or some other name, three main reasons exist for the longevity of this concept: (1) The process has proven to be a powerful tool for teacher professional development (Zeichner, 2003); (2) The process has become an important vehicle to raise teachers' voices in educational reform (Meyers & Rust, 2003); and (3) The process is a mechanism for expanding the knowledge base for teaching in important ways (Cochran-Smith & Lytle, 1993, 2009).

Because systematic, intentional study by teachers of their own classroom practice is critical for powerful professional development, raising teachers' voices in educational reform, and capturing and sharing the knowledge generated by teachers from within the four walls of their classrooms, engagement in teacher inquiry continues to be a process that is built into the fabric of many teacher education programs across the nation (see, for example, Cochran-Smith & Lytle, 2003, 2009; Crocco, Bayard & Schwartz, 2003; Dana, Silva, & Snow-Gerono, 2002). An important role the teacher educator plays is coaching action research, as the quality of any piece of action research completed by a practitioner inquirer is directly related to the coaching s/he receives in the process (Drennon & Cervero, 2002; Dana & Yendol-Hoppey, 2008; Poekert, 2010; Smeets & Ponte, 2009).

The scope and reach of teacher educators' action research coaching is often constrained by limitations of time and space. It is difficult to find time to bring teacher inquirers together to support one another in the action research cycle. It is also difficult to bring student teachers or practicing teachers in different schools or districts together because of geographic distance between places of practice that can inhibit face-to-face meeting time. Extending the coaching of action research to online environments may provide possibilities for negotiating challenges of time and space and enhance both the quantity and quality of the teacher educator's action research coaching opportunities.

In addition, Appana (2008) summarized some of the benefits of online learning that include increased learning, improved student interaction, increased satisfaction, availability of more learning resources, opportunity to bring in experts from different locations, promotion of life-long learning, flexibility, and opportunities for reflection. All of these benefits of online learning have the potential to be actualized in relationship to coaching the action research process online as pre-service and/or in-service teachers interact within a virtual classroom space to learn with and from each other through the investigation of their own classroom practice.

Because of the many benefits of online learning, a number of virtual teacher professional development initiatives are being introduced and studied, some of which contain an action research component (Dede, 2006). For example, Project WIDE (Wide-scale Interactive Development for Educators) World Online Professional Development at Harvard University is a job-embedded model of online professional development that contains an action research course. While the benefits of engagement in action research for teachers and online learning in general are being actualized in the creation of virtual teacher professional development programs like Project WIDE

World Online Professional Development, all of the details, particulars, and possibilities inherent in coaching action research online have not been fully articulated.

Hence, the purpose of this chapter is to explore online tools that can facilitate action research coaching. Following a background overview of action research and coaching, we will name and discuss five critical junctures in the action research process, including an overview of how each critical juncture has historically been addressed in face-to-face environments and the ways these strategies can be translated into the online context utilizing various synchronous and asynchronous technology tools such as Elluminate, web and telephone conferencing, and learning management systems and social networks such as Schoology and ARTI, an online tool created by Dawson (2009) that is further explored in this book in chapter 20.

BACKGROUND: ACTION RESEARCH AND COACHING

Action research provides pre-service and in-service teachers with a form of professional learning that allows them to focus on some specific aspect of their classroom or teaching that will contribute to improved outcomes for student learning. This process, unlike one-shot workshops, happens over an extended period of time and follows a cycle of inquiring, research, implementing, and reflecting that occurs over a period of months, an entire year, or even across several school years. The cycle is an active process for teachers and provides them space to critically examine their own practices and beliefs while also allowing them to interact with colleagues and external resources, such as teacher educators, in order to gain additional insights into their own practices. In fact, action research provides a reconceptualization of professional development that, in line with Webster-Wright (2009), focuses on professional learning as a holistic process rather than a string of potentially unrelated events aimed at helping teachers develop professionally. It initiates and reinforces a process of experiential learning and reflection that are tightly interwoven with pre-service and in-service teachers' immediate contexts (Webster-Wright, 2009). Moreover, action research can, in effect, either be a step toward developing or part of an established inquiry stance, embracing Cochran-Smith and Lytle's (2001) concept of "knowledge-*of*-practice" which assumes that

> *the knowledge teachers need to teach well is generated when teachers treat their own classrooms and schools as sites for intentional investigation at the same time that they treat the knowledge and theory produced by others as generative material for interrogation and interpretation. (p. 48)*

Teacher research affords the opportunity for in-service teachers to engage in professional development meaningful to their direct needs and contexts (classrooms, schools, content knowledge, pedagogy, students) (Cochran-Smith & Lytle, 1993, 2009; Crockett, 2002; Dana & Yendol-Hoppey, 2008, 2009; King, 2002; Poekert, 2010, Wolkenhauer, Boynton, & Dana, 2011) and for pre-service teachers to become socialized into the profession as inquirers (Dana & Yendol-Silva, 2001). They are allowed to "become students of their craft" (King, 2002), investigating a specific area of concern and acting in response to that investigation, ultimately to benefit student learning (Cochran-Smith & Lytle, 1993, 2009; Dana & Yendol-Hoppey, 2008, 2009; King, 2002; Poekert, 2010). Furthermore, action research pulls educators out of isolation in order to work collaboratively with other pre-service and in-service teachers to share knowledge and experiences, valuing the individual expertise and insights each can provide (King, 2002; Smeets & Ponte, 2009; Snow-Gerono, 2005).

In order to help cultivate the conditions for inquiry, the action research coach plays a vital and delicate role that ultimately impacts the quality of the experience for teacher researchers (Drennon & Cervero, 2002). Coaches have a unique function in this process that requires them to develop relationships with teachers, negotiate power dynamics between administrators and teachers, challenge teachers to think outside their norms, keep work sessions focused, and make relevant resources available to them (Aubusson et al., 2009; Drennon & Cervero, 2002; Dana & Yendol-Hoppey, 2008; Gallimore et al., 2009; Nelson & Slavit, 2007; Poekert, 2010; Smeets & Ponte, 2009). Some action research coaches may come from universities, especially when action research is embedded as part of a teacher education program, while others may come from within schools or districts. Thus, teacher educators may have to employ extra time and energy into developing relationships and fostering trust with teachers in order to allay fears or concerns teachers may have about the coach coming from outside the immediate school context (Aubusson et al., 2009; Drennon & Cervero, 2002). This may be particularly crucial when translating coaching to the virtual world in which there is a lack of the face-to-face interactions that tend to be more prevalent in the realm of education.

Although some pre-service or in-service teachers may be inherently motivated to engage in action research or collaborate with colleagues, an action research coach helps to foster and maintain an environment for effective collaboration and inquiry to take place (Gallimore et al., 2009). As a coach, teacher educators essentially assume the role of a "critical friend:"

[T]he concept of "friend" implies someone who will listen and is trusted enough by colleagues for them to take risks... while the concept of "critical" implies that the relationship is sufficiently robust to cope with questions and differing viewpoint. Such a person can confront issues that have the potential to be taken for granted or unnoticed by the school community. (Aubusson et al., 2009, p. 76)

Teacher educators may be able to bring a different voice to the table when working with in-service teachers and an understanding of the professional world of teaching when working with pre-service teachers. Within the role as coach, teacher educators play devil's advocate, mediate conflict, encourage individuals, raise issues, share expertise and resources, act as team players (thereby relinquishing authority), and support action research as a process (Aubusson et al., 2009; Dana & Yendol-Hoppey, 2008; Drennon & Cervero, 2002; Gallimore et al., 2009; Nelson & Slavit, 2007; Poekert, 2010; Smeets & Ponte, 2009). The coach brings his/her own knowledge to the table while also encouraging others to contribute their own knowledge—and valuing both equally (Aubusson et al., 2009). Teacher educators knowledgeable in the action research process and equipped to play the role of critical friend can be a vessel for introducing action research effectively into new contexts.

In addition to the work of building relationships and negotiating the process of action research, the role of coach further requires that teacher educators guide teacher researchers through the five critical junctures of the action research process (Dana & Yendol-Hoppey, 2008), discussed in detail below; the support of the coach through each of these junctures sets the tone and can make or break the process for teacher researchers (Krell & Dana, 2011).

CRITICAL JUNCTURES IN THE ACTION RESEARCH PROCESS

There are five critical junctures in the action research process where coaching is essential: (1) introducing the action research process, (2) developing a wondering/research question, (3) developing a plan for research, (4) analyzing data, and (5) sharing work with others (Dana & Yendol-Hoppey, 2008). We define critical junctures as times in the action research process where the coaching a practitioner researcher receives is critical to the ultimate outcome and quality of the action research endeavor. In the next sections of this chapter, we will review each of these critical junctures, why they are important, how these critical junctures can be addressed in face-to-face environments, and finally, how the face-to-face strategies can be translated to the online context.

Introducing the Action Research Process

While the process of action research has been around for ages, there are many teachers who are still not familiar with the process or have misconceptions about what the process entails and the ways it differs from traditional university research. Because teachers are inundated with so many mandates and encouraged to try the latest educational innovations one after another (Killion & Pinata, 2011), they often meet the invitation to engage in practitioner research as a powerful mechanism for their own professional learning with a fair amount of skepticism. For these reasons, the introduction teachers receive to the action research cycle is the first critical juncture in the coaching process.

The introduction teachers receive to action research must provide a solid overview of the process, help teachers unpack their prior conceptions of educational research and explore the ways practitioner inquiry differs from large scale educational research, excite teachers about the possibilities inherent in studying their own classrooms, and place teachers at ease that indeed they are capable of seamlessly integrating the act of research into their everyday practice and teaching lives. The most common way to accomplish these goals in a face-to-face environment is through a workshop where teachers have the opportunity to explore action research with an expert teacher researcher or other educational consultant knowledgeable about the action research process. Additionally, having the opportunity to hear from teachers themselves as they talk about and share their own classroom-based research and reflect on the meaning engagement in practitioner research has meant to their teaching practice is an extremely valuable way to introduce teacher research to new inquirers.

The face-to-face workshop(s) to introduce teachers to the process of action research can be translated into the online context easily by utilizing Elluminate, or another similar online collaboration tool. This particular web conferencing instrument supports the facilitation of highly interactive teaching and learning. Participants are able to view and share presentations, talk in real time, use text-based chat, share files, and break into small group sessions. Elluminate's interactive whiteboard and collaboration tools ensure teachers have ample opportunity to explore action research while being supported by other teacher researchers. The web-based tool also stores recordings of the sessions for participants to use for future reference or to watch if they are unable to attend one of the synchronous workshops.

We have successfully utilized Elluminate to virtually introduce the process of action research to numerous first time teacher inquirers. We began by uploading a PowerPoint presentation into a designated Elluminate room. This first presentation serves to define action research and provide an overview of each component of the action research cycle (Figure 1).

Utilizing some of the unique features of Elluminate that allow for interaction, we created the space for session participants to share their thinking with others as they made sense of the action research process during the PowerPoint presentation. For example, when we reviewed the many terms that are often utilized interchangeably to describe the process (including action research, teacher research, practitioner inquiry, and classroom research), we shared that our preferred term is simply "inquiry" and proceeded to demonstrate why by asking all session participants to take a minute to search the internet and use the screen

Figure 1. PowerPoint presentation introducing action research via Elluminate

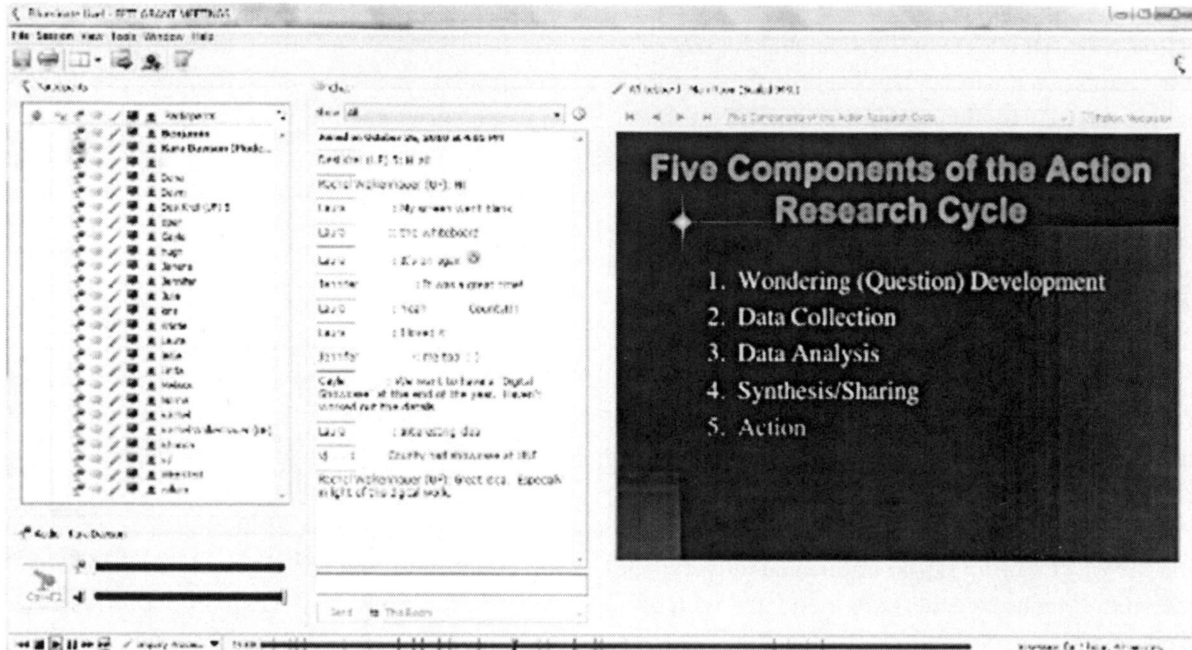

capture software Snagit to find an image that encapsulated what came to their minds when they heard the word "research." As participants found their images, they were able to paste them directly into the presentation on the Elluminate whiteboard and a powerful collage began to take shape with pictures of scientists in white lab coats, piles of books, and long numerical equations. We were able to utilize these images to alleviate participant anxiety about engaging in practitioner research by sharing that the images they selected were associated with large-scale research and antithetical to what the process of action research was all about. This short activity translated a very common face-to-face discussion facilitation strategy into this online context.

We were able to facilitate more informal interaction that typically takes place in face-to-face workshops as well. Participants were able to give virtual "thumbs up" or smiles when they understood content, were excited about upcoming activities, or were agreeing to support one another. They could raise virtual hands or frown to let workshop facilitators know their confusion, disagreement, or uneasiness with the process. Participants also had access to a continuous chat feature, allowing more specific questions and comments to be voiced to the whole group, a small group, or only the workshop facilitators. It worked well for us to have a co-facilitator in charge of the chat box while the others presented content and facilitated learning.

Following the introductory interactive presentation, we have also utilized Elluminate for "guest speaker" teacher inquirers to share their research with the participants during the introductory session. Illustrating with real life cases exemplifies each component of the action research cycle we defined in our opening PowerPoint. Hearing about real teacher research conducted by teachers in similar contexts to the participants' is perhaps the most powerful way to introduce the process to new teacher inquirers, and Elluminate is a fine mechanism to transcend the geographic distance that may exist between guest speaker teacher inquirers and those learning about the action research process for the first time. Elluminate facilitates this sharing particularly well, as the guest teacher researcher can easily show participants artifacts from his or her inquiry, share files that may be useful for future inquiries, and answer questions in real time.

While Elluminate can be a very effective way to translate a face-to-face workshop into the virtual environment, a potential challenge is that all of the participants may not be familiar with Elluminate and may not be knowledgeable with or have access to the tools necessary to participate in the workshop (i.e., the software Snagit in our example). To address this challenge, we learned that holding a brief 30-minute orientation to Elluminate and doing an equipment check prior to utilizing this web-conferencing tool to introduce the process of action research is essential.

Depending on the context of the action research coaching, another potential challenge can be scheduling a synchronous time for inquirers to meet for the action research introduction and other synchronous activities associated with coaching the AR process online described later in this chapter. If the coaching of action research is being done outside of a university course with a designated meeting time, it can be difficult to find a mutually agreed upon time for every inquirer one is coaching to meet virtually. To address this challenge, we learned that advanced planning and flexibility are essential. Clearly articulating the synchronous time meeting commitments early in the process and scheduling each synchronous meeting long before they occur so inquirers can place these commitments into their calendars is critical to attendance at and participation in synchronous online sessions. In addition, if we had difficulty finding one single time all of our inquirers could meet, sometimes we would offer two different times for attendance and offer the same session twice, giving our coachees time options for synchronous session attendance. Finally, if only one or two inquirers were unable to make a synchronous meeting time everyone else could agree to, we capitalized on an Elluminate feature that provides an advantage over face-to-face workshop sessions. Elluminate allows sessions to be recorded and shared so that individuals could view a session at an alternative time if unable to attend the initial live session. While we do not believe watching a recording of a session at a later date is optimal, it provided a viable way to troubleshoot the challenge of finding a mutually agreed upon meeting time across a number of teacher inquirers.

Developing a Wondering/ Research Question

Once teachers have been introduced to the action research process, igniting their own individual research journey begins by articulating a burning question they have about their practice. Burning questions, often referred to as "wonderings," emerge from issues, tensions, problems, and/or dilemmas teachers face when confronted each day with the complexities inherent in the daily act of teaching. As new teacher researchers articulate wonderings, it's important to note that

Rarely does any teacher researcher eloquently state his or her wondering immediately. It takes time, brainstorming, and actually 'playing' with the question...By playing with the wording of a wondering, teachers often fine-tune and discover more detail about the subject they are really passionate about understanding. (Dana & Yendol-Hoppey, 2009, p. 57-58)

Hence, the second critical component of coaching action research is creating a space for teachers to play with the possibilities for their research question(s).

Coaches often establishing a "wondering playground" by creating a space for new action researchers to discover, share, and reflect upon their felt difficulties or real world dilemmas. A series of exercises designed to help teachers reflect on their practice are available in a text we utilize in our online action research courses and in-service professional development work with districts. Chapter 2 of *The Reflective Educator's Guide to Classroom Research* (Dana & Yendol-Hoppey, 2009) provides numerous prompts to help teacher reflect on their passions for exploration through inquiry. It is in this reflective space that healthy, meaningful wonderings are born. In addition, coaches often create a space for new action researchers to "try out" the initial articulation of a wondering with other teachers and to enlist their colleagues' help in fine-tuning the wording and direction of the wondering. The most common way to accomplish these goals in the face-to-face environment is to have a small group of teacher researchers meet with one another to give and receive feedback on the questions they wish to explore for their own personal action research endeavors. These small group feedback meetings are facilitated by the action research coach as a part of a class session at the university or as an after-school meeting as a part of a district professional development program.

Just as Elluminate can be a viable online tool to introduce the process of action research to new teacher inquirers, Elluminate can be utilized to create the space for small groups of teacher researchers to meet and give and provide feedback to one another on emerging wonderings.

We have effectively utilized Elluminate to create a virtual space for small groups of first time teacher inquirers to meet with one another as a follow up to our Introduction to Action Research Elluminate session to work on the development of their wonderings. We began by offering a PowerPoint presentation that provided a large number of new teacher inquirers some tips for wondering development and for providing feedback on wonderings to their colleagues. These tips are available in another text we frequently use in our work about coaching inquiry. Chapter 3 of *The Reflective Educator's Guide to Professional Development: Coaching Inquiry-Oriented Learning Communities* (Dana & Yendol-Hoppey, 2008) focuses specifically on coaching others in the development of strong wonderings. Utilizing the breakout room feature of Elluminate, which provides private spaces for small groups of participants to collaborate, we then divided the large group of teacher inquirers into groups of four or five by dragging and dropping participants' names into designated rooms. In these rooms, each small group member had ten minutes to present their initial wondering and discuss it with members of their group. All of the same interactive tools afforded the whole group (i.e. "thumbs-up," audio, chat window, interactive whiteboard) were also available for small groups in breakout rooms. As the action research coaches and Elluminate moderators, we could visit each small group to listen in on their discussion and offer our suggestions for shaping their research questions.

Once again, the online tool Elluminate could be utilized to emulate small group discussion time that is often a component of face-to-face action research workshops or university action research class meetings. An added bonus through utilizing Elluminate is that groups of teacher researchers from different schools and districts could be placed in the same small groups without the time of physically travelling to one central meeting location, saving teacher researchers precious time and providing outside perspectives for each teacher researcher on his/her projected work.

Developing a Plan for Research

Once the process of action research is ignited with the birth of a wondering, a crucial next step is the development of a research plan. In the absence of a well-developed plan for inquiry, "teacher researchers risk making little or no progress in their work, getting lost, or even returning to the comfort of the ways their teaching has always been done without the benefits and insights that inquiry can bring" (Dana & Yendol-Hoppey, 2008, p. 95-96). For this reason, the third critical juncture teacher inquirers face is articulating a doable plan for their research that will provide a roadmap for the inquiry journey.

The development of a road map may take the form of an "inquiry brief," defined by Hubbard & Power (1999) as "a detailed outline completed before the research study begins" (p. 47). In general, a research brief is a one to two page summary that covers such aspects as the purpose of the study, a statement of the wondering(s), a plan for how the teacher researcher will collect and analyze data, and a timeline for the study to unfold (Dana & Yendol-Hoppey, 2009). Through the process of developing a brief, teacher inquirers commit their energies to one idea. The process also helps members gain insights into their wondering(s) and the "do-ability" of action research becomes apparent. Through the development of an inquiry brief, teacher researchers develop a sense of direction and know where to go next.

Just as it takes time and play for teacher researchers to articulate their wonderings, it takes time and playing with each component of the inquiry brief for teacher researchers to design a solid plan of attack for their research. The most common way to create opportunities for teacher researchers to design their studies is to create a space for coach and teacher researchers to give and provide feedback with and for one another in a similar fashion to the ways they provided feedback for one another on the initial wondering development.

Although the inquiry brief is, as its name entails, a relatively short document, it still takes significantly longer to read through and think about than the statement of a wondering. For this reason, we have found in our coaching work that creating an online space for teachers to articulate a research plan and receive feedback is best done in an asynchronous fashion.

We have successfully created the context for teachers to articulate and give/receive feedback on their plans for research in an asynchronous fashion via Schoology, a learning management system (LMS) and social network. On this site, or sites like it, teacher researchers can be placed in small groups, post their inquiry briefs, and fine-tune their action research plans by discussing each inquiry brief using the following questions as a guide:

- What match seems to exist (or not exist) between the proposed data collection plan and inquiry question?
- Are there additional types of data that would give the teacher researcher insights into his/her question?
- Rate the "do-ability" of this plan for inquiry. In what ways is the teacher researcher's plan meshed with the everyday work of a teacher?
- Is what ways does the teacher researcher's proposed timeline for study align with each step in the action research process?
- What possible disconnects and problems do you see?

In addition, after teacher researchers have fine-tuned their plans and collected data over time, they can return to the Schoology site and utilize it to articulate what they have learned from initially looking at their data in order to prepare for a synchronous data analysis meeting, the next critical juncture in the action research process (Figure 2).

Analyzing Data

Teacher researchers often feel overwhelmed when they get to the data analysis phase of their studies and face making sense of a huge pile of data. Often teacher researchers might think, "OK, I've collected all of this stuff. Now, what do I do with it?" Hence, the fourth critical juncture in the action research process is data analysis.

To help teacher researchers dig deeper into their data, many coaches continue the process of organizing small groups of teacher researchers to give and provide feedback to one another on

Figure 2. Participant discussions of inquiry briefs via Schoology

the sense making process of each individual's data utilizing a protocol designed by Dana and Yendol-Hoppey (2008) specifically for this purpose (Figure 3).

This protocol can be adapted for the online context easily by utilizing web and telephone conferencing tools (such as www.FreeConferenceCall.com). We have successfully utilized these tools to virtually connect teacher researchers to one another to discuss their data by following a modified version of Dana and Yendol-Hoppey's (2008) Action Research Data Analysis Protocol illustrated in Figure 3, originally designed for face-to-face use.

To begin, prior to the synchronous phone conference meeting, each inquirer we coached virtually was asked to complete the following steps to create a post on the Schoology site:

Step One: Gather all of your collected data into one place and organize it chronologically or in some other fashion that makes sense for your inquiry.

Step Two: Read through your entire data set one time to provide a sense of the entirety of your data set.

Step Three: Read through your entire data set a second time. As you read through your data set a second time, ask yourself "What am I noticing about my data?" Construct a list entitled "Inquiry – What I'm Noticing"

Step Four: Complete the following open-ended sentences:

The issue/tension/dilemma/problem/interest that led me to my inquiry was:

Therefore, the purpose of my inquiry was to:

My wondering was:

I collected data by:

So far, three discoveries I've made by reading through my data are:

(1)

(2)

(3)

Step Five: Post your open-ended sentence completion responses on the Schoology site.

Prior to the synchronous phone/web conference meeting, teacher researchers were asked to visit the Schoology site to refamiliarize themselves with their small group members' inquiries as well as become acquainted with what each teacher researcher had learned so far from their data by

Figure 3. Data analysis protocol (Adapted from Dana & Yendol-Hoppey, 2008)

Data Analysis Protocol:
Helping Your Colleagues Make Sense of What They Learned

Suggested Group Size: 4
Suggested Time Frame: 25 – 30 MINUTES PER GROUP MEMBER

Step One: Presenter Shares His/Her Inquiry (4 Minutes) – Presenter briefly shares with his/her group members the focus/purpose of his/her inquiry, what his/her wondering(s) were, how data were collected, and the initial sense that the presenter has made of his/her data. Completing the following sentences prior to discussion may help presenter organize his/her thoughts prior to sharing:

- The issue/dilemma/problem/interest that led me to my inquiry was . . .
- Therefore, the purpose of my inquiry was to . . .
- My wondering(s) was . . .
- I collected data by . . .
- So far, three discoveries I've made from reading through my data are . . .

Step Two: Group Members Ask Clarifying Questions (3 Minutes) – Group members ask questions that have factual answers to clarify their understanding of the inquiry, such as, "How long did you collect data for?" "How many students did you work with?"

Step Three: Group Members Ask Probing Questions (7 - 10 Minutes) - The group then asks probing questions of the presenter. These questions are worded so that they help the presenter clarify and expand his/her thinking about what he/she is learning from the data. During this 10-minute time frame, the presenter may respond to the group's questions, **but there is no discussion by the group of the presenter's responses.** Every member of the group should pose at least one question of the presenter. Some examples of probing questions might include:

a. What are some ways you might organize your data?
b. What might be some powerful ways to present your data?
c. Do you have any data that doesn't seem to "fit?"
d. Based on your data, what are you learning about yourself as a teacher?
e. What is your data telling you about the students you teach?
f. What are the implications of your findings for the content you teach?
g. What have you learned about the larger context of schools and schooling?
h. What are the implications of what you have learned for your teaching?
i. What changes might you make in your own practice?
j. What new wonderings do you have?

Step Four: Group Members Discuss The Data Analysis (6 Minutes) - The group talks with each other about the data analysis presented, discussing such questions as, "What did we hear?," "What didn't we hear that we think might be relevant?," "What assumptions seem to be operating?," "Does any data not seem to fit with the presenter's analysis?," "What might be some additional ways to look at the presenter's data?" During this discussion, members of the group work to deepen the data analysis. **The presenter doesn't speak during this discussion, but instead listens and takes notes.**

Step Five: Presenter Reflection (3 Minutes) - The presenter reflects on what s/he heard and what s/he is now thinking, sharing with the group anything that particularly resonated for him or her during any part of the group members' data analysis discussion.

Step Six: Reflection on the Process (2 Minutes) - Group shares thoughts about how the discussion worked for the group.

reading each others' postings. Furthermore, each teacher researcher was asked to have the Schoology site opened on their computer during the synchronous phone conference meeting so they could glance at each person's written summary of his/her inquiry as they provided feedback to one another.

At the synchronous meeting, one by one, each small group member had the opportunity to share his/her inquiry and receive feedback on the data analysis process by following these steps of the adapted protocol:

Step One: Brief Review of Inquiry (2-3 minutes)—Presenting inquirer briefly reminds the group about his/her inquiry by summarizing and referring to the post made on the Schoology site.

Step Two: Probing Questions (4-5 minutes)—Members of the group each take a turn posing one probing question of the presenter. Some examples of probing questions include:

- What are some ways you might organize your data?
- Do you have any data that doesn't seem to fit?
- Based on your data, what are you learning about yourself as a teacher?
- What is your data telling you about the students you teach?
- What changes might you make in your practice?

The presenting inquirer may or may not choose to answer the probing questions.

Step Three: Group Discussion (5 - 6 minutes)—Members of the group talk about things they have noticed or heard about the presenting inquirer's research and what he/she has been learning from data analysis. Group members discuss the inquiry as if the presenter was not present. During this discussion, group members may make suggestions and/or share thoughts that have occurred to them in relationship to the presenter's research. The presenting inquirer does not participate in this discussion, but listens and takes notes.

Step Four: Reflection (1-2 minutes)—Presenting inquirer reflects on what he or she heard and what he or she is now thinking, sharing with the group anything that particularly resonated for him or her during any part of the group members' data analysis discussion in Step Three.

Sharing Work with Others

An important way to bring closure to a cycle of inquiry for action researchers is to make their work public by sharing it with other professionals. Not only is this important to bring closure to one action research cycle, but the process of preparing to share one's action research with others itself helps teacher researchers clarify their own thinking about their work. In addition to clarifying their own thinking, in the actual sharing of their work, teacher researchers give other professionals access to their thinking so they can question, discuss, debate, and relate. The sharing process helps teacher researchers and their colleagues push and extend thinking about practice as well, enabling a teacher researcher's colleagues to learn from the research he/she conducted. For these reasons, sharing work with others is the fifth and final critical juncture in the action research process.

Those who coach action research both create and help structure a space for teacher researchers to share their work with one another, carefully scaffolding their preparation for the actual sharing event. A common way to accomplish these goals in the face-to-face environment is through an "Inquiry Showcase," a conference-like event that brings together numerous teacher researchers into one space and time to share their work with one another through poster sessions, round table discussions, and more traditional PowerPoint presentations (Dana & Yendol-Hoppey, 2008). The showcase may be a part of the last class session at the university if action research is being conducted as the culmination to a course on the topic or as an after-school meeting as a part of a school or district professional development program. The face-to-face "Inquiry Showcase" can be translated into the online context easily by once again utilizing Elluminate. We have successfully utilized Elluminate to create a virtual sharing space for teacher researchers by grouping three to four teacher researchers with related topics together to share their work in a 60-75 minute timeframe (depending on the number of teacher researchers presenting in one session). Different sessions were given a title and were advertised broadly to other teacher researchers and their colleagues, extending an invitation for others to attend and participate in the live webinar-style sessions.

During the session itself, each of the three to four action researchers presented a 10-12 minute PowerPoint presentation that was uploaded on Elluminate to guide the sharing of their work. The following directions (which included a PowerPoint template) were provided to the action researchers so there would be consistency across the three presentations:

To prepare for your Elluminate session, please create 5 – 10 PowerPoint slides that overview your study. The following format is suggested:

- **Slide One:** Title Slide
- **Slide Two:** Background (what led to your inquiry question)
- **Slide Three:** Statement of Your Wondering
- **Two-Three Slides:** What You Did and How You Collected Data
- **Two-Three Slides:** What Your Learned (supported by data)
- **Final Slide:** Next Steps – Where You Are Headed in the Future

A session facilitator welcomed all participants, introduced each teacher researcher, helped each teacher researcher stay within their allotted time by signaling presenters when they had five minutes and one minute left to their personal presentation time, helped each teacher researcher field questions, and led a discussion to synthesize all three action research presentations during the last fifteen minutes of each session.

In addition to Elluminate Action Research Sharing Sessions, we have also utilized a tool called Action Research for Technology Integration (ARTI) that was created by Dawson (2009) to allow educators to upload their inquiry work to a larger database, walking them step-by-step through the elements of producing a write-up of their work that could be shared with others. This sharing tool will be described in detail in chapter 19 of this text entitled "ARTI: An Online Tool To Support Teacher Action Research for Technology Integration," by Kara Dawson, Cathy Cavanaugh, and Albert Ritzhaupt.

FUTURE RESEARCH DIRECTIONS

Currently, research into virtual professional development is in its infancy (Dede et. al, 2009). This field is ripe for future research on coaching action research virtually, opening up possibilities in a number of areas.

In this chapter we have presented some of the tools utilized in our experiences with coaching action research virtually. Within the range of tools with which we have already experimented, further research could explore the effectiveness of these tools within virtual coaching. For example, how do virtual action research coaching experiences enhance (if at all) teacher researchers' work and/or their efficacy through the process? Additionally, there is room for investigating other possible tools—or creating new tools or spaces—for which coaching might be done effectively in virtual environments.

As more and more aspects of education move to online spaces, it is also important to consider the quality of the virtual coaching. Because online environments allow for coaching to cross some of the constraints of time and space that transpire in face-to-face settings, it is important to understand how the quality of face-to-face coaching is translated online, exploring how virtual coaches are building relationships with and among teachers and guiding them through the critical junctures of the inquiry process in meaningful ways. This would also be a space in which to explore some of the challenges that occur in translating face-to-face coaching online for both coaches and those with whom they work, including some of the gaps that might exist in knowledge of technology tools.

By further exploring what is already being done in inquiry communities and understanding how the quality of coaching transfers from face-to-face to online settings, we might also begin to see how virtual coaching might impact sustainability of inquiry. For example, how might the development of an inquiry stance vary (if at all) depending upon whether the inquiry process is coached in a face-to-face or an online environment? By better understanding some of these factors, we can continue to develop in our professional role as coaches and to broaden the reach of inquiry's transformative power.

CONCLUSION

Online action research coaching has the potential to help teacher educators who facilitate teacher professional development reach more teachers than they ever could in face-to-face workshop settings, and connect practicing teachers to one another across schools and districts, creating powerful networks of teacher researchers. Online action research coaching also expands the possibilities for teacher educators to support student teachers who engage in action research as a capstone experience in their teacher education programs, a common practice across the nation. By eliminating distance between university and various student teaching field placements, online action research coaching creates opportunities for student teachers to network with one another across schools. In sum, coaching action research online allows educators to reach out to each other and connect with professionals beyond their local context, broadening their professional community and support. In this chapter, we have described numerous mechanisms for online action research coaching and the potential these mechanisms hold for enhancing the promise and power of practitioner research.

REFERENCES

Adelman, C. (1993). Kurt Lewin and the origins of action research. *Educational Action Research*, *1*(1), 7–24. doi:10.1080/0965079930010102

Appana, S. (2008). A review of benefits and limitations of online learning in the context of the student, the instructor and the tenured faculty. *International Journal on E-Learning*, *7*(1), 5–22.

Aubusson, P. J., Ewing, R., & Hoban, G. (2009). *Action learning in schools: Reframing teachers' professional learning and development*. London, UK: Routledge.

Caro-Bruce, C., Klehr, M., Zeichner, K., & Sierra-Piedrahita, A. M. (2009). A school district-based action research program in the United States. In Noffke, S. E., & Somekh, B. (Eds.), *The SAGE handbook of educational action research* (pp. 104–117). Los Angeles, CA: SAGE Publications.

Cochran-Smith, M., & Lytle, S. (1993). *Inside/outside: Teacher research and knowledge*. New York, NY: Teachers College Press.

Cochran-Smith, M., & Lytle, S. (1999). Relationships of knowledge and practice: Teacher learning in communities. *Review of Research in Education*, *24*, 249–305.

Cochran-Smith, M., & Lytle, S. (2009). *Inquiry as stance: Practitioner research for the next generation*. New York, NY: Teachers College Press.

Cochran-Smith, M., & Lytle, S. L. (2001). Beyond certainty: Taking an inquiry stance on practice. In Lieberman, A., & Miller, L. (Eds.), *Teachers caught in the action: Professional development that matters* (pp. 45–58). New York, NY: Teachers College Press.

Corey, S. M. (1953). *Action research to improve school practice*. New York, NY: Teachers College Press.

Crocco, M., Bayard, F., & Schwartz, S. (2003). Inquiring minds want to know: Action research at a New York City professional development school. *Journal of Teacher Education*, *54*(1), 19–30. doi:10.1177/0022487102238655

Crockett, M. D. (2002). Inquiry as professional development: Creating dilemmas through teachers' work. *Teaching and Teacher Education*, *18*, 609–624. doi:10.1016/S0742-051X(02)00019-7

Dana, N. F., & Silva, D. Y. (2001). Student teachers as researchers: Developing an inquiry stance towards teaching. In Rainer, J., & Guyton, E. M. (Eds.), *Research on the effects of teacher education on teacher performance: Teacher education yearbook IX*. New York, NY: Kendall-Hunt Press.

Dana, N. F., Silva, D. Y., & Snow-Gerono, J. (2002). Building a culture of inquiry in a professional development school. *Teacher Education and Practice*, *15*(4), 71–89.

Dana, N. F., & Yendol-Hoppey, D. (2008). *The reflective educator's guide to professional development: Coaching inquiry-oriented learning communities*. Thousand Oaks, CA: Corwin Press.

Dana, N. F., & Yendol-Hoppey, D. (2009). *The reflective educator's guide to classroom research: Learning to teach and teaching to learn through practitioner inquiry*. Thousand Oaks, CA: Corwin Press.

Dawson, K., Cavanaugh, C., & Ritzhaupt, A. (2009, March). *The evolution of ARTI: An online tool to promote classroom-based technology outcomes via teacher inquiry*. Paper presented at the Society for Technology and Teacher Education International Conference, Charleston, SC.

Dede, C. (2006). *Online professional development for teachers: Emerging models and methods*. Harvard, MA: Harvard Education Press.

Dede, C., Jass Ketelhut, D., Whitehouse, P., Breit, L., & McCloskey, E. M. (2009). A research agenda for online teacher professional development. *Journal of Teacher Education*, *60*(1), 8–19. doi:10.1177/0022487108327554

Dewey, J. (1933). *Democracy and education*. New York, NY: The Free Company.

Drennon, C. E., & Cervero, R. M. (2002). The politics of facilitation in practitioner inquiry groups. *Adult Education Quarterly*, *52*(3), 193–209. doi:10.1177/0741713602052003003

Gallimore, R., Ermeling, B. A., Saunders, W. M., & Goldenberg, C. (2009). Moving the learning of teaching closer to practice: Teacher education implications of school-based inquiry teams. *The Elementary School Journal*, *109*(5), 537–553. doi:10.1086/597001

Hubbard, R. S., & Power, B. M. (1999). *Living the questions: A guide for teacher researchers*. York, ME: Stenhouse.

Killion, J., & Pinata, R. (2011). *Recalibrating professional development for teacher success*. Education Week Webinar.

King, M. B. (2002). Professional development to promote schoolwide inquiry. *Teaching and Teacher Education*, *18*, 243–257. doi:10.1016/S0742-051X(01)00067-1

Krell, D., & Dana, N. F. (2011, April). *Facilitating action research: A study of coaches, their experiences, and their reflections on leading teachers in the process of practitioner inquiry*. Paper presented at the meeting of the American Educational Research Association Annual Conference, New Orleans, LA.

Meyers, E., & Rust, F. (2003). *Taking action with teacher research*. Portsmouth, NH: Heinemann.

Nelson, T. H., & Slavit, D. (2007). Collaborative inquiry among science and mathematics teachers in the USA: Professional learning experiences through cross-grade, cross discipline dialogue. *Journal of In-service Education*, *33*(1), 23–39. doi:10.1080/13674580601157620

Poekert, P. (2010). The pedagogy of facilitation: Teacher inquiry as professional development in a Florida elementary school. *Professional Development in Education*, *37*(1), 19–38. doi:10.1080/19415251003737309

Smeets, K., & Ponte, P. (2009). Action research and teacher leadership. *Professional Development in Education*, *35*(2), 175–193. doi:10.1080/13674580802102102

Snow-Gerono, J. L. (2005). Professional development in a culture of inquiry: PDS teachers identify the benefits of professional learning communities. *Teaching and Teacher Education*, *21*, 241–256. doi:10.1016/j.tate.2004.06.008

Stokes, L. (2001). Lessons from an inquiring school: Forms of inquiry and conditions for teacher learning. In Lieberman, A., & Miller, L. (Eds.), *Teachers caught in the action: Professional development that matters* (pp. 141–158). New York, NY: Teachers College Press.

Webster-Wright, A. (2009). Reframing professional development through understanding authentic professional learning. *Review of Educational Research*, *79*(2), 702–739. doi:10.3102/0034654308330970

Wolkenhauer, R., Boynton, S., & Dana, N. F. (2011, February). *The power of practitioner research and development of an inquiry stance in teacher education programs*. Paper presented at the meeting for the Association of Teacher Educators, Orlando, FL.

Zeichner, K. (2003). Teacher research as professional development for P-12 educators in the USA. *Educational Action Research*, *2*(2), 301–326. doi:10.1080/09650790300200211

ADDITIONAL READING

Blackboard Collaborate. (2011). *Transform the teaching and learning experience*. Retrieved from http://www.blackboard.com/ Platforms/ Collaborate /Products/Blackboard-Collaborate/ Web-Conferencing.aspx

Carr, W., & Kemmis, S. (1986). *Becoming critical: Knowing through action research*. Geelong, Australia: Deakin University Press.

Check, J. (1997). Teacher research as powerful professional development. *Harvard Education Letter*, *13*(3), 6–8.

Copland, M. A. (2003). Leadership of inquiry: Building and sustaining capacity for school improvement. *Educational Evaluation and Policy Analysis*, *25*(4), 375–395. doi:10.3102/01623737025004375

Cushman, K. (1999). *The cycle of inquiry and action: Essential learning communities*. Oakland, CA: Coalition of Essential Schools. Retrieved from http://www.essentialschools.org/resources/72

Dana, N. F., Silva, D. Y., & Snow-Gerono, J. (2002). Building a culture of inquiry in a professional development school. *Teacher Education and Practice*, *15*(4), 71–89.

Dana, N. F., Thomas, C., & Boynton, S. (2011). *Inquiry: A districtwide approach to staff and student learning*. Thousand Oaks, CA: Corwin Press.

Dawson, K., & Dana, N. F. (2007). When curriculum-based, technology enhanced field experiences and teacher inquiry coalesce: An opportunity for conceptual change? *British Journal of Educational Technology*, *38*(4), 656–667. doi:10.1111/j.1467-8535.2006.00648.x

Easton, L. B. (2004). *Powerful designs for professional learning*. Oxford, OH: National Staff Development Council.

Elmore, R. F. (2007). Let's act like professionals. *Journal of Staff Development*, *28*(3), 31–32.

Florida Center for Instructional Technology. (2010). *Technology integration matrix*. Retrieved from http://fcit.usf.edu/matrix/

Glogowski, K., & Sessums, C. D. (2007). *Personal learning environments: Exploring professional development in a networked world*. Paper presented at the meeting of Webheads In Action Online Conference. Retrieved from http://www.webheadsinaction.org/node/168

Kern, A., & Levin, B. B. (2009). How National Board certified teachers are learning, doing, and sharing action research online! *Delta Kappa Gamma Bulletin*, *76*(1), 20–23.

Kincheloe, J. (1991). *Teachers as researchers: Qualitative inquiry as a path to empowerment*. London, UK: Falmer Press.

Mills, G. E. (2003). *Action research: A guide for the teacher researcher*. Saddle River, NJ: Pearson Education, Inc.

Moore, M., & Kearsley, G. (2005). *Distance education: A systems view* (2nd ed.). Belmont, CA: Wadsworth.

Schlechty, P. C. (2007). Move staff development into the digital world. *Journal of Staff Development*, *28*(3), 41–42.

Schmoker, M. (2004). The tipping point: From feckless reform to substantive instructional improvement. *Phi Delta Kappan*, *85*(6), 424–432.

Schwandt, T. A. (1997). *Qualitative inquiry: A dictionary of terms*. Thousand Oaks, CA: Sage Publications.

Shulman, L. (1986). Knowledge and teaching: Foundations of the new reform. *Harvard Educational Review*, *57*(1), 1–22.

Zeichner, K. M., & Liston, D. P. (1996). *Reflective teaching: An introduction*. Mahwah, NJ: Lawrence Erlbaum.

KEY TERMS AND DEFINITIONS

Action Research: A form of professional development in which educators engage in the systematic, intentional study of their own professional practice that will ultimately lead to improved outcomes for student learning. Although this term is often used synonymously with terms such as classroom research, teacher research, teacher inquiry, practitioner inquiry, etc., each form has its own nuanced meaning and history.

Action Research Coach: An educator who supports and facilitates the process of action research for other educators, guiding them through each of the critical junctures of the process.

ARTI (Action Research for Technology Integration): An inquiry sharing tool that walks teachers step-by-step through the elements of producing a write-up of their work that can be shared with others in ARTI's larger database.

Asynchronous Technology: A type of technology that allows individuals to communicate or collaborate outside the constraints of synchronous or face-to-face environments, such as email, online discussion boards, wikis, or blogs.

Inquiry Brief: A detailed outline completed before the action research project begins summarizing the purpose of the study, the statement of wondering(s), how the teacher researcher will collect and analyze data, and a timeline for the study's progression.

Inquiry Showcase: A conference-like event that brings together numerous teacher researchers into one space and time to share their inquiry work with one another through poster sessions, round table discussions, and more traditional professional presentations.

Learning Management System: A web-based or software application for carrying out tasks associated with web-based or blended instructional programs, allowing instructors to create course calendars, grade and provide feedback, upload learning materials, view and manage rosters, and engage in a variety of teaching tools, such as discussion boards, journals/blogs, or tests.

Social Networks: Online social structures that bring together groups of people with similar interests who use the website to communicate with one another, collaborate, and share resources.

Synchronous Technology: A type of technology that supports simultaneous communication among individuals, such as instant messaging, chat rooms, or Skype.

Web Conferencing: A tool used to conduct synchronous meetings, workshops, or presentations online for participants across geographic distances who connect with one another from their own computers through the Internet.

Chapter 20
ARTI:
An Online Tool to Support Teacher Action Research for Technology Integration

Kara Dawson
University of Florida, USA

Cathy Cavanaugh
Abu Dhabi Women's College, UAE

Albert Ritzhaupt
University of Florida, USA

ABSTRACT

Action research is recognized as a powerful tool for professional development and teacher preparation (Cochran-Smith & Lytle, 2009; Zeichner, 2003) and teachers require powerful professional development to effectively integrate technology (Hew & Brush, 2007). ARTI (Action Research for Technology Integration) is an online tool designed to support the merger of action research and technology integration. This chapter provides an introduction to ARTI followed by a discussion of its theoretical foundations. Next, the conceptual design of ARTI is described in terms of the three main purposes for its development which are to provide: (1) an online scaffold for teachers to inquire about their technology integration practices, (2) a mechanism to synthesize action research information from multiple teachers, and (3) a mechanism to capture evidence of student learning within technology integration inquiries. Finally, examples of ARTI implementation, implications and future possibilities for the tool in teacher preparation are discussed.

INTRODUCTION

Action research (also known as teacher inquiry) is widely recognized as a powerful tool for professional development and teacher preparation (Cochran-Smith & Lytle, 2009; Zeichner, 2003). It involves teachers systematically and intentionally studying their practices (Dana & Yendol-Hoppey, 2009) and has been shown to improve teacher practice, heighten teacher professionalism, lead to positive educational change, expand the knowledge base for teaching and provide a platform for teachers' voices in educational reform (Carr & Kemmis, 1986; Cochran-Smith & Lytle,

DOI: 10.4018/978-1-4666-1906-7.ch020

1993; Meyer & Rust, 2003). Technology integration refers to the ways teachers use technology to support and enhance teaching and learning. Technology integration is an important aspect of teaching (National Educational Technology Plan, 2010) and appropriate use of technology can support students in content area learning (Howland, Jonassen & Marra, 2011), provide students with essential workforce skills such as the ability to communicate and collaborate digitally (President's Report, 2010), prepare students for participation in an increasingly digital democracy (Partnership for 21st Century Skills, 2011) and enhance student motivation and engagement (Dawson, Cavanaugh & Ritzhaupt, 2008).

Successful technology integration requires opportunities for professional development (Hew & Brush, 2007) and, given the positive results associated with action research (Dana & Yendol-Hoppey, 2009), recent research has attempted to merge the two. This research suggests that action research is a vehicle through which teachers can systematically and intentionally study the ways that technology integration impacts student learning and as a lens through which teachers may experience conceptual change regarding their beliefs about technology integration practices (Dawson, 2006, Dawson, 2007; Dawson & Dana, 2007).

In part as a result of this research, action research has been used with hundreds of teachers involved in statewide technology integration efforts across the state of Florida. An online tool, known as ARTI (Action Research for Technology Integration), was developed to support these teachers. The chapter describes the theoretical and conceptual foundations of ARTI's design, examples of ARTI implementation, ARTI implications, and future possibilities for the tool in teacher preparation.

BACKGROUND

This section provides an overview of the theoretical and conceptual design of ARTI.

ARTI: Theoretical Framework

Cochran-Smith and Lytle's seminal work on teachers knowledge domains (1999) informed ARTI development. These domains described different types of knowledge teacher acquire through different types of professional experiences. In particular, they discuss three domains: knowledge for practice, knowledge in practice and knowledge of practice. This section provides an overview of these domains and explains why ARTI is built on the knowledge of practice domain.

Knowledge for Practice

Teacher preparation and inservice professional development programs have historically done an inadequate job of preparing effective technology-using educators (Milken Exchange on Educational Technology, 1999; Hew & Brush, 2007; Lawless & Pellegrino, 2007). The primary way prospective teachers learn about technology integration is through stand-alone technology integration courses or possibly through content-specific lessons offered as a part of methods coursework (Pierson & Thompson, 2005) while inservice teachers primarily learn in after school workshop formats (Broughman, 2006). These approaches emphasize the development of a certain type of knowledge referred to as *knowledge for practice* (Cochran-Smith & Lytle, 1999). This type of knowledge helps prospective and practicing teachers understand definitions, theories and concepts associated with technology integration. However, these experiences are rarely transferable to a classroom context because they negate the complexities of classroom technology integration and often teach isolated skills (Wei, Darling-Hammond, Andree, Richardson & Orphasno, 2009).

Knowledge in Practice

Recognizing the shortcomings of relying solely on workshops or university-based courses, many teacher education and inservice programs have experimented with a variety of ways to help prospective and practicing teachers use technology in real classrooms or other authentic contexts. Some teacher education programs explicitly provide opportunities for prospective teachers to use technology within traditional components of their programs such as the internship experience or methods courses (Strudler & Grove, 2004; Glazewski, Berg, & Brush, 2002; Dawson, Pringle & Adams, 2003; Hernandez-Ramos & Giancarlo, 2004). Other programs have sought to facilitate learning communities focusing on technology integration (Yendol-Hoppey, Dawson, Dana, League, Jacobs & Malik, 2007; O'Bannon & Nonis, 2002; Jacobsen & Lock, 2004). Others have experimented with virtual technology-based experiences (Knight, Pederson & Peters, 2004; Davis & Roblyer, 2005) or adding innovative technology-based experiences to programs (Dawson & Nonis, 2000; Schmidt, 2001).

Inservice strategies that have been described as effective for technology integration in practice include curriculum planning, grade level or vertical team planning, teamwork to analyze student data, teacher leadership academies, and collaboration with school based coaches (Wei et. al., 2009). The nature of these strategies emphasizes a certain type of knowledge referred to as *knowledge in practice* (Cochran-Smith & Lytle, 1999). *Knowledge in practice* enables prospective and practicing teachers to use technology in authentic contexts and synthesize their experiences via reflective activities and/or via dialogue with colleagues (Posner, 2005; Hudson, 2005). However, even these activities are not ideally suited to prepare prospective and practicing teachers as effective technology-using teachers because they tend to get caught up in logistical, technical, and managerial issues while ignoring important aspects of technology integration including its impact on student learning (Dawson, 2006; Dawson & Dana, 2007).

Knowledge of Practice

Knowledge of practice is the most transformational type of knowledge educators can gain (Cochran-Smith & Lytle, 1999). This knowledge emerges from teacher questions about their practice and results from the systematic study of their teaching (Cochran-Smith & Lytle, 1999), typically through action research. Merging action research with technology integration provides a scaffold for teachers to focus on student learning and contextual factors (Dawson, 2006; Dawson & Dana, 2007). As previously mentioned, action research also yields many other benefits for teachers and, thus, as ARTI scaffolds teachers through cycles of action research it supports the knowledge of practice domain.

ARTI: Conceptual Design

ARTI was developed for three purposes:

- To provide an online scaffold for teachers to inquire about their technology integration practices.
- To provide a mechanism to synthesize action research information from multiple teachers.
- To provide a mechanism to capture evidence of student learning within technology integration inquiries.

This section describes each purpose and the ways in which ARTI was designed to help meet each.

Purpose 1: To Provide an Online Scaffold for Teachers to Inquire about their Technology Integration Practices

ARTI was designed to scaffold teachers through The Reflective Educators' Action Research Cycle (Dana & Yendol-Hoppey, 2009). This cycle includes five stages: (1) developing an inquiry question, (2) identifying context, (3) planning data collection, (4) conducting data analysis and deriving findings and (5) determining the implications of the work. Here, we will briefly describe each stage in the cycle and how ARTI was designed to support them. However, it is important to note that while we will be presenting action research as a singular cycle, ideally teachers participate in multiple cycles throughout their teacher careers to support continual improvement of teaching practice and to examine new practices that build on previous action research findings. In addition, while we will present the cycle in a linear fashion, it is important to note that an action research cycle rarely follows a neat linear sequence. Rather, it is an iterative cycle. Given the focus of this chapter is on the ARTI tool, general information about action research is necessarily brief. However, a thorough overview of action research and its importance is provided in Chapter 19 of this book. To help readers better visualize ARTI, a screenshot is provided in Figure 1. The five stages of the Reflective Educators' Action Research Cycle (Dana & Yendol-Hoppey, 2009) are visible in the left-hand menu bar. When users click on each stage content in the main window adjusts accordingly.

Stage 1: Developing an Inquiry Question

Teachers begin the action research process by specifying a question about their practice. This question, often referred to as a "wondering" or "passion," might be about a teaching practice, a single lesson, an individual child or a small group of children. Within ARTI, teachers are guided to develop a question related to how their technology integration efforts influence student outcomes in the classroom. In particular, teachers are encouraged to include three components within their question: participants (i.e. 5th graders, low achieving 3rd graders, etc.), knowledge, skill or ability to be measured (i.e. science content, critical thinking, phonemic awareness, social skills, etc.) and intervention/strategy (i.e. collaborative learning strategies, reading buddies, lab simulations, etc.).

Stage 2: Identifying Context

In the second stage, teachers describe the specific context or setting for their action research. This enables teachers to think consciously about their classroom design, students, schedule and a host of other issues that may help inform the action research question. Within ARTI, teachers are scaffolded to think about items related to technology integration including, but not limited to, technology implementation strategies, student activities, technology configurations and types of technology used.

Stage 3: Planning Data Collection

In the third stage, teachers consider what forms of data will help inform their action research question. Ideally, data comes from activities already taking place in the classroom. For example, teachers might keep anecdotal records on student behavior or conduct informal interviews with students during a center time. Other strategies might include student created media, pre/post tests, samples of student work, interviews or surveys. Within ARTI, teachers are provided a list of potential data collection strategies and encouraged to consider which ones make the most sense for answering their particular question.

Figure 1. ARTI screenshot

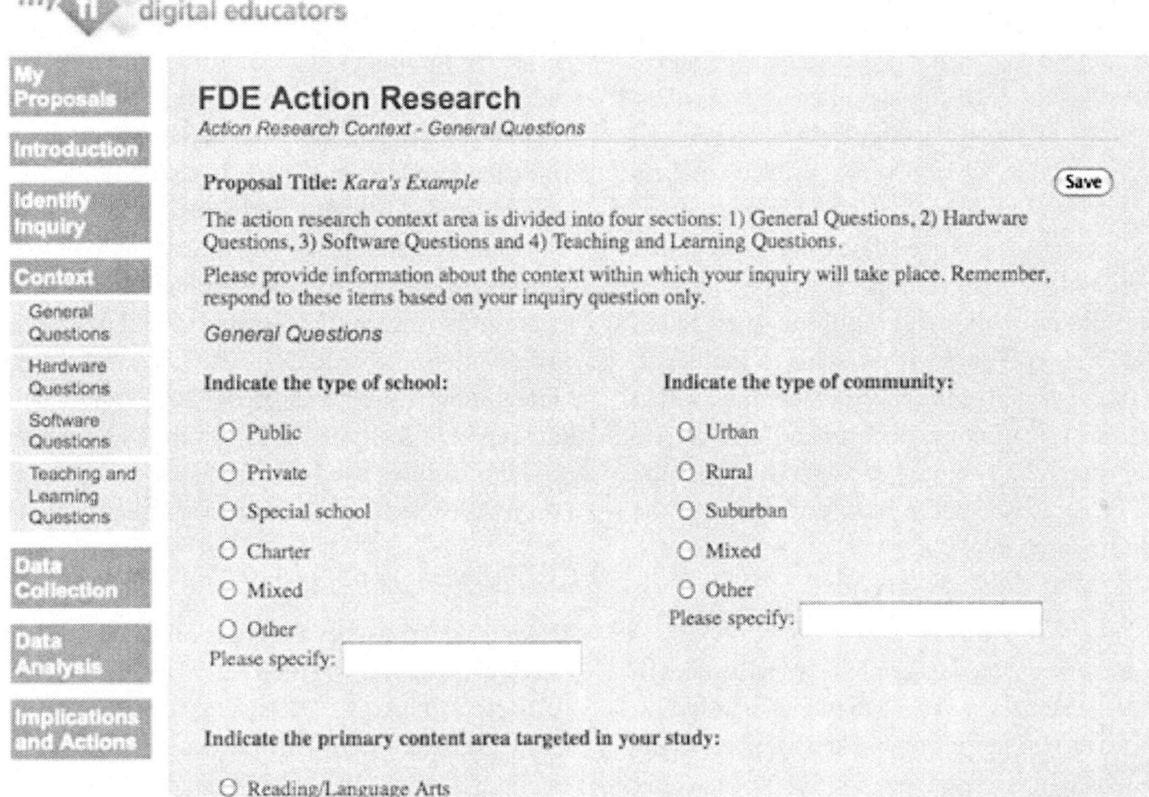

Stage 4: Conducting Data Analysis and Deriving Findings

In the fourth stage, teachers analyze data collected, synthesize it, and interpret it for insight into the action research question. Within ARTI, teachers are asked to submit their findings and associated artifacts to support them.

Stage 5: Determining the Implications of the Work

The action research cycle concludes when teachers consider broader implications of their findings and describe the actions that have or will result from their efforts. In most cases, this work is also shared with others either formally through some sort of showcase or informally through outlets such as grade level meetings. In ARTI, teachers are provided with spaces to share the implications resulting from their action research effort.

Purpose 2: To Provide a Mechanism to Synthesize Action Research Information from Multiple Teachers

Action research work typically occurs in a single classroom and action research results are seldom shared beyond the school level (Dana & Yendol-Hoppey, 2009). In fact, the isolated nature of most action research work has drawn criticism about usability of action research beyond a classroom context (Clausen, 2006). Thus, another purpose of

ARTI is to provide a mechanism to synthesize action research information from multiple teachers. This synthesis may provide a more generalizable pool of data to make implications beyond a single classroom. To do this, ARTI is designed, whenever possible, to collect easily aggregated data within each stage of the Reflective Educators' Action Research Cycle. Whenever possible, teachers input data using radio buttons, checkboxes or drop-down menus. This streamlines the aggregation process and provides teachers with a certain level of scaffolding while working through an action cycle. Decisions about what items to include in ARTI were strongly influenced by literature within educational technology, teaching and learning and action research. The next sections overview the types of data collected within each stage of the action research cycle.

Stage 1: Developing an Inquiry Question

Given the personal and unique nature of each question, teachers are asked to type their question into a textbox. They are also asked to specify their years of teaching experience and years of teaching with technology.

Stage 2: Identifying Context

In this stage teachers enter information about their context. Context is particularly important when aggregating information and, whenever possible, information collected here aligned with other research in the field. For example, teachers enter information about their community, school and classroom and this information is based on what was used in a well known technology and learning meta-analysis (Waxman, Linn & Michko, 2003). Teachers specify whether their school is private, public, charter or mixed and whether their school is urban, rural, suburban or mixed. They also specify information about the socioeconomic status, ethnicity and grade level of their students.

Teachers also specify information about their implementation strategies and student activities. For example, teachers identify the types of activities their students were engaged in throughout the action research cycle. Activity choices were derived from research-based categories related to effective teaching (Ross, Smith & Alberg, 1999) such as student discussion, experiential/hands-on learning, seatwork, individualized instruction and independent inquiry. Space is also provided for teachers to add student activities.

Finally, teachers specify the technology used during their action research cycles. Lists of hardware and software choices were modified from an observation protocol designed to study technology integration (Lowther & Ross, 2001) and teachers reported the types of hardware and software used, quantities used and frequency with which they were used during the action research cycle.

Stage 3: Planning Data Collection

Here teachers specify the types of data collection strategies used during their action research cycles using checkboxes. The list of data collection strategies were derived from Dana and Yendol-Hoppey's book on action research (2009) and included strategies such as test scores, pre/post artifacts, field notes, anecdotal records, surveys, journals and informal interviews. Blank textboxes are also provided for teachers to add strategies not included in the list. Here teachers also recorded the duration of their inquiries and the number of times students used technology during the action research cycle.

Stage 4: Conducting Data Analysis and Deriving Findings

In this section of ARTI, teachers write their findings in a textbox and upload an artifact with data supporting the finding. This is discussed in more detail later in the chapter.

Stage 5: Determining the Implications of the Work

In this section of ARTI, teachers record the implications of their work during their action research cycles. Teachers are provided with descriptions of implications that commonly result from action research efforts such as changes in teaching practices, changes of among school or grade level colleagues as a result of the action research work and feelings of professionalism. Then, they are asked to describe what they perceive to be the implications of their action research work.

Purpose 3: To Provide a Mechanism to Capture Evidence of Student Learning within Technology Integration Inquiries

There is a clear need for more school-based research related to technology integration (Bull, Knezek, Roblyer, Schrum & Thompson, 2005) and a need to document classroom-based student outcomes when technology is integrated (Schrum, Thompson, Sprague, Maddux, McAnear, Bell & Bull, 2005; Dawson & Ferdig, 2006). Such information could be derived from action research conducted by teachers in their classroom. However, the process of deriving findings often causes teachers "some discomfort or uncertainly" (Dana & Yendol-Hoppey, 2009, p. 52). In particular, they often struggle to ensure their findings align with the data they collected. This is a natural part of the action research process and scaffolding can help teachers make sense of their data and derive findings based on them (Dana & Yendol-Hoppey, 2009).

Several features are built into ARTI to support teachers in deriving findings aligned with the data they collected. First, ARTI provides a list of common categories from past action research efforts. These categories were generated from hundreds of action research cycles conducted using ARTI (Cavanaugh, Dawson, Nash & Ritzhaupt, 2008) and provide starting points from which teachers can consider their data (See Figure 2). For example, a teacher may look at her data and consider whether it demonstrates student achievement, evidence of conditions supporting learning, impact on different types of learners or a category not currently listed.

Once a teacher has created a general category for a finding, she writes the finding and provides supporting data. An important feature of ARTI is the fact that a finding cannot be submitted without an artifact displaying data supporting it. This helps to ensure teachers are aligning their findings with the data they collected. Figure 3 shows the screen to which teachers are taken when they submit a finding. Artifacts submitted by teachers range from spreadsheets of test scores to pre-post rubrics to student-created digital products.

ARTI IMPLEMENTATION

The following sections describe how the three purposes described above have played in practice.

Purpose 1: To Provide an Online Scaffold for Teachers to Inquire about their Technology Integration Practices

While American K-12 education has a decades-long history of action research, for the most part teachers have learned action research in face-to-face experiences from college instructors, professional development facilitators, school leaders, and peer mentors (Cochran-Smith & Lytle, 2009). The ARTI system represents a step forward in connecting greater numbers of teachers to mentors who share their interests and an understanding of their teaching context. The majority of US teachers work in schools with enrollments below 600 students (US Department of Education, 2010), making the chance small that they have local access to action research mentors in their subject, grade level or area of interest. ARTI provides a

Figure 2. ARTI screenshot: List of common categories

Cancel

FDE Action Research

Data Analysis - Create Finding

To create a finding, complete this form (keywords and description) and click on the **Create Finding** button. You will be returned to the Findings List where links will be made available to add artifacts and to further edit the entries on this form if you choose.

Keywords

Select the keyword/phrase that best relates to your finding. In none of the keywords/phrases relate to your finding, select other and type in your own keywords.

- ○ Conditions that lead to learning (e.g. Enjoyment, motivation, engagement, on-task behavior, positive school experience, etc.)
- ○ Instructional benefits of using technology (supporting individual differences, providing multimodal instruction, supporting repeated practice, providing instant feedback, use as a data collection device, tool for independent learning, etc.)
- ○ Twenty-first century/information skills (e.g. collaboration, computer skills, work force skills, students as producers, communication skills, leadership, innovation and creativity)
- ○ Student achievement (e.g. higher level thinking skills, retention, transfer, knowledge acquisition, etc.)
- ○ Impact on different learners (e.g. high achieving, low achieving, etc.)
- ○ Other

Please specify:

Finding Description

Please describe your finding in the text box below:

Create Finding

platform in which distributed teachers and mentors can share their action research expertise.

ARTI has been used for this purpose in a variety of ways. At the University of Florida, prospective teachers used the system to document their inquiries during their field experiences, thus adopting action research as a professional habit of mind even before their induction as professionals (Dawson, 2007). A multi-year statewide university-led research program used ARTI with hundreds of practicing teachers who documented their classroom technology implementation. Their action research work illuminated the impacts of the state's educational technology initiatives on student learning (Dawson, Ritzhaupt, Liu, Drexler, Barron, Kersaint & Cavanaugh, 2011). In addition, a group of virtual school teachers used ARTI

Figure 3. Submit a finding screen in ARTI

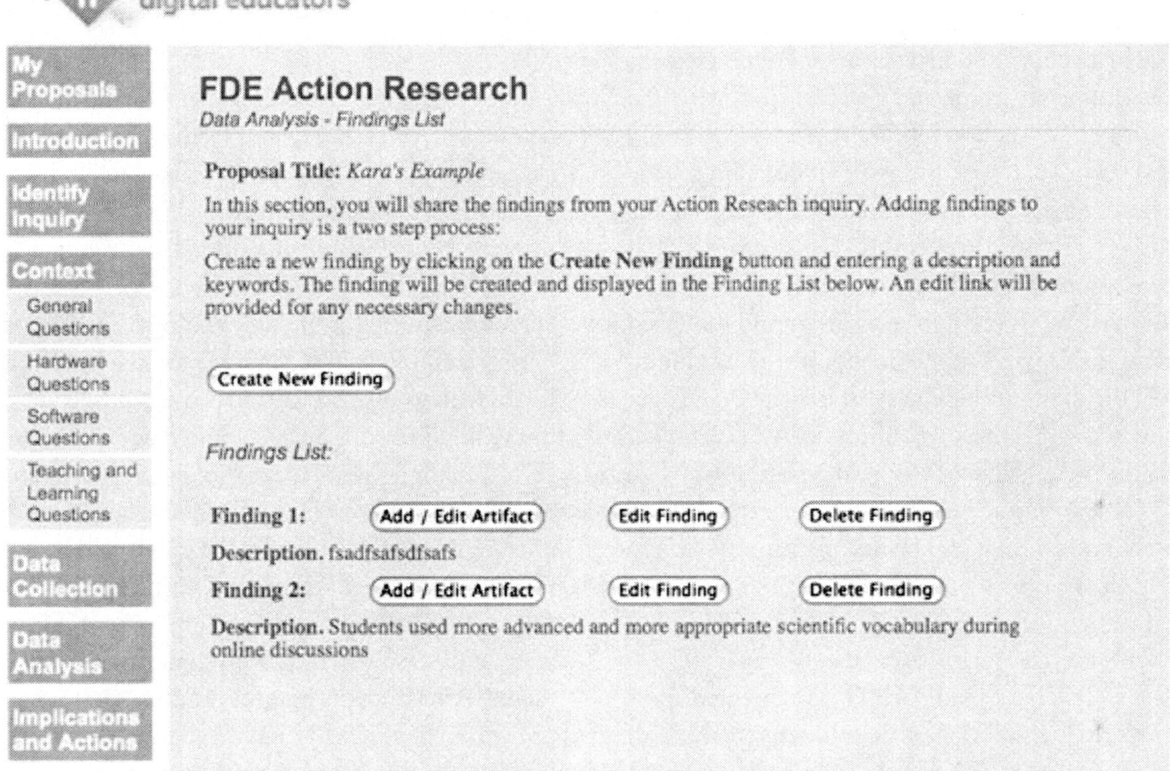

to examine their online teaching practice (Dana, Dawson, Krell & Wolkenhauer, 2012).

In each use of ARTI, the success of the action research teachers depended on the guidance they received from their mentors. Leaders at the university coached the distributed network of mentors. The web-based ARTI system included accounts for teachers who logged in to add their individual action research materials, accounts for their mentors who logged in to view the materials of their mentoring groups and to send feedback to teachers, and for the mentor coaches whose accounts allowed them to view all teacher materials and feedback. Each mentor was provided with synchronous online training sessions conducted by university educators. Dana, Krell and Wolkenhauer provide additional details about this in Chapter 18. Each mentor had access to the training materials throughout the year in which they mentored teachers and was able to contact the university educators for support at any time. Thus, while ARTI has served as an online scaffold for teachers to study their technology integration practices, the role of an action research mentor or coach is still critical to its success.

Purpose 2: To Provide a Mechanism to Synthesize Action Research Information from Multiple Teachers

During one year of the statewide study in which ARTI was used, over 350 action research efforts were collected in the system and together they formed a comprehensive view of teaching and learning in classrooms where teachers had access to new technology (Dawson & Cavanaugh, 2010; Dawson, in press). ARTI provided a demographic picture of the teachers that showed their experience in teaching and in teaching with technology. Data collected from ARTI also showed that the majority of the action research efforts occurred in rural school settings and with students in a wide range of ethnic and socioeconomic backgrounds. This suggests the power ARTI as a professional development platform for teachers who might otherwise be isolated in rural areas and it offered examples of ways that teachers were working to level the digital divide in rural districts. The data collected from these inquiries in ARTI showed that teachers had success in integrating technology for learning across all content areas and grade levels and with widely varying group sizes in their classrooms.

Data collected from ARTI has also provided insight into the nature of the action research efforts. Teachers using ARTI documented their efforts to achieve specific objectives through strategies that they had reason to believe would improve outcomes. In particular, the vast majority of inquiries occurred within the core content areas of mathematics, language arts or science with the primary goal of helping students master that content. Instructional strategies implemented by teachers varied but included a large number of project-based learning efforts in which a range of student-centered technology tools such as digital video/audio, productivity tools and the Internet were used during projects lasting a week or more. The information teachers provide through ARTI enables researchers, administrators and professional development providers to identify patterns of technology integration and shifts in technology integration over time which can guide improvements in teacher education and professional development programs. (Dawson, in press)

Purpose 3: To Provide a Mechanism to Capture Evidence of Student Learning within Technology Integration Inquiries

As previously mentioned, teachers document their findings in ARTI. For example, they might upload student samples of work, spreadsheets showing pre and post data, or results from other instruments they used to measure outcomes. This documentation provides insights into how teachers support student learning through technology integration.

For example, one teacher explored whether the use of technology-supported project-based learning would increase her 5th grade students' higher level thinking skills. This particular project related to a science unit on insects and students were to create a documentary about the plants and insects found on the school grounds. During her inquiry, she analyzed student work and one of her findings was that: "Sophistication of student knowledge increased as students progressed through stages of project development." Figure 4 shows the students' initial notes related to a particular plant on campus. The notes are factually based and do not exhibit much in terms of higher- level thinking. Figure 5 is a screen shot taken from the documentary and shows evidence of higher-level thinking as the students' had opportunities to engage with and think more deeply about the assigned topic.

In addition to artifacts documenting student learning, teachers also submitted artifacts that demonstrate changes in conditions that support learning, including time on task, perseverance, tasks attempted and completed, student success rate in learning activities, positive interactions in learning activities, attitude, self-concept or motivation. In addition, at least half of the teachers using the ARTI system so far have reported that the action research experience coupled with their broader technology integration efforts resulted in changed teaching practices. This suggests that teachers may rethink their instructional practices

Figure 4. Example of student notes on plants (lower order thinking illustration)

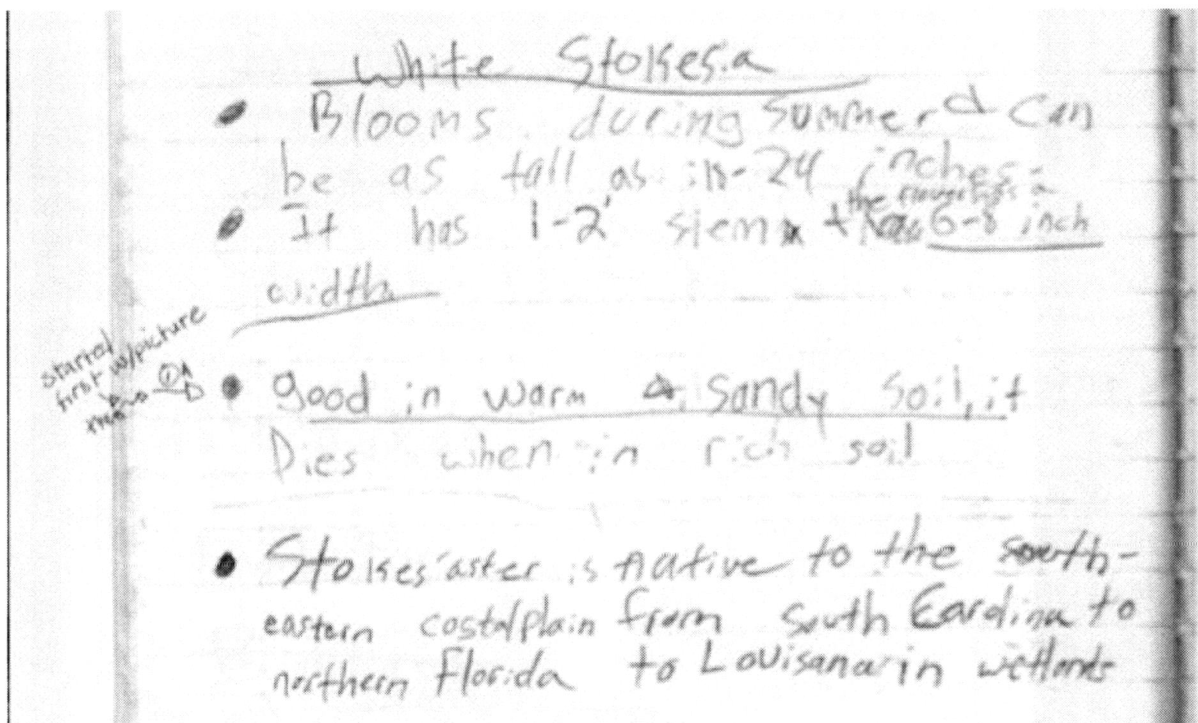

when their reflection is scaffolded using a system like ARTI.

IMPLICATIONS FOR ARTI

As a support system for teacher inquiry, professional development, research, and evaluation, ARTI has potential in a range of contexts. Initial teacher preparation programs can integrate ARTI into their courses and internship experiences in order to document inquiry efforts of Education majors at all stages of an education degree program. Application of ARTI in teacher education serves several purposes for students and faculty. Student growth over time in their technology integration practices could be documented at the level of the individual student, as well as at the level of cohort or program groups. Individual student data could be a valuable addition to a student's professional portfolio. Group and program level data could be useful in internal program evaluation efforts and to support program accreditation reviews. Faculty could also use ARTI to document their own faculty development initiatives as a stage in the scholarship of teaching (Austin & McDaniels, 2006).

Practicing teachers can use ARTI in their own individual professional development plans, as a gathering place for collaborative teams and lesson study groups, and for broader communities of practice. A district could award continuing education credits for completed inquires that would enable teachers to use their inquiries toward their recertification, additional licensure, and merit reviews. For professional development programs, ARTI could offer a repository for schools and districts to collect and share examples of innovation that may not only spur improvements at the classroom level, but also may support school improvement plans and school strategic goals. Districts and larger education agencies could use aggregated ARTI findings for technology planning, detection

Figure 5. Screenshot from student documentary (higher order thinking illustration)

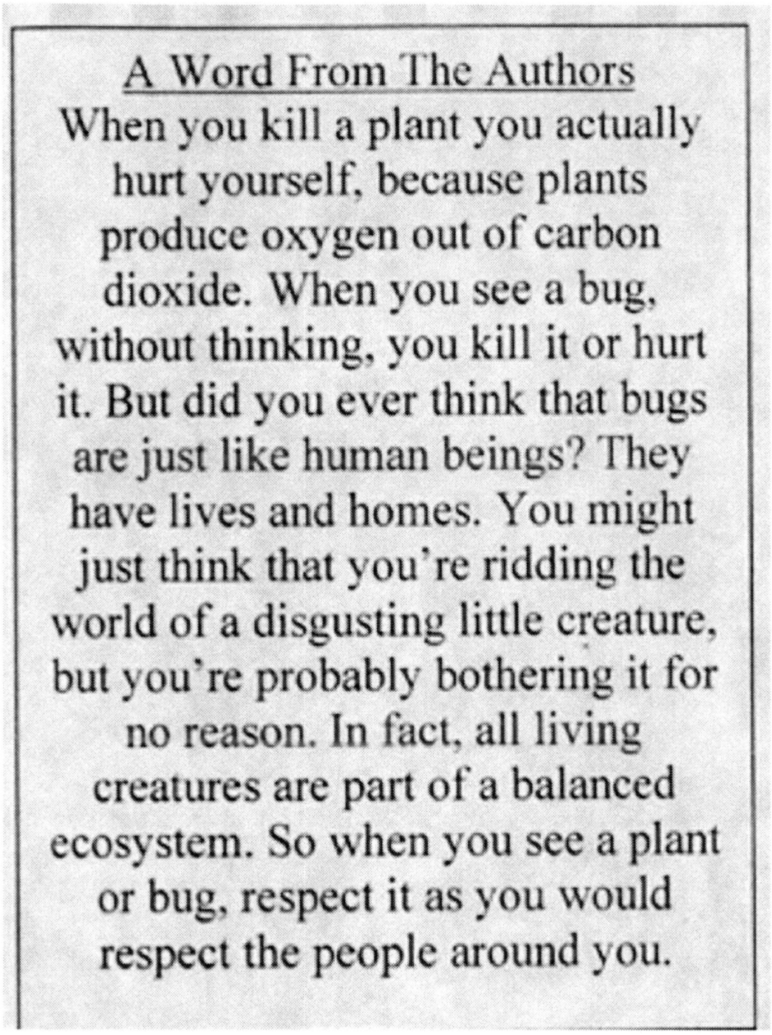

A Word From The Authors

When you kill a plant you actually hurt yourself, because plants produce oxygen out of carbon dioxide. When you see a bug, without thinking, you kill it or hurt it. But did you ever think that bugs are just like human beings? They have lives and homes. You might just think that you're ridding the world of a disgusting little creature, but you're probably bothering it for no reason. In fact, all living creatures are part of a balanced ecosystem. So when you see a plant or bug, respect it as you would respect the people around you.

of needs, and insights into trends that could be magnified with additional support and resources.

Researchers have in ARTI a comprehensive and rich data collection system that is low-impact and non-invasive in that researchers need no specialized instruments, training, or time in classrooms in order to gather information for a variety of analyses; although teachers must receive adequate coaching or mentoring to use ARTI effectively. The data elements in ARTI allow comparisons to be made among schools and other jurisdictions, grade levels, content areas, technology type, teaching strategy or any other category. ARTI is flexible enough that new categories can be added to accommodate shifting needs.

FUTURE DIRECTIONS

The ARTI system has been an effective tool for supporting individual teacher inquiry efforts in K-12 classroom technology integration, for understanding the uses of technology in a broad range of classroom contexts, and for capturing impacts of technology on learning on a broad scale. The system has promise for supporting similar inquiry and understanding in education beyond K-12 classroom technology integration. With modification to the data elements, ARTI can be transformed into a multi-purpose educator and researcher support system. In K-12 schools, an ARTI-based system could be developed that allows users to select a purpose, such as technology, a specific content area or areas, 21st century skills, addressing special student needs, etc., and the system could adaptively present contextual data elements and standards for the needs of the user. ARTI could also be modified to support group inquiries by allowing group accounts and data entry by multiple users on a team.

Because the affordances and demands of learning environments vary, teachers in online or blended courses would benefit from a version of ARTI tailored to practices known to be effective in those environments. Similarly, adapted ARTI systems could support home school and informal education instructors. Not only is the nature of teaching and learning different in adult and higher education, but also the nature of the scholarship of teaching varies in those contexts. Thus an ARTI-based system for adult learning environments could be developed to reflect principles of effective practice there.

Action research is an effective professional development approach in part due to the interactions among teachers and mentors. The ARTI system accommodates action research mentors by giving them access to teacher inquiry information and providing them an integrated channel for feedback to the teachers. But there is great potential for overlaying ARTI with more robust community capabilities and supports for mentors and coaches in the form of synchronous and asynchronous chat, videoconferencing, and discussion, for example. At this time, support and development for mentors occurs outside of ARTI, but could be interwoven throughout ARTI. Further, embedded prompts in text and multimedia forms can be added to enable more independent use of ARTI by teachers.

CONCLUSION

In its current form ARTI provides an online scaffold for teachers to inquire about how their technology integration practices affect student outcomes in their classroom, provides a mechanism to synthesize action research information from multiple teachers and provides a mechanism to capture evidence of student learning within technology integration inquiries. It has been used successfully with prospective teachers, practicing teachers and virtual school teachers.

In the future, ARTI can become a full-featured professional development and research support environment and can be developed in a form that seamlessly embeds it within a school's learning management system (LMS). Such embedding could include connection of the ARTI data with the LMS data to enable more detailed analysis of teaching and learning. A data system of this type that will further advance teaching and learning by illuminating the effect of technology integration practices on a broad scale.

REFERENCES

Austin, A., & McDaniels, M. (2006). Using doctoral education to prepare faculty to work within Boyer's four domains of scholarship. *New Directions for Institutional Research*, *129*, 51–65. doi:10.1002/ir.171

Broughman, S. P. (2006). *Teacher professional development in 1999–2000*. National Center for Education Statistics.

Bull, G., Knezek, G., Roblyer, M. D., Schrum, L., & Thompson, A. (2005). A proactive approach to a research agenda for educational technology. *Journal of Research on Technology in Education*, *37*(3), 217–220.

Carr, W., & Kemmis, S. (1986). *Becoming critical: Knowing through action research*. Geelong, Australia: Deakin University Press.

Cavanaugh, C., Dawson, K., Nash, R., & Ritzhaupt, A. (2008). *Florida digital educator program*. Final research report presented to the Florida Department of Education.

Clausen, K. W. (2006). It there meta in the madness?: Action research and the use of meta-analysis. *The Ontario Action Researcher, 8*(3). Retrieved from http://www.nipissingu.ca/oar/archive-V83E.htm

Cochran-Smith, M., & Lytle, S. (1993). *Inside/outside: Teacher research and knowledge*. New York, NY: Teachers College Press.

Cochran-Smith, M., & Lytle, S. (2009). *Inquiry as stance: Practitioner research for the next generation*. New York, NY: Teachers College Press.

Cochran-Smith, M., & Lytle, S. L. (1999). The teacher research movement: A decade later. *Educational Researcher, 28*(7), 15–25.

Dana, N., Dawson, K., Krell, D., & Wolkenhauer, R. (2012). *Using action research in professional development for virtual school educators: Exploring an established strategy in a new context.* Paper presented at the American Educational Research Conference. Vancouver, Canada.

Dana, N. F., & Silva, D. Y. (2009). *The reflective educator's guide to classroom research: Learning to teach and teaching to learn through practitioner inquiry* (2nd ed.). Thousand Oaks, CA: Corwin Press.

Davis, N., & Roblyer, M. D. (2005). Preparing teachers for the "schools that technology built": Evaluation of a program to train teachers for virtual schooling. *Journal of Research on Technology in Education, 37*(4), 399–409.

Dawson, K. (2006). Teacher inquiry: A vehicle to merge prospective teachers' experience and reflection during curriculum-based, technology-enhanced field experiences. *Journal of Research on Technology in Education, 38*(3), 265–292.

Dawson, K. (2007). The role of teacher inquiry in helping prospective teachers untangle the complexities of technology use in classrooms. *Journal of Computing in Teacher Education, 24*(1), 5–14.

Dawson, K. (2012). Using action research projects to examine teacher technology integration practices. *Journal of Digital Learning in Teacher Education, 28*(12), 117–124.

Dawson, K., & Cavanaugh, C. (2010). *Insights into classroom technology integration through action research: For whom, in what ways and with what outcomes and implications?* Paper presented at the American Educational Research Association. Denver, CO.

Dawson, K., Cavanaugh, C., & Ritzhaupt, A. (2008). Florida's Leveraging Laptops initiative and its impact on teaching practices. *Journal of Research on Technology in Education, 41*(2), 143–159.

Dawson, K., & Dana, N. (2007). When curriculum-based, technology-enhanced field experiences and teacher inquiry coalesce: An opportunity for conceptual change? *British Journal of Educational Technology, 38*(4), 656–667. doi:10.1111/j.1467-8535.2006.00648.x

Dawson, K., & Ferdig, R. E. (2006). Commentary: Expanding notions of acceptable research evidence in educational technology: A Response to Schrum et al. *Contemporary Issues in Technology and Teacher Education, 6*(1). Retrieved from http://www.citejournal.org/vol6/iss1/general/article2.cf

Dawson, K., Pringle, R., & Adams, T. (2003). Providing links between technology integration, methods courses and traditional field experiences: Implementing a model of curriculum-based and technology-enhanced microteaching. *Journal of Computing in Teacher Education, 20*(1), 41–47.

Dawson, K., Ritzhaupt, A., Liu, M., Drexler, W., Barron, A., Kersaint, G., & Cavanaugh, C. (2011). *Charting a course for the digital science, technology, engineering and mathematics (STEM) classroom: Research and evaluation report*. Florida Department of Education.

Glazewski, K., Berg, K., & Brush, T. (2002, March). *Integrating technology into preservice teacher education: Comparing a field-based model with a traditional approach*. Paper presented at the Society for Information Technology and Teacher Education, Nashville, TN.

Hernández-Ramos, P., & Giancarlo, C. A. (2004). Situating teacher education: From the university classroom to the real classroom. *Journal of Computing in Teacher Education*, *20*(4), 121–128.

Hew, K. F., & Brush, T. (2007). Integrating technology in K-12 teaching and learning: current knowledge gaps and recommendations for future research. *Educational Technology Research and Development*, *55*, 223–252. doi:10.1007/s11423-006-9022-5

Howland, J., Jonassen, D. H., & Marra, R. M. (2011). *Meaningful learning with technology* (4th ed.). Columbus, OH: Merrill/Prentice-Hall.

Hudson, M. (2005). The links between collaboration, agency, professional community and learning for teachers in a contemporary secondary school in England. *Educate*, *5*(2), 42–62.

Jacobsen, D. M., & Lock, J. V. (2004). Technology and teacher education for a knowledge era: Mentoring for student futures, not our past. *Journal of Technology and Teacher Education*, *12*(1), 75–100.

Knight, S. L., Pedersen, S., & Peters, W. (2004). Connecting the university with a professional development school: Pre-service teachers' attitudes toward the use of compressed video. *Journal of Technology and Teacher Education*, *12*(1), 139–154.

Lawless, K. A., & Pellegrino, J. W. (2007). Professional development in integrating technology into teaching and learning: Knowns, unknowns and ways to pursue better questions and answers. *Review of Educational Research*, *77*(4), 575–614. doi:10.3102/003465430730992l

Lowther, D. L., & Ross, S. M. (2001). *Observation of computer use: Reliability analysis*. Memphis, TN: Center for Research in Educational Policy, The University of Memphis.

Meyers, E., & Rust, F. (2003). *Taking action with teacher research*. Portsmouth, NH: Heinemann.

Milken Family Foundation. (2001). *Information technology underused in teacher education*. Retrieved from http://www.mff.org/edtech/article.taf?_function=detail&Content_uid1=131

National Educational Technology Plan. (2010). *Transforming American education: Learning powered by technology*. Washington, DC. Retrieved from http://www.ed.gov/technology/netp-2010

O'Bannon, B., & Nonis, A. (2002). A field-based initiative for integrating technology in the content areas: Using a team approach to preparing preservice teachers use technology. In D. Willis, et al. (Eds.), *Proceedings of Society for Information Technology & Teacher Education International Conference 2002* (pp. 1394-1397). Chesapeake, VA: AACE.

Partnership for 21st Century Skills. (2011). *Framework for 21st century learning*. Washington, DC. Retrieved from http://www.p21.org/tools-and-resources/publications/1017-educators#defining

Pierson, M. E. (2001). Technology integration practices as function of pedagogical expertise. *Journal of Research on Computing in Education*, *33*(4), 413–429.

President's Council of Advisors on Science and Technology. (2010). *Prepare and inspire: K-12 education in science, technology, engineering, and math (STEM) for America's future*. Retrieved from http://www.whitehouse.gov/sites/default/files/microsites/ostp/pcast-stem-edfinal.Pdf

Ross, S. M., Smith, L. J., & Alberg, M. (1999). *The school observation measure (SOM)*. Memphis, TN: Center for Research in Educational Policy, The University of Memphis.

Schmidt, D. A. (2001). *Simultaneous renewal in teacher education: Strategies for success.* Paper presented at the Society for Information Technology and Teacher Education, Orlando, FL.

Schrum, L., Thompson, A., Sprague, D., Maddux, C., McAnear, A., Bell, L., & Bull, G. (2005). Advancing the field: Considering acceptable evidence in educational technology research. *Contemporary Issues in Technology & Teacher Education*, *5*(3/4), 202–209.

Strudler, N., & Grove, K. (2002). Integrating technology into teacher candidates' field experiences: A two-pronged approach. *Journal of Computing in Teacher Education*, *19*(2), 33–38.

U.S. Department of Education, National Center for Education Statistics, Common Core of Data (CCD). (2010). *Public elementary/secondary school universe survey: 2006-07, 2007-08, and 2008-09.*

Waxman, H. C., Lin, M. F., & Michko, G. M. (2003). *A meta-analysis of the effectiveness of teaching and learning with technology on student outcomes*. Retrieved from the North Central Regional Educational Laboratory Web site: http://www.ncrel.org/tech/effects2/waxman.pdf

Wei, R. C., Darling-Hammond, L., Andree, A., Richardson, N., & Orphanos, S. (2009). *Professional learning in the learning profession: A status report on teacher development in the United States and abroad.* National Staff Development Council. Retrieved from http://www.learningforward.org/news/NSDCstudytechnicalreport2009.pdf

Yendol-Hoppey, D., Dawson, K., Dana, N. F., League, M., Jacobs, J., & Malik, D. (2006). Professional development communities: Vehicles for re-shaping field experiences to support school improvement. *Florida Journal of Teacher Education*, *4*(1), 37–48.

Zeichner, K. (2003). Teacher research as professional development for P-12 educators in the USA. *Educational Action Research*, *2*(2), 301–326. doi:10.1080/09650790300200211

ADDITIONAL READING

Dana, N. F. (2009). *Leading with passion and knowledge: The principal as action researcher*. Thousand Oaks, CA: Corwin Press, a Joint Publication with the American Association of School Administrators.

Dana, N. F., Thomas, C. H., & Boynton, S. (2011). *Inquiry: A district-wide approach to staff and student learning*. Thousand Oaks, CA: Corwin; a Joint Publication with Learning Forward.

Dana, N. F., & Yendol-Hoppey, D. (2009). *The reflective educator's guide to classroom research: Learning to teach and teaching to learn through practitioner inquiry*. Thousand Oaks, CA: Corwin Press.

Dana, N. F., Yendol-Silva, D., & National Staff Development Council (U.S.). (2008). *The reflective educator's guide to professional development: Coaching inquiry-oriented learning communities*. London, UK: SAGE.

Meyers, E., Paul, P. A., Kirkland, D. E., & Dana, N. F. (2009). *The power of teacher networks*. Thousand Oaks, CA: Corwin Press.

Noffke, S. E., & Somekh, B. (2009). *The Sage handbook of educational action research*. London, UK: SAGE Publications.

Yendol-Hoppey, D., & Dana, N. F. (2010). *Powerful professional development: Building expertise within the four walls of your school*. Thousand Oaks, CA: Corwin Press.

KEY TERMS AND DEFINITIONS

ARTI (Action Research for Technology Integration): A unique online tool that provides an online scaffold for teachers to inquire about how their technology integration practices affect student outcomes in their classroom, provides a mechanism to synthesize action research information from multiple teachers and provides a mechanism to capture evidence of student learning within technology integration inquiries.

Action Research (also known as Teacher Inquiry): The process by which an educator systematically and intentionally studies her practice in order to make improvements and share what was learned with others.

Knowledge for Practice: Knowledge teachers develop when they learn educational theories, concepts and definitions typically within university-based courses or professional development workshops.

Knowledge in Practice: Knowledge teachers developed from the experience of teaching.

Knowledge of Practice: Knowledge teachers develop when they study their own practices, make improvements based on what they learn and share with others.

Technology Integration: The process of using technology to enhance teaching and learning.

Virtual School: A school that offers courses completely or primarily through online methods

Chapter 21
Improving Student Learning in a Fully Online Teacher Leadership Program:
A Design-Based Approach

Scott L. Day
University of Illinois Springfield, USA

Leonard Bogle
University of Illinois Springfield, USA

Karen Swan
University of Illinois Springfield, USA

Daniel Matthews
University of Illinois Springfield, USA

Emily Boles
University of Illinois Springfield, USA

ABSTRACT

This chapter describes how faculty in a fully online Master's program in teacher leadership are using a design-based approach, grounded in theory and informed by data, to iteratively improve core courses and student learning from them. Specifically, the authors revised their courses to meet Quality Matters (QM) standards for online course design, and then made incremental and ongoing revisions focused on course implementation and based on student responses to the Community of Inquiry (CoI) survey. The first part of the chapter describes the online program in which course improvements are taking place, and the QM and CoI theoretical frameworks. In the main body of the chapter, specific course revisions are discussed and initial findings reported which show significant improvements in student outcomes as a result of these revisions. This section also describes the design-based approach the authors adopted and provides recommendations for others who might want to similarly improve individual courses or program offerings as a whole. The chapter closes with a brief discussion of directions for future research and conclusions, which highlight what the authors believe are the most important aspects of this work.

DOI: 10.4018/978-1-4666-1906-7.ch021

INTRODUCTION

In the past ten years, the number of higher education institutions offering online courses has increased dramatically; in fall 2010, for example, 31% of all higher education students, over 6.1 million students, took at least one online course (Allen & Seaman, 2011). Given the rapid growth of online education and its importance for postsecondary institutions, it is imperative that institutions of higher education provide quality online programs (Kim & Bonk, 2006). At the same time, state requirements for the improvement of teacher education programs are also increasing. In response to the U.S. Department of Education Title II guidelines and state regulations for teacher certification, teacher preparation programs are embracing the use of technology more readily. Online instructional delivery is one model with which schools of education have experimented in response to this mandate for change (Ragan, Lacey & Nagy, 2002).

This chapter will demonstrate how the Quality Matters (QM) and Community of Inquiry (CoI) frameworks can be applied at the program level to enhance curriculum development and maintain programmatic rigor when evaluating and updating course material for a teacher leadership program. First, we summarize the QM and CoI frameworks. We then describe how the QM standards were applied to individual core courses in the Teacher Leadership program to ensure they had high quality designs. We then describe how the CoI survey was employed across subsequent semesters to improve the implementation of these courses and ensure the development of social, cognitive, and teaching presence within them. We conclude with practical advice for others desiring to make substantive improvements in online learning at the programmatic level.

BACKGROUND

Teacher Leadership Program

The Masters Degree in Teacher Leadership was developed in 1999. Since the 1970's the educational leadership (EDL) department had offered one Master's degree focused on the Principalship. By the early 1990's, professors discussed the need to take the program in an additional direction since many students admitted to enrolling in this master's program when they had no interest in becoming school principals. However, some course content and assessments had less value for students who did not want to become administrative leaders. This collaborative discussion was a precursor to large-scale change in our department.

With a growing literature calling for the rethinking of organizational structures that facilitate teacher leadership from within the classroom (Silva, Gimbert & Nolan, 2000), a discussion about changing our program was begun. The faculty wanted to clearly separate the curriculum and develop two distinct master's degree programs. Our guiding question was to determine what was needed to affect student achievement in schools and to better prepare educators for their work with families, the community and social service agencies. To this end, courses, content, and assessments were designed to develop teacher leaders with instructional, curricular, action research, and assessment experiences.

We invited professors from two other programs in the College of Education and Human Services – teacher education and social work -- to join the development team, bringing together experts in school leadership, K-12 education, and the social work fields. Developers focused on what types of content, training, and assessments would result in the greatest amount of change in schools. We were also fortunate to receive a degree development grant from the Sloan Foundation, which was interested in assisting universities who were trying to offer teacher-oriented degrees online. This online option appealed to us, as we believed it would offer K-12 educators in rural America greater access to a teacher-focused advanced degree.

Indeed, once we opened the first two courses in the program, enrollments grew rapidly. In five years time, they grew from thirty to nearly three hundred students in the program. Today, our students represent 26 states and 6 countries outside the United States, and the EDL department has grown to become one of the largest graduate student departments on campus. With increased enrollments in the department, support came from the university administration for additional tenure-track faculty lines, an endowed distinguished professorship, and other faculty experts, such as a scholar-in-residence and clinical faculty positions. However, ten years passed quickly, and courses originally developed at the turn of the century were due for review and revision. It was particularly important that core courses (see Table 1) which had evolved into quite different offerings; be reoriented to common course goals and objectives. To guide this work, we turned to two theoretical frameworks specifically developed for online learning – Quality Matters (QM) and the Community of Inquiry (CoI). These are described in the sections which follow.

Quality Matters Framework

Quality Matters (QM) is a faculty-oriented, peer review process designed to assure quality in online and blended courses. The QM review process is centered on a rubric that was originally developed collaboratively by online educators working with MarylandOnline, Inc., to provide a replicable pathway for inter-institutional quality assurance and course improvements in online learning. These pioneers developed the initial rubric but also created and implemented a process to certify the quality of online courses and online components (Shattuck, 2007). The initial work on the Quality Matters framework was funded through FIPSE (Fund for the Improvement of Postsecondary Education), but the need for such a rubric was so widespread that it took on a life of its own. Today, it is a subscription-based service which also offers a variety of trainings and Quality Matters course reviews. QM subscribers include community and technical colleges, colleges and universities, K-12 schools and systems, and other academic institutions. Most recently, Blackboard became the first learning management system to become a QM subscriber.

In the QM framework, quality in online courses is assured through a peer-review process in which trained faculty review the design and organiza-

Table 1. Core course titles and descriptions

Course Title	Course Description
Education Research Methods	Basic research methods in education that have relevance to classroom practice. Goals include developing a general understanding of quantitative and qualitative methodologies, skills needed to make sense of the research literature, the ability to write a research proposal, and a basic understanding of tools for supporting data-based decision making
Teacher Leadership	Examination of leadership characteristics and applications of processes and strategies for teacher leadership. The goal is the ability to understand and apply problem-solving tools with people in educational settings.
Instructional Design	Examination of the role of standards and curriculum at district and school levels in shaping educational change and reform. The goal is to examine and understand curriculum and instructional issues within the context of reform efforts.
Foundations of Teacher Leadership	Provides an examination of learning to become a teacher leader and being a member of an online community. Included are strategies for e-learning, teacher leaders as a part of a virtual learning community, and graduate level research and writing.

tion of their peers' courses. QM assumes that this review is formative and that courses will undergo a processes of continuous improvement through certification (see Figure 1 below). It is important to note, however, that the QM review focuses on the design and organization of online courses and not on their implementation.

Quality Matters peer reviews are guided by a rubric designed to assess the quality of online courses consisting of 41 items describing criteria to be met (Appendix A). Items are assigned point values of 1, 2 or 3 depending on their perceived importance. To meet QM review expectations, courses must meet all 3 point criteria on the entire evaluation measure. Items in the rubric are organized into eight categories – course overview and introduction, learning objectives, assessment and measurement, resources and materials, learner engagement, course technology, learner support, and accessibility. These are described below.

Course overview and introduction is the first category in the rubric. It consists of 8 criterion items, totaling a possible 14 points, the first two of which must be met. This category is centered on a clear introduction to the course and course expectations. Courses that meet QM expectations must contain clear instructions on how to get started in the course and a clear statement of course purposes. Less important criteria in this category include netiquette and institutional expectations (2 points each), personal introductions by the instructor and students, and prerequisite content knowledge and technology skills (1 point each).

Learner objectives, the second category in the QM rubric, consists of 5 items totaling a possible 15 points, all of which must be met, making it a critical category. Criteria that must be met focus on course and unit objectives that are measurable, consistent, clearly stated, and which include adequate instructions on how students can meet them and are appropriate to the course level.

Figure 1. QM continuous course review model (Shattuck, 2007)

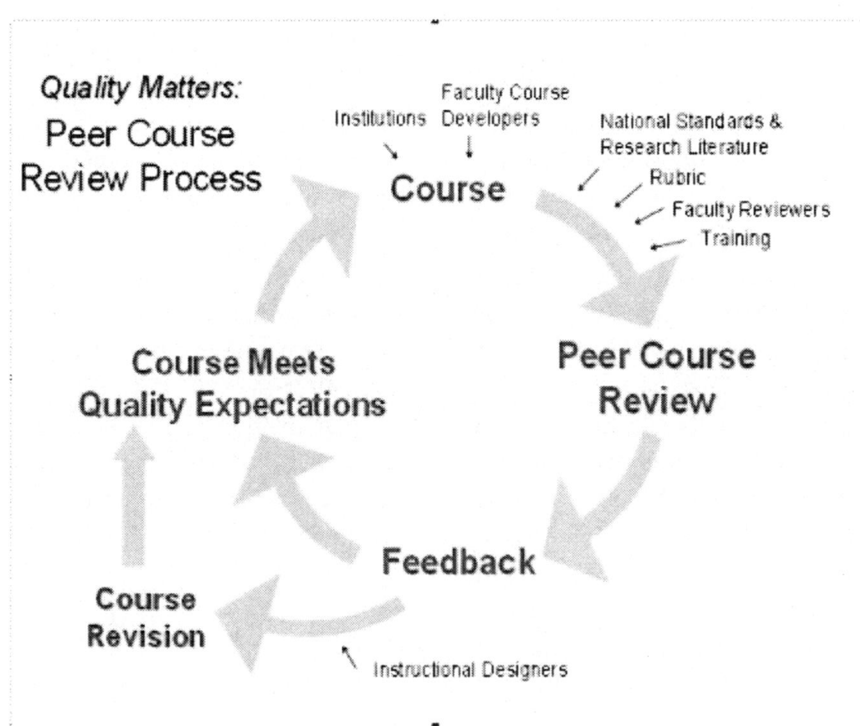

Assessment and measurement is the third category in the QM rubric. It also consists of 5 items, all but two of which must be met, making it an important category. Items in this category add up to a possible total of 13 points. Criteria that must be met include assessments that match objectives, a clearly stated grading policy, and specific evaluation criteria. Items that are not critical but still important (counting 2 points each) include the provision of self-check or other practice assignments, as well as "sequenced, varied and appropriate" assessments.

Instructional materials, the next category, consists of 6 items with a possible total of 12 points. Two of these criterion items must be met. They are that instructional materials contribute to the achievement of course and unit objectives and that the relationship between these materials and course activities are clearly explained. Criteria of less importance are that resources and materials are appropriately cited and current (2 points each) and that materials present a variety of perspectives and differences between required and optional materials are clearly stated (1 point each).

Learner interaction and engagement is the next category. It consists of 4 items, three of which must be met, for a possible total of 11 points. The three critical criteria are that learning activities promote the achievement of the course objectives, that learning activities provide opportunities for interaction "that support active learning," and clear statements concerning instructor availability and feedback. Of slightly lesser importance (2 points) is clearly articulated requirements for student interaction.

Course technology consists of 5 items with a possible total of 12 points. Three of the course technology criteria must be met. They focus on technology tools and media that contribute to the achievement of the course objectives and support active student engagement, as well as clear and transparent course navigation. Other criteria in this category include use of technologies that are readily accessible to students (2 points) and compatible with current delivery standards (1 point).

Learner support consists of 4 items with a possible total of 9 points. Two criteria, clear descriptions of technical support and accessibility policies, must be met. Other criteria include explanations or links to academic (2 points) and student support (1 point).

Accessibility is the final category in the QM rubric. It consists of 4 items with a possible total of 9 points. Only one criterion item in this category must be met, but it is an important one – conformance to ADA standards and institutional policies regarding accessibility. Other criteria in this category include screen readability, course pages and materials which provide equivalent auditory and visual content, and course design which accommodates the use of assistive technologies (2 points each).

Although the Quality Matters framework is relatively new, preliminary research on its effectiveness is promising. For example, Legon, Runyon and Aman (2007) surveyed students enrolled in QM certified courses (62 responses), in non-certified courses at QM institutions (33 responses), and in courses at non-QM institutions (77 responses). They found that students enrolled in QM certified courses and in non-certified courses at QM institutions were significantly more satisfied than students enrolled in courses at non-QM institutions.

Legon, Runyon, and Aman (2007) explored the relationship between course design and learner interaction with course content in a large enrollment class at the College of Southern Maryland. As part of the Quality Matters review process (Figure 1), each learning module in this information technology course was revised in 3 ways: 1) creation of a Learning Guide (explicit roadmap), 2) reorganized presentation and design, and 3) addition of classroom assessment techniques (CATs) in each course module. Runyon compared student grades before and after the QM redesign and found they were higher after the redesign (more "A"s, fewer "F"s). She also found greater learner interaction with course materials among students in the redesigned version of the course.

Community of Inquiry Framework

The Community of Inquiry (CoI) framework (Garrison, Anderson & Archer, 2000) is a process model of online learning. It is grounded in a social constructivist view of higher education which assumes that effective online learning requires the development of a course community (Rovai, 2002; Shea, 2006) that supports meaningful inquiry and deep learning. The CoI framework has been quite widely used to inform both research and practice in the online learning community, and an increasing body of research supports its efficacy for both describing and informing online learning (Arbaugh, et al., 2008; Swan, Garrison, & Richardson, 2009).

Building from the notion of social presence in online discussion, the CoI framework represents the online learning experience as a function of the relationship among three presences: social presence, teaching presence, and cognitive presence (see Figure 2). The CoI framework suggests that online learning is located at the intersection of these three presences; that is, it views all three presences as working together to support deep and meaningful learning processes.

Social Presence

Social presence refers to the degree to which learners feel socially and emotionally connected with others in an online environment. A number of research studies have found that the perception of interpersonal connections with virtual others is an important factor in the success of online learning (Picciano, 2002; Richardson & Swan, 2003; Swan, 2002; Swan & Shih, 2005; Tu, 2000). Garrison and Anderson (2003) identified three elements that contribute to the development of social presence in online courses –affective expression, open communication, and group cohesion–which research suggests are affected by both instructor behaviors (Shea, Li, Swan & Pickett, 2005; Shea & Bidjeramo, 2008) and course design (Swan & Shih, 2005; Tu & McIssac, 2002). Social presence has also been shown to predict 21% of retention in courses and programs (Boston, Diaz, Gibson, Ice, Richardson, & Swan, 2009).

Affective expression refers to participants' ability to express their personalities in virtual environments through the use of affective indicators (e.g., emotions), self-revelation, humor, the sharing of personal experiences, and the expression of personal beliefs and values. Course activities that provide opportunities for affective expression are obviously a necessary condition for its development, but it is also important for instructors to value and encourage these behaviors in students and model their use themselves.

It is also very important that course designers and instructors establish and maintain open communication, a climate in which students feel free to express themselves. Some strategies for establishing open communication include explicitly introducing students to the unique nature and learning potential of online discourse, establishing rules of Netiquette, and developing low stress "ice-breaker" activities to get students interacting at the beginning of a course. To maintain open communication, instructors should be supportive of all student contributions to discussions at the start of a course, but gradually reduce their participation to allow student voices to emerge (Vandergrift, 2003).

Finally, and perhaps most importantly, course designers and instructors can facilitate the development of social presence by providing opportunities and support for the development of group cohesion. Group cohesion is a sense of group commitment, a feeling that the class (or a smaller group within the class) is a community in which participants interact around shared intellectual activities and tasks. Obviously, this cannot happen unless course designers and instructors develop

Figure 2. CoI framework (Garrison, Anderson & Archer, 2000)

interactive and collaborative activities. Research also suggests that such activities are more successful when instructors provide explicit facilitation and direction (Meyer, 2003; Murphy, 2004; Shea & Bidjermo, 2008).

Teaching Presence

Teaching presence is defined as the design, facilitation, and direction of cognitive and social processes for the realization of personally meaningful and educationally worthwhile learning outcomes (Anderson, Rourke, Anderson, Garrison & Archer, 2001). Researchers have documented strong correlations between learners' perceived and actual interactions with instructors and their perceived learning (Jiang & Ting, 2000; Richardson & Swan, 2003; Swan, Shea, Fredericksen, Pickett, Pelz & Maher, 2000) and between teaching presence and student satisfaction, perceived learning, and development of a sense of community in online courses (Shea, Li, Swan & Pickett, 2005). In fact, the body of evidence attesting to the critical importance of teaching presence for successful online learning continues to grow (Garrison & Cleveland-Innes, 2005; Murphy, 2004; Swan & Shih, 2005; Vaughn & Garrison, 2006; Wu & Hiltz, 2004), with the most recent research suggesting it is the key to developing online communities of inquiry (Shea & Bidjeramo, 2008; Garrison, Cleveland-Innes, & Fung, 2010).

Garrison and Anderson (2003) identified three elements that contribute to the development of teaching presence in online courses – design and organization, facilitating discourse, and direct instruction – all of which deserve careful attention. The first category, *design and organization*, cannot be neglected in an online learning environment, especially as regards the clarity and consistency of course organization and clear statement of goals and objectives (Swan, et al., 2000). The selection of worthwhile collaborative and other learning activities is also especially important.

Facilitating discourse is particularly focused on facilitating online discussion, but it is also concerned with facilitating and directing collaborative activities (Meyer, 2003; Murphy, 2004; Shea & Bidjermo, 2008). In both cases, it is important to be supportive and present, but not overly so (Vandergrift, 2003). It is also very important to value student discourse by making it count for a significant portion of their grades (Jiang & Ting, 2000) and to provide diverse activities with ample opportunities for learner choice and control.

There will, of course, be times when it is necessary to intervene directly in online discussions to correct misconceptions, provide relevant information, summarize the discussion, and/or provide some meta-cognitive awareness. This is the third category of teaching presence, *direct instruction*, which also includes any lecture-like material presented in the course, as well as instruction included in feedback to students. As regards the former, teaching presence is enhanced when lecture materials are written in the first person and adopt a colloquial tone. As regards the latter, teaching presence is enhanced when feedback is timely and supportive (Jiang & Ting, 2000; Swan, et al., 2000).

Cognitive Presence

Cognitive presence describes the extent to which learners are able to construct and confirm meaning through course activities, sustained reflection, and discourse (Garrison, Anderson & Archer, 2001). In the CoI framework, cognitive presence is seen as consisting of the four phases of the Practical Inquiry Model, which begins with a triggering event and extends through exploration and integration to culminate in resolution (Figure 3).

Practical inquiry begins with a *triggering event* in the form of an issue, problem or dilemma that needs resolution. As a result of this event, there is a natural shift to *exploration*, the search for relevant information that can provide insight into the challenge at hand. As ideas crystallize, there is a move into the third phase – *integration*–in which connections are made, and there is a search for a viable explanation. Finally, there is a selection and testing (through vicarious or direct application) of the most viable solution and *resolution*. At each of the stages, there may be a need to return to a previous stage for new direction or information. The four phases described in the model are a telescoping of Dewey's phases of reflective thinking (Dewey, 1933) for the purposes of parsimony and understanding. Consistent with Dewey's rejection of dualism, the phases should not be seen as discrete or linear. In an actual educational experience, they would be very difficult to label, as those that have used this model to code transcripts will attest (Garrison, Anderson & Archer, 2001).

Indeed, while researchers have been able to find evidence of practical inquiry in online discussion, several studies have found that online discussion rarely moves beyond the exploration phase where participants share information and brainstorm ideas (Garrison & Arbaugh, 2007; Kanuka & Anderson, 1998; Luebeck & Bice, 2005; Murphy, 2004). While various explanations have been explored, it is most likely that much of this has to do with the nature of the assignments and instructional direction (teaching presence) provided (Garrison & Arbaugh, 2007). In studies in which students were challenged to resolve a problem and explicit facilitation and direction provided, students did progress to resolution (Akyol & Garrison, 2008; Meyer, 2003; Murphy, 2004; Shea & Bidjermo, 2008; Wang & Chang, 2008). In a study focusing on deep and meaningful learning, for example, Akyol and Garrison (2011) found that students were able to reach high levels of both cognitive presence and learning outcomes, which suggests that cognitive presence in a community of inquiry is associated with perceived and actual learning outcomes. It should be noted that one would ex-

Figure 3. Practical inquiry model (Garrison, Anderson & Archer, 2001)

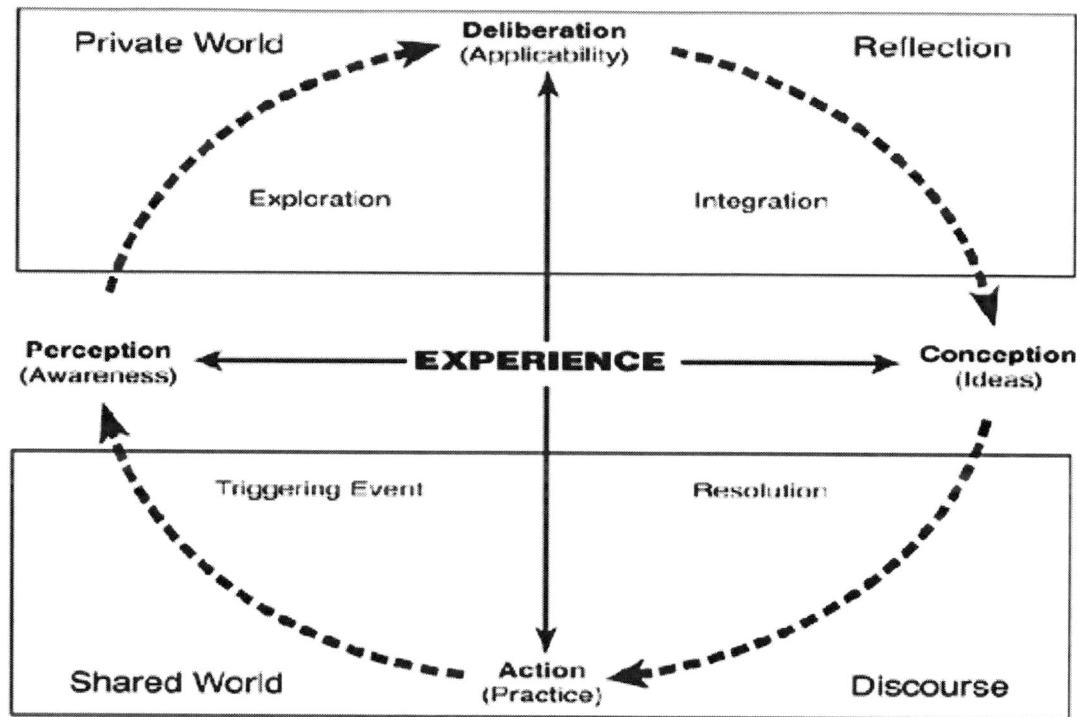

pect there to be more discussion postings during exploration than during the integration phase of inquiry and more postings during integration than in resolution. It could also be that ideas explored in online discussion are integrated and resolved through other course assignments (Archer, 2010).

The Community of Inquiry Survey

To begin to address some of the issues surrounding the cognitive presence construct, to introduce a means for researching all three presences at once as is consistent with the framework, and to move CoI research beyond single courses and institutions, CoI researchers came together and jointly created a survey to measure student perceptions of teaching, social, and cognitive presence in online courses (Arbaugh et al., 2008; Swan, Richardson, Ice, Garrison, Cleveland-Innes, & Arbaugh, 2008) The resulting instrument (Appendix B) was adapted from measures group members had successfully used to measure individual presences in the CoI framework. It includes twelve items designed to measure cognitive presence (3 for triggering events, 3 for exploration, 3 for integration, and 3 for resolution), nine items designed to measure social presence (3 for affective expression, 3 for open communication, and 3 for group cohesion), and thirteen items designed to measure teaching presence (4 for design and organization, 6 for facilitation of discourse, and 3 for direct instruction). In the summer of 2007, the survey was tested in graduate courses at four institutions located in the United States and Canada using principal component factor analysis, and the three factor (presences) construct predicted by the CoI framework was supported (Swan, et al., 2008).

Results of the factor analysis provide evidence that, as currently defined and operationalized, an online community of inquiry emerges out of social, cognitive, and teaching presence. Student responses to the survey's statements about their online experience clustered around items as defined by the theory. The results thus validate the survey as a measurement tool of agreed upon and statistically confirmed items that operationalize the concepts in the CoI model (Swan, et al., 2008). This measurement tool may be used for continued explication of concepts in the model and can serve as a ground for more qualitative investigations in mixed methods studies.

For example, Arbaugh, Bangert, and Cleveland-Innes (2010) used responses to the CoI survey to uncover differences is the development of teaching, social, and cognitive presence between online courses they identify as "hard" (eg., mathematics, sciences) and "soft" (eg., humanities, social sciences) disciplines. Akyol, Vaughan and Garrison (2011) employed the CoI survey to study changes in student perceptions of the presences over time. The CoI survey was also used in a large scale study of factors relating to retention of students in online programs (Boston, Diaz, Gibson, Ice, Richardson, & Swan, 2009. Shea and Bidjermo (2008) used Structural Equation Modeling (SEM) to study the impacts of teaching and social presence on cognitive presence, all as measured by the CoI survey. The results of their analysis are shown graphically in Figure 4. They reveal that teaching and social presence together account for 70% of the variation in students' reported level of cognitive presence. However, the authors also found that the development of social presence was contingent on the establishment of teaching presence; that is, social presence did not in itself directly affect cognitive presence but rather served as a mediating variable between teaching presence and cognitive presence.

The CoI framework has also been used to guide course design and implementation. Baber (2011), for example, use the CoI model in tandem with the design process to develop a blended approach which enhanced critical thinking in a ten-week graphic design foundations course. Ice, Gibson, Boston, and Becher (2011) used the CoI survey and a data mining approach to examine course-level retention through the lens of student perceptions of teaching, social and cognitive presence. In comparing courses in the highest and lowest retention quartiles of all courses at American Public University (APU), the value of effective *instructional design and organization*, and initiation of the *triggering event* phase of cognitive presence were found to be significant predictors of student satisfaction in the highest retention quartile. For the lowest retention quartile, the lack of follow-through vis-à-vis *facilitation of discourse* and *cognitive integration* were found to be negative predictors of student satisfaction. Faculty and staff at APU are using this information to improve course implementation.

COURSE REDESIGN

We began the redesign of the MTL program with the redesign of a single section of a single course in Educational Research Methods. Initially, we viewed the process as guided solely by a QM review, with the CoI survey being a secondary measure of learning processes. That is, we assumed that improved course design would result in improved learning processes, which in turn would enhance student learning outcomes. Primary outcome measures of learning included final course grades as well as scores on the final exam and the major course paper, all normalized to percentage correct.

QM Revisions

In the fall of 2009, an informal review of the Educational Research Methods course was undertaken.

The review process for Educational Research Methods involved two peer reviewers from the university, one from the department and one instructional designer from the Center for Online Learning, Research and Service (COLRS), who completed an informal review led by a Quality Matters expert. In addition, a forty-five minute

Figure 4. SEM analysis of effects of social & teaching presence on cognitive presence; Detail adapted from Shea & Bidjeramo, 2008; © 2008, Peter Shea, used with permission

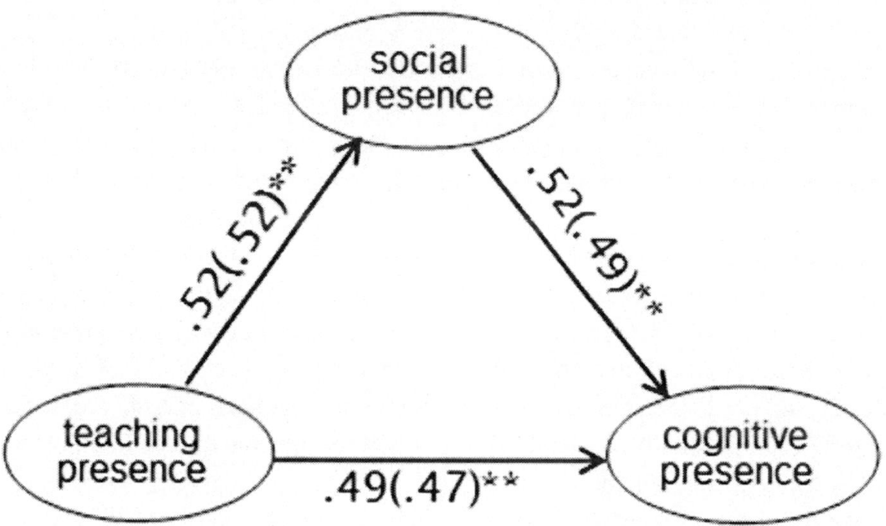

in-service provided guidance regarding the QM rubric and the application of the rubric to the assigned course. The instructor submitted a five-page narrative about the course design's relationship to the location of information by the students. The review process took approximately three hours per reviewer resulting in a score for each rubric item, a total score and detailed narrative for each item in the rubric that did not meet the QM standards. The detailed narrative served as a guide for the instructor during the revision process.

The course failed to meet 5 of the essential (3 point) standards, all of which had to do with learning objectives. These deficits were addressed in the spring 2010 version of the course, which was reviewed again by the Quality Matters expert and received a passing score.

CoI Revisions

Outcomes from the spring 2010 semester showed an increase in student performance (learning outcomes) but an actual reduction in student perceptions of learning processes (CoI scores) after the QM redesign. While we believe the unexpected initial reduction in student perceptions of learning process was attributable to the instructor's attention being focused on the revisions and having less time to dedicate to facilitating the learning process measured by the CoI, we were nonetheless struck by the unexpected differences in those early findings. This led us to see what we should have known from the start: that the QM and CoI frames view learning from differing perspectives and so measure different things. Because scores on the CoI survey went down after the QM redesign, we began exploring a second design-based notion; namely, whether iterative changes to the course based on CoI responses could lead to further increases in student performance.

We reviewed average scores for all items on the Spring 2010 CoI surveys and made changes based on the lowest of these, which for this semester were those below 3.75 (on a 5-point Likert scale). These turned out to be related to course discussions and collaborative learning tasks, so changes were made in those areas. The new version of Educational Research Methods was implemented in the Summer of 2010, and CoI scores improved. Outcome scores declined slightly but not meaningfully and perhaps understandably considering the reduced (eight week) semester. We continued to revise the course based on CoI scores (although in the second iteration this became scores below 4.0) for the Fall 2010 version of the course, and, in that semester, both CoI and outcome scores improved. Most importantly, outcome measures improved significantly from the Fall of 2009 to the Fall of 2010 (Figure 5). The results suggest that the two-phase (QM revisions followed by iterative CoI-based improvements) redesign process was highly effective.

Findings

Specifically, the data show increases in all learning outcome scores after the QM revision; then, after slight decreases in the summer semester (which we attributed to students having fewer weeks to learn the challenging material), further gains were made in the fall semester. The combination of both the QM and CoI revisions across all four semesters brought average scores on the research proposal from a 91 to a 97 and on the final exam from an 82 to a 90, while overall course grades went from a 90 to a 99. Analysis of variance found these differences were significant for the final exam scores at the p=.05 level and for overall course grades at the p=.001 level (Table 2). Post hoc analyses show these are only significant when comparing Fall 2009 and Fall 2010 scores. Differences in scores on the research proposal were not significant.

Cohen's (1992) analysis of eta squared results (Table 2) was used to calculate effect sizes for all results. Effect sizes of the cumulative QM/CoI revisions were small for the research proposal (.11) and the final exam (.16) but moderate for overall course grades (.29).

The findings suggest that revising Educational Research Methods around stated objectives (QM) and presence deficits identified by CoI scores resulted in better student performance, especially in terms of overall course grades (Swan, Matthews, Bogle, Boles, & Day, in press). The results thus indicate that ongoing course redesign, guided by the Quality Matters (QM) and Community of Inquiry (CoI) frameworks, can result in improved learning. Future research will explore

Figure 5. Comparison of learning outcomes across four semesters

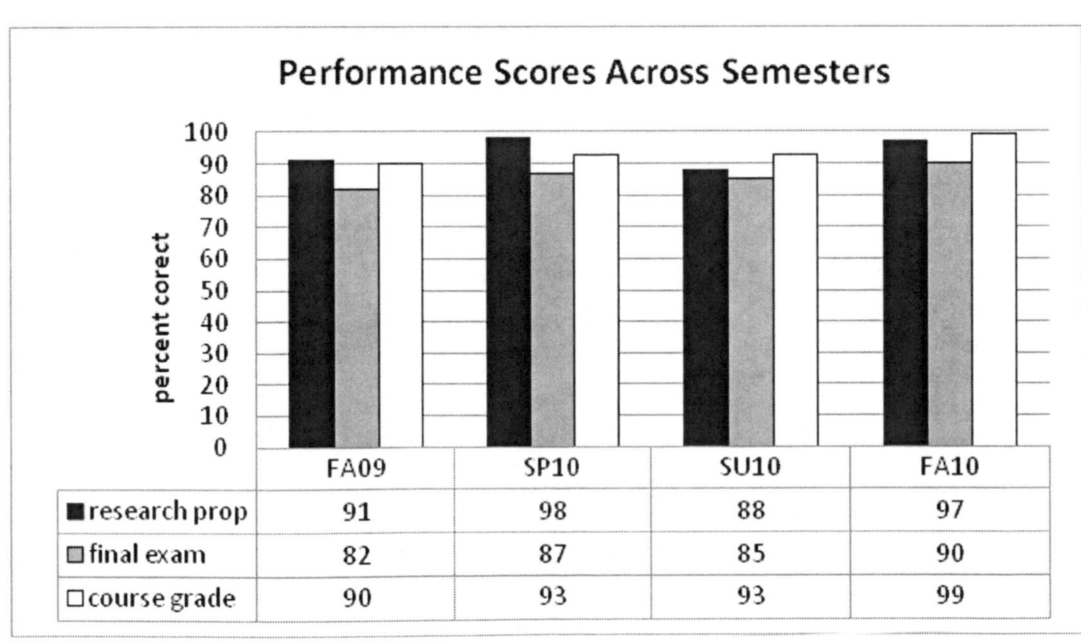

	FA09	SP10	SU10	FA10
■research prop	91	98	88	97
□final exam	82	87	85	90
□course grade	90	93	93	99

Table 2. Analysis of variance comparing learning outcomes across semesters

ANOVA TABLE	Mean Square	F	Sig
Research Proposal % * Semester Between Groups Within Groups Total	242.956 134.952	1.800	1.61
Final Exam % * Semester Between Groups Within Groups Total	503.526 179.503	2.805	0.05
Course Grade % * Semester Between Groups Within Groups Total	191.473 30.814	6.214	.001

whether such approach can work in other courses. Our initial results are quite promising. We note several limitations to the results to date that should be addressed in future research. The data reported here are from a relatively small sample comprised of multiple sections of a one professor's course. Further research, such as we are in the process of conducting, will be needed to confirm the effects are similar other courses and other instructors. Also, the current research is on graduate level courses in teacher leadership; future research will be needed to determine if the results generalize to undergraduate education and to other fields. Another methodological concern is that the baseline data for both the Community of Inquiry survey and the learning outcomes were rather high, creating the problem of ceiling effects.

A Design-Based Approach

As we continued to revise the course beyond our initial test of the effects of QM- and CoI-guided revisions, we realized our work had evolved into a design research, or design experiment, model. Design research blends empirical research with the theory-based design of learning environments. It centers on the systematic investigation of innovations designed to improve educational practice through an iterative process of design, development, implementation, and analysis in real-world settings (Cobb, Confrey, diSessa, Lehrer, & Schauble, 2003; Wang & Hannafin, 2005). Design-based research helps us understand "how, when, and why educational innovations work in practice" (Design-based Research Collective, 2003, p. 5), because the innovations it explores are grounded in educational theory. We have been able to capitalize on the potential of the design research approach to generate generalizable research from contextualized, formative evaluations in a manner that can provide guidance to others involved in the design process (Collins, Joseph, & Bielaczye, 2004). The words of Middleton, Gorard, Taylor, and Bannan-Ritland (2006) resonated with us: "Concerning diffusion of innovation, the issue of 'research-to-practice' should not be problematic, if the research *is* practice" (p. 26, italics theirs). Recognizing the value of the design research approach, we, like Anderson (2005) before us, decided to adopt it to our in-progress and ongoing improvements. Design research is now the overarching methodological perspective guiding our analysis of our program improvement activities.

On seeing the positive effects that our improvement efforts were having on the original course,

Table 3. Effect sizes for pre/post revisions learning outcomes

	Eta	*Eta Squared*
Research Proposal % * Semester	.327	.107
Final Exam % * Semester	.397	.158
Course Grade % * Semester	.541	.293

we recognized there was an opportunity to extend this iterative, model-guided improvement effort to other courses within the program. We immediately recognized two challenges to expanding our revision efforts to additional courses: getting other faculty to commit and identifying quantifiable outcome data that could be compared across courses, even though the courses' goals and assessments differed considerably.

We were fortunate in that, from the beginning, our research team had included faculty who taught other courses within the program, so the initial conversations on extending this work to those other courses grew spontaneously and informally out of our research group's meetings. At the same time, we quickly realized a major challenge of scaling up the reform efforts: the program relies on a healthy mix of adjunct faculty, faculty from other programs, and full-time program faculty to deliver a wide array of required and elective courses. We determined that by focusing solely on four core, required courses rather than on attempting to revise all the program courses, we would have a manageable task that would have a much greater likelihood of success. We were able to get the faculty for the four core courses agree to take part in this course improvement work because several of the faculty for those courses were already on our research team, and they knew the benefits and expectations of the proposed improvement efforts. The result was a non-hierarchical, collaborative process that led to faculty committing time and energy to an ongoing, iterative, data-driven, framework-guided course improvement process that can have program-wide effects.

The other challenge consisted of identifying quantifiable outcome data for the courses. That is, the four core courses have different goals, learning objectives, and assessments, yet we would need baseline and multiple semesters of data to inform the iterative revisions to the courses. In addition, the data would need to be interpretable within the frameworks (QM and CoI) that were guiding our revision efforts. For the outcome data, we came up with a flexible approach: we asked instructors to provide data on two to four key outcome assessments, converted to percentages correct for the baseline and ongoing semesters. We also asked instructors to offer their students the opportunity to take the CoI survey each semester during their courses. We then set up an Excel database designed to collect and store the outcome and processes (CoI) data in a way that allows us to analyze trends in outcomes and processes for each instructor, which he or she can then use to modify a specific course. The database is also designed to allow us to quantitatively assess the effects revisions have at the macro (program) level. In sum, at both the course and program level, the data will allow continuous assessment of the success of the interventions while also providing actionable data for the course instructors each semester.

Recommendations

Our ongoing work has led us to believe that the re-design of courses should be collaborative, data informed, continuous, iterative, and framework guided. The first and most important aspect of any change process is to effectively communicate with stakeholders and to get their acceptance before introducing new programs (Bolman & Deal, 2008). In our experience, a bottom-up approach, initiated by two faculty members working on a single course was particularly effective as other faculty were attracted by initial success. We also found addressing course redesign as a multi-stage process to be especially useful. Although forced on us through necessity, this approach allowed us to fully focus on differing aspects of our courses one step at a time, first design and then implementation.

While a single instructor can improve his or her own courses, programmatic improvements require collaborative efforts. We collaborated on all aspects of the redesign process. For example, we utilized internal reviewers for the QM process.

Although it may appear threatening to review a peer's course and/or to be reviewed by one's peers, it has helped us know more about the overall program and how our individual courses can work together to contribute to program goals A safe environment must be created for those willing to take such risks and work in a collaborative manner. In our case, we share more within our department and the project has become a professional development opportunity for us. You learn more about improving your own course by reviewing courses developed by others, and this, in turn, leads to clearer continuity among the same courses.

It is very important that any redesign efforts be data-driven. Programs should collectively decide what they particularly want to address in their redesign efforts and then find or develop instruments to measure it. In our case we were particularly concerned with learning outcomes, but other foci could include learning processes, student satisfaction, retention/progression toward a degree, or engagement. A program might also want to focus on improving specific skills such as writing, or critical thinking. Once a focus and measures have been decided, baseline data should be collected to which future data can be compared to determine if the design changes actually resulted in the desired improvements. Data collection should be ongoing and it is important to track the design and implementation changes made semester to semester and to associate these with the data collected so that progress can be assessed and related to specific changes. In the process of continuous/iterative improvement, one needs to recognize the time demands on faculty for making course adjustments. In our view, not all changes will lead to immediate improvements and data may initially show a drop in the chosen measures of student success. Benefits will emerge over time. Data collection over multiple semesters is thus necessary in order to obtain an accurate measure of the efficacy of the design changes.

Finally, although we used the QM and CoI frameworks to guide our course redesign efforts, and we believe they are the most appropriate for similar redesign efforts, we believe our approach can be used with other frameworks as well. What is most important is that a theoretical model be used to guide programmatic improvements. Regardless of the design analysis choice, course designs should be adjusted based on the chosen analysis instruments to keep the ongoing process true to its original vision and to explore the efficacy of the models chosen. The selection of a research-based guide for the design of online teacher education programs in particular is an excellent opportunity for professors to demonstrate that specific models or course design frameworks can be used to enhance the online learning experience.

FUTURE RESEARCH DIRECTIONS

We are currently investigating whether similar results can be replicated with other courses and other instructors, starting with our own core courses. We plan to test whether the combination of changes made to online courses, based on an initial Quality Matters review (QM revisions) and followed by iterative changes based on deficiencies identified in CoI surveys (CoI revisions) can lead to improved student learning. We would also like to explore the effects of adding process (especially social collaborative) objectives to courses to see how this will affect CoI survey scores. In addition, we will scrutinize our program redesign process itself by studying the effects of collaboration around course revisions on program faculty and will investigate how collaborative program redesign proceeds when it is grounded in theory-based analytics.

We have provided a design-based model for making improvements to online courses and programs that does not rely on the intuition of instructors and designers for quality improvements. We believe that this model could also be used to guide the transition from face-to-face to online delivery. We hope that others will also use it, either specifically or more generally, to do research that will improve our understanding of programmatic improvements in online teacher education.

CONCLUSION

Making connections between online course design and implementation and learning outcomes is long overdue in online education. This ongoing study is not only a first step in that direction, but it employs what are probably the two most commonly used theoretical frameworks in online education in the process. It also provides an important outcomes-based model for improvement in online teacher preparation programs. Findings suggest that, taken together, QM and CoI revisions can be linked to improved outcomes but unfortunately not to each other. However, they do suggest a trajectory – QM review and revision of courses and incremental "tweaking" of course implementation relative to deficiencies revealed by the CoI survey – for incremental improvement of online courses that might be of use to other institutions and programs. In addition, even if programs are wedded to other improvement frameworks, we believe a design-based approach could be useful.

Demonstrating a link between the QM framework and student outcomes and the CoI framework and student outcomes has great theoretical merit, especially as regards the latter and the cognitive presence construct in particular. More importantly, the efficacy of our efforts demonstrates the usefulness of design-based approaches to program improvement in online teacher education.

REFERENCES

Akyol, Z., & Garrison, D. R. (2008). The development of a community of inquiry over time in an online course: Understanding the progression and integration of social, cognitive and teaching presence. *Journal of Asynchronous Learning Networks*, *12*(2-3), 3–22.

Akyol, Z., & Garrison, D. R. (2011). Understanding cognitive presence in an online and blended community of inquiry: Assessing outcomes and processes for deep approaches to learning. *British Journal of Educational Technology*, *42*(2), 233–250. doi:10.1111/j.1467-8535.2009.01029.x

Akyol, Z., Vaughan, N., & Garrison, D. R. (2011). The impact of course duration on the development of a community of inquiry. *Interactive Learning Environments*, *19*(3), 231–246. doi:10.1080/10494820902809147

Allen, E. I., & Seaman, J. (2011). *Going the distance: Online education in the United States 2011*. Needham, MA: Babson Survey Research Group.

Anderson, T. (2005). Design-based research and its application to call centre innovation in distance education. *Canadian Journal of Learning and Technology*, *31*(2).

Anderson, T., Rourke, L., Garrison, D. R., & Archer, W. (2001). Assessing teaching presence in a computer conferencing context. *Journal of Asynchronous Learning Networks*, *5*(2), 1–17.

Arbaugh, B., Bangert, A., & Cleveland-Innes, M. (2010). Subject matter effects and the community of inquiry framework. *The Internet and Higher Education*, *13*(1-2). doi:10.1016/j.iheduc.2009.10.006

Arbaugh, J. B., Cleveland-Innes, M., Diaz, S., Garrison, D. R., Ice, P., & Richardson, J. C. (2008). Developing a community of inquiry instrument: Testing a measure of the community of inquiry framework using a multi-institutional sample. *The Internet and Higher Education*, *11*(3-4), 133–136. doi:10.1016/j.iheduc.2008.06.003

Archer, W. (2010). Beyond online discussions: Extending the community of inquiry framework to entire courses. *The Internet and Higher Education*, *13*(1-2), 69. doi:10.1016/j.iheduc.2009.10.005

Baber, T. C. (2011). The online crit: The community of inquiry meets design education. *Journal of Distance Education*, *25*(1).

Bolman, L., & Deal, T. (2008). *Reframing organizations: Artistry, choice, and leadership* (4th ed.). San Francisco, CA: Wiley.

Boston, W., Diaz, S. R., Gibson, A. M., Ice, P., Richardson, J., & Swan, K. (2009). An exploration of the relationship between indicators of the community of inquiry framework and retention in online programs. *Journal of Asynchronous Learning Networks*, *13*(3), 67–83.

Cobb, P., Confrey, J., diSessa, A., Lehrer, R., & Schauble, L. (2003). Design experiments in educational research. *Educational Researcher*, *32*(1), 9–13. doi:10.3102/0013189X032001009

Cohen, J. (2002). Statistics: A power primer. *Psychological Bulletin*, *112*(1), 155–159. doi:10.1037/0033-2909.112.1.155

Collins, A., Joseph, D., & Bielaczye, K. (2004). Design research: Theoretical and methodological issues. *Journal of the Learning Sciences*, *13*(1), 15–42. doi:10.1207/s15327809jls1301_2

Design-Based Research Collective. (2003). Design-based research: An emerging paradigm for educational inquiry. *Educational Researcher*, *32*(1), 5–8. doi:10.3102/0013189X032001005

Dewey, J. (1933). *How we think* (rev. ed.). Boston, MA: D. C. Heath.

Garrison, D. R., & Anderson, T. (2003). *E-learning in the 21st century: A framework for research and practice*. London, UK: Routledge/Falmer. doi:10.4324/9780203166093

Garrison, D. R., Anderson, T., & Archer, W. (2000). Critical inquiry in a text-based environment: Computer conferencing in higher education. *The Internet and Higher Education*, *2*, 87–105. doi:10.1016/S1096-7516(00)00016-6

Garrison, D. R., Anderson, T., & Archer, W. (2001). Critical thinking, cognitive presence, and computer conferencing in distance education. *American Journal of Distance Education*, *15*(1), 7–23. doi:10.1080/08923640109527071

Garrison, D. R., & Arbaugh, J. B. (2007). Researching the community of inquiry framework: Review, issues, and future directions. *The Internet and Higher Education*, *10*(3), 157–172. doi:10.1016/j.iheduc.2007.04.001

Garrison, D. R., & Cleveland-Innes, M. (2005). Facilitating cognitive presence in online learning: Interaction is not enough. *American Journal of Distance Education*, *19*(3), 133–148. doi:10.1207/s15389286ajde1903_2

Garrison, D. R., Cleveland-Innes, M., & Fung, T. S. (2010). Exploring relationships among teaching, cognitive and social presence: Student perceptions of the community of inquiry framework. *The Internet and Higher Education*, *13*(1-2), 31–36. doi:10.1016/j.iheduc.2009.10.002

Ice, P., Gibson, A. M., Boston, W., & Becher, D. (2011). An exploration of differences between community of indicators in low and high disenrollment online courses. *Journal of Asynchronous Learning Networks*, *15*(2).

Jiang, M., & Ting, E. (2000). A study of factors influencing students' perceived learning in a web-based course environment. *International Journal of Educational Telecommunications*, *6*(4), 317–338.

Kanuka, H., & Anderson, T. (1998). Online social interchange, discord, and knowledge construction. *Journal of Distance Education*, *13*(1), 57–75.

Kim, K., & Bonk, C. J. (2006). The future of online teaching and learning in higher education: The survey says… *Educause Quarterly Magazine*, *29*(4). Retrieved from http://www.educause.edu/EDUCAUSE+Quarterly/EDUCAUSEQuarterlyMagazineVolum/TheFutureofOnlineTeachingandLe/157426

Legon, R., Runyon, J., & Aman, R. (2007, October). *The impact of "Quality Matters" standards on courses: Research opportunities and results*. Paper presented at the 13th International Sloan-C Conference on Online Learning, Orlando, FL.

Lowenthal, P. R., & Dunlap, J. C. (2010). From pixel on a screen to real person in your students' lives: Establishing social presence using digital storytelling. *The Internet and Higher Education*, *13*(1), 70–72. doi:10.1016/j.iheduc.2009.10.004

Luebeck, J. L., & Bice, L. R. (2005). Online discussion as a mechanism of conceptual change among mathematics and science teachers. *Journal of Distance Education*, *20*(2), 21–39.

Meyer, K. A. (2003). Face-to-face versus threaded discussions: The role of time and higher-order thinking. *Journal of Asynchronous Learning Networks*, *7*(3), 55–65.

Middleton, J., Gorard, S., Taylor, C., & Bannan-Ritland, B. (2006). *The 'Compleat' design experiment: From soup to nuts*. Department of Educational Studies Research Paper 2006/5. York, PA: Department of Educational Studies, University of York. Retrieved from http://www.york.ac.uk/media/educationalstudies/documents/research/Paper18Thecompleatdesignexperiment.pdf

Murphy, E. (2004). Identifying and measuring ill-structured problem formulation and resolution in online asynchronous discussions. *Canadian Journal of Learning and Technology*, *30*(1), 5–20.

Picciano, A. G. (2002). Beyond student perceptions: Issues of interaction, presence and performance in an online course. *Journal of Asynchronous Learning Networks*, *6*(1), 21–40.

Quality Matters. (2005). *Research literature and standards sets support for Quality Matters review standards*. Retrieved from http://www.qualitymatters.org/Documents/Matrix%20of%20Research%20Standards%20FY0506.pdf

Ragan, P., Lacey, A., & Nagy, R. (2002). Web-based learning and teacher preparation: Stumbling blocks and stepping stones. *Teaching with Technology Today, 8*(5). Retrieved from http://www.wisconsin.edu/ttt/articles/ragan.htm

Richardson, J. C., & Swan, K. (2003). Examining social presence in online courses in relation to students' perceived learning and satisfaction. *Journal of Asynchronous Learning Networks*, *7*(1), 68–88.

Rovai, A. P. (2002). Development of an instrument to measure classroom community. *The Internet and Higher Education*, *5*(3), 197–211. doi:10.1016/S1096-7516(02)00102-1

Shattuck, K. (2007) Quality matters: Collaborative program planning at a state level. *Online Journal of Distance Learning Adminstration, 10*(3). Retrieved from http://www.westga.edu/~distance/ojdla/fall103/shattuck103.htm

Shea, P. (2006). A study of students' sense of learning community in online environments. *Journal of Asynchronous Learning Networks*, *10*(1), 35–44.

Shea, P., & Bidjerano, T. (2009). Cognitive presence and online learner engagement: A cluster analysis of the community of inquiry framework. *Journal of Computing in Higher Education*, *21*, 199–217. doi:10.1007/s12528-009-9024-5

Shea, P., Li, C., Swan, K., & Pickett, A. (2005). Developing learning community in online asynchronous college courses: The role of teaching presence. *Journal of Asynchronous Learning Networks*, *9*(4). Retrieved from http://www.sloan-c.org/publications/jaln/v9n4/v9n4_shea.asp

Silva, D. Y., Gimbert, B., & Nolan, J. (2000). Sliding the doors: Locking and unlocking possibilities for teacher leadership. *Teachers College Record*, *102*(4), 779–804. doi:10.1111/0161-4681.00077

Swan, K. (2002). Building communities in online courses: The importance of interaction. *Education Communication and Information*, *2*(1), 23–49. doi:10.1080/1463631022000005016

Swan, K., Garrison, D. R., & Richardson, J. C. (2009). A constructivist approach to online learning: The Community of Inquiry framework. In Payne, C. R. (Ed.), *Information technology and constructivism in higher education: Progressive learning frameworks* (pp. 43–57). Hershey, PA: IGI Global. doi:10.4018/978-1-60566-654-9.ch004

Swan, K., Matthews, D., Bogle, L., Boles, E., & Day, S. (2012). Linking online course design and implementation to learning outcomes: A design experiment. *The Internet and Higher Education*, *15*(2), 81–88. doi:10.1016/j.iheduc.2011.07.002

Swan, K., Shea, P., Fredericksen, E., Pickett, A., Pelz, W., & Maher, G. (2000). Building knowledge building communities: Consistency, contact and communication in the virtual classroom. *Journal of Educational Computing Research*, *23*(4), 389–413.

Swan, K., & Shih, L.-F. (2005). On the nature and development of social presence in online course discussions. *Journal of Asynchronous Learning Networks*, *9*(3), 115–136.

Swan, K. P., Richardson, J. C., Ice, P., Garrison, D. R., Cleveland-Innes, M., & Arbaugh, J. B. (2008). Validating a measurement tool of presence in online communities of inquiry. *E-mentor*, *2*(24). Retrieved from http://www.e-mentor.edu.pl/artykul_v2.php?numer=24&id=543

Tu, C. H. (2000). On-line learning migration: From social learning theory to social presence theory in CMC environment. *Journal of Network and Computer Applications*, *23*(1), 27–37. doi:10.1006/jnca.1999.0099

Tu, C. H., & McIsaac, M. S. (2002). The relationship of social presence and interaction in online classes. *American Journal of Distance Education*, *16*(3), 131–150. doi:10.1207/S15389286AJDE1603_2

Vandergrift, L. (2003). Orchestrating strategy use: Toward a model of the skilled second language listener. *Language Learning*, *53*(3), 463–496. doi:10.1111/1467-9922.00232

Vaughn, N., & Garrison, D. R. (2006). How blended learning can support a faculty development community of inquiry. *Journal of Asynchronous Learning Networks*, *10*(4), 139–152.

Wang, F., & Hannafin, M. J. (2005). Design-based research and technology enhanced learning environments. *Educational Technology Research and Development*, *53*(4), 5–23. doi:10.1007/BF02504682

Wang, Y.-M., & Chang, V. D.-T. (2008). Essential elements in designing online discussions to promote cognitive presence—A practical experience. *Journal of Asynchronous Learning Networks*, *12*(3-4), 157–177.

Wu, D., & Hiltz, S. R. (2004). Predicting learning from asynchronous online discussions. *Journal of Asynchronous Learning Networks*, *8*(2), 139–152.

ADDITIONAL READINGS

Acker-Hocevar, M., Pisapia, J., & Coukas-Semmel, E. (2002, April). *Bridging the abyss: Adding value and validity to leadership development through action learning--cases-in-point*. Paper presented at the Annual Meeting of the American Educational Research Association, New Orleans, LA.

Arbaugh, J. B. (2007). An empirical verification of the Community of Inquiry framework. *Journal of Asynchronous Learning Networks*, *11*(1), 73–85.

Barab, S., & Squire, K. (Eds.). (2004). Design-based research: Clarifying the terms. Introduction to the learning sciences methodology strand. *Journal of the Learning Sciences*, *13*(1), 115–128.

Bellanca, J., & Brandt, R. (2010). *21st century skills: Rethinking how students learn*. Bloomington, IN: Solution Tree Press.

Collopy, R., & Arnold, J. (2009). To blend or not to blend: Online and blended learning environments in undergraduate teacher education. *Issues in Teacher Education*, *18*(2), 85–101.

Dexter, S. (2011). School technology leadership: Artifacts in systems of practice. *Journal of School Leadership*, *21*(2), 166–189.

Dozier, T. (2007). Turning good teachers into great leaders. *Educational Leadership*, *65*(1), 54–59.

Franz, D., Hopper, P., & Kritsonis, W. (2007). National impact: Creating teacher leaders through the use of problem-based learning. *National Forum of Applied Educational Research Journal*, *20*(3), 1–9.

Harrigan, A. (2010). *Social presence and interactivity in online courses: Enhancing the online learning environment through discussion and writing*. Doctoral dissertation, University of Wyoming.

Jones, I. (2011). Can you see me now? Defining teaching presence in the online classroom through building a learning community. *Journal of Legal Studies Education*, *28*(1), 67–116. doi:10.1111/j.1744-1722.2010.01085.x

Kanuka, H. (2011). Interaction and the online distance classroom: Do instructional methods effect the quality of interaction? *Journal of Computing in Higher Education*, *23*, 143–156. doi:10.1007/s12528-011-9049-4

Ke, F. (2010). Examining online teaching, cognitive, and social presence for adult students. *Computers & Education*, *55*(2), 808–820. doi:10.1016/j.compedu.2010.03.013

Kelly, A. E. (Ed.). (2003). Theme issue: The role of design in educational research. *Educational Researcher*, *32*(1). doi:10.3102/0013189X032001003

Lehmann, C. (2009). Shifting ground. *Principal Leadership*, *10*(4), 18–21.

Luthra, S., & Fochtam, P. (2011). The road to lasting tech leadership. *Learning and Leading with Technology*, *38*(7), 16–20.

Maor, D. (2008). Changing relationship: Who is the learner and who is the teacher in the online educational landscape? *Australasian Journal of Educational Technology*, *24*(5), 627–638.

Matkin, G. (2010). The distance educator's opportunity for institutional leadership. *Continuing Higher Education Review*, *74*, 32–39.

Pape, L., & Wicks, M. (2009). *National standards for quality online programs*. International Association for K-12 Online Learning.

Swan, K., & Ice, P. (Eds.). (2010). Special issue on the Community of Inquiry framework: Ten years later. *The Internet and Higher Education, 13*(1-2). doi:10.1016/j.iheduc.2009.11.003

Young, M., Fuller, E., Carpenter, B., & Mansfield K. (2007). *Quality leadership matters. Policy brief series,* University Council for Educational Administration, 1(1).

KEY TERMS AND DEFINITIONS

Cognitive Presence: The extent to which learners are able to construct and confirm meaning through sustained reflection and discourse in a critical community of inquiry.

Community of Inquiry (CoI) Framework: A social-constructivist, process model of learning in online and blended educational environments which assumes that effective learning in higher education requires the development of a community of learners that supports meaningful inquiry.

Community of Inquiry Survey: A 34 question Likert Scale survey designed to determine the level of the three presences identified in the CoI framework as supporting online learning, teaching (13), social (9) and cognitive (12), as measured by student responses. The intersection of these presences is where student learning in an online environment occurs and the higher the score the greater the opportunity for students to acquire the core material for the class.

Educational Research Methods: The online course designed to provide students in the Master Teacher Leader (MTL) degree program with the basic understanding of research techniques and terminology in order to: 1) make them better consumers of research as related to their profession and; 2) provide them with the basic understanding of what good research entails in order to develop a Capstone project.

Practical Inquiry Model: A four-phase model adapted from John Dewey designed represent the development of cognitive presence in online courses. It begins with a triggering event designed by the instructor to provoke discussion and a desire to learn more about the identified topic. This triggering event leads to exploration of how to resolve the problem by integrating what has been learned with what the student already knows in order to resolve the problem. This resolution can lead to another triggering event resulting in the ongoing learning cycle.

Quality Matters (QM): A peer-review rating system designed to determine the quality of an online or blended class by rating the design of the class as it appears online. There are 41 items within the 8 categories consisting of 1, 2 and 3 point value levels. All 3 point items must be met for a course to meet the QM standards. This is a formative process and the instructor will make adjustments based upon the peer review ratings and comments.

Social Presence: The ability of participants in a virtual community of inquiry to project themselves socially and emotionally–as 'real' people.

Teaching Presence: The design, facilitation and direction of cognitive and social processes for the purpose of realizing personally meaningful and educationally worthwhile learning outcomes.

APPENDIX A: QUALITY MATTERS RUBRIC

Table A1.

	STANDARDS	PTS
Course Overview and Introduction	1.1 Instructions make clear how to get started and where to find various course components.	3
	1.2 Students are introduced to the purpose and structure of the course.	3
	1.3 Etiquette expectations (sometimes called "netiquette") for online discussions, email, and other forms of communication are stated clearly.	2
	1.4 Course and/or institutional policies with which the student is expected to comply are clearly stated, or a link to current policies is provided.	2
	1.5 Prerequisite knowledge in the discipline and/or any required competencies are clearly stated.	1
	1.6 Minimum technical skills expected of the student are clearly stated.	1
	1.7 The self-introduction by the instructor is appropriate and available online.	1
	1.8 Students are asked to introduce themselves to the class.	1
Learning Objectives (Competencies)	2.1 The course learning objectives describe outcomes that are measurable.	3
	2.2 The module/unit learning objectives describe outcomes that are measurable and consistent with the course-level objectives.	3
	2.3 All learning objectives are stated clearly and written from the students' perspective.	3
	2.4 Instructions to students on how to meet the learning objectives are adequate and stated clearly.	3
	2.5 The learning objectives are appropriately designed for the level of the course.	3
Assessment and Measurement	3.1 The types of assessments selected measure the stated learning objectives and are consistent with course activities and resources.	3
	3.2 The course grading policy is stated clearly.	3
	3.3 Specific and descriptive criteria are provided for the evaluation of students' work and participation and are tied to the course grading policy.	3
	3.4 The assessment instruments selected are sequenced, varied, and appropriate to the student work being assessed.	2
	3.5 Students have multiple opportunities to measure their own learning progress.	2
Instructional Materials	4.1 The instructional materials contribute to the achievement of the stated course and module/unit learning objectives.	3
	4.2 The purpose of instructional materials and how the materials are to be used for learning activities are clearly explained.	3
	4.3 All resources and materials used in the course are appropriately cited.	2
	4.4 The instructional materials are current.	2
	4.5 The instructional materials present a variety of perspectives on the course content.	1
	4.6 The distinction between required and optional materials is clearly explained.	1
Learner Interaction and Engagement	5.1 The learning activities promote the achievement of the stated learning objectives.	3
	5.2 Learning activities provide opportunities for interaction that support active learning.	3
	5.3 The instructor's plan for classroom response time and feedback on assignments is clearly stated.	3
	5.4 The requirements for student interaction are clearly articulated.	2
Course Technology	6.1 The tools and media support the course learning objectives.	3
	6.2 Course tools and media support student engagement and guide the student to become an active learner.	3
	6.3 Navigation throughout the online components of the course is logical, consistent, and efficient.	3
	6.4 Students can readily access the technologies required in the course.	2
	6.5 The course technologies are current.	1
Learner Support	7.1 The course instructions articulate or link to a clear description of the technical support offered and how to access it.	3
	7.2 Course instructions articulate or link to the institution's accessibility policies and services.	3
	7.3 Course instructions articulate or link to an explanation of how the institution's academic support services and resources can help students succeed in the course and how students can access the services.	2
	7.4 Course instructions articulate or link to an explanation of how the institution's student support services can help students succeed and how students can access the services.	1
Accessibility	8.1 The course employs accessible technologies and provides guidance on how to obtain accommodation.	3
	8.2 The course contains equivalent alternatives to auditory and visual content.	2
	8.3 The course design facilitates readability and minimizes distractions.	2
	8.4 The course design accommodates the use of assistive technologies.	2

For more information see: http://www.QMprogram.org

APPENDIX B: COMMUNITY OF INQUIRY SURVEY

Table A2.

The following statements relate to your perceptions of "**Teaching Presence**" – the design of this course and your instructor's facilitation of discussion and direct instruction within it. Please indicate your agreement or disagreement with each statement.						
#	**statement**	**Agreement** 1 = strongly disagree; 5 = strongly agree				
1	The instructor clearly communicated important course topics.	1	2	3	4	5
2	The instructor clearly communicated important course goals.	1	2	3	4	5
3	The instructor provided clear instructions on how to participate in course learning activities	1	2	3	4	5
4	The instructor clearly communicated important due dates/time frames for learning activities.	1	2	3	4	5
5	The instructor was helpful in identifying areas of agreement and disagreement on course topics that helped me to learn.	1	2	3	4	5
6	The instructor was helpful in guiding the class towards understanding course topics in a way that helped me clarify my thinking.	1	2	3	4	5
7	The instructor helped to keep course participants engaged and participating in productive dialogue.	1	2	3	4	5
8	The instructor helped keep the course participants on task in a way that helped me to learn.	1	2	3	4	5
9	The instructor encouraged course participants to explore new concepts in this course.	1	2	3	4	5
10	Instructor actions reinforced the development of a sense of community among course participants	1	2	3	4	5
11	The instructor helped to focus discussion on relevant issues in a way that helped me to learn.	1	2	3	4	5
12	The instructor provided feedback that helped me understand my strengths and weaknesses relative to the course's goals and objectives.	1	2	3	4	5
13	The instructor provided feedback in a timely fashion.	1	2	3	4	5
The following statements refer to your perceptions of "**Social Presence**" -- the degree to which you feel socially and emotionally connected with others in this course. Please indicate your agreement or disagreement with each statement.						
#	**statement**	**Agreement** 1 = strongly disagree; 5 = strongly agree				
14	Getting to know other course participants gave me a sense of belonging in the course.	1	2	3	4	5
15	I was able to form distinct impressions of some course participants.	1	2	3	4	5
16	Online or web-based communication is an excellent medium for social interaction.	1	2	3	4	5
17	I felt comfortable conversing through the online medium.	1	2	3	4	5
18	I felt comfortable participating in the course discussions.	1	2	3	4	5
19	I felt comfortable interacting with other course participants.	1	2	3	4	5
20	I felt comfortable disagreeing with other course participants while still maintaining a sense of trust.	1	2	3	4	5
21	I felt that my point of view was acknowledged by other course participants.	1	2	3	4	5
22	Online discussions help me to develop a sense of collaboration.	1	2	3	4	5

continued on following page

Table A2. Continued

The following statements relate to your perceptions of "**Cognitive Presence**" -- the extent to which you were able to develop a good understanding of course topics. Please indicate your agreement or disagreement with each statement.						
#	**statement**	**Agreement** 1 = strongly disagree; 5 = strongly agree				
23	Problems posed increased my interest in course issues.	**1**	**2**	**3**	**4**	**5**
24	Course activities piqued my curiosity.	**1**	**2**	**3**	**4**	**5**
25	I felt motivated to explore content related questions.	**1**	**2**	**3**	**4**	**5**
26	I utilized a variety of information sources to explore problems posed in this course.	**1**	**2**	**3**	**4**	**5**
27	Brainstorming and finding relevant information helped me resolve content related questions.	**1**	**2**	**3**	**4**	**5**
28	Online discussions were valuable in helping me appreciate different perspectives.	**1**	**2**	**3**	**4**	**5**
29	Combining new information helped me answer questions raised in course activities.	**1**	**2**	**3**	**4**	**5**
30	Learning activities helped me construct explanations/solutions.	**1**	**2**	**3**	**4**	**5**
31	Reflection on course content and discussions helped me understand fundamental concepts in this class.	**1**	**2**	**3**	**4**	**5**
32	I can describe ways to test and apply the knowledge created in this course.	**1**	**2**	**3**	**4**	**5**
33	I have developed solutions to course problems that can be applied in practice.	**1**	**2**	**3**	**4**	**5**
34	I can apply the knowledge created in this course to my work or other non-class related activities.	**1**	**2**	**3**	**4**	**5**

See also: http://communitiesofinquiry.com/methodology

Section 7
Moving Forward

Chapter 22
Puttering, Tinkering, Building, and Making:
A Constructionist Approach to Online Instructional Simulation Games

Joseph R. Feinberg
Georgia State University, USA

Audrey H. Schewe
Georgia State University, USA

Christopher D. Moore
Georgia State University, USA

Kevin R. Wood
Gwinnett County Public Schools, USA

ABSTRACT

Instructional simulation games are models of the real world that allow students to interact with events and objects that are normally inaccessible within a classroom setting. Yet, simply using an instructional simulation ignores powerful learning opportunities. Papert advocates going beyond simply using models. He promotes a fundamental change in how children learn through his theory of constructionism. Instead of constructivism with a "v," Papert advocates a theory of learning called constructionism with an "n." Constructionism aligns with constructivist theory with learners actively constructing knowledge from their experiences. But constructionism adds that new ideas are more likely to emerge when learners are actively engaged in designing or building an artifact or physical model that can be reflected upon and shared with others. Papert's theoretical approach to learning is relevant to teacher education and should be applied to instruction via interactive, multimedia, and computer-aided simulations.

DOI: 10.4018/978-1-4666-1906-7.ch022

INTRODUCTION

The prevailing assumption in education is that more content knowledge is better, which is based on the positivist assumption that the knowledge a learner must know is clearly identified (Rieber, 1996). The pressure to have students perform well on high-stakes tests continue to proliferate the tell and test methods of instruction (Federation of American Scientists, 2006). This traditional pedagogy naturally propagates students who are disengaged, disinterested, and discouraged to think at higher levels. The transmission of knowledge approach is contradictory to higher order thinking because it ignores the constructive nature of knowledge and fails to encourage rational questioning of evidence to enhance student understanding of the world. Constructivism provides a counter approach to the transmission of knowledge with simulation games offering a particular pedagogical toolset to promote higher levels of thinking. Teacher educators as gatekeepers have an important opportunity and obligation to share the research, theory, and practice regarding simulation games. In particular, this chapter details the complementary nature of simulations and constructivist learning, but greater theoretical emphasis is placed on Papert's (1991) theory of "constructionism" and the potential implications of a constructionist approach for teacher education and instruction that involves simulation games.

Papert (1991) describes how computers provide students and teachers with an excellent platform for learning. Used effectively, computers allow students and teachers to move about in a nearly endless virtual space where they can create meaning through their virtual interactions. Instruction utilizing digital mediums affords students the opportunity to manipulate objects, events, or processes and provides learners an opportunity to learn technology by doing technology. Papert (1998) writes that students disengage from school not because it is too hard but because they believe school is boring. According to Papert (1991), children enjoy computer games because they are challenging and because computer games force the child to engage in meaningful learning experiences. The point, according to Papert (1991), is that students are not afraid of challenges, but they hate boring work. Papert's (1991, 1998) assertions are echoed by Resnick (2007) who writes that digital mediums provide students with an instructional environment that is more dynamic and interactive than the traditional classroom allowing students to create powerful and lasting meanings out of their learning. However, learning is very complex and not all computer games are meaningful and not all traditional education is boring. According to educational theorists like Resnick (2007) and Papert (1991), simulation games provide teachers and students with the opportunity to engage in the meaningful creation of knowledge in the classroom that is authentic and lasting.

THEORETICAL BACKGROUND

Simulations as Authentic Learning Experiences

For nearly a century, education reformers have advocated for the immersion of students in authentic learning experiences, where content and skills are embedded in real-world contexts (Dewey, 1938). Constructivist teaching and learning strategies seek to shift the role of the student from passive recipient of content to active participant in constructing knowledge. Simulations model real world events or processes and provide students with authentic learning experiences. However, reviews of a half-century of research on the ability of simulations to produce intellectual gains reveal inconsistent findings and inconclusive data (Clegg, 1991; Feinberg, 2011). Simply promoting active learning is not the solution. For example, "Even highly active students can produce work that is intellectually shallow and weak" (Newmann, Marks, & Gamoran, 1996, p. 281). To be accepted as authentic learning experiences, constructivist or active learning approaches need both to be grounded in high intellectual standards, and also to enhance student academic performance (Newmann, Marks, & Gamoran, 1996).

Simulation games potentially offer students the opportunity to participate in activities designed to bring about authentic and meaningful learning experiences with higher levels of thinking. According to contemporary researchers, learning that occurs in meaningful and relevant contexts is more effective than learning that occurs outside of those contexts, as is the case with formal instruction (Eck, 2006). Gee (2007) makes the following case:

One good way to make people look stupid is to ask them to learn and think in terms of words and abstractions that they cannot connect in any useful way to images or situations in their embodied experiences in the world. Unfortunately, we do this regularly in schools (p. 72).

Learning through simulations allows students to apply and practice what they have learned in a real context that is easily transferable to the real world (Harmon, 2008). A growing body of literature supports the use of instructional or "serious" games (Kirkley, Duffy, Kirkley, & Kremer, 2011), but contemporary empirical research is not well developed and nascent relative to the observable behavior of students learning through simulations and games (Tobias, Fletcher, Dai, & Wind, 2011).

Although the integration of a variety of instructional support measures used with simulation games shows improved or more effective learning, the results vary depending upon the instructional goals for using each simulation model and how the goals are evaluated (de Jong & van Joolingen, 1998). Historically, the reviews of simulation game research highlight concerns about the overall findings and inconsistent research results (Brant, Hooper, & Sugrue, 1991; Feinberg, 2011; VanSickle, 1986). In a research review of scientific discovery learning with computer simulations, de Jong & van Joolingen (1998) concluded that findings are contradictory and do not clearly favor simulations (p.181). Rather than provide valid measures of learning outcomes, research on simulation use typically provides anecdotal evidence such as "kids who participate in game- and simulation-like learning are very excited, they're motivated, they're immersed, and they seem to do better" (Federation of American Scientists, 2006, p. 43). Thus, despite enthusiasm for simulations in some education circles, the degree of use by teachers as an instructional strategy is far from clear.

Simulation Game Theory

Digital simulation games are instructional tools that enable students to become participants in a virtual digital world that represents a real world event. This allows students to experience a cognitive domain and learn knowledge through virtual experience (Gee, 2005; Rice, 2007). Gee emphasized that when a person learns to play a simulation game he or she is learning a "semiotic" domain. In other words, the gamer/player is becoming literate in the rules, requirements, symbols, images, graphs, diagrams, artifacts, language, and culture of the game involved. Furthermore, the learning of one semiotic domain connects the students learning to other semiotic domains, which permit the learner to construct meaningful understandings of the new domain. The learning of a semiotic domain enables learners to connect their new understanding to their perception of the physical and virtual world. Gee (2007) compares this type of semiotic literacy learning with the traditional education view of content learning where content is often taught without meaningful context thus confusing the learner and creating a fragmented understanding of the subject. In an instructional simulation game, the player is engaged in an experience that will facilitate the learning of the games rules, choices, and moves that incorporate the structure of the game as well as any relevant content required to navigate within the simulation. The learner is constructing his or her understanding of the simulation game through his experiences because his or her experiences encompass knowledge construction within the semiotic domain of the game much as the learner would construct his or her understanding of the physical world through his experiences thus the learner is engaged in authentic learning.

In an instructional simulation game, the participant or learner has the opportunity to immerse him or herself in the role constructed by the game designers, but not all simulation games are ideally suited to take on the role of instructional simulation games. Facilitating a lesson using an instructional simulation game does not mean that the instructor can turn students loose in the simulation game and expect meaningful learning to take place. Scaffolding is essential to instructional practice. The producers of simulation games are prone to make broad claims that their simulation games provide educational scaffolding and students will learn as long as they play the game. According to Gee (2005), the responsibility of the instructor in a lesson incorporating the use of instructional simulation games is to provide a pathway for students to be able to navigate the many variables that make up the semiotic domain encapsulated by the game. Furthermore, instructors use scaffolding in the lesson to point out the links to other semiotic domains thus engaging the learners in meaningful learning experiences. In short, teachers matter and instruction matters even in our technology driven and ever changing world.

Simulation Design and Use for Teacher Education

Teacher educators now consider technological resources as an invaluable part of instruction. Bolick, Berson, Friedman, and Porfeli (2007) found that social studies professors, who prepare preservice teachers, are incorporating technology into their instructional practices. Furthermore, the results of the study conducted by Bolick et al. (2007) indicate that the type of technology incorporated by social studies teachers has changed to reflect the incorporation of new technologies such as computers and presentation software programs. Moreover, they found that institutional barriers to the incorporation of technology by teachers have decreased.

Although increased use of technology in education affords greater access to simulations, a contemporary focus on learning products detracts from the learning process and benefits of simulation games. In other words, the constraints of the university, college, and school environments inhibit focusing on the learning process. Learning products are more valued by society (consider high-stakes testing) and simulation games are evaluated based upon traditional outcomes rather than the benefits of learning through non-traditional methods that promote higher order thinking. Educators are reluctant to utilize simulation games because higher levels of learning, such as strategic thinking, problem-solving and decision-making, are not easily measured forms of knowledge. According to White (1985), "Orthodox evaluation methods are simply not appropriate for appraising the experiences drawn from simulation gaming" (26). The National Summit on Educational Games (2006) concludes, "If assessments are not measuring the right skills and knowledge – the higher-order skills that games may be able to develop – then the use of educational games and simulations may be viewed as having poor efficacy" (Federation of American Scientists, 2006, p. 43). Furthermore, time constraints may hinder the effectiveness of simulation games because teachers may be overly anxious to *cover* the next concept or topic rather than spending the time necessary to promote higher levels of thinking. Many simulation games, especially video games, are not compatible with the traditional rigid 45 or 50-minute class schedule. *Civilization*, for example, takes several hours to learn and can take more than 20 hours to play (Federation of American Scientists, 2006).

Some critics of simulations note the deficiencies and inflexibility of simulation models. White (1985), for example, suggests it is possible that student participants "will construct alternatives" that are different from those considered or created by simulation model designers. Therefore, according to White, simulation games fail to satisfy Dewey's ideals embodied in reflective inquiry (White, 1985, p. 27). As the following discussion shows in much greater detail, Gee's (2007) suggestion that participants construct their understanding is consistent with constructivist theories. However, simulation games limit participants to a particular model that is constructed and constrained by the designer and teacher.

Students should not only be given opportunities to alter existing models underlying the simulation, but they must also be encouraged to construct their own responses to simulation challenges (DeLeon, 2008). In a study of the simulations *Skyjack* and *House Design*, DeLeon (2008) argues that these simulations create and sustain dominant ideologies. "These limited choices point to the texts' binary construction of leadership decisions and simplistic portrayal of the decision-making process, a real problem when dealing with modeling to students the multiple perspectives needed to make informed decisions that affect communities and people" (DeLeon, 2008, p. 265). DeLeon emphasizes that for students to construct their own critical knowledge from simulations, they should be allowed to alter and/or discuss the underlying features and structures. Thus, simulations situated in the theories of constructivism may provide students with more authentic and meaningful learning experiences.

Constructivism emphasizes the active role of learners in acquiring knowledge that has multiple interpretations. This type of learning veers away from the more traditional positivistic approaches to teaching. Doolittle and Hicks (2003) note, "Constructivism involves the active creation and modification of thoughts, ideas and understanding as the result of experiences that occur within socio-cultural contexts" (p. 76). De Jong and van Joolingen (1998) explored the application of constructivism to computer simulations used in science education. However, they found that students using computer simulations without instructional support failed to understand the underlying models and thus construct their own meaning.

De Jong and van Joolingen speculate that students were unsuccessful for a variety of reasons. Learners may have lacked sufficient prior knowledge and skills to state a hypothesis, to interpret data, and to experiment in an unsystematic manner (de Jong & van Joolingen, 1998). Research findings indicate that support measures enable learners to activate prior knowledge without overwhelming them and not undermining the full complexity of the simulation. Yet, students are still working within the ("intolerable") limits of the prefabricated simulation models that they are not allowed to change or even create (Penner, 2001, p. 29). White (1985) recognized this problem prior to computer based video simulations and argued that an individual's creative ability is limited or confined to the particular simulation game being used. If students can be more involved in the creation and modification of thoughts and ideas, then the traditional limits of games and simulations do not pose such constrictive limits (Doolittle & Hicks, 2003).

Constructionism vs. Instructionism

Papert (1991) would argue that the findings reviewed by de Jong and van Joolingen (1998) fall short of going beyond "instructionism," which is Papert's term to describe a false belief that improving instruction is the key to a better education. In other words, Papert believes that traditional instruction is not the best means to improve education. According to Papert (1991), "It does not follow that the route to better performance is necessarily the invention by researchers of more powerful and effective means of instruction" (p. 4). Papert does not argue that instruction is bad and he clearly claims "That would be silly" (p.7). Instead, he promotes a radical change in our understanding of how children learn. Papert expressed in an interview, "Everybody knows that you learn much better when you're really interested in what you're doing, and it can't be the same for everybody" (de Pommereau, 1997, p. 11). According to Papert, technology provides instructional choices that were not available in the past when "the only way we could give knowledge was to put kids in a classroom and give them the knowledge bit by bit" (de Pommereau, 1997, p. 11). Papert connects his theoretical approach to educational practice through a computer programming language called Logo.

With his colleagues at MIT, Papert (1991) examined how children learned Logo, a simple computer programming language. They found that children learned through their exploration of the software, rather than through a linear and sequential process described in most developmental models. Instead of focusing on the cognitive potential of learners, Papert emphasizes that knowledge is constructed "felicitously in a context where the learner is engaged in the construction of something external or at least shareable…a sand castle, a machine, a computer program, a book" (Papert, 1991, p. 3). Thus, building on constructi*v*ism with a V, Papert advocates a theory of learning called constructio*n*ism with an N. Papert's theory of constructionism borrows from constructivist theory, but adds that new ideas are more likely to emerge when learners are actively engaged in designing or building an artifact or physical model that can be reflected upon and shared with others. Kafai and Resnick (1996) emphasize the point: "Children don't get ideas; they make ideas" (p.1). Constructionism also focuses on the affective components of learning. For example, proponents argue that personally meaningful activities or projects are more likely to intellectually engage students (Kafai & Resnick, 1996, p. 2). The foundation of constructionism is based upon two basic building blocks: the construction of knowledge and the building of personally meaningful objects (Kafai & Resnick, 1996, p. 1). Thus, the transition from a constructivist to constructionist platform suggests a task-based and product-oriented approach toward learning.

Papert's "constructionism" as a theory of learning must not be confused with Crotty's discussion of constructionism as an epistemology (Crotty, 1998). Constructionism for Crotty is one of three epistemologies, or ways of knowing, along with "objectivism" and "subjectivism." For Crotty, constructionism differs from constructivism in that constructivism focuses exclusively on the "meaning-making of the individual mind," whereas constructionism includes the "collective generation [and transmission] of meaning" (p. 58).

Solutions and Recommendations

The implications of Papert's theory of constructionism for the use of technology in teacher education, specifically for computer-based simulations, are abundant. As Doolittle and Hicks (2003) point out, while the promotion of technology integration into the classroom is steadfast, "the application of technology for the social studies has been theoretically underdeveloped" (p. 72). They position technology integration in the social studies within a framework of philosophical, theoretical and pedagogical constructivism; however, Doolittle and Hicks (2003) open the door to a constructionist approach with respect to computer-based simulations. To fully integrate technology in the social studies, Doolittle and Hicks advocate the implementation of technology as a tool for inquiry, to create authenticity, to foster global social interaction so students gain multiple perspectives, to build on their prior knowledge and interest, to provide timely and meaningful feedback and to foster autonomous, creative, and intellectual thinking (pp. 88-92). These six pedagogical strategies when adjusted for the constructionist assertion that learning occurs through designing, building and making an object, provide a theoretical argument for computer-assisted simulation games.

Constructionism is already being used as a theory that underlies how children learn through computer programming. Willett (2007) examined student activities in a computer games making class and analyzed the data in the context of a constructionist-learning model. The class instructor engaged the students (10 boys aged 9 to 13) in a critical analysis of computer games, provided a step-by-step walk-through of the game-making software and modeled the appropriate use of the software. However, despite the instructor's use of "good teaching," Willett noted that the boys did not learn as much as the instructor had hoped. Constructionism positions children as "innately inquisitive," making computers an ideal vehicle for exploring and developing their minds (Willett, 2007).

In order to examine the impact of computer programming activities on girls, Plass et al. (2007) used the theory of constructionism to develop *The RAPUNSEL* (Real-time, Applied Programming for Underrepresented Students' Early Literacy) Project, a game world called *Peeps* that encourages girls to design parts of a game and, in the process, develop computer programming skills. Students actively participated in authentic real-world tasks to solve a specific problem. The players must conceptualize and execute programming code to create an avatar (character) and function successfully in the game world. A series of tutorials and challenges provided scaffolds for the players as they developed their programming skills. The girls were "learning by doing, not only learning about programming, but learning how to think, to tinker, to putter, to make mistakes and to learn from them" (Plass et al., 2007, p. 3). The researchers found that as a result of playing the *Peeps* computer game, the girls' general and programming-related self-efficacy and self-esteem increased with medium to large effect sizes (p. 11).

Constructionism has also been suggested as a theory to support technology-enhanced language learning (Ruschoff & Ritter, 2001). Though transmission models for language learning are still dominant, calls for constructionist theories to inform the acquisition of language are surfacing. Ruschoff and Ritter (2001) propose the use of play and experimentation through self-structured and self-motivated processes of learning, and envision the classroom as a *workshop* where software tools such as word-processors, presentation tools, et cetera are available for students to construct physical objects. "Translated into language learning," state Rushoff and Ritter, "such an approach favors project-based, process-oriented, product-centered learning within a rich and facilitative multimodal learning environment" (p. 231).

Constructionism and Computer Simulations

A primary shortcoming of simulation models has been the inability of students to manipulate or alter their underlying design and programming (Colella, Klopfer, & Resnick, 2001; Resnick, 1994—cited in Penner, 2001, p. 16). Prior to constructionism, Shirts (1976) recommended that students assist with designing classroom simulations to enable a greater understanding of the simulated system. Shirts recognized that students are not given the opportunity to change a simulation model or create their own design to understand the underlying system or create meaning in a constructionist manner. Thus, Shirts appears to be an early advocate of constructionist principles in the social sciences and advances the idea that pre-computer simulation designs should be accessible to student alterations. Yet, as the examples provided above indicate, computers have radically changed simulations and modeling.

Computer technology on the surface provides many avenues to simplify instruction through simulations or demonstration models. Beyond the surface, technology can add a great deal of complexity to model design. Computers are powerful tools that allow teachers to display objects that are normally invisible and hide other things that are ordinarily visible (Horwitz, 1999). The autonomy of students in designing projects is unclear when analyzing exactly what level or degree of freedom they have in modeling and programming. Constructionists are willing to criticize simulations for being inflexible and incapable of providing students access to the underlying models. Yet, modeling language can be complex and requires learning that may detract from the content to be learned. Furthermore, there are "primitive elements" in modeling language that remain "black boxes" (Wilensky, 1999, 168). Wilensky (1999), a constructionist, emphasizes that it would be absurd to reduce model building to binary digits or building the hardware that supports the modeling language (p.168). According to Wilensky, even die-hard constructionists accept that all pieces of the model do not need construction.

For example, in order to make certain concepts more salient, Resnick, Berg, and Eisenberg (2000) explicitly decided to control the accessibility of certain design processes and mechanisms within "Crickets" or their approachable version of black boxes. The inner workings of black boxes are typically hidden and the devices are used to gather data or make measurements. In contrast, Crickets are "fully-programmable" computers that enable students the novel ability to program scientific instruments. Although Crickets are programmable, the inner mechanisms remain hidden or inaccessible for students to manipulate. Thus, the complexity of computers and computer programming forces educators to find the right "construction materials," such as the most appropriate software, that will allow exploration and application of previous knowledge (Willett, 2007). The incumbent role of the instructor and/or designer in deciding what to present and what not to present thus shows its influence in constructionist approaches to model building. The instructor understandably decides what type of modeling will take place in the classroom. The decision should not be an exclusionary choice.

The extreme gap between model using (pre-made simulations) and model building or construction (student created or student altered computer simulations) is filled with hybrid models that provide avenues for the integration of both approaches. Models that purely demonstrate phenomenon and do not allow students to modify the structures, processes, and operation of the model are called "demonstration models" (Wilensky, 1999, p. 168). Demonstration models become hybrids and verge upon model construction when they provide changeable parameters and allow students to actively explore the effects of changes on the model. In contrast, model construction actively engages students to formulate their own questions and seek tentative solutions and theories through an iterative process of reformulation and redesigning (Wilensky, 1999, p. 167). Complex interaction between model using and model building makes differentiation difficult because each approach offers relative benefits. The following discussion highlights hybrid simulation examples within various categorizations of simulation games, such as edutainment and augmented reality.

Edutainment Simulation Games

Edutainment software packages such as *SimCity*, *Colonization*, and *Civilization* are used in classrooms as demonstration models, but it is possible to classify them as hybrids because students alter the parameters and question the effects of changes. However, according to Colella, Klopfer, and Resnick (2001), such computer simulation models do not provide a "deeper understanding" that comes from programming or manipulating the underlying structure. The challenge is to determine what levels of activity, richness, motivation, and understanding are acceptable for social science learning that incorporates constructionist ideals. For example, *Civilization* users are given the opportunity to make decisions involving technological, political, and cultural developments of a society. Students analyze the effects of their decisions on the society they create and personalize (Frye & Frager, 1996). *SimCity*, *Colonization*, and *Civilization* provide students opportunities to develop new solutions to new problems as city manager, mayor, or governor. All three programs allow computer modeling of geographical concepts, general mathematical models and complex individual decisions.

Squire, DeVane, and Durga (2008) report findings of a recent study in which researchers introduced *Civilization III* in an after-school program to 12 5th and 6th graders who were described as having little interest in traditional school instruction and little background with computational technology. The researchers' goal was to help the students develop fluency in world history and advanced problem-solving skills. The participant students met once per week for 2 hours and during the summer and the students attended *Civilization* summer camp. According to Squire et al. (2008), after one year in the program, the students experimented and altered the simulation by changing the pacing, resource allocations and the starting point for the civilizations. One student even used the modifying tools to model the current war in Iraq by changing the civilizations and their leaders to reflect current geopolitical conditions. "For these students, the desire to modify games was not a distant and abstract goal, but rather a natural outgrowth of their desire to entertain friends, express themselves, and achieve status in their community" (Squire, DeVane, & Durga, 2008, p. 246). Though this study reflects the experiences of a small group of boys, it illustrates how constructionism as a learning theory can be put into practice through simulation games. The students construct knowledge through designing civilizations and sharing them with others in an engaging and interactive learning community. Similar to edutainment applications of constructionism, augmented reality simulation games afford relevant simulation game opportunities to build models and promote learning.

Augmented Reality Simulation Games

The pervasiveness of digital technologies and handheld computers provides another opportunity for students to construct knowledge by building models to share with others. Klopfer and Squire (2007) describe augmented-reality (AR) simulation games as "games that are played in the real world, in locations such as neighborhoods, historical sites, or watersheds, but using technologies to layer data over the real world" (Klopfer & Squire, 2007, p. 3). Data may include video, text or images, which players manipulate to create fictional characters and events. AR games allow students to participate as workers in a simulated world and provide opportunities to learn from the consequences of their work (Klopfer & Squire, 2007).

Consider the example of *Dow Day*, an AR game played on handheld devices that takes students back to the anti-Dow Chemical protests, which transpired on the University of Wisconsin-Madison campus in October of 1967 (Squire et al., 2007). Players role-play as journalists investigating the causes of the protests and why they turned violent. The game, which only lasts one and a half hours, is part of a larger inquiry-based unit, during which students read and analyze documents, develop questions, travel to the actual location of the events at the UWM campus, write newspaper articles and conduct independent research. In another AR game, *The Greenbush Game*, students investigate a multiethnic neighborhood in Madison, Wisconsin to gather research for the design of their own AR game. Students present their ideas and their cardboard models of historical Greenbush buildings to authentic audiences such as the city council. Through handheld devices, students participate in both model using and model building, as they wrestle with content knowledge, construct new and meaningful objects and share them with authentic audiences (Squire el al., 2007).

The trappings of the traditional transmission approach to simulations are avoided by actively engaging students in models that they alter, name, and control (Frye & Frager, 1996). Student interest, enjoyment and engagement noted by teachers and researchers (Frye & Frager, 1996; Rieber, Smith, & Noah, 1998) are comparable to the engagement and enjoyment of students participating in design. Higher levels of thinking are also stimulated through evaluating the strengths and weaknesses of cities, colonies, and civilizations (Frye & Frager, 1996). Through *SimCity* and teacher encouragement, students learn to evaluate, compare, and critique the simulated cities they create with real cities and the software links to real cities (Adams, 1998). Thus, students are quite active and motivated to learn through rich simulation models that may not fulfill constructionist ideals in the truest sense, but go well beyond demonstration models. For instance, although students do not alter the underlying assumptions and theories of the software, they learn about various complex systems by manipulating elaborate system variables and *building* their own societies. Within the context of constructionism, the societies built by students can be considered public artifacts with ideas, theories, or questions that are shared, displayed and reflected on and revised through new models. Thus, there are many facets of constructionism that benefit the educational value of AR and edutainment software. Simulations that integrate primary documents, such as diaries, writings, and pictures "provide real-world context and comprise two equally important components, authentic social studies material and authentic social studies inquiry" (Doolittle & Hicks, 2003, p. 89). The multifaceted constructionist benefits of learning through AR and edutainment are further enhanced through the advantages of social interaction that occurs through multiplayer simulation games.

Multiplayer Simulation Games

Constructionism is also relevant to learning when students are engaged in a multiplayer game. Through multiplayer games, players are absorbed in intensive social learning as each player negotiates to make meaning out of the space provided by the video game (Squire, 2005). Players are learning to make social sense out of their collective virtual world and learning how to navigate in the semiotic domain crafted by the game designers. Well-designed multiplayer simulation games compel players to navigate in the virtual world, to become literate in the semiotic domains of social practice, and to solve social or societal problems (Shaffer, Halverson, Squire, & Gee, 2004). In the multiplayer game River City, students adopt the role of scientists investigating a virtual outbreak of a disease. As citizens of a virtual community, the players work together using scientific principles to understand the causes of the simulated disease (Games & Squire, 2011). In addition to the learning opportunities that take place in the virtual world, multiplayer simulation games encourage participants to involve themselves in online chat rooms and messages boards maintained by their fellow game players. Participation in these online simulation game communities fosters the civic engagement that many pundits argue is lacking in our society (Steinkuehler, 2008). Steinkuehler believes that multiplayer games foster the development of online communities that encourage civic participation, much like coffee shops and other real world hangouts.

Simulation games foster the development of community learning among gamers according to Squire and Steinkuehler (2005). In addition, gamers involved in multiplayer simulation games share information, blur the distinction between the production and consumption of knowledge, and promote international communities. According to Gee (2007), gamers often prefer to play single player simulation games in groups and take turns playing the game and sharing knowledge of how to play the game. Single and multiplayer simulation games provide areas of shared community interest among the gamers that lead to the development of authentic communities. Students participating in a shared semiotic domain develop a shared understanding of that experience that typifies authentic classroom communities. Gee speculates that if educators use a video game as a classroom learning tool then the students' shared experience of playing the simulation game will help to create an authentic classroom community.

Single player simulation games foster the development of community practice among their players as illustrated in mediums such as player created Frequently Asked Questions ("FAQS") (Squire, 2006). FAQS are online spaces where players engage in online social practice in order to assist one another with the playing of a particular game. Games also allow players to adopt different identities in the game and coerce the player to think critically about identity including gender roles (Hayes, 2005). Hayes examined how women experience a single player simulation game and reached the conclusion that traditional gender stereotypes of men and women and simulation game play are simplistic and incorrect. Hayes reaches the conclusion that simulation games can potentially allow players of both genders to explore their identity in a critical format enhancing their ability to understand the social underpinnings of the real world. Multiplayer and single player games can inspire the player to engage in challenging acts of cognition that inspire the participants to create social networks to problem solve. Well-designed simulation games potentially offer players the opportunity to participate meaningfully in virtual social networks that are essential for 21st century citizens. In a world where Twitter, Facebook, political blogs, and numerous other digital media environments are as important as the traditional print or broadcast media for tech savvy citizens, teacher education via simulation games offer a method to link content, meaningful learning, and the technological skills essential for modern day citizenship involvement.

Teacher Education Simulations

The research on teacher education simulations is extremely limited by the paucity of relevant simulation games used in practice (Fischler, 2007). Indeed, researchers recently designed and created simulation games in the absence of commercially produced models. In early literacy education, an international team of researchers designed and investigated the implementation of a virtual kindergarten simulation (Ferry et al., 2005). The researchers utilized a case study approach and purposive sampling for three 90 minute trials of the simulation software with over 200 preservice teachers as participants. According to Ferry et al. (2005), data was collected through researcher observations, entries in the embedded simulation reflection tool ("thinking spaces"), and semi-structured interviews. However, only ten participants were interviewed.

Virtual kindergarten users reported that their experience with the simulation helped them to make their practicum experience more focused. The simulation provided classroom knowledge and experience to more fully appreciate the impact of subtle changes that experienced teachers made during lessons. Initial findings showed that "use of the simulation develops pre-service teacher awareness of the challenges they will face as beginning teachers" (p. 29). Moreover, the majority of participants "valued" the ability to critically reflect within the "thinking spaces," which provided greater opportunities to evaluate and justify decisions without consequences that would affect real children. In a real classroom, the rapid pace and continuous demands of kindergarten children can easily overwhelm a new teacher and inhibit reflection. In connection with constructionist modeling, the simulation classroom afforded an ability to "revisit and reflect on critical decision points and replay events in the light of new understandings" (p. 30).

In another innovative and original teacher education simulation, Fischler (2006, 2007) programmed SimTeacher to establish and research a classroom simulation. As the only programmer of SimTeacher, Fischler experienced technical problems, such as software bugs, and restricted media delivery. In particular, "There were no animated images, video, or sound" (Fischler, 2006, p. 150). Regardless of SimTeacher's limitations, the implications and results of Fischler's study provide powerful contributions to simulation game research and theory.

The application of Fischler's SimTeacher research involved three university instructors within a four-semester timeframe (Fischler, 2006). The three instructors taught 11 undergraduate teacher education courses with a total of 265 students. The instructors actually determined the extent of SimTeacher use within their teacher education courses. While Fischler's qualitative data collection focused on the instructors' use and perceptions of SimTeacher, the student participants also provided additional feedback through survey data. Fischler's research focus on instructors is related to the unique programming feature of SimTeacher that allowed the teacher educators to create their own content for the SimTeacher simulations. In other words, the instructors were afforded the opportunity to create content, choices, and the flow of assignments for their students (Fischler, 2006, 2007).

Within the simulation, participants assume the role of "SimTeachers" who interact with fictional students and other relevant characters.

The simulation came alive with seemingly real yet fictional characters. The daily tasks and story-like situations involved an animated and diverse crew, including students, a school counselor, nurse, secretary, parents of students, the principal, other teachers at the virtual school, and more. The instructor of a course occasionally played the role of the school principal or a parent of a student (though the instructor's students were usually not aware of this). Furthermore, some simulated characters made remarks and held opinions about other characters. Lastly, SimTeachers interacted with other SimTeachers with the semblance of being fellow teachers in the simulated environment (Fischler, 2006, p. 84).

The daily activities modeled real-world teaching and included tasks such as attendance, responding to parents' emails, grades, creating lesson plans, and even filling out an individualized education plan (IEP).

In addition to daily activities, "The interactive stories within SimTeacher were called situations and contained narratives, multimedia, and question and answer sessions. The instructor created the content for all the storyline branches or repurposed content another instructor had previously created" (Fischler, 2006, p. 79). Some interactive stories required SimTeacher participants to potentially think at higher levels by analyzing a problem, synthesizing relevant course concepts, and evaluate their chosen solution after observing the consequences. Furthermore, instructors could add content-related problems to assess knowledge-based learning. For example, an educational psychology student may encounter activities or situations "that emphasize classroom management, cognitive learning styles, student motivation, child development issues, and related topics" (Fischler, 2006, p. 81).

SimTeacher affords students multiple opportunities to apply concepts they were learning in teacher education classes to real world scenarios. However, Fischler's (2006) research results showed that integration of the educational simulation within the curriculum is very important to have a positive effect on learning. The teacher education instructors who reported the most success with simulation-based learning used "structurally rich storylines" and blended "simulation use with classroom discussion" (p. 164). One instructor did not promote interactivity so his/her students did not reflect with their peers. Papert would argue that sharing ideas and socially reflecting on ideas and products enhances learning. SimTeacher did not overwrite previous attempts and the history of student work was archived. Thus, SimTeacher contained a strong platform for constructionist learning and sharing. However, SimTeacher options were not limitless and most cases only offered two to four decisions, which leaves the door open for more sophisticated simulation games that provide an opportunity for more options. Why not involve participants in creating the scenarios and learning from their conceptions of the simulation and real-world classrooms?

FUTURE RESEARCH DIRECTIONS

Tobias et al. (2011) caution that research is still at an "early stage of development" for computer simulation games, and "any evaluation of the effectiveness of games used for instruction must be tentative rather than conclusive" (p. 159). The nascent field of simulation games recently evolved beyond inquiring whether educational games are simply possible (Games & Squire, 2011). The next step is to determine the most effective contexts and designs for simulation games that satisfy the dynamic and potential learning benefits of simulation games. For example, Gamestar Mechanic appears to synthesize Papert's constructionist vision with Gee's semiotic literacy by involving "children's language and literacy skill development in the context of an online community of game designers" (p. 36). Gamestar Mechanic players attempt to fix broken games and learn how to build their own games and share them online with others. Yet, higher levels of learning from this and similar simulations are complex and difficult to measure with standardized tests.

The research on teacher education simulation games offers considerable promise (Ferry et al., 2005; Fischler, 2006), but falls short of realizing the full technological and theoretical possibilities that are found in mass consumer and edutainment simulation games. According to Fischler (2006),

> *A more advanced educational simulation would let students come up with their own choices before continuing the storyline, perhaps relying on artificial intelligence to unfold the plot. It would be interesting to research the benefits of having students create their storyline paths rather than forcing them to choose from a limited list of options (pp. 150-151).*

Such an approach to research embraces model designing and many of the essential components to constructionist theory.

The larger obstacle for simulation game research is the difficulty in assessing and valuing the situated and socially constructed forms of learning in an environment that favors narrow standardized tests of knowledge (Games & Squire, 2011). Thus, much more research on the new and evolving simulations is necessary prior to making any global conclusions. Designing teacher education simulations that accurately model the complex interaction and learning processes of multiple students may seem impossible, but designers and researchers are making impressive progress (see Ferry et al., 2005; Fischler, 2006).

CONCLUSION

Model using and model designing offer educational opportunities to break from the traditional transmission approaches of teaching and learning. The merits of constructionism should be combined with model using to ensure deeper learning and active student engagement in creating models of complex systems. Unhealthy visions that promote one approach over the other will not benefit learning. Instead, researchers, designers, and educators should focus on integrating the benefits of both model using and model designing.

The constructionism models that are presented above are much more interactive and, therefore, not similar to the passive demonstration models used to exhibit a specific phenomenon. Teachers using simulation models should consider ways to encourage the designing or building of public artifacts to promote reflection and discussion. *SimCity*, *Colonization*, and *Civilization* allow students to build and exhibit cities, colonies, and civilizations they create. SimTeacher affords model design for the instructor that could include students or preservice teachers as well. Discussion concerning the models should invite questions that challenge students to reflect on their decisions and reformulate or redesign their models. If it is possible, teacher educators should encourage autonomous model creation to reflect the thoughts, questions, connections and theories of each individual modeler. Model artifacts exhibit the level of a student's understanding and enable the student to change or address any inadequacies. The ultimate goal is to ensure a thoughtful process of model using and building that considers the underlying dynamics of the system(s) being modeled.

Students learn to explore complex systems through modeling and using higher order thinking skills. Problems are evaluated through questions and theories that offer greater depth and go beyond simple solutions (Zaraza & Fisher, 1999, p. 50). The inherent complexity of human interaction with the environment requires students to become better thinkers and tinkerers. The goal of the social sciences is to promote informed, active, and reflective citizens who are capable of making rational decisions. The role of modeling in promoting citizenship is advanced by Wilensky (1999): "Our science, our social policy, and the importance of an engaged citizenry require an understanding of the dynamics of complex systems and the use of sophisticated modeling tools to display and analyze such problems" (p.174).

Modeling addresses the demands of citizenship, because higher order thinking is necessary to develop explanations of complex systems and the constructive process of model building requires students to make informed and rational decisions. Through model building, students and preservice teachers search for information, determine the quality of information, instantiate a position, and justify their position or model with evidence. In addition, it is essential that model users incorporate critical thinking that questions and examines the underlying system of the model(s) being used. Thus, teacher educators interested in promoting deeper learning should consider how to integrate model building with model using and at the very least question the underlying systems incorporated in simulations. For example, preservice teachers should learn to guide student users of *SimCity* to evaluate and reflect on its oversimplifications of urban systems, such as mayors who do not negotiate with city councils.

The role of the teacher is critical in determining the success or failure of modeling in a classroom environment. Thus, teacher educators must show or model (ideally through simulation games) how to implement instructional simulation games. The hurdles to integrating model design with model use will vary considerably depending upon the purposes of instruction and the type of system being modeled. Zaraza and Fisher (1999) caution that higher-order skills are needed to build models and it is not necessary for everyone to do it (p. 68). It is possible that a pure demonstration model satisfies the instructional goals or objectives to learn about a particular system. However, regardless of the type of modeling approach, teachers must scaffold learning and recognize the importance of providing individualized instruction. Scaffolding or instructional support measures safeguard effective learning and prevent learners from being overwhelmed by the full complexity of modeling. For example, Resnick (1999) works directly with students as they build models with StarLogo. He suggests projects, asks questions, challenges assumptions, helps with programming, and encourages students to reflect on their experiences. Through scaffolding efforts, teachers are in a better position to help students make personal connections and reflect on their modeling.

Even with the instructional support of teachers, it is difficult to motivate all students. Modeling provides opportunities to intrinsically motivate students. As Squire (2005) notes, our traditional secondary curriculum is characterized by the mastery of a pre-defined set of objectives, learned mostly through listening or participating in structured activities. "Those who prefer to develop understandings through building, tinkering, or more direct experience are left behind" (Squire, 2005). The aesthetic, interactive, personalized, and tangible process of model building supports various learning abilities and preferences.

Some of the benefits of modeling are lost if students are not attached or motivated to learn about their models. For example, Resnick, Berg, and Eisenberg (2000) found that students must truly care about their projects to follow through with data analysis. Allowing students to address the aesthetic nature of learning ensures that students feel and express a personalized connection to models. Overall, constructionism offers a great deal of promise to reshape how social science educators and researchers approach model using or model building. The merits of incorporating model building into teacher education are beneficial to the ultimate goal of promoting higher order thinking and informed citizens.

REFERENCES

Adams, P. C. (1998). Teaching and learning with SimCity 2000. *The Journal of Geography*, *97*(2), 47–55. doi:10.1080/00221349808978827

Bolick, C. M., Berson, M. J., Friedman, A. M., & Porfeli, E. J. (2007). Diffusion of technology innovation in the preservice social studies experience: Results of a national survey. *Theory and Research in Social Education*, *35*(2), 174–195. doi:10.1080/00933104.2007.10473332

Brant, G., Hooper, E., & Sugrue, B. (1991). Which comes first the simulation or the lecture? *Journal of Educational Computing Research*, *7*(4), 469–481. doi:10.2190/PWDP-45L8-LHL5-2VX7

Clegg, A. A. (1991). Games and simulations in social studies education. In Shaver, J. P. (Ed.), *Handbook of research on social studies teaching and learning* (pp. 523–529). New York, NY: Macmillan.

Colella, V. S., Klopfer, E., & Resnick, M. (2001). *Adventures in modeling: Exploring complex, dynamic systems with StarLogo*. New York, NY: Teachers College Press.

Crotty, M. (1998). *The foundations of social research: Meaning and perspective in the research process*. Thousand Oaks, CA: SAGE Publications.

de Jong, T., & van Joolingen, W. R. (1998). Scientific discovery learning with computer simulations of conceptual domains. *Review of Educational Research*, *68*(2), 179–201.

de Pommereau, I. (1997, April 21). Computers give the key to learning (an interview with Seymour Papert. *Christian Science Monitor, 89*(101), 11. Retrieved from http://www.csmonitor.com

DeLeon, A. P. (2008). Are we simulating the status quo? Ideology and social studies simulations. *Theory and Research in Social Education*, *36*(3), 256–277. doi:10.1080/00933104.2008.10473375

Dewey, J. (1938). *Experience and education*. New York, NY: Simon & Schuster.

Doolitle, P. E., & Hicks, D. (2003). Constructivism as a theoretical foundation for the use of techology in social studies. *Theory and Research in Social Education*, *31*(1), 72–104. doi:10.1080/00933104.2003.10473216

Eck, R. V. (2006). Digital game-based learning: It's not just the digital natives who are restless. *EDUCAUSE Review*, (March/April): 17–30.

Federation of American Scientists. (2006). *Harnessing the power of video games for learning.* Paper presented at the National Summit on Educational Games. Retrieved from http://www.fas.org/gamesummit/index.html

Feinberg, J. R. (2011). *Debriefing in simulation games: An examination of reflection on cognitive and affective learning outcomes*. Saarbrücken, Germany: LAP Lambert Academic Publishing.

Ferry, B., Kervin, L., Cambourne, B., Turbill, J., Hedberg, J., & Jonassen, D. (2005). Incorporating real experience into the development of a classroom-based simulation. *Journal of Learning Design, 1*(1), 22–32. Retrieved from www.jld.qut.edu.au/Vol_1_No_1

Fischler, R. (2007). SimTeacher.com: An online simulation tool for teacher education. *Techtrends: Linking Research and Practice to Improve Learning, 51*(1), 44–47. doi:10.1007/s11528-007-0011-2

Fischler, R. B. (2006). *SimTeacher: Simulation-based learning in teacher education*. (Doctoral dissertation). Retrieved from ProQuest Dissertations and Theses.(Accession Order No. AAT 3210046)

Frye, B., & Frager, A. M. (1996). Civilization, Colonization, SimCity: Simulations for the social studies classroom. *Learning and Leading with Technology, 24*(2), 21–23, 32.

Games, A., & Squire, K. D. (2011). Searching for the fun in learning: A historical perspective on the evolution of educational video games. In Tobias, S., & Fletcher, J. D. (Eds.), *Computer games and instruction* (pp. 371–394). Charlotte, NC: Information Age Publishers.

Gee, J. P. (2005). It's theories all the way down: A response to scientific research in education. *Teachers College Record, 107*(1), 10–18. doi:10.1111/j.1467-9620.2005.00452.x

Gee, J. P. (2007). *What video games have to teach us about learning and literacy* (Rev. and updated ed.). New York, NY: Palgrave Macmillan.

Harmon, S. W. (2008). A theoretical basis for learning in massive multiplayer virtual worlds. *Journal of Educational Technology Development and Exchange, 1*(1), 29–40.

Hayes, E. (2005). *Women and video gaming: Gendered identities at play*. Paper presented at the Games, Learning, & Society Conference. Madison, WI

Horwitz, P. (1999). Designing computer models that teach. In Fuerzeig, W., & Roberts, N. (Eds.), *Modeling and simulation in precollege science and mathematics Education* (pp. 179–196). New York, NY: Springer-Verlag. doi:10.1007/978-1-4612-1414-4_8

Kafai, Y. B., & Resnick, M. (1996). Introduction. In Kafai, Y. B., & Resnick, M. (Eds.), *Constructionism in practice* (pp. 1–8). Mahwah, NJ: Erlbaum.

Kirkley, J. R., Duffy, T. M., Kirkley, S. E., & Kremer, L. H. (2011). Implications of constructivism for the design and use of serious games. In Tobias, S., & Fletcher, J. D. (Eds.), *Computer games and instruction* (pp. 371–394). Charlotte, NC: Information Age Publishers.

Klopfer, E., & Squire, K. (2007). Case study analysis of augmented reality simulations on handheld computers. *Journal of the Learning Sciences, 16*(3), 371–413. doi:10.1080/10508400701413435

National Summit on Educational Games. (2006). *Harnessing the power of games*. Washington, DC: Federation of American Scientists. Retrieved from http://www.fas.org/gamesummit/Resources/Summit%20on%20Educational%20Games.pdf

Newmann, F. M., Marks, H. M., & Gamoran, A. (1996, August). Authentic pedagogy and student performance. *American Journal of Education, 104*(4), 280–312. doi:10.1086/444136

Papert, S. (1991). Situating constructionism. In Harel, I., & Papert, S. (Eds.), *Constructionism* (pp. 1–11). Norwood, NJ: Ablex.

Papert, S. (1998). Does easy do it? Children, games, and learning. *Game Developer Magazine*, *4*, 88–97.

Penner, D. E. (2001). Cognition, computers, and synthetic science: Building knowledge and meaning through modeling. *Review of Research in Education*, *25*, 1–35.

Plass, J. L., Goldman, R., Flanagan, M., Diamond, J. P., Dong, C., & Looui, S. … Perlin, K. (2007). *RAPUNSEL: How a computer game design based on educational theory can improve girls' self-efficacy and self-esteem*. Paper presented at the American Educational Research Association.

Resnick, M. (1994). *Turtles, termites, and traffic jams*. Cambridge, MA: MIT Press.

Resnick, M. (1999). Decentralized modeling and decentralized thinking. In Fuerzeig, W., & Roberts, N. (Eds.), *Modeling and simulation in precollege science and mathematics education* (pp. 114–137). New York, NY: Springer-Verlag. doi:10.1007/978-1-4612-1414-4_5

Resnick, M. (2007). Sowing the seeds for a more creative society. *Learning and Leading with Technology*, *7*, 18–22.

Resnick, M., Berg, R., & Eisenberg, M. (2000). Beyond black boxes: Bringing transparency and aesthetics back to scientific investigation. *Journal of the Learning Sciences*, *9*(1), 7–30. doi:10.1207/s15327809jls0901_3

Rice, J. (2007). Assessing higher order thinking in video games. *Journal of Technology and Teacher Education*, *15*(1), 87–100.

Rieber, L. P. (1996). Seriously considering play: Designing interactive learning environments based on the blending of microworlds, simulations, and games. *Educational Technology Research and Development*, *44*(2), 43–58. doi:10.1007/BF02300540

Rieber, L. P., Smith, L., & Noah, D. (1998). The value of serious play. *Educational Technology*, *38*(6), 29–37.

Ruschoff, B., & Ritter, M. (2001). Technology-enhanced language learning: Construction of knowledge and template-based learning in the foreign language classroom. *Computer Assisted Language Learning*, *14*(3-4), 219–232. doi:10.1076/call.14.3.219.5789

Shaffer, D. W., Halverson, R., Squire, K. R., & Gee, J. P. (2004). *Video games and the future of learning*. Retrieved from The University of Madison Wisconsin, Advanced Academic Distributed Learning Co-Laboratory http://www.academic-colab.org/resources/gappspaper1.pdf

Shirts, R. G. (1976). Simulation/gaming for the past 10 years: What has and what hasn't happened. *Simulation & Gaming*, *3*(5), 5–9.

Squire, K. (2005). Changing the game: What happens when video games enter the classroom? *Innovate, 1*(6). Retrieved from http://innovateonline.info/pdf/vol1_issue6/Changing_the_Game-__What_Happens_When_Video_Games_Enter_the_Classroom_.pdf

Squire, K. (2006). From content to context: Videogames as designed experience. *Educational Researcher*, *35*(8), 19–29. doi:10.3102/0013189X035008019

Squire, K. (2007). Games, learning, and society: Building a field. *Educational Technology*, *4*(5), 51–54.

Squire, K., DeVane, B., & Durga, S. (2008). Designing centers of expertise for academic learning through video games. *Theory into Practice*, *47*(3), 240–251. doi:10.1080/00405840802153973

Squire, K., Jan, M., Matthews, J., Wagler, M., Martin, J., DeVane, B., & Holden, C. (2007). Wherever you go, there you are: Place-based augmented reality games for learning. In Shelton, B. E., & Wiley, D. A. (Eds.), *The educational design and use of computer simulation games*. Rotterdam, The Netherlands: Sense Press.

Squire, K., & Steinkuehler, C. (2005). Meet the gamers: They research, teach, learn, and collaborate. So far, without libraries. *Library Journal*, *130*(7), 38–41.

Steinkuehler, C. (2005). The new third place: Massively multiplayer online gaming in American youth culture. *Tidskrift Journal of Research in Teacher Education*, *3*, 135–150.

Steinkuehler, C. (2008). Massively multiplayer online games as an educational technology: An outline for research. *Educational Technology*, *48*(1), 10–21.

Tobias, S., Fletcher, J. D., Dai, D. Y., & Wind, A. P. (2011). Review of research on computer games. In Tobias, S., & Fletcher, J. D. (Eds.), *Computer games and instruction* (pp. 371–394). Charlotte, NC: Information Age Publishers.

VanSickle, R. L. (1986). A quantitative review of research on instructional simulation gaming: A twenty-year perspective. *Theory and Research in Social Education*, *14*, 245–264.

White, C. S. (1985). Citizen decision making, reflective thinking and simulation gaming: A marriage of purpose, method and strategy. *Journal of Social Studies Research*, Monograph 2 (Summer).

Wilensky, U. (1999). GasLab—An extensible modeling toolkit for connecting micro- and macro-properties of gases. In Fuerzeig, W., & Roberts, N. (Eds.), *Modeling and simulation in precollege science and mathematics education* (pp. 151–178). New York, NY: Springer-Verlag. doi:10.1007/978-1-4612-1414-4_7

Willett, R. (2007). Technology, pedagogy and digital production: A case study of children learning new media skills. *Learning, Media and Technology*, *32*(2), 167–181. doi:10.1080/17439880701343352

Zaraza, R., & Fisher, D. M. (1999). Training system modelers: The NSF CC-STADUS and CC-SUSTAIN projects. In Fuerzeig, W., & Roberts, N. (Eds.), *Modeling and simulation in precollege science and mathematics education* (pp. 38–69). New York, NY: Springer-Verlag. doi:10.1007/978-1-4612-1414-4_2

ADDITIONAL READING

Brant, G., Hooper, E., & Sugrue, B. (1991). Which comes first the simulation or the lecture? *Journal of Educational Computing Research*, *7*(4), 469–481. doi:10.2190/PWDP-45L8-LHL5-2VX7

Colella, V. S., Klopfer, E., & Resnick, M. (2001). *Adventures in modeling: Exploring complex, dynamic systems with StarLogo*. New York, NY: Teachers College Press.

Doolitle, P. E., & Hicks, D. (2003). Constructivism as a theoretical foundation for the use of techology in social studies. *Theory and Research in Social Education*, *31*(1), 72–104. doi:10.1080/00933104.2003.10473216

Eck, R. V. (2006). Digital game-based learning: It's not just the digital natives who are restless. *EDUCAUSE Review*, (March/April): 17–30.

Federation of American Scientists. (2006). *Harnessing the power of video games for learning*. Paper presented at the National Summit on Educational Games. Retrieved from http://www.fas.org/gamesummit/index.html

Fuerzeig, W., & Roberts, N. (1999). *Modeling and simulation in precollege science and mathematics education*. New York, NY: Springer-Verlag. doi:10.1007/978-1-4612-1414-4

Gee, J. P. (2007). *What video games have to teach us about learning and literacy* (Rev. and updated ed.). New York, NY: Palgrave Macmillan.

Harel, I., & Papert, S. (1991). *Constructionism*. Norwood, NJ: Ablex.

Harmon, S. W. (2008). A theoretical basis for learning in massive multiplayer virtual worlds. *Journal of Educational Technology Development and Exchange*, *1*(1), 29–40.

Kafai, Y. B., & Resnick, M. (1996). *Constructionism in practice*. Mahwah, NJ: Erlbaum.

Klopfer, E., & Squire, K. (2007). Case study analysis of augmented reality simulations on handheld computers. *Journal of the Learning Sciences*, *16*(3), 371–413. doi:10.1080/10508400701413435

Newmann, F. M., Marks, H. M., & Gamoran, A. (1996). Authentic pedagogy and student performance. *American Journal of Education*, *104*(4), 280–312. doi:10.1086/444136

Papert, S. (1998). Does easy do it? Children, games, and learning. *Game Developer Magazine*, *4*, 88–97.

Resnick, M. (2007). Sowing the seeds for a more creative society. *Learning and Leading with Technology*, *7*, 18–22.

Rice, J. (2007). Assessing higher order thinking in video games. *Journal of Technology and Teacher Education*, *15*(1), 87–100.

Rieber, L. P., Smith, L., & Noah, D. (1998). The value of serious play. *Educational Technology*, *38*(6), 29–37.

Shaffer, D. W., Halverson, R., Squire, K. R., & Gee, J. P. (2004). *Video games and the future of learning*. Retrieved from The University of Madison Wisconsin, Advanced Academic Distributed Learning Co-Laboratory http://www.academic-colab.org/resources/gappspaper1.pdf

Shelton, B. E., & Wiley, D. A. (2007). *The educational design and use of computer simulation games*. Rotterdam, The Netherlands: Sense Press.

Squire, K. (2005). Changing the game: What happens when video games enter the classroom? *Innovate*, *1*(6). Retrieved from http://innovateonline.info/pdf/vol1_issue6/Changing_the_Game-__What_Happens_When_Video_Games_Enter_the_Classroom_.pdf

Squire, K., DeVane, B., & Durga, S. (2008). Designing centers of expertise for academic learning through video games. *Theory into Practice*, *47*(3), 240–251. doi:10.1080/00405840802153973

Steinkuehler, C. (2005). The new third place: Massively multiplayer online gaming in American youth culture. *Tidskrift Journal of Research in Teacher Education*, *3*, 135–150.

Tobias, S., & Fletcher, J. D. (Eds.). (2011). *Computer games and instruction*. Charlotte, NC: Information Age Publishers.

KEY TERMS AND DEFINITIONS

Augmented-Reality (AR): Real world simulation games that use technology to layer digital data over the real objects, e.g. *Dow Day*.

Constructionism: Papert's (1991) theory of learning borrows from constructivist theory, but adds that new ideas are more likely to emerge when learners are actively engaged in designing or building an artifact that can be reflected upon and shared with others.

Constructivism: Theory of learning with an active role of learners in acquiring knowledge that has multiple socio-cultural interpretations.

Edutainment: Video or computer games that offer educational processes and/or events, but this type of game is not necessarily developed for education (see serious games).

Instructional Simulation Games: Real world models that allow students to interact with events and object that are normally inaccessible and include an element of competition (i.e., game) within a classroom setting.

Semiotic Domain: Gee's (2007) theoretical type of simulation game literacy where the gamer or player becomes literate in the rules, requirements, symbols, images, graphs, diagrams, artifacts, language, and culture of the game involved.

Serious Games: Simulation games developed for educational purposes or offer relevant educational applications.

Simulation: Model of real world events, mechanisms, or processes that are usually safer, less costly, more convenient, and repeatable.

Simulation Game Computers or Consoles: Electronic devices that process software or online content and allow a player to become a participant in a digital virtual world that represents a real world event or process.

Chapter 23
Teacher Education with simSchool

David Gibson
Curveshift, USA

ABSTRACT

This chapter introduces an innovative online learning platform for the preparation of teachers through simulations, which addresses some of the systemic challenges of teacher education in the US. The chapter contrasts traditional course-based online learning experiences with a simulation approach to four areas of teacher preparation: conceptions of teaching & learning, the organization of knowledge, assessment practices and results, and the engagement of communities of practice. The chapter outlines a rationale for the new approach based in self-direction and personal validation in a complex but repeatable practice environment, supported by emergent interdisciplinary knowledge concerning the unique affordances of digital media assessment and social media. The online simulation simSchool is used as an example model that embodies the new paradigm.

INTRODUCTION

The plan of the chapter is to briefly outline some of the key problems with teacher education in the U.S., including preparing teachers in both traditional face-to-face and online courses and programs. Then the narrative presents the characteristics of a new model of self-directed teacher education supported in a game-like simulation context. The plan is to demonstrate that such a context has unique affordances compared with the alternatives, which entails describing how such an environment teaches as well as how it offers evidence for assessing whether someone has learned anything. The chapter is supported by a concrete example – simSchool - a flight simulator for teachers.

DOI: 10.4018/978-1-4666-1906-7.ch023

Some Problems with Teacher Education

About Half of All New Teachers Quit by the End of their Third Year

The National Commission on Teaching and America's Future (NCTAF) estimates that the national cost of public school teacher turnover could be over $7.3 billion a year. In addition to the nation losing billions of dollars, the constant churn of teachers drains resources, diminishes teaching quality, and undermines our ability to close global and even local student achievement gaps (Carroll, 2007). In 2004 most U.S. teachers were 52 years old and the average age was 43, but by 2008 most teachers were 28 years old (Carroll & Foster, 2010). This dramatic change in the demographics of teaching in the U.S. implies two things: 1) there is a need to prepare, mentor and support a much larger percentage of the teaching workforce for a longer period of time than at any other time in history and longer than formal education has traditionally been prepared to address, and 2) the current teaching workforce has grown up in an environment where technology, including access to the Internet and digital games, has been ubiquitous, informal and embedded in their lives (Beck & Wade, 2004). These implications help make a case for the potential value of a scalable, informal online support environment that is available at all times to augment human performance in education. How are the current formal and informal teacher education systems set up to accept, respond to and deliver this capability? To examine the situation, we'll briefly review gaps in formal program quality, policies and online learning.

Quality of Programs and Experiences (and thus the Teachers Produced) Varies Widely

When Title 2 (Sections 205 through 208) of the Higher Education Opportunity Act was passed in 2008, reauthorizing the Higher Education Act of 1965, it required institutional and state report cards on teacher preparation quality as well as reforms that included "implementing teacher preparation program curriculum changes that improve, evaluate, and assess how well all prospective and new teachers develop teaching skills." (Higher Education Opportunity Act, 2008, Section 202). Among the data reported each year are the pass rates of programs and the number of teachers working without certification in each state. The pass rates generally fall between 80% to 100% of students who achieve certification, while the states report that in certain areas such as special education, up to 20% of teachers are working on waivers, meaning that they are not certified for the area they are teaching.

This data indicates that in spite of the self-reported high pass rates, the actual quality and effectiveness of programs for ensuring the quality of teaching to meet existing needs varies widely, putting students at risk in various ways depending on where they live. Formal teacher preparation programs have come under attacks for many years and are seen as unprotected by the higher education establishment because, according to some analysts, they are "basement" offerings that were fit historically for lower status women, and are seen as cash cows at most institutions (Maher, 2002). Until fundamental changes in the status of teaching take hold, for example, with changes in incentives, career opportunities, school resource availability in low-income locations, and the cessation of waiver-based policies, the wide variance in teacher quality is likely to continue in the U.S. As these problems have been evident for much of the history of U.S. education, this argues for the need for dramatic shifts in the way educators are identified, recruited into the profession, made ready to teach, and supported in the field of service.

Policies and Practices in the U.S. Compare Poorly with Other Countries, as do Student Results

The U.S. is not the global leader in student results, teacher support, and technology vision for education. Poor student results are well-known; news and national reports since the early 1980's have lamented the situation, warned and chided policymakers, and painted a picture of dire circumstances (Broad, 2005; COSEPUP, 2006; NCEE, 1983; TIMSS, 2003). International comparisons of teacher support are just as dire; in some other countries teachers enter teaching being paid as much as beginning doctors, that are supported by mentor teachers and have fifteen or more hours a week to work and learn together (Darling-Hammond, 2011). A global comparison of technology leadership fares no better; South Korea announced in early July 2011 that it plans to spend over $2 billion developing digital textbooks, replacing paper in all of its schools by 2015 (Honig, 2011). These examples highlight the gaps, lack of vision and the impact of both the de-professionalized status of teaching in America and the lack of investment and focus on the role of advanced technologies such as games and simulations in informal and formal education, including teacher education, which is out of step with the strategies pursued by the world's educational leaders, a point made clear in 2011 at the first International Summit on Teaching (Stewart, 2011). These gaps suggest some policy areas in need of attention.

One needy policy area concerns how and where student teachers practice what they have learned. Teacher educators have trouble finding enough appropriate placements with mentor teachers who can provide effective critical feedback about a student teacher's instructional skills and effectiveness (Grossman, 2010). In our work with teacher education programs in nearly 100 institutions, we've learned that in certain parts of the country, student teachers do not have field placements within driving distance of their college that provide experiences with diversity of race, socioeconomic differences, and special education conditions. In addition, student teachers cannot "experiment" with students during these field experiences; mentor teachers sometimes only allow a small amount of control, or small amounts of time, for the student teacher to be in total control (Gibson & Kruse, 2011). Many student teachers thus only get a handful of hours in highly constrained conditions, which may be contributing to the high rate of turnover and low productivity of teacher preparation programs (Carroll, 2007).

Another policy area concerns the relationship of self-direction to motivation and engagement, that is present in typical professional development systems and is mirrored in K-12 systems. The literature on self-directed professional development has noted the value and necessity of choice for building personal relevance (Mezirow, 1997; NSDC, 2001). In order to develop the habits of self-direction and to carry those habits forward from professional development into K-12 settings, student teachers need experiences with setting professional development goals and self-monitoring progress toward those goals, as well as in developing modules and lesson plans that utilize similar approaches with their own future students (Zimmerman, 2000). The culture of games and simulations has a natural fit with this sort of empowered decision-making (Prensky, 2001) and thus has great potential to help programs fill gaps in this policy area.

Online Learning: Has it Made a Difference so Far?

Higher education online courses have soared in recent years. Approximately 5.6 million students were enrolled in at least one higher education online course in fall 2009 (Allen & Seaman, 2010), up from 2.35 million reported in 2004 (Allen & Seaman, 2005). In K-12, the move to online teaching, which started small with roughly 45,000 students taking an online course in 2000 (Staker, 2011) grew to an estimated 1,030,000 in 2008 (Picciano & Seaman, 2009) and is now over 4 million.

Has this online rush improved learning? Some critics wonder whether online educators are doing enough to use the media's possibilities; they point out that e-learning often consists of taped lectures being posted online (Lueg & Siebert, 2010). Bored students drop out of online classes pleading for richer and more engaging online learning experiences (Bonk, 2002). One touted remedy is "blended" learning, a hybrid of face-to-face and online learning, which ranges from offline to online "content delivery" crossed with supervised to remote "geographic location" (Staker, 2011). An all-offline-teacher-supervised learning environment is the traditional brick-and-mortar education system; while an all-online-remote-learning environment is as far as one can get on the other end of the spectrum. Neither of these endpoints is "blended," but note that the point of blending in the Staker panoply concerns "delivery and place" not "learning, media and mediation."

What is clearly needed in online and blended learning teacher education programs are solutions that improve the impacts of teaching, not simply change when, where and how it takes place. Not only do teachers need to be well versed in using online learning tools and resources in teaching because many of their students will expect it (USDOE, 2010), but also because that online digitized world offers powerful new ways for teachers to practice theories of instruction and build theory-to-practice experience (Gibson, 2006). Teachers thus need to be experienced and skilled in taking advantage of technology affordances for honing their skills as well as promoting learning.

The Paradigm Needs Shifting

Online learning environments (in higher education as well as K-12) have been said to be facing a "perfect e-storm," linking pedagogy, technology, and learner needs (Bonk, 2004). The national technology plan of 2010 puts a fine point on the problem and offers concrete insights for the needed paradigm shift, which we will outline below. The next section discusses the characteristics of the changed viewpoint, which is consistent with the need for a new online method of teacher education that is innovatively online, scalable, personalized, data-rich, well-suited to the gamer teacher generation, and adaptable for many different levels of expertise that are acquired over a lifetime of learning to teach.

Formalizing Informal Online Experiences and the Future of Learning

Why Students Will Attend a School of their Own Devising

Imagine that you'd like to brush up on Algebra, or go over the causes of the Vietnam War. Do you first go to a bookstore or library and find a book, or look up and contact a teacher to talk about a class you might like to take? Do you scour a local college's catalog of courses? Or, are you one of the 300 million people who use Google daily to see what might be available for free online? If you do the latter, you will probably come across the Khan Academy (www.khanacademy.org), which is a nonprofit organization on a mission to help you learn whatever you want, whenever you want, at your own pace. The site has over 2400 lesson videos integrated with a content map and assessments, and it's free. As you interact with the materials, you can choose to take tests, get scores, and track your progress. If you are a teacher of one of the many subjects offered, while your students work on their own time, you can use the site to monitor their individual progress over time, freeing your time to teach different material (hopefully in new ways). The site has several characteristics that drive people to learn online now and into the future, characteristics that are part of the reason students will devise their own "schools."

Characteristics of Learner-Driven Online Learning

How many of these characteristics appeal to you? Exploring, dabbling and learning in spare moments. The choice and pace is yours. Your interests and control determine what is learned. The materials, resources and feedback methods are responsive to you. The learning is without pressure, yet is guided by an exacting external standard of performance in a highly specific area of knowledge. These features concern performance, adaptability, and cost-effectiveness, but are more importantly about personalization. What other method of going to school, watching an expert, practicing on your own, and testing your learning has these characteristics? Compare them with typical formal education: not free, not your choice (except to drop out if you don't like what is offered), not your pace, and so forth. The differences could hardly be starker.

However, missing essential elements in this example include experiencing social relationships, having a live mentor, and gaining formal external validity needed for certification. If these are not provided, then there is no course meeting time, nor a cohort of other people studying "with you" on the same content, following the same curriculum path. Finally, the record of achievement in your personal file does not guarantee an external reviewer that it was really you doing that work. Some would argue that these limitations are not serious enough to warrant ignoring this form of learning. Since there is no pressure to attend, watch, practice and learn, what harm can it do to partake of it as little or as much as you'd like? After all, missing elements can be added in a new form of blended learning that counts individual and social learning experiences more than teacher supervision and geography.

How the U.S. National Technology Plan of 2010 Views Teaching, Learning, and Assessment

The vision of teaching in the U.S. national technology plan (USDOE, 2010) is "connected" in three ways; teams rather than solo teachers guide learners, data on student performance and analysis tools are always available, and there are online resources always at hand to help act on insights. Learning is "powered by technology" to be engaging, relevant and empowering. Finally, assessment is viewed as diagnostic and designed for multi-stakeholder collaboration for continuous improvement. This vision of teaching, learning and assessment is congruent with informal education and a simulation-based vision of teacher education, but at the same time raises several challenges for formal education.

The challenges to formal education arise as a result of several conditions in the U.S system. For example, teaching is primarily a solo job in which data on student performance from standardized tests arrives infrequently, long after the fact, and is usually disconnected from the local curriculum. As a result, the data often has little formative value to students and teachers (Wiggins, 1999). Digital learning innovators who want to address this isolation and lack of timely performance information need to create formally valid yet formative and responsive performance opportunities with embedded analytics and reporting that allows and encourages team-based learning among teachers, demonstration, guidance and feedback.

A second example comes from the lack of daily learner choice that is common in much of formal education, and the problem of learner engagement that is highly dependent upon a model of the teacher-as-entertainer (Garoian, 1999). The challenge in this case is to provide learner choice based on current interests, motivations and aspirations as the heart of daily activities for every student – a job that no one can do alone, without technology, and without integrating the classroom and the real world (USDOE, 2010).

A third example is the problem of curriculum relevance. Formal learning has the potential to be relevant to the learner, but is often more relevant to the ideals (both stated and hidden) of curriculum experts (Giroux 1994). The challenge of relevance is to connect the strengths and aspirations of learners to the next-best step within their reach in the world they understand.

Finally, formal learning has the potential to be empowering; but that potential is realized primarily by learners who do things well aligned with what experts including teachers say are needed for success (Lindsay & Breen, 2002). The challenge of empowerment is to widen the limits of authority and validation while maintaining a thread of connectivity from the informal to the formal system – from the self to peers to the larger world community.

Addressing these challenges with technology-empowered solutions will build bridges between formal and informal learning that will change the paradigm of education (Gibson, 2010). A key example of the shift can be found in simulation based learning, where the technology of a computational model and the web-based freedoms of informal education can be combined with the needs of formal education to provide concrete solutions to the challenges just outlined. However, more examples are needed in all subject areas and these need to be included in teacher education experiences, because future teachers need to experience it to believe it (Hancock, Knezek, & Christensen, 2007; Knezek & Christensen, 2009).

Educating a Teacher who Says "I Teach the Way that I Wish I Was Taught"

To meet the kinds of challenges just outlined – addressing the problems facing teacher education and building the needed bridges from informal to formal education – we need teachers who can teach using methods other than the ones they've experienced in K-12 schools and colleges of education. The premise of teacher education with simSchool is that a flight simulator for teachers can provide a new way of learning how to teach. To make the case, the next section describes what sets simulation-based learning apart, how a simulation works, and in particular how simSchool works and what evidence exists for it as a model of teaching and learning.

Can Someone Learn to Teach by Practicing with a Classroom Simulation?

What Makes Game and Simulation-Based Learning More Powerful than Reading Online Texts and Writing Papers

When the concept of creating a simulation to train teachers was first introduced, an early challenge was convincing people that the complexities of teaching and learning could be adequately represented. We briefly review that argument again, to assure the reader that these considerations have been addressed. In addition, since the creation of simSchool in 2003, we've discovered that powerful learning in a simulator is unlike learning in other modalities, and these differences are noted.

A simulation of a classroom with enough complexity to educate teachers stretches the imagination, so it is important to begin with an illustration of some of the affordances and limitations of all simulation models. Imagine that an elementary teacher is teaching a lesson in earth science. As a constructivist and inquiry-oriented teacher, she gives the students a bare lamp bulb on a stand as a light source, a basketball and a softball, and challenges the students to turn down the lights and experiment with these objects until they find a way to show the phases of the moon.

After several minutes of playing with these objects, the students discover several facts about the sun-earth-moon system in spite of the fact that the light source is too small to represent the true size of the sun and the room is too small to represent the distances involved among the three bodies. Even though the relative positions, sizes and shapes of the balls are not an accurate representation of the earth or moon, the students discover important truths about some of the dynamics, geomechanics and perspectives from the earth that lead to the observable phases of the moon.

This vignette illustrates that the affordances and limitations of a model such as *simSchool* offer benefits for learning. One benefit is *shearing away details in a simplification of a real system*. Models allow us to hold, in our hands and minds, some aspects of a system that cannot otherwise be experienced. Connected to and entailed by the characteristic of simplification is *increased safety* (e.g. a pilot in training can crash a virtual plane and a beginning teacher can crash a student or a class), *decreased costs* (e.g. virtual materials are more easily built and shared), and *enhanced focus on the relationships* among the simplified features (e.g. making a theory operational and amenable to manipulation). Simulations also provide *multiple chances to practice*, including making attempts with higher risks and causing spectacular failures, and to learn, retry and master new skills more rapidly and with less effort than through experiences that are not mediated by computers (Holland, 1995; Wolfram, 2002). This is part of the reason that simulations are used in aviation, medicine and the military with increasing frequency and effectiveness (Prensky, 2001). In teacher preparation, simulations that provide targeted feedback can develop teachers' understanding and practice (Grossman, 2010) and may be as effective as field experience (Christensen, Knezek, Tyler-Wood, & Gibson, 2011).

Limitations of models include the fact that *they are not full substitutes for real experience*, and as simplifications, there is *a danger that something vital may have been left out*. However, the progress of science attests to the fact that despite these limitations, models are vital parts of the advancement of knowledge. Happily, recent policy recommendations for teacher professional development now mention simulations among the promising new tools (Carroll, 2000, 2009; Dede, 2009; Grossman, 2010). This recognition leads to an important principle in the underlying theory of learning in simSchool; practice in a variety of settings builds expertise – and virtual settings may be as good as real ones in certain circumstances, due to the characteristics and benefits of models.

How a Simulation of a Classroom Can Entail Enough Realism to be a Worthwhile Learning Environment for Teachers

What teachers do in the classroom matters a great deal and is part of a causal network that brings about student learning as evidenced in their skill- and knowledge-based performances (Darling-Hammond, 1997; Darling-Hammond & Youngs, 2002; Rice, 2003). Teacher decisions can be thought of as independent variables in an ongoing experiment in their own classrooms that builds expertise over time (G. Girod, Girod, & Denton, 2006; M. Girod & Girod, 2006). A simulated classroom like simSchool provides cycles of experimentation and practice with few of the dangers associated with mistakes made on real students in real classrooms.

The "classroom experiment" metaphor is represented with four kinds of variables in *simSchool*: Observable, Hidden, Independent and Dependent variables (Figure 1).

1. The **observable variables** are what the teacher can see, which includes a typical student record passed down from teacher to teacher and kept on file in the school data system (e.g. grades, comments, psychological profile), behaviors in class including

Figure 1. An experimental framework for developing teaching expertise

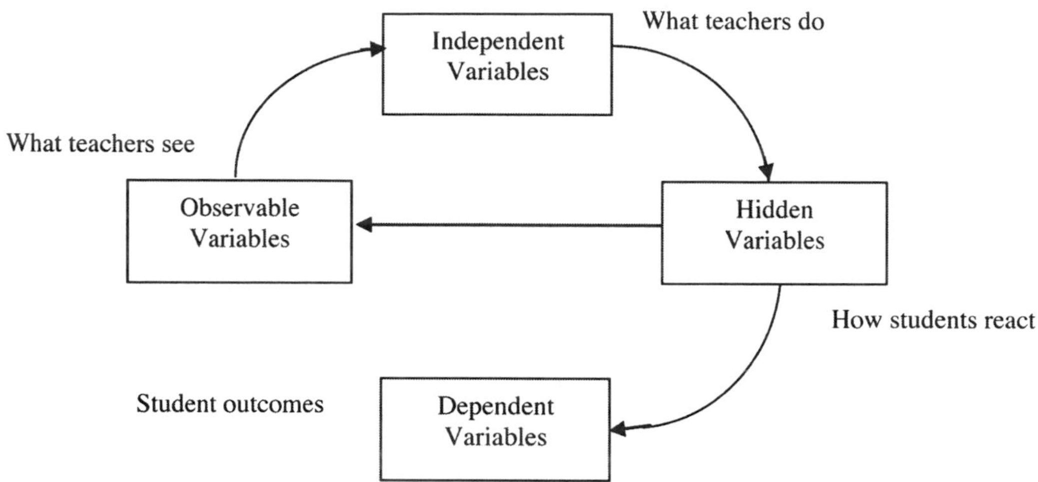

students' talking and body positions. In the simulator, unlike in real classrooms, the teacher can also see dynamic trailing indicator clusters that show the immediate result of the interaction of a current task or a teacher's talking behavior on the student's learning, happiness and self-efficacy/sense of power.

2. The **hidden variables** include detailed factor-level views of the individual variables of the student profile that are changing every instant in response to tasks and teacher talk. These variables, are also hidden in real classrooms, but are revealed in *simSchool* during the after-action reflection to help analyze why the student learned and behaved in a particular way.
3. The **independent variables** are the teacher's selection and timing of tasks and decisions about whether to talk and how to say things. This is the area where a teacher's knowledge and practice repertoire make a difference in student learning.
4. The **dependent variables** are the trailing indicators, which are revealed at the end of a session to allow a detailed reflection and analysis of the moment-by-moment interactions and effects caused by the teacher's decisions.

How simSchool Works and what Research Has Shown are its Main Effects

People who practice with a simulator develop *heuristic knowledge* of the underlying theories because the immersive multimedia experience taps into physical, emotional as well as cognitive pathways, heightening the sense of importance of the experience (Dede, Gibson, & Damon, 2007; Gee, 2004; Shaffer, 2007; Squire, 2006). Heuristic knowledge is readily accessible and probabilistic in nature; it develops from experience and while not guaranteeing a result, does make a desired result more likely. A heuristic approach to solving a problem is one that is followed when there are too many choices to exhaustively search for the best solution; examples include educated guesses, rules of thumb, and intuitive judgments.

Independent research teams have found initial evidence indicating the development of heuristic knowledge of teaching. In one study for example, preservice teachers could not articulate what they had learned, but outscored non-players on teaching skills that had been developed through exposure to simSchool (Christensen, Knezek, Tyler-Wood, & Gibson, 2011). To help illustrate how heuristic knowledge can develop absent the metacognitive awareness of a simulation user, we next provide an example of a simulated student, whose profile is based on a real student. The example demonstrates how the simSchool experimental framework variables dynamically interact and work to influence the development of a heuristic skill of teaching.

THOMAS

Thomas is a simulated seventh grader who has a severe, post-lingual hearing loss. His hearing loss impacts his oral and written language skills and makes it difficult for him to work independently and integrate socially. A user playing the role of Thomas' teacher in simSchool has two choices to make while teaching him: which tasks should Thomas do now, and what should the teacher say to him? Figure 2 is a graphic representation of a few moments of a simulation with Thomas showing one of the variables accessible in simSchool.

Observable Variables

Thomas' profile describes him as functioning below grade level academically, with low self-esteem and difficulty respecting boundaries and classroom expectations. These characteristics imply that certain kinds of tasks and certain verbal approaches are likely to work better than others. How the user learns these is a combination of reading Thomas' school file and working with him in class over time to see the impacts of task choice and talking attempts.

Hidden Variables

Thomas' hidden characteristics relate to his cognitive ability, emotional stability, other personality dimensions and his slower learning pace. These affect his interactions at every instant in response to tasks and teacher talk. As the user attempts to

Figure 2. Task expectations

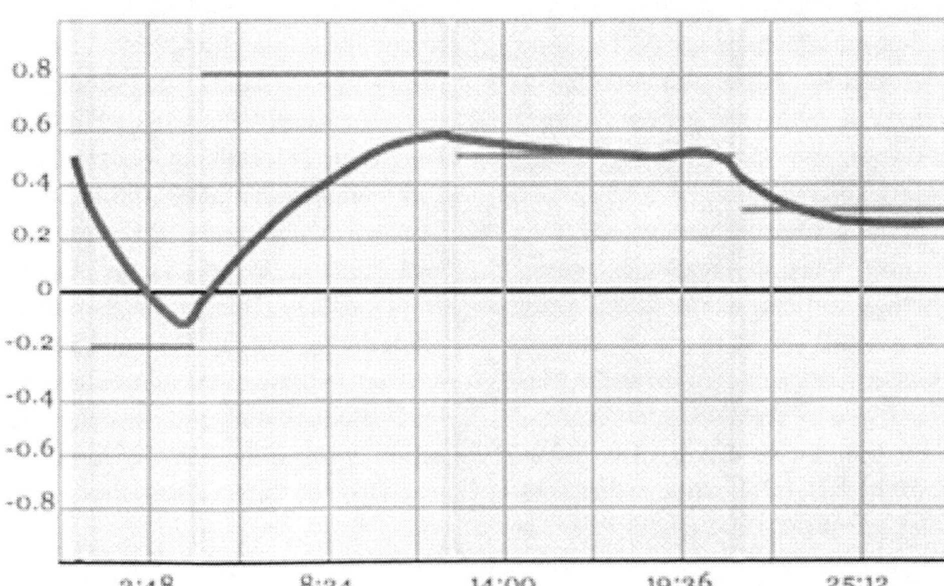

teach Thomas, these hidden variables determine how he will behave in class; his body position, his attitude in talking back, and his academic performance.

Independent Variables

A user/teacher selects a task such as "do silent reading" that interacts with Thomas' hidden variables. In figure 2, about thirty seconds after the task is assigned, the user talks to Thomas, which also impacts his hidden variables.

Dependent Variables

The impact of a teacher's decisions, such as to assign "silent reading" is presented as a simple visual graph for reflection and discussion.

Looking at this example (Figure 2), which represents about four minutes of real time and simulates about 26 minutes of class time, provides a way to discuss the learning opportunity for developing habits of teaching. The constant lines during task time represent the tasks' academic expectation level. Note that while "nothing" is the task, there is no change in the learner and no chance of acquiring any academic content. But when a task is given with a certain academic requirement, then the simulated learner begins to respond. Tasks with low expectations elicit low performance and tasks above the students' current performance level require time to struggle with the task in order to make gains.

What are the possible heuristic lessons from the choice of a task? The task's academic requirement was too high for this learner at this point in time and as a result, the learner's exhibited a slight deterioration in academic performance. This might be rationalized by the user in terms of the difficulty of the task, or the timing of introducing this task, or the lack of effective preparatory tasks or talking, all of these are possibilities for interpretation in part or in some combination, since they are all equally true.

Later during this one minute of simulation, the user talks to the student and that interaction improves the student's academic prospects, but the distance to the task requirement is still too far away, so again, slight academic deterioration occurs. Note, however, that if the simulation stopped immediately after the teacher talk, it would appear that the student learned from the talking behavior. What are the heuristic possibilities here? That teacher talk imparted critical but insufficient information, and that given more time, the teacher may have succeeded in raising the student's readiness to achieve this task, and perhaps other strong possibilities. While this talking interaction was going on, what were other students in the class doing? The point being made here is that teaching is complex, layered and multicausal and that actual user performance and detailed documentation in a simulation provides a window into these subtleties.

While there are no easy completely "right" answers, there are hosts of partially right ones, and this facet of simulation is evidence for the development of heuristic knowledge. What has to be learned, instead of verbalized pat answers, is a set of estimations, educated guesses, and rules of thumb for similar situations. A user can actually re-run this simulation many times to try out alternatives and build up a history of interaction-based knowledge. Does talking sooner help more? Would a different task better set Thomas up for success? Of the available tasks, which one would be the best "first" task for Thomas? Of the available ways to talk to Thomas, which ones have the best effect during this particular task? There are sets of answers to these questions that are clearly better than other answers, and since the sets overlap in complex ways, the knowledge that builds up as one develops them is heuristic.

How Teachers Can Be Part of a Continuous Design Team Helping to Build the Future of Research on Teaching and Learning

The infrastructure of simSchool includes a conception of teaching as a socially constructed field of knowledge. So, a web site hosts a social hub of users, some of whom are learning about the theories, others using the simulation in classes and others who are authors of new challenges. Still others are undertaking a variety of independent research studies and writing reflective articles on the challenges, opportunities and limitations of learning to teach with a simulator (see www.simschool.org/simPartners). Since one of the most important issues in educational research is assessing what people learn, the next section outlines the main issues in digital media assessment related to a simulation like simSchool.

Knowing What Online Students Know

Affordances of Digital Media for Assessment of Learning and Performance

A currently emerging strand of inquiry in digital media and learning concerns the data-mining of diverse sources of information about the learner. There are two primary types of development of these analytics; data gathered across many learning experiences (e.g. courses, programs, over time) to predict program performance such as graduation, certification, and dropping out, and data gathered within a single learning experience (e.g. during a game or simulation) to predict, respond to and adapt to the immediate needs o the learner. The first type is called *learning analytics* and the second is referred to as *learner* or *academic analytics.*

A number of writers concur that the development of *learning analytics*

"...responds to calls for accountability on campuses across the country, and leverages the vast amount of data produced by students in day-to-day academic activities. While learning analytics has already been used in admissions and fund-raising efforts on several campuses, "academic analytics" is just beginning to take shape" (Johnson, Smith, Willis, Levine, & Haywood, 2011).

In contrast with this type, learner-centered *academic analytics* is gathered from unobtrusive or stealth assessment of the impact of a particular digital media learning experience on a learner. An effective academic analytics analysis comments on one or more of the digital media affordances for learning (Gibson, 2006):

1. *Complex performances supported and documented* in network-based assessments via multimedia, multileveled, and multiply connected bases of knowledge.
2. The many instances of the learner interacting with applications in different times, places and contexts, leading to network-based assessments that build a long-term record of documentation, *showing how learners change over time.*
3. *Analysis of expert-novice differences* facilitated across groups, across space and time, drawing from an evolving common knowledge store.
4. The interactive potential of network-based assessments for *fostering and determining the metacognitive skills* of the learner.
5. Emerging capabilities in metadata generation for *identifying the problem-solving strategies* of learners.
6. Network-based assessments that include *statistical analysis and displays* of information to assist learners and teachers in making inferences about performance.
7. Unobtrusive observation techniques combined with libraries of evidence and tasks which make possible *timely feedback to*

learners and teachers and matching of current needs with best "next step" materials, tasks and challenges, including tasks that involve transfer of learning to new contexts.

Assessing Informal Online Learning and Performance Using Digital Media Assessment Theories and Tools

Assessing teaching with simSchool is founded on the above listed digital media affordances for learning shaped by three approaches to digital media assessment: evidence-centered design, portfolios, and complex systems modeling.

First, the Evidence-Centered Design (ECD) framework (Almond, Steinberg, & Mislevy, 2002; Mislevy, Steinberg, & Almond, 1999, 2003) stems from a program of research and application development first carried out at Educational Testing Service. ECD is a principled framework for designing, producing, and delivering educational assessments. Special attention is given to assessments that incorporate interactive and adaptive features such as those found in digital media learning applications. The design process includes domain analysis and modeling, two stages that are engaged to meet the product requirements of a specific assessment (e.g. an assessment of problem-solving skill in chemistry), which are then made operational in a three-part structure: a student model, task model and evidence model. The objects and specifications in that structure provide a blueprint for 1. Creating tasks and the computational models used to infer from performance of the tasks, 2. Implementing the assessment (operating it and collecting data from it) and 3. Analyzing data and reporting results.

Second, Gibson (2004) proposed a three-dimensional model of the decision-making space that includes "Audiences," "Reflective Purposes" and "Artifacts"- areas that intersect to form key questions in the creation of an assessment. The three-dimensional model follows closely the Activity Theory of Leontev (1896 – 1934), which depicts all purposeful human activity as interactions among subjects (audiences), artifacts (tools), and objects (purposes). The model provides a source for principled inquiry into the nature of the interaction of an assessment's creators and users, the artifacts that stand as evidence of what people know and can do, and the inferences one might make about learners. The answers to the key questions in the model also contain a number of irresolvable dilemmas that designers need to live with and maximize or minimize, as their needs demand.

The intersection of *audience* (self, trusted others, the public) with *purposes* (mirror, map, sonnet) and *artifacts* (configuration complexity, intentional content, affordances and constraints of the media) suggests a multidimensional analysis space for considering design and implementation issues (Gibson, 2004, 2006). More research is needed that applies this framework to digital media assessment planning processes and specific embedded implementations. Negotiating among the questions inherent in the three-part structure, an assessment designer or design team must settle on sets of decisions that change for different audiences, purposes and artifact types. These negotiated configurations form an institutional design fingerprint that shapes the assessments in the system. The challenge of multidimensional assessments extends to teachers designing assessments for their own classrooms, and to professors designing courses about assessment in teacher preparation programs.

Third and finally, since the problem of assessing performances in digital media learning environments involves several types of knowledge and action, often dynamically integrated to accomplish a nontrivial task, it is a complex matter. Complexity science offers new tools of theory and analysis for digital media assessments, as it encompasses several interdisciplinary theoretical frameworks in the search for patterns and causes in adaptive, changeable systems (Bar-Yam, 1997; Bechtel, 1993; Holland, 1995). Some of complexity science's analysis frameworks are coalescing around network theory, which has been found useful in a wide variety of fields from biology, e.g. (Kauffman, 2000) space and physical sciences, e.g. (Prigogine, 1980), to social and economic sciences, e.g. (Beinhocker, 2006), to neurosciences, e.g. (Sporns, 2011). Complexity and network approaches have also been noted for their potential in evaluation and educational assessment (Dexter, 2003; Gibson, 2003; Patton, 2011). The new methods include network analysis, artificial neural nets, Bayesian nets, coherence nets, ontologies and semantic nets, data mining techniques and methods, and knowledge engineering (Gibson & Jakl, in preparation). In simSchool, for example, the crowd-sourced performance of various groups is data mined to form network images of the knowledge-in-action or performance representation of the groups. These can then be compared with an individual performance to make adaptations in the digital experience as well as make inferences about what the user knows and can do.

SUMMARY

When using a simulation platform such as simSchool to prepare teachers, the conceptions of teaching and learning, the organization of knowledge, assessment practices and results, and the engagement communities of practice stand in dramatic contrast with traditional course-based online learning experiences. The new paradigm focuses on the development of heuristic knowledge through direct experience, guided by self-direction and personal validation in a complex but repeatable environment. New methods of automated data gathering and analysis help address some of the important unmet challenges of teacher education by supporting emergent interdisciplinary knowledge and leveraging the unique affordances of digital and social media.

REFERENCES

Allen, I., & Seaman, J. (2005). *Growing by degrees: Online education in the United States, 2005*. Babson Park, MA: Sloan Consortium: Babson Survey Research Group.

Allen, I., & Seaman, J. (2010). *Class differences: Online education in the United States, 2010*. Babson Park, MA: Sloan Consortium: Babson Survey Research Group.

Almond, R., Steinberg, L., & Mislevy, R. (2002). Enhancing the design and delivery of assessment systems: A four process architecture. *The Journal of Technology, Learning, and Assessment*, *1*(5), 3–63.

Bar-Yam, Y. (1997). *Dynamics of complex systems*. Reading, MA: Addison-Wesley.

Bechtel, W. (1993). *Discovering complexity: Decomposition and localization as strategies in scientific research*. Princeton, NJ: Princeton University Press.

Beck, J., & Wade, M. (2004). *Got game: How the gamer generation is reshaping business forever*. Boston, MA: Harvard Business School Press.

Beinhocker, E. (2006). *The origin of wealth: Evolution, complexity and the radical remaking of economics*. Boston, MA: Harvard Business School Press.

Bonk, C. (2002). Online teaching in an online world (executive summary). *USDLA Journal*, *16*(1). Retrieved from http://www.usdla.org/html/journal/JAN02_Issue/article02.html

Bonk, C. (2004). *The perfect e-storm: Emerging technologies, enhanced pedagogy, enormous learner demand, and erased budgets*. London, UK: The Observatory on Borderless Higher Education.

Broad, W. (2005, October 13). Top advisory panel warns of an erosion of the U.S. competitive edge in science. *New York Times*.

Carroll, T. (2000, March). *If we didn't have the schools we have today, would we create the schools we have today?* Paper presented at the Society for Information Technology and Teacher Education International Conference, San Diego, CA.

Carroll, T. (2007). *The high cost of teacher turnover*. National Commission on Teaching and America's Future.

Carroll, T. (2009, September 12). Press release: Transforming schools into 21st century learning environments. *eSchool News*.

Carroll, T., & Foster, E. (2010). *Who will teach? Experience matters*. Washington, DC: National Commission on Teaching and America's Future.

Christensen, R., Knezek, G., Tyler-Wood, T., & Gibson, D. (2011). SimSchool: An online dynamic simulator for enhancing teacher preparation. *International Journal of Learning technology, 6*(2), 201-220.

COSEPUP. (2006). *Rising above the gathering storm: Energizing and employing America for a brighter economic future*. The National Academy of Sciences, The National Academy of Engineering, The Institute of Medicine.

Darling-Hammond, L. (1997). Quality teaching: The critical key to learning. *Principal, 77*, 5–6.

Darling-Hammond, L., & Youngs, P. (2002). Defining "highly qualified teachers": What does "scientifically-based research" actually tell us? *Educational Researcher, 31*(9), 13–25. doi:10.3102/0013189X031009013

Dede, C. (2009). *Address to the ITEST Summit 2009. Unpublished Lecture - slide show*. NSF.

Dede, C., Gibson, D., & Damon, T. (2007). Learning games and simulations (video podcast). *National Educational Computing Conference*. Atlanta, GA: International Society for Technology in Education.

Dexter, S. (2003, March). *The promise of network-based assessment for supporting the development of teachers' technology integration and implementation skills*. Paper presented at the Society for Information Technology and Teacher Education International Conference, New Orleans, LA.

Garoian, C. (1999). *Performing pedagogy: Toward an art of politics*. Albany, NY: SUNY Press.

Gee, J. (2004). *What video games have to teach us about learning and literacy*. New York, NY: Palgrave Macmillan.

Gibson, D. (2003). Network-based assessment in education. *Contemporary Issues in Technology & Teacher Education, 3*(3), 310–333.

Gibson, D. (2004). *E-portfolio decisions and dilemmas*. Paper presented at the Society for Information Technology in Teacher Education International Conference, Atlanta, GA.

Gibson, D. (2006). Elements of network-based assessment. In Jonson, D., & Knogrith, K. (Eds.), *Teaching teachers to use technology* (pp. 131–150). New York, NY: Haworth Press.

Gibson, D. (2010). Bridging informal and formal learning: Experiences with participatory media. In Baek, Y. K. (Ed.), *Gaming for classroom-based learning: Digital role playing as a motivator of study* (pp. 84–99). Hershey, PA: IGI Global. doi:10.4018/978-1-61520-713-8.ch005

Gibson, D., & Jakl, P. (in preparation). Measuring teaching skills with a simulation: simSchool with Leverage. *Journal of Educational Data Mining*.

Gibson, D., & Kruse, S. (2011). *Personal communications with LEVEL 3 partnership colleges. simSchool Modules EDUCAUSE Project*. VT: Stowe.

Girod, G., Girod, M., & Denton, J. (2006). Lessons learned modeling "connecting teaching and learning". In Gibson, D., Aldrich, C., & Prensky, M. (Eds.), *Games and simulations in online learning: Research & development frameworks*. Hershey, PA: Idea Group. doi:10.4018/978-1-59904-304-3.ch010

Girod, M., & Girod, J. (2006). *Simulation and the need for quality practice in teacher preparation*. Paper presented at the American Association of Colleges for Teacher Education. Retrieved from http://www.allacademic.com/meta/p36279_index.html

Giroux, H. (1994). Doing cultural studies: youth and the challenge of pedagogy. *Harvard Educational Review*, *64*(3), 278–308.

Grossman, P. (2010). *Learning to practice: The design of clinical experience in teacher preparation*. Washington, DC: Partnership for Teacher Quality (AACTE and NEA).

Hancock, R., Knezek, G., & Christensen, R. (2007). Cross-validating measures of technology integration: A first step toward examining potential relationships between technology integration and student achievement. *Journal of Computing in Teacher Education*, *24*(1), 15–21.

Higher Education Opportunity Act, H.R. 4137 1125 (2008).

Holland, J. (1995). *Hidden order: How adaptation builds complexity*. Cambridge, MA: Perseus Books.

Honig, Z. (2011, July 3). South Korea plans to convert all textbooks to digital, swap backpacks for tablets by 2015. *Engadget*. Retrieved from http://www.engadget.com/2011/07/03/south-korea-plans-to-convert-all-textbooks-to-digital-swap-back/

Johnson, L., Smith, R., Willis, H., Levine, A., & Haywood, K. (2011). *The 2011 horizon report*. Austin, TX: New Media Consortium.

Kauffman, S. (2000). *Investigations*. New York, NY: Oxford University Press.

Knezek, G., & Christensen, C. (2009). Preservice educator learning in a simulated teaching environment. In Maddux, C. (Ed.), *Research highlights in technology and teacher education* (*Vol. 1*, pp. 161–170). Chesapeake, VA: Society for Information Technology and Teacher Education.

Lindsay, R., & Breen, R. (2002). Different disciplines require different motivations for student success. *Research in Higher Education*, *43*(6), 693–725. doi:10.1023/A:1020940615784

Lueg, A., & Siebert, D. (2010, March 16). *E-learning's potential is hampered by misuse, critics say*. Deutsche Welle DW-WORLD.DE. Retrieved from http://www.dw-world.de/dw/article/0,5359341,00.html

Maher, F. (2002). The attack on teacher education and teachers. *Radical Teacher*, *64*(5), 5–8.

Mezirow, J. (1997). Transformative learning: Theory to practice. *New Directions for Adult and Continuing Education*, *74*, 5–12. doi:10.1002/ace.7401

Mislevy, R., Steinberg, L., & Almond, R. (1999). *Evidence-centered assessment design*. Educational Testing Service.

Mislevy, R., Steinberg, L., & Almond, R. (2003). On the structure of educational assessments. *Interdisciplinary Research and Perspectives*, *1*, 3–67. doi:10.1207/S15366359MEA0101_02

NCEE. (1983). *A nation at risk*. Washington, DC: National Commission on Excellence in Education.

NSDC. (2001). *National standards for online learning*. National Staff Development Council.

Patton, M. (2011). *Developmental evaluation: Applying complexity concepts to enhance innovation and use*. New York, NY: The Guilford Press.

Picciano, A., & Seaman, J. (2009). *K-12 online learning: A 2008 follow-up of the survey of U.S. school district administrators*. Babson Park, MA: The Sloan Consortium: Babson Survey Research Group.

Prensky, M. (2001). *Digital game-based learning*. New York, NY: McGraw-Hill.

Prigogine, I. (1980). *From being to becoming: time and complexity in the physical sciences*. San Francisco, CA: W.H. Freeman.

Rice, J. (2003). *Teacher quality: Understanding the effectiveness of teacher attributes*. Washington, DC: Economic Policy Institute.

Shaffer, D. (2007). Epistemic games. *Innovate, 1*. Retrieved from http://innovateonline.info/pdf/vol1_issue6/Epistemic_Games.pdf

Sporns, O. (2011). *Networks of the brain*. Cambridge, MA: MIT Press.

Squire, K. (2006). From content to context: Videogames as designed experience. *Educational Researcher*, *35*(8), 19–29. doi:10.3102/0013189X035008019

Staker, H. (2011). *The rise of K-12 blended learning: Profiles of emerging models*. Mountain View, CA: Innosight Institute.

Stewart, V. (2011). *Improving teacher quality around the world: The international summit on the teaching profession*. New York, NY: Asia Society and U.S. Department of Education.

Strauss, V. (2011, March 23). Darling-Hammond, L.: U.S. vs highest-achieving nations in education. *The Answer Sheet*. Retrieved from http://www.washingtonpost.com/blogs/answer-sheet/post/darling-hammond-us-vs-highest-achieving-nations-in-education/2011/03/22/ABkNeaCB_blog.html

TIMSS. (2003). *International report on achievement in the mathematics cognitive domains*. Retrieved from http://timss.bc.edu/timss2003i/mcgdm.html

USDOE. (2010). *Learning powered by technology: Transforming American education*. Washington, DC: U.S. Department of Education.

Wiggins, G. (1999). *Educative assessment: Designing assessments to inform and improve student performance*. San Francisco, CA: Jossey Bass Publishers.

Wolfram, S. (2002). *A new kind of science*. Champaign, IL: Wolfram Media.

Zimmerman, B. (2000). Attaining self-regulation: A social-cognitive perspective. In Boekaerts, M., Pintrich, P., & Zeichner, M. (Eds.), *Handbook of self-regulation* (pp. 13–39). Orlando, FL: Academic Press.

ADDITIONAL READING

Aldrich, C. (2005). *Learning by doing: The essential guide to simulations, computer games, and pedagogy in e-learning and other educational experiences*. San Francisco, CA: Jossey-Bass.

Bransford, J., Brown, A., & Cocking, R. (2000). *How people learn: Brain, mind, experience and school*. Washington, DC: National Academy Press.

Carroll, T. (2007). *The high cost of teacher turnover*. National Commission on Teaching and America's Future.

Gibson, D. (2010). Bridging informal and formal learning: Experiences with participatory media. In Baek, Y. K. (Ed.), *Gaming for classroom-based learning: Digital role playing as a motivator of study* (pp. 84–99). Hershey, PA: IGI Global. doi:10.4018/978-1-61520-713-8.ch005

Mozilla. (2012). Badges. *Mozilla Wiki*. Retrieved from https://wiki.mozilla.org/Badges

Pellegrino, J., Chudowsky, N., & Glaser, R. (2001). *Knowing what students know: The science and design of educational assessment*. Washington, DC: National Academy Press.

Prensky, M. (2001). *Digital game-based learning*. New York, NY: McGraw-Hill.

KEY TERMS AND DEFINITIONS

Analytics (Learning, Learner, Academic): These terms are emerging and not fully defined or agreed to in the literature or practice communities. However, the core idea is that visual and statistical representations are assembled automatically, using computational models and artificial intelligence, in near real-time, integrating data from a variety of sources to create a complex representation. The target user of the data gives rise to the various kinds of analytics; those related to progress of achievement, the personal achievement, goals and progress of the user, and the achievement pathways and levels as evidenced by the data sets and their integration.

Digital Media Assessment: Gibson (2010) outlined three elements for digital media assessment: evidence-centered design (Mislevy, Stenberg, Almond, 1999), network-based affordances, and complexity theory. The core idea is that performance in digital spaces (with digital tools, resources and affordances including hypertext and social affordances) allows unobtrusive assessment that produces data that requires new complex, time-dependent and network-based analyses to represent as evidence of the knowledge and action capabilities of a user.

Learning Powered by Technology: This term is the key idea of the 2010 U.S. national technology plan. It includes three key ideas that are made more possible and more powerful with networked technologies: engagement, relevance and personal empowerment.

Unobtrusive Assessment, Stealth Assessment: Unobtrusive assessment is a method that collects data without interrupting the natural actions of a learner or performer, by being embedded within the media of performance. For example, the choices and actions of someone who is working with digital media can be recorded by the media without stopping or interrupting the action of the user with the media.

Chapter 24
Windows into Teaching and Learning:
Uncovering the Potential for Meaningful Remote Field Experiences in Distance Teacher Education

Tina L. Heafner
University of North Carolina at Charlotte, USA

Michelle Plaisance
University of North Carolina at Charlotte, USA

ABSTRACT

Windows into Teaching and Learning (WiTL), a project conceived and actualized by authors situated in a large urban university in the southeastern region of the United States, captures the nuisance of online learning as a method for transforming school-based clinical experiences in teacher preparation programs. This chapter introduces and describes the theoretical context in which the project was developed in hopes to convey the potential for uncomplicated and intuitive innovations in teacher education to recalibrate current practices to the demands of the 21st Century classroom. An overview of the challenges facing colleges of education in providing meaningful and relevant clinical experiences to pre-service teachers enrolled in online distance education courses is discussed and serves as the impetus of WiTL. In the chapter, the authors explain the methods and technology used by the researchers to demonstrate the project's practical duplicability in almost any course with clinical requirements. Furthermore, the authors provide a glimpse into the potential impact of WiTL as a means of facilitating meaningful field experiences in distance education and traditional coursework, as well as corollary benefits realized for student participants and mentor teachers.

DOI: 10.4018/978-1-4666-1906-7.ch024

INTRODUCTION

Technology has changed the nature of teaching and learning in the 21st Century. This is especially prevalent in higher education. With the advent of online learning and Web 2.0 tools, how institutions of higher education serve their student populations is evolving. The possibilities of technology-mediated learning along with economic contractions have led to administrative decisions that have shifted investments in infrastructure renovations and new construction to low overhead options, specifically online learning. As a result, universities and colleges have experienced significant fiscal expenditures on information technologies and a philosophical transition from the "brick and mortar era" of schooling to virtual learning environments (Schulken, 2008, p. 1). As expansion in online education continues in higher education, innovative strategies for addressing traditional requirements of classroom-based teaching and learning have been sought.

This chapter introduces and describes the theoretical context in which Windows into Teaching and Learning (WiTL), designed to capture the nuisances of online learning as a method for transforming school-based clinical experiences in teacher preparation programs, was conceptualized and developed. An overview of the challenges facing colleges of education in providing meaningful and relevant clinical experiences to pre-service teachers enrolled in online distance education courses is discussed and serves as the impetus of WiTL. In the chapter, the authors explain the methods and technology used by the researchers. Furthermore, they provide a glimpse into the potential impact of WiTL as a means of facilitating meaningful field experiences in distance education and traditional coursework, as well as corollary benefits realized for student participants and mentor teachers.

BACKGROUND

In teacher education, field experiences in clinical settings (e.g. observations in PK-12 schools) are considered essential licensure requirements for preparing preservice and lateral entry teachers. These classroom experiences bridge theoretically-based university coursework in content and pedagogy with practical applications of teaching and learning with PK-12 learners. The importance of these authentic experiences is affirmed by the national accrediting body, NCATE (National Council for Accreditation in Teacher Education), who identifies field experiences and clinical practices as one of six *Professional Standards for the Accreditation of Teacher Preparation Institutions*. In the words of NCATE (2007), field experiences allow teacher candidates to observe and reflect on content, professional, and pedagogical knowledge, skills, and dispositions in a variety of settings with diverse students and teachers. Both field experiences and clinical practice extend the institutional goals into PK-12 setting through modeling by practicing teachers, systematic reflective analysis, and well-designed opportunities to learn about methods and content applications (p. 29-30). Thus, the need for university managed clinical experiences poses a unique opportunity for remote distance education teacher licensure programs.

Field Experiences in Distance Education-Defining Challenges

Traditionally, in most colleges of education, all clinical experiences are completed onsite in PK-12 schools. In the case of the University of North Carolina in Charlotte (UNC Charlotte), clinical experiences are required in most, if not all, teacher education classes leading to initial licensure, whether the candidate seeks a Master of Arts in Teaching (MAT) or a Graduate Certificate in Teaching (GCT). Clinical experiences range from 10 to 30 hours per education course. These hours are intended to be an extension of class expectations to assist candidates in bridging pedagogy and practice as well as providing authentic settings for applied learning. Given that most university teacher candidates take multiple education classes during the same semester, the possibility arises that clinical requirements in a single semester could be as great as 90 hours. While these requirements traditionally have not posed a barrier for university students, new challenges have emerged with online licensure programs.

In the College of Education within UNC Charlotte, field experiences are arranged by the Office of Field Experiences (OFE) for all teacher education candidates. OFE has partnerships with the largest school system in the state, Charlotte-Mecklenburg School System, and twelve surrounding counties in which schools have agreed to place UNC Charlotte teacher candidates in classrooms and to support candidates' licensure training in their field-based placements. With the growth and expansion of online licensure programs, teacher candidates are no longer limited to residence within the greater Charlotte region traditionally served by the University. For example, the Department of Middle, Secondary, and K-12 Education's GCT licensure program enrolls candidates from over a third of the counties across the state. The licensure candidates who reside outside of the aforementioned boundaries (defined henceforth as proximity parameters) are required to find their own clinical placements. This can pose challenges for teacher candidates who do not have relationships with local school systems. Moreover, UNC Charlotte cannot guarantee the quality of placement for these candidates since they are not involved in these clinical arrangements and do not know firsthand the teachers candidates observe.

Not only are teacher candidates outside the proximity parameters expected to gain placements for one semester, they are expected to arrange three diverse placements with three different schools (e.g. three semester placements in an urban, suburban, and rural school) over the duration of their program of study. The expectation for diverse clinical placements is a licensure requirement established by the North Carolina Department of Public Instruction and informed by NCATE who posit standards such as: "Field experiences allow candidates to apply and reflect on their content, professional, and pedagogical knowledge, skills, and professional dispositions in a variety of settings" (NCATE, Standard 3, p. 29). The rationale for this requirement is to prepare future teachers for the complexity of working in multifaceted school settings that pose unique challenges prevalent in metropolitan regions. For example, needs of suburban schools and the students they serve are quite different from schools within urban centers or in rural communities. For teacher candidates to understand these differences, opportunities to learn in diverse schools are necessitated. Students who reside outside of the proximity parameters of urban areas often have limited access to schools within their own geographic boundaries. It is highly unlikely for a candidate in a rural county to experience schooling from an urban context, such as CMS, through observation alone. Living in these communities and working in urban schools are essential experiences to addressing geographical limitations. Overcoming these barriers is a unique challenge encountered in online programs that recruit candidates from underserved counties in rural locales such as those of a geographically-diverse state like North Carolina.

In tough economic times and when unemployment is on the rise, the general populous seeks ways of improving or diversifying their skill-base and employability. Given past trends of teacher shortages in this southern state, and continued demand for qualified professionals in STEM subjects, many working professionals have sought second career opportunities as teachers. However, economic constraints, while attracting an interested population, have posed limitations for access to careers in teacher education. Specifically, clinical requirements are frequently problematic for working professionals who are taking classes during evening hours while maintaining a full-time job during the day or for individuals who cannot afford the added costs of travel. The traditional approach to teacher education classes has been to conduct clinicals during the working hours and in reality this is when schools operate. Requiring second career professionals to take a day (or a week) off from work or travel significant distances to conduct clinical observations places financial burdens on individuals resulting in reductions of teacher education applicants. Finding viable alternatives that might allow career professionals to take off an hour or two out of the work day or time outside of the daytime work hours would be a potential recruitment tool for underserved areas. These options could open, not close, doors to increasing the diversity of teacher education enrollments in graduate licensure and degree programs.

Finally, summer university classes are especially attractive to lateral entry teachers and to unemployed individuals seeking the most cost-effective pathway to licensure and future employment as a teacher. As a point of reference, summer school classes are more affordable options than courses offered during traditional semesters. Distance education classes also are offered at UNC Charlotte for approximately half the cost of on-campus classes. For these reasons, many teacher candidates in licensure programs take a significant number of summer classes and most of these are online courses. Recently the university made a cost-saving decision and migrated all summer courses in teacher licensure to online platforms. Not only are online classes more affordable to students, they do not bear infrastructure expenditures of campus-based classes.

A new challenge restricting field placement accessibility during summer sessions poses another economic constriction. Public schools that have typically hosted clinicals are no longer available in the summer months. In prior years, public summer school programs for remediation were available, but due to state and local education budget cuts, most PK-12 schools have eliminated these programs. These economically driven decisions result in teacher candidates enrolled in summer classes being asked to find their own clinical placements in very creative settings, such as summer camps, churches, or YMCAs. While these venues may serve as meaningful experiences for understanding young children, they are far removed from teaching subject area content in middle and high schools. Thus, seasonal access poses the fourth identified barrier to online programmatic outreach and the quality of online teacher education.

In summary, four major challenges related to providing meaningful clinical experiences in distance education licensure programs have been identified: accessibility, flexibility, quality and relevance. These potential barriers evoke a need to re-conceptualize clinical requirements, both in terms of *what* they entail and *how* they are experienced. While perceived as barriers, a possible solution to these challenges creates opportunities for transforming licensure training.

In response to these needs, Drs. Tina Heafner, Associate Professor and Teresa Petty, Assistant Professor, both with the University of North Carolina in Charlotte embarked on a project entitled Windows into Teaching and Learning (WiTL). Through a Scholarship of Teaching and Learning (SoTL) grant funded by the Center for Teaching and Learning, Drs. Heafner and Petty designed a project utilizing technological advancements to facilitate remote observations that might potentially address the previously-described barriers and transform distance education for preservice teachers, as well as those seeking licensure through alternative paths.

While WiTL possesses a myriad of corollary opportunities and avenues for expansion, it is best understood when grounded in current literature which positions the project into three major interwoven and recurrent themes: the potential benefits and pitfalls of technologically supported remote versus face-to-face observations, the exploration and potential of synchronous and asynchronous learning environments, and the use of video as a tool for growth in the classroom.

Theoretical Framework

Technologically Supported Remote versus Face-to-Face Observations: A Fundamental Debate

Current research in technology integration is beginning to reflect the previously described evolution of teacher education programs. Technology as a medium for learning in higher education is becoming institutionalized and as a result, questions about the effectiveness of methods in online instructional platforms surface. Focus is shifting from *whether or not* to utilize technology that supports distance education to, rather, *how* to most effectively engage students through this technology and evaluate the learning outcomes of participants in these non-traditional settings (Ajayi, 2009). The migration of research from *if* to *how* is significant in documenting the manner in which technology permeates learning experiences in all careers. It is especially intriguing as to the role technology should play in the preparation of teachers.

There is no known project which duplicates exactly the procedures and framework of WiTL; however, much can be gained by looking to other computer-supported innovations in teacher education. In a study involving 85 student teachers and 14 supervisors from the Netherlands, Admiraal, Lockhorst, Wubbels, Kothagen and Veen (1997) examined online interaction within established discussion forums, both in terms of supervisor's mediation and peer interaction. They found that participants felt equally supported in an online environment versus traditional settings. Fifty percent reported that the same type of writing required by WiTL participants caused them to reflect more deeply on their experiences and supervisors reported that the online platform allowed them to more closely monitor their students' learning.

Technologically Supported Experiences

Hixon and So (2009) define five specific benefits to the use of technology to support remote observations: exposure to diverse teaching and settings, increased reflection, cognitive development, technology integration and the shared experience that results from collective viewing of videotaped clinical experiences. The researchers delineate three types of technology-supported clinical experiences, with remote observations via video conferencing being one that appears to provide all of the described benefits. And while they present four potential drawbacks to the use of technology-supported clinicals, only two apply to remote observations of real classrooms. The first, participant perception of reduced interaction with the mentoring teacher, is addressed in the design and implementation of WiTL's interactive observations. The second drawback, technical difficulties and frustration for participants, remains to be addressed and will be an ongoing focus in the continuing development of WiTL.

In one study with similar objectives, Kent (2007) used interactive video conferencing (IVC) to supply teacher candidates with access to classrooms with exemplary teaching models. She found that the integration of the clinical experience within the university course assisted candidates in bridging the gap between theory and practice. In addition, increased reflection as evidenced in candidates written responses to the clinical experiences was noted. While many of the procedures of the Kent study closely resemble WiTL, a major difference is seen in the fact that the IVC technology used is quite costly and requires a dedicated classroom in the PK-12 setting. In contrast, the technology utilized by WiTL, described below, is fairly inexpensive and quite portable, allowing these types of clinical experiences to be incorporated into university coursework with relative ease and facilitating observations of a diverse ranges of classrooms.

Greene (2005) conducted a somewhat similar pilot study of online learning communities comprised of preservice teachers and mentor teachers who shared in the viewing of virtual recorded teaching episodes. She found that a model of interaction with active teachers, in conjunction with class lectures enhanced student understanding of the teaching they viewed. Moreover, Ziechner (2010) argues for the need for a "third space" (p. 92), or a common area, between knowledge acquired in university classrooms and what is experienced by preservice teachers during their clinical placements. He suggests the need for the mentoring teacher to be a presence in the university setting, and similarly, that the teacher educator share in the clinical experience. The way in which the observations facilitated through WiTL were collectively experienced and explored by the student participants, mentor teachers and university faculty suggests the potential to host such a space.

Inclusive Learning Environments

The use of online communication tools to support distance education initiatives in teacher preparation programs holds the potential to attract a more diverse pool of candidates than traditional programs in an on-campus, university setting (Ajayi, 2009). In addition to making such programs available to students who might have otherwise been excluded due to geographical and financial considerations, online learning appeals to a different range of learning styles than face-to-face instruction. In a study focused on communication and dialogue in an online setting, Jackobsson (2006) found that students from non-academic backgrounds were able to effectively utilize online collaborative environments to achieve beneficial critical reflection through the use of computer mediated communication, even more so than their academic-oriented peers. In addition, Im and Lee (2003), in a study comparing over 3000 synchronous and asynchronous comments posted by 40 preservice teachers in a teacher education program in South Korea, suggest that asynchronous learning may allow females to interact in a more equitable environment than face-to-face classrooms, where dialogue is often dominated by male participants. Thus, as technology redefines learning, new possibilities of a more inclusive and diverse learning experience emerge.

THE PROJECT

The overarching goal of the WiTL project focused on the exploration of possibilities for technology-supported alternatives to traditional clinical experiences in distance education programs for teacher education candidates. However, the researchers conceptualize potential benefits far outside the scope of expected outcomes in traditional field experiences. The foundation of WiTL is built upon the integration of both synchronous and asynchronous interactions among mentor practicing teachers, teacher candidates and university instructors to facilitate an interactive learning community designed to provide a forum for a deeper understanding of the wide range of considerations that teachers face on a day-to-day basis and to link theory and practice to real classrooms. The WiTL project presents a practical technology mediated alternative to face-to-face field-based experiences while scaffolding participant learning through shared experiences and collective pedagogical growth (Figure 1).

Teacher candidates enrolled in a university methods course with clinical observation requirements are given the opportunity to view the teaching of multiple expert teachers in their area of specialty, across a range of grade levels in diverse school settings. Archived teaching sessions allow flexibility and convenience for geographically diverse students and students enrolled in summer programs, while synchronous sessions provided

Figure 1. The WiTL framework

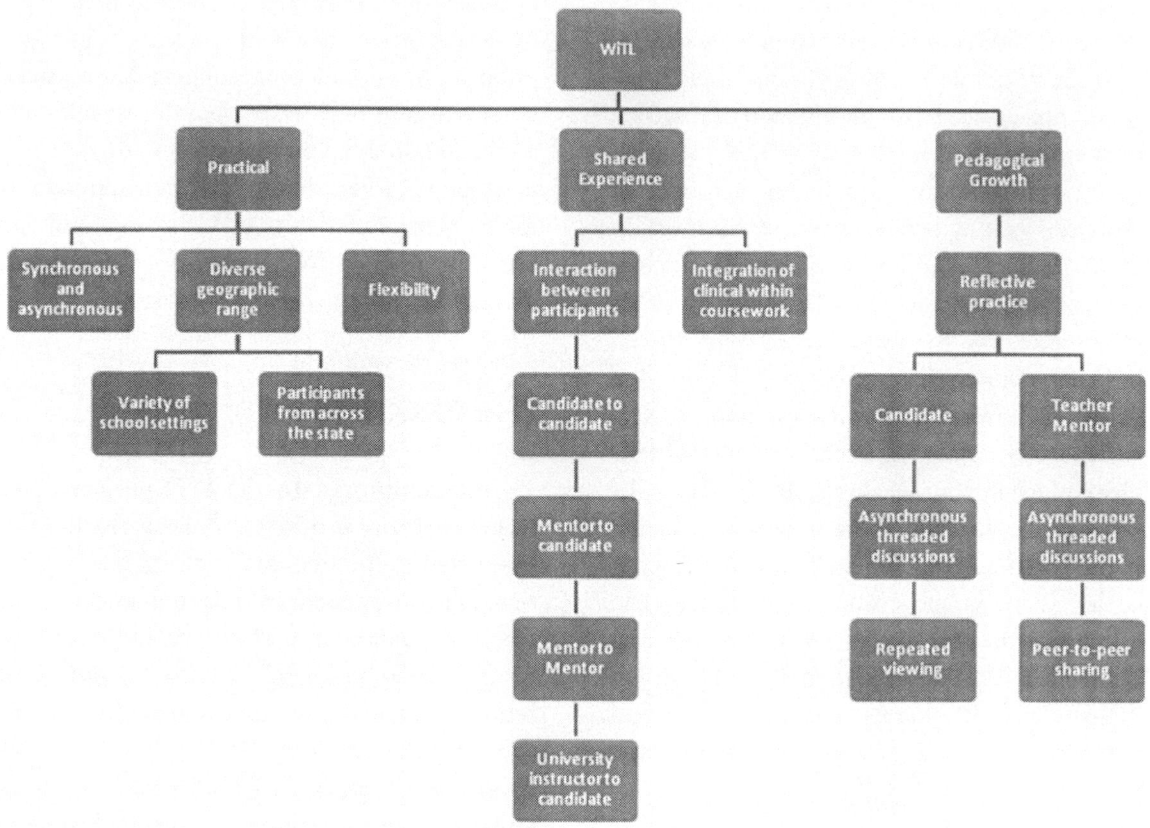

participants a unique opportunity to collectively share in the viewing experience. The incorporation of post-observation debriefings, either in the form of real-time, live discussions or extended threaded discussions, allows participants (candidates, teacher mentors, and university methods faculty) to receive the benefit of both spontaneous and systematic reflection, while simultaneously allowing the candidates to dissect pedagogical thinking by exploring the *why* and *how* of teaching. Because the context of WiTL promotes interaction among all participants, teacher candidates, teacher mentors and university supervisors, a collective understanding of shared clinical experiences becomes available as a tool for further exploration, learning, community building, and even professional development.

Evolving Technology

Technology, with all of its innovations and at time, frustrations, factors heavily into the implementation and ongoing development of WiTL. For asynchronous teaching sessions, WiTL utilizes webcams to transmit the video footage to a laptop computer, while wireless microphones are worn by the teacher mentors (Figure 2). Software such as Camtasia Screen Recording Software (http://www.techsmith.com/camtasia/) captures the images on the computer screen, which includes an open Microsoft Word document that allows the observation facilitator to annotate the lesson with both comments and instructional input about the pedagogy being observed. Synchronous observations are scheduled to align with 90-minute blocks of instruction in the teacher mentor's schedule.

During these live viewings, students log into the university course's online platform, Wimba. At the scheduled time, the observation facilitator opens the "window" into the classroom, allowing student participants to view the classroom activities and teaching from their remote locations. During these synchronous sessions, the observation facilitator is able to supply comments and supplemental information via the text chat feature available in Wimba. Students also have access to this feature and used it to communicate with one another regarding the lesson they are observing.

WiTL is in the initial stages of development, and consequently, technical challenges are anticipated and will be used as a means of honing future procedures. For example, the researchers initially anticipated using the Wimba technology to record both the synchronous and asynchronous observations. The clarity of image and familiarity of the Wimba technology for both the researchers and student participants were major benefits. In addition, the natural connection between Wimba and Moodle could have eliminated extensive amounts of time spent uploading the video files. However, data file size became a major hurdle, as did the restriction that would be posed in terms of accessing the videos with future courses. Ultimately, images are captured using Logitech Webcams. There is some diminishment in image clarity, in addition to a small viewing pane for student participants, which make this a somewhat unfortunate compromise. However, value is gained in the low-cost, portable and inconspicuous nature of these cameras. An unanticipated drawback is the exorbitant amount of time required to convert the file formats and to prepare them to be uploaded to Moodle. Moodle, itself, presents minor challenges in that files cannot be migrated from course to course, but rather, must be downloaded to an external drive, converted to MP4 files, and re-uploaded to a new course if needed. Internet provider bandwidth restrictions on video uploads and downloads make this a time-consuming process. Teacher candidates also experience time delays in accessing large video files in Moodle.

Using the Webcam means that there was a mildly restricted view of the classroom at times. While the observation facilitator is able to reposition the camera to capture the movement of the teacher or students engaged in various activities, zooming in for close up views has been found to distort the image beyond the benefit provided by an up-close perspective. However, this obstacle is easily overcome through the use of typed annotation provided in the text chat by the observation facilitator who is able to easily draw candidate participants' attention to elements that might be otherwise been overlooked. An additional challenge presents itself in the fact that the teacher mentors, in some cases, made extensive use of their electronic SmartBoards. The images displayed on these boards were not visible to the teacher candidates. The researchers seek to address

Figure 2. Screenshot of typical WiTL observation using text chat

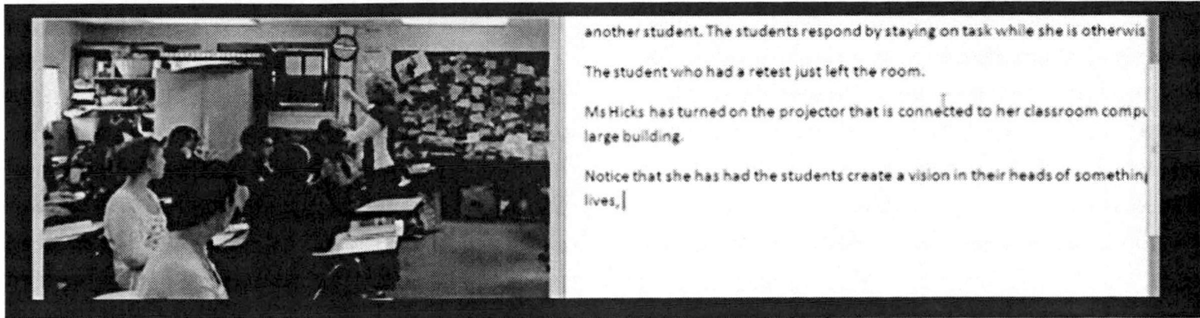

this obstacle in future versions of the project by obtaining software that will allow the candidate participants to view the SmartBoard input, potentially in a separate window displayed in tandem with the video footage.

In addition to video issues, audio recording presents a minor challenge at times. Currently, teacher mentors wear high-quality wireless microphones that provided superb audio recordings of their speech and dialogue within the classroom. However, at times it may be difficult to hear student discourse or other audio occurrences in the room. One final unavoidable obstacle is the limited amount of access to the Internet available in some classrooms. Most classrooms are equipped with only one internet port, which frequently is needed by the teacher mentor in the delivery of their lesson. The researchers have been able to address this issue through the use of a portable internet access card, which, while costly and inconsistently reliable, achieves the desired result.

Synchronous versus Asynchronous Environments: Tools for Learning

The use of asynchronous and synchronous teaching platforms brings with it several factors that should be considered. Asynchronous distance learning is typically characterized by the use of course materials organized in a chronological fashion that allow the students to move through the coursework at a pace outlined in terms of windows of opportunity (Figure 3). Class lectures are written by the instructor, or delivered via archived video sessions with tools such as Camtasia and PowerPoint. Students submit assignments via email or upload them through programs specifically designed to manage distance education courses. Interaction between both the instructor and student and among the students themselves occurs in a threaded discussion format where students post questions and comments and others post replies.

Conversely, synchronous learning in distance education most closely resembles that of a traditional face-to-face classroom in that course material is delivered in a "real-time" format. Lectures are delivered via a teleconferencing tool such as Wimba or Centra, both of which support an interactive whiteboard and PowerPoint. Students "attend" class at a given date and time. Interaction between peers and between students and the instructor occur via a face-to-face video session or in a format similar to that of an online chat room, where typed messages are delivered and replied to instantly (Motiwalla & Tello, 2000; Song, Singleton, Hill & Koh, 2004). Additionally, interactive tools such as texting, polling, question-answer checks, whiteboard notes and drawing tools, and feeling response button icons (laughing, smiley face, confusion, and applause) are utilized to engage the lecture participants and conduct informal assessment measures. Unless clickers are used in on campus classes, these tools make the online learning experience more engaging, accountable, and immediate than traditional on campus courses. WiTL capitalizes on technology's ability to engage the learner by incorporating all of the elements described above, adding the unique ability to collectively view and discuss actual teaching as it occurs.

When synchronous learning opportunities are incorporated into distance education programs, Hastie, Chen and Kuo (2007), argue that the immediacy of such communication increases learning outcomes. They describe a six-year study of a trial investigating an online distance education project that utilized synchronous sessions between instructors and students to deliver content to geographically-diverse 5-8 year-old students. During this trial, students and teachers logged into a web-supported, interactive whiteboard where instruction took place via webcam, microphone and the use of a mouse or graphic tablet. The teacher delivered instructional lessons and then worked interactively with the student in practicing and

Figure 3. Asynchronous threaded discussion between candidates and teacher mentors

Thanks for allowing us in your class today. Is your class level, I mean is this an advanced or regular class?
Is this an inclusion class?
So if it is below level...are there 10th and 11th graders??
What are some of the stations specifically that you used today?
could you explain how you grouped the students today?
How are they advancing through the school year? are they increasing their performing level at this point of the school year?
Do you use stations often during the school year, or is this something special for the end of the year?
Is the leader in each group rotating or are the same all the time?
Are they having a test soon? Was this activity a review for an upcoming test?
Were today's activities differentiated for different learning levels?
Can you explain some of the stations that you used today?
Do you teach only geometry or do you teach other classes?
Do you add words to the word wall throughout the year?
How long you have being teaching math?
At your school, do you work with a team of teachers or other teachers in different curriculums?
What do you think is the most challenging part of teaching math?
Does your school has tutoring after school time?
Do you have any advice for getting students more interest in math or helping them build confidence?
Is the classroom always set up in a group seating arrangement?

applying the material that had been delivered. The synchronous technology afforded the teacher the opportunity to provide students with continuous and immediate feedback throughout the lesson. The authors attribute the successful outcomes for these students to this increased interaction between student and teacher. And, while their study focuses on young children, it would seem the results could be generalized to adult learners in that all students learning through synchronous platforms were found to spend more time on task, have greater interaction with the instructor and demonstrate higher levels of concentration. The use of texting in the virtual classroom is one such example that creates a level of discourse with the instructor that would not occur in a face to face setting. Thus, the virtual classroom has the potential to become a more interactive and personal learning space.

Benefits of Synchronous Dialogue

Levin, He, and Robbins (2006) also suggest that there are benefits to using synchronous components. In their study they examined participant preference for synchronous versus asynchronous online discussions of cases in a post baccalaureate class of 36 preservice teachers enrolled in a web-supported classroom management course. Students submitted an initial analysis for each of six assigned cases, followed by online synchronous and asynchronous discussions with the instructor and other members of the course. Through a pre- and post-preference survey and comparison of the content of eight purposefully selected case analyses, Levin, et al. were able to conclude that there were significant changes in the type of responses that followed the synchronous online discussion. Specifically, they found that students showed increased evidence of critical reflection, as defined by Dewey (1933, as cited in Levin, et al., 2006), when they experienced synchronous learning opportunities versus those students who engaged in asynchronous online discussions. In addition, through comparison of the pre- and post-survey responses, they were able to conclude a significant change in preference for synchronous discussions over the asynchronous formats that were used in the study.

Outcomes of Asynchronous Dialogue

Asynchronous learning environments carry distinct advantages as well. Im and Lee (2003), in the previously described study, found that this mode of communication lends itself to more task-oriented discussions. Their study also revealed that the level of communication, in terms of approaching meaningful learning and knowledge acquisition, was present more in asynchronous, versus synchronous environments. Specifically, the authors found that synchronous communication was used primarily and most effectively for social applications, such as establishing rapport. Conversely, collaborative efforts, such as the sharing of ideas and resources, as well as cognitively demanding processes such as evaluating and analyzing occurred almost entirely in asynchronous settings. In addition, analysis of the content of the discussions revealed that asynchronous posts are more likely to be task-oriented and on topic. Im and Lee conclude that asynchronous models of communication should be used to promote knowledge acquisition and learning, while synchronous modes can be employed to establish and improve relationships between participants.

Communicative, Community Learning

Wise, Padmanabhan, and Duffy (2009) learned that when students in online courses analyze classroom practice by reviewing videos, the students can develop a shared context and common experience that frames their discussions. Online discussions among students can facilitate knowledge construction and shared meaning because the experiences in the course serve as a shared context that provides a common experience that they may not share outside of the course (Wise, Padmanabhan, & Duffy, 2009). In a project focused on defining the types of communication essential to online learning, Soo and Bonk (1998) surveyed eight veteran distance educators and concluded that their preference for asynchronous settings stemmed, in part, from the ability of asynchronous platforms to support the most important type of communication for effective online learning- learner to learner interaction. This type of interaction was consistently ranked highest in terms of importance, closely followed by other asynchronous settings, such as learner-teacher and learner-material. MaKinster, Barab, Harwood and Anderson (2006), in their study examining the impact of online social context on the reflective practices of student teachers enrolled in a university methods course, found that students who were given the opportunity to reflect on their teaching experiences via online discussion achieved a more in-depth analysis of personal experience that those students charged with journaling through more traditional, private settings. Systematic reflection, a byproduct of threaded discussions like those incorporated in WiTL, enabled richer and more frequent instructor-student or student-student interaction in relation to examining practicum applications. Viewing these interactions as essential for learning as per a social constructivist perspective, the authors found that the targeted level of reflection was seen most often in student responses to other students' posts, reflecting an opportunity to develop a discourse that simply does not exist in traditional journaling tasks. Many other advantages to asynchronous interaction have been cited. In an ongoing, online survey being conducted by Branon and Essex (2001), asynchronous instruction is found to be useful for the flexibility it affords students with challenging schedules, the capability of archiving discussions for future use, the ability to increase participation by allowing multiple responses to a single question, and the perception that it provides a means of eliciting more in-depth, reflective responses from participants.

Pitfalls of Online Learning

While there are distinct advantages to both, there are potential pitfalls that must be considered as well. For example Brannon and Essex (2001) found that synchronous interaction, in certain instances, may present sizeable challenges in terms of managing large groups of students, allowing adequate time for response to questioning, and negotiating the schedules of many students at the same time. Similarly, they found that asynchronous learning opportunities may be characterized by delayed feedback, poor participation and feelings of disconnect on the part of the students. Though the results of a study by Hawkes and Romiszowski (2001) that examined interaction among 28 teachers in Chicago schools suggest that the disadvantage of less interaction could potentially be outweighed by the benefits of increased collaboration, as well as the essential critical reflection that present themselves in online versus face-to-face formats.

A Blended Approach

Given both the advantages and disadvantages of each type of online communication, it appears that a blended approach to program design may be best to promote a strong sense of community while allowing for the depth of discussion necessary for learning to occur. Im and Lee (2003) suggest that different types of technology be used for different objectives. Synchronous methods are more appropriate for establishing social interaction, while asynchronous is better for efficiently and effectively tackling tasks. Levin, He and Robbins (2006) also suggest that there are certain circumstances that merit the use of one format or the other, while certain learners perform better depending on the type of interaction facilitated by the online technology. It is reasonable to assume then, a carefully considered program that utilizes both synchronous and asynchronous methods based on their strengths would likely produce the best results. Likewise, program decisions about technological platforms for learning should weigh benefits and limitations of each process. There may be justification for following one format but it is not without tradeoffs. The framework of WiTL addresses this need for a balance between synchronous and asynchronous learning opportunities, capitalizing on the unique benefits afforded by both while simultaneously offsetting the potential drawbacks that have just been described.

Use of Video in the Classroom

To Notice

The nature of the WiTL project introduces the use of video in the classroom and brings with it a host of intended and unintended potential outcomes for both the professional educators and preservice teachers involved in the project. Sharin and van Es (2005) presented the findings from two studies that involve the use of video in a learning environment, asserting that in both cases, the use of video impacted the way in which teachers or students notice various occurrences inside the classroom. The theoretical framework upon which they designed these studies asserts that an educator's ability to notice in the classroom implies teacher expertise in that it allows the teacher to navigate and prioritize the multitude of complex interactions occurring simultaneously within the classroom. In addition, proficiency in noticing affords teachers the ability to conceptualize large concepts from small details, as well as allows them to apply increased logic and reason to their teaching contexts through their acquired expertise in their content area. Based upon these assumptions, the authors argued that video affords teachers the opportunity to notice more and different factors related to their teaching because they are freed from the constraints of memory and forced to view things from a varied perspective. This type of consideration is a form of reflection, but one that extends beyond traditional afterthought of lessons and practice. The nature of the WiTL framework calls for participants to explore the perspectives of others- teacher mentors see one another through the eyes of the students they teach as well as the teacher candidates who they mentor.

In the second of the studies, which relates most closely to WiTL, Sharin and van Es (2005) employed video as a tool in a teacher education program with six preservice teachers, requiring them to write reflections of the type of details they noticed when watching themselves teach. Over time, when compared with other preservice teachers not utilizing this tool, those who used video were able to hone their noticing skills and discern significant from insignificant events in the classroom. In addition, they found that the content of their analysis elevated from being simply evaluative to interpretive in nature. These findings are affirmed by practices utilized in the National Board Professional Teaching Standards Certification Process (http://www.nbpts.org/), although NBPTS use of video is not through online learning. Such benefits, embodied in the potential of WiTL, can be surmised as the importance of video as a tool for systematic reflection in ongoing professional growth.

Caution: Decomposed Teaching

While many benefits have been described and innovative teaching methods may be enhanced, Hatch and Grossman (2009) caution that there are limitations and shortcomings that arise as a result of using video in teacher education programs. In an article describing a project that explored the use of what they coined *representations of teaching*, they caution teacher educators to be cautious of isolated videotaped teaching segments that are broken down, or "decomposed" (p. 71) in order to illustrate specific points in the curriculum. And while their project focused specifically on instructing teachers in the facilitation of rich classroom discussions, the results are easily generalized to working with preservice teachers on any relevant subject matter. Fearful that these segments will over-simplify the act of teaching, Hatch and Grossman further warn that videotaped lessons fail to capture the planning as well as the strategic thinking on the part of the teacher. Furthermore, using video representations of teaching does not allow for the illustration of overarching themes or concepts, but rather isolated lessons that appear in a decontextualized setting. The authors suggest that the transmittal or posting of peripheral artifacts, such as lesson plans or unit overviews, might assist in overcoming these obstacles. In addition, the authors admit that the shortcomings they outlined are offset somewhat by the undeniable advantage of using video, including the ability to replay a segment multiple times, the furnishing of a "near peer" (p. 73) when one is otherwise unavailable and the unique ability to provide a shared viewing experience of clinical experiences between students and instructor.

Real Teaching

An additional benefit that accompanies the use of video in the classroom, in lieu of observers being physically present for the observation, is the opportunity to witness real teaching and student behavior. Johnson, Sullivan and Williams (2009) conducted a study that examined the use of video as a vehicle to observe daily classroom routines from a "naturalistic orientation" (p. 36). They found that, despite technical challenges, by placing cameras in classrooms, observations were less intrusive and a more authentic representation of the true learning environment was obtained. While a researcher was present during the WiTL observations, the use of an inconspicuous camera allowed for real teaching to be observed because the passing of time permitted students and teachers to acclimate to its placement without the disruption of the actual presence of another adult or multiple teacher candidates within the classroom.

LOOKING FORWARD WITH WITL: SOLUTIONS AND RECOMMENDATIONS

Citing the importance of effective clinical experiences in the overall improvement of teacher preparation programs in the United States, Darling-Hammond (2006) calls for "a radical overhaul of the status quo" (p.307) in terms of how field experiences are conceptualized and enacted. She strongly asserts that these changes are necessary to continue the age-old struggle to connect theory and practice in teacher preparation programs.

This means that the enterprise of teacher education must venture out further and further from the university and engage ever more closely with schools in a mutual transformation agenda, with all of the struggle and messiness that implies (Darling-Hammond, 2006, p. 302).

WiTL presents itself as a step in this direction. While the primary purpose of the WiTL project was to provide teacher candidates seeking licensure through distance education meaningful, authentic and relevant clinical experiences, a host of additional potential in terms of possible positive outcomes has emerged.

In traditional teacher education programs, assigned clinical hours are typically spent with one teacher, in one school, in one district. WiTL opens a window of opportunity for the teacher candidates to experience a diverse range of school settings and multiple teaching contexts. Over the course of a 5-week online summer school class, student participants were able to witness teaching in a rural and an urban district. In addition, they were afforded the opportunity to interact and view expert teachers from two different licensure levels (6-9 and 9-12), providing a more thorough understanding of the long-term and developmental needs of adolescent learners and vertical curricular alignments. Unlike traditional clinical experiences, where candidates might only view their content area for a small portion of the day, all of the lessons they attended related directly to their specialization. More importantly, observations such as those facilitated by WiTL allowed the university instructors to control for quality to ensure that candidates observe sound pedagogy and exemplary teaching practices rooted in content area research.

Another positive potential that lies within the framework of WiTL is the ability of candidate participants to collectively share in each clinical experience, both amongst one another as well as with their university instructor and teacher mentors. Ordinary clinical placements leave preservice teachers in isolated settings to make sense of the teaching they see with little or no guidance. Within the typical methods course, there might be twenty to thirty teacher candidates completing their required hours in twenty to thirty very distinct settings. Given the overwhelming quantity of anecdotal tales that inevitably surface, it becomes impossible to bring the clinical experience into the university setting. Course instructors must compromise by discussing what is *typical* or *noteworthy*, rather than the *reality* and *immediacy* of what candidates have just witnessed. Likewise, students are unable to adequately contribute to one another's construction of understanding, because their individual clinical experiences typically differ significantly. By providing a text chat feature during the synchronous clinical observations and text narrated asynchronous observations, WiTL exposes teacher candidates to an integrative learning community- one where they can share thoughts and ideas as the teaching episode unfolds. Shared viewing experiences such as these, allow the instructor to point out the methods and strategies that are being discussed in the course literature as they are actually unfolding in the context of the classroom. Further enhancing this powerful tool is the fact that the teacher mentor can speak directly to the novice candidates immediately before, during, and after the teaching session- providing an opportunity to explore teacher thinking and instructional planning. This type of three-way collaboration is simply not possible in traditional clinical settings.

Asynchronous remote observations, as facilitated by WiTL, also provide substantial opportunities for collaboration, extending the benefit even further to the teams of teacher mentors who participate. The archived teaching sessions can be viewed in a variety of ways. First, candidate participants and teacher mentors are able to view the sessions at their leisure, from remote locations, as many times as is necessary to master the objective or primary focus of the teaching episode. Archived synchronous sessions provide yet another opportunity to revisit content observed in real-time teaching sessions. Furthermore, the asynchronous discussion board permits candidate participants to reflect on what they have seen and pose thoughtful and constructive questions and comments. Rather than participating in a one-on-one discussion or an individualized written reflection, which often results in more summary that authentic reactive thinking, candidates are presented with a team of teacher mentors who provide a variety of perspectives, greatly enriching the discussion and enhances the level of collective discourse.

The second use for the archived teaching sessions is peer-to-peer viewing. Peer observations are "perhaps the simplest way to break down professional isolation" (Darling-Hammond, 2010), but rarely occur due to the time demands of the typical school day or the physical restraints of classrooms. WiTL's asynchronous observations allow teacher mentors to view the pedagogy of their peers outside of the instructional day, even from the comfort of their own homes. In addition, collectively participating in the asynchronous threaded discussions allows the mentor teachers to learn more about the teaching philosophies, strategies and methods being employed by others, both in their own school and other districts in diverse settings. Moreover questions posed by teacher candidates challenged teacher mentors to question the *why*, undergirding theoretical rationale for their pedagogical decision-making. Rather than overreliance on personal affirmations of instructional success, forced reflection as a byproduct of university teacher candidates' questions challenged mentor teachers to articulate formative and data-based assessments of practice and teaching efficacy.

Professional development presents a final use for the asynchronous, archived videotaped teaching sessions. Reflection is a powerful tool in promoting growth in professional educators (Darling-Hammond, 2010). Teacher mentors are given access to their contributions to the WiTL project, and thus, the opportunity to reflect deeply on their teaching skills and practices. In addition, comments presented by candidate participants, via both the synchronous interviews and they asynchronous threaded discussion, provide the teacher mentor with both direct and indirect feedback on their pedagogy. As previously stated, peer observations, while inarguably beneficial, are challenging due to scheduling constraints and may posed barriers due to an unwillingness to openly criticize a colleagues methods. Even more so is the notion of cross-curricular exposure, where, for example, math teachers can watch the delivery of social studies to gain a better understanding of the entire educational environment in which students find themselves. Much is to be gained by increasing the exposure of teachers to colleagues in other licensure levels, other schools, and even other districts. WiTL possesses the potential to simplify these types of experiences, facilitating them at the convenience of the viewer and within restricted economic budgets (figure 4). Teacher mentors are given the opportunity to assume leadership roles within their schools by sharing their archived videos in staff development or team planning meeting when they feel there is benefit to their peers. Teachers as leaders serves to align ongoing practice with exemplary expectations for 21st Century teaching professionals (North Carolina Department of Public Instruction, 2010).

FUTURE DIRECTIONS

Field experiences are a vital resource within teacher preparation programs, there is little dispute as to the necessity of this requirement for preservice professional educators. Traditionally, these placements have been problematic for distance education students, lateral entry, second career, and summer school university students. WiTL opens widows to the possibilities of providing an alternative to traditional, face-to-face observation. The goal of WiTL is not to replace field-based clinicals altogether but to seek unique ways of providing instruction in online programs. There is no substitute for sitting in a classroom and absorbing its culture and ambience. But this type of

Figure 4. Features of synchronous and asynchronous sessions

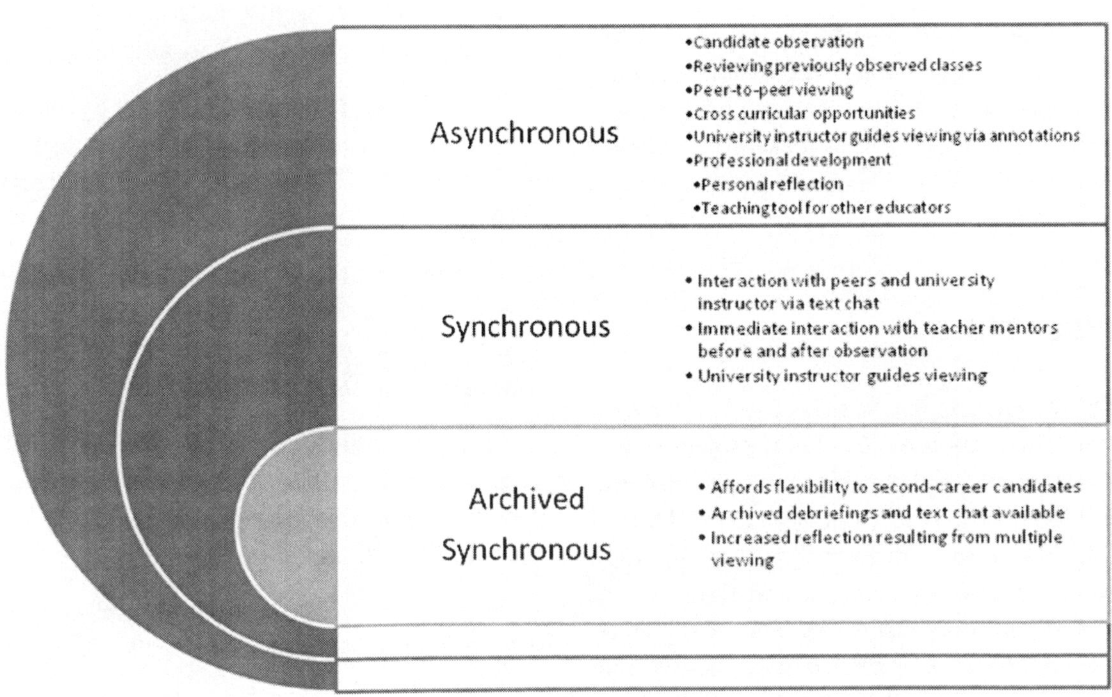

experiential learning has fallen short of expectations; candidates are not consistently gaining the intended knowledge or experience that is critical in developing effective professional educators.

The full evolution of WiTL is still in process, with its designers conceptualizing ways in which to tap its full potential for the betterment of distance education programs, traditional preparation programs, as well as university-community relationships. Within this vision is the idea of prolonged, collaborative mentoring communities- ones where teams of experienced and expert teachers come together in support of novice teachers. This community may be born though clinical experience, but continue throughout the teacher candidate's licensure program, providing a multifaceted bank of resources upon which to draw. The communal relationship should be comprised of synchronous, asynchronous and face-to-face interaction, so as to capitalize on the benefits of each and provide the diversity that is essential to true understanding. When properly supported by university faculty and school administrators, this relationship holds the potential to serve as a portal into the world of professional teaching. The teacher mentors stand much to gain as well. The opportunity to interact with the academic knowledge that is so often kept out of schools is a means of professional growth. In addition, WiTL presents seemingly endless possibilities for leadership, reflection and collaboration.

CONCLUSION

The potential that has been uncovered through the implementation of the WiTL project stands to make substantial contributions in bridging what Grossman, Hammerman and McDonald (2008) refer to as a "historic separation between university-based course work and fieldwork in local K-12 schools" (p. 275). They call for an active collaboration between practitioners and researchers to ensure that novice teachers gain a sense of coherence between what they hear in the lecture halls and what they are asked to enact in their classrooms. The multi-faceted elements that comprise the essence of WiTL, where university instructors work in tandem with expert teacher mentors to support the teacher candidate's journey from theory to reality, are just the types of changes that will enact the transformation necessary to ensure that teacher education programs produce professionals prepared to successfully navigate diversity of 21st Century schools.

REFERENCES

Admiraal, W. F., Lockhorst, D., Wubbels, T., Korthagen, F. A. J., & Veen, W. (1998). Computer-mediated communication environments in teacher education: Computer conferencing and the supervision of student teachers. *Learning Environments Research*, *1*(1), 59–74. doi:10.1023/A:1009936715640

Ajayi, L. (2009). An exploration of pre-service teachers' perceptions of learning to teach while using asynchronous discussion board. *Journal of Educational Technology & Society*, *12*(2), 86–100.

Branon, R. F., & Essex, C. (2001). Synchronous and asynchronous communication tools in distance education. *TechTrends*, *45*(1), 3642. doi:10.1007/BF02763377

Darling-Hammond, L. (2006). Constructing 21st-century teacher education. *Journal of Teacher Education*, *57*(3), 300–314. doi:10.1177/0022487105285962

Darling-Hammond, L. (2010). *The flat world and education: How America's commitment to equity will determine our future*. New York, NY: Teachers College Press.

Dewey, J. (1933). *How we think, a restatement of the relation of reflective thinking to the educative process*. Boston, MA: D.C. Heath and Co.

Greene, C. H. (2005). Creating connections: A pilot study on an online community of learners. *Journal of Interactive and Online Learning*, *3*(3), 1–21.

Hastie, M., Chen, N. S., & Kuo, Y. H. (2007). Instructional design for best practice in the synchronous cyber classroom. *Journal of Educational Technology & Society*, *10*(4), 281–294.

Hatch, T., & Grossman, P. (2009). Learning to look beyond the boundaries of representation. *Journal of Teacher Education*, *60*(1), 70–85. doi:10.1177/0022487108328533

Hawkes, M., & Romiszowski, A. (2001). Examining the reflective outcomes of asynchronous computer-mediated communication on inservice teacher development. *Journal of Technology and Teacher Education*, *9*(2), 285–308.

Hixon, E., & So, H. J. (2009). Technology's role in field experiences for preservice teacher training. *Journal of Educational Technology & Society*, *12*(4), 294–304.

Im, Y., & Lee, O. (2003). Pedagogical implications of online discussion for preservice teacher training. *Journal of Research on Technology in Education*, *36*(2), 155–170.

Jakobsson, A. (2006). Students' self-confidence and learning through dialogues in a net-based environment. *Journal of Technology and Teacher Education*, *14*(2), 387–405.

Johnson, B., Sullivan, A. M., & Williams, D. (2009). A one-eyed look at classroom life: Using new technologies to enrich classroom-based research. *Issues in Educational Research*, *19*(1), 34–47.

Kent, A. M. (2007). Powerful preparation of preservice teachers using interactive video conferencing. *Journal of Literacy and Technology*, *8*(2), 42–59.

Levin, B. B., He, Y., & Robbins, H. H. (2006). Comparative analysis of preservice teachers' reflective thinking in synchronous versus asynchronous online case discussions. *Journal of Technology and Teacher Education*, *14*(3), 439–460.

Makinster, J. G., Barab, S. A., & Harwood, W. (2006). The effect of social context on the reflective practice of preservice science teachers: Incorporating a web-supported community of teachers. *Journal of Technology and Teacher Education*, *14*(3), 543–579.

Motiwalla, L., & Tello, S. (2000). Distance learning on the Internet: An exploratory study. *The Internet and Higher Education*, *2*(4), 253–264. doi:10.1016/S1096-7516(00)00026-9

National Board for Professional Teaching. (2011). *National Board for Professional Teaching standards*. Retrieved from http://www.nbpts.org

National Council for the Accredidation of Teacher Education. (2010). *NCATE Unit Standards*. Retrieved from http://www.ncate.org/Standards/NCATEUnitStandards/UnitStandardsinEffect2008/tabid/476/Default.aspx

North Carolina Department of Public Instruction. (2010). *Professional development*. Retrieved from http://www.dpi.state.nc.us/profdev/training/teacher/

Partnership for 21st Century Skills (2009). *Framework for 21st century learning*. Retrieved from http://www.21stcenturyskills.org/documents/framework_flyer_updated_jan_09_final-1.pdf

Schulken, M. (2008, September 25). The end of brick and mortar era. *The Charlotte Observer*.

Sherin, M. G., & Van Es, E. A. (2005). Using video to support teachers' ability to notice classroom interactions. *Journal of Technology and Teacher Education*, *13*(3), 475–491.

Song, L., Singleton, E. S., Hill, J. R., & Koh, M. H. (2004). Improving online learning: Student perceptions of useful and challenging characteristics. *The Internet and Higher Education*, *7*(1), 59–70. doi:10.1016/j.iheduc.2003.11.003

Soo, K. S., & Bonk, C. J. (June, 1998). *Interaction: What does it mean in online distance education?* Paper presented at The World Conference on Educational Multimedia and Hypermedia & World Conference on Educational Telecommunications. Freiburg, Germany.

Wise, A. F., Padmanabhan, P., & Duffy, T. M. (2009). Connecting online learners with diverse local practices: The design of effective common reference points for conversation. *Distance Education*, *30*(3), 317–338. doi:10.1080/01587910903236320

Zeichner, K. (2010). Rethinking the connections between campus courses and field experiences in college- and university-based teacher education. *Journal of Teacher Education*, *61*(1-2), 89–99. doi:10.1177/0022487109347671

ADDITIONAL READING

Admiraal, W. F., Lockhorst, D., Wubbels, T., Korthagen, F. A. J., & Veen, W. (1998). Computer-mediated communication environments in teacher education: Computer conferencing and the supervision of student teachers. *Learning Environments Research*, *1*(1), 59–74. doi:10.1023/A:1009936715640

Ajayi, L. (2009). An exploration of pre-service teachers' perceptions of learning to teach while using asynchronous discussion board. *Journal of Educational Technology & Society*, *12*(2), 86–100.

Alexander, G. C. (2003). Reaching out to rural schools: University-practitioner linkage through the Internet. *Journal of Technology and Teacher Education*, *11*(2), 321–330.

Annetta, L., Murray, M., Laird, S. G., Bohr, S., & Park, J. (2008). Investigating student attitudes toward a synchronous, online graduate course in a multi-user virtual learning environment. *Journal of Technology and Teacher Education*, *16*(1), 5–34.

Ballantyne, R., & Mylonas, A. (2001). Improving student learning during 'remote' school-based teaching experience using flexible delivery of teacher mentor and student preparation programmes. *Asia-Pacific Journal of Teacher Education*, *29*(3), 263–273. doi:10.1080/13598660120091865

Bonk, C. (2003). I should have known this was coming: Computer-mediated discussions in teacher education. *Journal of Research on Technology in Education*, *36*(2), 95–102.

Branon, R. F., & Essex, C. (2001). Synchronous and asynchronous communication tools in distance education. *TechTrends, 45*(1), 36, 42.

Broadbent, C. (1998). Preservice students' perceptions and level of satisfaction with their field experiences. *Asia-Pacific Journal of Teacher Education*, *26*(1), 27. doi:10.1080/1359866980260103

Brophy, J. (Ed.). (2010). *Using video in teacher education*. United Kingdom: Emerald JAI.

Brzycki, D., & Dudt, K. (2005). Overcoming barriers to technology use in teacher preparation programs. *Journal of Technology and Teacher Education*, *13*(4), 619–641.

Compton, L., Davis, N., & Correia, A.-P. (2010). Pre-service teachers' preconceptions, misconceptions, and concerns about virtual schooling. *Distance Education*, *31*(1), 37–54. doi:10.1080/01587911003725006

Dai, Chifeng, Sindelar, P. T., Denslow, D., Dewey, J., & Rosenberg, M. S. (2007). Economic analysis and the design of alternative-route teacher education programs. *Journal of Teacher Education*, *58*(5), 422–439. doi:10.1177/0022487107306395

Ferdig, R. E., & Roehler, L. R. (2003). Student uptake in electronic discussions: Examining online discourse in literacy preservice classrooms. *Journal of Research on Technology in Education*, *36*(2), 119–136.

Garrison, D. R., & Kanuka, H. (2004). Blended learning: Uncovering its transformative potential in higher education. *The Internet and Higher Education*, *7*(2), 95–105. doi:10.1016/j.iheduc.2004.02.001

Gillies, D. (2008). Student perspectives on videoconferencing in teacher education at a distance. *Distance Education*, *29*(1), 107–118. doi:10.1080/01587910802004878

Gomez, L. M., Sherin, M. G., Griesdorn, J., & Finn, L.-E. (2008). Creating social relationships. *Journal of Teacher Education*, *59*(2), 117–131. doi:10.1177/0022487107314001

Grant, M. M. (2004). Learning to teach with the web: Factors influencing teacher education faculty. *The Internet and Higher Education*, *7*(4), 329–341. doi:10.1016/j.iheduc.2004.09.005

Hastie, M., Chen, N.-S., & Kuo, Y.-H. (2007). Instructional design for best practice in the synchronous cyber classroom. *Journal of Educational Technology & Society*, *10*(4), 281–294.

Hatch, T., & Grossman, P. (2009). Learning to look beyond the boundaries of representation. *Journal of Teacher Education*, *60*(1), 70–85. doi:10.1177/0022487108328533

Hawkes, M., & Romiszowski, A. (2001). Examining the reflective outcomes of asynchronous computer-mediated communication on inservice teacher development. *Journal of Technology and Teacher Education*, *9*(2), 285–308.

Hew, K., & Hara, N. (2007). Empirical study of motivators and barriers of teacher online knowledge sharing. *Educational Technology Research and Development*, *55*(6), 573–595. doi:10.1007/s11423-007-9049-2

Hixon, E., & So, H.-J. (2009). Technology's role in field experiences for preservice teacher training. *Journal of Educational Technology & Society*, *12*(4), 294–304.

Im, Y., & Lee, O. (2003). Pedagogical implications of online discussion for preservice teacher training. *Journal of Research on Technology in Education*, *36*(2), 155–170.

Jakobsson, A. (2006). Students' self-confidence and learning through dialogues in a net-based environment. *Journal of Technology and Teacher Education*, *14*(2), 387–405.

Jetton, T. L. (2003). Using computer-mediated discussion to facilitate preservice teachers' understanding of literacy assessment and instruction. *Journal of Research on Technology in Education*, *36*(2), 171–191.

Johnson, G. M., & Howell, A. J. (2005). Attitude toward instructional technology following required versus optional WebCT usage. *Journal of Technology and Teacher Education*, *13*(4), 643–654.

Jung, I. (2005). Cost□effectiveness of online teacher training. *Open Learning*, *20*(2), 131–146. doi:10.1080/02680510500094140

Lao, T., & Gonzales, C. (2005). Understanding online learning through a qualitative description of professors and students' experiences. *Journal of Technology and Teacher Education*, *13*(3), 459–474.

Levin, B. B., He, Y., & Robbins, H. H. (2006). Comparative analysis of preservice teachers' reflective thinking in synchronous versus asynchronous online case discussions. *Journal of Technology and Teacher Education*, *14*(3), 439–460.

Maher, M., & Jacob, E. (2006). Peer computer conferencing to support teachers' reflection during action research. *Journal of Technology and Teacher Education*, *14*(1), 127–150.

Makinster, J. G., Barab, S. A., & Harwood, W. (2006). The effect of social context on the reflective practice of preservice science teachers: Incorporating a web-supported community of teachers. *Journal of Technology and Teacher Education*, *14*(3), 543–579.

Motiwalla, L., & Tello, S. (2000). Distance learning on the Internet: An exploratory study. *The Internet and Higher Education*, *2*(4), 253–264. doi:10.1016/S1096-7516(00)00026-9

Oslington, P. (2004). The impact of uncertainty and irreversibility on investments in online learning. *Distance Education*, *25*(2), 233–242. doi:10.1080/0158791042000262120

Rich, P. J., & Hannafin, M. (2009). Video annotation tools. *Journal of Teacher Education*, *60*(1), 52–67. doi:10.1177/0022487108328486

Rosaen, C. L., Lundeberg, M., Cooper, M., Fritzen, A., & Terpstra, M. (2008). Noticing noticing. *Journal of Teacher Education*, *59*(4), 347–360. doi:10.1177/0022487108322128

Sherin, M. G., & Van Es, E. A. (2005). Using video to support teachers' ability to notice classroom interactions. *Journal of Technology and Teacher Education*, *13*(3), 475–491.

Simpson, M. (2006). Field experience in distance delivered initial teacher education programmes. *Journal of Technology and Teacher Education*, *14*(2), 241–254.

Skylar, A. A., Higgins, K., & Boone, R. (2005). Distance education: An exploration of alternative methods and types of instructional media in teacher education. *Journal of Special Education Technology*, *20*(3), 25–33.

Song, L., Singleton, E. S., Hill, J. R., & Koh, M. H. (2004). Improving online learning: Student perceptions of useful and challenging characteristics. *The Internet and Higher Education*, *7*(1), 59–70. doi:10.1016/j.iheduc.2003.11.003

Swinglehurst, D., Russell, J., & Greenhalgh, T. (2008). Peer observation of teaching in the online environment: An action research approach. *Journal of Computer Assisted Learning*, *24*(5), 383–393. doi:10.1111/j.1365-2729.2007.00274.x

Winitzky, N., & Arends, R. (1991). Translating research into practice: The effects of various forms of training and clinical experience on preservice students' knowledge, skill, and reflectiveness. *Journal of Teacher Education*, *42*(1), 52–65. doi:10.1177/002248719104200108

Wise, A. F., Padmanabhan, P., & Duffy, T. M. (2009). Connecting online learners with diverse local practices: The design of effective common reference points for conversation. *Distance Education*, *30*(3), 317–338. doi:10.1080/01587910903236320

Zeichner, K. (2010). Rethinking the connections between campus courses and field experiences in college- and university-based teacher education. *Journal of Teacher Education*, *61*(1-2), 89–99. doi:10.1177/0022487109347671

KEY TERMS AND DEFINITIONS

Asynchronous Learning/Instruction: Not occurring at the same time.

Clinicals: Common course requirement that teachers spend a fixed amount of time in a classroom setting, synonymous with field experiences.

Distance Education: Instructional programs that utilize the Internet to provide communication between students and instructors.

Remote Field Experience: Clinical requirements of time spent in a classroom setting are provided in a technology-supported manner that allows the teacher candidate to remain physically outside of the classroom setting.

Synchronous Learning/Instruction: Occurring real-time.

Video in the Classroom: The presence of technology that allows teaching and learning to be recorded for a variety of uses.

Web-Supported Learning: Any combination of synchronous, asynchronous and face to face learning that is enhanced or facilitated through use of the Internet.

Compilation of References

Aaronson, D., Barrow, L., & Sander, W. (2007). *Teachers and student achievement in the Chicago Public High Schools*. Retrieved from http://www.csa.com

Adams, P. C. (1998). Teaching and learning with SimCity 2000. *The Journal of Geography*, *97*(2), 47–55. doi:10.1080/00221349808978827

Adelman, C. (1993). Kurt Lewin and the origins of action research. *Educational Action Research*, *1*(1), 7–24. doi:10.1080/0965079930010102

Admiraal, W. F., Lockhorst, D., Wubbels, T., Korthagen, F. A. J., & Veen, W. (1998). Computer-mediated communication environments in teacher education: Computer conferencing and the supervision of student teachers. *Learning Environments Research*, *1*(1), 59–74. doi:10.1023/A:1009936715640

Aiken, I. P., & Day, B. D. (1999). Early field experiences in preservice teacher education: Research and student perspectives. *Action in Teacher Education*, *21*(3), 7–12. doi:10.1080/01626620.1999.10462965

Ajayi, L. (2009). An exploration of pre-service teachers' perceptions of learning to teach while using asynchronous discussion board. *Journal of Educational Technology & Society*, *12*(2), 86–100.

Ajjan, H., & Hartshorne, R. (2008). Investigating faculty decisions to adopt Web 2.0 technologies: Theory and empirical tests. *The Internet and Higher Education*, *11*(2), 71–80. doi:10.1016/j.iheduc.2008.05.002

Akyol, Z., & Garrison, D. R. (2008). The development of a community of inquiry over time in an online course: Understanding the progression and integration of social, cognitive and teaching presence. *Journal of Asynchronous Learning Networks*, *12*(2-3), 3–22.

Akyol, Z., & Garrison, D. R. (2011). Understanding cognitive presence in an online and blended community of inquiry: Assessing outcomes and processes for deep approaches to learning. *British Journal of Educational Technology*, *42*(2), 233–250. doi:10.1111/j.1467-8535.2009.01029.x

Akyol, Z., Vaughan, N., & Garrison, D. R. (2011). The impact of course duration on the development of a community of inquiry. *Interactive Learning Environments*, *19*(3), 231–246. doi:10.1080/10494820902809147

Alavi, M., & Gallupe, R. (2003). Using information technology in learning: Case studies in business and management education programs. *Academy of Management Learning & Education*, *2*, 139–153. doi:10.5465/AMLE.2003.9901667

Alban, T., Proffitt, T. D., & SySantos, C. (1998). *Defining performance based assessment within a community of learners: The challenge & the promise*. Retrieved from http://www.csa.com

Al-Bataineh, A., Brooks, S. L., & Bassoppo-Moyo, T. C. (2005). Implications of online teaching and learning. *International Journal of Instructional Media*, *32*(3), 285–294.

Allen, E. I., & Seaman, J. (2011). *Going the distance: Online education in the United States, 2011*. Needham, MA: Sloan Consortium. Retrieved from http://sloanconsortium.org/publications/survey/going_distance_2011

Allen, E., & Seaman, J. (2011). *Class differences: Online education in the United States, 2010*. Newburyport, MA: Sloan Consortium (Sloan-C).

Allen, I. E., & Seaman, J. (2010). *Class differences: Online education in the United States, 2010.* The Alfred P. Sloan Foundation: Author. Retrieved from http://sloanconsortium.org/publications/survey/class_differences

Allen, I., & Seaman, J. (2005). *Growing by degrees: Online education in the United States, 2005.* Babson Park, MA: Sloan Consortium: Babson Survey Research Group.

Allen, E., & Seaman, J. (2007). *Online nation five years of growth in online learning.* Needham, MA: The Sloan Consortium.

Allington, R. (2002). *Big brother and the national reading curriculum: How ideology trumped evidence.* Portsmouth, NH: Heinemann.

Almond, R., Steinberg, L., & Mislevy, R. (2002). Enhancing the design and delivery of assessment systems: A four process architecture. *The Journal of Technology, Learning, and Assessment, 1*(5), 3–63.

Althaus, S. L. (1997). Computer-mediated communication in the university classroom: An experiment with on-line discussions. *Communication Education, 46*, 158–174. doi:10.1080/03634529709379088

Ambient Insight. (2011). *2011 learning technology research taxonomy: Research methodology, buyer segmentation, product definitions, and licensing model.* Monroe, WA: Author. Retrieved from http://www.ambientinsight.com/Resources/Documents/AmbientInsight_Learning_Technology_Taxonomy.pdf

American Association of Colleges for Teacher Education and The Partnership for 21st Century Skills. (2010). *21st century knowledge and skills in educator preparation.* Upper Saddle River, NJ: Pearson.

American Association of Colleges of Teacher Education and the Partnership for 21st Century Skills. (2010). *Educator preparation: A vision for the 21st century.* Retrieved from http://aacte.org/email_blast/president_e-letter/files/02-16-2010/Educator%20Preparation%20and%2021st%20Century%20Skills%20DRAFT%20021510.pdf

American Association of Colleges of Teacher Education and the Partnership for 21st Century. (2010). *Evidence of teacher effectiveness by pathway to entry into teaching.* Retrieved from http://aacte.org/pdf/Publications/Reports%20_Studies/Evidence%20of%20Teacher%20Effectiveness%20by%20Pathway.pdf

American Education Research Association. (2005). *The impact of teacher education: What do we know?* Retrieved from http://www.aera.net/uploadedFiles/News_Media/News_Releases/2005/STE-WhatWeKnow1.pdf

Anderson, B., & Simpson, M. (2005). *From distance teacher education to beginning teaching: What impacts on practice?* Paper presented at the Open and Distance Learning Association of Australia Conference, Adelaide. Retrieved from http://www.odlaa.org/events/2005conf/ref/ODLAA2005Anderson-Simpson.pdf

Anderson, K., Allen, L., Brooks, K., DiBiase, W., Finke, J., & Gallagher, S. Calhoun, M. (2004). *Rising to the challenge: Preparing excellent professionals: The conceptual framework for professional education programs at UNC Charlotte* (2nd ed.). Retrieved from http://education.uncc.edu/coe/Conceptual_Framework/conceptual%20framework.pdf

Anderson, L., & Krathwohl, D. (Eds.). (2001). *A taxonomy for learning, teaching, and assessing: A revision of Bloom's taxonomy of educational objectives.* New York, NY: Longman Publishers.

Anderson, T. (2005). Design-based research and its application to call centre innovation in distance education. *Canadian Journal of Learning and Technology, 31*(2).

Anderson, T., Rourke, L., Garrison, D. R., & Archer, W. (2001). Assessing teaching presence in a computer conferencing context. *Journal of Asynchronous Learning Networks, 5*(2), 1–17.

Anders, P., Hoffman, J., & Duffy, G. (2000). Teaching teachers to teach reading: Paradigm shifts, persistent problems, and challenges. In Kamil, M. L., Mosenthal, P. B., Pearson, P. D., & Barr, R. (Eds.), *Handbook of reading research* (*Vol. III*, pp. 719–742). Mahwah, NJ: Lawrence Erlbaum Associates.

Anstey, M., & Bull, G. (2006). *Teaching and learning multiliteracies: Changing times, changing literacies.* Newark, DE: International Reading Association.

Anthony, G., & Kane, R. (2008). *Making a difference: The role of initial teacher education and induction in the preparation of secondary teachers*. New Zealand: Crown.

Appana, S. (2008). A review of benefits and limitations of online learning in the context of the student, the instructor and the tenured faculty. *International Journal on E-Learning*, *7*(1), 5–22.

Aragon, S. R. (2003). Creating social presence in online environments. *New Directions for Adult and Continuing Education*, *100*, 57–68. doi:10.1002/ace.119

Arbaugh, B., Bangert, A., & Cleveland-Innes, M. (2010). Subject matter effects and the community of inquiry framework. *The Internet and Higher Education*, *13*(1-2). doi:10.1016/j.iheduc.2009.10.006

Arbaugh, J. B., & Benbunan-Fich, R. (2005). Contextual factors that influence ALNs. In Hiltz, S. R., & Goldman, R. (Eds.), *Learning together online: Research on asynchronous learning networks* (pp. 123–144). Mahwah, NJ: Lawrence Erlbaum Associates.

Arbaugh, J. B., Cleveland-Innes, M., Diaz, S., Garrison, D. R., Ice, P., & Richardson, J. C. (2008). Developing a community of inquiry instrument: Testing a measure of the community of inquiry framework using a multi-institutional sample. *The Internet and Higher Education*, *11*(3-4), 133–136. doi:10.1016/j.iheduc.2008.06.003

Archambault, L., & Crippen, K. (2009). K-12 distance educators at work: who's teaching online across the United States. *Journal of Research on Technology in Education*, *41*(4), 363–376.

Archer, W. (2010). Beyond online discussions: Extending the community of inquiry framework to entire courses. *The Internet and Higher Education*, *13*(1-2), 69. doi:10.1016/j.iheduc.2009.10.005

Aronson, J. Z., & Timms, M. J. (2003). *Net choices, net gains: Supplementing the high school curriculum with online courses*. San Francisco, CA: WestEd. Retrieved from www.wested.org/online_pubs/KN-03-02.pdf

Ashton, P. (1984). Teacher efficacy: A motivational paradigm for effective teacher education. *Journal of Teacher Education*, *35*(5), 28–32. doi:10.1177/002248718403500507

Atkinson, T. A. (2009). *Syllabus for READ 6422 clinical procedures in the identification and evaluation of reading disabilities*. East Carolina University.

Aubusson, P. J., Ewing, R., & Hoban, G. (2009). *Action learning in schools: Reframing teachers' professional learning and development*. London, UK: Routledge.

Aud, S., Hussar, W., Kena, G., Bianco, K., Frohlich, L., Kemp, J., & National Center for Education Statistics. (2011). *The condition of education 2011*. NCES 2011-033, National Center for Education Statistics.

Aufderheide, P. (1997). Media literacy: From a report on the National Leadership Conference on media literacy. In R. Kubey (Ed.), *Media literacy in the information age: Current perspectives* (pp. 79-86). New Brunswick, NJ: Transaction Publishers.

Austin, A., & McDaniels, M. (2006). Using doctoral education to prepare faculty to work within Boyer's four domains of scholarship. *New Directions for Institutional Research*, *129*, 51–65. doi:10.1002/ir.171

Avidov-Ungar, O., & Eshet-Alkakay, Y. (2011). Teachers in a world of change: Teachers' knowledge and attitudes towards the implementation of innovative technologies in schools. *Interdisciplinary Journal of E-Learning and Learning Objects*, *7*, 291–303.

Baber, T. C. (2011). The online crit: The community of inquiry meets design education. *Journal of Distance Education*, *25*(1).

Baldwin, T. T., & Ford, J. K. (1988). Transfer of training: A review and directions for future research. *Personnel Psychology*, *41*, 63–105. doi:10.1111/j.1744-6570.1988.tb00632.x

Ball, D. L., Thames, M. H., & Phelps, G. (2008). Content knowledge for teaching: What makes it special? *Journal of Teacher Education*, *59*(5), 389–407. doi:10.1177/0022487108324554

Banks, J. A. (2000). *Cultural diversity and education: Foundations, curriculum, and teaching*. Boston, MA: Allyn & Bacon.

Banks, J. A., & McGee Banks, C. A. (Eds.). (2006). *Multicultural education: Issues and perspectives*. New York, NY: John Wiley.

Barab, S. A., MaKinster, J. G., Moore, J. A., & Cunningham, D. J. (2001). Designing and building an on line community: The struggle to support sociability in the Inquiry Learning Forum. *Educational Technology Research and Development*, *49*(4), 71–96. doi:10.1007/BF02504948

Barab, S., Barnett, M., & Squire, K. (2002). Developing an empirical account of a community of practice: Characterizing the essential tensions. *Journal of the Learning Sciences*, *11*(4), 489–542. doi:10.1207/S15327809JLS1104_3

Baran, E., & Correia, A. (2009). Student-led facilitation strategies in online discussions. *Distance Education*, *30*(3), 339–361. doi:10.1080/01587910903236510

Barbour, M. (2010). Researching K-12 online learning. *Distance Learning*, *7*(2), 6–12.

Barbour, M. K. (2007). Principles of effective web-based content for secondary school students: Teacher and developer perceptions. *Journal of Distance Education*, *21*(3), 93–114.

Barbour, M. K. (2009). Today's student and virtual schooling: The reality, the challenges, the promise.... *Journal of Distance Learning*, *13*(1), 5–25.

Barbour, M. K. (2011). Training teachers for a virtual school system: A call to action. In Polly, D., Mims, C., & Persichitte, K. (Eds.), *Creating technology-rich teacher education programs: Key issues* (pp. 499–517). Hershey, PA: IGI Global.

Barbour, M. K., & Reeves, T. C. (2009). The reality of virtual schools: A review of the literature. *Computers & Education*, *52*(2), 402–416. doi:10.1016/j.compedu.2008.09.009

Barkley, E. F., Cross, P. K., & Major, C. H. (2005). *Collaborative learning techniques. A handbook for college faculty*. San Francisco, CA: Jossey-Bass.

Barnett, M. (2002). *Issues and trends concerning electronic networking technologies for teacher professional development: A critical review of the literature*. Paper presented at the Annual Meeting of the American Educational Research Association, New Orleans, LA.

Barone, D., & Morrell, E. (2007). Multiple perspectives on preparing teachers to teach reading. *Reading Research Quarterly*, *42*, 167–180. doi:10.1598/RRQ.42.1.10

Barron, B. (2003). When smart groups fail. *Journal of the Learning Sciences*, *12*(3), 307–359. doi:10.1207/S15327809JLS1203_1

Bartell, C. (2005). *Cultivating high-quality teaching through induction and mentoring*. Thousand Oaks, CA: Corwin Press.

Barton, D. (2007). *Literacy: An introduction to the ecology of written language*. Oxford, UK: Blackwell.

Barton, D., & Hamilton, M. (1998). *Local literacies: Reading and writing in one community*. New York, NY: Routledge.

Bar-Yam, Y. (1997). *Dynamics of complex systems*. Reading, MA: Addison-Wesley.

Bawane, J., & Spector, M. (2009). Prioritization of online instructor roles: Implications for competency-based teacher education programs. *Distance Education*, *30*(3), 383–397. doi:10.1080/01587910903236536

Beattie, J., Spooner, F., Jordan, L., Algozzine, B., & Spooner, M. (2002). Evaluating instruction in distance learning classes. *Teacher Education and Special Education*, *25*(2), 124–132. doi:10.1177/088840640202500204

Bechtel, W. (1993). *Discovering complexity: Decomposition and localization as strategies in scientific research*. Princeton, NJ: Princeton University Press.

Beck, J., & Wade, M. (2004). *Got game: How the gamer generation is reshaping business forever*. Boston, MA: Harvard Business School Press.

Beile, P. M., & Boote, D. N. (2002). Library instruction and graduate professional development: Exploring the effect of learning environments on self-efficacy and learning outcomes. *The Alberta Journal of Educational Research*, *48*, 364–367.

Beinhocker, E. (2006). *The origin of wealth: Evolution, complexity and the radical remaking of economics*. Boston, MA: Harvard Business School Press.

Benbunan-Fich, R., Hiltz, S. R., & Turoff, M. (2003). A comparative content analysis of face-to-face vs. asynchronous group decision making. *Decision Support Systems*, *34*(4), 457–469. doi:10.1016/S0167-9236(02)00072-6

Benbunan-Fich, R., & Hilz, S. R. (2003). Mediators of the effectiveness of online courses. *IEEE Transactions on Professional Communication*, *46*(4), 298–312. doi:10.1109/TPC.2003.819639

Bender, T. (2003). *Discussion-based online teaching to enhance student learning*. Sterling, VA: Stylis.

Bennett, G., & Green, F. (2001). Promoting service learning via online instruction. *College Student Journal*, *35*(4), 491–497.

Berg, B. L. (2004). *Qualitative research methods for the social sciences* (5th ed.). Boston, MA: Pearson.

Berge, Z. L. (2002). Active, interactive, and reflective elearning. *Quarterly Review of Distance Education*, *3*, 181–190.

Bernard, R. M., Abrami, P. C., Lou, Y., Borokhovski, E., Wade, A., & Wozney, L. (2004). How does distance education compare with classroom instruction? A meta-analysis of the empirical literature. *Review of Educational Research*, *74*(3), 379–439. doi:10.3102/00346543074003379

Berry, B., Daughtrey, A., & Wieder, A. (2010). *Preparing to lead an effective classroom: The role of teacher training and professional development programs.* Center for Teaching Quality. Retrieved from http://www.eric.ed.gov/PDFS/ED509718.pdf

Blake, D. (2009, December). What I learned from teaching adult learners online. *Elearn Magazine.* Retrieved from http://elearningmag.acm.org/archive.cfm?aid+1692866

Bloom, B. S., & Krathwohl, D. R. (1956). *Taxonomy of educational objectives: The classification of educational goals*. New York, NY: Longmans, Green.

Bloome, D., Carter, S. P., Christian, B. M., Otto, S., & Shuart-Faris, N. (2005). *Discourse analysis and the study of classroom language and literacy events: A mircoethnographic perspective*. Mahwah, NJ: Lawrence Erlbaum Associates.

Bogden, R. R., & Biklen, S. K. (2003). *Qualitative research in education: An introduction to theories and methods* (4th ed.). Boston, MA: Allyn & Bacon.

Boler, M. (Ed.). (2004). *Democratic dialogue in education: Troubling speech, disturbing silence*. New York, NY: Peter Lang.

Bolick, C. M., Berson, M. J., Friedman, A. M., & Porfeli, E. J. (2007). Diffusion of technology innovation in the preservice social studies experience: Results of a national survey. *Theory and Research in Social Education*, *35*(2), 174–195. doi:10.1080/00933104.2007.10473332

Bolman, L., & Deal, T. (2008). *Reframing organizations: Artistry, choice, and leadership* (4th ed.). San Francisco, CA: Wiley.

Bonk, C. (2002). Online teaching in an online world (executive summary). *USDLA Journal*, *16*(1). Retrieved from http://www.usdla.org/html/journal/JAN02_Issue/article02.html

Bonk, C. (2004). *The perfect e-storm: Emerging technologies, enhanced pedagogy, enormous learner demand, and erased budgets*. London, UK: The Observatory on Borderless Higher Education.

Bonk, C. J. (2004). *The perfect E-storm: Emerging technologies, enhanced pedagogy, enormous learner demand, and erased budgets*. London, UK: The Observatory on Borderless Higher Education.

Bonk, C. J., Wisher, R. A., & Lee, J. Y. (2003). Moderating learner-centered e-learning: Problems and solutions, benefits and implications. In Roberts, T. S. (Ed.), *Online collaborative learning: Theory and practice* (pp. 54–85). Hershey, PA: Information Science Publishing. doi:10.4018/978-1-59140-174-2.ch003

Boston, W., Diaz, S. R., Gibson, A. M., Ice, P., Richardson, J., & Swan, K. (2009). An exploration of the relationship between indicators of the community of inquiry framework and retention in online programs. *Journal of Asynchronous Learning Networks*, *13*(3), 67–83.

Boswell, C. (2003). Well-planned online discussions promote critical thinking. *Online Cl@ssroom*, 1–8.

Bowers, C., Vasquez, M., & Roaf, M. (2000). Native people and the challenge of computers: Reservation schools, individualism and consumerism. *American Indian Quarterly*, *24*(2), 182–199.

Brand, G. (1997). What research says: Training teachers for using technology. *Journal of Staff Development*, *19*(1), 10–13.

Branon, R. F., & Essex, C. (2001). Synchronous and asynchronous communication tools in distance education. *TechTrends*, *45*(1), 3642. doi:10.1007/BF02763377

Branon, R. F., & Essex, C. (2001). Synchronous and asynchronous communication tools in distance education. *TechTrends*, *45*(1), 3642. doi:10.1007/BF02763377

Brant, G., Hooper, E., & Sugrue, B. (1991). Which comes first the simulation or the lecture? *Journal of Educational Computing Research*, *7*(4), 469–481. doi:10.2190/PWDP-45L8-LHL5-2VX7

Brescia, W., Swartz, J., Pearman, C., Williams, D., & Balkin, R. (2004). Peer teaching in web based threaded discussions. *Journal of Interactive Online Learning*, *3*(2). Retrieved from http://www.ncolr.org/jiol/issues/pdf/3.2.1.pdf

Bridgstock, R., Dawson, S., & Hearn, G. (2011). Innovation through social relationships: A qualitative study of outstanding Australian innovators in science, technology, and the creative industries. In Mesquita, A. (Ed.), *Technology for creativity and innovation: Tools, techniques and applications*. Hershey, PA: IGI Global. doi:10.4018/978-1-60960-519-3.ch005

Broad, W. (2005, October 13). Top advisory panel warns of an erosion of the U.S. competitive edge in science. *New York Times*.

Brocklesby, J., & Sandford, M. (2006). *Evaluation of the graduate teacher training programme*. Prepared for Victoria University of Wellington College of Education. Wellington, New Zealand: Research New Zealand. A Powerpoint presentation.

Brookfield, S. D., & Preskill, S. (2005). *Discussion as a way of teaching: Tools and techniques for democratic classrooms* (2nd ed.). San Francisco, CA: Jossey-Bass.

Brooks, D. M., & Kopp, T. W. (1989). Technology in teacher education. *Journal of Teacher Education*, *40*, 2–8. doi:10.1177/002248718904000402

Broughman, S. P. (2006). *Teacher professional development in 1999–2000*. National Center for Education Statistics.

Brown, K. (2000). *Diploma in teleLearning and rural school teaching*. A presentation at Hook Line & Net, Clarenville, NL. Retrieved from http://www.snn-rdr.ca/snn/hln2000/telelearning.html

Brown, B. W., & Liedholm, C. E. (2002). Can web courses replace the classroom in principles of microeconomics? *The American Economic Review*, *92*(2), 1–12. doi:10.1257/000282802320191778

Brown, J. S., Collins, A., & Duguid, P. (1989). Situated cognition and the culture of learning. *Educational Researcher*, *18*(1), 32–42.

Buckingham, D. (2003). *Media education: Literacy, learning and contemporary culture*. Malden, MA: Polity Press.

Bull, G., Knezek, G., Roblyer, M. D., Schrum, L., & Thompson, A. (2005). A proactive approach to a research agenda for educational technology. *Journal of Research on Technology in Education*, *37*(3), 217–220.

Burbules, N. C. (2004). Navigating the advantages and disadvantages of online pedagogy. In Haythornthwaite, C., & Kazmer, M. M. (Eds.), *Learning, culture and community in online education: Research and practice* (pp. 3–17). New York, NY: Peter Lang.

Burbules, N. C., & Callister, T. A. Jr. (2000). Universities in transition: The promise and the challenge of new technologies. *Teachers College Record*, *102*(2), 271–293. doi:10.1111/0161-4681.00056

Burstein, N., Huddleston, C., Gonda, E., & Roberts, J. (Producers), & Burstein, N. (Director). (2008). *American teen*. United States: Paramount Vantage.

Calvani, A., Fini, A., Molino, M., & Ranieri, M. (2010). Visualizing and monitoring effective interactions in online collaborative groups. *British Journal of Educational Technology*, *41*(2), 213–226. doi:10.1111/j.1467-8535.2008.00911.x

Cameron, M., & Baker, R. (2004). *Research on initial teacher education in New Zealand: 1993-2004 literature review and annotated bibliography*. Wellington, New Zealand: New Zealand Council for Educational Research.

Cannings, T., & Talley, S. (2002). Multimedia and online video-based studies for preservice teacher preparation. *Education and Information Technologies*, *7*(4), 359–367. doi:10.1023/A:1020969723060

Capraro, M., Cpraro, R., & Helfeldt, J. (2010). Do differing types of field experiences make a difference in teacher candidates' perceived level of competence? *Teacher Education Quarterly*, *37*(1), 131–154.

Caro-Bruce, C., Klehr, M., Zeichner, K., & Sierra-Piedrahita, A. M. (2009). A school district-based action research program in the United States. In Noffke, S. E., & Somekh, B. (Eds.), *The SAGE handbook of educational action research* (pp. 104–117). Los Angeles, CA: SAGE Publications.

Carr-Chellman, A. (2005). Stealing our smarts: Indigenous knowledge in on-line learning. *International Journal of Media. Technology and Lifelong Learning*, *1*(2), 1–10.

Carroll, A. E., Rivera, F. P., Ebel, B., Zimmerman, F. J., & Christakis, D. A. (2005). Household computer and Internet access: The digital divide in a pediatric clinical population. Proceedings AMIA Annual Symposium, (pp. 111-115).

Carroll, T. (2000, March). *If we didn't have the schools we have today, would we create the schools we have today?* Paper presented at the Society for Information Technology and Teacher Education International Conference, San Diego, CA.

Carroll, T. (2009, September 12). Press release: Transforming schools into 21st century learning environments. *eSchool News*.

Carroll, T., & Fulton, K. (2004). The true cost of teacher turnover. *Threshold,* Spring, 16-17. Retrieved August 14, 2011, from http://www.nctaf.org/resources/news/nctaf_in_the_news/

Carroll, T. (2007). *The high cost of teacher turnover*. National Commission on Teaching and America's Future.

Carroll, T., & Foster, E. (2010). *Who will teach? Experience matters*. Washington, DC: National Commission on Teaching and America's Future.

Carr, W., & Kemmis, S. (1986). *Becoming critical: Knowing through action research*. Geelong, Australia: Deakin University Press.

Carter, J. F. (2009). Lines of communication: Using a wiki in a mathematics course. *Problems, Resources, and Issues in Mathematics Undergraduate Studies*, *19*(1), 1–17.

Cavanaugh, C., Dawson, K., Nash, R., & Ritzhaupt, A. (2008). *Florida digital educator program*. Final research report presented to the Florida Department of Education.

Caywood, K., & Duckett, J. (2003). Online vs. on-campus learning in teacher education. *Teacher Education and Special Education*, *26*(2), 98–105. doi:10.1177/088840640302600203

Cervetti, G., Damico, J., & Pearson, P. D. (2006). Multiple literacies, new literacies, and teacher education. *Theory into Practice*, *45*(4), 378–386. doi:10.1207/s15430421tip4504_12

Chamberlin, M., & Powers, R. (2010). The promise of differentiated instruction for enhancing the mathematical understandings of college students. *Teaching Mathematics and Its Applications*, *29*, 113–139. doi:10.1093/teamat/hrq006

Chant, R. H. (2002). The impact of personal theorizing on beginning teaching: Experiences of three social studies teachers. *Theory and Research in Social Education*, *30*, 516–540. doi:10.1080/00933104.2002.10473209

Chapman, C., Ramondt, L., & Smiley, G. (2005). Strong community, deep learning: Exploring the link. *Innovations in Education and Teaching International*, *47*(3), 217–230. doi:10.1080/01587910500167910

Charmaz, K. (2006). *Constructing grounded theory. A practical guide through qualitative analysis*. London, UK: Sage.

Chastain, T., & Elliot, A. (2001). Cultivating design competence: Online support for beginning design studio. In *Proceedings of Association of Computer Aided Design in Architecture* (pp. 130-139). Quebec City, Canada.

Chen, N., & Ko, H., Kinshuk, & Lin, T. (2005). A model for synchronous learning using the Internet. *Innovations in Education and Teaching International*, *42*, 181–194. doi:10.1080/14703290500062599

Cherubini, L. (2009). Exploring prospective teachers' critical thinking: Case-based pedagogy and the standards of professional practice. *Teaching and Teacher Education*, *25*, 228–234. doi:10.1016/j.tate.2008.10.007

Chiseri-Strater, E., & Sunstein, B. (2006). *What works? A practical guide for teacher research*. Portsmouth, NH: Heinemann.

Christensen, R., Knezek, G., Tyler-Wood, T., & Gibson, D. (2011). SimSchool: An online dynamic simulator for enhancing teacher preparation. *International Journal of Learning technology, 6*(2), 201-220.

Christensen, C. M., Horn, M. B., & Johnson, C. W. (2008). *Disrupting class: How disruptive innovation will change the way the world learns*. New York, NY: McGraw-Hill.

Cisco Systems. (2006). *Technology in schools: What the research says*. Retrieved from http://www.cisco.com/web/strategy/docs/education/TechnologyinSchoolsReport.pdf

Clandinin, D. J. (1986). *Classroom practice*. London, UK: Falmer.

Clark, T. (2001). *Virtual schools: Trends and issues—A study of virtual schools in the United States*. San Francisco, CA: Western Regional Educational Laboratories. Retrieved from http://www.wested.org/online_pubs/virtualschools.pdf

Clark, R. C., & Mayer, R. E. (2008). *E-learning and the science of instruction: Proven guidelines for consumers and designers of multimedia learning* (2nd ed.). San Francisco, CA: Wiley.

Clark, T. (2007). Virtual and distance education in North American schools. In Moore, M. G. (Ed.), *Handbook of distance education* (2nd ed., pp. 473–490). Mahwah, NJ: Lawrence Erlbaum Associates.

Clausen, K. W. (2006). It there meta in the madness?: Action research and the use of meta-analysis. *The Ontario Action Researcher, 8*(3). Retrieved from http://www.nipissingu.ca/oar/archive-V83E.htm

Clegg, A. A. (1991). Games and simulations in social studies education. In Shaver, J. P. (Ed.), *Handbook of research on social studies teaching and learning* (pp. 523–529). New York, NY: Macmillan.

Cleveland, D. (2003). A semester in the life of alternatively certified teachers: Implications for alternative routes to teaching. *High School Journal, 86*, 17–34. doi:10.1353/hsj.2003.0002

Cobb, P., Confrey, J., diSessa, A., Lehrer, R., & Schauble, L. (2003). Design experiments in educational research. *Educational Researcher, 32*(1), 9–13. doi:10.3102/0013189X032001009

Cochran-Smith, M. (1999). Learning to teach for social justice. In Griffin, G. (Ed.), *The education of teachers: Ninety-eighth yearbook of the National Society for the Study of Education* (pp. 114–144). Chicago, IL: University of Chicago Press.

Cochran-Smith, M. (2004). *Walking the road: Race, diversity and social justice in teacher education*. New York, NY: Teacher College Press.

Cochran-Smith, M., & Lytle, S. (1993). *Inside/outside: Teacher research and knowledge*. New York, NY: Teachers College Press.

Cochran-Smith, M., & Lytle, S. (1999). Relationships of knowledge and practice: Teacher learning in communities. *Review of Research in Education, 24*, 249–305.

Cochran-Smith, M., & Lytle, S. (2009). *Inquiry as stance: Practitioner research for the next generation*. New York, NY: Teachers College Press.

Cochran-Smith, M., & Lytle, S. L. (1999). The teacher research movement: A decade later. *Educational Researcher, 28*(7), 15–25.

Cochran-Smith, M., & Lytle, S. L. (2001). Beyond certainty: Taking an inquiry stance on practice. In Lieberman, A., & Miller, L. (Eds.), *Teachers caught in the action: Professional development that matters* (pp. 45–58). New York, NY: Teachers College Press.

Cohen, E. G., Lotan, R. A., Abram, P. L., Scarloss, B. A., & Schultz, S. E. (2002). Can groups learn? *Teachers College Record, 104*(6), 1045–1068. doi:10.1111/1467-9620.00196

Cohen, J. (2002). Statistics: A power primer. *Psychological Bulletin, 112*(1), 155–159. doi:10.1037/0033-2909.112.1.155

Cohen, S. (2005). *Teachers' professional development and the elementary mathematics classroom: Bringing understandings to light*. Mahwaw, NJ: Lawrence Erlbaum Associates, Inc.

Coiro, J. (2009). Rethinking online reading assessment. *Educational Leadership, 66*(6), 59–63.

Colella, V. S., Klopfer, E., & Resnick, M. (2001). *Adventures in modeling: Exploring complex, dynamic systems with StarLogo*. New York, NY: Teachers College Press.

Collazos, C. A., Guerrero, L. A., Pino, J. A., Renzi, S., Klobas, J., & Ortega, M. (2007). Evaluating collaborative learning processes using system-based measurement. *Journal of Educational Technology & Society*, *10*(3), 257–274.

College of Education. (2009). *Annual data report: Introduction*. Retrieved from http://education.uncc.edu/assessment/secure/2009_Data_Report.htm

Collins, A., & Halverson, R. (2009). *Rethinking education in the age of technology: The digital revolution and schooling in America*. New York, NY: Teachers College Press.

Collins, A., Joseph, D., & Bielaczye, K. (2004). Design research: Theoretical and methodological issues. *Journal of the Learning Sciences*, *13*(1), 15–42. doi:10.1207/s15327809jls1301_2

Collis, B. (1999). Designing for differences: Cultural issues in the design of WWW-based course-support sites. *British Journal of Educational Technology*, *30*(3), 201–215. doi:10.1111/1467-8535.00110

Collison, G., Elbaum, B., Haavind, S., & Tinker, R. (2000). *Facilitating online learning*. Madison, WI: Atwood Publishing.

Conceicao, S., & Drummond, S. B. (2005). Online learning in secondary education: A new frontier. *Educational Considerations*, *33*, 31–37.

Conner, M. (2004). *Pedagogy + andragogy.* Ageless Learner. Retrieved from http://agelesslearner.com/intros/andragogy.html

Conrad, R. M., & Donaldson, J. A. (2004). *Engaging the online learner: Activities and resources for creative instruction*. San Francisco, CA: Jossey-Bass.

Cope, B., & Kalantzis, M. (2000). *Multiliteracies: Literacy learning and the design of social futures*. New York, NY: Routledge.

Corbell, K., Booth, S., & Reiman, A. J. (2010). The commitment and retention intentions of traditionally and alternatively licensed math and science beginning teachers. *Journal of Curriculum and Instruction*, *4*, 50–69. doi:10.3776/joci.2010.v4n1p50-69

Corbin, J., & Strauss, A. (1990). Grounded theory research: Procedures, cannons, and evaluative criteria. *Qualitative Sociology*, *13*, 3–31. doi:10.1007/BF00988593

Corey, S. M. (1953). *Action research to improve school practice*. New York, NY: Teachers College Press.

Cortés, C. (2005). How the media teach. In Schwarz, G., & Brown, P. U. (Eds.), *Media literacy: Transforming curriculum and teaching. The 104th yearbook of the National Society for the Study of Education, Part I* (pp. 55–73). Malden, MA: Blackwell Publishing.

Coryell, J. E., & Chulp, D. T. (2007). Implementing e-learning components with adult English language learners: Vital factors and lessons learned. *Computer Assisted Language Learning*, *20*, 263–278. doi:10.1080/09588220701489333

COSEPUP. (2006). *Rising above the gathering storm: Energizing and employing America for a brighter economic future*. The National Academy of Sciences, The National Academy of Engineering, The Institute of Medicine.

Council of Chief State School Officers. (2011, April). *Interstate Teacher Assessment and Support Consortium (INTASC) model core teaching standards: A resource for state dialogue*. Washington, DC: Author. Retrieved from http://www.ccsso.org/Documents/2011/_Model_Core_Teaching_Standards_2011.pdf

Creswell, J. (2003). *Research design: Qualitative, quantitative, and mixed methods approaches* (2nd ed.). Thousand Oaks, CA: Sage.

Crocco, M., Bayard, F., & Schwartz, S. (2003). Inquiring minds want to know: Action research at a New York City professional development school. *Journal of Teacher Education*, *54*(1), 19–30. doi:10.1177/0022487102238655

Crockett, M. D. (2002). Inquiry as professional development: Creating dilemmas through teachers' work. *Teaching and Teacher Education*, *18*, 609–624. doi:10.1016/S0742-051X(02)00019-7

Crotty, M. (1998). *The foundations of social research: Meaning and perspective in the research process*. Thousand Oaks, CA: SAGE Publications.

Cruickshank, D. R., & Armaline, W. D. (1986). Field experiences in teacher education: Considerations and recommendations. *Journal of Teacher Education*, *37*(3), 34–40. doi:10.1177/002248718603700307

Cuban, L. (1986). *Teachers and machines*. New York, NY: Teachers College Press.

Cuban, L. (2001). *Oversold and underused: Computers in the classroom*. Cambridge, MA: Harvard University Press.

Cyrs, T. E. (1997). Competence in teaching at a distance. In Cyrs, T. E. (Ed.), *Teaching and learning at a distance: What it takes to effectively design, deliver, and evaluate programs* (pp. 15–18). San Francisco, CA: Jossey-Bass Publishers.

Dabbagh, N. (2007). The online learner: Characteristics and pedagogical implications. *Contemporary Issues in Technology & Teacher Education*, *7*(3), 217–226.

Dabbagh, N., & Bannan-Ritland, B. (2005). *Online learning: Concepts, strategies, and application*. Upper Saddle River, NJ: Prentice Hall.

Dana, N., Dawson, K., Krell, D., & Wolkenhauer, R. (2012). *Using action research in professional development for virtual school educators: Exploring an established strategy in a new context*. Paper presented at the American Educational Research Conference. Vancouver, Canada.

Dana, N. F., & Silva, D. Y. (2001). Student teachers as researchers: Developing an inquiry stance towards teaching. In Rainer, J., & Guyton, E. M. (Eds.), *Research on the effects of teacher education on teacher performance: Teacher education yearbook IX*. New York, NY: Kendall-Hunt Press.

Dana, N. F., & Silva, D. Y. (2009). *The reflective educator's guide to classroom research: Learning to teach and teaching to learn through practitioner inquiry* (2nd ed.). Thousand Oaks, CA: Corwin Press.

Dana, N. F., Silva, D. Y., & Snow-Gerono, J. (2002). Building a culture of inquiry in a professional development school. *Teacher Education and Practice*, *15*(4), 71–89.

Dana, N. F., & Yendol-Hoppey, D. (2008). *The reflective educator's guide to professional development: Coaching inquiry-oriented learning communities*. Thousand Oaks, CA: Corwin Press.

Dana, N. F., & Yendol-Hoppey, D. (2009). *The reflective educator's guide to classroom research: Learning to teach and teaching to learn through practitioner inquiry*. Thousand Oaks, CA: Corwin Press.

Daniels, G., Schur, M., Klein, H., & Miner, D. (Producers). (2009). *Parks and recreation*. Van Nuys, CA: Universal Media Studios.

Darling-Hammond, L. (1997). Quality teaching: The critical key to learning. *Principal*, *77*, 5–6.

Darling-Hammond, L. (2000). How teacher education matters. *Journal of Teacher Education*, *51*(3), 166–173. doi:10.1177/002248710005100302

Darling-Hammond, L. (2003). Keeping good teachers. *Educational Leadership*, *60*(8), 6–13.

Darling-Hammond, L. (2003). Keeping good teachers: Why it matters, what leaders can do. *Educational Leadership*, *60*(8), 6–13.

Darling-Hammond, L. (2006). Constructing 21st-century teacher education. *Journal of Teacher Education*, *57*(3), 300–314. doi:10.1177/0022487105285962

Darling-Hammond, L. (2006). *Powerful teacher education: Lessons from exemplary programs*. San Francisco, CA: John Wiley & Sons.

Darling-Hammond, L. (2010). *The flat world and education: How America's commitment to equity will determine our future*. New York, NY: Teachers College Press.

Darling-Hammond, L., & Bransford, J. (Eds.). (2005). *Preparing teachers for a changing world: What teachers should learn and be able to do*. San Francisco, CA: Jossey-Bass.

Darling-Hammond, L., Chung, R., & Frelow, F. (2002). Variation in teacher preparation: How well do different pathways prepare teachers to teach. *Journal of Teacher Education*, *53*(4), 286–302. doi:10.1177/0022487102053004002

Darling-Hammond, L., Hammerness, K., Grossman, P., Rust, F., & Shulman, L. (2005). The design of teacher education programs. In Darling-Hammond, L., & Bransford, K. (Eds.), *Preparing teachers for a changing world: What teachers should learn and be able to do* (pp. 390–440). San Francisco, CA: Jossey-Bass.

Darling-Hammond, L., & Youngs, P. (2002). Defining "highly qualified teachers": What does "scientifically-based research" actually tell us? *Educational Researcher*, *31*(9), 13–25. doi:10.3102/0013189X031009013

Daves, D. P., & Roberts, J. G. (2010). Online teacher education programs: Social connectedness and the learning experience. *Journal of Instructional Pedagogies, 4*. Retrieved from http://www.aabri.com/jip.html

Davis, N., Demiraslan, Y., & Wortmann, K. (2007, October). *Preparing to support online learning in K-12*. A presentation at the Iowa Educational Technology Conference, Des Moines, IA. Retrieved from http://ctlt.iastate.edu/~tegivirtual school/TEGIVIRTUAL SCHOOL/publications/ITEC2007-presentations.pdf

Davis, B. (1993). *Tools for teaching*. San Francisco, CA: Jossey-Bass.

Davis, N. E., & Niederhauser, D. S. (2007). Virtual schooling. *Learning and Leading with Technology, 34*(7), 10–15.

Davis, N. E., & Roblyer, M. D. (2005). Preparing teachers for the "schools that technology built": Evaluation of a program to train teachers for virtual schooling. *Journal of Research on Technology in Education, 37*, 399–409.

Davis, N. E., Roblyer, M. D., Charania, A., Ferdig, R., Harms, C., Compton, L. K. L., & Cho, M. O. (2007). Illustrating the "virtual" in virtual schooling: Challenges and strategies for creating real tools to prepare virtual teachers. *The Internet and Higher Education, 10*(1), 27–39. Retrieved from http://ctlt.iastate.edu/~tegivs/TEGIVS/publications/JP2007%20davis&roblyer.pdfdoi:10.1016/j.iheduc.2006.11.001

Davis, N., & Roblyer, M. D. (2005). Preparing teachers for the "schools that technology built": Evaluation of a program to train teachers for virtual schooling. *Journal of Research on Technology in Education, 37*(4), 399–409.

Dawley, L., Rice, K., & Hinck, G. (2010). *Going Virtual! 2010: The status of professional development and unique needs of K-12 online teachers*. Boise, ID: Boise State University. Retrieved from http://edtech.boisestate.edu/goingvirtual/goingvirtual3.pdf

Dawson, K., & Cavanaugh, C. (2010). *Insights into classroom technology integration through action research: For whom, in what ways and with what outcomes and implications?* Paper presented at the American Educational Research Association. Denver, CO.

Dawson, K., & Ferdig, R. E. (2006). Commentary: Expanding notions of acceptable research evidence in educational technology: A Response to Schrum et al. *Contemporary Issues in Technology and Teacher Education, 6*(1). Retrieved from http://www.citejournal.org/vol6/iss1/general/article2.cf

Dawson, K., Cavanaugh, C., & Ritzhaupt, A. (2009, March). *The evolution of ARTI: An online tool to promote classroom-based technology outcomes via teacher inquiry*. Paper presented at the Society for Technology and Teacher Education International Conference, Charleston, SC.

Dawson, K. (2006). Teacher inquiry: A vehicle to merge prospective teachers' experience and reflection during curriculum-based, technology-enhanced field experiences. *Journal of Research on Technology in Education, 38*(3), 265–292.

Dawson, K. (2007). The role of teacher inquiry in helping prospective teachers untangle the complexities of technology use in classrooms. *Journal of Computing in Teacher Education, 24*(1), 5–14.

Dawson, K. (2012). Using action research projects to examine teacher technology integration practices. *Journal of Digital Learning in Teacher Education, 28*(12), 117–124.

Dawson, K., Cavanaugh, C., & Ritzhaupt, A. (2008). Florida's Leveraging Laptops initiative and its impact on teaching practices. *Journal of Research on Technology in Education, 41*(2), 143–159.

Dawson, K., & Dana, N. (2007). When curriculum-based, technology-enhanced field experiences and teacher inquiry coalesce: An opportunity for conceptual change? *British Journal of Educational Technology, 38*(4), 656–667. doi:10.1111/j.1467-8535.2006.00648.x

Dawson, K., Pringle, R., & Adams, T. (2003). Providing links between technology integration, methods courses and traditional field experiences: Implementing a model of curriculum-based and technology-enhanced microteaching. *Journal of Computing in Teacher Education, 20*(1), 41–47.

Dawson, K., Ritzhaupt, A., Liu, M., Drexler, W., Barron, A., Kersaint, G., & Cavanaugh, C. (2011). *Charting a course for the digital science, technology, engineering and mathematics (STEM) classroom: Research and evaluation report*. Florida Department of Education.

de Jong, T., & van Joolingen, W. R. (1998). Scientific discovery learning with computer simulations of conceptual domains. *Review of Educational Research, 68*(2), 179–201.

de Pommereau, I. (1997, April 21). Computers give the key to learning (an interview with Seymour Papert. *Christian Science Monitor, 89*(101), 11. Retrieved from http://www.csmonitor.com

Debuse, J. C. W., Hede, A., & Lawley, M. (2009). Learning efficacy of simultaneous audio and on-screen text in online lectures. *Australasian Journal of Educational Technology, 25*, 748–762.

DeCandido, G. A. (2006). On my mind. *American Libraries, 37*, 23.

Dede, C., Gibson, D., & Damon, T. (2007). Learning games and simulations (video podcast). *National Educational Computing Conference*. Atlanta, GA: International Society for Technology in Education.

Dede, C. (2006). *Online professional development for teachers: Emerging models and methods*. Cambridge, MA: Harvard Education Press.

Dede, C. (2009). *Address to the ITEST Summit 2009. Unpublished Lecture - slide show*. NSF.

Dede, C., Jass Ketelhut, D., Whitehouse, P., Breit, L., & McCloskey, E. M. (2009). A research agenda for online teacher professional development. *Journal of Teacher Education, 60*(1), 8–19. doi:10.1177/0022487108327554

Dede, C., Ketelhut, D., Whitehouse, P., Breit, L., & McCloskey, E. (2009). A research agenda for online teacher professional development. *Journal of Teacher Education, 60*, 8–19. doi:10.1177/0022487108327554

DeLeon, A. P. (2008). Are we simulating the status quo? Ideology and social studies simulations. *Theory and Research in Social Education, 36*(3), 256–277. doi:10.1080/00933104.2008.10473375

Delfino, M., & Persico, D. (2007). Online or face-to-face? Experimenting with different techniques in teacher training. *Journal of Computer Assisted Learning, 23*, 351–365. doi:10.1111/j.1365-2729.2007.00220.x

Dell, C. A., Low, C., & Wilker, J. F. (2010). Comparing student achievement in online and face-to-face class formats. *MERLOT Journal of Online Learning and Teaching, 6*(1), 30–42.

Demiraslan-Cevik, Y. (2008). *Final report to FIPSE for P116B040216 – TEGIVS: Teacher education goes into virtual schooling*. Ames, IA: Iowa State University. Retrieved from http://yunus.hacettepe.edu.tr/~yasemind/HCIPortfolio/TEGIVSPerformanceNarrative.pdf

Dennis, T., El-Gayar, O. F., & Zhou, Z. (2002). A conceptual framework for hybrid distance delivery for information system programs. *Issues in Information Systems, 3*, 137–143.

Department of Education, Science and Training. (DEST). (2002). *Universities online: A survey of online education and services in Australia*. Canberra, Australia: Author.

Desai, M., Hart, J., & Richards, T. (2009). E-learning: Paradigm shift in education. *Education, 129*(2), 327–334.

Design-Based Research Collective. (2003). Design-based research: An emerging paradigm for educational inquiry. *Educational Researcher, 32*(1), 5–8. doi:10.3102/0013189X032001005

Dewey, J. (1891, 1980). *School and society*. Carbondale, IL: Southern Illinois University Press.

Dewey, J. (1897). My pedagogic creed. *The School Journal, 55*(3), 77–80.

Dewey, J. (1916). *How we think*. Boston, MA: Houghton Mifflin Company.

Dewey, J. (1933). *Democracy and education*. New York, NY: The Free Company.

Dewey, J. (1933). *How we think* (rev. ed.). Boston, MA: D. C. Heath.

Dewey, J. (1933). *How we think, a restatement of the relation of reflective thinking to the educative process*. Boston, MA: D.C. Heath and Co.

Dewey, J. (1933). *How we think*. New York, NY: D.C. Heath.

Dewey, J. (1938). *Experience and education*. New York, NY: Simon & Schuster.

Dewey, J. (1997). *Democracy and education: An introduction to the philosophy of education*. New York, NY: The Free Press. (Original work published 1916)

Dexter, S. (2003, March). *The promise of network-based assessment for supporting the development of teachers' technology integration and implementation skills*. Paper presented at the Society for Information Technology and Teacher Education International Conference, New Orleans, LA.

DiGiulio, R. C. (2004). *Great teaching: What matters most in helping students succeed*. Thousand Oaks, CA: Corwin Press.

Dillenbourg, P. (1999). What do you mean by 'collaborative learning? In Dillenbourg, P. (Ed.), *Collaborative-learning: Cognitive and computational approaches* (pp. 1–19). Oxford, UK: Elsevier.

Dinkelman, T. (2003). Self-study in teacher education: A means and end tool for promoting reflective teaching. *Journal of Teacher Education*, *54*, 6–18. doi:10.1177/0022487102238654

DiPietro, M. (2010). Virtual school pedagogy: The instructional practices of K-12 virtual school teachers. *Journal of Educational Computing Research*, *42*(3), 327–354. doi:10.2190/EC.42.3.e

Dirkx, J., & Smith, R. (2004). Thinking out of a bowl of spaghetti: Learning to learn in online collaborative groups. In Roberts, T. S. (Ed.), *Online collaborative learning: Theory and practice* (pp. 132–159). Hersey, PA: Information Science Publishing.

Dodge, B. (1995). WebQuests: A technique for internet-based learning. *Distance Education*, *1*(2), 10–13.

Dollisso, A., & Martin, R. (1999). Perceptions regarding adult learners' motivation to participate in educational programs. *Journal of Agricultural Education*, *40*, 38–46. doi:10.5032/jae.1999.04038

Dönmez, P., Rosé, C. P., Stegmann, K., Weinberger, A., & Fischer, F. (2005). Supporting CSCL with automatic corpus analysis technology. In T. Koschmann, D. Suthers & T. W. Chan (Eds.), *Proceedings of the International Conference on Computer Supported Collaborative Learning – CSCL 2005* (pp. 125–134). Taipei, Taiwan: Erlbaum.

Dooley, K., Lindner, J., & Dooley, L. (2005). Engaging learners and fostering self-directedness. In Dooley, K., Lindner, J., & Dooley, L. (Eds.), *Advanced methods in distance education: Applications and practices* (pp. 76–97). Hershey, PA: Information Science Publishing. doi:10.4018/978-1-59140-485-9.ch005

Dooley, K., Lindner, J., Dooley, L., & Wilson, S. (2005). Adult learning principles and learner differences. In Dooley, K., Lindner, J., & Dooley, L. (Eds.), *Advanced methods in distance education: Applications and practices* (pp. 56–75). Hershey, PA: Information Science Publishing. doi:10.4018/978-1-59140-485-9.ch004

Doolitle, P. E., & Hicks, D. (2003). Constructivism as a theoretical foundation for the use of techology in social studies. *Theory and Research in Social Education*, *31*(1), 72–104. doi:10.1080/00933104.2003.10473216

Doolittle, P. (1999). Constructivist pedagogy. *Educational Psychology*.

Drennon, C. E., & Cervero, R. M. (2002). The politics of facilitation in practitioner inquiry groups. *Adult Education Quarterly*, *52*(3), 193–209. doi:10.1177/0741713602052003003

Drops, G. (2003). Assessing online chat sessions. *Online Cl@ssroom*, 1–8.

Duffy, G. G., Miller, S. D., Kear, K. A., Parsons, S. A., Davis, S. G., & Williams, J. B. (2008). Teachers' instructional adaptations during literacy instruction. In Y. Kim, V. J. Risko, D.L. Compton, D. K. Dickinson, M. K., Hundley, R. T. Jimenez, K. M. Leander, & D. W. Rowe (Eds.), *57th yearbook of the National Reading Conference* (pp. 160-171). Oak Creek, WI: National Reading Conference.

Duffy, G. G., Webb, S., & Davis, S. G. (2009). Literacy education at a crossroad: A strategy for countering the trend to marginalize quality teacher education. In J. V. Hoffman & Y. Goodman (Eds.), *Changing literacies for changing times* (pp. 189-197). New York, NY: Routledge, Taylor and Francis.

Durham, J. (2005). Bowles plans to help ECU meet goals. *Pieces of Eight*. Retrieved August 17, 2011, from http://www.ecu.edu/cs-admin/news/poe/0015/bowles.cfm

Duffy, G. (1993). Teachers' progress toward becoming expert strategy teachers. *The Elementary School Journal*, *94*(2), 109–120. doi:10.1086/461754

Duffy, G. (1994). How teachers think of themselves: A key to mindfulness. In Mangieri, J., & Collins-Block, C. (Eds.), *Creating powerful thinking in teachers and students: Diverse perspectives* (pp. 3–26). Fort Worth, TX: Holt, Rinehart & Winston.

Duffy, G. (1998). Teaching and the balancing of round stones. *Phi Delta Kappan*, *79*, 777–780.

Duffy, G. (2002). Visioning and the development of outstanding teachers. *Reading Research and Instruction*, *41*, 331–344. doi:10.1080/19388070209558375

Duffy, G. G., Webb, S., & Davis, S. G. (2009). Literacy education at a crossroad: A strategy for countering the trend to marginalize quality teacher education. In Hoffman, J. V., & Goodman, Y. (Eds.), *Changing literacies for changing times* (pp. 189–197). New York, NY: Routledge, Taylor and Francis.

Duffy, T., Kirkley, J., del Valle, R., Malopinsky, L., Scholten, C., & Neely, G. (2006). Online teacher professional development: A learning architecture. In Dede, C. (Ed.), *Online professional development for teachers* (pp. 175–197). Cambridge, MA: Harvard Education Press.

Duncan, A. (March 3, 2010). *Using technology to transform schools: Remarks by Secretary Arne Duncan at the Association of American Publishers Annual meeting*. Retrieved from http://www.ed.gov/news/speeches/using-technology-transform-schools%E2%80%94remarks-secretary-arne-duncan-association-american-

Duncan, H. E. (2005). Online education for practicing professionals: A case study. *Canadian Journal of Education*, *28*, 874–896. doi:10.2307/4126459

Duncan, H. E., & Barnett, J. (2010). Experiencing online pedagogy: A Canadian case study. *Teaching Education*, *21*, 247–262. doi:10.1080/10476210903480340

Dykman, C., & Davis, C. (2008). Online education forum: Part two--Teaching online versus teaching conventionally. *Journal of Information Systems Education*, *19*, 157–164.

Dziuban, C. D., Moskal, P. D., & Hartman, J. (2005). Higher education, blended learning, and the generations: Knowledge is power: No more. In Bourne, J., & Moore, J. C. (Eds.), *Elements of quality online education: Engaging communities*. Needham, MA: Sloan Center for Online Education.

Eastman, J. K., & Swift, C. (2001). New horizons in distance education: The online learner-centered marketing class. *Journal of Marketing Education*, *23*(1), 25–34. doi:10.1177/0273475301231004

Easton, S. (2003). Clarifying the instructor's role in online distance learning. *Communication Education*, *52*(2), 87–105. doi:10.1080/03634520302470

Eck, R. V. (2006). Digital game-based learning: It's not just the digital natives who are restless. *EDUCAUSE Review*, (March/April): 17–30.

Edelson, D. C. (2002). Design research: What we learn when we engage in design. *Journal of the Learning Sciences*, *11*(1), 105–121. doi:10.1207/S15327809JLS1101_4

Edelstein, S., & Edwards, J. (2002). If you build it, they will come: Building learning communities through learning discussions. *Online Journal of Distance Learning Administration*, *5*(1). Retrieved from http://www.westga.edu/~distance/ojdla/spring51/edelstein51.html

Edwards, B., Flowers, C., & Stephenson-Green, E. (2005). *Comprehensive assessment system*. UNC Charlotte College of Education. Retrieved from http://education.uncc.edu/assessment/

E-Learning Advisory Group. (2002). *Highways and pathways: Exploring New Zealand's e-learning opportunities*. The Report of the E-Learning Advisory Group, March 2002.

Ellis, C., & Bochner, A. (2000). Autoethnography, personal narrative, reflexivity. In Denzin, N. K., & Lincoln, Y. S. (Eds.), *Sage handbook of qualitative research* (2nd ed., pp. 733–768). Thousand Oaks, CA: Sage.

Elmore, R. (2000). *Building a new structure for school leadership*. Washington, DC: Alberta Shanker Institute.

Emihovich, C. (2008). Preparing global educators: New challenges for teacher education. *Teacher Education and Practice*, *21*(4), 446–448.

Erkens, G., Janssen, J., Jaspers, J., & Kanselaar, G. (2006). Visualizing participation to facilitate argumentation. In S. A. Barab, K. E. Hay, & D. T. Hickey (Eds.), *Proceedings of the 7th International Conference of the Learning Sciences (ICLS)* (Vol. 2, pp. 1095–1096). Mahwah, NJ: Lawrence Erlbaum Associates.

Ernst. J. V. (2008). A comparison of traditional and hybrid online instructional presentations in communication technology. *Journal of Technology Education, 19*(2). Retrieved from http://scholar.lib.vt.edu/ejournals/JTE/v19n2/pdf/ernst.pdf

Ertmer, P., & Ottenbreit-Leftwich, A. (2010). Teacher technology change: How knowledge, confidence, beliefs, and culture intersect. *Journal of Research on Technology in Education, 42*(3), 255–284.

Evans, R. (1996). *The human side of school change: Reform, resistance, and the real-life problems of innovation.* San Francisco, CA: Jossey-Bass Publishers.

Faulkner, W. (1990). *As I lay dying: The corrected text.* New York, NY: Vintage Books.

Faux, T. L., & Black-Hughes, C. (2000). A comparison of using the Internet versus lectures to teach social work history. *Research on Social Work Practice, 10*, 454–466.

Federation of American Scientists. (2006). *Harnessing the power of video games for learning.* Paper presented at the National Summit on Educational Games. Retrieved from http://www.fas.org/gamesummit/index.html

Feiman-Nemser, S. (2003). Keeping good teachers: What new teachers need to learn. *Educational Leadership, 60*(8), 25–29.

Feiman-Nemser, S., & Buchmann, M. (1985). Pitfalls of experience in teacher preparation. *Teachers College Record, 87*(1), 53–65.

Feinberg, J. R. (2011). *Debriefing in simulation games: An examination of reflection on cognitive and affective learning outcomes.* Saarbrücken, Germany: LAP Lambert Academic Publishing.

Feistritzer, C. (2005). *Profile of alternative route teachers.* National Center for Education Information. Retrieved from http://www.ncei.com/PART.pdf

Feldman, A. (2003). Validity and quality in self-study. *Educational Researcher, 32*(4), 26–28. doi:10.3102/0013189X032003026

Fenstermacher, G. D., & Richardson, V. (2005). On making determinations of quality in teaching. *Teachers College Record, 107*(1), 186–213. doi:10.1111/j.1467-9620.2005.00462.x

Ferguson, S. (2008). Key elements for a Māori e-learning framework. *MAI Review, 3*(3), 1–7.

Ferry, B., Kervin, L., Cambourne, B., Turbill, J., Hedberg, J., & Jonassen, D. (2005). Incorporating real experience into the development of a classroom-based simulation. *Journal of Learning Design, 1*(1), 22–32. Retrieved from www.jld.qut.edu.au/Vol_1_No_1

Fischler, R. B. (2006). *SimTeacher: Simulation-based learning in teacher education.* (Doctoral dissertation). Retrieved from ProQuest Dissertations and Theses.(Accession Order No. AAT 3210046)

Fischler, R. (2007). SimTeacher.com: An online simulation tool for teacher education. *Techtrends: Linking Research and Practice to Improve Learning, 51*(1), 44–47. doi:10.1007/s11528-007-0011-2

Fisher, E. A., & Wright, V. H. (2010). Improving online course design through usability testing. *MERLOT Journal of Online Learning and Teaching, 6*(1), 228–245.

Fishman, B. (2007). Fostering community knowledge sharing using ubiquitous records of practice. In Goldman, R., Pea, R. D., Barron, B., & Derry, S. J. (Eds.), *Video research in the learning sciences* (pp. 495–506). Mahwah, NJ: Erlbaum.

Fishman, B. J., Marx, R. W., Best, S., & Tal, R. T. (2003). Linking teachers and student learning to improve professional development in systemic reform. *Teaching and Teacher Education, 19*(6), 643–658. doi:10.1016/S0742-051X(03)00059-3

Forcheri, P. (2011). Editorial: Reimagining schools: The potential of virtual education. *British Journal of Educational Technology, 42*(3), 363–372. doi:10.1111/j.1467-8535.2011.01178.x

Freebody, P., & Luke, A. (1990). Literacies programs: Debates and demands in cultural context. *Prospect: Australian Journal of TESOL, 5*(7), 7–16.

Freese, A. (2006). Reframing one's teaching: Discovering our teacher selves through reflection and inquiry. *Teaching and Teacher Education*, *22*, 100–119. doi:10.1016/j.tate.2005.07.003

Frey, T. (2008). Determining the impact of online practicum facilitation for inservice teachers. *Journal of Technology and Teacher Education*, *16*(2), 181–210.

Frieberg, H. J. (1995). Promoting reflective practices. In Slick, G. A. (Ed.), *Emerging trends in teacher preparation: The future of field experiences* (pp. 25–42). Thousand Oaks, CA: Corwin Press, Inc.

Friedman, T. L. (2006). *The world is flat [updated and expanded]: A brief history of the twenty-first century*. New York, NY: Farrar, Straus and Giroux.

Friend, B., & Johnston, S. (2005). Florida virtual school: A choice for all students. In Berge, Z. L., & Clark, T. (Eds.), *Virtual schools: Planning for success* (pp. 97–117). New York, NY: Teachers College Press.

Frye, B., & Frager, A. M. (1996). Civilization, Colonization, SimCity: Simulations for the social studies classroom. *Learning and Leading with Technology*, *24*(2), 21–23, 32.

Fullan, M. (2001). *Leading in a culture of change*. San Francisco, CA: Jossey Bass.

Gallimore, R., Ermeling, B. A., Saunders, W. M., & Goldenberg, C. (2009). Moving the learning of teaching closer to practice: Teacher education implications of school-based inquiry teams. *The Elementary School Journal*, *109*(5), 537–553. doi:10.1086/597001

Gallo, R., & Little, E. (2003). Classroom behavior problems: The relationship between preparedness, classroom experiences, and self-efficacy in graduate and student teachers. *Australian Journal of Educational & Developmental Psychology* (3), 21-34.

Games, A., & Squire, K. D. (2011). Searching for the fun in learning: A historical perspective on the evolution of educational video games. In Tobias, S., & Fletcher, J. D. (Eds.), *Computer games and instruction* (pp. 371–394). Charlotte, NC: Information Age Publishers.

Garet, M., Porter, A., Desimone, L., Briman, B., & Yoon, K. (2001). What makes professional development effective? Analysis of a national sample of teachers. *American Educational Research Journal*, *38*(4), 915–945. doi:10.3102/00028312038004915

Garland, K., & Pace, B. G. (2011, July). *Shifting concepts of literacy: How media literacy education can serve as transformative pedagogy for secondary students and pre-service teachers*. Paper presented at the meeting of the National Association for Media Literacy Education, Philadelphia, PA.

Garoian, C. (1999). *Performing pedagogy: Toward an art of politics*. Albany, NY: SUNY Press.

Garrison, D. R. (1989). *Understanding distance education: A framework for the future*. London, UK: Routledge.

Garrison, D. R., & Anderson, T. (2003). *E-learning in the 21st century: A framework for research and practice*. London, UK: Routledge/Falmer. doi:10.4324/9780203166093

Garrison, D. R., Anderson, T., & Archer, W. (2000). Critical inquiry in a text-based environment: Computer conferencing in higher education. *The Internet and Higher Education*, *2*, 87–105. doi:10.1016/S1096-7516(00)00016-6

Garrison, D. R., Anderson, T., & Archer, W. (2001). Critical thinking, cognitive presence, and computer conferencing in distance education. *American Journal of Distance Education*, *15*(1), 7–23. doi:10.1080/08923640109527071

Garrison, D. R., & Arbaugh, J. B. (2007). Researching the community of inquiry framework: Review, issues, and future directions. *The Internet and Higher Education*, *10*(3), 157–172. doi:10.1016/j.iheduc.2007.04.001

Garrison, D. R., & Cleveland-Innes, M. (2005). Facilitating cognitive presence in online learning: Interaction is not enough. *American Journal of Distance Education*, *19*(3), 133–148. doi:10.1207/s15389286ajde1903_2

Garrison, D. R., Cleveland-Innes, M., & Fung, T. S. (2010). Exploring relationships among teaching, cognitive and social presence: Student perceptions of the community of inquiry framework. *The Internet and Higher Education*, *13*(1-2), 31–36. doi:10.1016/j.iheduc.2009.10.002

Garrison, D. R., & Cleveland-Innis, M. (2005). Facilitating cognitive presence in online learning: Interaction is not enough. *American Journal of Distance Education*, *19*, 133–148. doi:10.1207/s15389286ajde1903_2

Gay, G. (2002). Preparing for culturally responsive teaching. *Journal of Teacher Education*, *53*(2), 106–116. doi:10.1177/0022487102053002003

Gaytan, J., & McEwen, B. C. (2007). Effective online instructional and assessment strategies. *American Journal of Distance Education*, *21*, 117–132. doi:10.1080/08923640701341653

Gee, J. P. (2007). *What video games have to teach us about learning and literacy* (Rev. and updated ed.). New York, NY: Palgrave Macmillan.

Gee, J. (2004). *What video games have to teach us about learning and literacy*. New York, NY: Palgrave Macmillan.

Gee, J. P. (1996). *Social linguistics and literacies: Ideology in discourses* (2nd ed.). New York, NY: Routledge.

Gee, J. P. (2003). *What video games have to teach us about literacy and learning*. New York, NY: Palgrave MacMillan Press.

Gee, J. P. (2005). It's theories all the way down: A response to scientific research in education. *Teachers College Record*, *107*(1), 10–18. doi:10.1111/j.1467-9620.2005.00452.x

Gibson, D. (2004). *E-portfolio decisions and dilemmas*. Paper presented at the Society for Information Technology in Teacher Education International Conference, Atlanta, GA.

Gibson, D. (2003). Network-based assessment in education. *Contemporary Issues in Technology & Teacher Education*, *3*(3), 310–333.

Gibson, D. (2006). Elements of network-based assessment. In Jonson, D., & Knogrith, K. (Eds.), *Teaching teachers to use technology* (pp. 131–150). New York, NY: Haworth Press.

Gibson, D. (2010). Bridging informal and formal learning: Experiences with participatory media. In Baek, Y. K. (Ed.), *Gaming for classroom-based learning: Digital role playing as a motivator of study* (pp. 84–99). Hershey, PA: IGI Global. doi:10.4018/978-1-61520-713-8.ch005

Gibson, D., & Jakl, P. (in preparation). Measuring teaching skills with a simulation: simSchool with Leverage. *Journal of Educational Data Mining*.

Gibson, D., & Kruse, S. (2011). *Personal communications with LEVEL 3 partnership colleges. simSchool Modules EDUCAUSE Project*. VT: Stowe.

Gibson, W. (1984). *Neuromancer*. New York, NY: Ace Books.

Gillies, R., & Boyle, M. (2010). Teachers' reflections on cooperative learning: Issues of implementation. *Teaching and Teacher Education*, *26*, 933–940. doi:10.1016/j.tate.2009.10.034

Gillis, D. (2008). Student perspectives on video conferencing in teacher education at a distance. *Distance Education*, *29*(1), 107–118. doi:10.1080/01587910802004878

Gilman, T. (2010). Designing effective online assignments. *The Chronicle of Higher Education*, *56*, 44–45.

Girod, M., & Girod, J. (2006). *Simulation and the need for quality practice in teacher preparation*. Paper presented at the American Association of Colleges for Teacher Education. Retrieved from http://www.allacademic.com/meta/p36279_index.html

Girod, G., Girod, M., & Denton, J. (2006). Lessons learned modeling "connecting teaching and learning". In Gibson, D., Aldrich, C., & Prensky, M. (Eds.), *Games and simulations in online learning: Research & development frameworks*. Hershey, PA: Idea Group. doi:10.4018/978-1-59904-304-3.ch010

Girod, M., & Girod, G. (2006). Exploring the efficacy of the Cook school district simulation. *Journal of Teacher Education*, *57*(5), 481–497. doi:10.1177/0022487106293742

Giroux, H. (1994). Doing cultural studies: youth and the challenge of pedagogy. *Harvard Educational Review*, *64*(3), 278–308.

Givens Generett, G., & Hicks, M. A. (2004). Beyond reflective competency: Teaching for audacious hope-in-action. *Journal of Transformative Education*, *2*, 187–203. doi:10.1177/1541344604265169

Glazer, E. M., & Hannafin, M. J. (2006). The collaborative apprenticeship model: Situated professional development within school settings. *Teaching and Teacher Education*, *22*(2), 179–193. doi:10.1016/j.tate.2005.09.004

Glazewski, K., Berg, K., & Brush, T. (2002, March). *Integrating technology into preservice teacher education: Comparing a field-based model with a traditional approach*. Paper presented at the Society for Information Technology and Teacher Education, Nashville, TN.

Gliem, J. A., & Gliem, J. R. (2003). *Calculating, interpreting, and reporting Cronbach's alpha reliability coefficient for Likert-type scales.* Midwest Research to Practice Conference in Adult, Continuing, and Community Education. Retrieved from https://scholarworks.iupui.edu/bitstream/handle/1805/344/Gliem+&+Gliem.pdf? sequence=1

Goddard, R. D. (2003). The impact of schools on teacher beliefs, influence, and student achievement: The role of collective efficacy beliefs. In Raths, J. D., & McAninch, A. R. (Eds.), *Teacher beliefs and classroom performance: The impact of teacher education. Advances in teacher education* (*Vol. 6*, pp. 183–202). Greenwich, CT: Information Age Pub.

Goddard, R. D., Hoy, W. K., & Hoy, A. W. (2000). Collective teacher efficacy: Its meaning, measure, and impact on student achievement. *American Educational Research Journal*, *37*(2), 479–507.

Goe, L. (2007). Linking teacher quality and student outcomes. In C. A. Dwyer (Ed.), *America's challenge: Effective teachers for at-risk schools and students.* National Comprehensive Center for Teacher Quality. Retrieved from http://www.tqsource.org/publications/NCCTQBiennialReport.pdf

Goe, L. (2007). *The link between teacher quality and student outcomes: A research synthesis*. National Comprehensive Center for Teacher Quality. Retrieved from http://www.tqsource.org/publications/LinkBetweenTQandStudentOutcomes.pdf

Goe, L., & Stickler, L. M. (2008). *Teacher quality and student achievement: Making the most of recent research.* National Comprehensive Center for Teacher Quality. Retrieved from http://www.tqsource.org/publications/March2008Brief.pdf

Goldman, S., Booker, A., & McDermott, M. (2008). Mixing the digital, social, and cultural: Learning, identity and agency in youth participation. In Buckingham, D. (Ed.), *Youth, identity, and digital media* (pp. 185–206). Cambridge, MA: The MIT Press.

Gold, S. (2001). A constructivist approach to online training for online teachers. *Journal of Asynchronous Learning Networks*, *5*, 35–57.

Good, A. J., O'Connor, K. A., Greene, H. C., & Luce, E. F. (2005). Collaborating across the miles: Telecollaboration in a social studies methods course. *Contemporary Issues in Technology & Teacher Education*, *5*(3/4), 300–317.

Goodman, J. (1986). Making early field experience meaningful: a critical approach. *Journal of Education for Teaching*, *12*(2), 109–125. doi:10.1080/0260747860120201

Goodson, I. F. (2003). *Professional knowledge, professional lives: Studies in educational change*. Philadelphia, PA: Open University Press.

Graham, C. R., & Misanchuk, M. (2004). Computer-mediated learning groups: Benefits and challenges to using group work in online learning environments. In Roberts, T. S. (Ed.), *Online collaborative learning: Theory and practice* (pp. 181–214). Hershey, PA: Information Science Publishing.

Grant, M. (1996). Development of a model using information technology for support of rural Aboriginal students off-campus learning. *Australian Journal of Educational Technology*, *12*(2), 94–108.

Greene, H. C. (2009). Multimedia observations: Examining the roles and learning outcomes of traditional, CD-ROM based, and videoconference observations in pre-service teacher education [Electronic Version]. *Current Issues in Education, 11*. Retrieved from http://cie.ed.asu.edu/volume11/number3/

Greene, B. A., & Land, S. M. (2000). A qualitative analysis of scaffolding use in a resource-based learning environment involving the World Wide Web. *Journal of Educational Computing Research*, *23*(2), 151–180. doi:10.2190/1GUB-8UE9-NW80-CQAD

Greene, C. H. (2005). Creating connections: A pilot study on an online community of learners. *Journal of Interactive and Online Learning*, *3*(3), 1–21.

Greenlaw, S. A., & DeLoch, S. B. (2003). Teaching critical thinking with electronic discussion. *The Journal of Economic Education*, *34*(1), 36–52. doi:10.1080/00220480309595199

Grossman, P. (2010). *Learning to practice: The design of clinical experience in teacher preparation.* Washington, DC: Partnership for Teacher Quality (AACTE and NEA).

Grossman, P., Wineburg, S., & Woolworth, S. (2001). Toward a theory of teacher community. *Teachers College Record*, *103*, 942–1012. doi:10.1111/0161-4681.00140

Gulati, S. (2004). *Constructivism and emerging online learning pedagogy: A discussion for formal to acknowledge and promote the informal*. London, UK: University of Glamorgan.

Gunawardena, C. N. (2005). *Social presence and implications for designing online learning communities*. Paper presented at the Fourth International Conference on Educational Technology, Nanchang, China. Retrieved from http://www.edu.cn/include/new_jiaoyuxxh/xiazai/gunawardena.ppt

Gunawardena, C. N., & Zittle, F. J. (1997). Social presence as a predictor of satisfaction within a computer-mediated conferencing environment. *American Journal of Distance Education*, *11*(3), 8–26. doi:10.1080/08923649709526970

Guri-Rosenblit, S. (2005). Eight paradoxes in the implementation process of e-learning in higher education. *Higher Education Policy*, *18*, 5–29. doi:10.1057/palgrave.hep.8300069

Guskey, T. R. (2000). *Evaluating professional development*. Thousand Oaks, CA: Corwin Press.

Haberman, M., & Post, L. (1992). Does direct experience change education students' perceptions of low-income minority students? *Mid-Western Educational Researcher*, *5*(2), 29–31.

Hagie, C., Hughes, M., & Smith, S. J. (2005). The positive and challenging aspects of learning online and in traditional face-to-face classrooms: A student perspective. *Journal of Special Education Technology*, *20*(2), 52–59.

Hall, A., Yates, R., & Campbell, N. (1998). *Teacher education at a distance: Reflections on the first year of a mixed media teacher education programme*. Unpublished paper presented to VITAL day at the University of Waikato, 17 February, 1998.

Ham, V., & Wenmoth, D. (2007). *Evaluation of the e-Learning collaborative development fund.* Final Report to Tertiary Education Commission, Wellington. Retrieved from www.tec.govt.nz/templates/standard.aspx?id=755

Hammerness, K., Darling-Hammond, L., Bransford, J., Berliner, D., Cochran-Smith, M., McDonald, M., & Zeichner, K. (2005). How teacher learn and develop. In Darling-Hammond, L., & Bransford, J. (Eds.), *Preparing teachers for a changing world: What teachers should learn and be able to do* (pp. 358–389). San Francisco, CA: Jossey-Bass.

Hammerness, K., Darling-Hammond, L., Grossman, P., Rust, F., & Shulman, L. (2005). The design of teacher education programs. In Darling-Hammond, L., & Bransford, J. (Eds.), *Preparing teachers for a changing world: What teachers should learn and be able to do* (pp. 390–441). San Francisco, CA: Jossey-Bass.

Hammond, M. (2000). Communication within online forums: The opportunities, the constraints and the value of communicative approach. *Computers & Education*, *35*, 251–262. doi:10.1016/S0360-1315(00)00037-3

Hammond, M., Reynolds, L., & Ingram, J. (2011). How and why do student teachers use ICT? *Journal of Computer Assisted Learning*, *27*(3), 191–203. doi:10.1111/j.1365-2729.2010.00389.x

Hampel, R., & Baber, E. (2003). Using internet-based audio-graphic and video conferencing for language teaching and learning. In Felix, U. (Ed.), *Learning language online towards best practice* (pp. 171–191). The Netherlands: Dripps, Meppel.

Hancock, R., Knezek, G., & Christensen, R. (2007). Cross-validating measures of technology integration: A first step toward examining potential relationships between technology integration and student achievement. *Journal of Computing in Teacher Education*, *24*(1), 15–21.

Hara, N., Bonk, C. J., & Angeli, C. (2002). Content analysis of online discussion in an applied educational psychology course. *Instructional Science*, *28*, 115–152. doi:10.1023/A:1003764722829

Harasim, L. (2006). A history of e-learning: Shift happened. In Weiss, J., Nolan, J., & Trifonas, P. (Eds.), *International handbook of virtual learning environments* (pp. 25–60). Dordrecht, The Netherlands: A A Dordrecht. doi:10.1007/978-1-4020-3803-7_2

Harasim, L. (2006). A history of e-learning: shift happened. In Weiss, J., Nolan, J., & Trifonas, P. (Eds.), *International handbook of virtual learning environments* (pp. 25–60). Dordrecht, The Netherlands: Kluwer. doi:10.1007/978-1-4020-3803-7_2

Harasim, L., Hiltz, S., Teles, L., & Turoff, M. (1995). *Learning networks: A field guide to teaching and learning online*. Cambridge, MA: MIT Press.

Hargreaves, A., Earl, L., Moore, S., & Manning, S. (2001). *Learning to change: Teaching beyond subject and standards*. San Francisco, CA: Jossey Bass.

Harmon, S. W. (2008). A theoretical basis for learning in massive multiplayer virtual worlds. *Journal of Educational Technology Development and Exchange*, *1*(1), 29–40.

Harms, C. M., Niederhauser, D. S., Davis, N. E., Roblyer, M. D., & Gilbert, S. B. (2006). Educating educators for virtual schooling: Communicating roles and responsibilities. *The Electronic Journal of Communication, 16*(1-2). Retrieved from http://ctlt.iastate.edu/~tegivs/TEGIVS/publications/JP2007%20harms&niederhauser.pdf

Harrell, P. E., & Harris, M. (2006). Teacher preparation without boundaries: A two-year study of an online teacher certification program. *Journal of Technology and Teacher Education*, *14*(4), 755–774.

Harris, A., Leithwood, K., Day, C., Sammons, P., & Hopkins, D. (2007). Distributed leadership and organizational change: Reviewing the evidence. *Journal of Educational Change*, *8*(4), 337–347. doi:10.1007/s10833-007-9048-4

Harrison, S., & Dourish, P. (1996). Re-place-ing space: The roles of place and space in collaborative systems. *Proceedings of the 1996 ACM Conference on Computer Supported Collaborative Work*.

Hart, P. D. (2010). *Career changes in the classroom: A national portrait*. The Woodrow Wilson National Fellowship Foundation. Retrieved from http://www.woodrow.org/images/pdf/policy/CareerChangersClassroom_0210.pdf

Hartshorne, R., Heafner, T., & Petty, T. (2011). Examining the effectiveness of the remote observation of graduate interns. *Journal of Technology and Teacher Education*, *19*(4), 395–422.

Harvard, B., Du, J., & Xu, J. (2008). Online collaborative learning and communication media. *Journal of Interactive Learning Research*, *19*(1), 37–50.

Hassel, B., & Terrell, M. (2004). *How can virtual schools be a vibrant part of meeting the choice provisions of the No Child Left Behind Act?* Virtual School Report, [White Paper], Summer 2004. Retrieved from http://www.connectionsacademy.com/virtualreport.asp

Hastie, M., Chen, N. S., & Kuo, Y. H. (2007). Instructional design for best practice in the synchronous cyber classroom. *Journal of Educational Technology & Society*, *10*(4), 281–294.

Hatch, T., & Grossman, P. (2009). Learning to look beyond the boundaries of representation. *Journal of Teacher Education*, *60*(1), 70–85. doi:10.1177/0022487108328533

Hattie, J. (2009). *Visible learning*. London, UK: Routledge.

Hawkes, M., & Romiszowski, A. (2001). Examining the reflective outcomes of asynchronous computer-mediated communication on inservice teacher development. *Journal of Technology and Teacher Education*, *9*(2), 285–308.

Hayes, E. (2005). *Women and video gaming: Gendered identities at play*. Paper presented at the Games, Learning, & Society Conference. Madison, WI

Haythornthwaite, C., & Kazmer, M. M. (Eds.). (2004). *Learning, culture and community in online education: Research and practice*. New York, NY: Peter Lang.

Haythornthwaite, C., Kazmer, M. M., Robins, J., & Shoemaker, S. (2004). Community development among distance learners: Temporal and technological dimensions. In Haythornthwaite, C., & Kazmer, M. M. (Eds.), *Learning, culture and community in online education: Research and practice* (pp. 35–57). New York, NY: Peter Lang. doi:10.1111/j.1083-6101.2000.tb00114.x

Heafner, T. L. (2011). *Windows into teaching and learning [WiTL]: Exploring online clinicals for a distance education social studies methods course*. Paper presented at the Annual Meeting of the College and University Faculty Assembly of the National Council for the Social Studies, Washington, DC.

Heafner, T. L., & Petty, T. (2010). Observing graduate interns remotely. *Kappa Delta Pi Record*, *47*, 39–43.

Heafner, T. L., Petty, T. M., & Hartshorne, R. (2011). Evaluating modes of teacher preparation: A comparison of face-to-face and remote observations of graduate interns. *Journal of Digital Learning in Teacher Education*, *27*(4), 154–164.

Heafner, T. L., Petty, T. M., & Hartshorne, R. (2012). Moving beyond four walls: Qualitative evaluation of ROGI (Remote Observation of Graduate Interns) for the expanding online teacher preparation classroom. In Alias, N. A., & Hashim, S. (Eds.), *Instructional technology research, design and development: Lessons from the field* (pp. 370–400). Hershey, PA: IGI Global Publishing.

Heath, S. B. (2001). Protean shapes in literacy events: Ever-shifting oral and literate traditions. In E. Cushman, et al. (Eds.), *Literacy: A critical sourcebook* (pp. 443-466). Boston, MA: Bedford/St. Martin's. (Reprinted from *Spoken and written language: Exploring orality and literacy*, pp. 91-117, by D. Tannen, Ed., 1982, Norwood, NJ: Ablex)

Heath, S. B. (1991). The sense of being literate: Historical and cross-cultural features. In Barr, R., Kamil, M. L., Mosenthal, P., & Pearson, P. D. (Eds.), *Handbook of reading research* (*Vol. II*, pp. 3–25). New York, NY: Longman Publishing Group.

Heck, D. J., Banilower, E. R., Weiss, I. R., & Rosenberg, S. L. (2008). Studying the effects of professional development: The case of the NSF's local systemic change through teacher enhancement initiative. *Journal for Research in Mathematics Education*, *39*(2), 113–152.

Heider, K. L. (2005). Teacher isolation: How mentoring programs can help. *Current Issues in Education, 8*(14). Retrieved August 18, 2011, from http://cie.ed.asu.edu/volume8/number14/

Hemingway, E. (1997). *A farewell to arms*. New York, NY: Scribner Classics.

Henn, M., Weinstein, M., & Foard, N. (2006). *A short introduction to social research*. London, UK: Sage Publications.

Henri, F. (1992). Computer conferencing and content analysis. In Kaye, A. R. (Ed.), *Collaborative learning through computer conferencing: The Najaden Papers* (pp. 116–136). Berlin, Germany: Springer-Verlag. doi:10.1007/978-3-642-77684-7_8

Herie, M. (2005). Theoretical perspectives in online pedagogy. *Journal of Technology in Human Services*, *2*(1), 29–52. doi:10.1300/J017v23n01_03

Heritage Foundation. (2011). *2011 index of economic freedom*. Retrieved from http://www.heritage.org/index/Ranking

Hernández-Ramos, P., & Giancarlo, C. A. (2004). Situating teacher education: From the university classroom to the real classroom. *Journal of Computing in Teacher Education*, *20*(4), 121–128.

Hertzog, N. (1998). Open-ended activities: Differentiation through learner responses. *Gifted Child Quarterly*, *42*, 212–227. doi:10.1177/001698629804200405

Hess, F. M., Rotherham, A. J., & Walsh, K. (2004). *A qualified teacher in every classroom? Appraising old answers and new ideas*. Cambridge, MA: Harvard Education Press.

Heward, W. (2006). *Exceptional children: An introduction to special education* (8th ed.). Upper Saddle River, NJ: Prentice Hall.

Hew, K. F., & Brush, T. (2007). Integrating technology in K-12 teaching and learning: current knowledge gaps and recommendations for future research. *Educational Technology Research and Development*, *55*, 223–252. doi:10.1007/s11423-006-9022-5

Heyman, G. D. (2008). Children's critical thinking when learning from others. *Current Directions in Psychological Science*, *17*(5), 344–347. doi:10.1111/j.1467-8721.2008.00603.x

Higher Education Opportunity Act, H.R. 4137 1125 (2008).

Hiltz, S. R., Coppola, N., Rotter, N., & Turoff, M. (2001). Measuring the importance of learning for the effectiveness of ALN: A multi-measure, multi-method. *Journal of Asynchronous Learning Networks*, *4*(2), 103–125.

Hiltz, S. R., Turoff, M., & Harasim, L. (2007). Development and philosophy of the field of asynchronous learning networks. In Andrews, R., & Haythornthwaite, C. (Eds.), *The sage handbook of e-learning research* (pp. 55–72). London, UK: Sage, Ltd.

Hixon, E., & So, H. J. (2009). Technology's role in field experiences for preservice teacher training. *Journal of Educational Technology & Society*, *12*(4), 294–304.

Ho, S. (2002). *Evaluating students' participation in on-line discussions.* Retrieved from http://ausweb.scu.edu.au/aw02/papers/refereed/ho/paper.html

Hoadley, C., & Kilner, P. G. (2005). Using technology to transform communities of practice into knowledge-building communities. *SIGGROUP Bulletin*, *25*(1), 31–40.

Hobbs, R. (1997). Expanding the concept of literacy. In Kubey, R. (Ed.), *Media literacy in the information age: Current perspectives* (pp. 163–183). New Brunswick, NJ: Transaction Publishers.

Hodges, C. B., & Hunger, C. M. (2011). Communicating mathematics on the Internet: Synchronous and asynchronous tools. *TechTrends*, *55*(5), 39–44. doi:10.1007/s11528-011-0526-4

Hodson, J. (2004). Aboriginal learning and healing in a virtual world. *Canadian Journal of Native Education*, *28*(1/2), 111–122.

Holland, J. (1995). *Hidden order: How adaptation builds complexity*. Cambridge, MA: Perseus Books.

Hollister, C. D., McGahey, L., & Mehrotra, C. (2001). *Distance learning: Principles for effective design, delivery, and evaluation*. Thousand Oaks, CA: Sage Publications.

Holmberg, B. (2005). *The evolution, principles and practices of distance education*. Oldenburg, Germany: BIS-Verlag Carl von Ossietzky Universitat.

Honig, Z. (2011, July 3). South Korea plans to convert all textbooks to digital, swap backpacks for tablets by 2015 [Electronic Version]. *Engadget*. Retrieved from http://www.engadget.com/2011/07/03/south-korea-plans-to-convert-all-textbooks-to-digital-swap-back/

Hopkins, S. (1995). Using the past; guiding the future. In Slick, G. A. (Ed.), *Emerging trends in teacher preparation: The future of field experiences* (pp. 1–9). Thousand Oaks, CA: Corwin Press, Inc.

Horwitz, P. (1999). Designing computer models that teach. In Fuerzeig, W., & Roberts, N. (Eds.), *Modeling and simulation in precollege science and mathematics Education* (pp. 179–196). New York, NY: Springer-Verlag. doi:10.1007/978-1-4612-1414-4_8

Hou, H., & Wu, S. (2011). Analyzing the social knowledge construction behavioral patterns of an online synchronous collaborative discussion instructional activity using an instant messaging tool: A case study. *Computers & Education*, *57*, 1459–1468. doi:10.1016/j.compedu.2011.02.012

Howard, G. (2006). *We can't teach what we don't know: White teachers, multiracial schools* (2nd ed.). New York, NY: Teachers College Press.

Howland, J., Jonassen, D. H., & Marra, R. M. (2011). *Meaningful learning with technology* (4th ed.). Columbus, OH: Merrill/Prentice-Hall.

Hoy, W. K., & Woolfolk, A. E. (1990). Socialization of student teachers. *American Educational Research Journal*, *27*(2), 279–300.

Hraslinski, S. (2008). Asynchronous and synchronous e-learing. *EDUCAUSE Quarterly*, *31*(4).

Hubbard, R. S., & Power, B. M. (1999). *Living the questions: A guide for teacher researchers*. York, ME: Stenhouse.

Huberman, A. M., & Miles, M. B. (2002). *The qualitative researcher's companion*. Thousand Oaks, CA: Sage Publications.

Hudson, M. (2005). The links between collaboration, agency, professional community and learning for teachers in a contemporary secondary school in England. *Educate*, *5*(2), 42–62.

Huebner, T. A. (2010). Differentiated instruction. *Educational Leadership*, *67*, 79–81.

Humphries, S. (2010). Five challenges for new online teachers. *Journal of Technology Integration*, *2*(1), 15–24.

Hurston, Z. N. (1990). *Their eyes were watching God: A novel*. New York, NY: Perennial Library.

Ice, P., Gibson, A. M., Boston, W., & Becher, D. (2011). An exploration of differences between community of indicators in low and high disenrollment online courses. *Journal of Asynchronous Learning Networks*, *15*(2).

Im, Y., & Lee, 0. (2004). Pedagogical implications of online discussion for preservice teacher training. *Journal of Research on Technology in Education*, *36*, 155–170.

Im, Y., & Lee, O. (2003). Pedagogical implications of online discussion for preservice teacher training. *Journal of Research on Technology in Education*, *36*(2), 155–170.

Institute for Higher Education Policy. (1999). *What's the difference? A review of contemporary research on the effectiveness of distance learning in higher education*. Washington, DC: National Education Association. Retrieved from http://www.ihep.org/assets/files/publications/s-z/WhatDifference.pdf

Institutes of Technology and Polytechnics of New Zealand. (2004). *Critical success factors for effective use of e-learning with Māori learners*. Retrieved from http://elearning.itpnz.ac.nz/index.htm

International Society for Technology in Education. (2007). *NETS for students*. Retrieved from http://www.iste.org/Content/NavigationMenu/NETS/ForStudents/NETS_for_Students.htm

International Society for Technology in Education. (2008). *NETS for teachers 2008*. Retrieved from http://www.iste.org/Content/NavigationMenu/NETS/ForTeachers/2008Standards/NETS_for_Teachers_2008.htm

Interstate New Teacher Assessment and Support Consortium. (1992). *Model standards for beginning teacher licensing, assessment and development: A resource for state dialogue*. Washington, DC: Council of Chief State School Officers.

Interstate Teacher Assessment and Support Consortium. (2010). *Official website*. Retrieved from http://www.ccsso.org/Resources/Programs/ Interstate_Teacher_Assessment_Consortium_(InTASC).html

Inyega, H., & Ratliff, J. (2007). Teaching online course: Lessons learned. In Sampson, M. B., Szabo, S., Falk-Ross, F., Foote, M., & Linder, P. (Eds.), *Multiple literacies in the 21st century* (pp. 344–363). Commerce, TX: Texas A & M University Commerce.

ISTE. (2008). *The ISTE NETS and performance indicators for teachers*. International Society for Technology in Education.

Jacobsen, D. M., & Lock, J. V. (2004). Technology and teacher education for a knowledge era: Mentoring for student futures, not our past. *Journal of Technology and Teacher Education*, *12*(1), 75–100.

Jakobsson, A. (2006). Students' self-confidence and learning through dialogues in a net-based environment. *Journal of Technology and Teacher Education*, *14*(2), 387–405.

Jeong, H., & Hmelo-Silver, C. E. (2010). Productive use of learning resources in an online problem-based learning environment. *Computers in Human Behavior*, *26*, 84–99. doi:10.1016/j.chb.2009.08.001

Jiang, M., & Ting, E. (2000). A study of factors influencing students' perceived learning in a web-based course environment. *International Journal of Educational Telecommunications*, *6*(4), 317–338.

Johnson, B., Sullivan, A. M., & Williams, D. (2009). A one-eyed look at classroom life: Using new technologies to enrich classroom-based research. *Issues in Educational Research*, *19*(1), 34–47.

Johnson, C. (2005). Lessons learned from teaching web-based courses: The 7-year itch. *Nursing Forum*, *40*(1), 11–17. doi:10.1111/j.1744-6198.2005.00002.x

Johnson, D. W., & Johnson, R. T. (1996). Cooperation and the use of technology. In Jonassen, D. H. (Ed.), *Handbook of research for educational communications and technology* (pp. 1017–1044). New York, NY: Simon and Shuster Macmillan.

Johnson, D. W., Johnson, R. T., & Smith, K. A. (1998). *Active learning: Cooperation in the college classroom*. Edina, MN: Interaction Book Company.

Johnson, J. L. (2003). *Distance education: The complete guide to design, delivery, and improvement*. New York, NY: Teachers College Press.

Johnson, L., Levine, A., Scott, C., Smith, R., & Stone, S. (2009). *The Horizon Report: 2009 economic development edition*. Austin, TX: The New Media Consortium.

Johnson, L., Smith, R., Willis, H., Levine, A., & Haywood, K. (2011). *The 2011 horizon report*. Austin, TX: New Media Consortium.

Johnson, S., & Aragon, S. (2003). An instructional strategy framework for online learning environments. *New Directions for Adult and Continuing Education*, (100): 31–43. doi:10.1002/ace.117

Johnson, T., Maring, G., Doty, J., & Fickle, M. (2006). Cybormentoring: Evolving high-end video conferencing practices to support preservice teacher training. *Journal of Interactive Online Learning*, *5*(1), 59–74.

Johnston, S. (1994). Experience is the best teacher; or is it? An analysis of the role of experience in learning to teach. *Journal of Teacher Education*, *45*(3), 199–208. doi:10.1177/0022487194045003006

Jonassen, D. H. (1999). *Computers in the classroom: Mindtools for critical thinking*. Englewood Cliffs, NJ: Prentice Hall.

Jones, R., Fox, C., & Levin, D. (2011). *National educational technology trends: 2011. Transforming education to ensure all students are successful in the 21st century.* State Educational Technology Directors Association. (ERIC: ED522777).

Journell, W. (2008). Facilitating historical discussions using asynchronous communication: The role of the teacher. *Theory and Research in Social Education*, *36*, 317–355. doi:10.1080/00933104.2008.10473379

Journell, W. (2010). Perceptions of e-learning in secondary education: A viable alternative to classroom instruction or a way to bypass engaged learning? *Educational Media International*, *47*, 69–81. doi:10.1080/09523981003654985

Journell, W. (in press). Proceeding with caution: Online learning in K-12 education. *Phi Delta Kappan*.

Joyce, B., & Showers, B. (1980). Improving inservice training: The messages of research. *Educational Leadership*, *37*(5), 379–385.

Jun, J. (2005). Understanding e-dropout. *International Journal on E-Learning*, *4*, 229–240.

Kafai, Y. B., & Resnick, M. (1996). Introduction. In Kafai, Y. B., & Resnick, M. (Eds.), *Constructionism in practice* (pp. 1–8). Mahwah, NJ: Erlbaum.

Kajder, S. (2003). Plugging in: What technology brings to the English/language arts classroom. *Voices from the Middle*, *11*(3), 6–9.

Kane, R. (2005). *Initial teacher education policy and practice. (A research report to the Ministry of Education and the New Zealand Teachers Council)*. Wellington, New Zealand: Ministry of Education.

Kanuka, H., & Anderson, T. (1998). Online social interchange, discord, and knowledge construction. *Journal of Distance Education*, *13*(1), 57–75.

Kapitzke, C., & Pendergast, D. (2005). Virtual schooling service: Productive pedagogies or pedagogical possibilities? *Teachers College Record*, *107*, 1626–1651. doi:10.1111/j.1467-9620.2005.00536.x

Karber, D. (2001). Comparisons and contrasts in traditional versus on-line teaching in management. *Higher Education in Europe*, *16*, 533–536. doi:10.1080/03797720220141852

Kaseman, L., & Kaseman, S. (2000). How will virtual schools effect homeschooling? *Home Education Magazine* (November-December), 16-19. Retrieved from http://homeedmag.com/HEM/176/ndtch.html

Kauffman, J., Mock, D., Tankersley, M., & Landrum, T. (2008). Effective service delivery models. In Morris, R. J., & Mather, N. (Eds.), *Evidence-based interventions for students with learning and behavioral challenges* (pp. 359–378). Mahwah, NJ: Lawrence Erlbaum Associates.

Kauffman, S. (2000). *Investigations*. New York, NY: Oxford University Press.

Kearsley, G., & Lynch, W. (1991). Computer networks for teaching and research: Changing the nature of educational practice and theory. *DEOSNEWS – The Distance Education Online Symposium*, *1*(18). BITNET.

Kearsley, G. (2000). *Online education cyberspace learning and teaching*. Belmont, CA: Wadsworth.

Keeler, C. G., & Horney, M. (2007). Online course designs: Are special needs being met? *American Journal of Distance Education*, *21*, 61–75. doi:10.1080/08923640701298985

Ke, F., & Xie, K. (2009). Toward deep learning for adult students in online courses. *The Internet and Higher Education*, *12*(3/4), 136–145. doi:10.1016/j.iheduc.2009.08.001

Kellner, D., & Share, J. (2005). Toward critical media literacy: Core concepts, debates, organizations, and policy. *Discourse: Studies in the Cultural Politics of Education*, *26*(3), 369–386. doi:10.1080/01596300500200169

Kelly, P. (2011). What is teacher learning? A socio-cultural perspective. *Oxford Review of Education*, *32*(4), 505–519. doi:10.1080/03054980600884227

Kennedy, K., & Archambault, L. (2011). The current state of field experiences in K-12 online learning programs in the U.S. In M. Koehler & P. Mishra (Eds.), *Proceedings of Society for Information Technology & Teacher Education International Conference 2011* (pp. 3454-3461). Chesapeake, VA: AACE.

Kent, A. M. (2007). Powerful preparation of preservice teachers using interactive video conferencing. *Journal of Literacy and Technology*, *8*(2), 42–59.

Kickul, G., & Kickul, J. (2006). Closing the gap: Impact of student productivity and learning goal orientation on e-learning outcomes. *International Journal on E-Learning*, *5*, 361–372.

Killion, J., & Pinata, R. (2011). *Recalibrating professional development for teacher success*. Education Week Webinar.

Kim, K., & Bonk, C. J. (2006). The future of online teaching and learning in higher education: The survey says... *Educause Quarterly Magazine*, *29*(4). Retrieved from http://www.educause.edu/EDUCAUSE+Quarterly/EDUCAUSEQuarterlyMagazineVolum/TheFutureofOnlineTeachingandLe/157426

Kim, K.-J., Bonk, C. J., & Zeng, T. (2005, June). Surveying the future of workplace e-learning: The rise of blending, interactivity, and authentic learning. *E-Learn Magazine*, *6*. Retrieved from http://elearnmag.acm.org/archive.cfm?aid=1073202

Kim, K. J., & Bonk, C. J. (2006). The future of online learning: The survey says.... *EDUCAUSE Quarterly*, *4*, 22–30.

Kim, K.-J., Liu, S., & Bonk, C. J. (2005). Online MBA students' perceptions of online learning: Benefits, challenges and suggestions. *The Internet and Higher Education*, *8*(4), 335–344. doi:10.1016/j.iheduc.2005.09.005

King, M. B. (2002). Professional development to promote schoolwide inquiry. *Teaching and Teacher Education*, *18*, 243–257. doi:10.1016/S0742-051X(01)00067-1

Kinzer, C. K., & Leander, K. M. (2003). Technology and the language arts: Implications of an expanded definition of literacy. In Flood, J., Lapp, D., Squire, J., & Jensen, J. (Eds.), *Handbook of research on teaching the English language arts* (pp. 546–566). Mahwah, NJ: Erlbaum.

Kirkley, J. R., Duffy, T. M., Kirkley, S. E., & Kremer, L. H. (2011). Implications of constructivism for the design and use of serious games. In Tobias, S., & Fletcher, J. D. (Eds.), *Computer games and instruction* (pp. 371–394). Charlotte, NC: Information Age Publishers.

Kirschner, P. A., Strijbos, J. W., Kreijns, K., & Beers, P. J. (2004). Designing electronic collaborative learning environments. *Educational Technology Research and Development*, *52*(3), 47–66. doi:10.1007/BF02504675

Kist, W. (2005). *New literacies in action: Teaching and learning in multiple media*. New York, NY: Teachers College Press.

Klopfer, E., & Squire, K. (2007). Case study analysis of augmented reality simulations on handheld computers. *Journal of the Learning Sciences*, *16*(3), 371–413. doi:10.1080/10508400701413435

Knezek, G., & Christensen, C. (2009). Preservice educator learning in a simulated teaching environment. In Maddux, C. (Ed.), *Research highlights in technology and teacher education* (*Vol. 1*, pp. 161–170). Chesapeake, VA: Society for Information Technology and Teacher Education.

Knight, S. L., Pedersen, S., & Peters, W. (2004). Connecting the university with a professional development school: Pre-service teachers' attitudes toward the use of compressed video. *Journal of Technology and Teacher Education*, *12*(1), 139–154.

Knowles, M. (1980). *The modern practice of adult education*. Chicago, IL: Association Press.

Knowles, M. (1984). *The adult learner: A neglected species* (3rd ed.). Houston, TX: Gulf Publishing Co.

Knowles, M. (1986). *Using learning contracts: Practical approaches to individualizing and structuring learning*. San Francisco, CA: Jossey-Bass.

Knowles, M. S. (1990). *The adult learner: A neglected species*. Houston, TX: Gulf.

Knowles, M. S., Holton, E. F., & Swanson, R. A. (1998). *The adult learner: The definitive classic in adult education and human resource development*. Woburn, MA: Butterworth-Heinemann.

Knowles, M. S., Swanson, R. A., & Holton, E. F. III. (2005). *The adult learner: The definitive classic in adult education and human resource development* (6th ed.). CA: Elsevier Science and Technology Books.

Koehler, M. J., & Mishra, P. (2005). Teachers learning technology by design. *Journal of Computing in Teacher Education*, *21*(3), 94–102.

Koehler, M. J., & Mishra, P. (2009). What is technological pedagogical content knowledge? *Contemporary Issues in Technology & Teacher Education*, *9*(1), 60–70. Retrieved from http://www.citejournal.org/vol9/iss1/general/article1.cfm

Korir Bore, J. C. (2008). Perceptions of graduate students on the use of web-based instruction in special education personnel preparation. *Teacher Education and Special Education*, *31*, 1–11. doi:10.1177/088840640803100101

Ko, S., & Rosen, T. (2010). *Teaching online: A practical guide*. New York, NY: Routledge.

Kramarski, B., & Mevarech, Z. R. (2003). Enhancing mathematical reasoning in the classroom: The effects of cooperative learning and metacognitive training. *American Educational Research Journal*, *40*(1), 281–310. doi:10.3102/00028312040001281

Krathwohl, D. R. (2002). A revision of Bloom's taxonomy: An overview. *Theory into Practice*, *41*(4), 212–218. doi:10.1207/s15430421tip4104_2

Kreijns, K., Kirschner, P. A., & Jochems, W. (2003). Identifying the pitfalls for social interaction in computer-supported collaborative learning environments: A review of the research. *Computers in Human Behavior*, *19*, 335–353. doi:10.1016/S0747-5632(02)00057-2

Krell, D., & Dana, N. F. (2011, April). *Facilitating action research: A study of coaches, their experiences, and their reflections on leading teachers in the process of practitioner inquiry*. Paper presented at the meeting of the American Educational Research Association Annual Conference, New Orleans, LA.

Kress, G. (2003). *Literacy in the new media age*. New York, NY: Routledge. doi:10.4324/9780203164754

Kucer, S. B. (2005). *Dimensions of literacy: A conceptual base for teaching reading and writing in school settings* (2nd ed.). Mahwah, NJ: Lawrence Erlbaum Associates.

Kuhn, D., & Dean, D. (2004). A bridge between cognitive psychology and educational practice. *Theory into Practice*, *43*(4), 268–273. doi:10.1207/s15430421tip4304_4

Kumar, S., & Vigil, K. (2011). The Net generation as preservice teachers: Transferring familiarity with new technologies to educational environments. *Journal of Digital Learning in Teacher Education*, *27*(4), 144–153.

Kyong-Jee, K., & Bonk, C. J. (2006). The future of online teaching and learning in higher education: The survey says.... *EDUCAUSE Quarterly*, *29*(4). Retrieved from http://www.educause.edu/apps/eq/eqm06/eqm0644.asp?bhcp=1

Labaree, D. F. (2004). *The trouble with ed schools*. New Haven, CT: Yale University Press.

Lacoss, J., & Chylack, J. (1998). What makes a discussion section productive? *Teaching Concerns*, Fall.

Ladson-Billings, G. (2000). Fighting for our lives: Preparing teacher to teach African-American students. *Journal of Teacher Education*, *51*, 206–213. doi:10.1177/0022487100051003008

Lambdin, D. V., Duffy, T. M., & Moore, J. A. (1996). *A hypermedia system to aid in preservice teacher education: Instructional design and evaluation*. Paper presented at the National Convention of the Association for Educational Communications and Technology, Indianapolis, IN.

Landrum, T. J., & McDuffie, K. A. (2010). Learning styles in the age of differentiated instruction. *Exceptionality*, *18*, 6–17. doi:10.1080/09362830903462441

Lang, W., & Wilkinson, J. (2008, March). *Measuring teacher dispositions with different item structures: An application of the Rasch model to a complex accreditation requirement*. Paper presented at Annual Meeting of the American Educational Research Association, New York, NY.

Lankshear, C., & Knobel, M. (2003). *New literacies: Changing knowledge and classroom teaching*. Philadelphia, PA: Open University Press.

Lawless, K. A., & Pellegrino, J. W. (2007). Professional development in integrating technology into teaching and learning: Knowns, unknowns and ways to pursue better questions and answers. *Review of Educational Research*, *77*(4), 575–614. doi:10.3102/0034654307309921

Lawrence-Brown, D. (2004). Differentiated instruction. *American Secondary Education*, *32*, 34–62.

Lawrence, S. A., & Mongillo, G. (2010). Multi-media literacy projects: Strategies for using technology with K to 20 learners. *The Language and Literacy Spectrum. Journal of New York State Reading Association*, *20*, 39–49.

Leander, K., & Boldt, G. (2008). *New literacies in old literacy skins*. Paper presented at the Annual Meeting of the American Educational Research Association, New York.

LeCompte, M., & Schensul, J. (Eds.). (1999). *The ethnographers' toolkit*. Walnut Creek, CA: AltaMira Press.

Legon, R., Runyon, J., & Aman, R. (2007, October). *The impact of "Quality Matters" standards on courses: Research opportunities and results*. Paper presented at the 13th International Sloan-C Conference on Online Learning, Orlando, FL.

Lehman, J. D., & Richardson, J. (2007, April). *Linking teacher preparation program with k-12 schools via video conferencing: Benefits and limitations*. Paper presented at the Annual Meeting of the American Educational Research Association, Chicago, IL. Retrieved from http://p3t3.education.purdue.edu/AERA2007_Videoconf_Paper.pdf.

Lehman, J., & Richardson, J. (2003). Virtual field experiences: Helping pre-service teachers learn about diverse classrooms through video conferencing connections with K-12 classrooms. In D. Lassner & C. McNaught (Eds.), *Proceedings of World Conference on Educational Multimedia, Hypermedia and Telecommunications 2003*, Chesapeake, VA: AACE, 1727-1728.

Leontiev, A. N. (1978). *Activity, consciousness, and personality*. Hillsdale, NJ: Prentice-Hall.

Leu, D. J. Jr. (2000). Literacy and technology: Deictic consequences for literacy education in an information age. In Kamil, M. L., Mosenthal, P. B., Pearson, P. D., & Barr, R. (Eds.), *Handbook of reading research* (*Vol. 3*, pp. 743–770). Mahwah, NJ: Erlbaum.

Leu, D. J. Jr. (2002). The new literacies: Research on reading instruction with the Internet and other digital technologies. In Samuels, J., & Farstrup, A. E. (Eds.), *What research has to say about reading instruction* (pp. 310–336). Newark, DE: International Reading Association.

Levin, S. R., Waddoups, G. L., Levin, J., & Buell, J. (2001). Highly interactive and effective online learning environments for teacher professional development (electronic version). *International Journal of Educational Technology, 2*. Retrieved from http://smi.curtin.edu.au/ijet/v2n2/slevin/index.html

Levin, B. B., He, Y., & Robbins, H. H. (2006). Comparative analysis of preservice teachers' reflective thinking in synchronous versus asynchronous online case discussions. *Journal of Technology and Teacher Education*, *14*, 439–460.

Levin, B., & Fullan, M. (2008). Learning about system renewal. *Educational Management Administration & Leadership*, *36*(2), 289–303. doi:10.1177/1741143207087778

Levitan, S., & Lloyd, C. (Producers). (2009). *Modern family* [Television series]. Los Angeles, CA: 20th Century Fox.

Lewin, L. (2006). *Teaching resources from Larry Lewin*. Retrieved from http://www.larrylewin.com/teachingresources/checbrics.html

Lewin, K. (1935). *A dynamic theory of personality*. New York, NY: McGraw Hill.

Li, C., & Irby, B. (2008). An overview of online education: Attractiveness, benefits, challenges, concerns and recommendations. *College Student Journal*, *42*, 449–458.

Lieberman, A., & Hoody, L. L. (1998). *Closing the achievement gap: Using the environment as an integrating context for learning*. San Diego, CA: State Environment and Education Roundtable.

Lincoln, Y. S., & Guba, E. G. (1985). *Naturalistic inquiry*. Newbury Park, CA: Sage.

Lindblom-Ylanne, S., & Pihlajamaki, H. (2003). Can a collaborative network environment enhance essay-writing processes? *British Journal of Educational Technology*, *34*(1), 17–30. doi:10.1111/1467-8535.00301

Lindner, J. R., Dooley, K. E., & Williams, J. R. (2003). Teaching, coaching, mentoring, facilitating, motivating, directing … What is a teacher to do? *The Agricultural Education Magazine*, *76*, 26–27.

Lindsay, R., & Breen, R. (2002). Different disciplines require different motivations for student success. *Research in Higher Education*, *43*(6), 693–725. doi:10.1023/A:1020940615784

Loucks-Horsley, S., Love, N., Stiles, K. E., Mundry, S., & Hewson, P. W. (2009). *Designing professional development for teachers of science and mathematics* (3rd ed.). Thousand Oaks, CA: Corwin Press.

Loveland, E. (2002). Connecting communities and classrooms. *Rural Roots Newsletter*, *3*(4). Retrieved from http://www.seer.org/pages/rsct.html

Lowell, N. (2006). Collaborating online: Learning together in community. *Quarterly Review of Distance Education*, *7*(2), 211–214.

Lowenthal, P. R., & Dunlap, J. C. (2010). From pixel on a screen to real person in your students' lives: Establishing social presence using digital storytelling. *The Internet and Higher Education*, *13*(1), 70–72. doi:10.1016/j.iheduc.2009.10.004

Lowther, D. L., & Ross, S. M. (2001). *Observation of computer use: Reliability analysis*. Memphis, TN: Center for Research in Educational Policy, The University of Memphis.

Lucas, T., Villegas, A. M., & Freedson-Gonzalez, M. (2008). Linguistically responsive teacher education: Preparing classroom teachers to teach English language learners. *Journal of Teacher Education*, *59*, 361–373. doi:10.1177/0022487108322110

Ludlow, B. L. (2006). Overview of online instruction. In Ludlow, B., Collins, B. C., & Menlove, R. (Eds.), *Online instruction for distance education: Preparing special educators in and for rural areas. Special monograph of the American council on rural special education* (pp. 2–42). Victoria, BC: Trafford Publishing.

Ludvig, M., Kirshstein, R., Sidana, A., Ardila-Rey, A., & Bae, Y. (2010). *An emerging picture of the teacher preparation pipeline: A report by the American Association of Colleges for Teacher Education and the American Institutes for Research*. Retrieved from http://aacte.org/pdf/Publications/Resources/PEDS%20Report%20-%20An%20Emerging%20Picture%20of%20the%20Teacher%20Preparation%20Pipeline.pdf

Luebeck, J. L., & Bice, L. R. (2005). Online discussion as a mechanism of conceptual change among mathematics and science teachers. *Journal of Distance Education*, *20*(2), 21–39.

Lueg, A., & Siebert, D. (2010, March 16). *E-learning's potential is hampered by misuse, critics say*. Deutsche Welle DW-WORLD.DE. Retrieved from http://www.dw-world.de/dw/article/0,5359341,00.html

Luke, A., & Freebody, P. (1999, October 26). Further notes on the four resources model. *Reading Online*. Retrieved from http://www.readingonline.org/research/lukefreebody.html#hasan

Luke, A., & Elkins, J. (1998). Reinventing literacy in "New Times". *Journal of Adolescent & Adult Literacy*, *42*(1), 4–7.

Lund, C. P., & Volet, S. (1998). Barriers to studying online for the first time: Students' perceptions. In C. McBeath & R. Atkinson (Eds.), *Proceedings EdTech'98: Planning for Progress, Partnership and Profit*: Perth, Australia: Australian Society for Educational Technology. Retrieved from http://www.aset.org.au/confs/edtech98/pubs/articles/lund.html

Mabrito, M. (2006). A study of synchronous versus asynchronous collaboration in an online business writing class. *American Journal of Distance Education, 20*, 93–107. doi:10.1207/s15389286ajde2002_4

MacGregor, J. (1990). Collaborative learning: Shared inquiry as a process of reform. In Svinicki, M. (Ed.), *The changing face of college teaching, New Directions for Teaching and Learning No. 42*. San Francisco, CA: Jossey-Bass. doi:10.1002/tl.37219904204

Macrorie, K. (1988). *The I-search paper: Revised edition of searching writing*. Portsmouth, NH: Boynton/Cook Publishers.

Maeroff, G. I. (2003). *A classroom of one: How online learning is changing our schools and colleges*. New York, NY: Palgrave Macmillan.

Maher, F. (2002). The attack on teacher education and teachers. *Radical Teacher, 64*(5), 5–8.

Major, J. (2005). Teacher education for cultural diversity: Online and at a distance. *Journal of Distance Learning, 9*(1), 15–26.

Makinster, J. G., Barab, S. A., & Harwood, W. (2006). The effect of social context on the reflective practice of preservice science teachers: Incorporating a web-supported community of teachers. *Journal of Technology and Teacher Education, 14*(3), 543–579.

Maor, D. (2006). Using reflective diagrams in professional development with university lecturers: A developmental tool in online teaching. *The Internet and Higher Education, 9*, 133–145. doi:10.1016/j.iheduc.2006.03.005

MarylandOnline. (2010). *Quality matters (QM)*. Retrieved August 16, 2011, from http://www.qmprogram.org

Mason, C. (2000). Online teacher education: An analysis of student teachers' use of computer-mediated communication. *The International Journal of Social Education, 15*(1), 19–38.

Massey, D. D. (2009). Teacher research: What's the point and who's it for? *Journal of Curriculum and Instruction, 3*. Retrieved from http://www.joci.ecu.edu/index.php/JoCI/issue/view/20

Mayer, R. E., & Wittrock, M. C. (2006). Problem solving. In Alexander, P. A., & Winne, P. H. (Eds.), *Handbook of educational psychology* (2nd ed.). Mahwah, NJ: Lawrence Erlbaum Associates.

McClam, S., & Sevier, B. (2010). Troubles with grades, grading, and change: Learning from adventures in alternative assessment practices in teacher education. *Teaching and Teacher Education, 26*, 1460–1470. doi:10.1016/j.tate.2010.06.002

McClay, J. M., & Mackey, M. (2009). Distributed assessment in OurSpace: This is not a rubric. In Burke, A., & Hammett, R. F. (Eds.), *Assessing new literacies: Perspectives from the classroom*. New York, NY: Peter Lang.

McCombs, B. L. (2001). What do we know about learner and learning? The learner-centered framework: Bringing the educational system into balance. *Educational Horizons, 53*, 82–191.

McGreal, R., & Elliott, M. (2004). Technologies of online learning. In Anderson, T., & Elloumi, F. (Eds.), *Theory and practice of online learning* (pp. 115–135). Athabasca, Canada: Athabasca University.

McKenzie, W., & Murphy, D. (2000). "I hope this goes somewhere": Evaluation of an online discussion group. *Australian Journal of Educational Technology, 16*(3), 239–257. Retrieved from http://cleo.murdoch.edu.au/ajet/ajet16/mckenzie.html

McLoughlin, C., & Lee, M. J. W. (2008). Future learning landscapes: Transforming pedagogy through social software. *Innovate: Journal of Online Education, 4*(5). Retrieved from http://innovateonline.info/?view=article&id=539

McLuhan, M. (1964). *Understanding media: The extensions of man*. New York, NY: Mentor.

McNamara, J. M., Swalm, R. L., Stearne, D. J., & Covassin, T. M. (2008). Online weight training. *Journal of Strength and Conditioning Research, 22*, 1164–1168. doi:10.1519/JSC.0b013e31816eb4e0

Means, B., Toyama, Y., Murphy, R., Bakia, M., & Jones, K. (2009). *Evaluation of evidence-based practices in online learning: A meta-analysis and review of online learning studies*. Washington, DC: U.S. Department of Education. Retrieved from http://www.ed.gov/rschstat/eval/tech/evidence-basedpractices/finalreport.pdf

Means, B. (Ed.). (1994). *Technology and education reform*. Sand Francisco, CA: Jossey-Bass Publishers.

Mehrotra, C. M., Hollister, C. D., & McGahey, L. (2001). *Distance learning: Principles for effective design, delivery, and evaluation*. Thousand Oaks, CA: Sage Publications.

Memorial University of Newfoundland. (1999). *Courses in telelearning and rural school teaching*. St. John's, NL: Author. Retrieved from http://www.mun.ca/regoff/cal99_00/EducationTeleLearningandRuralSchoolTeachingCourses.htm

Merriam, S. B., & Caffarella, R. (1999). *Learning in adulthood: A comprehensive guide*. San Francisco, CA: Jossey-Bass Publishers.

Merton, A. G. (2002). *Improving education outcomes: In colleges, universities, and beyond*. Panel Discussion, Conference Series 47, Education in the 21st Century: Meeting the Challenges of a Changing World. Retrieved from http://www.bos.frb.org/economic/conf/conf47/conf47u.pdf

MetLife. (2010). *Survey of the American teacher: Collaborating for student success*. Retrieved from http://www.metlife.com/assets/cao/contributions/foundation/american-teacher/MetLife_Teacher_Survey_2009.pdf

Meyer, K. A. (2003). Face-to-face versus threaded discussions: The role of time and higher-order thinking. *Journal of Asynchronous Learning Networks*, *7*(3), 55–65. Retrieved from http://www.sloan-c.org/publications/jaln/v7n3/pdf/v7n3_meyer

Meyer, K. A. (2005). The ebb and flow of online discussions: What Bloom can tell us about our students' conversations. *Journal of Asynchronous Learning Networks*, *9*(1), 53–63.

Meyers, E., & Rust, F. (2003). *Taking action with teacher research*. Portsmouth, NH: Heinemann.

Mezirow, J. (1997). Transformative learning: Theory to practice. *New Directions for Adult and Continuing Education*, *74*, 5–12. doi:10.1002/ace.7401

Michigan Department of Education. (2006). *Michigan merit curriculum guidelines: Online experience*. Retrieved from http://www.michigan.gov/documents/mde/Online10.06_final_175750_7.pdf

Michigan Department of Education. (2008). *Standards for the preparation of teachers: Educational technology*. Lansing, MI: Author. Retrieved from http://www.michigan.gov/documents/mde/EducTech_NP_SBEApprvl.5-13-08.A_236954_7.doc

Middleton, J., Gorard, S., Taylor, C., & Bannan-Ritland, B. (2006). *The 'Compleat' design experiment: From soup to nuts*. Department of Educational Studies Research Paper 2006/5. York, PA: Department of Educational Studies, University of York. Retrieved from http://www.york.ac.uk/media/educationalstudies/documents/research/Paper18Thecompleatdesignexperiment.pdf

Midobuche, E., & Benavides, A. H. (2006). Preparing teachers to teach English language learners: Best practices for school and after-school programs. In Cowart, M., & Dam, P. (Eds.), *Cultural and linguistic issues for English language learners* (pp. 83–107). Texas Woman's University: Cahn Nam Publishers, Inc.

Milken Family Foundation. (2001). *Information technology underused in teacher education*. Retrieved from http://www.mff.org/edtech/article.taf?_function=detail&Content_uid1=131

Ministry of Education. (2010a). *OECD review on evaluation and assessment frameworks for improving school outcomes: New Zealand country background report*. Retrieved from http://www.educationcounts.govt.nz/__data/assets/pdf_file/0009/90729/966_OECD-report.pdf

Ministry of Education. (2010b). *Ngā haeata mātauranga: The annual report on Māori education, 2008/09*. Wellington, New Zealand: Author. Retrieved from http://www.educationcounts.govt.nz/publications/series/5851/75954/4#discussion

Mishra, P., & Koehler, M. J. (2006). Technological pedagogical content knowledge: A framework for teacher knowledge. *Teachers College Record*, *108*(6), 1017–1054. doi:10.1111/j.1467-9620.2006.00684.x

Mislevy, R., Steinberg, L., & Almond, R. (1999). *Evidence-centered assessment design*. Educational Testing Service.

Mislevy, R., Steinberg, L., & Almond, R. (2003). On the structure of educational assessments. *Interdisciplinary Research and Perspectives*, *1*, 3–67. doi:10.1207/S15366359MEA0101_02

Moallem, M. (2003). An interactive online course: a collaborative design model. *Educational Technology Research and Development*, *51*(4), 85–103. doi:10.1007/BF02504545

Moje, E. B., Ciechanowski, K. M. I., Kramer, K., Ellis, L., Carrillo, R., & Collazo, T. (2004). Working toward third space in content area literacy: An examination of everyday funds of knowledge and discourse. *Reading Research Quarterly*, *39*(1), 38–70. doi:10.1598/RRQ.39.1.4

Moller, L., Foshay, W., & Huett, J. (2008). The evolution of distance education: Implications for instructional design on the potential of the Web. *TechTrends*, *52*(4), 66–70. doi:10.1007/s11528-008-0179-0

Moore, S. A. (2001). Technology, place, and nonmodern thesis. *Journal of Architectural Education*, *54*(3), 130-139.

Moore, M. (2007). The theory of transactional distance. In Moore, M. (Ed.), *The handbook of distance education* (2nd ed., pp. 89–108). Mahwah, NJ: Lawrence Erlbaum.

Moore, R. (2003). Reexamining the field experiences of preservice teachers. *Journal of Teacher Education*, *54*(1), 31–42. doi:10.1177/0022487102238656

Moreno, R., & Mayer, R. E. (2002). Verbal redundancy in multimedia learning: When reading helps listening. *Journal of Educational Psychology*, *94*, 156–163. doi:10.1037/0022-0663.94.1.156

Morrell, E. (2004). *Linking literacy and popular culture: Finding connections for lifelong learning*. Norwood, MA: Christopher Gordon Publishers.

Morris, S. (2002). *Teaching and learning online: A step-by-step guide for designing an online K-12 school program*. Lanham, MD: Scarecrow Press Inc.

Motiwalla, L., & Tello, S. (2000). Distance learning on the Internet: An exploratory study. *The Internet and Higher Education*, *2*(4), 253–264. doi:10.1016/S1096-7516(00)00026-9

Moursund, D., & Bielefeldt, T. (1999). *Will new teachers be prepared to teach in a digital age? A national survey on information technology in teacher education*. Santa Monica, CA: Milken Exchange on Education Technology.

Muijs, D. (2004). *Doing quantitative research in education with SPSS*. London, UK: Sage Publications.

Muirhead, B. (2002). *Salmon's e-tivities: The key to active online learning*. Sterling, VA: Stylus.

Murphy, E. (2004). Identifying and measuring ill-structured problem formulation and resolution in online asynchronous discussions. *Canadian Journal of Learning and Technology*, *30*(1), 5–20.

Murphy, E., Rodriguez-Manzanares, M. A., & Barbour, M. (2011). Asynchronous and synchronous online teaching: Perspectives of Canadian high school distance education teachers. *British Journal of Educational Technology*, *42*, 583–591. doi:10.1111/j.1467-8535.2010.01112.x

National Board for Professional Teaching. (2011). *National Board for Professional Teaching standards*. Retrieved from http://www.nbpts.org

National Center for Alternative Licensure. (2010). *A state by state analysis: Introduction*. Retrieved from http://www.teach-now.org/intro.cfm

National Commission on Teaching and America's Future. (2002). *Unraveling the "teacher shortage" problem: Teacher retention is the key*. Washington, DC: National Commission on Teaching and America's Future.

National Commission on Teaching and America's Future. (2003). *No dream denied: A pledge to America's children: A summary*. Retrieved from http://www.ecs.org/html/Document.asp?chouseid=4269

National Council for Accreditation of Teacher Education. (2008). *Unit standards*. Retrieved from http://www.ncate.org/standards/ncateunitstandards/unitstandardsineffect2008/tabid/476/default.aspx

National Council for Accreditation of Teacher Education. (2010). *NCATE unit standards*. Retrieved from http://www.ncate.org/standards/ncateunitstandards/unitstandardsineffect2008/tabid/476/default.aspx

National Council for the Accredidation of Teacher Education. (2010). *NCATE Unit Standards*. Retrieved from http://www.ncate.org/Standards/NCATEUnitStandards/UnitStandardsinEffect2008/tabid/476/Default.aspx

National Council of Teachers of English, & International Reading Association. (1996). *Standards for the English language arts*. Urbana, IL: NCTE.

National Council of Teachers of English. (2007). *21st-century literacies: A policy research brief.* Retrieved from http://www.ncte.org/library/NCTEFiles/Resources/PolicyResearch/21stCenturyResearchBrief.pdf

National Education Association. (2006). *Guide to teaching online courses*. Retrieved from http://www.nea.org/assets/docs/onlineteachguide.pdf

National Educational Technology Plan. (2010). *Transforming American education: Learning powered by technology*. Washington, DC. Retrieved from http://www.ed.gov/technology/netp-2010

National Partnership for Excellence and Accountability in Teaching (NPEAT). (2000). *Revisioning professional development: What learner-centered professional development looks like*. Oxford, OH: Author. Retrieved from http://www.nsdc.org/library/policy/npeat213.pdf

National Summit on Educational Games. (2006). *Harnessing the power of games*. Washington, DC: Federation of American Scientists. Retrieved from http://www.fas.org/gamesummit/Resources/Summit%20on%20Educational%20Games.pdf

NCEE. (1983). *A nation at risk*. Washington, DC: National Commission on Excellence in Education.

Nelson, E., Beggs, K., Campbell, B., Fairclough, P., Meditch, A., Ortenberg, T., & Palovak, J. (Producers), & Herzog, W. (Director). (2005). *Grizzly man* [Motion picture]. United States: Lions Gate Films.

Nelson, J., Christopher, A., & Mims, C. (2009). TPACK and web 2.0: Transformation of teaching and learning. *TechTrends*, *53*(5), 80–85. doi:10.1007/s11528-009-0329-z

Nelson, T. H., & Slavit, D. (2007). Collaborative inquiry among science and mathematics teachers in the USA: Professional learning experiences through cross-grade, cross discipline dialogue. *Journal of In-service Education*, *33*(1), 23–39. doi:10.1080/13674580601157620

New London Group. (1996). A pedagogy of multiliteracies: Designing social futures. *Harvard Educational Review*, *66*, 60–92.

New Zealand Council for Educational Research. (2004). *Statistical profile of Māori in tertiary education and engagement in e-learning*. Retrieved from http://www.itpnz.ac.nz/

New Zealand Teachers Council. (2011). *Graduating teacher standards*. Retrieved from http://www.teacherscouncil.govt.nz/te/gts/index.stm

Newmann, F. M., Marks, H. M., & Gamoran, A. (1996, August). Authentic pedagogy and student performance. *American Journal of Education*, *104*(4), 280–312. doi:10.1086/444136

Ng, C. F. (2006). Academics telecommuting in open and distance education universities: Issues, challenges, and opportunities. *International Review of Research in Open and Distance Learning*, *7*(2), 1–16.

Nielsen, L. E. (2003). *Online discussion rubric*. Retrieved from http://www2.uwstout.edu/content/profdev/rubrics/discussionrubric.html

Nielsen, H. D. (1997). Quality assessment and quality assurance in distance teacher education. *Distance Education*, *18*(2), 284–317. doi:10.1080/0158791970180207

Nieto, S., & Bode, P. (2008). *Affirming diversity: The sociopolitical context of multicultural education*. New York, NY: Addison-Wesley.

Njenga, J. K., & Fourie, L. C. H. (2010). The myths about e-learning in higher education. *British Journal of Educational Technology*, *41*, 199–212. doi:10.1111/j.1467-8535.2008.00910.x

Noble, D. F. (2001). *Digital diploma mills: The automation of higher education*. New York, NY: Monthly Review Press.

Norlander-Case, K. A., Reagan, T. G., & Case, C. W. (1999). *The professional teacher: The preparation and nurturance of the reflective practitioner*. San Francisco, CA: Jossey-Bass.

North American Council for Online Learning. (2011). *iNACOL national standards of quality for online courses*. Retrieved August 8, 2011, from http://www.inacol.org/research/nationalstandards/index.php

North Carolina Department of Public Instruction. (2010). *Professional development*. Retrieved from http://www.dpi.state.nc.us/profdev/training/teacher/

North Carolina school executive principal and assistant principal evaluation process. (2008). *North Carolina Department of Public Instruction*. Retrieved from http://www.dpi.state.nc.us/docs/profdev/training/principal/evaluationprocess.pdf

North Carolina State Board of Education, Department of Public Instruction. (1997). *A strategic plan for reading literacy*. Retrieved from http://www.ncpublicschools.org/docs/curriculum/languagearts/elementary/strategicplanforreadingliteracy.pdf

North Carolina teacher evaluation process. (2009). *North Carolina Department of Public Instruction*. Retrieved from http://www.dpi.state.nc.us/docs/profdev/training/teacher/teacher-eval.pdf

Northwest Educational Technology Consortium. (2005). *Videoconferencing to enhance instruction*. Retrieved from http://www.netc.org/digitalbridges/uses/useei.php

Norton, P., & Hathaway, D. (2008). On its way to classrooms, Web 2.0 goes to graduate school. *Computers in the Schools*, *25*(3), 163–180. doi:10.1080/07380560802368116

NSDC. (2001). *National standards for online learning*. National Staff Development Council.

O'Brien, D. G., & Bauer, E. (2005). New literacies and the institution of old learning. *Reading Research Quarterly. Essay Book Review*, *40*, 120–131.

O'Connor, E. A. (2011). The effect on learning, communication, and assessment when student-centered Youtubes of microteaching were used in an online teacher-education course. *Journal of Educational Technology Systems*, *39*(2), 135–154. doi:10.2190/ET.39.2.d

O'Connor, K. A., Good, A. J., & Greene, H. C. (2006). Lead by example: The impact of teleobservation on social studies methods courses. *Social Studies Research and Practice*, *1*, 165–178.

O'Hanlon, N., & Diaz, K. R. (2010). Techniques for enhancing reflection and learning in an online course. *MERLOT Journal of Online Learning and Teaching*, *6*(1), 43–54.

O'Neil, H. F. Jr, Chung, G. K. W. K., & Brown, R. (1997). Use of networked simulations as a context to measure team competencies. In O'Neil, H. F. Jr., (Ed.), *Workforce readiness: Competencies and assessment* (pp. 411–452). Mahwah, NJ: Erlbaum.

O'Bannon, B., & Nonis, A. (2002). A field-based initiative for integrating technology in the content areas: Using a team approach to preparing preservice teachers use technology. In D. Willis, et al. (Eds.), *Proceedings of Society for Information Technology & Teacher Education International Conference 2002* (pp. 1394-1397). Chesapeake, VA: AACE.

OECD. (2009a). *Creating effective teaching and learning environments: First results from TALIS*. Paris, France: OECD.

OECD. (2009b). *PIAAC problem solving in technology rich environments: Conceptual framework*. Paris, France: OECD.

Ogobonna, E., & Harris, L. C. (2003). Innovation organizational structure and performance. *Journal of Organizational Change Management*, *16*(5), 512–533. doi:10.1108/09534810310494919

Oliveira, J., & Orivel, F. (2003). The cost of distance education for training teachers. In Robinson, B., & Latchem, C. (Eds.), *Teacher education through open and distance learning*. London, UK: Routledge-Falmer.

Oliver, K. (2010). Integrating Web 2.0 across the curriculum. *TechTrends*, *54*(2), 50–60. doi:10.1007/s11528-010-0382-7

Oliver, R., & Herrington, J. (2003). Exploring technology-mediated learning from a pedagogical perspective. *Interactive Learning Environments*, *11*(2), 111–126. doi:10.1076/ilee.11.2.111.14136

Olson, G. M., & Olson, J. S. (2000). Distance matters. *Human-Computer Interaction*, *15*(2), 139–178. doi:10.1207/S15327051HCI1523_4

Olson, G. M., Zimmerman, A., & Bos, N. (Eds.). (2008). *Scientific research on the Internet*. Cambridge, MA: MIT Press.

Olson, S. J., & Werhan, C. (2005). Teacher preparation via online learning: A growing alternative for many. *Action in Teacher Education*, *27*(3), 76–84. doi:10.1080/01626620.2005.10463392

Orrill, C. H. (2001). Building learner-centered classrooms: A professional development framework for supporting critical thinking. *Educational Technology Research and Development*, *49*(1), 15–34. doi:10.1007/BF02504504

Orvis, K. L., & Lassiter, A. L. R. (2008). *Computer-supported collaborative learning: Best practices and principles for instructors*. Hersey, PA: Information Science Publishing. doi:10.4018/978-1-59904-753-9

Ota, C. DiCarlo, C., Burts, D., Laird, R., & Gioe, C. (2006, December). Training and the needs of adult learners. *Journal of Extension*, *44*(6). Article Number: 6TOT5. Retrieved from www.joe.org/joe/2006december/tt5.shtml

oz-TeacherNet. (2001). *oz-TeacherNet: Teachers helping teachers: Revised Bloom's Taxonomy*. Retrieved from http://rite.ed.qut.edu.au/oz-teachernet/index.php?module=ContentExpress&func=display&ceid=29

Pace, B. G. (2006). *Teaching narrative media online and in other social spaces*. Unpublished manuscript.

Pace, B. G. (2010, November). *Embellishing conceptions of literacy with pre-service English teachers*. Paper presented at the meeting of the National Council of Teachers of English, Orlando, FL.

Pace, B. G., Rodesiler, L. B., & Tripp, L. (2010). Pre-service English teachers and Web 2.0: Teaching and learning literacy with digital applications. In Maddux, C. D., Gibson, D., & Dodge, B. (Eds.), *Research highlights in technology and teacher education 2010* (pp. 177–184). Chesapeake, VA: Society for Information Technology and Teacher Education.

Paese, P. C. (1996). Contexts: Overview and framework. In McIntyre, J., & Byrd, D. M. (Eds.), *Preparing tomorrow's teachers: The field experience* (pp. 1–7). Thousand Oaks, CA: Corwin Press, Inc.

Pahl, K., & Rowsell, J. (2005). *Literacy and education: Understanding the new literacy studies in the classroom*. Thousand Oaks, CA: Sage.

Palloff, R. M., & Pratt, K. (2001). *Lessons from the cyberspace classroom: The realities of online teaching*. San Francisco, CA: Jossey-Bass.

Palloff, R. M., & Pratt, K. (2005). *Collaborating online: Learning together in community*. San Francisco, CA: Jossey-Bass.

Palloff, R. M., & Pratt, K. (2009). *Assessing the online learner: Resources and strategies for faculty*. San Francisco, CA: Jossey-Bass.

Palloff, R., & Pratt, K. (2007). *Building online learning communities: Effective strategies for the virtual classroom*. San Francisco, CA: Jossey-Bass.

Paloff, R., & Pratt, K. (2003). *The virtual student: A profile and guide to working with online learners*. San Francisco, CA: Jossey-Bass.

Pankowski, M. M. (2003). *How do undergraduate mathematics faculty learn to teach online?* Unpublished Doctoral dissertation, Duquesne University.

Papert, S. (1991). Situating constructionism. In Harel, I., & Papert, S. (Eds.), *Constructionism* (pp. 1–11). Norwood, NJ: Ablex.

Papert, S. (1998). Does easy do it? Children, games, and learning. *Game Developer Magazine*, *4*, 88–97.

Paquette, G. (2004). *Instructional engineering in networked environments. Instructional technology & training series*. San Francisco, CA: Pfeiffer.

Parr, F. W. (1929). *A remedial program for the inefficient silent reader*. An unpublished Doctoral dissertation, State University of Iowa.

Parsad, B., Lewis, L., & Tice, P. (2008). *Distance education at degree-granting postsecondary institutions: 2006-07: First look*. National Center for Education Statistics. Retrieved from http://nces.ed.gov/pubs2009/2009044.pdf

Parsons, S., & Burrowbridge, S. (2011). *Thoughtfully adaptive teaching: An overlooked form of differentiated instruction*. Manuscript submitted for publication.

Parsons, S. (2008). Providing all students ACCESS to self-regulated literacy learning. *The Reading Teacher*, *61*, 628–635. doi:10.1598/RT.61.8.4

Parsons, S. A., Massey, D., Vaughn, M., Scales, R. Q., Faircloth, B. S., & Howerton, S. (2011). Developing teachers' reflective thinking and adaptability in graduate courses. *Journal of School Connections*, *3*(1), 91–111.

Partnership for 21st Century Skills (2009). *Framework for 21st century learning*. Retrieved from http://www.21stcenturyskills.org/documents/framework_flyer_updated_jan_09_final-1.pdf

Partnership for 21st Century Skills. (2004). *Framework for 21st century learning*. Retrieved August 2, 2011, from http://www.p21.org/index.php?option=com_content&task=view&id=254&Itemid=120

Partnership for 21st Century Skills. (2011). *Framework for 21st century learning*. Washington, DC. Retrieved from http://www.p21.org/tools-and-resources/publications/1017-educators#defining

Pascarella, E. T., & Terenzini, P. T. (2005). *How college affects students: A third decade of research* (*Vol. 2*). San Francisco, CA: Jossey-Bass.

Patton, M. (2011). *Developmental evaluation: Applying complexity concepts to enhance innovation and use*. New York, NY: The Guilford Press.

Patton, M. Q. (1990). *Qualitative evaluation and research methods* (2nd ed.). Newbury Park, CA: Sage.

Patton, M. Q. (2002). *Qualitative research and evaluation methods* (3rd ed.). Thousand Oaks, CA: Sage.

Paul, R., & Elder, L. (2008). *The miniature guide to critical thinking: Concepts and tools*. Dillon Beach, CA: The Foundation for Critical Thinking Press.

Payr, S. (2005). Not quite an editorial: Educational agents and (e-)learning. *Applied Artificial Intelligence*, *19*(3/4), 199–213. doi:10.1080/08839510590910147

Pearson, P. D., & Gallagher, M. C. (1983). The instruction of reading comprehension. *Contemporary Educational Psychology*, *8*, 317–344. doi:10.1016/0361-476X(83)90019-X

Penner, D. E. (2001). Cognition, computers, and synthetic science: Building knowledge and meaning through modeling. *Review of Research in Education*, *25*, 1–35.

Perna, L. (2010). Understanding the working college student. *Academe Online, 96*. Retrieved from http://www.aaup.org/AAUP/pubsres/academe/2010/JA/

Perraton, H., & Potashnik, M. (1997). Teacher education at a distance. *Education and Technology Series, 2*(2).

Perraton, H., Creed, C., & Robinson, B. (2002). *Teacher education guidelines: Using open and distance learning*. Paris, France: UNESCO.

Peterson, C. L., & Bond, N. (2004). Online compared to face-to-face teacher preparation for learning standards-based planning skills. *Journal of Research on Technology in Education*, *36*(4), 345–360.

Peterson, E. (2009). Using a wiki to enhance cooperative learning in a real analysis course. *PRIMUS (Terre Haute, Ind.)*, *19*(1), 18–28. doi:10.1080/10511970802475132

Petty, T., Heafner, T., & Hartshorne, R. (2009, March). Examining a pilot program for the remote observation of graduate interns. In R. Weber, K. McFerrin, R. Carlsen, & D. A. Willis, (Eds.), *2009 Society for Information Technology and Teacher Education Annual: Proceedings of SITE2009* (pp. 2658-2660). Norfolk, VA: Association for the Advancement of Computing in Education (AACE).

Petty, T., & Heafner, T. (2009). What is ROGI? *Journal of Technology Integration in the Classroom*, *1*(1), 21–27.

Piaget, J. (1969). *The mechanisms of perception*. London, UK: Routledge and Kegan Paul.

Piaget, J. (1972). *The psychology of the child*. New York, NY: Basic Books.

Picciano, A. G., & Seaman, J. (2009). *K-12 online learning survey: A survey of U.S. school district administrators*. Retrieved August 2, 2011, from http://sloanconsortium.org/publications/survey/K-12_06

Picciano, A., & Seaman, J. (2009). *K-12 online learning: A 2008 follow-up of the survey of U.S. school district administrators*. Babson Park, MA: The Sloan Consortium: Babson Survey Research Group.

Picciano, A. G. (2002). Beyond student perceptions: Issues of interaction, presence and performance in an online course. *Journal of Asynchronous Learning Networks*, *6*(1), 21–40.

Picciano, A. G., & Seaman, J. (2009). *K-12 online learning: A 2008 follow-up of the survey of U.S. school district administrators*. Needham, MA: Alfred P. Sloan Foundation.

Pierson, M. E. (2001). Technology integration practices as function of pedagogical expertise. *Journal of Research on Computing in Education, 33*(4), 413–429.

Plass, J. L., Goldman, R., Flanagan, M., Diamond, J. P., Dong, C., & Looui, S. … Perlin, K. (2007). *RAPUNSEL: How a computer game design based on educatinal theory can improve girls' self-efficacy and self-esteem*. Paper presented at the American Educational Research Association.

Poekert, P. (2010). The pedagogy of facilitation: Teacher inquiry as professional development in a Florida elementary school. *Professional Development in Education, 37*(1), 19–38. doi:10.1080/19415251003737309

Polly, D., & Hannafin, M. J. (2010). Reexamining technology's role in learner-centered professional development. *Educational Technology Research and Development, 58*(5), 557–571. doi:10.1007/s11423-009-9146-5

Poole, D. M. (2000). Student participation in a discussion-oriented online course: A case study. *Journal of Research on Computing in Education, 33*(2), 162–177.

Pope, C., & Golub, J. (2000). Preparing tomorrow's English language arts teachers today: Principles and practices for infusing technology. *Contemporary Issues in Technology & Teacher Education, 1*(1), 89–97. Retrieved from http://www.citejournal.org/vol1/iss1/currentissues/english/article1.htm

Porima, L. (2005). *Understanding the needs of Māori learners for the effective use of elearning*. Retrieved from http://www.itpnz.ac.nz/

Posey, L. (2007). Critical thinking and collaboration in post-secondary online education. In T. Bastiaens & S. Carliner (Eds.), *Proceedings of World Conference on E-Learning in Corporate, Government, Healthcare, and Higher Education 2007* (pp. 1770-1774). Chesapeake, VA: AACE.

Posner, G. J. (2005). *Field experience: A guide to reflective teaching* (6th ed.). White Plains, NY: Allyn and Bacon.

Prensky, M. (2001). *Digital game-based learning*. New York, NY: McGraw-Hill.

Prensky, M. (2001). Digital natives, digital immigrants. *Horizon, 9*(5), 1–6. Retrieved from http://www.marcprensky.com/writing/Prensky%20%20Digital%20Natives,%20Digital%20Immigrants%20-%20Part1.pdfdoi:10.1108/10748120110424816

President's Council of Advisors on Science and Technology. (2010). *Prepare and inspire: K-12 education in science, technology, engineering, and math (STEM) for America's future*. Retrieved from http://www.whitehouse.gov/sites/default/files/microsites/ostp/pcast-stem-edfinal.Pdf

Prigogine, I. (1980). *From being to becoming: time and complexity in the physical sciences*. San Francisco, CA: W.H. Freeman.

Pugach, M. (2009). *Because teaching matters* (2nd ed.). Hoboken, NJ: Wiley and Sons, Inc.

Putnam, R. T., & Borko, H. (2000). What do new views of knowledge and thinking have to say about research on teacher learning? *Educational Researcher, 29*, 4–15.

Quality Matters. (2005). *Research literature and standards sets support for Quality Matters review standards*. Retrieved from http://www.qualitymatters.org/Documents/Matrix%20of%20Research%20Standards%20FY0506.pdf

Quinlan, A. M. (2011). 12 tips for the online teacher. *Phi Delta Kappan, 92*(4), 28–31.

Ragan, P., Lacey, A., & Nagy, R. (2002). Web-based learning and teacher preparation: Stumbling blocks and stepping stones. *Teaching with Technology Today, 8*(5). Retrieved from http://www.wisconsin.edu/ttt/articles/ragan.htm

Raths, J. (2001). Teachers' beliefs and teaching beliefs. *Early Childhood Research & Practice, 3*(1). Retrieved from http://ecrp.uiuc.edu/v3n1/raths.html

Reeves, T. C., Herrington, J., & Oliver, R. (2005). Design research: A socially responsible approach to instructional technology research in higher education. *Journal of Computing in Higher Education, 16*(2), 97–116. doi:10.1007/BF02961476

Reiser, R. A., & Dempsey, J. V. (2007). *Trends and issues in instructional design* (2nd ed.). Upper Saddle River, NJ: Pearson Education, Inc.

Report, H. (2008). *The Horizon Report: 2008 edition.* A collaboration between The New Media Consortium and the EDUCAUSE Learning Initiative (ELI), an EDUCAUSE program. Retrieved from http://www.nmc.org/pdf/2008-Horizon-Report.pdf

Resnick, D. P., & Resnick, L. B. (1977). The nature of literacy: An historical exploration. *Harvard Educational Review*, *47*(3), 370385.

Resnick, M. (1994). *Turtles, termites, and traffic jams*. Cambridge, MA: MIT Press.

Resnick, M. (1999). Decentralized modeling and decentralized thinking. In Fuerzeig, W., & Roberts, N. (Eds.), *Modeling and simulation in precollege science and mathematics education* (pp. 114–137). New York, NY: Springer-Verlag. doi:10.1007/978-1-4612-1414-4_5

Resnick, M. (2007). Sowing the seeds for a more creative society. *Learning and Leading with Technology*, *7*, 18–22.

Resnick, M., Berg, R., & Eisenberg, M. (2000). Beyond black boxes: Bringing transparency and aesthetics back to scientific investigation. *Journal of the Learning Sciences*, *9*(1), 7–30. doi:10.1207/s15327809jls0901_3

Rhine, S., & Bryant, J. (2007). Enhancing pre-service teachers' reflective practice with digital video-based dialogue. *Reflective Practice*, *8*(3), 345–358. doi:10.1080/14623940701424884

Rice, K., & Dawley, L. (2007). *Going virtual! The status of professional development for K-12 online teachers*. Boise, ID: Boise State University. Retrieved from http://edtech.boisestate.edu/goingvirtual/goingvirtual1.pdf

Rice, J. (2003). *Teacher quality: Understanding the effectiveness of teacher attributes*. Washington, DC: Economic Policy Institute.

Rice, J. (2007). Assessing higher order thinking in video games. *Journal of Technology and Teacher Education*, *15*(1), 87–100.

Richardson, J. C., & Swan, K. (2003). Examining social presence in online courses in relation to students' perceived learning and satisfaction. *Journal of Asynchronous Learning Networks*, *7*(1), 68–88.

Richardson, L. (2000). Writing: A method of inquiry. In Denzin, N. K., & Lincoln, Y. S. (Eds.), *Sage handbook of qualitative research* (2nd ed., pp. 923–948). Thousand Oaks, CA: Sage.

Richardson, V., & Roosevelt, D. (2004). Teacher preparation and the improvement of teacher education. In Smylie, M. A., & Miretzky, D. (Eds.), *Developing the teacher workforce*. Chicago, IL: National Society for the Study of Education. doi:10.1111/j.1744-7984.2004.tb00032.x

Richardson, W. (2006). *Blogs, wikis, podcasts, and other powerful Web tools for classrooms*. Thousand Oaks, CA: Corwin Press.

Rideout, V. J., Foehr, U. G., & Roberts, D. F. (2010). *Generation M^2: Media in the lives of 8- to 18-year olds*. Kaiser Foundation. Retrieved from the http://www.kff.org/entmedia/8010.cfm

Rieber, L. P. (1996). Seriously considering play: Designing interactive learning environments based on the blending of microworlds, simulations, and games. *Educational Technology Research and Development*, *44*(2), 43–58. doi:10.1007/BF02300540

Rieber, L. P., Smith, L., & Noah, D. (1998). The value of serious play. *Educational Technology*, *38*(6), 29–37.

Riel, M., & Polin, L. (2004). Online learning communities: Common ground and critical differences in designing technical environments. In Barab, S. (Eds.), *Designing for virtual communities* (pp. 16–50). Cambridge, UK: Cambridge University Press.

Riffel, S., & Sibley, D. (2005). Using web-based instruction to improve large undergraduate biology courses: An evaluation of a hybrid course format. *Computers & Education*, *44*, 217–235. doi:10.1016/j.compedu.2004.01.005

Riis, J. (1890). *How the other half lives*. New York, NY: Charles Scribner's Sons. doi:10.1037/12986-000

Rimor, R. (2002). *From search for information to construction of knowledge: Organization and construction of knowledge in database environment*. Unpublished Doctoral Dissertation, Ben-Gurion University of the Negev, Israel.

Rimor, R., Rosen, Y., & Naser, K. (2010). Complexity of social interactions in collaborative learning: The case of online database environment. *Interdisciplinary Journal of E-Learning and Learning Objects*, *6*, 355–365.

Rivkin, S. G., Hanushek, E. A., & Kain, J. F. (2005). Teachers, schools, and academic achievement. *Econometrica: Journal of the Econometric Society*, *73*(2), 417–458. doi:10.1111/j.1468-0262.2005.00584.x

Roberts, T. S., & McInnerney, J. M. (2007). Seven problems of online group learning (and their solutions). *Journal of Educational Technology & Society*, *10*(4), 257–268.

Robinson, B., & Latchem, C. (Eds.). (2003). *Teacher education through open and distance learning*. London, UK: Routledge-Falmer.

Roblyer, M. D., Davis, L., Mills, S. C., Marshall, J., & Pape, L. (2008). Toward practical procedures for predicting and promoting success in virtual school students. *American Journal of Distance Education*, *22*, 90–109. doi:10.1080/08923640802039040

Roblyer, M. D., Freeman, J., Stabler, M., & Schniedmiller, J. (2007). *External evaluation of the Alabama ACCESS initiative phase 3 report*. Eugene, OR: International Society for Technology in Education.

Roblyer, M. D., & McKenzie, B. (2000). Distant but not out-of-touch: What makes an effective distance learning instructor? *Learning and Leading with Technology*, *27*(6), 50–53.

Robyler, M. D. (1999). Is choice important in distance learning? A study of student motives for taking internet-based courses at the high school and community college levels. *Journal of Research on Computing in Education*, *32*, 157–171.

Rock, M. L., Gregg, M., Thead, B. K., Acker, S. E., Gable, R. A., & Zigmond, N. P. (2009). Can you hear me now? Evaluation of an online wireless technology to provide real-time feedback to special education teachers in-training. *Teacher Education and Special Education*, *32*, 64–82. doi:10.1177/0888406408330872

Rock, M., Gregg, M., Ellis, E., & Gable, R. (2008). REACH: A framework for differentiating classroom instruction. *Preventing School Failure*, *52*, 31–47. doi:10.3200/PSFL.52.2.31-47

Rodgers, C. (2002). Defining reflection: Another look at John Dewey and reflective thinking. *Teachers College Record*, *104*, 842–866. doi:10.1111/1467-9620.00181

Rogoff, B. (2003). *The cultural nature of human development*. Oxford, UK: Oxford University Press.

Romano, T. (1990). The multigenre research paper: Melding fact, interpretation, and imagination. In Daiker, D., & Morenberg, M. (Eds.), *The writing teacher as researcher: Essays in the theory and practice of class-based research* (pp. 123–141). New Hampshire: Boyton/Cook Publishers.

Romano, T. (1995). *Writing with passion: Life stories, multiple genres*. Portsmouth, NH: Boyton/Cook.

Rosaen, C. L., Lundeberg, M., Cooper, M., Fritzen, A., & Terpstra, M. (2008). Noticing noticing: How does investigation of video-records change how teachers reflect on their experiences? *Journal of Teacher Education*, *59*(4), 347–360. doi:10.1177/0022487108322128

Roschelle, J., & Teasley, S. D. (1995). The construction of shared knowledge in collaborative problem-solving. In O'Malley, C. E. (Ed.), *Computer-supported collaborative learning* (pp. 69–97). Berlin, Germany: Springer-Verlag. doi:10.1007/978-3-642-85098-1_5

Rosenthal, P., Romano, R., Smiley, S., Rosegarten, R., Schneider, L., & Cawley, T. … Stevens, J. (Producers). (1996). *Everybody loves Raymond* [Television series]. Burbank, CA: Warner Bros. Studios.

Rosen, Y., & Rimor, R. (2009). Using collaborative database to enhance students' knowledge construction. *Interdisciplinary Journal of E-Learning and Learning Objects*, *5*, 187–195.

Ross, S. M., Smith, L. J., & Alberg, M. (1999). *The school observation measure (SOM)*. Memphis, TN: Center for Research in Educational Policy, The University of Memphis.

Rourke, L., & Kanuka, H. (2009). Learning in communities of inquiry: A review of the literature. *Journal of Distance Education*, *23*(1), 19–48.

Rovai, A. P. (2000). Online and traditional assessments: What is the difference? *The Internet and Higher Education*, *3*(3), 141–151. doi:10.1016/S1096-7516(01)00028-8

Rovai, A. P. (2000a). Building and sustaining community in asynchronous learning networks. *The Internet and Higher Education, 3*, 285–297. doi:10.1016/S1096-7516(01)00037-9

Rovai, A. P. (2000b). Online and traditional assessments: What is the difference? *The Internet and Higher Education, 3*, 141–151. doi:10.1016/S1096-7516(01)00028-8

Rovai, A. P. (2001). Building classroom community at a distance: A case study. *Educational Technology Research and Development, 49*(4), 33–48. doi:10.1007/BF02504946

Rovai, A. P. (2002). Building sense of community at a distance. *International Review of Research in Open and Distance Learning, 3*(1), 1–16.

Rovai, A. P. (2002). Development of an instrument to measure classroom community. *The Internet and Higher Education, 5*(3), 197–211. doi:10.1016/S1096-7516(02)00102-1

Rovai, A. P., & Barnum, K. T. (2003). On-line course effectiveness: An analysis of student interactions and perceptions of learning. *Journal of Distance Education, 18*(1), 57–73. Retrieved from http://proquest.umi.com.helicon.vuw.ac.nz/pqdweb?RQT=318&pmid=57387&cfc=1

Ruschoff, B., & Ritter, M. (2001). Technology-enhanced language learning: Construction of knowledge and template-based learning in the foreign language classroom. *Computer Assisted Language Learning, 14*(3-4), 219–232. doi:10.1076/call.14.3.219.5789

Russell, S. (2006). *An overview of adult learning processes: Adult learning principles*. Retrieved from http://www.medscape.com/viewarticle/547417_2

Russell, M., Carey, R., Kleiman, G., & Venable, J. D. (2009). Face-to-face and online professional development for mathematics teachers: A comparative study. *Journal of Asynchronous Learning Networks, 13*(2), 71–87.

Russell, T. (2001). *The no significant difference phenomenon: A comparative research annotated bibliography on technology for distance education* (5th ed.). Littleton, CO: IDECC.

Salmon, G. (2004). *E-moderating: The key to teaching and learning online* (2nd ed.). London, UK: Routledge.

Santagata, R., Zannoni, C., & Stigler, J. W. (2007). The role of lesson analysis in pre-service teacher education: an empirical investigation of teacher learning from a virtual video-based field experience. *Journal of Mathematics Teacher Education, 10*, 123–140. doi:10.1007/s10857-007-9029-9

Santamaria, L. (2009). Culturally responsive differentiated instruction: Narrowing gaps between best pedagogical practices benefiting all learners. *Teachers College Record, 111*, 214–247.

Saroyan, A., & Amundsen, C. (Eds.). (2004). *Rethinking teaching in higher education*. Sterling, VA: Stylus Press.

Savery, J. R., & Duffy, T. M. (1996). Problem-based learning: An instructional model and its constructivist design. In Wilson, B. G. (Ed.), *Constructivist learning environments: Case study in instructional design* (pp. 135–148). Englewood Cliffs, NJ: Educational Tech Pubs.

Scheetz, N. A., & Gunter, P. L. (2004). Online versus traditional classroom delivery of a course in manual communication. *Exceptional Children, 71*, 109.

Schifter, D., Bastable, V., & Russell, S. J. (2008). *Developing mathematical ideas*. Parsippany, NJ: Dale Seymour.

Schiro, M. (2008). *Curriculum theory: Conflicting visions and enduring concerns*. Los Angeles, CA: Sage.

Schlager, M. S., Farooq, U., Fusco, J., Schank, P., & Dwyer, N. (2009). Analyzing online teacher networks: Cyber networks require cyber research tools. *Journal of Teacher Education, 60*, 86–100. doi:10.1177/0022487108328487

Schlager, M. S., & Fusco, J. (2004). Teacher professional development, technology, and communities of practice: Are we putting the cart before the horse? In Barab, S., Kling, R., & Gray, J. (Eds.), *Designing virtual communities in the service of learning* (pp. 120–153). Cambridge, MA: Cambridge University Press. doi:10.1080/01972240309464

Schmidt, D. A. (2001). *Simultaneous renewal in teacher education: Strategies for success.* Paper presented at the Society for Information Technology and Teacher Education, Orlando, FL.

Schmoker, M. (2006). *Results now*. Alexandria, VA: Association for Supervision and Curriculum Development.

Schmuck, R. A., & Schmuck, P. A. (2004). *Group processes in the classroom* (8th ed.). Madison, WI: Brown & Benchmark.

Schon, D. A. (1983). *The reflective practitioner: How professionals think in action*. New York, NY: Basic Books.

Schön, D. A. (1984). *The reflective practitioner: How professionals think in action*. Aldershot, UK: Arena.

Schon, D. A. (1987). *Educating for reflective practitioner: Toward a new design for teaching and learning in professions*. San Francisco, CA: Jossey-Bass.

Schon, D. A. (1991). *The reflective turn: Case studies in and on educational practice*. New York, NY: Teachers College.

Schraw, G., Crippen, K. J., & Hartley, K. (2006). Promoting self-regulation in science education: Metacognition as part of a broader perspective on learning. *Research in Science Education*, *36*, 111–139. doi:10.1007/s11165-005-3917-8

Schrum, L. (2004). The web and virtual schools. *Computers in the Schools*, *21*, 81–89. doi:10.1300/J025v21n03_09

Schrum, L., Burbank, M. D., & Capps, R. (2007). Preparing future teachers for diverse schools in an online learning community: Perceptions and practice. *The Internet and Higher Education*, *10*(3), 204–211. doi:10.1016/j.iheduc.2007.06.002

Schrum, L., Thompson, A., Sprague, D., Maddux, C., McAnear, A., Bell, L., & Bull, G. (2005). Advancing the field: Considering acceptable evidence in educational technology research. *Contemporary Issues in Technology & Teacher Education*, *5*(3/4), 202–209.

Schulken, M. (2008, September 25). The end of brick and mortar era. *The Charlotte Observer*.

Schuyler, L., Stohn, S., & Yorke, B. (Producers). (2001). *Degrassi* [Television series]. Toronto, Canada: Epitome Pictures.

Schwartz, C. S. (2012). Counting to 20: Online implementation of a face-to-face, elementary mathematics methods problem-solving activity. *TechTrends*, *56*(1), 34–39. doi:10.1007/s11528-011-0551-3

Scoville, S. A., & Buskirk, T. D. (2007). Traditional and virtual microscopy compared experimentally in a classroom setting. *Clinical Anatomy (New York, N.Y.)*, *20*, 565–570. doi:10.1002/ca.20440

Scribner, S., & Cole, M. (2001). Unpacking literacy. In E. Cushman, et al. (Eds.), *Literacy: A critical sourcebook* (pp. 123-137). Boston, MA: Bedford/St. Martin's. (Reprinted from *Writing: The nature, development, and teaching of written communication*, pp. 71-87, by Marcia Farr Whiteman, Ed., 1981, Mahwah, NJ: Lawrence Erlbaum Associates)

Selby, M. (2006). Language, matauranga Māori and technology? *He Puna Korero: Journal of Māori and Pacific Development*, *7*(2), 79–86.

Sener, J. (2010). Why online education will attain full scale. [JALN]. *Journal of Asynchronous Learning Networks*, *14*(4), 3–16.

Sewell, J. P., Frith, K. H., & Colvin, M. M. (2010). Online assessment strategies: A primer. *MERLOT Journal of Online Learning and Teaching*, *6*(1), 297–305.

Shaffer, D. (2007). Epistemic games [Electronic Version]. *Innovate*, *1*. Retrieved from http://innovateonline.info/pdf/vol1_issue6/Epistemic_Games.pdf

Shaffer, D. W., Halverson, R., Squire, K. R., & Gee, J. P. (2004). *Video games and the future of learning*. Retrieved from The University of Madison Wisconsin, Advanced Academic Distributed Learning Co-Laboratory http://www.academiccolab.org/resources/gappspaper1.pdf

Shakespeare, W. (1963). *The tragedy of Hamlet, prince of Denmark*. New York, NY: Signet Classics.

Sharpe, L., Hu, C., Crawford, L., Saravanan, G., Khine, M. S., Moo, S. N., & Wong, A. (2003). Enhancing multipoint desktop video conferencing (MDVC) with lesson video clips: Recent developments in pre-service teaching practice in Singapore. *Teaching and Teacher Education*, *19*(3), 529–541. doi:10.1016/S0742-051X(03)00050-7

Shattuck, K. (2007) Quality matters: Collaborative program planning at a state level. *Online Journal of Distance Learning Adminstration*, *10*(3). Retrieved from http://www.westga.edu/~distance/ojdla/fall103/shattuck103.htm

Shea, P. (2006). A study of students' sense of learning community in online environments. *Journal of Asynchronous Learning Networks*, *10*(1), 35–44.

Shea, P., & Bidjerano, T. (2009). Cognitive presence and online learner engagement: A cluster analysis of the community of inquiry framework. *Journal of Computing in Higher Education*, *21*, 199–217. doi:10.1007/s12528-009-9024-5

Shea, P., Li, C., Swan, K., & Pickett, A. (2005). Developing learning community in online asynchronous college courses: The role of teaching presence. *Journal of Asynchronous Learning Networks*, *9*(4). Retrieved from http://www.sloan-c.org/publications/jaln/v9n4/v9n4_shea.asp

Sheridan, J. (1989). Rethinking andragogy: The case for collaborative learning in continuing higher education. *Journal of Continuing Higher Education*, *37*(2), 2–6. doi:10.1080/07377366.1989.10401167

Sherin, M. G., & Van Es, E. A. (2005). Using video to support teachers' ability to notice classroom interactions. *Journal of Technology and Teacher Education*, *13*(3), 475–491.

Shirts, R. G. (1976). Simulation/gaming for the past 10 years: What has and what hasn't happened. *Simulation & Gaming*, *3*(5), 5–9.

Short, J. E., Williams, E., & Christie, B. (1976). *The social psychology of telecommunications*. New York, NY: Wiley.

Shulman, L. S. (1986). Those who understand: Knowledge growth in teaching. *Educational Researcher*, *15*(2), 4–14.

Shulman, L. S., & Shulman, J. (2004). How and what teachers learn: A shifting perspective. *Journal of Curriculum Studies*, *36*(2), 257–271. doi:10.1080/0022027032000148298

Signer, B. (2008). Online professional development: Combining best practices from teacher, technology and distance education. *Journal of In-service Education*, *34*(2), 205–218. doi:10.1080/13674580801951079

Silva, D. Y., Gimbert, B., & Nolan, J. (2000). Sliding the doors: Locking and unlocking possibilities for teacher leadership. *Teachers College Record*, *102*(4), 779–804. doi:10.1111/0161-4681.00077

Silverman, B., Daniels, G., Gervais, R., Merchant, S., Klein, H., & Lieberstein, P. (Producers). (2005). *The office* [Television series]. Van Nuys, CA: Universal Media Studios.

Simpson, M. (2003). *Distance delivery of pre-service teacher education: Lessons for good practice from twenty-one international programs*. Unpublished Doctoral dissertation, The Pennsylvania State University, State College, PA.

Simpson, M., & Kehrwald, B. (2010). Educational principles and policies framing teacher education through open and distance learning. In P. Danaher & A. Umar (Eds.), *Teacher education through open and distance learning* (pp. 23-34). Vancouver, Canada: Commonwealth of Learning.

Simpson, M. (2006). Field experience in distance delivered initial teacher education programmes. *Journal of Technology and Teacher Education*, *14*(2), 241–254.

Singer, N., Catapano, S., & Huisman, S. (2010). The university's role in preparing teachers for urban schools. *Teaching Education*, *21*(2), 119–130. doi:10.1080/10476210903215027

Sirkin, R. M. (2005). *Statistics for the social sciences* (3rd ed.). Thousand Oaks, CA: Sage Publications.

Slagter van Tryon, P., & Bishop, M. J. (2009). Theoretical foundations for enhancing social connectedness in online learning environments. *Distance Education*, *30*(3), 291–315. doi:10.1080/01587910903236312

Slattery, P. (2006). *Curriculum development in the postmodern era* (2nd ed.). New York, NY: Rutledge.

Slavin, R. E. (1997). *Educational psychology: Theory and practice* (5th ed.). Needham Heights, MA: Allyn & Bacon.

Smeets, K., & Ponte, P. (2009). Action research and teacher leadership. *Professional Development in Education*, *35*(2), 175–193. doi:10.1080/13674580802102102

Smith, R., Clark, T., & Blomeyer, R. L. (2005). *A synthesis of new research on K-12 online learning*. Naperville, IL: Learning Point Associates. Retrieved from http://www.ncrel.org/tech/synthesis/synthesis.pdf

Smith, S., Smith, S., & Boone, R. (2000). Increasing access to teacher preparation: The effectiveness of traditional instructional methods in an online learning environment. *Journal of Special Education Technology*, *15*(2), 37–46.

Snow, C. E., Griffin, P., & Burns, M. (Eds.). (2005). *Knowledge to support the teaching of reading: Preparing teachers for a changing word*. San Francisco, CA: Jossey-Bass.

Snow-Gerono, J. L. (2005). Professional development in a culture of inquiry: PDS teachers identify the benefits of professional learning communities. *Teaching and Teacher Education*, *21*, 241–256. doi:10.1016/j.tate.2004.06.008

Sobel, D. (2004). *Place based education: Connecting classrooms and communities*. Great Barrington, MA: Orion Society.

So, H. J., Lossman, H., Lim, W. Y., & Jacobson, J. M. (2009). Designing an online video-based platform for teacher learning in Singapore. *Australasian Journal of Educational Technology*, *25*(3), 440–457.

So, H.-J., & Kim, B. (2009). Learning about problem based learning: Student teachers integrating technology, pedagogy and content knowledge. *Australasian Journal of Educational Technology*, *25*(1), 101–116.

Song, L., Singleton, E. S., Hill, J. R., & Koh, M. H. (2004). Improving online learning: Student perceptions of useful and challenging characteristics. *The Internet and Higher Education*, *7*(1), 59–70. doi:10.1016/j.iheduc.2003.11.003

Soo, K. S., & Bonk, C. J. (June, 1998). *Interaction: What does it mean in online distance education?* Paper presented at The World Conference on Educational Multimedia and Hypermedia & World Conference on Educational Telecommunications. Freiburg, Germany.

Sorensen, E. K., & Takle, E. S. (1998). *Collaborative knowledge building in web-based learning: assessing the quality of dialogue*. Paper presented at the Annual Meeting of ED-MEDIA 1998 – World Conference on Educational Multimedia, Hypermedia and Telecommunications Finland: Tampere.

Southern Association of Colleges and Schools. (2000). *Best practices for electronically offered degree and certificate programs*. Retrieved March 15, 2005 from http://www.sacscoc.org/pdf/commadap.pdf

Spicer, J. (2002). Even better than face-to-face? In Thorson, A. (Ed.), *By your own design: A teacher's professional learning guide* (pp. 32–33). Columbus, OH: Eisenhower National Clearinghouse for Mathematics and Science Education.

Spooner, F., Jordan, L., Algozzine, B., & Spooner, M. (1999). Evaluating instruction in distance learning classes. *The Journal of Educational Research*, *92*, 132–140. doi:10.1080/00220679909597588

Sporns, O. (2011). *Networks of the brain*. Cambridge, MA: MIT Press.

Spurlock, M. (Producer & Director). (2004). *Super size me* [Motion picture]. United States: Samuel Goldwyn Films.

Squire, K. (2005). Changing the game: What happens when video games enter the classroom? *Innovate*, *1*(6). Retrieved from http://innovateonline.info/pdf/vol1_issue6/Changing_the_Game-__What_Happens_When_Video_Games_Enter_the_Classroom_.pdf

Squire, K. (2006). From content to context: Videogames as designed experience. *Educational Researcher*, *35*(8), 19–29. doi:10.3102/0013189X035008019

Squire, K. (2007). Games, learning, and society: Building a field. *Educational Technology*, *4*(5), 51–54.

Squire, K., DeVane, B., & Durga, S. (2008). Designing centers of expertise for academic learning through video games. *Theory into Practice*, *47*(3), 240–251. doi:10.1080/00405840802153973

Squire, K., Jan, M., Matthews, J., Wagler, M., Martin, J., DeVane, B., & Holden, C. (2007). Wherever you go, there you are: Place-based augmented reality games for learning. In Shelton, B. E., & Wiley, D. A. (Eds.), *The educational design and use of computer simulation games*. Rotterdam, The Netherlands: Sense Press.

Squire, K., & Steinkuehler, C. (2005). Meet the gamers: They research, teach, learn, and collaborate. So far, without libraries. *Library Journal*, *130*(7), 38–41.

Stahl, G. (2006). *Group cognition: Computer support for building collaborative knowledge*. Cambridge, MA: MIT Press.

Stahl, N., & Smith-Burke, M. T. (1999). National Reading Conference: The college and adult reading years. *Journal of Literacy Research, 31*(1), 47–66. Retrieved from http://findarticles.com/p/articles/mi_qa3785/is_199903/ai_n8848968/?tag=content;col1 doi:10.1080/10862969909548036

Stahl, S. (1999). Different strokes for different folks? *American Educator, 23*, 27–31.

Staker, H. (2011). *The rise of K-12 blended learning: Profiles of emerging models*. Mountain View, CA: Innosight Institute.

State of Wisconsin. (2010). *Guidance on the 30 hours of professional development for teaching online courses*. Madison, WI: Author. Retrieved from http://dpi.wi.gov/imt/pdf/online_course_pd.pdf

Statistics New Zealand. (2011). *National population projections: 2009 to 2061*. Retrieved from http://www.stats.govt.nz/tools_and_services/tools/TableBuilder/population-projections-tables.aspx

Steinkuehler, C. (2005). The new third place: Massively multiplayer online gaming in American youth culture. *Tidskrift Journal of Research in Teacher Education, 3*, 135–150.

Steinkuehler, C. (2008). Massively multiplayer online games as an educational technology: An outline for research. *Educational Technology, 48*(1), 10–21.

Stein, M. K., Grover, B. W., & Henningsen, M. (1996). Building student capacity for mathematical thinking and reasoning: An analysis of mathematical tasks used in reform classrooms. *American Educational Research Journal, 33*, 455–488.

Steinweg, S. B., Davis, M. L., & Thomson, W. S. (2005). A comparison of traditional and online instruction in an introduction to special education course. *Teacher Education and Special Education, 28*, 62–73. doi:10.1177/088840640502800107

Stevens, C. J., & Dial, M. (1993). A qualitative study of alternatively certified teachers. *Education and Urban Society, 26*, 63–77. doi:10.1177/0013124593026001006

Stewart, V. (2011). *Improving teacher quality around the world: The international summit on the teaching profession*. New York, NY: Asia Society and U.S. Department of Education.

Stobaugh, R. R., & Tassell, J. L. (2011). Analyzing the degree of technology use occurring in pre-service teacher education. *Educational Assessment, Evaluation and Accountability, 23*(2), 143–157. doi:10.1007/s11092-011-9118-2

Stodel, E. J., Thompson, T. L., & MacDonald, C. J. (2006). Learners' perspectives on what is missing from online learning: Interpretations through the community of inquiry framework. *International Review of Research in Open and Distance Learning, 7*(3), 1–24.

Stokes, L. (2001). Lessons from an inquiring school: Forms of inquiry and conditions for teacher learning. In Lieberman, A., & Miller, L. (Eds.), *Teachers caught in the action: Professional development that matters* (pp. 141–158). New York, NY: Teachers College Press.

Storeygard, J. (2009). *My kids can*. Boston, MA: Heinemann.

Strauss, V. (2011, March 23). Darling-Hammond, L.: U.S. vs highest-achieving nations in education. *The Answer Sheet*. Retrieved from http://www.washingtonpost.com/blogs/answer-sheet/post/darling-hammond-us-vs-highest-achieving-nations-in-education/2011/03/22/ABkNeaCB_blog.html

Street, B. (2001). The new literacy studies. In E. Cushman, et al. (Eds.), *Literacy: A critical sourcebook* (pp. 430-442). Boston, MA: Bedford/St. Martin's. (Reprinted from *Cross-cultural approaches to literacy*, pp. 1-21, by B. Street, Ed., 1993, London: Cambridge University Press)

Strudler, N., & Grove, K. (2002). Integrating technology into teacher candidates' field experiences: A two-pronged approach. *Journal of Computing in Teacher Education, 19*(2), 33–38.

Summerville, J., & Reid-Griffin, A. (2008). Technology integration and instructional design. *TechTrends, 52*(5), 45–51. doi:10.1007/s11528-008-0196-z

Swan, K. (2005). *Threaded discussion*. Retrieved from www.oln.org/conferences/ODCE2006/papers/Swan_Threaded_Discussion.pdf

Swan, K. P., Richardson, J. C., Ice, P., Garrison, D. R., Cleveland-Innes, M., & Arbaugh, J. B. (2008). Validating a measurement tool of presence in online communities of inquiry. *E-mentor, 2*(24). Retrieved from http://www.e-mentor.edu.pl/artykul_v2.php?numer=24&id=543

Swan, K. (2002). Building communities in online courses: The importance of interaction. *Education Communication and Information, 2*(1), 23–49. doi:10.1080/146363 1022000005016

Swan, K., Garrison, D. R., & Richardson, J. C. (2009). A constructivist approach to online learning: The Community of Inquiry framework. In Payne, C. R. (Ed.), *Information technology and constructivism in higher education: Progressive learning frameworks* (pp. 43–57). Hershey, PA: IGI Global. doi:10.4018/978-1-60566-654-9.ch004

Swan, K., Garrison, D. R., & Richardson, J. C. (2009). A constructivist approach to online learning: The Community of Inquiry framework. In Payne, C. R. (Ed.), *Information technology and constructivism in higher education: Progressive learning frameworks* (pp. 43–57). Hershey, PA: Information Science Publishing. doi:10.4018/978-1-60566-654-9.ch004

Swan, K., & Ice, P. (2010). The community of inquiry framework ten years later: Introduction to the special issue. *The Internet and Higher Education, 13*(1-2), 1–4. doi:10.1016/j.iheduc.2009.11.003

Swan, K., Matthews, D., Bogle, L., Boles, E., & Day, S. (2012). Linking online course design and implementation to learning outcomes: A design experiment. *The Internet and Higher Education, 15*(2), 81–88. doi:10.1016/j.iheduc.2011.07.002

Swan, K., Shea, P., Fredericksen, E., Pickett, A., Pelz, W., & Maher, G. (2000). Building knowledge building communities: Consistency, contact and communication in the virtual classroom. *Journal of Educational Computing Research, 23*(4), 389–413.

Swan, K., & Shih, L. F. (2005). On the nature and development of social presence in online course discussions. *Journal of Asynchronous Learning Networks, 9*, 115–136.

Swenson, J., Rozema, R., Young, C. A., McGrail, E., & Whitin, P. (2005). Beliefs about technology and the preparation of English teachers: Beginning the conversation. *Contemporary Issues in Technology & Teacher Education, 5*(3/4), 210–236. Retrieved from http://www.citejournal.org/vol5/iss3/languagearts/article1.cfm

Szwed, J. F. (2001). The ethnography of literacy. In E. Cushman, et al. (Eds.), *Literacy: A critical sourcebook* (pp. 421-429). Boston, MA: Bedford/St. Martin's. (Reprinted from *Writing: The nature, development, and teaching of written communication*, pp. 13-23, by Marcia Farr Whiteman, Ed., 1981, Mahwah, NJ: Lawrence Erlbaum Associates)

Tallent-Runnels, M. K., Thomas, J. A., Lan, W. Y., Cooper, S., Ahern, T. C., Shaw, S. A., & Liu, X. (2006). Teaching courses online: A review of the research. *Review of Educational Research, 76*, 93–135. doi:10.3102/00346543076001093

Taylor, D. B. (2012). Multiliteracies: Moving from theory to practice in teacher education. In Polly, A. B., Mims, C., & Persichitte, K. (Eds.), *Creating technology-rich teacher education programs: Key issues* (pp. 266–287). Hershey, PA: IGI Global. doi:10.4018/978-1-4666-0014-0.ch018

Teclehaimanot, B., Mentzer, G., & Hickman, T. (2011). A mixed methods comparison of teacher education faculty perceptions of the integration of technology into their courses and student feedback on technology proficiency. *Journal of Technology and Teacher Education, 19*(1), 5–21.

Thames, M. H., & Ball, D. L. (2010). What mathematical knowledge does teaching require? Knowing mathematics in and for teaching. *Teaching Children Mathematics, 17*(4), 220–225.

Theissen, J., & Ambrock, V. (2008). Value added: The editor in design and development of online courses. In T. Anderson (Ed.), *The theory and practice of online learning* (2nd ed.) (pp. 265-276). Edmonton, Canada: Athabasca Press.

Thomson, D. L. (2010). Beyond the classroom walls: teachers' and students' perspectives on how online learning can meet the needs of gifted students. *Journal of Advanced Academics, 21*, 662–712. doi:10.1177/1932202X1002100405

Tiakiwai, S.-J., & Tiakiwai, H. (2010). *A literature review focused on virtual learning environments (VLEs) and e-learning in the context of te reo Māori and kaupapa Māori education: Report to the Ministry of Education.* Wellington, New Zealand: Ministry of Education. Retrieved from http://www.educationcounts.govt.nz/__data/assets/pdf_file/0004/72670/936_LitRev-VLEs-FINALv2.pdf

Tiezzi, L., & Cross, B. (1997). Utilizing research on prospective teachers' beliefs to inform urban field experiences. *The Urban Review*, *29*(2), 113–125. doi:10.1023/A:1024634623688

TIMSS. (2003). *International report on achievement in the mathematics cognitive domains*. Retrieved from http://timss.bc.edu/timss2003i/mcgdm.html

Tobias, S., Fletcher, J. D., Dai, D. Y., & Wind, A. P. (2011). Review of research on computer games. In Tobias, S., & Fletcher, J. D. (Eds.), *Computer games and instruction* (pp. 371–394). Charlotte, NC: Information Age Publishers.

Tomlinson, C. (1999). *The differentiated classroom: Responding to the needs of all learners*. Alexandria, VA: Association for Supervision and Curriculum Development.

Tomlinson, C. (2000). Reconcilable differences: Standards-based teaching and differentiation. *Educational Leadership*, *58*, 6–11.

Treacy, B., Kleiman, G., & Peterson, K. (2002). Successful online professional development. *Leading & Learning with Technology*, *30*(1), 42–47.

Trilling, B., & Fadel, C. (2009). *21st century skills: Learning for life in our times*. San Francisco, CA: Jossey-Bass.

Troutwine, C., Romano, C., & O'Meara, D. (Producers), & Ewing, H., Gibney, A., Gordon, S., Grady, R., Jarecki, E., & Spurlock, M. (Directors). (2010). *Freakonomics* [Motion picture]. United States: Magnolia Pictures.

Tschannen-Moran, M., & Barr, M. (2004). Fostering student learning: the relationship of collective teacher efficacy and student achievement. *Leadership and Policy in Schools*, *3*(3), 189–209. doi:10.1080/15700760490503706

Tschannen-Moran, M., & Woolfolk Hoy, A. (2001). Teacher efficacy: Capturing an elusive construct. *Teaching and Teacher Education*, *17*, 783–805. doi:10.1016/S0742-051X(01)00036-1

Tu, C. H. (2000). On-line learning migration: From social learning theory to social presence theory in CMC environment. *Journal of Network and Computer Applications*, *23*(1), 27–37. doi:10.1006/jnca.1999.0099

Tu, C. H., & McIsaac, M. S. (2002). The relationship of social presence and interaction in online classes. *American Journal of Distance Education*, *16*(3), 131–150. doi:10.1207/S15389286AJDE1603_2

Tu, C., & McIsaac, M. (2002). The relationship of social presence and interaction in online classes. *American Journal of Distance Education*, *16*, 131–150. doi:10.1207/S15389286AJDE1603_2

Tunison, S., & Noonan, B. (2001). On-line learning: Secondary students' first experience. *Canadian Journal of Education*, *26*, 495–514. doi:10.2307/1602179

Tyack, D., & Cuban, L. (1995). *Tinkering toward utopia: A century of public school reform*. Cambridge, MA: Harvard University Press.

U. S. Department of Education. (2009). *Evaluation of evidence-based practices in online learning: A meta-analysis and review of online learning studies*. Retrieved August 10, 2011, from http://www.ed.gov/about/offices/list/opepd/ppss/reports.html

U.S. Department of Education, National Center for Education Statistics, Common Core of Data (CCD). (2010). *Public elementary/secondary school universe survey: 2006-07, 2007-08, and 2008-09*.

UNESCO. (2008). *ICT competency standards for teachers*. Paris, France: UNESCO.

United Nations Development Programme. (2010). *Human development report: 20th anniversary edition: The real wealth of nations: Pathways to human development*. New York, NY: Palgrave. Retrieved from http://www.beta.undp.org/content/dam/undp/library/corporate/HDR/HDR_2010_EN_Complete_reprint-1.pdf

University of North Carolina System. (2007). *The University of North Carolina online*. Retrieved August 15, 2011, from http://online.northcarolina.edu/

US Department of Education. (2009). *Race to the top application for initial funding*. Retrieved from http://www2.ed.gov/programs/racetothetop/phase1-applications/minnesota.pdf

USDOE. (2010). *Learning powered by technology: Transforming American education*. Washington, DC: U.S. Department of Education.

Van Laarhoven, T. R., Munk, D. D., Lynch, K., Bosma, J., & Rouse, J. (2007). A model for preparing special and general education preservice teachers for inclusive education. *Journal of Teacher Education*, *58*, 440–455. doi:10.1177/0022487107306803

Vandergrift, L. (2003). Orchestrating strategy use: Toward a model of the skilled second language listener. *Language Learning*, *53*(3), 463–496. doi:10.1111/1467-9922.00232

VanSickle, R. L. (1986). A quantitative review of research on instructional simulation gaming: A twenty-year perspective. *Theory and Research in Social Education*, *14*, 245–264.

Vaughn, N., & Garrison, D. R. (2006). How blended learning can support a faculty development community of inquiry. *Journal of Asynchronous Learning Networks*, *10*(4), 139–152.

Venn, M. L., Moore, R. L., & Gunter, P. L. (2001). Using audio/video conferencing to observe field-based practices of rural teachers. *Rural Educator*, *22*(2), 24–27.

Verduin, J. R., & Clark, T. A. (1991). *Distance education: The foundations of effective practice*. San Francisco, CA: Jossey-Bass.

Vonnegut, K. (2005). *Slaughterhouse-five, or, The children's crusade: A duty-dance with death*. New York, NY: Dial Press.

Vygotsky, L. (1930/1978). *Mind in society*. Cambridge, MA: Harvard University Press.

Vygotsky, L. (1978). *Mind and society: The development of higher mental processes*. Cambridge, MA: Harvard University Press.

Vygotsky, L. S. (1978). *Mind in society: The development of higher psychological processes*. Cambridge, MA: Harvard University Press.

Wadmany, R., Rimor, R., & Rosner, E. (2011). The relationship between attitude, thinking and activity of students in an e-learning course. *Research on Education and Media*, *3*(1).

Waldron, N., & McClesky, J. (1998). The effects of an inclusive school program on students with mild and severe learning disabilities. *Exceptional Children*, *64*, 395–405.

Walizer, B. R., Jacobs, S. L., & Danner-Kuhn, C. L. (2007). The effectiveness of face-to-face vs. web camera candidate observation evaluations. *Academic Leadership*, *5*(3), 1–9.

Wall, K. (2008). Reinventing the wheel? Designing an Aboriginal recreation and community development program. *Canadian Journal of Native Education*, *31*(2), 70–93.

Wang, F., & Hannafin, M. J. (2005). Design-based research and technology enhanced learning environments. *Educational Technology Research and Development*, *53*(4), 5–23. doi:10.1007/BF02504682

Wang, Y.-M., & Chang, V. D.-T. (2008). Essential elements in designing online discussions to promote cognitive presence — A practical experience. *Journal of Asynchronous Learning Networks*, *12*(3-4), 157–177.

Ward, B. (2004). The best of both worlds: A hybrid statistics course. *Journal of Statistics Education*, *12*(3), 74–79.

Watson, J., Murin, A., Vashaw, L., Gemin, B., & Rapp, C. (2010). *Keeping pace with K–12 online learning: An annual review of policy and practice*. Evergreen, CO: Evergreen Education Group. Retrieved from http://www.kpk12.com/wp-content/uploads/KeepingPaceK12_2010.pdf

Waxman, H. C., Lin, M. F., & Michko, G. M. (2003). *A meta-analysis of the effectiveness of teaching and learning with technology on student outcomes*. Retrieved from the North Central Regional Educational Laboratory Web site: http://www.ncrel.org/tech/effects2/waxman.pdf

Webster-Wright, A. (2009). Reframing professional development through understanding authentic professional learning. *Review of Educational Research*, *79*(2), 702–739. doi:10.3102/0034654308330970

Wegerif, R. (2006). Towards a dialogic understanding of the relationship between teaching thinking and CSCL. *International Journal of Computer-Supported Collaborative Learning*, *1*(1), 143–157. doi:10.1007/s11412-006-6840-8

Wei, R. C., Darling-Hammond, L., Andree, A., Richardson, N., & Orphanos, S. (2009). *Professional learning in the learning profession: A status report on teacher development in the United States and abroad.* National Staff Development Council. Retrieved from http://www.learningforward.org/news/NSDCstudytechnical-report2009.pdf

Weigel, V. (2005). From course management to curricular capabilities: A capabilities approach for the next-Generation CMS. *EDUCAUSE Review*, *40*(3), 54–67.

Weimer, M. (2002). *Learner-centered teaching*. San Francisco, CA: Jossey-Bass.

Weinberger, A., & Fischer, F. (2006). A framework to analyze argumentative knowledge construction in computer-supported collaborative learning. *Computers & Education*, *46*, 71–95. doi:10.1016/j.compedu.2005.04.003

Weinberger, A., Stegmann, K., Fischer, F., & Mandl, H. (2007). Scripting argumentative knowledge construction in computer-supported learning environments. In Fischer, F., Kollar, I., Mandl, H., & Haake, J. (Eds.), *Scripting computer-supported communication of knowledge—Cognitive, computational and educational perspectives* (pp. 191–211). New York, NY: Springer. doi:10.1007/978-0-387-36949-5_12

Weiner, C. (2003). Key ingredients to online learning: Adolescent students study in cyberspace—The nature of the study. *International Journal on E-Learning*, *2*, 44–50.

Weiner, L. (1999). *Urban teaching: The essentials*. New York, NY: Teachers College Press.

Wellington College of Education. (2001). *Approval documentation, secondary online*. Wellington, New Zealand: Author.

Wenglinsky, H. (2000). *How teaching matters: Bringing the classroom back into discussions of teacher quality*. Policy Information Center, Mail Stop 04-R, Educational Testing Service. Retrieved from http://www.csa.com

Wenglinsky, H. (2000). *Teaching the teachers: Different settings, different results. policy information report.* Policy Information Center. Retrieved from http://www.csa.com

Wenglinsky, H. (2002). How schools matter: The link between teacher classroom practices and student academic performance. *Education Policy Analysis Archives*, *10*(12). Retrieved from http://www.csa.com

White, C. S. (1985). Citizen decision making, reflective thinking and simulation gaming: A marriage of purpose, method and strategy. *Journal of Social Studies Research*, Monograph 2 (Summer).

Wiggins, G. (1999). *Educative assessment: Designing assessments to inform and improve student performance*. San Francisco, CA: Jossey Bass Publishers.

Wiggins, R. A., & Follo, E. J. (1999). Development of knowledge, attitudes, and commitment to teach diverse student populations. *Journal of Teacher Education*, *50*, 94–105. doi:10.1177/002248719905000203

Wilensky, U. (1999). GasLab—An extensible modeling toolkit for connecting micro- and macro-properties of gases. In Fuerzeig, W., & Roberts, N. (Eds.), *Modeling and simulation in precollege science and mathematics education* (pp. 151–178). New York, NY: Springer-Verlag. doi:10.1007/978-1-4612-1414-4_7

Willett, R. (2007). Technology, pedagogy and digital production: A case study of children learning new media skills. *Learning, Media and Technology*, *32*(2), 167–181. doi:10.1080/17439880701343352

Williams, P. (2008). Leading schools in the digital age: a clash of cultures. *School Leadership & Management*, *28*(3), 213–228. doi:10.1080/13632430802145779

Wilson, G., & Stacey, E. (2003). Online interaction impacts on learning: Teaching the teachers to teach online. In G. Crisp, D. Thiele, I. Scholten, S. Baker, & J. Baron (Eds.), *Interact, integrate, impact: Proceedings of the 20th Annual Conference of the Australasian Society for Computers in Learning in Tertiary Education*. Adelaide, Australia: ASCILITE.

Wilson, S. M., Floden, R. E., & Ferrini-Mundy, J. (2002). Teacher preparation research: An insider's view from the outside. *Journal of Teacher Education*, *53*(3), 190–204. Retrieved from http://jte.sagepub.com/content/53/3/190.full.pdf+htmldoi:10.1177/0022487102053003002

Wise, A. F., Padmanabhan, P., & Duffy, T. M. (2009). Connecting online learners with diverse local practices: The design of effective common reference points for conversation. *Distance Education, 30*(3), 317–338. doi:10.1080/01587910903236320

Wise, A., Chang, J., Duffy, T., & del Valle, R. (2004). The effects of teacher social presence on student satisfaction, engagement, and learning. *Journal of Educational Computing Research, 31*(3), 247–271. doi:10.2190/V0LB-1M37-RNR8-Y2U1

Wiske, M. S., Perkins, D., & Spicer, D. E. (2006). Piaget goes digital: Negotiating accommodation of practice to principles. In Dede, C. (Ed.), *Online professional development for teachers* (pp. 49–67). Cambridge, MA: Harvard Education Press.

Wolf, P. D. (2006). Best practices in the training of faculty to teach online. *Journal of Computing in Higher Education, 17*, 47–78. doi:10.1007/BF03032698

Wolfram, S. (2002). *A new kind of science*. Champaign, IL: Wolfram Media.

Wolkenhauer, R., Boynton, S., & Dana, N. F. (2011, February). *The power of practitioner research and development of an inquiry stance inteacher education programs*. Paper presented at the meeting for the Association of Teacher Educators, Orlando, FL.

Wood, C. (2005). Highschool.com: The virtual classroom redefines education. *Edutopia, 1*(4), 31-44. Retrieved from http://www.edutopia.org/high-school-dot-com

Work, A. P. A. Group of the Board of Educational Affairs. (1997). *Learner-centered psychological principles: A framework for school reform and redesign*. Washington, DC: Author.

Wright, S. P. Horn, S. P., & Sanders, W. L. (1997). Teacher and classroom context effects on student achievement: Implications for teacher evaluation. *Journal of Personnel Evaluation in Education, 11*(1), 57-67. Retrieved from http://www.sas.com/ govedu/edu/teacher_eval.pdf

Wu, C. C., & Kao, H. C. (2008). Streaming videos in peer assessment to support training pre-service teachers. *Journal of Educational Technology & Society, 11*(1), 45–55.

Wu, D., & Hiltz, S. R. (2004). Predicting learning from asynchronous online discussions. *Journal of Asynchronous Learning Networks, 8*(2), 139–152.

Yang, S. C., & Liu, S. F. (2004). Case study of online workshop for the professional development of teachers. *Computers in Human Behavior, 20*, 733–761. doi:10.1016/j.chb.2004.02.005

Yeh, H. (2005). *The use of instructor's feedback and grading in enhancing students' participation in asynchronous online discussion*. Advanced Learning Technologies (ICALT 2005).

Yendol-Hoppey, D., Dawson, K., Dana, N. F., League, M., Jacobs, J., & Malik, D. (2006). Professional development communities: Vehicles for re-shaping field experiences to support school improvement. *Florida Journal of Teacher Education, 4*(1), 37–48.

Young, A., & Lewis, C. W. (2008). Teacher education programmes delivered at a distance: An examination of distance student perceptions. *Teaching and Teacher Education, 24*(3), 601–609. doi:10.1016/j.tate.2007.03.003

Young, C. A., & Bush, J. (2004). Teaching the English language arts with technology: A critical approach and pedagogical framework. *Contemporary Issues in Technology & Teacher Education, 4*(1), 1–22. Retrieved from http://www.citejournal.org/vol4/iss1/languagearts/article1.cfm

Young, E. E., Grant, P. A., Montbriand, C., & Therriault, D. J. (2001). *Educating preservice teachers: The state of affairs*. Naperville, IL: North Central Regional Educational Laboratory.

Young, S. (2006). Student views of effective online teaching in higher education. *American Journal of Distance Education, 20*(2), 65–77. doi:10.1207/s15389286ajde2002_2

Zaraza, R., & Fisher, D. M. (1999). Training system modelers: The NSF CC-STADUS and CC-SUSTAIN projects. In Fuerzeig, W., & Roberts, N. (Eds.), *Modeling and simulation in precollege science and mathematics education* (pp. 38–69). New York, NY: Springer-Verlag. doi:10.1007/978-1-4612-1414-4_2

Zeichner, K. (2003). Teacher research as professional development for P-12 educators in the USA. *Educational Action Research*, *2*(2), 301–326. doi:10.1080/09650790300200211

Zeichner, K. (2010). Rethinking the connections between campus courses and field experiences in college- and university-based teacher education. *Journal of Teacher Education*, *61*(1-2), 89–99. doi:10.1177/0022487109347671

Zeichner, K. M. (2007). Accumulating knowledge across self-studies in teacher education. *Journal of Teacher Education*, *58*, 36–46. doi:10.1177/0022487106296219

Zeichner, K. M., & Schulte, A. K. (2001). What we know and don't know from peer-reviewed research about alternative teacher certification programs. *Journal of Teacher Education*, *52*, 266–282. doi:10.1177/0022487101052004002

Zhu, E. (2006). Interaction and cognitive engagement: An analysis of four asynchronous online discussions. *Instructional Science*, *34*(6), 451–480. doi:10.1007/s11251-006-0004-0

Zhu, P. (1998). Learning and mentoring: Electronic discussion in a distance learning course. In Bonk, C. J., & King, K. S. (Eds.), *Electronic collaborators: Learner-centered technologies for literacy, apprenticeship, and discourse* (pp. 233–259). Mahwah, NJ: Erlbaum.

Zimmerman, B. (2000). Attaining self-regulation: A social-cognitive perspective. In Boekaerts, M., Pintrich, P., & Zeichner, M. (Eds.), *Handbook of self-regulation* (pp. 13–39). Orlando, FL: Academic Press.

Zimmerman, J. (2006). Why some teachers resist change and what principals can do about it. *NASSP Bulletin*, *90*(3), 238–249. doi:10.1177/0192636506291521

Zucker, A., & Kozma, R. (2003). *The virtual high school: Teaching generation V*. New York, NY: Teachers College Press.

Zuckerbrod, N. (2011). From readers' theater to math dances: Bright ideas to make differentiation happen. *Instructor*, *120*, 33–38.

Zygouris-Coe, V., & Swan, B. (2010). Challenges of online teacher professional development communities: A statewide case study in the United States. In Lindberg, J. O., & Oloffson, A. D. (Eds.), *Online learning communities and teacher professional development: Methods for improved education delivery* (pp. 114–133). Hershey, PA: Information Science Publishing.

About the Contributors

Richard Hartshorne is an Associate Professor of Educational Technology at the University of Central Florida. He earned his Ph.D. in Curriculum and Instruction from the University of Florida. At the University of Central Florida, his teaching focuses on the integration of technology into the educational landscape, as well as instructional design and development. His research interests primarily involve the production and effective integration of instructional technology into the teaching and learning environment. The major areas of his research interest are rooted in online teaching and learning, technology and teacher education, and the integration of emerging technology into the K-Post-secondary curriculum. His articles have appeared in such publications as *Journal of Technology and Teacher Education, Journal of Computing in Higher Education, Journal of Digital Learning in Teacher Education, The Internet and Higher Education, The Journal of Educational Computing Research,* and others. He has also authored numerous book chapters and serves in editorial capacities for a number of journals in the field of educational technology.

Tina L. Heafner, Ph.D., is an Associate Professor of Social Studies in the Department of Middle, Secondary, and K-12 Education at the University of North Carolina at Charlotte. Her administrative responsibilities include coordinating the M.Ed. in Secondary Education and the Minor in Secondary Education. Tina's teaching and research focus on promoting effective practices in social studies and teacher education such as technology integration, content literacy development, and service learning. Other research interests include policy and curriculum issues in social studies, online teaching and learning, and the impact of content-based applications of emerging technologies for K-12 learners. Her articles have appeared in such publications as *Journal of Technology and Teacher Education, Journal of Computing in Higher Education, Journal of Digital Learning in Teacher Education Journal of Social Studies Research, The Journal of Adolescent and Adult Literacy,* and *Theory and Research in Social Studies Education*. She is the co-author of *Targeted Vocabulary Strategies for Secondary Social Studies, Strategic Reading in World History,* and *Strategic Reading in U.S. History.*

Teresa M. Petty is an Assistant Professor in the Department of Middle, Secondary, and K-12 Education at the University of North Carolina at Charlotte. She also serves as the department's coordinator of online programs. Teresa earned her Ed.D. in Curriculum and Instruction from the University of North Carolina at Chapel Hill. At the University of North Carolina at Charlotte, her teaching focuses on instructional design, teacher leadership, and instructional methods in the middle and secondary mathematics classroom. Her research interests include teacher attraction/retention in high-need schools, online teaching/learning, and National Board Certification.

* * *

Terry S. Atkinson is an Associate Professor of Reading Education in the College of Education at East Carolina University. Terry is a former elementary school teacher, reading specialist, and principal in both public and independent school settings. Her research interests include the new literacies of the Internet, and effective practices for technology integration and professional development in both face-to-face and online teaching. She is currently involved in a collaborative research project to identify exemplary teachers who integrate technology effectively in local schools. Her work appears in venues such as *Teacher Education Quarterly, The Reading Teacher, Language Arts, Reading Research, and Instruction, Journal of School Connections, and the Journal of Curriculum and Instruction.*

Michael Barbour is an Assistant Professor at Wayne State University in Detroit, Michigan, where he teaches Instructional Technology and Qualitative Research Methodology. Michael's interest in K-12 distance education began after accepting his first teaching position in a rural high school. Having been educated in an urban area, Michael was troubled by the inequity of opportunity provided to his rural students and began a program to offer Advanced Placement social studies courses online at his own school and other schools in the district. For more than a decade now, Michael worked with numerous K-12 online learning programs in Canada, the United States, New Zealand, and around the world as a teacher, course developer, administrator, evaluator, and researcher. His current research interests focus on the effective design, delivery, and support of online learning to K-12 students in virtual school environments, particularly those in rural jurisdictions. Michael currently resides in Windsor, Ontario, Canada.

John Beattie is an Assistant Professor in the Department of Special Education and Child Development at the University of North Carolina at Charlotte. Dr. Beattie's current areas of interest and research include alternative algorithms that promote mathematics achievement of all students with high-incidence disabilities, mathematics interventions designed to meet the guidelines of RTI tier-model, the use of distance education or online instructional technologies to bridge the research-to-practice gap, and strategies to enhance the performance of students with high-incidence disabilities on the SAT/ACT. Dr. Beattie recently published *Teaching Students with Special Needs: A Guide for Future Educators* with chapter co-author Chris O'Brien in an effort to improve teacher preparation for inclusive instruction, particularly for aspiring general education teachers.

Leonard Bogle, Ed. D., Associate Professor of Educational Leadership, has 34 years experience as an educator in the public schools of Illinois with the last 28 years served as an administrator to include Assistant and Head Middle School Principal, Elementary Principal, Director of Adult Education, Grant Writer, and School Superintendent. Dr. Bogle currently teaches courses in the areas of Supervision of Instruction, Introduction to Research (online), Teacher Leadership (online), Instructional Design (online), Master's Project, Organizational Dynamics, and the Capstone Project (online). National presentations and publications focused on the delivery of on-line and blended classes to graduate student, the identification of quality on-line programs, effective delivery of online instruction using a variety of tools and the development and delivery of school-wide cooperative instructional programs.

Emily Boles, M.S., is an Instructional Developer in the Center for Online Learning, Research and Service (COLRS) at the University of Illinois Springfield. At COLRS, she mentors faculty members in effective practices in online learning and technology integration. She coordinates several grant projects for the Center and is leading new UIS Continuing Education Online initiative. Emily teaches online courses for UIS and facilitates online workshops and webinars for the Sloan Consortium. Her experience in the field spans community colleges and four-year institutions. She has presented at regional, national, and international conferences on online teaching, learning, and technologies.

M. Joyce Brigman currently occupies the position of Assistant Clinical Professor in the UNC Charlotte's College of Education. She was among the first cohort of Nationally Certified Teachers and has since recertified. With past experience from thirty years in K-12 public schools, she is now involved in preparing educators for the 21st century classroom. She also directs Middle Grades University, a mentoring program which introduces middle level students to the possibilities of college in their own future. Her research interests include alternative licensure, instructional design, distance education, adolescent development, diverse learners, middle level education as well as preparation of next generations of effective classroom teachers.

Victoria M. Cardullo received her undergraduate degree from the University of Central Florida in Elementary Education. She went on to further her education and received a Master's degree in Reading Education with a focus on Phonics. She is currently completing her Ed.D. in curriculum with a concentration in reading. She has worked at the University of Central Florida as an instructor in Elementary Education since 2006. Her research interest focuses on teacher preparation, online learning, adolescent reading, and New Literacies.

Cathy Cavanaugh is currently Associate Director of the Abu Dhabi Women's College, a campus of the Higher Colleges of Technology in the United Arab Emirates. Previously, she was an Associate Professor of Educational Technology at the University of Florida specializing in instructional design and online/blended learning environments. She served as a Fulbright Senior Scholar to Nepal in 2010. She currently serves an advisor to the European Union Lifelong Learning Programme and as a panelist for the National Center of Science and Technology of the Republic of Kazakhstan. She has over 100 publications in educational technology, including books of virtual schools and online education research. Cathy's funded research focused on classroom technology and professional development in schools, effective practices in virtual schools, online professional development, and design of online and blended courses. Her primary research interests are in indicators of quality in online and blended education, and equitable access to quality education through technology.

Jerad J. Crave is a Doctoral student at the University of North Carolina at Greensboro in Teacher Education. He earned his M.Ed. from Harvard University and B.S. from Cornell University. He taught for ten years at several elementary schools and is studying the impact of technology on student motivation to learn literacy content as well as effective online instruction for pre-service teachers. At UNCG, he teaches undergraduate methods classes and supervises student teachers.

Nancy Fichtman Dana is currently Professor of Education in the School of Teaching and Learning at the University of Florida, Gainesville. Since earning her Ph.D. from Florida State University in 1991, she has been a passionate advocate for practitioner inquiry and the promise this movement holds to enable all educators to experience powerful, job-embedded professional learning. She has coached the practitioner research of numerous educators from various districts across the nation, as well as published seven books and over 50 articles in professional journals and books focused on teacher and principal professional development and practitioner inquiry. Most recently, her research interests have focused on the ways action research plays out in virtual environments.

Kara Dawson is an Associate Professor of Educational Technology in the School of Teaching and Learning at the University of Florida where she serves as Program Coordinator for traditional and online graduate programs. She has secured nearly three million dollars in funding since 2002, much of which focuses on the impact of professional development, classroom technologies and K-12/university partnerships on teaching practices, student achievement, and school culture. She has published refereed articles in journals such as the *British Journal of Educational Technology* and the *Journal of Research on Technology in Education,* over a dozen editor-reviewed articles in outlets such as the *Chronicle of Higher Education and Educational Leadership,* numerous book chapters, and an edited book. Prior to working at the University of Florida, Dr. Dawson worked as a post-doctoral fellow at the Center for Technology and Teacher Education within the Curry School at the University of Virginia.

Scott L. Day, Ed. D., is an Associate Professor and Chair in the Department of Educational Leadership at the University of Illinois at Springfield (UIS). He earned his Doctorate from the University of Illinois at Urbana-Champaign in Education Organization and Leadership. He began his career in teaching courses for the principalship program and later taught one of the first courses in the online Master of Arts in Teacher Leadership (MTL) program at UIS. His current teaching assignment includes courses in Instructional Design (MTL) and Instructional Leadership at the Post-Master's level. His research interests include school district reorganization, school personnel issues, and improving online courses and programs through the use of Quality Matters and Community of Inquiry frameworks.

David M. Dunaway currently serves as an Assistant Professor of Educational Leadership at the University of North Carolina at Charlotte – a post he has held for the past seven years. He is a frequently published writer on school organizations, leadership, and improvement. With 35 years in public education at the K-12 level, he has served as a District Superintendent in Gibson County, Indiana, and as Deputy Superintendent for Instruction in the Owensboro Public Schools in Owensboro, Kentucky. Dr. Dunaway has taught at the middle and high school levels, served as an assistant principal of a large urban high school, and as a high school principal for a dozen years. Dr. Dunaway received his B.S. from Auburn University in 1965, his M.A. from the University of South Alabama in 1976, and his Ed.D. from Auburn University in 1985.

Joseph Feinberg is an Assistant Professor of Social Studies Education at Georgia State University. Joseph earned his Doctorate, Education Specialist, and Master degrees at The University of Georgia, and his Bachelor degree from UNC Chapel Hill. He previously taught social studies at Campbell High School in Smyrna, Georgia where he received a Martin Luther King Humanitarian Award. His research interests include simulation games, teacher education, and service learning.

Shaqwana M. Freeman, MAT, is a Doctoral student of Special Education at the University of North Carolina at Charlotte. Her current areas of interest include empirically-based teaching and learning strategies for youth with high incidence disabilities in middle and secondary schools with an emphasis on mathematics education in urban settings, and innovative use of web-based instruction to address the promotion of research-to-practice in special education.

David Gibson received his Ed.D. from the University of Vermont in Leadership and Policy Studies in 1999, and works as an educational researcher, professor, learning scientist, and innovator at local, state, and national levels of education. He consults with project and system leaders, formulates strategies, and helps people articulate their vision for innovation. Dr. Gibson's research focuses on complex systems analysis and modeling of education, games and simulations in teacher education, web applications and the future of learning, and the use of technology to personalize education via cognitive modeling, design, and implementation. His articles and books on games and simulations in learning follow from his role as creator of simSchool, a classroom flight simulator for preparing educators, and eFolio an online performance-based assessment system.

Miguel Gomez is a Ph.D. Student at The University of North Carolina at Greensboro. He received his B.A in History from the University of Georgia in 1999 and his Master's in Education from The University of North Carolina at Greensboro in 2009. He spent 13 years working in the information technology sector as a senior system engineer specializing in network security and system administration. He currently teaches middle school courses in technology education. His research interests include technology pedagogy, online learning in K-12 education, and social studies education, with a focus on social justice.

Elizabeth A. Gross is currently a Freelance Research Consultant and a Doctoral candidate at Wayne State University in Detroit, Michigan. Previously she taught in the public school system for 20 years. Elizabeth's extensive experience includes classroom and distance learning both as a student and as a teacher. As a result of her interests in the difficulties surrounding information search and cognitive reconciliation, her research centers on the effects of cognitive load on the information-seeking process for high school students. She capitalizes on this understanding by assisting both students and teachers in becoming more savvy and efficient researchers. Her current research efforts include analysis of the effects of reflective practice as it relates particularly to classroom teachers and their professional development. She is an All-American athlete who remains an avid runner. Elizabeth is married and currently resides in Houston, Texas.

Emily Hixon is an Assistant Professor of Educational Psychology and Instructional Technology in the Department of Teacher Preparation at Purdue University Calumet. She also serves as the Faculty Instructional Design Consultant in the university's Office of Instructional Technology and co-directs the institution's Digital Learning Faculty Certificate program. Her degrees are from Indiana University Bloomington in Instructional Systems Technology (M.S. and Ph.D.) and Educational Psychology (M.S). Her research interests involve effective technology integration and professional development and enhancement at both the K-12 and higher education levels. Through her work in faculty development, she has also been involved in many distance learning initiatives and is pursuing research in that area as well.

LuAnn Jordan is an Associate Professor in the Department of Special Education and Child Development at the University of North Carolina at Charlotte. Dr. Jordan's current areas of interest and research include written expression for students with disabilities and students at risk for academic failure, learning strategies and supports which increase the academic achievement of adolescents with high-incidence disabilities in middle and secondary schools, and inclusion of students with disabilities in general education settings. Dr. Jordan's experience at UNC Charlotte has included extensive distance education program planning, teaching, and advising beginning special education teachers, especially professionals entering special education as a second career.

Wayne Journell is an Assistant Professor and Secondary Social Studies Program Coordinator at The University of North Carolina at Greensboro. He received his Ph.D. from the University of Illinois at Urbana-Champaign in 2009. Wayne teaches undergraduate courses in social studies methods and graduate courses in social studies pedagogy, secondary education, and online instructional theory. His research interests include the civic development of adolescents, specifically the teaching of politics and political processes in secondary education, and online learning in K-12 education. Recent publications include articles in *Theory and Research in Social Education, Educational Leadership, Phi Delta Kappan, Educational Studies, Journal of Social Studies Research, The Clearing House, Educational Media International,* and *E-Learning*. Wayne also serves as the Interdisciplinary Education Feature editor for *Social Studies Research and Practice.*

Desi Krell is a Doctoral student in the School of Teaching and Learning at the University of Florida, Gainesville, where she is studying both Teacher Education and English Education. A former middle school teacher, she initially developed an interest in action research after understanding how it promoted teacher reflection and action in the classroom. Currently, one area of her research lies within practitioner inquiry, specifically the way in which coaches or mentors facilitate the process with educators, and has recently expanded into how the action research process is facilitated in virtual environments. In her work, she has coached both traditional public school educators and virtual educators through the action research process.

Salika A. Lawrence is Associate Professor of Literacy and Teacher Education at William Paterson University of New Jersey. Dr. Lawrence is a former middle and high school teacher. She currently serves as Director of the Master of Education in Literacy program. Her research interests include literacy instruction, adolescent literacy, and teacher education and professional development.

Jayme Nixon Linton currently serves as instructional technology facilitator and staff development coordinator for Newton-Conover City Schools in Catawba County, North Carolina. She began her career in education teaching elementary students for six years. After leaving the classroom, Jayme worked as an instructional coach for two-and-a-half years before moving into her current position. Jayme is a Doctoral student in the Teacher Education and Development program at the University of North Carolina at Greensboro. Her research interests include effective professional development for technology integration. She is particularly interested in exploring online tools to leverage the work of professional learning communities.

Dixie D. Massey teaches face-to-face courses at the University of Washington and online courses for The University of North Carolina at Greensboro and East Carolina University. Her research interests include literacy education and teacher research. Her articles have appeared in such publications as *The Reading Teacher, Reading Research and Instruction, The Journal of Adolescent and Adult Literacy*, and *The Journal of Literacy Research.* She is the co-author of the curriculum series, *Comprehension Strategies for World History and U.S. History in the Social Studies.* Currently, she serves as Historian for the Literacy Research Association.

Daniel Matthews, Ph.D., is an Associate Professor in the Department of Educational Leadership at the University of Illinois at Springfield (UIS). He earned his Doctorate from the University of Illinois at Urbana Champaign. He began his career in higher education teaching undergraduate students who were pre-service teachers and currently works with graduate students in the online Master of Arts in Teacher Leadership program at UIS. His current responsibilities include teaching on-campus and online educational research methods courses and guiding graduate students through their Master's research projects. His research interests include geographic diversity in online education, identifying factors related to success in online education, and improving online education through the use of Quality Matters and Community of Inquiry frameworks.

Christopher Moore is currently a Doctoral student at Georgia State University, concentrating on Teaching and Learning in Social Studies Education. He received his Master's degree from Piedmont College and his Bachelor of Arts degree from the University of Georgia. Currently, he is teaching social studies at the high school level for Gwinnett County Public Schools after teaching middle schools social studies for six years in private schools, where he helped write and implement social studies curriculum for grades 1-8. Christopher has presented at national conferences, including College and University Faculty Assembly (CUFA) and NCSS. His research interests include simulations and games, technology usage in social studies, and student engagement.

Christopher O'Brien is an Assistant Professor in the Department of Special Education and Child Development at the University of North Carolina at Charlotte. Dr. O'Brien's current areas of interest and research include learning strategies that promote the academic achievement of adolescents with high-incidence disabilities in middle and secondary schools (particularly in urban settings), the use of technology-enhanced reading performance for students with high-incidence disabilities, and technological innovations in the promotion of research-to-practice in special education. Most projects in the latter area have related to the use of distance education or online instructional technologies including extensive efforts to use video modeling and instructional exemplars to bridge the research-to-practice gap. Dr. O'Brien recently published *Teaching Students with Special Needs: A Guide for Future Educators* with chapter co-author John Beattie in an effort to improve teacher preparation for inclusive instruction, particularly for aspiring general education teachers.

Barbara G. Pace is an Associate Professor in the School of Teaching and Learning at the University of Florida where she serves as co-coordinator of the English Education Program. Her primary teaching responsibility is the preparation of English teachers and her research is focused on effective teaching practices in literacy and on learning from popular culture and multimodal texts in face-to-face and online environments. She has published in *English Journal*, the *International Journal of Qualitative Studies in Education, The Advocate*, *The Journal of Adolescent and Adult Literacy*, *English Education* and other scholarly journals in literacy education. She currently reviews research articles for *Contemporary Issues in Teaching and Technology Education* and has served on the Editorial Board of the *Journal of Adolescent and Adult Literacy*, the NCTE Commission on Media Literacy, and the Board of Directors for the Florida Council of Teachers of English.

Michelle Plaisance is a Doctoral student at the University of North Carolina at Charlotte. She is pursuing a PhD in Urban Education with an emphasis on Teaching English as a Second Language. As a former elementary ESL teacher, her research interests include differentiated instructional strategies and curricular access for emerging bilingual students.

Drew Polly (http://coedpages.uncc.edu/abpolly) is an Assistant Professor at the University of North Carolina at Charlotte. His research agenda focuses on examining how to best support teachers' use of learning technologies, learner-centered tasks, and standards-based mathematics pedagogies in elementary school classrooms.

Rikki Rimor is a member of the faculty of Education and Psychology at the Open University and at the Seminar Hakibutzim College of Education, Technology and Arts, in Israel. She teaches graduate programs concerning educational technologies in courses such as qualitative research methodologies, evaluation of online learning environments, and knowledge construction in database environments. Her major research and publications focus mainly on the cognitive and metacognitive aspects of learning in online environments. Her recent papers concern collaborative knowledge construction in Google Docs environments, as well as the study of social interaction patterns in online collaborative environments.

Albert D. Ritzhaupt is an Assistant Professor of Educational Technology at the University of Florida. His primary research areas focus on the development of technology-enhanced instruction, educational games and simulations, and technology integration in education. His publications have appeared in leading educational technology venues, including the *Journal of Research on Technology in Education*, *Computers & Education*, *Journal of Educational Computing Research*, *Journal of Computing in Higher Education*, *Behavior Research Methods*, *Journal of Interactive Learning Research*, and *Computers in Human Behavior*. Dr. Ritzhaupt regularly attends and presents at the American Educational Research Association, the International Society for Technology in Education, and the Association of Educational and Communication Technology.

Luke Rodesiler is a Doctoral fellow studying English education in the School of Teaching and Learning at the University of Florida and a teacher-consultant of the Red Cedar Writing Project at Michigan State University. His work has appeared in various publications, including *English Journal*, *Classroom Notes Plus*, *Screen Education*, and *Computers and Composition Online*. He has also made contributions to various edited collections, including IGI Global's *Virtual Professional Development and Informal Learning via Social Networks* and NCTE's *Lesson Plans for Developing Digital Literacies*. His scholarly interests include teacher education, self-directed professional learning, and the use of popular media and technology in the teaching of English.

Yigal Rosen is a member of the faculty of Education at the University of Haifa in Israel. His research interests are in promoting and assessing higher-order thinking skills in technology-rich learning environments, with special emphasis on collaborative problem-solving, social interactions, applying knowledge into new situations, reasoning, and critical thinking. He teaches graduate and undergraduate courses on educational technology, higher-order thinking skills and computer-based assessment. As an assessment expert he is involved in developing the new generation of the international computer-based assessment OECD PISA 2015. He obtained his Ph.D. degree from the Faculty of Education at the University of Haifa. Dr. Rosen was a post-doctoral fellow at Harvard University Graduate School of Education, and at Tel Aviv University School of Education.

Felicia Saffold is an Associate Professor in the Department of Curriculum and Instruction. In her current position, Professor Saffold teaches courses in urban education. Her primary research areas of interest include urban education, teacher preparation, online learning, and teacher development.

Audrey Schewe is currently a Doctoral student in the Middle-Secondary and Instructional Technology department in the College of Education at Georgia State University. Her concentration is on secondary social studies teaching and learning. Schewe earned her Master of Science in Education (Secondary Social Studies) and her Bachelor of Arts in History from the University of Pennsylvania in Philadelphia, Pennsylvania. Prior to pursuing her Doctorate, Schewe taught social studies at Albert Leonard Middle School in New Rochelle, New York and at Sweetwater Middle School in Lawrenceville, Georgia. For more than a decade, she served as the Senior Curriculum Development Manager at CNN/Turner Learning. In that role, Schewe developed a multitude of award-winning curriculum materials, including the educator guide to CNN's landmark Cold War series and Turner Adventure Learning, a series of interactive, electronic field trips. Her research interests include applications of motivation theory to social studies curriculum development, authentic education, and backwards design lesson planning.

Lindsay Sheronick Yearta is a Doctoral candidate in Curriculum and Instruction with a focus on Urban Literacy at the University of North Carolina at Charlotte. She also currently teaches fifth grade in Rock Hill, South Carolina. She is passionate about reading and learning as well as enabling others to do so. She seeks to inspire her students to become thoughtful, reflective learners who seek to enact positive change in their communities. Her research interests and teaching primarily focus on culturally responsive pedagogy, in kindergarten through twelfth grade as well as in higher education, and utilizing technology to improve the educational opportunities for all students.

Jason Siko is an Assistant Professor of Educational Technology at Grand Valley State University in Grand Rapids, Michigan. Previously, Jason was a high school science teacher at Clarkston High School, a suburb of Detroit, where he taught chemistry and biology for over 10 years. His research focuses on the use of game design as an instructional tool (i.e., the notion that students learn by constructing educational games as well as by playing them). Jason's other research interests include K-12 online learning and teacher professional development as it relates to technology and online learning. He also has a Master's degree in Strategic Foresight, and has worked as a consultant for non-profits and professional associations.

Hyo-Jeong So is an Assistant Professor in the Learning Sciences & Technologies Academic Group at the National Institute of Education (NIE). The Institute is housed in the Nanyang Technological University, Singapore. Dr. So is also a researcher in Singapore's Learning Sciences Laboratory. She received her Ph.D. in Instructional Systems Technology from Indiana University, Bloomington and worked as instructional designer and education consultant before joining NIE. She currently heads several research projects funded by the Singapore National Research Foundation as well as the Office for Educational Research at NIE. Her recent research focuses on teachers' epistemological beliefs about learning and technology, mobile learning for in-situ knowledge building, and using video annotation programs for teachers' meaning making. She has published and presented in the fields of distance education, teacher education, and computer-supported collaborative learning (CSCL).

Karen Swan, Ed.D., is the Stukel Professor of Educational Leadership and a Research Associate in the Center for Online Learning, Research and Service (COLRS) at the University of Illinois Springfield. Her research is in the area of media/technology and learning. She has authored over 100 publications and several hypermedia programs, and co-edited two books on educational technology topics. Her current interests include learning analytics and online learning. Dr. Swan has been involved with online teaching and learning for over a decade, both as an instructor and as a researcher. She helped develop one of the first fully online Master's degrees, is active in the online learning community, and is well known for her research on learning effectiveness in online environments. She is a Sloan-C Fellow and the 2006 recipient of the Sloan-C award for Outstanding Individual Achievement.

Bruce Taylor is an Associate Professor at the University of North Carolina at Charlotte in the Department of Reading and Elementary Education. He is Director of the Center for Adolescent Literacies. His research and teaching focus on the social and cultural aspects of literacy and learning of adolescents and, in particular, ways to meet the academic learning needs of diverse and often marginalized students. His work explores the role of diverse texts—including digital texts—in content-area classrooms and the role of discourse in the lives of adolescents. Recently, he has begun to explore the role of inquiry and service learning as ways to advance literacy and promote agency among marginalized adolescents.

Mary O. Taylor is a Doctoral candidate in Teacher Education and Development at The University of North Carolina at Greensboro. Mary also works as an Academically/Intellectually Gifted Teaching Specialist in a public elementary school. She earned National Board for Professional Teaching Standards certification in Early Adolescence/English Language Arts. In her more than fifteen years of teaching, Mary has had multiple opportunities to participate in the development of technology use in education at elementary and post-secondary levels. Her current studies emphasize the relationship among school culture, teacher beliefs, the process of change, and technology integration. Mary co-authored "D^3: Empowering Technology Integration in Education," published in *Ubiquitous Learning*.

Kecia Waddell is a Doctoral student in Wayne State University's Instructional Technology program, where she is an active broad member of the Michigan Chapter of Association for Educational Communications and Technology. Professionally, Kecia has been a Special Education teacher at the secondary level for over 15 years and has become conditioned to think in terms of alternative teaching and learning options for learners who struggle with exceptionalities that adversely impact their academic, social or emotional success. Kecia is passionate about technology integration to individualize instruction, monitor progress, support weaknesses, enhance strengths and motivate work production. Her current research interests focus on blended learning and individualized instructional design using learning analytics of at risk populations. Kecia is a happy wife of one supportive husband and proud mother of three dynamic daughters, two of whom are twins.

Melissa Walker Beeson is a full time Doctoral student at the University of North Carolina at Greensboro studying Teacher Education with an emphasis in Instructional Technology. She has taught undergraduate courses in Elementary Social Studies Methods and graduate courses in Instructional Technology and Principles of Effective Instruction. She also supervises teacher education students in their field placements. Her research interests include teacher planning for meaningful technology integration, specifically within the Technological Pedagogical Content Knowledge Framework, and effective technology uses in the content areas.

Rachel Wolkenhauer is a Doctoral student in the School of Teaching and Learning at the University of Florida, Gainesville, where she is studying Curriculum, Teaching, and Teacher Education. A former third and fourth grade teacher, she became interested in practitioner inquiry as a teacher researcher in her own classroom. Experiencing the impact of the empowering, job-embedded professional learning firsthand, Rachel developed a passion for the process and its potential impact on the teaching profession. She has coached many practitioner researchers, including those working in virtual school environments, and those working in some of Florida's highest need schools. Her current research interests focus on the ways in which inquiry is used as a tool for innovation and as a source of empowerment for teachers and students in both traditional and virtual schools.

Kevin Wood, an educator for over 10 years, currently serves as the Assistant Principal for curriculum and instruction at Central Gwinnett High School in Lawrenceville, Georgia. Kevin earned his Doctorate degree, Master's degree, and Bachelor's degree at Georgia State University. He also taught at Dacula High School in Dacula, Georgia and at Sharp Learning Center in Covington, Georgia where he was named Teacher of the Year and a finalist for Newton County Teacher of the Year. Kevin serves on the Gwinnett County Public Schools EClass School Advisory Board. Kevin's research interests include simulation video games, teacher education, and digital education.

Anne Yates is the Director the Graduate Diploma of Teaching (Secondary) program of learning for online student teachers at Victoria University of Wellington, New Zealand. She is the Course Coordinator and Principal Lecturer in EPSY 302 Teaching Models and Strategies, which is a core paper for both the Graduate Diploma of Teaching (Secondary) and the Graduate Diploma of Teaching (Primary): a course which is offered to student teachers in both the online and campus modes. Anne has strong research interests in the development of learning through online interaction. She is also interested creating effective initial teacher education for both campus and online student teachers and is currently using self-study to determine the effectiveness of lecturer modeling in initial teacher education.

Vassiliki Zygouris-Coe is an Associate Professor of Education at the University of Central Florida, College of Education. Her research focuses in literacy in the content areas, online learning, and professional development. Dr. Zygouris-Coe has impacted reading instruction in the state of Florida through the *Florida Online Reading Professional Development* project—Florida's first online large-scale project that has serviced over 45,000 preK-12 educators since 2003. Her work has been published in *The Reading Teacher, Reading & Writing Quarterly, Reading Horizons, Childhood Education, Early Childhood Education Journal, The International Journal of Qualitative Studies in Education, Focus in the Middle, Journal of Technology and Teacher Education, The International Journal of E-Learning, Florida Educational Leadership Journal,* and *Florida Reading Quarterly,* among others. She serves in several editorial roles, including Co-Editor of the *Literacy Research and Instruction* journal, Associate Editor of *Florida Educational Leadership*, and the *Florida Association of Teacher Educators Journal.*

Index

F

G

H

I

K

L

M

N

O

P

Q

R

S

T

U

V

W

X

CPSIA information can be obtained at www.ICGtesting.com
Printed in the USA
BVOW050710050313
314493BV00013BB/211/P